MOLECULES IN NATURAL SCIENCE AND MEDICINE
An Encomium for Linus Pauling

MOLECULES IN NATURAL SCIENCE AND MEDICINE

An Encomium for Linus Pauling

ZVONIMIR B. MAKSIĆ B.Sc., Ph.D.
Professor of Theoretical Chemistry, Faculty of Sciences and Mathematics
University of Zagreb, Croatia, Yugoslavia

MIRJANA ECKERT-MAKSIĆ B.Sc., M.Sc., Ph.D.
Department of Organic Chemistry and Biochemistry
The Rudjer Bosković Institute, Zagreb, Croatia, Yugoslavia

ELLIS HORWOOD
NEW YORK LONDON TORONTO SYDNEY TOKYO SINGAPORE

First published in 1991 by
ELLIS HORWOOD LIMITED
Market Cross House, Cooper Street,
Chichester, West Sussex, PO19 1EB, England

A division of
Simon & Schuster International Group
A Paramount Communications Company

Printed and bound in Great Britain
by Redwood Press Ltd, Melksham, Wiltshire

British Library Cataloguing-in-Publication Data

Maksić, Zvonimir B.
Molecules in natural science and medicine: An enconium for Linus Pauling.
I. Title. II. Eckert-Maksić, Mirjana.
574.8
ISBN 0–13–561598–4

Library of Congress Cataloging-in-Publication Data available

Table of contents

Foreword

M. F. Perutz

As a chemistry student in Vienna, I was made to memorize the 759 pages of Karl Hoffman's *Inorganic Chemistry* and the 866 pages of Paul Karrer's *Organic Chemistry*. I looked upon such tasks as feats of endurance which gave me a certain sporting satisfaction, like walking from Land's End to John o'Groat's, but they gave me little intellectual satisfaction because the books did not explain the properties of matter. Why does water freeze at 0 °C and methane at −184 °C? Why does one form of selenium melt at a temperature 76 °C higher than the other? Why is sulphur soft and diamond hard? Why is one form of silica, quartz, optically active, while the two others, tridymite and crystoballite, are not? Why is salicylic acid stronger than benzoic acid? No such questions were answered for me.

Later I became a research student in X-ray crystallography at Cambridge. For Christmas 1939 a girl-friend gave me a book token which I used to buy Linus Pauling's recently published *Nature of the Chemical Bond*. His book transformed the chemical flatland of my earlier textbooks into a world of three-dimensional structures. It stated that 'the properties of a substance depend in part upon the type of bonds between its atoms and in part upon the atomic arrangement and the distribution of bonds', and it proceeded to illustrate this theme with many examples. For instance, Pauling discusses the cause of a discontinuity in the melting points of the fluorides of the second row elements thus:

An abrupt change in properties in a series of compounds, such as in the melting points or boiling points of metal halides, has sometimes been considered to indicate an abrupt change in bond type. Thus of the fluorides of the second-row elements, those of high melting points have been described as salts, and the others as covalent compounds; and the drop in melting point of 1100 °C in going from aluminium fluoride to silicon fluoride has been interpreted as showing that the bonds change sharply from the extreme ionic type to the extreme covalent type.[1] I consider the bonds in aluminium fluoride to be only slightly different in character from those in silicon fluoride, and attribute the abrupt change in properties to a change in the nature of the atomic arrangement.[2] In NaF, MgF_2 and AlF_3 each of the metal atoms is surrounded by an octahedron of fluorine atoms, and the stoichiometric relations then require that each fluorine atom be held jointly by several metal atoms. In each of these crystals the molecules are thus combined into giant polymers, and the processes of fusion and vaporization can take place only by breaking the strong chemical bonds between metal and non-metal atoms; in consequence the substances have high melting points and boiling points. The stable coordination number of silicon relative to fluorine is, on the other hand, four, so that the SiF_4 molecule has little tendency to form polymers. The crystal of silicon fluoride consists of SiF_4 molecules piled together and held together only by weak van der Waals forces.

Characteristically for Pauling's showmanship, reference 1 is to N. V. Sidgwick's classic *The Electronic Theory of Valency*, and reference 2 to one of his own papers. By such examples Pauling's book fortified my belief, already inspired by Bernal, that knowledge of three-dimensional structure is all-important and that the functions of living cells will never be understood without knowing the structures of the large molecules composing them.

In my physical chemistry practical I had had to prove to myself that acetic acid in solution forms dimers, but it needed Pauling to drive home the importance of the hydrogen bonds that are responsible for their formation:

> Although the hydrogen bond is not strong it has great significance in determining the properties of substances. Because of its small bond energy and the small activation energy involved in its formation and rupture, the hydrogen bond is especially suited to play a part in reactions occurring at normal temperatures. It has been recognized that hydrogen bonds restrain protein molecules to their native configurations, and I believe that as the methods of structural chemistry are further applied to physiological problems it will be found that the significance of the hydrogen bond for physiology is greater than that of any other single structural feature.

This prophecy has come true.

In later years the valence bond and resonance theories which formed the theoretical backbone of Pauling's book were superseded by R. S. Mulliken's molecular orbital theory which provided a deeper understanding of chemical bonding. For instance, it allowed C. Longuet-Higgins and W. Lipscomb to predict and explain the structures of the boranes, which would not have been possible on the basis of Pauling's concepts.

Despite this shortcoming, Pauling's imaginative approach, his synthesis of structural, theoretical and practical chemistry, his capacity of drawing on a wide variety of observations to prove his generalizations, and his vivid writing drew the dry facts of chemistry together into a coherent intellectual fabric for me and thousands of other students for the first time.

Leafing through the book while writing this foreword I came across a passage headed 'Discontinuous change in bond type', which contains the following sentence printed in italics: 'If the two structures under consideration involve different numbers of unpaired electrons, then the transition between the two must be discontinuous, the discontinuity being associated with the pairing or unpairing of electrons'. This predicted the key to the understanding of oxygen transport by haemoglobin. Three years before publication of the book, Pauling himself forged that key when he and his student Charles Coryell discovered the spin change of the heme iron on combination of haemoglobin with oxygen. Michael Faraday had found blood to be diamagnetic, even though it contains iron. Coryell showed that this was because Faraday's blood had been saturated with oxygen; on removal of oxygen, haemoglobin becomes paramagnetic with a spin of its ferrous iron of $S = 2$.

I asked Linus Pauling at his 75th birthday what made him decide on this crucial experiment as the very first he ever did with haemoglobin. He replied that people had been uncertain whether oxygen formed a chemical bond with the iron or whether it was just adsorbed by the heme; it occurred to him then that the formation of a bond might be accompanied by a magnetic change. Having discovered the magnetic transition, he left it at that, even though he was to point out in *Nature of the Chemical Bond* that the transition of ferrous iron from low to high spin is accompanied by an increase in its ionic radius.

Until 1970, the significance of that passage was lost also on me. I then found that the high-spin iron in deoxyhaemoglobin is displaced from the porphyrin plane, while in low-spin iron-porphyrins it was known to lie in that plane. This made me realize that the low-spin iron only just fits into the hole between the four porphyrin nitrogens, while the high-spin iron is forced out because it is too large; the resulting movement of the iron relative to the porphyrin is the trigger for the change of structure that the entire haemoglobin molecule undergoes every time it takes up and releases oxygen. That change of structure underlies its cooperative reaction with oxygen on which the respiration of fast-moving animals depends.

I first met Pauling 10 years after I had read his book and was intrigued by his lectures, where he would reel off the top of his head atomic radii and interatomic distances with the gusto of an organist playing a fugue; afterwards he would look around for applause as I had seen Bertrand Russell do after quoting one of his own elegant metaphors. Pauling's lectures reinforced the chief message of his book: to understand the properties of molecules, not only must you know their structures but you must know them *accurately*.

At his 90th birthday celebrations in Pasadena on 28th February, Pauling gave a lecture in which he recalled the thrill of determining the first organic structures by X-ray analysis. It was as lively a performance as the first lecture I heard him give when he came to a haemoglobin symposium here in Cambridge in 1948.

Cambridge, 11 June 1991

Preface

Linus Pauling, the great American Nobel prizewinner, is one of the outstanding men of this century. His brilliant mind has revealed to the world a staggering range of scientific disclosures. In particular Linus Pauling is the grand old man of molecular sciences. He is the architect of modern chemistry who more than anyone else reconciled empirical knowledge with quantum theory. It was he who first realized that molecular structure is the central theme of chemistry. A deep understanding of molecular stereochemistry made possible many of his important discoveries in other branches of natural science, to mention only mineralogy, biochemistry, molecular biology and medicine.

The main characteristic of his scientific method is an understanding of the quintessence of phenomena and their interpretation in a fundamental but simple and transparent way. He is a grandmaster of modelling and his models define a scale of quality in interpretive quantum chemistry. One usually says that a model description of a studied system approaches the Pauling point if it gives a reasonable agreement with experiment in a most economical way. In other words, a model at the Pauling point best satisfies Occam's razor. The fruitfulness of this approach is best illustrated by Pauling's numerous discoveries which made historical breakthroughs in several widely different research areas. He unravelled the alpha-helix structure of proteins, identified abnormal haemoglobin structure in sickle-cell anaemia, discovered the 'molecular clock of evolution' etc. Some of his predictions were controversial; but as somebody nicely put it, it is generally true that 'the mainstream converges with Pauling's opinion twenty years later'.

His work in medicine deserves particular attention. Pauling put forward a theory of anaesthesia and founded orthomolecular medicine—a new branch of medicine where prevention and treatment of illnesses is achieved by varying the concentration of substances which are normally present in the body. He is a proponent of the 'mega-ascorbic' therapy of various diseases, ranging from the common cold to cancer.

However, Pauling is not only one of the greatest living scientists, he is also a giant public figure and a citizen of the world. Pauling's peace rallies are well known. His collection of 11,000 signatures from leading intellectuals from all over the world on a petition to ban atomic bomb explosions has eventually led to a treaty which put an end to atomic experiments in the atmosphere.

Linus Pauling is a great fighter for democracy, social justice and human rights. He was one of the main authors of the Cavtat Declaration (*International Journal of Quantum Chemistry* No. 1 (1989) and *Croatica Chemica Acta* No. 4 (1989)) coined during the scientific Symposium on the Electronic Structure of Molecules, Clusters and Crystals held in Cavtat (Dubrovnik), Croatia, August 1988. The Declaration issues a warning against the four apocalyptic dangers which threaten to destroy life on

Earth: nuclear annihilation, ecological catastrophe, economic collapse and the population explosion. It gives in addition some suggestions for concerted democratic efforts in their overcoming. A very strong emphasis was put on human rights. We cite here just a small part of the relevant section: 'As members first and foremost of the human race, we neither expect nor desire that one single political or social system should prevail on the whole Earth. A disappearance of cultural differences would make our world poorer. But we have to live with diversity in a manner that does not threaten the security and prosperity of all people. This requires above all tolerance and respect for human life, freedom from political, economic or religious oppression ...'. We could not have anticipated at that time that exactly these words would be so badly needed in Yugoslavia. We could not have imagined that precisely in Croatia, where the Cavtat Declaration was designed, human rights, including the most basic right to life, would be so ruthlessly violated by foreign aggression.

This book has been prepared under very unfavourable circumstances in one of the most dramatic moments in the history of the Croatian people. More than 10,000 Croats have been forced to leave their homes in Eastern and Central Croatia and become refugees in their own country, trying to save their lives. The last pages were written in days when up to 100 Croatian patriots were dying in defence of their right to live on their own land and in their own state, to live in a democratic country without all sorts of oppression which characterize totalitarian systems. But these pages were also completed in the hope that the Croatian dream of freedom, democracy and justice, so well expressed in the Cavtat Declaration, would finally come true. Pauling's life-long struggle for the same ideals was a source of permanent inspiration.

This book, dedicated to Linus Pauling, the greatest scientist, democrat and humanist, is prepared in the best spirit of his seminal ideas, thanks to valuable contributions from the world's leading experts in the field of the molecular sciences. It is at the same time a symbol of the small but not insignificant Croatian contribution to world science and international cooperation.

We would like to thank all the authors for their scholarly chapters which made this book possible. A part of the editorial work has been performed at the Organisch-chemisches Institut der Universität Heidelberg. Hence our special thanks go to the Alexander von Humboldt Stiftung for financial support and to Professor R. Gleiter for hospitality and advice.

Heidelberg, August 1991

Zvonimir B. Maksić

Mirjana Eckert-Maksić

I. Molecular Structure

1

Gas electron diffraction: its role in molecular structure determination

Jerome Karle

Naval Research Laboratory, Washington, DC, USA

INTRODUCTION

Linus Pauling's interests in, and contributions to, atomic, molecular and crystal structures have been expressed in many contexts. One of his early interests concerned the study of molecular structure by the use of the diffraction of electrons from gaseous molecules, a technique that he introduced into the United States very soon after it was first demonstrated in Europe. This chapter will review some of the history of gas electron diffraction and describe the developments and implications that have occurred over the years as a consequence of the dedicated talents of individuals in a discipline that has a relatively small number of members.

One of the well-refined techniques for investigating molecular structure and internal motion is electron diffraction by gases. Electron diffraction had its origins in major developments in physics that occurred early in this century. They were the concept of Louis de Broglie [1–3], that there is a wave associated with propagated material particles whose wavelength is defined in terms of the mass and velocity of the particles, the introduction of quantum mechanics by Erwin Schrödinger [4], Werner Heisenberg [5] and Max Born and Pascual Jordan [6, 7] which provided the theory for interpreting the diffraction patterns, and the impetus provided by the experiments of Davisson and Kunsman [8] and Davisson and Germer [9] whose electron scattering experiments on polycrystalline and single-crystal nickel surfaces, respectively, were later recognized as showing interference effects. These developments were followed by experiments with thin polycrystalline foils by Thomson and Reid [10], mica by Kikuchi [11] and gases by Mark and Wierl [12]. It had occurred to Herman Mark that the strong scattering of electrons by atoms would make an electron beam an attrative source for carrying out diffraction experiments on gases, and he stimulated Raimund Wierl in his laboratory to undertake the construction of an appropriate apparatus and to initiate experiments. Valuable structural results were soon forthcoming [13].

Toward the end of the 1920s, after receiving the Ph.D. degree, Linus Pauling visited and worked with some of the leading theorists in atomic and molecular structure in Europe, e.g. Arnold Sommerfeld, Erwin Schrödinger, Peter Debye and Niels Bohr. He also took the opportunity to visit the laboratory of Herman Mark, where he learned of the work being done on the structure of polymers by X-ray and electron diffraction. He was attracted to the new method of gas electron diffraction, particularly because of his own interests, and was afforded the details of the work that had been done. Soon after, Linus Pauling returned to the California Institute of Technology and brought the latter information with him.

Lawrence O. Brockway was one of the first graduate students of Linus Pauling. He told me that he was somewhat difficult to satisfy when he discussed possible research problems with Professor Pauling. He had been offered a number of possible problems, but could not make up his mind. Then Professor Pauling told him about the new technique of gas electron diffraction, which immediately caught his imagination and became his thesis problem. With the information made available by Herman Mark, Lawrence Brockway proceeded to put together a suitable apparatus, Fig. 1, and undertook a research program that reflected Linus Pauling's interests in the nature of the chemical bond. Numerous publications of Pauling and Brockway ensued in which configurations, conformations, bonding and multiple bonding were characterized. Much work ensued in Pauling's laboratory. This was also true of Wierl, who investigated many structures until his untimely fatal accident in 1932. A few other laboratories were soon also attracted to this field. The high degree of productivity and

Fig. 1.—Lawrence Brockway with the gas electron diffraction apparatus at the California Institute of Technology, *circa* 1935.

rapid development of the field of gas electron diffraction are recorded in a general review article by Lawrence Brockway [14], published in 1936, describing theory, experiment and analytical methodology in which the structures of approximately 150 molecules are tabulated.

In the latter part of the 1930s, Lawrence Brockway accepted a position in the Chemistry Department of the University of Michigan, where Isabella Lugoski (later my wife) and I were soon to become graduate students. After taking a physical chemistry course given by Lawrence Brockway during the school year from the Fall of 1940 through the Spring of 1941, we asked Professor Brockway if he would accept us as graduate students and be our thesis advisor. He agreed to do so, and just as he was one of Linus Pauling's first graduate students, we were among Lawrence Brockway's first graduate students. Our introduction to, and training in, gas electron diffraction was thus initiated and we had the benefit fairly directly of the developments in this technique from its inception.

In the course of time, we had the opportunity to add a bit of our own to the technique in a way that turned out to have a very profound effect on the course of our professional lives. In 1941, Peter J.W. Debye published a paper that concerned the mathematical treatment of idealized data in such a way that one could conceive of obtaining from electron diffraction data not only average interatomic distances but also information concerning internal motion [15]. The subject of this paper was a further development of the Fourier transform of the interference intensity, introduced by Pauling and Brockway [16]. As first used, interatomic distances were obtained from the positions of peaks in the Fourier transform, which is also known as a radial distribution function.

To fulfill the promise of this analysis would require highly accurate data and sophisticated data reduction techniques to put the data in a form suitable for application of the theory. We wanted to try to pursue this matter, and deferred it to a future opportunity to do so.

The opportunity came when we went to Washington, D.C. to join the Naval Research Laboratory. NRL had a post-war program to enhance its capacity for basic reseach, which provided an ideal setting for pursuing our interests in research on molecular structure. Our attempts to fulfill the promise of Debye's theoretical paper were very well supported, and the manner in which it was achieved has been a matter of interest to me because it provides a good example of a type of scientific activity that plays a key role in scientific research. The activity may be called 'bridging' because it affords connections where previously none existed and, in such circumstances, makes the difference between success and failure. Methods for overcoming barriers to progress may, at times, seem to be unattainable and then, by processes associated with bridging, the problems become solvable and the difficulties are replaced by routine procedures. After bridging techniques are applied, the methods by which awkward problems are overcome often seem to be self-evident. For interesting examples of the not so obvious in the scientific community, see 'Oliver Heaviside', by Paul J. Nahin, in the June *Scientific American* **262**, 122 (1990).

Problems arose in realizing the potential of Debye's theory because it was based on data in form and extent that are not available from experiment. In order to create a bridge between theory and experiment, use was made of some advanced techniques for collecting and measuring data that had been developed in European laboratories, and changes were introduced in the way diffraction data were measured and subsequently treated to make such data suitable for application of the theory. The theory, in turn, required some further treatment to make it suitable for use with the resulting data.

The pioneering method in structural research by gas electron diffraction initiated by Mark and Wierl and further developed by Pauling and Brockway was known as the 'visual method'. The name arose from the visual observation of the diffraction photographs which appeared to consist of light and dark rings, Fig. 2. Remarkably good structure determinations could be made from careful measurement of the diameters of the rings and approximate estimates of their intensities. Actually, the diffraction patterns are composed of an oscillating but relatively weak interference pattern superimposed on a steeply falling background intensity. The eye somehow largely corrects for the steeply falling background and extracts the interference pattern. If it is desired, however, to measure the interference pattern with high precision, a requirement for the relization of the full promise of the electron diffraction technique, it is apparent that the nature of the diffraction data presents considerable difficulties with respect to measurement and extraction of the interference pattern.

It would seem that microphotometry could replace the visual estimate of the diffraction intensities. Microphotometry is difficult to carry out accurately, however, when the objective is to measure a weak, oscillating signal superimposed on a steeply falling background. This problem was overcome by use of a rotating sector introduced into the diffraction apparatus independently by Christian Finbak [17] and P.P.

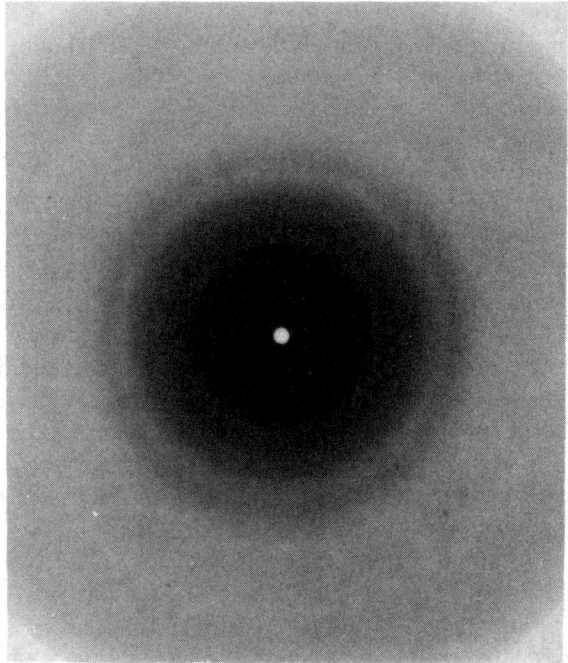

Fig. 2. — Reproduction of an electron diffraction photograph of the vapor of hexafluoropropene [71], showing some of the light and dark diffuse rings characteristic of such patterns. (The reproduction degrades the quality of the original photograph.)

Debye [18] (son of P.J.W.). The sectors could be designed to make the steeply falling background much more level. Finbak, Hassel and their collaborators at the University of Oslo pioneered in the use of the rotating sector and microphotometry [19–22]. The tradition of gas electron diffraction research in Norway has continued under the leadership of Otto Bastiansen and his students. The graininess of the photographic emulsions was minimized considerably by use of an apparatus for rotating the photographic plate [23] while it was being traced, thus improving the quality of the densitometer traces.

Another problem arose because in order for the Fourier transform of the interference scattering to be interpretable in terms of a probability distribution of interatomic distances, as proposed in Debye's theory, it is necessary for the interference scattering to correspond to that which would be obtained from point atoms. Actual atoms scatter electrons from their electron densities as well as their nuclei and therefore do not fit the requirement. This problem was overcome mainly by manipulation of the total intensities of scattering [24–26].

Still another problem concerned the fact that the mathematical analysis in Debye's theory implied the use of an effective infinite amount of data. In such cases when the maximum scattering angle accessible to measurement was less than that at which the damping of the interference scattering, as a consequence of the internal motion, made the amplitude of the oscillations negligible, something needed to be done to compensate for the absent data. This was achieved by the use of an additional methematical damping function and the use of Fourier transform theory to correct for its effect on the final results [25, 26]. The damping function was earlier introduced by Charles Degard [27] and Verner Schomaker [28], also a student of Pauling in the 1930s, in order to bring visually estimated data to negligible values at the largest experimental scattering angles.

Finally, it was necessary to determine the shape of the steeply falling background intensity on which the interference scattering was superimposed. Relatively small errors in the background intensity could introduce serious errors into the extraction of the weak signal that comprises the interference scattering. This problem was solved by applying the criteria of smoothness [25, 29–31] and the non-negativity [26] of the resulting Fourier transform of the interference scattering. The latter is required for a function that represents a probability distribution, in this case, the probability distribution of interatomic distances in a molecule.

These were the main considerations in fulfilling the promise of Debye's theory. Final parameter values and estimated errors were obtained by making comparisons with the experimental intensity values, for example, Fig. 3. At first, intensity curves based on the model from the radial distribution function, for example, Fig. 4, were calculated for making the comparisons. This was augmented or replaced by least squares techniques that benefited greatly by the advances in computing technology in the 1950s. Least squares analysis for electron diffraction was discussed by Cruickshank and Viervoll [32], studied further by Hamilton [33], applied by Bastiansen, Hedberg and Hedberg [34], and developed further by Bonham and Bartell [35], Hedberg and Iwasaki [36], Morino, Kuchitsu and Murata [37], Murata and Morino [38], Hilderbrandt and Bauer [39], and Seip, Strand and Stolevik [40].

With the increase in accuracy that ensued and the opportunities that arose from a number of additional technical and theoretical developments, electron diffraction of

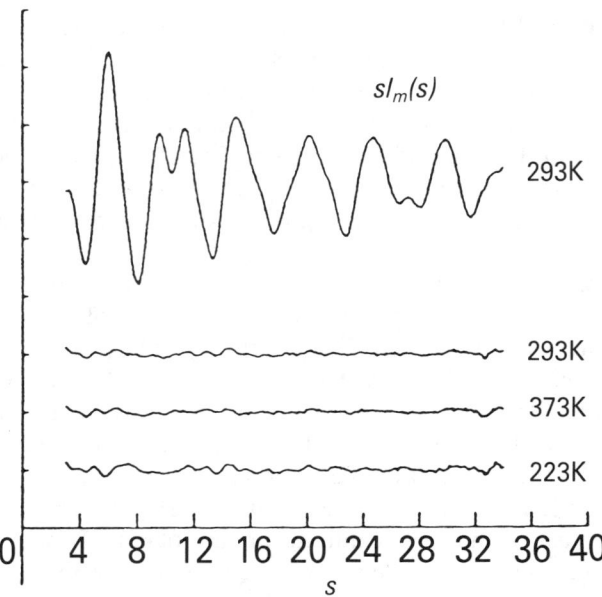

Fig. 3. — Experimental interference intensity (molecular intensity) curve at 293K for hexafluoro-propene [71] and the differences between experimental and theoretical curves at three different sample temperatures. The theoretical curves were computed with a threefold potential barrier having a height of 1.5 kcal mol^{-1}. The differences between the experimental intensity curves at different temperatures are larger than the differences between the experimental and model intensity functions for the corresponding temperatures.

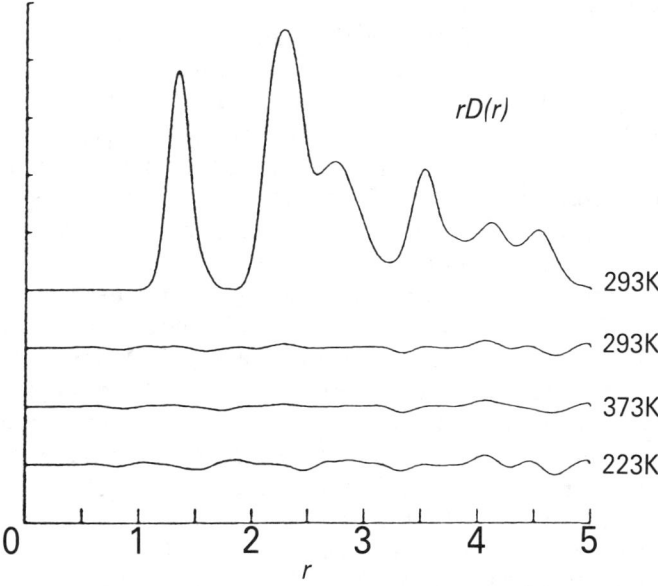

Fig. 4. — Experimental radial distribution curve for hexafluoropropene [71] and also Fourier transforms of the molecular intensity differences in Fig. 3. The value of the ordinate for each of the four curves when $r = 0$ is zero. The non-negativity of the radial distribution curve is apparent.

gases has made major contributions to the understanding of many aspects of molecular structure in the gaseous state. A broad variety of investigations followed.

Very many studies of configuration and conformation proceeded. They are described in many review articles, e.g. [41–46]. The higher accuracies permitted more detailed studies of bonding and the effects of attached moieties on bond distances. Of special interest have been steric effects, i.e. the crowding of large atoms or groups in molecules that can lead to distortion of the structure from commonly found configurtions. Along with this were studies of the internal motion of molecules which, for small amplitude motions, were expressed in terms of root mean square amplitudes of vibration between pairs of atoms [15, 25, 26]. Soon, Yonezo Morino, Kozo Kuchitsu and their collaborators developed a theory for computing root mean square amplitudes from spectroscopic data [47, 48]. Comparisons of results from spectroscopy and electron diffraction have been helpful in evaluating the accuracy of normal coordinate assignments, force models and analyses of diffraction data.

Associated with internal motion is the possibility that some of the longer interatomic distances may be observed to be different than that calculated from known bond distances and angles. When the calculated value is larger than the observed value, the effect of the internal motion is called shrinkage. This is observed in the linear molecule, CO_2, in which the O—O distance has been indicated to be smaller than twice the value of the C—O distance [25]. This may be explained in terms of a bending mode for which the O—O distance is equal to the sum of the C—O distances only when the molecule passes through the equilibrium position. Otherwise it is smaller. Internal rotations of large amplitude can cause large shrinkage effects on *trans* distances, but would enhance the observed values for *cis* distances [49, 50]. Additional studies of shrinkage were made by Morino and collaborators [51–53] and by Bastiansen and Collaborators [54–56]. An extensive treatise concering internal motion in molecules as deduced from both spectroscopic and electron diffraction methods has been written by Sven Cyvin [57]. For the purpose of combining interatomic distance information in an optimal way from spectroscopic and electron diffraction data, it has been necessary to analyze in detail the meaning of the information provided by each technique. Reviews of these matters have been given by Kuchitsu [58] and Kuchitsu and Cyvin [59].

A special type of internal motion that can attain large amplitudes is internal rotation about single bonds [16, 60]. Investigations of internal rotation have had, among their objectives, the evaluation of the barriers to the rotation. Both theoretical and experimental studies have been made [50, 51, 61–64] and have included multiple rotors [65].

Investigations of barrier heights or thermodynamic quantities or both by means of gas electron diffraction studies at different temperatures have been made on a number of substances, e.g. ethylene chlorohydrin by Almenningen *et al.* [66], dinitrogen tetrafluoride by Gilbert *et al.* [67], oxalyl chloride and 2,3 butanedione by Hagen and Hedberg [68, 69], formaldazine by Hagen *et al.* [70] and hexafluoropropene by Lowrey *et al.* [71].

The possibility of investigating the structures of substances that require very high temperatures to vaporize sufficiently for gas electron diffraction to be practicable was considered quite early in the history of this technique. Maxwell, Hendricks and Moseley performed structural investigations of alkali halide vapors in 1937 [72, 73].

Evidently, special apparatus is required to carry out the experiments. Not only is it necessary to achieve high temperatures, it is also necessary to protect photographic emulsions from the radiation. Akishin, Rambidi, Spiridonov and colleagues [74] have made extensive developments and investigated many types of low-volatility compounds such as halides, oxides, sulfides, molybdates, metaborates and perrhenates. A review has been written by Schäfer [75]. Various special interpretive problems are associated with this technique that arise from the fact that the vapors at very high temperatures are generally of mixed composition and are in various excited states, for example, states having anharmonic motions of large amplitude.

High-temperature experiments can also be applied to studies of reactive species such as the free radical CCl_3 investigated by Denis Kohl [76]. Reactions can also be studied, for example, the pyrolysis at 750° of [2,2]paracyclophane to form p-xylylene investigated by Mahaffy, Wieser and Montgomery [77]. Shen and Hedberg [78] investigated NO_2/N_2O_4, $GaBr_3/Ga_2Br_6$ and aluminum halide systems at various temperatures, observing the reaction equilibria. They determined the structures of the reactants and evaluated thermodynamic properties associated with the reactions.

Electron diffraction investigations of molecules containing atoms that have a considerable disparity in atomic number were observed to give results that were at considerable variance with those obtained from other techniques, for example, X-ray crystal structure analysis. The explanation for this was given by Verner Schomaker and Roy Glauber [79, 80], who pointed out that the Born approximation [81, 82], that was used for calculating the scattering obtained from individual atoms, was not adequate for the accelerating voltages normally used and that the discrepancy increased with atomic number. With use of the theoretical work of Schomaker and Glauber, Russell Bonham and collaborators have tabulated the scattering obtained from individual atoms (atomic scattering factors) [83, 84]. Some discrepancies have been found between the theory and experiment which have been thought to be explainable in terms of multiple scattering processes whose occurrence would be greater for atoms of high atomic number. Investigations concerning multiple scattering [85–88] have indicated that it probably is at least partly responsible for the discrepancies observed.

CONCLUDING REMARKS

I have presented here a brief summary of the variety of topics and the broad progress that characterize the discipline of gas electron diffraction. The details given bring the reader into the 1970s. For greater details and deeper insights that reach into the 1980s, there is the two-volume publication *Stereochemical Applications of Gas-phase Electron Diffraction* edited by Istvan and Magdolna Hargittai [89]. The articles in these volumes concern various aspects of the electron diffraction technique and structural studies of a variety of classes of compounds.

Electron diffraction of gases has blossomed from the pioneering work of Mark and Wierl and that of Pauling and Brockway into a technique of wide application to many aspects of molecular structure investigation. This vigorous area of activity complements and interacts with spectroscopic techniques and calculations in theoretical chemistry to provide the structural and related information with which other areas of science progress.

Postscript

One of the conditions for forming a suitable background intensity in gas diffraction, namely, the non-negativity of the resulting radial distribution function, stimulated other applications of the non-negativity criterion. Evidently. the non-negativity of the electron density distributions in crystals could be of interest, and we undertook a study of its implications. The appropriate mathematical description for this work is an infinite set of determinantal inequalities that represents the necessary and sufficient condition for a Fourier series to be non-negative [90, 91]. Work by David Harker (student of Pauling) and John Kasper based on Schwarz and Cauchy inequalities preceded ours [92]. Low-order determinants afforded practical restraints on the so-called phases whose values are needed in order to calculate the electron density distributions in crystals. These results stimulated further theoretical and experimental investigations over the years until in the early 1960s a general procedure for solving the structures of both centrosymmetric and noncentrosymmetric crystals — the symbolic addition procedure [93, 94] — was developed. This is one more example of a broad development that had its origins in gas electron diffraction [95].

REFERENCES

[1] L. de Broglie, *Dissertation*, Masson, Paris, 1924.
[2] L. de Broglie, *Phil. Mag.* **47** 446 (1924).
[3] L. de Broglie, *Ann. de Phys.* **3**, 22 (1925).
[4] E. Schrödinger, *Ann. d. Physik* **79**, 361 (1926); **79**, 489 (1926); **80**, 437 (1926); and **81**, 109 (1926).
[5] W. Heisenberg, *Zeits. f. Physik* **33**, 879 (1925).
[6] M. Born and P. Jordan, *Zeits. f. Physik* **34**, 858 (1925).
[7] M. Born, W. Heisenberg and P. Jordan, *Zeits. f. Physik* **35**, 557 (1925).
[8] C.J. Davisson and C.H. Kunsman, *Phys. Rev.* **22**, 242, (1923).
[9] C.J. Davisson and L.H. Germer, *Nature* **119**, 558 (1927).
[10] G.P. Thomson and A. Reid, *Nature* **119**, 890 (1927).
[11] S. Kikuchi, *Jap. Jour. Phys.* **5**, 83 (1928).
[12] H. Mark and R. Wierl, *Zeits. f. Physik* **60**, 741 (1930).
[13] R. Wierl, *Ann. d. Physik* **8**, 521 (1931).
[14] L.O. Brockway, *Rev. Mod. Phys.* **8**, 231 (1936).
[15] P.J.W. Debye, *J. Chem. Phys.* **9**, 55 (1941).
[16] L. Pauling and L.O. Brockway, *J. Am. Chem. Soc.* **57**, 2684 (1935).
[17] C. Finbak, *Avhandl. Norske Videnskaps—Akad. Oslo, I. Mat.-Naturv. Kl.* No. 13 (1938).
[18] P.P. Debye, *Physik, Zeits.* **40**, 66 and 404 (1939).
[19] C. Finbak, *Avhandl. Norske Videnskaps—Akad. Oslo, I. Mat.-Naturv. Kl.* No. 7 (1941).
[20] C. Finbak and O. Hassel, *Arch. Math. Naturvidenskab* **45**, No. 3 (1941).
[21] H. Viervoll, *Acta Chem. Scand.* **1**, 120 (1947).
[22] O. Hassel and H. Viervoll, *Acta Chem. Scand.* **1** 149 (1947).
[23] I.L. Karle, D. Hoober and J. Karle, *J. Chem. Phys.* **15**, 765 (1947).
[24] J. Karle and I.L. Karle, *J. Chem. Phys.* **15**, 764 (1947).

[25] I.L. Karle and J. Karle, *J. Chem. Phys.* **17**, 1052 (1949).

[26] J. Karle and I.L. Karle, *J. Chem. Phys.* **18**, 957 (1950).

[27] C. Degard, *Bull. Soc. Roy. Sci. Liege* **12**, 383 (1937).

[28] V. Schomaker, presented to American Chemical Society, Baltimore, April 1939.

[29] C. Finbak and O. Hassel, *Arch. Math. Naturvidenskab* **45**, No. 3 (1941).

[30] H. Viervoll, *Acta Chem. Scand.* **1**, 120 (1947).

[31] O. Hassel and H. Viervoll, *Acta Chem. Scand.* **1**, 149 (1947).

[32] D.W.J. Cruickshank and H. Viervoll, *Acta Chem. Scand.* **3**, 560 (1949).

[33] W.C. Hamilton, Thesis, California Institute of Technology (1954).

[34] O. Bastiansen, L. Hedberg and K. Hedberg, *J. Chem. Phys.* **27**, 1311 (1956).

[35] R.A. Bonham and L.S. Bartell, *J. Chem. Phys.* **31**, 702 (1959).

[36] K. Hedberg and M. Iwasaki, *Acta Cryst.* **17**, 529 (1964).

[37] Y. Morino, K. Kuchitsu and Y. Murata. *Acta Cryst.* **18**, 549 (1965).

[38] Y. Murata and Y. Morino, *Acta Cryst.* **20**, 605 (1966).

[39] R. Hilderbrandt and S.H. Bauer, *J. Mol. Struct.* **3**, 825 (1969).

[40] H.M. Seip, T.G. Strand and R. Stolevik, *Chem. Phys. Letters* **3**, 617 (1969).

[41] L.S. Bartell, in *Physical Methods of Chemistry*, Eds A. Weissberger and B.W. Rossiter, Wiley–Interscience, New York, 1971.

[42] O. Bastiansen, H.M. Seip and J.E. Boggs, in *Prespectives in Structural Chemistry*, Eds J.D. Dunitz and J.A. Ibers, Wiley–Interscience, New York, 1971.

[43] S.H. Bauer, in *Physical Chemistry, An Advanced Treatise*, Ed. D. Henderson, Academic Press, New York, 1970.

[44] A. Haaland, L. Vilkov, L.S. Khaikin, A. Yokozeki and S.H. Bauer, *Top. Curr. Chem.* **53**, 1 (1973).

[45] R.L. Hilderbrandt and R.A. Bonham, *Annu. Rev. Phys. Chem.* **22**, 279 (1971).

[46] J. Karle, in *Determination of Organic Structures by Physical Methods*, Eds F.C. Nachod and J.J. Zuckerman, Academic Press, New York, 1973.

[47] Y. Morino, K. Kuchitsu and T. Shimanouchi, *J. Chem. Phys.* **20**, 726 (1952).

[48] Y. Morino, K. Kuchitsu, A. Takahashi and K. Maeda, *J. Chem. Phys.* **21** 1927 (1953).

[49] J. Karle and H. Hauptman, *J. Chem. Phys.* **18**, 875 (1950).

[50] J. Ainsworth and J. Karle, *J. Chem. Phys.* **20**, 425 (1952).

[51] Y. Morino and E. Hirota, *J. Chem. Phys.* **23**, 737 (1955).

[52] Y. Morino, *Acta Cryst.* **13**, 1107 (1960).

[53] Y. Morino, S.J. Cyvin, K. Kuchitsu and T. Iijima, *J. Chem. Phys.* **36**, 1109 (1962).

[54] A. Almenningen, O. Bastiansen and T. Munthe-Kaas, *Acta Chem. Scand.* **10**, 261 (1956).

[55] A. Almenningen, O. Bastiansen and M. Traetteberg, *Acta Chem. Scand.* **13**, 1699 (1959).

[56] O. Bastiansen and M. Traetteberg, *Acta Cryst.* **13**, 1108 (1960).

[57] S.J. Cyvin, *Molecular Vibrations and Mean Square Amplitudes*, Elsevier, Amsterdam, 1968.

[58] K. Kuchitsu, in *MTP International Review of Science, Molecular Structure and Properties*, Ed. G. Allen, Medical and Technical Publ. Co., Oxford, 1972.

[59] K. Kuchitsu and S.J. Cyvin, in *Molecular Structures and Vibrations*, Ed. S.J. Cyvin, Elsevier, Amsterdam, 1972.

[60] J. Karle, *J. Chem. Phys.* **22**, 1246 (1954).

[61] Y. Morino and E. Hirota, *J. Chem. Phys.* **28**, 185 (1958).

[62] J. Karle, *J. Chem. Phys.* **45**, 4149 (1966).

[63] R.E. Kundsen, C.F. George and J. Karle, *J. Chem. Phys.* **44**, 2334 (1966).

[64] I. Hargittai and J. Brunvoll, *J. Mol. Struct.* **44**, 107 (1978).

[65] J. Karle, *J. Chem. Phys.* **59**, 3659 (1973).

[66] A. Almenningen, O. Bastiansen, L. Fernholt and K. Hedberg, *Acta Chem. Scand.* **25**, 1946 (1971).

[67] M.M. Gilbert, G. Gundersen and K. Hedberg, *J. Chem. Phys.* **56**, 1691 (1972).

[68] K. Hagen and K. Hedberg, *J. Am. Chem. Soc.* **95**, 1003 (1973).

[69] K. Hagen and K. Hedberg, *J. Am. Chem. Soc.* **95**, 8266 (1973).

[70] K. Hagen, V. Bondybey and K. Hedberg, *J. Am. Chem. Soc.* **99**, 1365 (1977).

[71] A.H. Lowrey, C.F. George, P. D'Antonio and J. Karle, *J. Mol. Struct.* **53**, 189 (1979).

[72] L.R. Maxwell, S.B. Hendricks and V.M. Mosley, *Phys. Rev.* **52**, 968 (1937).

[73] L.R. Maxwell and V.M. Mosley, *Phys. Rev.* **55**, 238 (1939).

[74] P.A. Akishin, N.G. Rambidi and V.P. Spiridonov, High temperature electron diffraction by gases, in *The Characterization of High Temperature Vapours*, Ed. J.L. Margrave, Wiley, New York, 1967.

[75] L. Schäfer, *Appl. Spectrosc.* **30**, 123 (1976).

[76] D. Kohl, *Transactions of American Crystallographic Association* **13**, 31 (1977).

[77] P.G. Mahaffy, J.D. Wieser and L.K. Montgomery, *J. Am. Chem. Soc.* **99**, 4514 (1977).

[78] Q. Shen and K. Hedberg, *5th Austin Symposium on Gas Phase Molecular Structure*, p. 25 (1974); Q. Shen, Ph.D. Thesis, Oregon State University (1974).

[79] V. Schomaker and R. Glauber, *Nature* **170**, 290 (1952).

[80] R. Glauber and V. Schomaker, *Phys. Rev.* **89**, 667 (1953).

[81] M. Born, *Zeits f. Physik* **38**, 803 (1926).

[82] N.F. Mott, *Proc. Roy. Soc. (London)* **A127**, 658 (1930).

[83] R.A. Bonham and L. Schäfer, Section 2.5 in *International Tables for X-Ray Crystallography*, Vol. IV, Ed. J.A. Ibers, Kynoch Press, Birmingham, 1974.

[84] H.L. Sellers, L. Schäfer and R.A. Bonham, *J. Mol. Struct.* **49**, 125 (1978).

[85] L.S. Bartell, *J. Chem. Phys.* **63**, 3750 (1975).

[86] E.J. Jacob, H.B. Thompson and L.S. Bartell, *J. Mol. Struct.* **8**, 383 (1971).

[87] E.J. Jacob and L.S. Bartell, *J. Chem. Phys.* **53**, 2231 (1970).

[88] G. Gundersen, K. Hedberg and G. Strand, *J. Chem. Phys.* **68**, 3548 (1978).

[89] *Stereochemical Applications of Gas-phase Electron Diffraction*, Parts A and B, Eds I. Hargittai and M. Hargittai, VCH Verlagsgesellschaft, Weinheim, 1988.

[90] O. Toeplitz, *Rend. Circ. Mat. Palermo*, 191 (1911).

[91] J. Karle and H. Hauptman, *Acta Cryst.* **3**, 181 (1950).

[92] D. Harker and J.S. Kasper, *Acta Cryst.* **1**, 70 (1948).

[93] I.L. Karle and J. Karle, *Acta Cryst.* **16**, 969 (1963).

[94] J. Karle and I.L. Karle, *Acta Cryst.* **21**, 849 (1966).

[95] J. Karle, *Angew. Chem. Int. Ed. Engl.* **25**, 614 (1986).

2

Some reflections on the modeling of molecules for structure analysis by gas-phase electron diffraction

Kenneth Hedberg

Oregon State University, Corvallis, OR, USA

INTRODUCTION

The gas-phase electron-diffraction experiment is inherently very simple. A well columnated beam of monochromatic electrons is caused to intersect a jet of gas at an angle of 90° in a chamber under high vacuum. The scattered electrons form a cone-like pattern symmetrically disposed about the undiffracted beam and are recorded, traditionally at any rate, by impinging on a photographic plate positioned perpendicular to the undiffracted electron beam. Thus, what is needed to carry out such an experiment is a vacuum chamber fitted with a mechanism for positioning photographic plates, a pumping system capable of maintaining a pressure of 10^{-5}–10^{-6} torr, an electron gun, a stable high-voltage (40–60 kV) power supply for electron acceleration, a suitable gas nozzle, and various auxiliary apparatus. Although construction of such a system is not a task to be undertaken lightly, once done, the experiments can be carried out rapidly and efficiently.

The diffraction pattern generated from the experiment just described comprises a set of diffuse rings that reflects the molecular structure of the scattering substance, superimposed on a largely featureless background that declines steeply with increasing scattering angle. The early years following discovery of the method, say 1930–1940, quite naturally saw it applied to simple molecules—ones with one to three shape parameters—whose structures could be reliably determined with the analytical procedures available at the time. These procedures were non-instrumental; they involved an interpretation of the appearance of the pattern based on caliper measurements of the apparent maxima and minima of the diffuse rings, and a comparison of the interpretation with molecular intensity functions calculated according to the so-called Wierl equation

$$I(s) = k \sum_{i \neq j} Z_i Z_j \frac{\sin r_{ij}s}{r_{ij}s} \tag{1}$$

In the equation the Zs are the atomic numbers of the scattering pair, $s = 4\pi\lambda^{-1}\sin(\theta/2)$ where θ is the scattering angle, and the rs are the interatomic distances.

The equation contains no recognition of molecular vibration, but it was adequate for the quality of the data. Models of the molecule, defined by parameters such as bond lengths and bond angles, were constructed and the corresponding set of interatomic distances calculated. The decision about the structure was made on the basis of the set of rs that led to the best agreement between the calculated function and the appearance of the pattern. Because the experiment was short, the calculations simple, and the results of unprecedented accuracy, electron diffraction was the method of choice for determination of the structures of free molecules. Indeed, the period immediately before the Second World War saw literature reports of the measurements of hundreds of molecules, a large number of which, incidentally, were authored by Linus Pauling and his students at the California Institute of Technology.

The modern electron-diffraction method differs from the one just described in many respects. Although the photographic plate is still the predominant medium for recording the diffraction pattern, other experimental improvements, such as use of a rotating sector that allows microdensitometric measurement instead of visual estimates of the scattered intensity, have greatly increased the precision of the measurements. This increased precision has made possible the investigation of other aspects of structure, such as intramolecular motion, that were effectively denied to the early workers. At the same time, the improved experiment has made necessary the use of a more precise scattering theory and more powerful computational procedures. Nowadays, microphotometry of the photographic plates yields digitized data consisting of a molecular-structure-sensitive part and a background modified by the rotating sector. Reduction of these data is designed to extract the molecular structure part of the total intensity in a form convenient for computational analysis. This form differs slightly from laboratory to laboratory. That used at Oregon State University is the following,

$$sI_m(s) = k \sum_{i \neq j} \frac{A_i(s)A_j(s)}{r_{ij}} \cos \Delta\eta_{ij}(s) e^{-l_{ij}^2 s^2/2} \sin s(r_{ij} - \kappa_{ij}s^2) \qquad (2)$$

where l is a root-mean-amplitude of vibration, η a phase factor, κ an anharmonicity constant, A a modified electron-scattering amplitude, and r and s are as described for equation (1). Although the process is much more involved than in the early days, the determination of a structure is still no more than the choice of a model, defined by a convenient set of internal coordinates, whose corresponding set of interatomic distances (the r_{ij}s) and vibrational amplitudes (the l_{ij}s) gives a theoretical scattered intensity in better agreement with the observed intensity than other models.

As is evident from equations (1) and (2), the scattered molecular intensity appears as a superposition of damped sine waves of different frequencies—the rs. The question of accuracy in the measurement of the distances is thus connected to the resolution of these terms in the recorded pattern, which in turn depends on (1) the angular range of the experimental data, (2) the phase differences ($\Delta\eta$), (3) the differences between the rs, and (4) the magnitudes of the associated vibrational amplitudes. The accuracy question is especially complicated by the interplay between the vibrational amplitudes that damp the individual terms of the intensity, and the distance differences. This interplay is such that it may not be possible to determine one without assumptions about the other. For example, based on data that extend to $s = 30\ \text{Å}^{-1}$, the intensity from two distances of equal weight differing by about 0.1 Å may be fit very well by a single term representing their average value modified by a damping factor. Obviously, since the experiment does not yield the answer in such a case, the conclusion one reaches about the structure would depend upon the assumption(s) made.

The electron-diffraction method of today is enormously more powerful than the pre-war one, and successful structure analyses are routinely carried out on molecules of complexity unimaginable to the early workers. The instrumental and computational advances mentioned above lead to results that are an order of magnitude greater both in the number of measurable parameters and in the precision of their values. These advances are not without increased difficulties. In general, only a fraction cf the total number of parameters in complex structures is measurable and assumptions about the remainder are required. It is this matter that comprises the subject of this chapter. As one whose research interest lies in this field, I have become increasingly concerned with the limitations inherent in the results reported from electron-diffraction investigations. It is not that these limitations are hidden; it is that they are often lost, forgotten, or at best are inadequately recognized in the report. The following discussion of these matters is aimed both at the investigator and at the reader. It is a plea to the former to explore more thoroughly the ramifications of assumptions about the models to which his results apply, and to make these clear not only in the main text but in the abstract—after all, it is the content of the abstract that draws the attention of most readers. It is also a warning to those who wish to make use of the reported results. Although most nonexperts cannot be expected to understand the structural consequences of erroneous or limiting assumptions in the investigation, they should be alert to the fact that assumptions play an important role in defining the results. Caution appropriate to the intended use of such results is indicated. In the remaining pages I will discuss some of these matters.

CONSEQUENCES OF ERRONEOUS ASSUMPTIONS

It is obvious that a model incorporating assumptions is to that extent restricted. It is not that assumptions are inherently bad—in fact they are necessary—it is rather the importance of the restrictions they impose on the model that should be recognized. Possible errors in some quantities implicitly assumed to be correct, such as the electron scattering amplitudes and phases used in the calculations, generally have only minor influence on the structural results. Errors in other assumptions, however, can have a substantial effect on the derived results, and some may be far enough off the mark to lead to grossly incorrect conclusions. Following are four examples that illustrate these points.

1. Assumption: the identity and purity of the sample are known

In most structure studies by electron diffraction it is implicitly assumed that the identity and purity of the sample are known. It is also usually assumed that the results of the investigation will reveal any problem arising from misidentification of the sample or the presence of a large amount of an impurity. This is not necessarily true. Suppose one knows nothing about the structure of nitrosyl fluoride (NOF) and does an experiment on a nitrogen dioxide (NO_2) sample thought to be nitrosyl fluoride. (In this example the sample may be regarded as grossly impure.) The data in this case [1], handled by the Oregon State procedures [2], yield a preliminary experimental radial distribution curve with two strong peaks whose positions provide estimates of the average length of a pair of bonds presumed to the be N—F and N=O, and of the F—N=O bond angle in the presumed sample of nitrosyl fluoride. The usual least squares refinement of this model based on intensity data leads to the results given in Table 1. They are, of course, quite different from the correct results [3], but nevertheless they give an excellent fit to the data as is seen in the comparison of the corresponding theoretical radial distribution curve (Fig. 1) with the 'experimental' one. None of this is surprising to the expert, who knows that the FNO and NO_2 molecules have similar scattering powers, and he might argue that the similarity of the bond lengths is unreasonable enough to awaken suspicion. Nevertheless, the point remains: it is possible to obtain an excellent fit for a wrong molecule and thus to deduce completely wrong parameter values for it. Clearly, in less extreme cases, such as a sample contaminated with an impurity, the results will be perturbed in an unpredictable way if the nature of the impurity goes unrecognized.

The lesson: *Electron diffraction is not a good analytical tool and thus impurities will not necessarily manifest themselves in the course of a structure analysis. Ignorance of impurities may lead to large errors in parameter values.*

2. Assumption: the most accurate parameter values of a structure correspond to the model yielding the smallest value of the quality-of-fit factor, R [4]

The determination of a structure usually proceeds through a series of least squares refinements of different models. It is common to select the best model, i.e. the most accurate result, by comparison of the magnitudes of R. Although this often gives a true result, when restrictions are imposed on some parameters the results for the refined parameters must be viewed with caution. An illustration of this point is

Table 1 — Results for FNO fitted to intensity from NO_2[a]

	$r_a/Å$	$l/Å$
$\langle N—F, O\rangle$[b]	1.197 (2)	
$\Delta(N—F, O)$[c]	−0.005 (11)	
$\angle F—N=O$[d]	134.3 (8)	
$N=O$[d]	1.199 (4)	0.030 (5)
$N—F$[d]	1.195 (7)	0.057 (10)
$F·O$[d]	2.206 (6)	0.59 (5)
R[e]	0.094	

[a] Quantities in parentheses are estimated 2σ uncertainties. [b] Number average of the bond lengths. [c] Equal to $r(N—F) − r(N=O)$. [d] Correct experimental values are $r(N=O) = 1.136$ (5) Å, $r(N—F) = 1.512$ (5) Å, $r(F·O) = 2.181$ Å, $\angle F—N=O = 110.1$ (5)°; see ref. [3]. [e] Ref. [4].

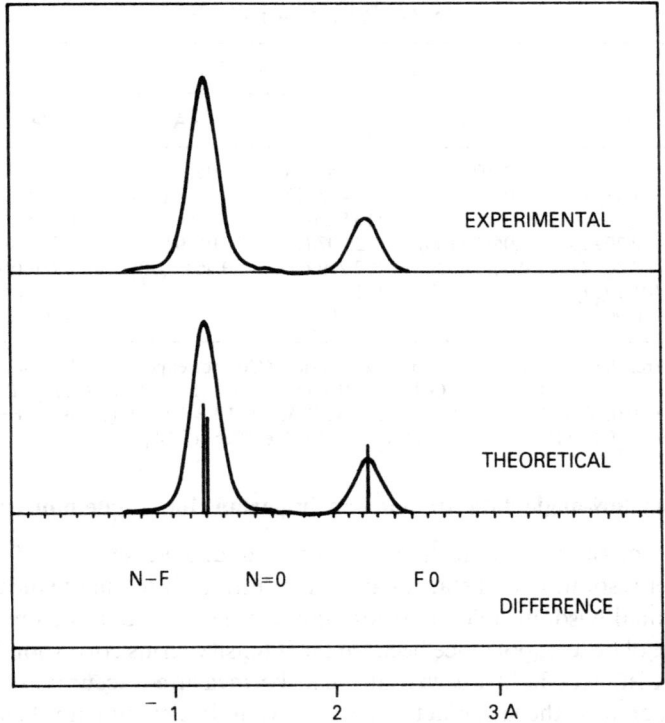

EXPERIMENTAL

THEORETICAL

N–F   N=0   F O

DIFFERENCE

-1 2   3 A

Fig. 1. — Radial distribution curves showing the fit obtainable from a wrong sample. The experimental curve is for NO_2. The theoretical curve is for nitrosyl flouride (FNO) refined to fit the NO_2 data. The FNO model is that of Table 1. The difference curve is experimental minus theoretical.

provided by some results for ONF_3. In this test an experimental intensity curve was constructed as described for the $FNO—NO_2$ system discussed above [5]. Least squares adjustment of all parameters leads to the results listed as model 1 in Table 2. Restriction of the $O \cdot F$ amplitude to a value near one of the extremes of the calculated uncertainty leads to the results given as model 2. Notable for model 2 is an $F \cdot F$ amplitude well outside the range obtained for it in the general refinement. If instead the $F \cdot F$ amplitude is restricted in like fashion, the resulting $O \cdot F$ amplitude is similarly changed (model 3). The quality of fit obtained with each of these models is essentially identical and in each case excellent, yet the models differ in important respects. If for some reason one of the restricted refinements were to have been the only one carried out, the values of the nonbond amplitudes would be substantially in error. It is worth pointing out that the uncertainties attached to the $O \cdot F$ and $F \cdot F$ amplitudes from the restricted refinements are *smaller* than those from the unrestricted one despite the fact that less has been learned about the structure. The implication is that the size of an uncertainty and the precision of the corresponding parameter determination may have nothing in common.

The lesson: *The uncertainties attached to refined parameters may be unrealistic if other parameters have been restricted to assumed values.*

Table 2 — Results for ONF_3[a]

	Model 1[b]		Model 2		Model 3	
	$r_a/Å$	$l/Å$	$r_a/Å$	$l/Å$	$r_a/Å$	$l/Å$
N=O	1.158 (3)	0.0307 (31)	1.157 (2)	0.0307 (31)	1.158 (2)	0.0308 (31)
N—F	1.430 (2)	0.0511 (26)	1.428 (2)	0.0510 (26)	1.430 (5)	0.0511 (26)
∠F—N=O	117.0 (3)		116.8 (6)		116.7 (5)	
O·F	2.209 (5)	0.0657 (100)	2.207 (8)	$[0.0500]^c$	2.207 (6)	0.0486 (28)
F·F	2.207 (4)	0.0497 (53)	2.209 (10)	0.0631 (39)	2.210 (9)	$[0.0650]^c$
∠F—N—F	101.1 (3)		101.3 (7)		101.3 (6)	
R^d	0.068		0.069		0.068	

[a] Quantities in parentheses are estimated 2σ uncertainties. [b] Correct experimental values are $r(N=O)$ = 1.158 (4) Å, $r(N—F)$ = 1.431 (3) Å, $r(O·F)$ = 2.214 (13) Å, $r(F·F)$ = 2.206 (16) Å, $l(N=O)$ = 0.0289 (44) Å, $l(N—F)$ = 0.0507 (32) Å, $l(O·F)$ = 0.0584 (89) Å, $l(F·F)$ = 0.0551 (77) Å, $\angle F—N=O$ = 117.1 (9)°, $\angle F—N—F$ = 100.8 (11)°; see ref. [5]. [c] Assumed value. [d] See ref. [4].

3. Assumption: low-angle data are not very important for simple molecules

The effect of missing data at higher scattering angles, say $s \geqslant 20 \text{ Å}^{-1}$, is well known — lower resolution of distances and larger uncertainties in the measured values. The conventional wisdom holds that for simple molecules, data at small angles, e.g. $s \leqslant 5 \text{ Å}^{-1}$, are of little importance because the intensity terms corresponding to bonds and geminal distances hold up well and can be measured accurately at the higher angles. Although true, the latter fact is to some extent irrelevant if it is believed that the molecule may be more complicated than it actually is. The following example is drawn from our own experience.

Chromium tetrafluoride, CrF_4, has the shape of a regular tetrahedron and a bond length of 1.706 (2) Å [6]. Because of the known tendency of chromyl fluorides to polymerize, the experiments were carried out at a high temperature to minimize the presence of possible polymers. The original set of data encompassed the range $6.00 \leqslant s/\text{Å}^{-1} \leqslant 33.75$, from which was obtained a radial distribution curve similar to the leveled one in Fig. 2; i.e. one with more peaks than the two expected for a tetrahedral molecule. As it turned out, these extra peaks could be fit rather well by a dimer of CrF_4 with a Cr—Cr bond, as is seen by the theoretical and difference curves of Fig. 2. How can peaks corresponding to a set of non-existent distances appear in an experimental radial distribution curve? The answer lies in the use of theoretical data from a dimer model for the missing low-angle experimental data; such data are routinely used to level the radial distribution curve and make its interpretation easier. Since the dimer of CRF_4 just described seemed most unlikely, an additional set of data over the low-angle range $2.00 \leqslant s/\text{Å}^{-1} \leqslant 13.75$ was collected and indeed ruled it out.

The lesson: *Prediction of a structure can be self-fulfilling if the analysis is not based on a full range of data.*

4. Assumption: the successful convergence of a least squares refinement leads to the correct structure

The process of structure refinement based on electron-diffraction data is termed 'non-linear least squares', a term that recognizes the non-linear dependence of the intensity

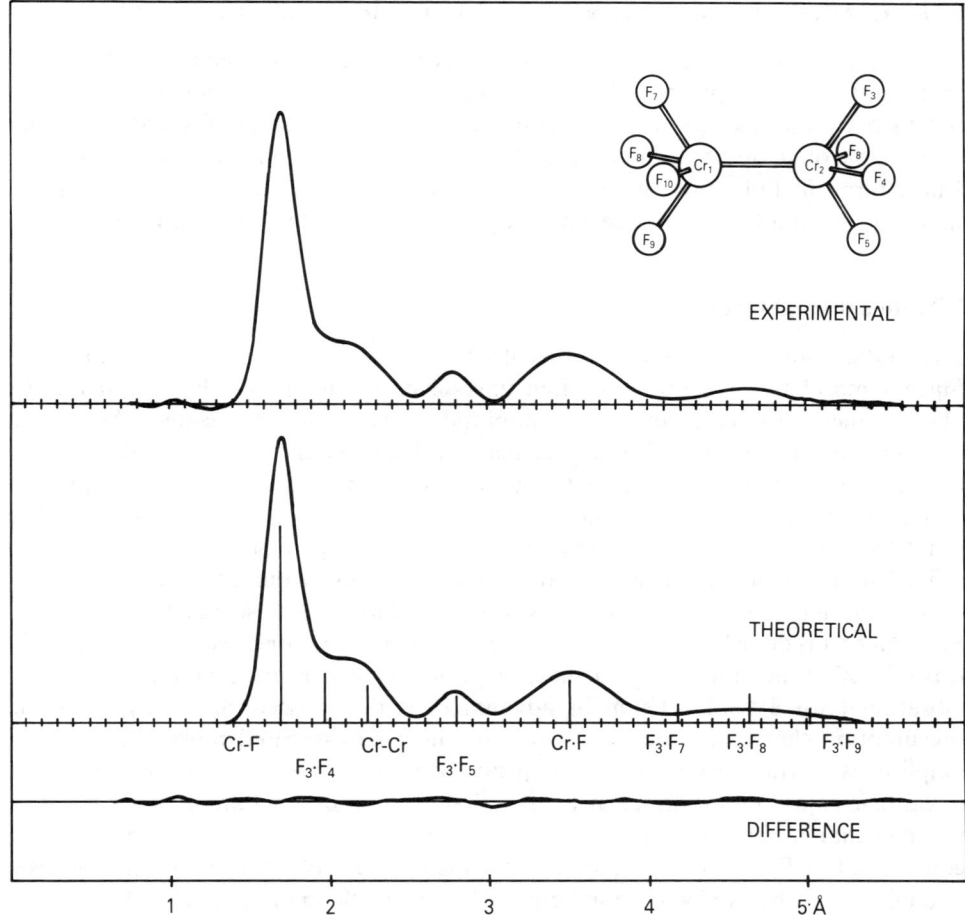

Fig. 2. — Radial distribution curves showing the fit obtainable with a wrong model when low-angle data are missing. (The correct structure is that of a monomeric CrF_4 molecule and has T_d symmetry.) The vertical bars indicate the weights and positions of distances in the hypothetical dimer.

function on the structural parameters. The mathematical process, however, is a linear one, made so by familiar procedures [7], that leads at each cycle to parameter shifts that should improve the agreement between the experimental and calculated intensities. It is clear that if the parameter space is multi-dimensional, there is a good chance for the existence of multiple minima. Accordingly, the result one obtains may depend upon the starting model and could well correspond to an incorrect structure—a so-called 'false minimum'.

The lesson: *An incomplete exploration of parameter space may lead to wrong conclusions about the structure.*

STRUCTURAL CONSEQUENCES OF COMMON ASSUMPTIONS

The examples of the previous section concern the consequences of obviously inappropriate assumptions. Although these examples may be thought unrealistic, having been simplified for the sake of illustrating the points, situations that are similar in principle often do occur in practice. A more common circumstance is that to which I now turn: the limitations imposed on structural results through use of reasonable, necessary assumptions about certain properties of the system being investigated.

Vibrational amplitudes

Each interatomic distance in a molecule has an associated vibrational amplitude. Since some of the distances are often unresolved (as in the ONF_3 case discussed above), these distances and their amplitudes are highly correlated. Attempted simultaneous refinements of these parameters will almost always fail. It is the practice in such cases, therefore, to restrict some of the parameters, usually amplitudes, to assumed values, which permits successful refinement of the remaining quantities. The point to be noted is that the difference between, say, a pair of unresolved distances refined under these circumstances, and thus the values of the distances themselves, depend intimately on the amplitude assumptions. How are these assumptions made? Sometimes it is done by adopting experimental values from other work. More recently it is becoming the practice to calculate amplitude values from an assumed quadratic vibrational force field. The main advantage of this procedure is based on the presumption that quadratic force constants may be estimated more reliably than amplitudes, particularly if they are adjusted to fit any observed wave numbers; this advantage gains in importance by the fact that most amplitudes are not very sensitive to small changes in force constants. (This is not the case, however, for some bending constants. For PF_5, two sets of force constants for the e′ block, each of which gives a good fit to the observed wave numbers, lead to very different calculated values for the nonbond amplitudes [8]. A somewhat similar situation exists for SOF_4 [8].) Another advantage is that a complete set of internally consistent amplitudes is obtained, which permits useful comparisons with experimentally evaluated ones. With this apparently sound approach one might expect reliable results for the interatomic distances affected by the amplitude assumptions. The difficulty lies in the observation that even in cases where distance resolution plays no role, i.e., CO_2, the calculated amplitude values may sometimes lie outside the uncertainty range of the measured ones. It is thus obvious that the imposition of calculated amplitude values on a system, ones that may be inconsistent with the 'natural' ones, will lead to wrong values for the associated distances.

Vibrational anharmonicity

Since electron diffraction leads to distance values averaged over vibrational state populations, the anharmonicity of molecular vibration must affect these values; the theory of their effect on the scattered intensity is seen in the quantities κ_{ij} in equation (2). For bonds exclusive of those involving a hydrogen atom, κ has a value of about 2×10^{-6} Å^3 and thus the correction to r, averaged over data in the range $0 \leqslant s/\text{Å}^{-1} \leqslant 30$, is negligible, When the upper part of the data range nears 40 Å^{-1}, however, the distance correction approaches the magnitude of the uncertainty returned by the refinement. (Bonds linking a hydrogen atom have much larger κs, but the small weights of hydrogen-containing terms in the scattered intensity make any corrections small compared to the uncertainties of the measurements.) What about the nonbond terms between heavy atoms? The anharmonicities associated with these terms have generally been ignored, partly because little is known about them, and partly because it was felt that the damping caused by their larger amplitudes tends to render them negligible. Although neglect of nonbond anharmonicity is usually justified, there are some cases when it may be important to take account of it. The reason may be seen in the formula $\kappa = al^4/6$. If the term has high multiplicity (several identical rs), with high consequent scattered intensity, a large l value can lead to appreciable corrections to distances. Cases in point are experiments done on molecules at high temperatures [9], and, occasionally, even experiments on room-temperature molecules [10].

Weighting of the data

The refinement of structures based on intensity data requires the use of suitable weights, the determination of which has not received adequate attention. It is usual to use a unit matrix (P in the expression $V'PV \rightarrow min$ where V is the matrix of residuals), but others designed to deemphasize data at the lower angles where inelastic scattering contributes importantly, or at high angles where the coherent signal becomes weak, have also been used. Other factors, such as the form of the intensity function being used, also influence the choice of weighting. How does this choice affect the structural results? Since the intensity terms for nonbond distances separated by several bonds tend to be damped quickly because they have large vibrational amplitudes, information about these distances is obtained mostly at small scattering angles. Conversely, although bond-distance information is found at small as well as large scattering angles, increased precision in the bond-distance measurements is obtained from the high-angle data. Weighting that emphasizes the small-angle region over the large tends to degrade the precision of the bond measurements, whereas emphasis of the high-angle data tends to degrade information about large amplitude motion found in the low-angle region. Further, the more precise values of bond lengths obtained by the latter weighting may require attention to the effects of bond anharmonicity.

'Shrinkage'

This term was coined by Bastiansen and Trætteberg [11] to describe the effect of vibrational averaging on interatomic distances. This effect causes distances from electron-diffraction measurements to differ from their equilibrium values so that, for example, for carbon tetrachloride with T_d symmetry the apparent bond angle is less than the tetrahedral value. The phenomenon may be easily understood by consideration of the molecular vibrations of the carbon dioxide molecule which has a linear equilibrium arrangement of the three atoms. In the harmonic approximation, the two stretching modes do not affect the prediction $\langle r(O \cdot O) \rangle = 2\langle r(C=O) \rangle$; however, all displacements of the degenerate bending mode, whether or not these are viewed as the usual linear movement perpendicular to the bonds or as a curvilinear motion, lead to $\langle r(O \cdot O) \rangle < 2\langle r(C=O) \rangle$. Typical values for the distances in carbon dioxide are

$r_a(C-O) = 1.165$ Å and $r_a(O \cdot O) = 2.324$ Å; in this case the r_a shrinkage is equal to $2 \times 1.165 - 2.324 = 0.006$ Å. Shrinkage is a general phenomenon, but is not evident when the number of interatomic distances is equal to the number of geometrical parameters required for model specification in the known symmetry. The bent triatomic molecule is such a case. Here, the shrinkage may be regarded as the difference between the nonbond distance predicted from the measured bond lengths with the equilibrium bond angle, and the measured nonbond distance; however, since the equilibrium angle is not known and all the distances can be adjusted as independent parameters, the measurements will reflect the best individual fits. Shrinkage does become important when the measured distances violate the known equilibrium molecular symmetry. The question then is, when a model is formulated, what is the consequence of simultaneously assuming a molecular symmetry and at the same time neglecting the effect of shrinkage? Since there are more distances than geometrical parameters, the adjustment will be a compromise that is determined by the weights of the distance terms. In the case of carbon tetrabromide where the dominant term is Br $\cdot$ Br, the main error will occur in the value of $r_a(C-Br)$, but with silicon tetrafluoride, the larger error would be in the value of $r_a(F \cdot F)$. The size of the errors in any given case is harder to estimate, since it depends on many things—for example, the relative weights of the terms, the weighting of the data, and magnitude of the vibrational amplitudes.

In order to take account of shrinkage effects, use is made of the fact that the time-average *positions* of the atoms in a molecule undergoing molecular vibration will conform to the equilibrium symmetry of the molecule even though the average interatomic distances do not. (In the carbon dioxide example cited above, for example, the average positions of all the atoms lie on a straight line.) The distances between the atomic positions are designated r_α, and of course conform to the equilibrium symmetry. Since the electron-diffraction analysis returns the r_a-type distance consistent with equation (2), specification of the structure in terms of r_α-type distances (i.e., making use of the equilibrium symmetry) requires the corrections $r_a - r_\alpha$ given by equation (3):

$$r_a - r_\alpha = \delta n + K - \frac{l^2}{r} \tag{3}$$

Unfortunately, although the amplitudes (l) may sometimes be estimated with good accuracy, the centrifugal distortions (δn) and the perpendicular amplitude corrections (K) in general cannot; they must be calculated from a harmonic force field. Moreover, there are some circumstances, such as when large-amplitude torsional motion occurs, where the calculated values of the Ks are inappropriate. A further complication is that the force fields themselves are not usually available and, as stated earlier, the force constants must be estimated from other work. The result of these considerations is again a structure that contains within it the uncertainties built into the force field used for the corrections.

Use of rotational constants as contraints

An important development in gas-phase electron-diffraction studies is the use of rotational constants, usually from microwave spectroscopy, as structural constraints.

Such a use most often occurs when the number of rotational constants from parent and isotopic species is insufficient for a complete structure determination from the microwave data alone. Since both the accuracy and the precision of the rotational constant measurements are very high, the electron-diffraction results must be consistent with them. The problem that arises is two-fold. First, the rotational constants are usually (ground state) B_0, so that the thermal average distances from electron-diffraction (r_a) need correction to a distance type that can be made consistent with the Bs. This turns out to be r_α^0, or r_α at 0 K. Second, extraction of distances from the B_0s leads to the quantities $\langle 1/r^2 \rangle^{-1/2}$, which are not the same as $\langle r \rangle$. In order to fit simultaneously both rotational constants and electron-diffraction intensities the B_0 rotational constants must be converted to B_z, from which are derived a distance type symbolized by r_z; r_z is essentially identical to r_α^0. The connection between the two types of r is given by the formula

$$r_a^T = r_\alpha^0 + (3a/2)[(l^2)^T - (l^2)^0] + \delta r^T + K^0 - \frac{(l^2)^T}{r} \tag{4}$$

and the two types of B by

$$B_z = B_0 + \sum_i \alpha_i^{har}/2 \tag{5}$$

All these quantities except a, the Morse anharmonicity constant, are readily obtained from calculations based on a harmonic vibrational force field. It is obvious that the quality of the results from this type of 'joint' analysis is, once again, affected by the choice of vibrational force field and, therefore, by the assumptions that figure in that choice.

A surprising feature of the use of rotational constants in the way just described is that the structure analysis is not always aided by use of more than one set of rotational constants. The reason lies in isotope effects that are not easily introduced. If this matter is ignored and rotational constants from several isotopic species are included as constraints, substantial errors may result. Another aspect of the problem concerns the relative weighting of the spectroscopic and diffraction data. Since this is connected to the confidence one has in the many necessary corrections outlined above, and to the different types, amounts, and precisions of the data, it is difficult to devise a weighting scheme appropriate to all occasions. We have found that a relative weighting $\Sigma w_i(B_{obsd,i})^2/\Sigma w_i(s_i I_{m,i,obsd})^2$ equal to 200–300 gives satisfactory results.

Multiple scattering

Equation (2) is based on the presumption that the participating scattering centers in the molecule need be considered only in pairs. Under certain circumstances, however, there is a substantial additional contribution from three-atom (multiple) scattering. The problem has been considered by Bartell [12], Kohl [13], and their coworkers. These authors assert that the effect on the values of distances and amplitudes derived from use of equation (2) ignoring multiple scattering is negligible except when three atoms of the molecule are arranged at an angle very close to 90° or 180°. However, when these conditions exist (examples are octahedrally coordinated molecules) the effect can be appreciable, particularly on the values of the vibrational amplitudes.

Large-amplitude motion

The problems that arise in connection with vibrations of reasonably small amplitude have been discussed in preceding sections. When the motion has large amplitude, however, the harmonic vibration approximation adequate for treatment of small-amplitude motion becomes unsatisfactory. Typical examples are molecules in which internal rotation plays an important role. It is the practice in such cases to view the system as a set of rotational pseudoconformers, each characterized by a different value of the torsion angle and each undergoing 'frame' vibration as if it were an otherwise normal molecule locked at its particular torsion angle. The pseudoconformers are usually defined at appropriate intervals of the torsion angle ϕ. In addition to the usual structure and amplitude parameters, the shape of the torsional potential $V(\phi)$ can be investigated. This is done by finding a weighting scheme for the pseudoconformers that yields the best fit to the intensity. The weights are assumed to follow a Boltzmann distribution $(w \sim \exp[-V(\phi)/RT])$ and, after assumption of a functional form for $V(\phi)$, may be obtained by adjusting the constants of the potential function. Several problems present themselves. One is the form of $V(\phi)$: experience shows that it is often possible to obtain satisfactory fits with different functions. Another is the necessity of adopting assumptions about the structures of the pseudoconformers: since the total number of distance and amplitude parameters is proportional to the number of pseudoconformers, the difficulties of resolution present with one conformer are many-fold greater. The matter is usually handled by the assumption that the structures of the pseudoconformers differ only in the values of their torsion angles. Still a third problem concerns the calcuated quantities, such as certain vibrational amplitudes that comprise a part of the frame vibrations mentioned above; these are in principle different for each pseudoconformer. Fortunately, estimates of the changes connected to vibration that occur with changes in the torsional angle may be had from normal coordinate calculations. In summary, the investigation of large-amplitude motion requires a number of assumptions that affect all aspects of the structural conclusions.

Symmetry

I have left to the last a topic that is arguably the most important of the many assumptions that figure in the process of structure analysis by electron diffraction: the assumption of molecular symmetry. It cannot be overemphasized that this process involves the fitting of a *model* of the structure *that has been formulated by the investigator*; the process does not itself produce the model. The fitting procedure, usually least squares based on intensity data, will return the best parameter values within the constraints imposed by the assumed model type. It is obvious, but nevertheless often overlooked, that *the successful fit of a model is not a proof of the assumed symmetry*. It is fortunate that in most cases there exist data from other sources that point toward, say, the most likely molecular symmetry and to approximate values for the bond lengths and bond angles; still, there are often cases in which the diffraction data are quite consistent with certain distortions, usually small, from a higher to a lower symmetry.

SUMMARY AND RECOMMENDATIONS

There are many who have no special expertise with the method of structure determination by gas-phase electron diffraction, but who make use of the results from it. For such people it would seem important to have some understanding of the limitations imposed on those results by the constraining assumptions that are always invoked by the investigator. The preceding discussion was meant to draw attention to the many assumptions that play a role in the derived values of the parameters and to give some hints about the effects that may result from errors in those assumptions.

There are also the experts. It has been my observation that although the structural limitations imposed by assumptions may be understood by experts in the method, the consequences of errors in assumptions are seldom mentioned. (Even worse, the true situation is often masked by careless or inappropriate statements such as, for example, one that suggests the molecular symmetry has been 'determined'.) The investigator has the responsibility to note in article abstracts or summaries the use of any assumptions that limit the structural conclusions in an important way. Further, if uncertainty attaches to some of these assumptions, the effect of changes in them should be investigated. Descriptions of these and other conceivably important assumptions should be available in the main text so that, even if their consequences have not been explored, other experts may draw on their experience for estimates of effects.

It is not possible to investigate a structure without the adoption of some assumptions about it. The question is, then, given this fact, how can such a study be carried out to provide the best possible estimates of parameter values and their associated uncertainties? The quick answer is to change the constraining quantities in a systematic way over an appropriate range and observe the effects on the measurements. Unfortunately, such an approach is not practicable for most systems—the number of assumptions is too large. It is possible, however, to test a few of the more important assumptions and this should be done whenever reasonable questions about them may be raised. Another tactic is to ensure that necessary assumptions are drawn from as firm a theoretical base as possible. Normal coordinate calculations provide better estimates of vibrational amplitudes, particularly nonbond amplitudes, than do guesses from experience. *Ab initio* calculations can be used to assign values to distances, or differences between distances, that cannot be measured; they can also be used to calculate vibrational force fields and to estimate the compositions of conformational mixtures. Molecular mechanics yields results similar to those from *ab initio* work. Even these methods, however, amount to the use of certain assumed constraints. In the end, one can do little more than to hope that the investigator has done his best and to be aware of uncertainties that may not appear in the list of results.

ACKNOWLEDGMENT

This work has grown out of many electron-diffraction investigations supported by the National Science Foundation. I am grateful to the many students and colleagues who have contributed to that work.

REFERENCES

[1] The data for this example are synthetic. The 'experimental' intensity was generated with use of equation (2) from the values $r(N{=}O) = 1.199$ Å, $r(O \cdot O) = 2.207$ Å, $l(N{=}O) = 0.0416$ Å, $l(O \cdot O) = 0.0530$ Å (Shen Q., Ph.D. Thesis, Oregon State University, 1974). Noise was added to the curve. The radial distribution curve was calculated from this curve from the equation $P(r)/r = k \sum_i (s_i I_i(s))(A_N A_O)^{-1} \exp(-0.0025) \Delta s$; data in the region $s \leqslant 2$ Å^{-1} normally unavailable experimentally were taken from a theoretical curve for FNO.

[2] G. Gundersen and K. Hedberg, J. Chem. Phys. 51 (1969) 2500.

[3] K.S. Buckton, A.C. Legon and D.J. Millen, Trans. Faraday Soc. 19 (1969) 1975.

[4] $R = [\sum w_i \Delta_i^2 / \sum w_i (s_i I_i(s)^{\text{obsd}})^2]^{1/2}$, where w is an element of the weight matrix and $\Delta_i = s_i I_i(s)^{\text{obsd}} - s_i I_i(s)^{\text{calcd}}$.

[5] The synthetic 'experimental' data (see reference [1]) were generated from the parameter values given in footnote b, Table 2. The source of the values is Plato, V., Hartford, W.D. and Hedberg, K., J. Chem. Phys. 53 (1970) 3488.

[6] L. Hedberg, K. Hedberg, G.L. Gard and J.O. Udeaja, Acta Chem. Scand. A 42 (1988) 318.

[7] See K. Hedberg and M. Iwasaki, Acta Crystallog. 17 (1964) 529 for an account of least squares applied to electron-diffraction data.

[8] L. Hedberg, J. Phys. Chem. 86 (1982) 593.

[9] L.S. Bartell and J.F. Stanton, J. Chem. Phys. 81 (1984) 3792.

[10] H. Thomassen and K. Hedberg, J. Mol. Struct. 240 (1990) 151.

[11] O. Bastiansen and M. Trætteberg, Acta Crystallog. 13 (1960) 1108.

[12] (a) L.S. Bartell, J. Chem. Phys. 63 (1975) 3750; (b) B.R. Miller and L.S. Bartell, J. Chem. Phys. 72 (1980) 800.

[13] D. Kohl and M. Arvedson, J. Chem. Phys. 73 (1980) 381.

3

The early crystallographic works of Linus Pauling

Richard E. Marsh and Verner Schomaker

The California Institute of Technology, Pasadena, CA, USA

INTRODUCTION

Pauling seems never to have had great interest in the methodology of X-ray diffraction: he did not design new equipment for recording or measuring diffraction patterns, nor did he make conscious attempts to develop new or improved methods for interpreting those patterns. Rather, he became intensely, almost totally, devoted to the structural knowledge that was gained by X-ray diffraction. In other words, he was immensely interested in the results afforded by the technique, but was quite willing to leave to others the technical improvements in applying it. His interests were in chemical structure and in the principles of chemical bonding. In fostering these interests he became expert in many areas — electron diffraction, quantum mechanics and thermodynamics as well as X-ray diffraction — but his love was in the productive (*and correct*) application of the techniques, not in their development.

It is, then, perhaps somewhat surprising that Pauling has in fact contributed enormously to the development of the methodology of X-ray diffraction. His contribution was indirect, and resulted from his superlatively brilliant use of diffraction results. The most characteristic feature here is the brilliance of his structural imagination, which he strengthened year after year by critically (this is a key word: '*critically*') incorporating the structural information being published throughout the world into his picture — his body of understanding — of the chemical (and physical) universe. Pauling acutely realized, in the early days, that X-ray diffraction was not very powerful (at that time, the 1920s) as a method for directly discovering structures [1]: there were no 'direct methods', no Patterson function. However, diffraction afforded a great amount of information that could be used to test whether any model of a crystal structure is the correct model (a situation that Pauling has often spoken and written of). One method to proceed, then, was to formulate *all* the structures that could possibly fit the observed size and symmetry of the unit cell, test them all, and reject all except the correct one. However, a crystal structure has to be quite unusually simple if it is to be solved by this method. So Pauling adopted another way: he would devise a reasonable model (in the light of his growing knowledge and intuition), and test whether it fit the diffraction data from the crystal; if it did fit, it almost certainly was correct.

Eventually, Pauling found a name for this method — *stochastic*, based on an etymology not usually given: *apt to divine the truth by conjecture* [2, 3].

He has applied this method throughout his career. Even after the introduction of the Patterson function and direct phasing methods — and the development of high-speed computers — Pauling preferred to develop structures merely by ruminating, an encyclopedia of structure in his mind and a six-inch slide rule in his hand, until he found the one obviously correct structure. Few, if any, others could apply the stochastic method with his amazing proficiency, for few others had comparable powers of three-dimensional visualization and of depth and ingenuity of insight into the principles of structure.

A large part of this chapter will be concerned with Pauling's *stochastic* method. Most of the rest will consist of a number of examples in which Pauling has used his great insight into the world of structural chemistry to make significant corrections to the crystallographic work of others. We shall concentrate on Pauling's early work — the period until about 1935 — not only to demonstrate the critical insight that he showed even then but also because of the joy we found in reviewing X-ray diffraction papers of that time. As it turns out, this comes down to commenting on nearly all the 'Papers on the Structure of Crystals' as categorized by Albrecht in his compilation of Pauling's bibliography [4] through 1935.

Some of these contributions represent substantial or even enormous sustained effort; others can properly be called incidental; but all, we feel, were important in contributing to or illustrating the fact that, one way or another, X-ray crystallography had become a much changed craft after the first decade of Pauling's post-Ph.D. research career.

At the end, we shall list Pauling's other major contributions to X-ray crystallography, to the extent that we recognise them, even though they may be much more extensively covered elsewhere in this volume.

THE STOCHASTIC METHOD

In tracing the evolution of the stochastic method as developed by Pauling, we may start with his Ph.D. thesis (1925). It has five chapters, each a reprint of a crystal-structure paper published [5–9] in the *Journal of the American Chemical Society*. Two of these papers are corrections of structures previously proposed by others; we shall dwell on them later. The remaining three are original determinations: molybdenite (MoS_2), with Roscoe Dickinson as co-author, and magnesium stannide and the isomorphous compounds ammonium fluoferrate, fluo-aluminate and oxyfluomolybdate, which he published alone. All of these determinations (in fact, one was not original: the compound he thought to be ammonium fluo-aluminate, $(NH_4)_3AlF_5$, was later shown [10, 11] to have been the fluosilicate, $(NH_4)_2SiF_6$, whose structure was already known) were carried out in the traditional fashion of the day: data were obtained from Laue and 'spectral' photographs (these were made by recording successive orders of reflection of $MoK\alpha$ radiation from a specific crystal plane — from a developed face or, if the crystal was small, 'by reflection during the transmission of the beam through the crystal'); powder photographs were often used to verify the cell dimensions. The structure was derived by placing the various atoms in sites consistent with the molecular formula, the crystal density, and the Laue symmetry; since a number of such sites were usually available, the correct choice was selected by considering the relative intensities of the various reflections. Sometimes — as in the case of Mg_2Sn, which has the fluorite structure — there were no parameters to adjust; the single parameter necessary to complete the structure of molybdenite (or of the fluo compounds) was derived by more extensive considerations of the relative diffraction intensities as a function of the parameter. These considerations were often complicated by uncertainties as to the relative scattering powers of different atoms — the 'form factors'; these were normally assumed to be proportional to the atomic numbers, but in the case of magnesium stannide, Pauling proceeded merely 'on the very safe assumption that an atom of tin scatters X-rays more strongly than an atom of magnesium'.

In this early work, Pauling clearly showed his remarkable understanding of three-dimensional patterns, and in particular showed his ability to relate these patterns to actual atom arrangements — here was the key to the stochastic method. As an example, in 1925 Hendricks and Pauling reported [12] the structures of NaN_3, KN_3 and KNCO; in arriving at the structure of NaN_3 he ruled out, on the basis of intensity inequalities, a number of arrangements of atoms, remarking: 'Since these arrangements are the only ones placing the three nitrogen atoms in a ring, such a structure cannot be assigned to the trinitride ion.' The structural implications of the atom arrangements were always foremost in his mind.

Pauling's application of the stochastic method for the derivation of a crystal structure can be traced to 1928, when he and Sturdivant published [13] their determination of the structure of brookite, the orthorhombic form of TiO_2. This was a difficult, nine-parameter problem, the atoms lying in three sets of general positions in space group $V_h^{15} - Pbca$. Establishing these parameters directly from diffraction photographs apparently seemed a hopeless task to Pauling and Sturdivant, so they attacked the problem on the basis of their accumulated knowledge of related structures.

Shortly before, W.L. Bragg, with G.B. Brown [14] and with J. West [15], had proposed that the structures of certain silicates could be understood and, in some instances, predicted on the premise that the oxygen ions, being the largest atoms in the structure, would form a close-packed array with the cations occupying interstices. Pauling and Sturdivant proposed a more restrictive 'coordination theory', based on their analysis of the known structures of rutile and anatase (tetragonal forms of TiO_2). In this theory the Ti atoms in brookite should lie at the centers of octahedra with Ti–O distances of 1.95 Å; these octahedra would 'share edges and corners with each other to such an extent as to give the crystal the correct chemical composition', and the shared edges would be shortened, to $O \cdots O = 2.50$ Å, so as to lengthen the $Ti \cdots Ti$ distance and reduce cationic repulsion.

With these conditions in mind, Pauling and Sturdivant started constructing models. They 'made no attempt to consider exhaustively the possible simple structures, but instead investigated the two which presented themselves first.' Both of these models were based on 'staggered strings' of octahedra, each octahedron sharing edges with the one above and below it in the string. In their first model, strings were joined by sharing octahedral vertices; however, this led to too small a unit cell, containing only four TiO_2 units. For their second structure, they joined strings by sharing edges;

such a model, they saw, would have approximately correct cell dimensions and would also have the symmetry of V_h^{15}.

In order to measure the atom coordinates corresponding to this model, they constructed octahedra from heavy paper; these octahedra were 'distorted' so that the O⋯O distances in the three shared edges (two edges within a string, one connecting the strings) were at 2.50 Å and the other edges 'distorted as little as possible' so as to maintain equal Ti–O distances of 1.95 Å. The octahedra were then glued together to form a full unit cell. (Pauling tells us that Mrs Pauling made the first models, sewing them together.) The dimensions of the cell were then carefully measured, and found to agree with the known dimensions within 0.04 Å. Further measurements on these models established coordinates for the three independent types of atoms; Pauling and Sturdivant noted that these coordinates differed by as much as 0.35 Å from values that would correspond to regular (undistorted) octahedra.

The next step was to test the structure; they did this by calculating structure factors ('with the assumption that the relative reflecting powers of titanium and oxygen atoms are proportional to their atomic numbers') for about 50 reflections whose intensities they had estimated from rotation photographs about the three principal axes. While the general agreement was satisfactory, they found they could improve it by adjusting the x coordinate of Ti by 0.004 — about 0.04 Å. They made no further changes. We note that the biggest difference between their coordinates and the latest ones we have found [16] is about 0.15 Å for the y coordinate of Ti; the oxygen coordinates agree within 0.07 Å. In other words, the distorted O_8 octahedra constructed by Pauling and Sturdivant were very nearly correct: the three shared edges are indeed short, all at about 2.5 Å, and the remaining edges are roughly equal at about 2.85 Å. But the Ti–O distances are *not* all equal to 1.95 Å; they range from 1.87 to 2.04 Å [16].

In this same year there appeared another prototypical paper [17] based on Pauling's stochastic method, reporting on the structure of topaz, $Al_2SiO_4(F,OH)_2$. In it, he describes how, apparently with great ease, he derived a structure that fit the composition, the unit cell dimensions, the space group, the characteristic features of the diffraction intensities (as already reported by Leonhardt [18]) — and, of course, his ideas concerning the most probable atom arrangements. This structure was based on SiO_4 tetrahedra with each vertex shared with two AlO_4F_2 octahedra, of which the F-vertices (Pauling spared himself any mention of OH substitution for F) are unshared, the anions in an approximate double hexagonal close-packed arrangement layered perpendicular to the orthorhombic b-axis. Pauling also noted that a structure for topaz 'suggested with some reserve' by Bragg and West [15] and based on Bragg's principle of close-packed anions was incorrect, 'for it involves the identification of the hexagonal axis of the close-packed structure with the a-axis rather than the b-axis of topaz, and is, moreover, simple rather than double hexagonal close packing'. Within the same year, Alston and West [19] published a revised structure, now in agreement with Pauling's and again derived from a rather laborious application of Bragg's principle of close-packing of anions.

At about this time, Pauling published a series of five water-shed papers [20-24] describing his deductions concerning the basic principles of structure of ionic compounds, including a listing of the sizes of the important ionic species and a discussion of the importance of these sizes in determining the structures and properties of crystals. The culmination was the paper [24] 'The Principles Determining the Structure of Complex Ionic Crystals'.

There were five principles:

1. 'A coordinated polyhedron of anions is formed about each cation, the cation–anion distance being determined by the radius sum and the coordination number of the cation by the radius ratio';
2. 'The electric charge of each anion tends to compensate the strength of the electrostatic valence bonds reaching to it from the cations at the centers of the polyhedra of which it forms a corner', this electrostatic bond strength being defined as the charge of the cation divided by its coordination number;
3. Edge-sharing, and particularly face-sharing, of the polyhedra decreases the stability of the structure;
4. For cations with large valence and small coordination number, even vertex-sharing is disadvantageous; and
5. 'The Rule of Parsimony', i.e. 'The number of essentially different kinds of constituents in a crystal tends to be small.'

It may seem that these principles are only a re-statement, in words, of the Born-Madelung treatment of the energy of formation of an ionic cystal from the constituent ions. They are much more than this, however, representing Pauling's structural and theoretical understanding and also his enormously rich experience in devising ionic structures to fit whatever special information was at hand — his effort of the time, we fancy, to set down a general guide for the ionic architect, the detective of structure. He didn't want merely to recite the best physical theory — he wanted a physically sound practical guide. As an eager competitor he also wanted, we are sure, an even more eloquent statement than the one so often applied by his illustrious predecessor-contemporary W.L. Bragg, the idea that the structures of ionic crystals, especially of the oxides, are determined by closest- or nearly closest-packing of the large anions, the cations occupying some of the resulting interstices. In the course of a splendid long article, 'The Structure of Silicates' [25], Bragg indeed graciously acknowledged Pauling's principles with a full exposition of them, but including a suggestion to the effect that Pauling's principles weren't so different from his own: 'In contrasting his method with that adopted by the author, Pauling has laid emphasis on the difference between them which is perhaps rather excessive; it is only a matter of convenience of description to replace the idea of regular groups of large anions around small cations with that of tetrahedra and octahedra, and of linking by sharing oxygen atoms with the sharing of corners and edges ...'. Bragg placed much the greatest emphasis on 2.), the electrostatic valence rule, and rather downplayed 1.), partly because the same polyhedron is not always formed by given anions around a given cation, even when the indication from the radius ratio would seem unambiguous. We believe, however, that Pauling had written what he wanted to write with good reason, and that he, at least, was prepared to recognize and interpret departures from the rules whenever they appeared.

It is curious that he didn't state as a principle or as a corollary to 3.) the fact that shared edges are notably shortened — to about 2.5 Å for O···O, by several tenths of an angstrom. Its illustrations in his structures are numerous. As for 5.), which of course isn't always obeyed, it says, in other words, that the most efficient arrangement of atoms tends to be repeated throughout the structure. While this is a simple concept, and seems almost intuitively obvious from the very fact that crystals exist, it was an important component of Pauling's method: once he had decided on the most probable motif for a grouping of atoms, he proceeded with confidence (and, usually, success) in repeating this motif as many times as needed to complete the structure. The Rule of Parsimony finds many references in his works.

The year 1930 brought a spate of structure determinations based on Pauling's use of the stochastic method. We review some of them:

Pseudobrookite

Here, Pauling [26] used his Principles to deduce the structure of pseudobrookite, Fe_2TiO_5 — an investigation with the special distinction that an uncertainty in the chemical composition had to be resolved as well as the structure. In the course of determining the unit cell and probable space group from Laue and oscillation photographs of a natural crystal from Aranyar Berg, Transylvania, Pauling found that his data required the composition to be Fe_2TiO_5, in agreement with the long-known composition of synthetic pseudobrookite, but in disagreement with the existing careful analyses of the mineral, both of the Transylvanian material and of the pseudobrookite from Havredal, Norway, which corresponded to the formula $Fe_4Ti_3O_{12}$. He first showed that the unit cell and space group are compatible with the formula Fe_2TiO_5 but not with $Fe_4Ti_3O_{12}$. In his report he continued, 'Since the large number of parameters precluded the rigorous deduction of the atomic arrangement from X-ray data, there was predicted with the aid of the coordination theory and our knowledge of interatomic distances a structure satisfying the previously formulated rules [24] determining the stability of ionic crystals.' He also noted, in passing, that the space group reported by Mark and Rosbaud [27] is incorrect — a result which 'is no doubt to be explained in the usual way as arising from errors in their assignment of indices to reflections on their photographs'.

From the systematic absences and the holohedral morphology, Pauling settled 'with some assurance' on space group V_h^{17}, in setting $Bbmm$. He then noted that the length of the c axis, 3.725 Å, suggested strings of TiO_5 octahedra sharing corners (rather than edges as in rutile, for the distance would then be only 3.0 Å); it was then 'seen at once that the ferric ions could be introduced in such a way that each was surrounded by four oxygen ions about 1.90 Å away.... No other structure was predicted for pseudobrookite in the course of our study, as this, the first, was found satisfactory.' Indeed, it was also found to have 'the space-group symmetry of V_h^{17}, with the correct distribution of symmetry elements among the axes.'

Pauling noted that the structure depends on eight parameters: one for Ti at $(u, \frac{1}{4}, 0)$, two for Fe at $(s, w, 0)$, one for O_I at $(u, \frac{1}{4}, 0)$ and two each for O_{II} and O_{III} at $(v, w, 0)$. He first assigned these eight parameters by assuming Ti–O = 1.91–1.95 Å, Fe–O = 1.93 Å and O⋯O $\geqslant$ 2.50 Å; these parameters were found to give good intensity agreement for 'reflections from simple planes, but not from complicated ones'. He accordingly adjusted the value of w_{Fe} on the basis of the $0k0$ reflections and then the values of u_{Ti} and v_{Fe} on the basis of the $h00$s; finally, he made small changes in the oxygen positions in order to keep the Ti–O and Fe–O distances at reasonable values. He then made 'an exacting test of the structure ... by calculating structure factors and comparing them with the estimated intensities of the numerous reflections on rotation and Laue photographs'. With obvious pride he noted that they paralleled one another 'with remarkable fidelity', making it 'highly probable that the correct structure of pseudobrookite has been found'.

The anomalous composition also was explained. The photographs showed some 'rather hazy' extra spots, which, Pauling saw, had their origin in 'a large number of small crystals with nearly but not quite the same orientation.' The interplanar distances indicated by the extra reflections showed them to be from rutile, 12–14% of which accounts for the chemical analysis.

Sodalite and helvite, $Na_4Al_3Si_3O_{12}Cl$ and $(Mn, Fe, Zn)_4Be_3Si_3O_{12}S$

The structure of sodalite contains the features of a vast class of aluminosilicate minerals, including the zeolites (both the mineral species and most of the synthetic Molecular Sieves of commerce). Indeed, the Type-A Molecular Sieve has essentially the same asymmetric unit of structure as sodalite and the Type-Y, enormously important as the basis of the vast majority of present-day petroleum cracking catalysts and structurally similar to the rare mineral faujasite, has a framework built up of the same four- and six-rings of AlO_4 and SiO_4 tetrahedra as occur in the sodalite structure.

Pauling [28] first worked out the structure of sodalite, noting three previous investigations [29–31] which agreed that sodalite crystals are cubic with cell edge a about 8.83 Å. In one of these papers, Jaeger [31] describes the structure of nosean (which is closely related to the ultramarines); while similar to sodalite, with S_x^{--} replacing Cl^-, it has a larger unit cell (9.13 Å) and, Jaeger emphasizes, a quite different diffraction pattern [31]. He even remarks, 'As will soon become clear, however, neither *sodalite*, nor the *garnets* belong, structurally, to the same group as *nosean* (*hauyne*) and the *ultramarines*.' Bragg, Claringbull, and Taylor [32] have the following introductory passage on 'Sodalite, Helvine, Ultramarine': 'The cubic crystals of these groups are all based on the structure of linked tetrahedra first discovered by Jaeger. Jaeger's analysis of ultramarine marked an important turning point in the analysis of silicates, both because it was the first framework structure to be determined, and because he showed that the framework imposed cubic symmetry upon the crystals although the strict requirements of space-group theory could not be satisfied. The latter point has now become familiar through many analyses, but Jaeger's demonstration that Al and Si replace each other to a variable extent, and that the framework includes variable groups which could hardly be supposed to have cubic symmetry, was an important step in its establishment.' At the same time, Bragg *et al.* give full, single credit to Pauling in their section on sodalite itself. Thus, although Pauling's discovery of the sodalite structure was surely original, Jaeger's work on *nosean* predated it, and Jaeger is given primary acknowledgment for the structure type.

In the sodalite framework, alternating silicate and aluminate tetrahedra share corners in such a way that the centers of the tetrahedra lie at the vertices of truncated octahedra; these octahedra are fused together by sharing square faces (oriented perpendicular to the cube axes) and also by sharing hexagonal faces (oriented perpendicular to the body diagonals). The unit cell contains two truncated octahedra, one centered at $(0, 0, 0)$ and one at $(\frac{1}{2}, \frac{1}{2}, \frac{1}{2})$, and each octahedron encapsulates a Cl^- ion surrounded by a tetrahedron of four Na^+ ions. (In Type-A zeolites the square faces are *linked*, through oxygen atoms, rather than fused; as a result, the cavity at $(\frac{1}{2}, \frac{1}{2}, \frac{1}{2})$ is much larger than in sodalite.) Pauling describes how he derived the structure of sodalite, adjusting it to fit his knowledge of ionic structural chemistry and the observed X-ray data. Pauling's further description of the sodalite structure is eloquent and instructive:

> The crystal provides a remarkable example of a framework structure. The forces between the highly charged cations Si^{4+} and Al^{3+} and the oxygen ions are by far the strongest forces in the crystal. They cause the joined tetrahedra to form a strong framework, of composition $Al_6Si_6O_{12}$, extending throughout the crystal and essentially determining its structure. Within the framework are rooms and passages, spaces which can be occupied by other ions or atoms or molecules, in this case sodium and chlorine ions. The framework, while strong, is not rigid, for there are no strong forces tending to hold it tautly expanded. In sodalite the framework collapses, the tetrahedra rotating about the two-fold axes until the oxygen ions come into contact with the sodium ions, which themselves are in contact with chloride ions. This partial collapse of the framework reduces the edge of the unit from its maximum value, about 9.4 Å, to about 8.87 Å.

Pauling went on to point out that hauynite $(Na_3CaAl_3Si_3O_{12} \cdot SO_4)$ and noselite $(Na_5Al_3Si_3O_{12} \cdot SO_4)$ must have analogous structures to sodalite, with evident substitutions, and that cation exchanges can be accomplished, e.g. by treating these minerals with fused NaCl to produce sodalite; further, that exchange with the larger cations of K, Rb, Cs causes the cell edge to increase [33] to 9.4 Å, his predicted maximum value. Helvite is a similar story, with the third-row metals replacing Na, the Be replacing Al, and S replacing the Cl of sodalite.

He did not mention Jaeger's *nosean* structure, which is essentially correct *and* similar to the sodalite structure, not even in connection with his remarks on how the structure of *nosean* (noselite) must be the sodalite structure somewhat expanded.

Nickel chlorostannate hexahydrate

Here [34], Pauling describes his determination of the structure of $NiSnCl_6 \cdot 6H_2O$ (which is better formulated as $Ni(H_2O)_6^{++} \cdot Sn\,Cl_6^=$) — one of dozens of isostructural crystals (he notes that Groth lists 35) of double halides of a bivalent and a quadrivalent element — and notes that Hassel and Salvesen had done X-ray work on several more [35], without determining the atomic arrangement. Three of these last, e.g. $Co(NH_3)_4(H_2O)_2Co(CN)_6$, are additions to the class of substances that give X-ray data indicating that non-identical atoms are sometimes to be considered as crystallographically equivalent. This phenomenon, Pauling notes, was first observed in the case of ammonium oxyfluomolybdate, $(NH_4)_3MoO_3F_3$, which he found, in his thesis work, to have a structure 'experimentally indistinguishable' from that of $(NH_4)_3AlF_5$.

Menzer [10], however, pointed out that Pauling's material must have been the fluosilicate, not the fluo-aluminate, as Pauling agreed [11]. (What Menzer noticed is that tervalent iron and aluminum have nearly the same radii, in strong discord with

Pauling's reported difference in M—F, 0.24 Å.) It seems that in writing of a striking new phenomenon, Pauling was ruling out as less remarkable the standard isomorphous substitutions in minerals, e.g. of Na^+ and Si^{+4} for Ca^{+2} and Al^{+3}.

The micas and chlorites

In two other papers published in 1930 in the Proceedings of the (United States) National Academy of Sciences, Pauling describes the structures of the related classes of aluminosilicates, the micas [36] and the chlorites [37].

For the micas, Pauling suggests the general formula $KX_nY_4O_{10}(OH, F)_2$, 'with $2 \leqslant n \leqslant 3$, in which X represents cations of coordination number 6 (Al^{+3}, Mg^+, Fe^{++}, Fe^{+3}, Mn^{++}, Mn^{+3}, Ti^{+4}, Li^+, etc. and Y cations of coordination number 4 (Si^{+4}, Al^{+3}, etc). The subscript n can have any value between 2 (hydrargillite layer) and 3 (complete octahedral layer) ... The distribution of the various ions X and Y must be such as to give general agreement with the electrostatic valence rule.' He notes that Mauguin [38] had previously reported that muscovite, $KSi_3Al_3O_{10}(OH, F)_2$, forms monoclinic, pseudohexagonal crystals with $a = 5.17$, $b = 8.94$, $c = 20.01$ Å, $\beta = 96°$, $Z = 4$, and Pauling finds nearly the same unit cell for fuchsite, a variety of muscovite. He then proceeds to derive the structure without ever mentioning a possible space group. (Later work by Jackson and West [39] reports the space group as $C2/c$.)

Pauling derived the structure by noting that the dimensions of the pseudohexagonal basal plane, 5.2×9.0 Å, are close to those found in hydrargillite (gibbsite; $Al(OH)_3$), which forms layers of edge-sharing $Al(OH)_5$ octahedra, and also to those found in β-tridymite and β-cristobalite, which contain layers of corner-sharing SiO_4 tetrahedra. He also noted that another type of tetrahedral SiO_4 layer with the same dimensions can be formed by pointing all the tetrahedra in the same direction, in which case the unshared vertices of these tetrahedra 'can be imposed on the hydrargillite layer with the tetrahedron corners coincident with two-thirds of the shared octahedron corners'; the remaining tetrahedron corners would remain unshared and would be occupied by hydroxide or fluoride ions. This proposed arrangement would extend 10.0 Å along the c direction, so two units would be needed.

To verify the structure, Pauling calculated the intensities of the '18 even orders of reflection from (001)', and compared them with the observed intensities obtained from photographs of fuchsite. The calculations depended on only three parameters: the z coordinates of the centers of the SiO_4 tetrahedra (a 3:1 mixture of Si and Al in fuchsite), of the $O^=$ and (OH^-, F^-) layers (which are common to the tetrahedra and the octahedra), and of the $O^=$ layers on the opposite side; the Al atoms in the centers of the octahedra were placed at $z = 0$ and the K^+ atoms at $z = \frac{1}{4}$, since these layers are at one-dimensional centers of symmetry. His initial estimate of the three parameters, based on the known thickness of the layers in hydrargillite and on an assumption of regular SiO_4 tetrahedra 2.60 Å on an edge, led to 'general agreement' with the observed intensities; the agreement became 'striking' when the parameter of the Si(Al) atom was shifted from 0.137 to 0.135. Thus, the question of space group never arose, nor were the other coordinates—x and y—considered; obviously, they could be immediately written down by analogy with hydrargillite and tridymite.

Pauling continues:

The physical properties of talc, pyrophillite, the micas, and the brittle micas are in agreement with the suggested structure. To tear apart one of the pseudohexagonal layers it is necessary to break the strong Si—O, Al—O, etc., bonds; as a consequence these individual layers are tough. But they can be easily separated from one another, giving rise to the pronounced basal cleavage shown by all these minerals. In talc and pyrophillite the layers are electrically neutral, and are held together only by stray electrical forces. These crystals are accordingly very soft, feeling soapy to the touch as do graphite crystals. To separate the layers in mica it is necessary to break the bonds of the univalent potassium ions, so that the micas are not so soft, thin plates being sufficiently elastic to straighten out after being bent. Separation of layers in the brittle micas involves breaking bonds of bivalent calcium ions; these minerals are hence harder, and brittle instead of elastic, but still show perfect basal cleavage.

As in some of Pauling's other publications in the Proceedings of the National Academy, he states that a further account of the X-ray investigations 'will be published in the *Zeitschrift für Kristallographie*', but we find no such publication.

The structure Pauling derived for the chlorites is quite similar, and he suggests the general chemical formula $X_mY_4O_{10}(OH)_8$ with m usually equal to 6 and X usually Mg^{++} or Al^{+3}. The unit cell has approximately the same basal dimensions as the micas, 5.2×9.0 Å, but the c axis is appreciably shorter, at about 14.5 Å; accordingly, Pauling presumed that, along the c direction, a single unit of the mica structure alternates with an octahedral unit of the brucite $(Mg(OH)_2)$ structure. There were now four parameters necessary to calculate the intensities of the $00l$ reflections, in this case the odd orders as well as the even. Once again the agreement with observed values (from a crystal of penninite, $Al_2Mg_5Si_3O_{16}(OH)_8$) was excellent, and 'deleteriously affected by changing any of the parameters by as much as 0.01 from its predicted value'. And once again we fail to find the promised follow-up article in *Zeitschrift für Kristallographie*.

Zunyite

Three years later, Pauling produced [2] what may be the type example of his use of his trademark method. It is here that he first used the designation *stochastic*, and it is here that we find the most impressive application of the method, at least among his determinations of mineral structures. It is a description of his derivation of the structure of Zunyite, $Al_{13}Si_5O_{20}(OH, F)_{18}Cl$.

About *stochastic*, Pauling wrote, 'The large size of the structure [$a_0 = 13.820 \pm 0.005$ Å] and the complexity of the chemical formula make the deduction of the atomic arrangement from X-ray data alone impractical if not impossible. We consequently make use also of arguments based on analogy with other structures, semi-empirical structural rules regarding ionic sizes and ionic environments, etc., with ultimate recourse to the stochastic method, which has already been applied to brookite, topaz, mica, natrolite, and many other crystals — a detailed structure being suggested, and then tested by a comparison of observed and calculated intensities of reflection.' In a footnote he adds the following:

I am indebted to Dr. Karl K. Darrow of the Bell Telephone Laboratories for acquainting me with this word and with its use by Alexander Smith, who wrote in his "Inorganic Chemisty", 1909, p. 142, the following: "When Mitscherlich discovered that Glauber's salt gave a definite pressure of water vapor, he at once formed the hypothesis, that is, supposition, that other hydrates would be found to do likewise. Experiments showed this supposition to be correct. The hypothesis was at once displaced by fact. This sort of hypothesis predicts the probable existence of certain facts or connections of facts, hence, reviving a disused word, we call it a stochastic hypothesis (Greek στοχάστιχος, apt to divine the truth by conjecture). It differs from the other kind in that it professes to be composed entirely of verifiable facts and is subjected to verification as quickly as possible ..." The method of treating very complex crystals which has been used recently, in which a plausible structure is guessed with the aid of hints provided by the observed size of unit and space-group symmetry, and the stochastic hypothesis that this is the actual structure of the crystal thereupon is either verified or disproved by intensity observations, may well be called the stochastic method, in contradistinction to the "rigorous" method, which latter involves the straightforward testing by intensity data of all of the possible arrangements provided by the theory of space groups.

In describing his work on zunyite, Pauling begins by noting an earlier worker's error: Gossner [40] had prepared powder, rotation, and Laue photographs of zunyite, reported a value for the cubic lattice constant, suggested a chemical formula, and 'assigned the crystal symmetry of space group T_d^1, although his data indicated a face-centered lattice.' From his own Laue photographs, and the tetrahedral face development of his crystals, Pauling confirmed the space group $T_d^2 - F_4^-$, $3m$.

Six published chemical analyses of zunyite specimens from three localities, carefully considered in view of possible isomorphous replacements and the space-group multiplicities available for Si, led Pauling to the formula $Al_{18-n}Si_nO_{20}(OH, F)_{18}Cl$, with $n \geqslant 5$. With almost no apparent difficulty, Pauling arrived at the complete structure. He first noted that, with at least five silicon atoms in the formula unit, there must be at least 20 silicon atoms in the unit cell — that is, 20 tetrahedra. He first investigated the 24-fold positions in T_d^2, and found that none could accommodate tetrahedra without leading to $O—O < 2.4$ Å or else requiring that three tetrahedra share a vertex, which contradicts the electrostatic valence rule. So he settled on a combination of a 16-fold and a 4-fold position which led to a grouping of five tetrahedra with T_d symmetry, the one at the center sharing vertices with the surrounding four. He then needed, for the aluminum atoms, four sets of 12 octahedra, again with point symmetry T_d. He found such units in spinel, $MgAl_2O_4$; however, he was unable to join the spinel group with the five silicon tetrahedra without violating his electrostatic valence rule. But he noted that the spinel group was based on four triangular Al_3O_{13} units (three AlO_5 octahedra sharing three edges and with one vertex common to all three, the oxide ions in two close-packed layers); by inverting these Al_3O_{13} units they can share vertices, rather than edges, and thus form new groupings $Al_{12}O_{46}$, still with T_d symmetry; these clusters could, in turn, share vertices with one another and also with the Si_5 clusters. Lo and behold, this arrangement 'leads to a value of 13.82 Å for a_o, in exact agreement with the observed value,' Moreover, a tetrahedral vacancy remained which, when occupied by the remaining aluminum atom, satisfied the electrostatic valence rule; the Cl^- atoms, placed in four octahedral holes surrounded by six O or F atoms, completed the structure. It surely seemed, to Pauling, 'highly probable' that he had found the structure of zunyite.

Ten position parameters were needed to describe the structure. Key elements in his derivation of these parameters were the bond lengths 'Si—O = 1.59 Å, Al—O = 1.89 Å in octahedra, 1.75 Å in tetrahedra,'. He confirmed the structure by comparing calculated intensities for 56 Bragg reflections with those measured on oscillation photographs, noting that the agreement was 'satisfactory'; he then notes: 'It is probable that complete agreement with the observed intensities could be obtained with the use of parameter values differing only slightly from those given above; but the large number of the parameters and the labor involved in the structure-factor calculations makes the determination of the parameter variations impracticable.'

In fact, the parameter values required only very slight changes when the structure was reinvestigated by Barclay Kamb [41] (who counted 72 intensities calculated by Pauling—we find 62), by Louisnathan and Gibbs [42], and by Baur and Ohta [43]. This last study, with electron microprobe analyses of four samples from three localities and X-ray diffraction studies of two crystals, from two of the localities, showed silicon numbers ranging from 4.95(6) down to 4.61(17), fluorine numbers of 3.6(3) up to 4.3(3), and unweighted R values, for the two crystals, of 0.016 and 0.020. It seems clear from the results for the two crystals that the silicon deficiency of the second—about 0.25 per formula unit—is concentrated in the central tetrahedron of the five-linked-tetrahedron unit $Si(OSiO_3)_4$ of the structure, causing the central Si$\cdots$O bond to be lengthened by just about the right amount (0.029 Å) by the replacement of Si by Al, and the bonds from the intermediate oxygens to the peripheral silicons to be shortened by 0.018 Å, because of the 'underbonding' of the intermediate oxygen. In comparison with other aluminosilicates, however, Baur finds discrepancies that cause him to remark, 'Clearly, nature does not read our empirical and theoretical papers, or else the zunyite structure would look differently.' For this and a host of other reasons, he concludes with, 'Half a century after the structure of zunyite was determined by Pauling it continues to be a fascinating subject of study.'

Zunyite is wonderfully complex, each element of its high crystallographic symmetry paradoxically presenting new prospects to entrance the student's perception just as it simplifies the essential description down to the location of ten key atoms. As we have noted, Pauling described the structure as built up from Al_3O_{13} triangular groups; four of these groups, each sharing three vertices, form octahedral structures that surround each of the lattice points of the face-centered unit cell, being joined together in at least three ways: by tetrahedral aluminum atoms whose oxygen atoms are already present at the centers of four of the faces of the just-mentioned octahedral units; by tetrahedral $Si(OSi)_4$ groups, the peripheral silicon atoms of which find the oxide ions to complete their tetrahedra already present in the remaining faces of the octahedral units; and by sharing pairs of oxide ions in the twelve [1, 1, 0] directions with others of the octahedral structures. The triply connected framework so constructed has large hollows at the center and each of the edge centers of the unit cell, where reside the chloride ions. But Pauling does not seem to have paused to enjoy—or be bewildered by—the numerous other prospects the structure affords.

Zagal'skaya and N.V. Belov [44] were delighted to notice that instead of with what they labeled as the 'Prominent octahedral hollows in the structure of zunyite'

— Pauling's octahedral structure around the lattice points, the 'hollows' being in fact just equal in size to the surrounding oxide ions — the structure could be built up from two kinds or complex ions modeled on the structure proposed by Pauling [45] for the anion of the heteropolyacid $H_3[PW_{12}O_{40}]$, one kind centered on the tetrahedral aluminum ions and the other on the Si_5 groups. The oxide-ion positions are nearly the same in both, but in the first kind the 'peripheral' tetrahedra are vacant while in the second the octahedra are vacant. So far, so good, but Zagal'skaya and Belov also proposed that Pauling's structure was wrong, that the tetrahedral aluminum and the central silicon should be interchanged. In this *they* were quite wrong: this assignment destroys the good agreement with the electrostatic valence rule and is incompatible with the observed interatomic distances, a point that became more obviously binding with the increased precision of the more recent refinements [42, 43].

Still further conceptual views of the structure were presented by Louisnathan and Gibbs [42], who show Pauling's octahedral units as schematic cubes — the corners of which are the positions of Belov's filled (Si) and empty tetrahedra — joined to large cubododecahedra by lines drawn to the positions of the tetrahedral Al and the central Si atoms. These diagrams do some justice to the environment of the enclosed chloride ion, which as Kamb [41] concluded and Bartl verified by his single-crystal neutron-diffraction study [46], receives hydrogen bonds along the cube-axis directions from six of the OH groups of the formula unit. Unfortunately, neither all the diagrams taken together nor Bartl's neutron study has provided a clear view of the linking of Pauling's octahedral groups by the sharing of (OH, F) atoms along the [1, 1, 0] directions or their hydrogen bonding, so that this aspect of the zunyite structure remains unclear. Kamb proposed an arrangement that Bartl's work seems not to have tested, and Bartl's B value (the displacement coefficient) for the second type of hydrogen atom (13; his value for the first is 1.9) is so large as to call his identification of the atom into question. Besides, this atom, as reported, would account for only six of the remaining 12 (OH, F) atoms of the formula. The analyses [43] reported by Baur and Ohta, as well as the ones originally cited by Pauling, all suggest a maximum F content of about four.

We have gone to this length about the hydrogen bonding in zunyite, not because Pauling might have been interested enough to speculate on it, even in 1933. Rather, it may be that he didn't because that would have carried him past the Pauling Point [47].

CORRECTIONS TO OTHER CRYSTALLOGRAPHIC WORK

Our decision to include a number of examples of Pauling's critical examination of the crystallographic literature was prompted by our ruminations over Robert John Paradowski's remark, in his quite wonderful Ph.D. thesis [48], 'The Structural Chemistry of Linus Pauling', 'Pauling seems to have developed a penchant for correcting the work of others during these early papers.' This remark may seem to have a somewhat wry flavor, and as such it deserves further comment. It is true enough, as Paradowski's fine account (seventy-four pages) of Pauling's crystal-structure work and our experience in compiling the following pages clearly show. However, it is clear that Pauling did not simply make it his business to go looking for errors, but rather that errors came to his attention in the course of his continual survey, digestion, and eventual understanding of essentially *all* the publications in the area of structural chemistry. When he noticed a report that didn't fit in with his understanding, he was compelled to determine the true facts. If the report was wrong, it had to be rectified; if it was not wrong, something in his system of structural chemistry would have to be corrected. He soon had acquired a facility for detecting errors that rivaled his facility for digesting and remembering *all* (it seems to us) of his scientific experiences and readings. His ability to detect errors is, perhaps, a part of his stochastic method: if a correct structure could be built from first principles, an incorrect structure could be detected. What Pauling was looking for, we believe, was the ability to *predict*; he sought such a deep understanding of structural concepts that he could explain, or be skeptical of, new observations. In this framework we can understand the phrase 'It might have been predicted that ...' which has occurred so often in Pauling's lectures. So what follows, although categorized under 'corrections', is an account of much or most of Pauling's crystallographic work, up to about 1935, that we have not already discussed.

Uranyl nitrate hexahydrate

This first 'correction' constitutes the third part of Pauling's thesis and was published with Roscoe Dickinson [5]. It was based on a paper [49] by G.L. Clark, who reported that $UO_2(NO_2)_2 \cdot 6H_2O$ forms orthorhombic crystals with the uranium atoms in face-centered positions. Clark had deduced the unit-cell dimensions ($11.45 \times 13.02 \times 7.93$ Å) on the basis of 'Z-radiation' reflections measured with an ionization spectrometer. This was an interesting phenomenon of the time, which warrants some description.

Z-radiation was first announced by Clark and Duane in 1923 [50]. They interpreted certain peaks recorded in their spectrometer as 'due to X-rays characteristic of elements in the crystals and produced by such elements under the influence of the impinging beam, and reflected from the crystals at angles given by the relation $n\lambda = 2d \sin \theta$.' In other words, they believed that some atoms in a crystal could, upon irradiation with the full spectrum from a tungsten tube, produce characteristic X-radiation which would obey the same Bragg's Law as the incident beam. This was recognized as a potentially important phenomenon; for example, Wyckoff [51], who (among others) was unable to repeat the experiment, comments: 'As yet the properties of this "characteristic reflection" are incompletely understood; but its very existence and the opportunity it offers of producing distinctive diffraction effects from only a part of the atoms of a crystal make it of immediate interest and probably great future value to crystal analysis.'

In the case of uranyl nitrate hexahydrate, Pauling and Dickinson noted that Clark [49] had interpreted one of the 'Z' peaks as due to first-order diffraction of the $L_{\gamma 1}$ radiation from uranium (wavelength 0.61283 Å) by the (010) face of the crystal, and they further noted that the first-order reflection should be absent if the uranium atoms form a face-centered array. Thus, reasoned Pauling and Dickinson, either the identification of the Z peaks was incorrect or the structure was incorrect; in fact, their results showed that Clark was wrong in both regards.

Pauling and Dickinson found the orthorhombic unit cell ($a = 13.15$, $b = 8.02$,

$c = 11.42$ Å) to be end-centered on (0 0 1), assigned the crystal to space group $V_h^{17} - Cmcm$, and placed the uranium atoms in the positions $0u\frac{1}{4}$; $0\bar{u}\frac{3}{4}$; $\frac{1}{2}\frac{1}{2} + u\frac{1}{4}$; $\frac{1}{2}\frac{1}{2} - u\frac{3}{4}$; with $u = 0.13$. Their cell and u value have subsequently been confirmed ($u = 0.1284(3)$ [52], and $u = 0.1292(5)$ [53]), but for the complete structure they passed by the correct space group $C_{2v}^{12} - Cmc2_1$ (which cannot be distinguished from V_h^{17} by the systematic absences). Perhaps they were misled by Groth's [54] identification of the crystals as 'rhombic bipyramidal'; perhaps they were thinking only of the partial structure, represented by the uranium atoms, for which the two space groups are equivalent. The latter explanation carries weight because of the title of the paper, 'The Crystal Structure of ...', when in fact uranium was the only atom considered. It would soon no longer be appropriate, having located only one atom of a moderately complex structure 'X', to title the report 'The Crystal Structure of X'.

Hematite and corundum (Fe_2O_3 and Al_2O_3)

For hematite and corundum the question was not of correcting an erroneous structure determination, but instead of improving an important aspect of the Braggs' marvelous, very early deduction of the structure [55–57].

The Braggs had assigned values to two parameters of the structure, one for the metal atoms and one for the oxygens, by analogy with the known structure of calcite ($CaCO_3$): they assumed that six oxygen atoms lay around each metal atom at the vertices of a regular octahedron, and they used a few X-ray reflections only as a guide. (Surely a fine example of the stochastic method.)

Pauling and Hendricks [58] proceeded to make a thorough X-ray study of crystals of hematite and corundum, remarking that 'an exact knowledge of the arrangement of the constituent atoms' would make much more convincing the arguments given in recent papers that included discussions of the birefringence [59] and the temperature dependence of the X-ray reflections [60] of ruby. Pauling and Hendricks prepared the usual Laue and spectral photographs; from them they carefully selected pairs of reflections whose relative intensities were found to be nearly insensitive to the relative scattering powers of the two different types of atoms. By this means, they managed to determine within narrow limits the two parameters needed to describe each structure, and they showed that Bragg's assumption of all-equal Al–O distances in substantially in error. In corundum, three of the distances, the ones involved in both face-sharing and edge-sharing with other octahedra, are longer (1.990 ± 0.020 Å) than the other three, which are involved only in the edge sharing and in corner sharing (1.845 ± 0.015 Å); there are similar differences for hematite. Moreover, the shortest $O\cdots O$ distances, those of the shared face, are only 2.495 Å (corundum) and 2.545 Å (hematite), much shorter than the mean $O\cdots O$ distance, approximately 2.71 and 2.86 Å. The disparity in M–O bond lengths, Pauling and Hendricks noted, is less in hematite than in corundum, and they remark 'an iron ion with 23 electrons within a volume only slightly greater than that of an aluminum ion, with 10 electrons, would approximate a sphere much more closely than the aluminum ion.' Pauling and Hendricks failed to note another point of a kind that Pauling would have emphasized a few years later: of the twelve edges of an AlO_6 octahedron, alike in threes, not only are the three $O\cdots O$ edges involved in face-sharing the shortest, but the three involved only in edge-sharing are shorter than the remaining six edges which are not shared.

The atom parameters in corundum and hematite were determined again nearly forty years later by Newnham and de Haan [61], by least-squares refinement of the fit to visually estimated intensities of reflection observed on zero-level Weissenberg photographs from crystals rotated about the direction [100]. For the metal atom parameter z of the positions $\pm[0, 0, z; 0, 0, z + \frac{1}{2}]$ (space group $D_{3d}^4 - P\bar{3}c1$), Newnham and de Haan found 0.3520(3) for corundum and 0.3550(3) for hematite, compared to Pauling and Hendricks's value of 0.3550 ± 0.0010 for both compounds; for the oxygen parameter x of the positions $\pm[x, 0, \frac{1}{4}; 0, x, \frac{1}{4}; \bar{x}, \bar{x}, \frac{1}{4}]$, Newnham and de Haan found 0.306(4) (corundum) and 0.300(4) (hematite) compared with Pauling and Hendricks's 0.303 ± 0.003 and 0.292 ± 0.007. The agreement is pretty good, but we have to wonder what the best modern study would yield. The four types of $O \cdots O$ distances given by Newnham and de Haan for corundum are 2.52, 2.62, 2.73, and 2.87 Å, the last being for the edges of the face opposite the shared face of the octahedron.

Barite, BaSO₄

Samuel K. Allison [62], evidently inspired by the work of Clark and Duane, chose to investigate barite because it 'is a crystal which contains an element whose K critical absorption frequency lies in a region easily accessible to the ionization spectrometer. Very great interest has been drawn to spectra from such crystals by the recent work of Clark and Duane.' [63]. While Allison found 'no good evidence' for the characteristic Ba spectrum, he reported the orthorhombic unit cell of barite as $a = 4.449$, $b = 5.448$, and $c = 7.170$ Å, with $Z = 2$ in space group $V_h^{13} - Pnma$. Pauling and Emmett [64] found reason to complain. They pointed out that Allison reported the spacing d_{obs} of 5.562 Å for the plane (102) compared with $d_{calc} = 2.792$ Å, further noting that the observed spacing could be smaller (by a factor $1/n$) than calculated but not larger. They assigned the cell dimensions as $d_{100} = 8.846$, $d_{010} = 5.430$, and $d_{001} = 7.10$ Å, concluding that the four Ba^{++} lay on the mirror planes at $(x, \frac{1}{4}, z)$, etc., but did not determine any other coordinates.

In the same year, 1925, four other X-ray diffraction studies of barite were published that are listed in Donnay and Ondik's *Crystal Data Determinative Tables*, 1973, and one of these [65] reports the complete structure determination. By careful consideration of all possibilities, James and Wood deduced that the barium and sulfur atoms lie on mirrors. Then, accepting that the sulfate ion is most likely a regular tetrahedron with S—O bond length 1.5 Å, as in LiKSO₄, the structure of which had just been determined by A.J, Bradley, they were left with only five parameters, two each for Ba and S and the angle of orientation of the sulfate group about an axis perpendicular to the mirror. These parameters were fixed by comparing observed integrated intensities of reflection from about ten crystal faces (some natural, some prepared by grinding with emery) with calculations from the crystal model based on Hartree's atom scattering factors, made available before publication and used with much agonizing over the possible roles of absorption, extinction, and thermal vibration. In the end, James and Wood took care to see that no oxygen atoms not in the same sulfate ion were nearer to each other than, nor any Ba–O distance shorter than, 2.7 Å. We have not seen that this structure determination has ever been repeated.

In their introduction, James and Wood make remarks that in considerable measure anticipate Pauling's later description of his stochastic method: 'This paper describes an attempt to analyse the structures of three crystals', $CaSO_4$, $SrSO_4$, and $PbSO_4$, members of a numerous, isomorphous series. 'The structures of such crystals must be determined by a process of trial and error. Atomic arrangements must be assumed, and then tested by comparing the intensities of the X-ray spectra predicted from them with those actually observed.' The difference is that James and Wood emphasize the process of adjusting their trial model to fit the X-ray data, while Pauling emphasizes deducing or guessing the structure in all detail, and having recourse to the data, in the ideal case, only in order to verify the model.

CsI_3 and $CsIBr_2$

The next correction of published error is part 5 of Pauling's thesis, also published in *J. Amer. Chem. Soc.*, with R.M. Bozorth [9]. Once again we encounter Clark and Duane's 'Z-radiation' phantasy. In this case, Clark and Duane had interpreted the presumed Z-peaks due to secondary radiation from Cs as having been formed by diffraction solely by planes of Cs atoms in the crystal, and the presumed iodine-radiation Z-peaks correspondingly from the planes of iodine atoms; both kinds of planes were observed parallel to each of the three faces of the orthorhombic unit cell, and the iodine planes had only 1/4 the separation of the cesium planes. Accordingly, Clark and Duane placed the Cs atom at the corners of the unit cell and the I atoms at positions 1/4, 1/2 and 3/4 along the body diagonal [63]. Which diagonal? Random diagonals, because 'There is no evidence that the diagonals are all unidirectional in a crystal.' It was apparently Bozorth, rather than Pauling, who first became suspicious, perhaps of the reality of Z-radiation, perhaps of the structure itself. In 1923 he measured the density of CsI_3 and re-determined the cell dimensions, finding them to be quite different (by factors of about 1.5) from those reported by Clark and Duane; his unit cell would contain four CsI_3 units rather than one. Recognizing the importance of the discrepancy between the two results, Pauling, in 1924, carried out a completely independent investigation, with results that matched closely with those of Bozorth. The two agreed that the unit cell is orthorhombic and primitive, with $a \approx 6.8$, $b \approx 9.9$, $c \approx 11.0$ and with four formula units per cell; on the basis of systematic absences and the holohedral morphology reported by Groth [66] they thought the probable space group was V_h^{16} (*Pmcn*), but did not rule out the hemihedral C_{2v}^9, $P2_1cn$. However, they did not attempt to assign atom positions 'because of the number of parameters involved'. The complete structure determination of CsI_3, showing an unsymmetrical tri-iodide ion, confirms the assignment of $V_h^{16} - Pmcn$ [67]. A footnote by Bozorth and Pauling, added after their paper was submitted for publication, references a paper [68] by Armstrong, Duane, and Havighurst which, without referring specifically to any of the previous papers on Z-radiation, disavows the new method, reporting that the anomalous observations on which it was based were due to (use quote B and P) 'reflection in the usual way from small crystals with axes slightly displaced from those of the main crystal.'

Potassium chloroplatinate

Here [69], Ewing and Pauling correct a report by Scherrer and Stoll [70] that potassium chloroplatinate has the potassium chlorostannate structure [71] but with the parameter value u for chlorine equal to 0.16, in contrast to Wyckoff and Posnjak's [72] carefully determined limits 0.22 to 0.24 for the corresponding value in $(NH_4)_2PtCl_6$. Wyckoff [73] had pointed out the discrepancy, noting that u is the same for the potassium and ammonium chlorostannates and estimating $u = \pm 0.24$ for K_2PtCl_6. Ewing and Pauling also remark that Lennard-Jones and Dent [74] had

> recently developed a theory designed to predict parameter values for crystals of this type, and have obtained results agreeing with Scherrer and Stoll's work on potassium chloroplatinate and disagreeing with Dickinson's on potassium chloro-stannate. ... The research described in the following pages leads definitely to the conclusion that Scherrer and Stoll's parameter value is incorrect, and hence that Lennard-Jones and Miss Dent's theory disagrees with experiment for potassium chloroplatinate as well as for potassium chlorostannate.

Ewing and Pauling's value is $u = 0.240 \pm 0.005$. It was confirmed in 1935 and again in 1955.

Cubic telluric acid, Te(OH)₆

Kirkpatrick and Pauling [75] prepared some of the cubic form of $Te(OH)_6$, and made spectral (molybdenum K-radiation) and Laue photographs with crystals of greatest edge length 0.3 mm. 'Es ist bemerkungswert, dass so kleine Kristalle sehr guten Aufnahmen gaben.' They reported the unit cell to be face-centered with edge 15.48 Å, containing 32 $Te(OH)_6$, and the space group to be O_h^8; there was also reason to conclude that the Te site symmetry is C_{3i} or D_3, which suggests the formula $Te(OH)_6$ rather than $(H_2Te)_4 \cdot 2H_2O$.

When Gossner and Kraus [76], working from oscillation photographs, reported that the cubic modification of telluric acid belongs to space group O_h^5 with the cell edge halved at $a = 7.83$ Å and Te—O bond length ~ 1.96 Å, Pauling was quick to respond [77]. Indeed, he not only presented new evidence in support of the original conclusions but also took the occasion to give a short lecture on an aspect of X-ray crystallography:

> In the investigation of crystals with X-rays the error has often been made of accepting as correct a unit of structure which is smaller than the actual unit; this error is usually the result of the consideration of insufficient experimental data, as, for example, the use of rotation photographs taken with exposure times too small to make the weaker layer lines evident. A much more egregious error is to assume a unit of structure larger than the actual unit, inasmuch as this error is the result of carelessness on the part of the investigator in overlooking the fact that a smaller unit can account for his data, or in otherwise interpreting his experimental results.
>
> Nine years ago L.M. Kirkpatrick and I prepared and investigated cubic crystals of telluric acid, $Te(OH)_6$, reporting the cubic unit of structure, with $a_0 = 15.48$ Å, to contain 32 molecules. Recently Gossner and Kraus have reinvestigated the crystal, coming to the conclusion that the unit of structure contains only four molecules,

and stating that Kirkpatrick and I had made the error of assuming too large a unit. As a matter of fact, the unit of structure which we reported is correct, and Gossner and Kraus are themselves guilty of accepting too small a unit of structure.

Pauling goes on to comment that the original paper had cited four Laue reflections that require the 32-atom unit, and includes a rotation photograph, prepared for the purpose by himself and Sturdivant, that clearly shows weak odd layer lines, confirming his 32-atom unit cell. He also remarks that 'the atomic arrangement suggested by Gossner and Kraus is of course incorrect.'

The most recent investigation of the structure of $Te(OH)_6$ that we have found [78], by both X-ray and neutron diffraction from single crystals, has confirmed Pauling's unit cell once again, but has reduced the space group from $O_h^8 - Fd3c$ to $O^4 - F4_132$. (In neutron diffraction a large number (46 out of 245) reflections were observed that violate the absences required by the glide planes of O_h^8.) The $Te(OH)_6$ molecules now have only C_3 symmetry with two Te–O distances, 1.81(1) and 1.98(1) Å, rather startlingly different in view of the symmetry of the hydrogen bonding situation. Each oxygen forms two hydrogen bonds to neighboring molecules, both showing two (disordered) half-hydrogens with equal occupancy. The hydrogen bonding network was not explored except to note that there is a considerable range of O—H···O distances that seems to correlate with the variation in Te–O bond lengths.

Pauling's second paper on $Te(OH)_6$ appeared in the same year as his beautiful paper [79] on the disorder of the hydrogen bonds and consequent residual entropy of ice; it would be nice to know more in this respect about the situation in $Te(OH)_6$.

Rubidium halides

In 'Note on the pressure transitions of the rubidium halides' [80], Pauling argues convincingly that P.W. Bridgman's high-pressure forms of rubidium chloride, rubidium bromide, and rubidium iodide indeed have the CsCl structure, formed by transition from the low-pressure NaCl structure, contrary to Bridgman's expressed 'extreme scepticism' on the ground that the observed volume contractions average only 12% instead of the 23% (Bridgman wrote 30%, a slip, as Pauling noted, of density increase for magnitude of volume decrease) that would hold if the MX distances were to remain constant and that the theory of Born and Hund would not account for the stability of the CsCl structure at the observed transition pressures. Pauling's argument, based on his recent publications (1927 and 1928) on ionic crystals in Z. Krist., J. Am. Chem. Soc., and Proc. Roy. Soc., shows that the MX distances would be expected to increase about 3%, not to remain constant, and that the polarizabilities of the ions should stabilize the high-coordination structure, relative to the low-, by more than had previously been allowed for. He acknowledges a letter from V.M. Goldschmidt referring to Geochemische Verteilungsgesetze der Elemente, VII, p 16, which indicates a 3% increase in MX distance on going from the NaCl to the CsCl type of structure.

Bridgman's concern was great. He wrote, 'When I began work on the transitions of the rubidium salts I had no question that the modification found by Slater at high pressures was of the CsCl type. All the a priori probability pointed in this direction ... It became at once obvious, however, on determining the change in volume that it is highly questionable whether the transition can be of this type,' and discussed his concern for two more pages, proposing (and eventually rejecting) a radical, less symmetrical structure that by his calculation would give the right volume contraction, and concluding with 'It is perhaps not impossible that we shall be driven ultimately to recognize that in at least some cases of polymorphism the atoms have different electronic configurations in the different modifications.' We do not believe that Pauling was unsympathetic to this anticipation — only to its mention in the context of the subject pressure transitions of the three rubidium halides.

Indeed, after he learned about a transition [81] with 10% volume decrease — and changes in paramagnetic susceptibility — occurring in metallic cerium under some conditions below 109 K, he suggested to Schuch and Sturdivant [82] that 'it is caused by the promotion of a 4f electron to a bond-forming orbital, and that the dense phase be studied by x-ray diffraction.' Schuch and Sturdivant found that at 90 K the new phase is face-centered cubic-closest-packed, like the room-temperature phase (which itself persists to a variable extent), but with a_0 reduced from 5.14 kX (about 5.12 kX at 90 K) to 4.82 kX. The dense phase did not appear unless the cerium 'had been quenched by an air blast from at least 300° C to room temperature', and in successive coolings of the same quenched specimen with liquid air 'a smaller and smaller proportion of the new structure appears.' Subsequent to these experiments the transition was produced by Lawson and Tang [83] at 15,000 atmospheres, who thanked their colleague W.H. Zachariasen for conversations that led to *their* proposal that the '4f electron [of cerium] is literally squeezed into a 5d state.' (Pauling caught Zachariasen, who was a wonderful crystallographer, out-of-bounds surprisingly many times; see below. This case shows, however, that they didn't always run on different wavelengths.)

The Trombe–Foex volume change was only 10%, apparently, because the transition to the dense form is not complete. However, this is a complicated story, and we make no pretense of following it beyond Schuch and Sturdivant's report.

The paper that follows Bridgman's 'Rubidium Halide' paper [84] is one of Pauling's monumental more theoretical papers, 'The sizes of ions and their influence on the properties of salt-like compounds', one of the three papers he referred to in his note.

Cyanite (Kyanite), Al_2SiO_5

In his paper '*The Principles Determining the Structure of Complex Ionic Crystals* [24], his fifth on the subject, Pauling takes the polymorphic silicates cyanite, andalusite, and sillimanite as examples to illustrate the application of some of the principles. Although he had just remarked

> So far as I know, Al^{+3} has the coordination number 6 in all its compounds with oxygen the structures of which have been determined. The coordination number 4 would also be expected for it, however; it is probable that it forms tetrahedra in some of its compounds, as, for, example, γ-alumina, the cubic form of Al_2O_3, and the feldspars, in which there occurs replacement of Na^+ and Si^{+4} by Ca^{++} and Al^{+3} [and he gives more discussion later in the paper].

Pauling based his discussion of the composition Al_2SiO_5 on the assumption of only 6-coordination for aluminum; this proved correct for cyanite but not for andalusite and sillimanite, which *also* have aluminum bipyramids and tetrahedra, respectively. Perhaps he would have been led to think of such possibilities if he had noticed Table 1 of 'The Structure of Certain Silicates' [85], which lists the volumes per oxygen atom (and refractive indices) for a number of silicates in order to illustrate something about the extent to which their structures are all based on closest packings of the oxygen atoms. The volumes per O atom in $Å^3$ are 13.94, 15.05, 17.15, and 16.30 for a 'close-packed assemblage', cyanite, andalusite, and sillimanite. This comparison, together with the already expressed prediction that aluminum might not always have 6-coordination with oxygen, might then have suggested that 6-coordination does not prevail in andalusite and sillimanite. The observed increases in volume per oxygen atom for andalusite and sillimanite might then have been attributed to either 4-coordination, the aluminum ions entering and expanding the oxygen tetrahedra in a closest-packed array of oxide ions, or to 5-coordination, which again would disrupt closest packing in order to form trigonal bipyramids. But having also adopted the principle of parsimony, his rule V, to restrict aluminum to only one kind of coordination in these polymorphs, Pauling drew several conclusions about the three structures, which he noted were

> in pronounced disagreement with the complete ionic arrangement proposed by Taylor and Jackson [86], whose suggested structure conflicts with most of our principles. Their structure is far from parsimonious, with four essentially different kinds of octahedra and two of tetrahedra. Each silicon tetrahedron shares a face with an octahedron, contrary to Rule III. The electrostatic valence principle is not even approximately satisfied; one oxygen ion, common to four aluminum octahedra and one silicon tetrahedron, has $\sum s_i = 3$, while another, common to two octahedra only, has $\sum s_i = 1$. For these reasons the atomic arrangement seems highly improbable.

A little later, Naray-Szabo, Taylor, and Jackson [87], referring to a communication from Pauling, wrote, 'A complete structure for cyanite was suggested by two of the authors [based on rotating-crystal photographs]. In a private communication Mr. L. Pauling pointed out that the structure proposed was open to certain serious objections, and a re-examination shewed the criticism to be well-founded. The present paper describes a structure which we believe to be correct.' This structure is consistent with Pauling's rules and was successfully refined by C.W. Burnham [88].

Note on the paper of A. Schröder: Beitrage zur Kenntnis des Feinbaues des Brookits usw [89]

Having determined the space group of brookite (orthorhombic TiO_2) to be V_h^{15} from Laue photographs only, having in their paper 'pointed out the desirability of basing space-group determinations on Laue photographs rather than on rotation photographs, on account of the small chance of error in assigning indices to the planes producing Laue spots and the much larger uncertainty in identifying planes on rotation photographs,' and being especially pleased with the derivation of their structure for brookite, Sturdivant and Pauling here take pains to demonstrate that Schröder's argument [90], purporting to show that V_h^{15} is not the space group of brookite and based on nine reflections observed on his 'rotation' photographs, is contrary to their Laue observations (some previously reported, some newly reported here) in eight cases, the ninth, plane (033), being Laue-indeterminate because it would fall on (022); further, they exhibit a newly prepared rotation photograph of their own, which proves that Schröder indeed mis-indexed his.

We were interested to see that Pauling called Schröder's 'Drehspektrogramm' an oscillation photograph, despite Schröder's term for it. The reason, evidently: Schröder rotated his crystal continuously through less than a whole revolution, e.g. by 20° during a 40-minute exposure.

The crystal structure of the A-modification of the rare earth sesquioxides

Zachariasen had suggested [91] a structure for the A-modification of the oxides La_2O_3, Ce_2O_3, Pr_2O_3, and Nd_2O_3 that Pauling found quite unreasonable. Based on powder photographs, it evoked the following introductory comment [92].

For several reasons this atomic arrangement cannot be accepted as correct.

1. Theoretical considerations as well as the examinations of known crystals containing oxygen anions and relatively small cations have shown [referring to paper in press] that stable structures of such crystals contain coordinated polyhedra of anions about the cations. Such polyhedra are not present in Zachariasen's structure; each cation is in contact with three oxygen ions only, which are not coplanar with the cation.

2. The structure is a layer structure with M^{+3} as the polarizable constituent, whose polarization in the field of the anions presumably stabilizes the structure. But the mole refraction of La^{+3} is only 2.64 and that of the other cations still smaller, so that no appreciable polarization of these anions can occur.

3. The minimum M–O distance is given as 2.10–2.15 Å, and the minimum O···O distance as about 1.70 Å. These are much smaller than the accepted ionic radii, which give

$$La^{+3}-O^= = 2.50 \text{ to } 2.55 \text{ Å}, \quad O^=\cdots O^= = 2.60 \text{ to } 2.80 \text{ Å}.$$

In particular is the oxygen–oxygen distance much too small; the closest approach of oxygen ions previously reported for an oxide is about 2.5 Å.

Pauling formulated an 'alternative and more satisfactory arrangement by applying the principles determining the structures of coordinated crystals.' This is based on two double layers of cations imbedded in six closest-packed layers of oxide ions, with distortions such that the coordination, first taken as octahedral, becomes 7-fold, one oxide from a 'foreign' double layer approaching along the hexagonal axis as close to the cation as the nearer three of the original six of the octahedron. Pauling points to the calculated line intensities for his structure as being in 'fully as good agreement with the experimental data as the structure proposed by Zachariasen', who, however, offered a rebuttal [93] in which he argued that his pyramidal LaO_3 coordination was *not* unreasonable, in view of the existence of such in chlorates, bromates, As_2O_3, and Sb_2O_3, disputed Pauling's discussion of the mole refraction of A-La_2O_3, and, using a single-crystal newly prepared 'in liebenswürdiger Weise' by Professor Goldschmidt, added Laue photographic evidence of holohedric symmetry, which is violated by Pauling's structure. He also scoffed at Pauling's statement that *his* structure makes it understandable that there should be excellent cleavage perpendicular to the z-axis in view of the presence in that structure of an La–O distance of only 2.4 Å perpendicular to the cleavage plane. Several decades later, Müller-Buschbaum and von Schnering [94] obtained their own single crystals, and with a disordered version of Pauling's model in $P6_3/mmm$, rather than Pauling's ordered model in $P\bar{3}m1$ (which they regarded as generally accepted) managed to reduce R to about 7%, only half the best they could R with Zachariasen's model similarly construed. Both original models gave about the same R value of about 30%, and for both a treatment assuming twinning instead of disorder brought R down to only about 14%. It appeared that Pauling's model was indeed generally correct, although there was still doubt about the details of and reasons for the disorder that is required to realize the observed holohedry of this high-temperature form of La_2O_3.

The most recent work we have found on the A-modification structure was done by Greis [95]. It strongly supports the Pauling structure. We quote from his Abstract: 'Both X-ray powder and electron single-crystal diffraction revealed the A-type sesquioxide structure. No difference could be found in samples either quenched or slowly cooled to room temperature. Two space groups, $P\bar{3}m1$ and $P6_3/mmc$, are reported for the A-type structure. X-ray powder studies seemed to support the latter on the evidence of extinctions. Electron diffraction from single crystals, however, indicated the space group $P\bar{3}m1$, confirming the so-called Pauling structure, while $P6_3/mmc$ can now be excluded unequivocally.'

Greis cites a slightly earlier neutron-diffraction study [96] as also showing that the A-type structure belongs to $P\bar{3}m1$, but that above 2030°C, 'the so-called H type is the stable form which is to be described with $P6_3/mmc$', likely statically or dynamically disordered, and he concludes with 'It is now also clear that the Müller-Buschbaum/ Schnering structure does not describe the A-type [but possibly] the H-type structure ... retained ... on quenching.' The M-B-S preparation included a final heating at 2200°C.

Bixbyite, $FeMnO_3$

Here [97], Pauling seems to have been motivated not by a perception of technical error, but again by his search for order in the structural world. He noticed in the structure found by Zachariasen [98] that 'not only are the interatomic distances reported abnormally small, but also the structure does not fall in line with the set of principles found to hold for coordinated structures in general.' [24]. Among these principles was 'The Rule of Parsimony', which states that 'the number of essentially different kinds of constituents in a crystal tends to be small' and, thus, 'the polyhedra circumscribed about all chemically identical cations should, if possible, be chemically similar'.

Zachariasen had shown that bixbyite, which had once been thought to be related to substances of the perovskite type, is instead isostructural with the long series of sesquioxides (the C-Modification sesquioxides) recognized by V.M. Goldschmidt, and that it must be formulated with the metal atoms equivalent — as $(Fe, Mn)_2O_3$ rather than, as had been thought, $Fe^{II}Mn^{IV}O_3$. The structure proposed by Zachariasen, in space group $T^5 - I2_13_1$ placed the 32 M atoms in 8-fold sites of the type (x, x, x) and in two sets of 12-fold sites $(u, 0, \frac{1}{4})$. Pauling and Shappell were studying the mineral braunite, $3 Mn_2O_3 \cdot MnSiO_3$, which seemed to be related to bixbyite in structure and led them to Zachariasen's work. They noted that the structure proposed by Zachariasen was very nearly compatible with space group $T_h^7 - Ia3$, in which the two 12-fold positions would be combined into a single set of 24-fold positions, and remarked that 'it is difficult to find a physical explanation of this distortion from a more symmetrical structure'. Accordingly they made a careful study, using Laue and oscillation photographs, of many reflections of the type $0kl$ with k (and l) odd, which would be extinguished in T_h^7. They found none, and they saw that Zachariasen had been forced to reject T_h^7 because he had incorrectly assumed the contribution of the oxygen atoms to the structure factors to be negligible, whereas it is in fact great enough to account for certain crucial observed inequalities; e.g. between {271} and {217}, for bixbyite and even for Tl_2O_3. 'Dr. Zachariasen,' they noted, 'has kindly informed us he now agrees with our choice of the space group T_h^7.'

In the revised structure, the M atoms in these M_2O_3 compounds form an approximate cubic-closest-packed array and, as Zachariasen had recognized, the oxygens occupy only six of the eight cubic coordination sites around each M, as if two of the anions around each calcium in fluorite were removed or if two were added to the four around each zinc in zinc blende (cubic ZnS). The two types of M sites show different vastly distorted octahedral coordinations. Thus, Pauling's Rule of Parsimony is violated while at the same time it is confirmed: after all, the metal atoms could have adopted more than one coordination number. Pauling and Shappell did not comment on this, noting only that 'the coordinated octahedra are deformed so that the anions are at six cube corners, and their radius ratio will accordingly tend to the range of values giving coordination number 8.'

By considering the intensities of reflections of the type ($h00$), Pauling and Shappell concluded that the coordinate of the atoms in the 24-fold positions ($u, 0, \frac{1}{4}$) was $|u| = 0.030$. They then made a very interesting observation:

Now there are two physically distinct arrangements of the metal atoms corresponding to $|u| = 0.030$, the first with $u = 0.030$, and the second with $u = -0.030$; and it is not possible to distinguish between them with the aid of the intensities of reflection of X-rays which they give. ... the value of the structure factor is the same for a given positive as for the same negative value of x, except for a difference in sign in some cases. But the positive and negative parameter values correspond to structures which are not identical, but are distinctly different, as can be seen when the attempt to bring them into coincidence is made. This is a case where *two distinct structures give the same intensity of X-ray reflection from all planes*, so that they could not be distinguished from one another by X-ray methods. The presence of atoms in $8e$ or $8i$ [now $8a$ or $8b$] does not change this result.

It would have been helpful if Pauling and Shappell had pointed out two fairly simple aspects of the situation, aspects that must have been obvious to them, but evidently were something of a puzzle to Glusker (*vide infra*) in her commentary, and certainly were puzzling to us — until we finally saw the light: (1) Positions $8e$ and $8i$ do not 'change this result', but only because, although quite different in general when the 24-fold positions are occupied, they do interchange roles precisely when the parameter u of the 24-fold is changed from $u \approx 0$ to $u \approx 0.25$; (2) that the 24-fold cases with $u = |u|$ and $u = -|u|$ are genuinely different becomes clear if, for example, one examines the six 24-fold points, for $u \approx 0$, around any of the $8e$ points, like the origin, each of which lies on a three-fold axis; a small positive displacement of u from $u = 0$ moves the six points *nearer* to the three-fold, a small negative, farther from it. Pauling and Shappell continued:

> In the case of bixbyite a knowledge of the position of the oxygen atoms would enable the decision between these alternatives to be made, but the rigorous evaluation of the three oxygen parameters from the X-ray data cannot be carried out.

In order to resolve the dilemma, Pauling and Shappell persisted with the oxygen atoms. With $u = +0.030$ (which approximated Zachariasen's structure), they could find no arrangement of oxygens that led to satisfactorily large interatomic distances; with $u = -0.03$, they obtained satisfactory distances with the oxygens in 48-fold general positions with $x \cong \frac{3}{8}$, $y \cong \frac{1}{8}$, and $z \cong \frac{3}{8}$. They settled on the values $x = 0.385$, $y = 0.145$, $z = 0.380$; with this structure, each metal atom is surrounded by six oxygen atoms at a satisfactory distance of 2.01 Å, and the minimum O$\cdots$O distance is 2.50 Å. They caution, though: 'It is probable that the various metal–oxygen distances are not exactly equal, but show variations of possibly ± 0.05 Å.'

With the bixbyite paper, Pauling and Shappell exposed a frightening gap in the conceptual underpinning of X-ray crystallography, one that has never been satisfactorily filled in. With more and more successes to their credit, X-ray workers had become more and more confident that they could determine *unambiguously* the structure of any crystal, if only they could discover it. The metal-atom structure in bixbyite is an absolute counter-example. Pauling, perhaps among few others, remained aware of it. In 1934, A.L. Patterson showed that his now famous 'F^2 series' could lead directly to structures, and in 1939 he thought he had found a proof that no two physically different structures made up of a given set of atoms could have the same Patterson diagram [99].

Pauling saw this abstract and wrote to Patterson, 'Shappell and I stated that Wyckoff's positions 24(e) [now 24(d)] for the space group T_h^7 gives the same values of F^2 for positive and negative values of the parameter, which however correspond to different atomic arrangements.' [Pauling and Shappell actually showed more: his statement holds for the whole 32 metal atoms.] Patterson eventually found many possible examples of such *Homometric* structures, and there is now a considerable literature on the problem, both crystallographically oriented and more purely mathematical, but to our knowledge it is true both that no full crystal structure has been found to be ambiguous in the exact Pauling–Patterson sense and that no useful practical test has ever been discovered for telling whether a given arrangement of atoms is possibly homometric with some unknown, different arrangement of the same set of atoms [100].

The structure of bixbyite has been re-examined by H. Dachs [101], who indeed found that the M–O distances deviate appreciably from the constancy assumed by Pauling and Shappell in the purely numerical estimation of their precise values for the oxygen parameters. Dachs's M–O lengths average 2.02 Å with estimated scatter standard deviation 0.16 Å, about three times the range estimated by Pauling and Shappell. His crystal was twinned, on (110); theirs apparently wasn't. Perhaps it still remains to be seen how regular the M–O bond lengths in bixbyite really are.

The cadmium chloride structure

In 'On the Crystal Structures of the Chlorides of Certain Bivalent Elements', [102] Pauling derived what it now often called the Cadmium Chloride structure to fit the data that had been reported by Bruni and Ferrari [103] for $MgCl_2$, Bruni and Ferrari [104] for $ZnCl_2$, $CdCl_2$, and $MnCl_2$, and Ferrari [105] for $CoCl_2$ and $NiCl_2$ all of which, according to these workers, had the same type of structure, with rhombohedral unit cells containing 2 or 16 MCl_2, a equal to about 5 or about 10 Å, and α about 90°. Pauling remarked, 'None of these investigators succeeded in deducing the atomic arrangement in these crystals.' He did, emphasizing $CdCl_2$. He indexed all the spots on the Bruni–Ferrari Laue pattern for $CdCl_2$ in terms of a rhombohedral cell with $a = 6.35$ Å, $\alpha = 36°40'$, and $Z = 1$, incidentally, therefore, correcting errors in unit-cell determination that had been made by all the cited authors and remarking of his unit cell in relation to the $Z = 16$ cell that it 'has lattice points at the face-centered positions of the rhombohedron which is itself formed by the face-centered positions of the original pseudocubic unit.' Goldschmidt [106] had assigned also $RuCl_2$, RhCls2, $PdCl_2$, $IrCl_2$, and $PtCl_2$ to this structure, according to Pauling [107]. However, as we read Goldschmidt [106], he had surmised that 'the dichlorides of the platinum metals' had the $NiCl_2$ structure, but that Oftedal had shown this not to be the case.

With his student J.L. Hoard he published a full study of $CdCl_2$, including the determination of the value of the parameter, u of the position uuu, etc., shortly thereafter [107].

The standard attribution of credit for these structures is somewhat unclear. Strukturbericht confirms that Bruni and Ferrari had wrong unit cells, but Crystal Data assigns all the structures except of $CdCl_2$ to Ferrari, presumably in view of a paper [108] by Ferrari *et al.*, a paper that we haven't seen. Anyhow, Ferrari later published a summary and refinements [109] from which, perhaps, one can understand what happened. The $LiCl-MgCl_2$ story is interesting.

Crystal Data lists **no** structures of the type for any of the five heavy-metal chlorides, nor, in fact, any mention at all of three of them.

Homochiral phenylaminoacetic acid

This is a strange case to which we shall devote quite a lot of attention. George L. Clark (whom Pauling and others had found cause to criticize on a number of occasions) published with his student G. Robert Yohe a paper [110] that claimed to demonstrate the long-accepted concept of molecular asymmetry in a new way — by determining the space group of crystals of an optically resolved substance. Pauling and Dickinson submitted their 'Note of objection' [111] hardly a month later, but it was not published for two years, not until after Clark had published another error-ridden paper, on biphenyl and five of its derivatives [112], to which Maurice L. Huggins had submitted *his* objections [113], and Clark had at last (24 August 1931) submitted his unconvincing rebuttal [114] to both complaints.

Pauling and Dickinson were correct and compelling, but they didn't address all of Clark and Yohe's errors. They could have: they were especially well oriented to the problem because Dickinson had worked out the structure of the crystals of sodium chlorate and sodium bromate [115], which were well known as optically active, although their solutions are optically *inactive* and despite the fact that even in the crystal structure the constituent ions have intrinsic symmetry incompatible with optical activity. (Their crystal environment is sufficiently less symmetrical, however.) Interesting details about the crystal structure and optical activity of sodium chlorate reach back to about 1850 and continue to be published still [116—118].

Clark and Yohe had measured the density and cell dimensions of *l*-phenylamino-acetic acid crystals, finding an orthorhombic cell with $Z = 4$, and on these grounds concluded that the space group must be one of the 14 in the orthorhombic system ($D_2^{1\cdots4}, C_{2v}^{1\cdots10}$) whose general positions are 4-fold. On the basis of systematic absences they then assigned space group $C_{2v}^5 - Pca2_1$, 'which by its purely geometric deriva-tion demands four asymmetric molecules per unit cell... Thus... the space group of the crystal requires the molecule to be asymmetric.'

Pauling and Dickinson caught the egregious error in this argument: that Clark and Yohe, 'entirely without argument or justification, eliminated from discussion the forty-two orthorhombic space groups which provide positions for four non-asymmet-ric equivalent molecules'. They also noted that 'the space-group criteria for C_{2v}^5 are identical with those for Q_h^{11}, so that Clark and Yohe's data alone could not possibly be used to indicate C_{2v}^5 uniquely' and that 'there is no adequate demonstration that the crystal is even orthorhombic,' Clark and Yohe having presented evidence only that it is metrically orthorhombic but not that it has orthorhombic symmetry. Pauling and Dickinson also noted that Clark and Yohe's space-group assignment is inconclusive in another way, even if the crystal is assumed to be orthorhombic, because they arbitrarily considered their axis of length 9.66 Å to be the *c*-axis of the Astbury–Yardley tables instead of considering all the possible identifications of observed axes with the axes of the tables.

Then another surprising error. Clark and Yohe had pointed out that 'C_{2v}^5 calls for halving $\{h\,0\,l\}$ if h is odd, and halving of $\{0\,k\,l\}$ if l is odd', and had presented two tables apparently indicating these systematic absences; Pauling and Dickinson noticed that the data in the two tables are identical, both obviously pertaining to the same zone. 'Accordingly, not more than one of the C_{2v}^5 conditions can be regarded as satisfied.' They pointed out also that the data used 'could not be made the basis of a space-group determination even if treated correctly', being restricted to only one prism zone, whereas in the absence of knowledge of the macroscopic point-group symmetry a 'knowledge of the systematic presence or absence of reflections from planes of all three prism zones would be necessary'.

They concluded by remarking, 'we do not contend that the *l*-phenylaminoacetic acid molecule is symmetrical or that the space-group symmetry is necessarily other than $C_{2v}^5;\ldots$'.

It is astonishing, however, that Pauling and Dickinson could have written that last remark, or that they could have accepted (as their article seems to suggest they had) the initial Clark and Yohe premise, p. 2797, 'there are a few [space groups] which will admit of no molecular symmetry. If, in crystallizing, an optically active compound builds its crystalline form according to one of these space groups which will admit of no symmetry of the molecule this may be taken as direct evidence that its molecules are asymmetric.' Can it be that Clark and Yohe were misled by the column headings in Astbury and Yardley that read 'Possible Molecular Symmetry', but which in more recent tables read simply 'Punktsymmetrie' (*Internationale Tabellen...*), 'point symmetry', (*International Tables for X-ray Crystallography*, I), or 'site symmetry' (*International Tables for Crystallography*, A)? It has become crystallographically commonplace for a molecule to show, to within the limits of experimental error, 'noncrystallographic' symmetry greater than the symmetry of the site it occupies in a crystal structure. If then, contrary to the already so firmly grounded chemical knowledge, going back to van't Hoff, Le Bel, and Pasteur, *l*-phenylaminoacetic acid somehow *had* intrinsic symmetry 'of the second kind', it might still occupy unsymmetrical sites in its crystal structure. It is doubtless true that non-crystallographic molecular symmetry is more often recognized nowadays than it was in 1929, but it is astonishing to us that Pauling and Dickinson didn't make a point of Dickinson's example from 1920: in sodium chlorate the chlorate site symmetry is only C_3, but the chlorate ion structure, as determined by Dickinson, has symmetry C_{3v}, as it automatically has, if it has symmetry C_3. And the knowledge that sodium chlorate crystals, all of them, rotate the plane of polarized light in one direction or the other, although the substance in solution is optically inactive, goes back to about 1850. The Clark and Yohe premise is indeed blatantly false.

We now come to Pauling and Dickinson's last remark, that they do not contend that the space group is necessarily other than C_{2v}^5. We can only assume that they were wholly engaged with presenting their technical points, to the exclusion of more general aspects, in effect were overcome by the unimaginable number of Clark and Yohe's errors. For *no* point symmetry including *improper* symmetry elements (i.e. inversion axes) can normally be adopted by a homochiral optically active substance. In C_{2v}^5, for example, reflection in any one of the glide planes ensures that half the molecules are of opposite configuration from the other half. Indeed, this point *was* made by Huggins [113], p. 3824, with the words 'The assignment of 3,3'-diaminodimesityl, which "has been resolved into two active forms," to C_{2h}^5, cannot be correct, for space groups containing symmetry planes are impossible for optically active crystals.' This quote may seem ambiguous, but in its context is not: Clark and Pickett worked on both the *d* and *l* forms, reporting that they gave identical X-ray data. And Clark, in his rebuttal, has an italicized paragraph insisting that the rotation photographs show a plane of symmetry, and arguing that the optical isomerism, being of (Roger Adam's) new type, is to be thought of as outside the realm of the classical stereochemical theory.

One more related bit, old but still new, has to be mentioned about optical activity and symmetry of crystals. It is an ancient and remarkable fact [119] that the system *d*-carvoxime–*l*-carvoxime forms an apparently continuous set of solid solutions with maximum melting point at the composition 50:50. Its basis was considerably clarified rather recently by the determination of its crystal structure [120], in space group $P2_1/c$ and with centrosymmetric cyclic-hydrogen-bonded dimers, and by the demonstration that when a moderate excess of one of the enantiomers is present the 'excess' molecules of the majority species occupy sites of the wrong chirality at random, changing the space group from $P2_1/c$ to $P2_1$ but not greatly changing other aspects of the structure [121]. (A 'wrong' molecule fits into the structure quite well, only the isopropenyl group causing any trouble.) Going further, the homochiral crystal is clearly similar to the racemic structure but has symmetry $P2_1$. No discontinuity in the phase diagram, corresponding to the change in space group, has been detected, but both racemic and homochiral samples have been obtained in two forms, one melting five or ten degrees below the other and having improbably a lesser ΔS of melting by about 2.3 or 4.6 cal deg^{-1} mol^{-1} and thus the greater S. The forms sometimes intertransform in the melting region [122]. Kroon *et al.* do not make it clear which of the forms the structure crystals belonged to. The high-temperature forms should have the *greater* entropies, corresponding to orientational disorder of the isopropenyl group at the higher temperature in both the racemic and the homochiral cystals and, in the racemic case, also to substitutional disorder of *d* vs *l*. If the crystal structures that have been determined were of the low-temperature forms, as seems likely, we could then understand the continuous liquidus and solidus of the phase diagram and would consider the transition in space group to be continuous also, for the high forms. But we are stumped by the report that the higher melting forms also have the higher heats of melting. However, all this suggests that someday a good approximation may turn up to the case ruled out by Huggins and our received crystallographic wisdom. Did Pauling and Dickinson have such an occurrence in mind?

As far as we are aware, there has still been no satisfactory assignment of space group to crystals of *l*-phenylaminoacetic acid.

Chalcopyrite, $CuFeS_2$

Interested in the interatomic distances in $CuFeS_2$, of accepted structure by Burdick and Ellis [123] (who, however, had not attempted a precise fix on the parameter *u* of the sulfur positions $\frac{1}{4}\frac{1}{4}u, \ldots$), Pauling and Brockway [124] found that 'the unit is twice as large as the pseudocubic unit described above [of the accepted structure], and the distribution of copper and iron atoms is completely different.' All except the pseudocubic reflections are of low intensity so that the 'evidence for the space-group is not overwhelming', but the newly derived structure fitted all the data, 'so we believe the space-group is correct.' The structure is still closely related to the familiar zinc-blende structure, but with one axis doubled, space-group $D_{2d}^{12} - I\bar{4}2d$, $Z = 4$, and with rather curious, unexplained rules of alternation of Cu and Fe. Each Fe or Cu is surrounded by 4 S, and each S by 2 Fe and 2 Cu; in the 'straight' zig-zag chains ... M–S–M–S ... in the basal plane, Cu and Fe alternate, but in the chains that angle up the sides of the unit cell, the M-sequence is ... Cu Cu Fe Fe ....

The Burdick and Ellis paper is No. 3 from the Chemical Laboratories of the Throop College of Technology, but the submission line at the end is 'Cambridge, Mass.,' an echo of the role of A.A. Noyes in establishing X-ray crystallography at what is now Caltech.

Sulvanite, Cu_3VS_4

This structure was first studied by W.F. de Jong [125], who based it on a face-centered cubic cell with $a_0 = 10.75$ Å. Pauling and Ralph Hultgren [126], noted that the powder diffraction pattern described by de Jong could be explained by a primitive unit cell with a_0 one-half as large (their value was 5.370 ± 0.005 Å), with space group $T_d^1 - P\bar{4}3m$ (de Jong's was $O_h^5 - Fm3m$); from a careful consideration of the relative intensities of Laue reflections they assigned a value $u = 0.235 \pm 0.004$ to the positions of the S atom (at *uuu* etc.). In this revised structure, each Cu is surrounded tetrahedrally by 4 S (and octahedrally by 6 V, more distant), whereas in de Jong's structure the 4 S atoms are coplanar with Cu. Pauling and Hultgren ignored this discrepancy, focussing their discussion on the very unusual coordination about the S atoms: 3 Cu atoms in tetrahedral directions, and a vanadium atom 'not at the fourth corner of the tetrahedron, but in the negative position to this: that is, in the pocket formed by the three copper atoms'. This feature (which was also shown by de Jong's structure) seems to have puzzled Pauling ever since.

Rubidium nitrate

Pauling and Sherman's 'Note on the Crystal Structure of Rubidium Nitrate [127] points out that Zachariasen's [128] unit cell and space group for the crystal form in question are incorrect, gives evidence for a larger unit cell ($Z = 9$ instead of $Z = 4$), which 'may be accepted with some confidence . . . , provided the crystals are considered to be hexagonal', deduces the approximate locations of the rubidium and oxygen atoms, and suggests a possible scheme for the nitrogen atoms. At least until 1973 (*Crystal Data*, 3rd edn) the full structure had not been found.

Note on undergraduate research

One of our oldest Caltech memories is of Pauling telling about the importance of undergraduate research, and giving as two examples of involvement with undergraduate crystallographic research at Caltech the names of Edwin McMillan and Ken Pitzer, whose later great success, mainly at the University of California, Berkeley, is at least not inconsistent with Pauling's strong belief in undergraduate research participation.

   Pitzer, besides being a coauthor with A.A. Noyes, J.L. Hoard, and Clarence L. Dunn of two articles on the higher valence Ag^{++} as produced by ozone oxidation of silver nitrate in nitric acid solution, determined the structure of tetramminocadmium perrhenate [129], short of evaluating all the parameters, concluding that analogous zinc and cobalt compounds 'undoubtedly have the same structure, which is probably that of the similar permanganate compounds.' His acknowledgment: 'The author wishes to express his appreciation to Dr. Linus Pauling for his interest and advice during this investigation.' Edwin McMillan (who later won the Nobel prize in Physics) and Pauling had studied the Pb–Tl system by powder diagrams [150], finding that $PbTl_2$ is not a phase, in agreement with certain other indications, but disproving the claim of Lewkonja [131] that it is. Tang and Pauling returned to the problem [132], finding no evidence of the sharp bend at 54.6 atomic percent thallium in the curve of cell-edge *vs* atom-fraction as reported by Ölander [133], but noting reasons to believe that the possible simple ordered cubic structures AB_3 and AB_7 are of some importance in the forms $Pb(Pb, Tl)_3$ and $PbTl(Pb, Tl)_6$.

Later crystallographic works

The crystallographic work that Pauling carried out during the decade 1925–1935 was truly prodigious, not only in terms of the quantity and variety of structural problems that he solved (or corrected) but, far more important, in terms of the insight into the world of structural chemistry that he gained and passed on to others. But he was doing other things as well. In 1935, he and E. Bright Wilson published [134] *Introduction to Quantum Mechanics* — 'a textbook of practical quantum mechanics', aimed primarily for chemistry students, which found wide use in graduate schools throughout the United States. He applied his knowledge of quantum mechanics to the field of crystallography: in 1932, obviously tired of having to make assumptions concerning the relative scattering powers of different atoms, he and Sherman [135] proposed a set of energy-level 'screening constants' and used them to calculate form factors for most of the neutral atoms and for a score of important cations and anions. They noted that, while their form factors were not as 'reliable' as those calculated by Hartree's method [136], 'they are obtained with much less labor'. We note that James and Brindley [137] had, a year previously, carried out more exact (and more laborious) Hartree-type calculations for a number of important atoms; it was not until the 1950s, when better computing facilities and more accurate wave functions became available, that the Pauling–Sherman and James–Brindley form factors were superseded.

In 1937–1938, as Baker Lecturer at Cornell University, Pauling collected much of his chemical and structural insight into his classic book *The Nature of the Chemical Bond* which forever changed the way that chemists and biologists looked at their worlds: molecules became truly three-dimensional, and their properties and reactions could be closely related to the bonds — covalent, or ionic, or hydrogen, or van der Waals — that they formed. But after about 1936, Pauling would have ever more responsibilities than his immediate researches: he became division chairman at Caltech; he participated in war-time research projects; he had more students to guide; and his non-crystallographic researches and his teaching and public activities increased. He was stricken with glomerular nephritis (1940–1941), was largely confined to bed rest for months, and fought the malady for several years. Indeed, his time in a diffraction laboratory was severely limited. Yet his interests in structural chemistry never waned, nor did his capacity for combining structural results with his great insight into structural principles. He became more and more interested in protein structure, later in the nucleic acids, and, all along, in the structures of intermetallic compounds. With driven, characteristic recurrent persistence he pursued the case of $NaCd_2$, which he had first encountered in 1923 [6]. (His Laue photographs were 'so complicated that it was not found possible to assign indices with certainty.') In the 1950s he more than once felt he was on the point of discovering the structure, especially with the theme of icosahedral coordination, which is more efficient than closest packing of equal spheres, 12 equal atoms being accommodated around not an equal sphere, but a sphere of 10% smaller radius. But although this theme led to notable successes, for example with $Mg_{32}(Al, Zn)_{49}$, by Bergman, Waugh, and Pauling [138], where icosahedra were correctly presumed to be major components of the structure, $NaCd_2$ continued elusive. He took obvious pride that the structure of $NaCd_2$ was eventually solved [139] by an associate of his at Caltech, Sten Samson; in the structure (which has over 1100 atoms in the unit cell), there are many fascinating polyhedra, some of them more important to Samson's derivation of the structure than the icosahedra. The concept of regular polyhedra was also the basis of Pauling's derivation of the structure of the hydrates of smaller gas molecules [140].

But surely the most notable success of Pauling's stochastic method was his derivation of the basic configurations of polypeptide chains, the alpha helix and the beta sheets [141]. Here, he started with the basic assumptions — deduced from his huge inventory of structural information — of a planar peptide group and of hydrogen bonds N—H···O between groups. And in deriving the alpha helix he came up with another crucial concept — that the polypeptide chain did not need to be periodic in the crystallographic sense, but instead could form helices with an irrational number of residues per turn while still giving rise to a perfectly good fiber diffraction pattern. While this concept seems obvious today, it was, as far as we are aware, entirely original with Pauling. Apparently, he constructed his first model of the alpha helix in much the same way that he constructed his earliest models of mineral structures — out of paper.

A non-integral helix is also, of course, the basis of the structure of DNA. In 1953, Pauling and Corey [142] proposed such a structure, very tentatively, for DNA; however, they improbably placed the phosphate groups, rather than the purine and pyrimidine bases, on the inside of the helix. This is one of the few failures of the stochastic method in Pauling's hands. (It was successful for Watson and Crick!) Pauling's use of the stochastic method has continued into the 1990s. In the latest example we find [143], he proposes — recalling the structure of basic beryllium acetate [144] but primarily invoking his principles — that the compound presumed to be $LiBeH_3$, for which Selvam and Yvon [145] had been unable to find a satisfactory structure, is actually $Li_6Be_4OH_{12}$; he proposes coordinates for all the atoms. However, although the title of the article, 'Determination of the Crystal Structure and Composition of $Li_6Be_4OH_{12}$ by the Stochastic Method', seems to promise a full application of the stochastic method, his standard, independent final confrontation of observed and calculated diffraction intensities is not reported, and the structure remains unverified. Pauling has also been fascinated — 'obsessed', he says — with the problem of 'quasicrystals', which give diffraction patterns with apparently exact or nearly exact five-fold symmetry; here, he has proposed [146] that (in contrast to the widely popular theory based on Penrose tiling) the five-fold symmetry is due to repeated twinning of discrete crystallites, the structure of each crystallite being based on icosahedral coordination perhaps similar to that found in $NaCd_2$. The detailed, atom-by-atom structures of these fascinating materials are still unknown.

We have written here that he has not been much interested in mechanical and computational aids to the working out of crystal structures, but he *has* made important contributions and has given advice. We have mentioned his polyhedral paper models of ionic structures and his famous paper model of the alpha-helix, but so far have neglected the beautiful, precisely machined stainless steel, 'ball-and-stick' models that he used in exploring the helices, pleated sheets, and the like, as well as the enormously useful space-filling models [147] that were developed over the years in the Caltech shops and evolved into the widely used 'CPK' models. As to computation, he supported the initial development of an analog machine for Fourier syntheses, but on the advice of W.J. Eckert was converted to punched cards, almost instantly designed a scheme for using them, and took part in its development [148]. It was a pioneering step and was important, at Caltech and elsewhere (especially as a demonstration of what could be done), until electronic computers came on the scene. Later, when almost all crystallographers were using substantial, if primitive, computers, he chided the community for its pedestrian style, which required more and more time to complete a structure determination, taking several or many times as long as their predecessors had in earlier days [149]. Perhaps he didn't make enough allowance for how the structures being solved were crystallographically much more complicated than those of the earlier days (though no more challenging to visualize). In any event, his admonition was followed by three prodigiously important advances: the maturation of 'direct methods', the coming of really powerful computers, and the rise and commercialization of the automatic, computer-controlled diffractometer. We have no desire to review all this except in the context of Pauling's very recent advice, at his ninetieth-birthday celebration at Caltech: Rejoicing in the ordinary great ease and speed of crystal-structure determinations, but dismayed by the frequency of error, he strongly deplored the execution and presentation of so much of this work 'without thought'.

During his career, Pauling has, we believe, contributed as much to the development of X-ray diffraction as any person since Laue and the Braggs. He has done this not by working *on* the method but by working *with* it. He recognized, very early in his career, the tremendous importance of the technique: he fully appreciated how knowledge of *structure* was vital to an understanding of other properties of matter; here was a technique to be used as vigorously and as imaginatively as possible. And this is how Pauling has used it — vigorously and imaginatively. Moreover, he has passed this vigor and imagination to others in the field, both through his copious contributions to the literature and through the large number of associates and students that he attracted to Caltech. This is his legacy to X-ray diffraction.

We are indebted to V. Waser for advice and criticism of an early partial draft, to W. P. Schaefer for reading the completed manuscript, and to Judy Schomaker for taking great careful pains with the proof.

NOTES AND REFERENCES

[1] Pauling has written many fascinating accounts of his work and surroundings, often with emphasis on X-ray crystallography and its history, especially at Caltech. One of his great contributions, which we shall not otherwise dwell on, was to grow into, take over, and greatly expand the school of structural chemistry that Arthur Amos Noyes had established almost at the very beginning of his association with the Pasadena institution, even before it got its present name. Noyes, through Burdick and Ellis (who appear briefly later in our narrative), began the work already in 1917, and brought Dickinson from MIT to become Caltech's first Ph.D. recipient (1920) and, later, Pauling's teacher in crystallography. Here is a listing of some of Pauling's scientific reminiscences:

[1a] Roscoe Gilkey Dickinson, *Science* **102** (1945) 216.

[1b] Problems of Inorganic Structures, In: *Fifty Years of X-ray Diffraction* (Dedicated to Max von Laue), P.P. Ewald (Ed.), N.V.A. Oosthoek's Uitgeversmaatschappij, Utrecht (1962), pp. 136–146.

[1c] Early Work on X-ray Diffraction in the California Institute of Technology, In: *Fifty Years of X-ray Diffraction, ibid*, 623–628.

[1d] Acceptance of the Roebling medal of the Mineralogical Society of America, *Am. Mineral.* **53** (1968) 521–530.

[1e] Fifty Years of Physical Chemistry in the California Institute of Technology, *Ann. Rev, Phys.* **16** (1965) 1–13.

 [1f] The Early Years of X-ray Crystallography in the United States, In: *Structural Studies on Molecules of Biological Interest: A Volume in Honour of Professor Dorothy Hodgkin*, Guy Dodson, Jenny P. Glusker, and David Sayre (Eds), Clarendon Press, Oxford (1981), pp. 35–42.

[1g] Foreword. In: *Crystallography in North America*, D. McLachlan, Jr. and J.P. Glusker (Eds), American Crystallographic Association, New York (1983), v–vi.

[1h] X-ray Crystallography in the California Institute of Technology, In: *Crystallography in North America*, D. McLachlan, Jr. and J.P. Glusker (Eds), American Crystallographic Association, New York (1983) pp. 27–30.

[1i] The Discovery of the Structure of the Clay Minerals, *CMS News* (A Publication of The Clay Minerals Society) (1990) 25–27.

[1j] Historical Perspective, In: *Structure and Bonding in Crystals*, volume 1, Michael O'Keeffe and Alexandra Navrotsky, (Eds.), Academic Press, New York (1981) pp. 1–12.

[2] L. Pauling, *Z. Krist.* **84** (1933) 442–452.

[3] L. Pauling, The Stochastic Method and the Structure of Proteins, In: *13th International Congress of Pure and Applied Chemistry: Plenary Lectures*, Stockholm (1954) 37–52 *Am. Scientist* **43** (1955) 285–297.

[4] G. Albrecht, In: *Structural Chemistry and Molecular Biology*, A. Rich and N. Davidson (Eds), Freeman and Company, San Francisco and London (1968) pp. 887–907.

[5] The Crystal Structure of Molybdenite, R.G. Dickinson and L. Pauling, *J. Am. Chem. Soc.* **45** (1923) 1466–1471.

[6] The Crystal Structure of Magnesium Stannide, L. Pauling, *J. Am. Chem. Soc.* **45** (1923) 2777–2780.

[7] The Crystal Structure of Uranyl Nitrate Hexahydrate, L. Pauling and R.G. Dickinson, *J. Am. Chem. Soc.* **46** (1924) 1615–1622.

[8] The Crystal Structures of Ammonium Fluoferrate, Fluo-aluminate and Oxy-fluomolybdate, L. Pauling, *J. Am. Chem. Soc.* **46** (1924) 2738–2751.

[9] The Crystal Structures of Cesium Tri-iodide and Cesium Dibromo Iodide, R.M. Bozorth and L. Pauling, *J. Am. Chem. Soc.* **47** (1925) 1561–1571.

[10] G. Menser, *Z. Krist.* **73** (1930) 113–150.

[11] L. Pauling, *Z. Krist.* **74** (1930) 104–105.

[12] S.B. Hendricks and L. Pauling, *J. Am. Chem. Soc.* **47** (1925) 2904–2920.

[13] L. Pauling and J.H. Sturdivant, *Z. Krist.* **68** (1928) 239–256.

[14] W.L. Bragg and G.B. Brown, *Proc. Roy. Soc.* **A 110** (1926) 34–63.

[15] W.L. Bragg and J. West, *Proc. Roy Soc.* **A 114** (1927) 450–473.

[16] W.H. Baur, *Acta Cryst.* **14** (1961) 214–216.

[17] L. Pauling, *Proc. Nat. Acad. Sci.* **14** (1928) 603–606.

[18] J. Leonhardt, *Z. Krist.* **59** (1924) 216–229.

[19] N.A. Alston and J. West, *Z. Krist.* **69** (1928) 149–167.

[20] The Sizes of Ions and the Structure of Ionic Crystals, L. Pauling, *J. Am. Chem. Soc.* **49** (1927) 765–790.

[21] The Sizes of Ions and Their Influence on the Properties of Salt-like Compounds, L. Pauling, *Z. Krist.* **67** (1928) 377–404.

[22] The Influence of Relative Ionic Sizes on the Properties of Salt-like Compounds, L. Pauling, *J. Am. Chem. Soc.* **50** (1928) 1036–1045.

[23] The Coordination Theory of the Structure of Ionic Crystals, L. Pauling, In *Festschrift zum 60. Geburtstage Arnold Sommerfelds*, Leipzig: Verlag Hirzel (1928), pp. 11–17.

[24] L. Pauling, *J. Am. Chem. Soc.* **51** (1929) 1010–1026.

[25] W.L. Bragg, *Z. Krist.* **74** (1930) 237–305.

[26] L. Pauling, *Z. Krist.* **73** (1930) 97–112.

[27] H. Mark and P. Rosbaud, *N. Jb. Min.* **54 A** (1926) 127.

[28] L. Pauling, *Z. Krist.* **74** (1930) 213–225.

[29] T. Barth, *Norsk Geol. Tidskr.* **9** (1926) 40.

[30] F.M. Jaeger, H.G.K. Westenbrink, and F.A. van Melle, *Proc. Acad. Amsterdam* **30** (1927) 249–267.

[31] F.M. Jaeger, *Trans. Faraday Soc.* **25** (1929) 320–345.

[32] *Crystal Structure of Minerals*, W.L. Bragg, G.F. Claringbull, and W.H. Taylor, Cornell University Press, Ithaca, New York (1965), p. 371.

[33] F.M. Jaeger and F.A. van Melle, *Proc. Acad. Amsterdam* **30** (1927) 479–498, 885–904.

[34] L. Pauling, *Z. Krist.* **72** (1930) 482–492.

[35] O. Hassel and R. Salvesen, *Z. physik. Chem.* **128** (1927) 345–361.

[36] L. Pauling, *Proc. Nat. Acad. Sci.* **16** (1930) 123–129.

[37] L. Pauling, *Proc. Nat. Acad. Sci.* **16** (1930) 578–582.

[38] Ch. Mauguin, *Comp. Rend.* **185** (1927) 288–291.

[39] W.W. Jackson and J. West, *Z. Krist.* **76** (1931) 211–227.

[40] B. Gossner, *N. Jb. Min., Beil.-Bd.* **55 A** (1927) 319.

[41] W. Barclay Kamb, *Acta Cryst.* **13** (1960) 15–24, 24–27.

[42] S. John Louisnathan and G.V. Gibbs, *Am. Mineralogist* **57** (1972) 1089–1108.

[43] W.H. Baur and T. Ohta, *Acta Cryst.* **B 38** (1982) 390–401.

[44] Yu. G. Zagal'skaya and N.V. Belov, *Soviet Physics—Crystallography* **8** (1964) 429–432.

[45] L. Pauling, *J. Am. Chem. Soc.* **51** (1929) 2868–2880.

[46] H. Bartl, *Neues Jahrb. Mineral. Monatsh.* (1970) 552–557.

[47] According to Per-Olov Löwdin (Lecture in Pasadena in the early 50s) the point to which Pauling carried the refinement of an approximate quantum-mechanical calculation. Up to the Pauling Point, increased refinement gives improved results; beyond it, further refinement gives poorer results.

[48] R.J. Paradowski, *Thesis*, University of Wisconsin (1972), p. 201.

[49] G.L. Clark, *J. Am. Chem. Soc.* **46** (1924) 372–384 (p. 379).

[50] G.L. Clark and W. Duane, *Proc. Nat. Acad. Sci.* **9** (1923) 126–130.

[51] R.W.G. Wyckoff, *The Structure of Crystals*, The Chemical Cataloges., New York (1924) p. 80.

[52] D. Hall, A.D. Rae, and T.N. Waters, *Acta Cryst.* **19** (1965) 389–395.

[53] J.C. Taylor and M.H. Mueller, *Acta Cryst.* **19** (1965) 536–543.

[54] P. Groth, *Chemische Kristallographie*, Engelmann, Leipzig, **2** (1908) p. 142.

[55] W.H. Bragg and W.L. Bragg, *X-rays and Crystal Structure*, G. Bell and Sons, London (1915), p. 171.

[56] W.L. Bragg, *Nature* **105** (1920) 646–648. A report on a Royal Institution

Discourse, this article shows photographs of models of the structures of NaCl, calcite ($CaCO_3$), zinc blende (ZnS), and ruby (Al_2O_3). Bragg describes the structure of ruby as a stacking of Al_2 pairs, each with a belt of three oxygens, and involving two parameters — the separation of the two Als and the radius of the oxygen belt.

[57] W.H. Bragg and W.L. Bragg, *X-rays and Crystal Structure*, 4th edn, G. Bell and Sons, London (1924), p. 183.

[58] L. Pauling and S.B. Hendricks, *J. Am. Chem. Soc.* **47** (1925) 781–790.

[59] W.L. Bragg, *Proc. Roy. Soc.*, London **106** (1924) 346–368.

[60] I. Backhurst, *Proc. Roy. Soc.*, *London* **102** (1920) 340–368.

[61] R.E. Newnham and Y.M. de Haan, *Z. Krist.* **117** (1962) 235–237.

[62] S.K. Allison, *The American Journal of Science* **8** (1924) 261–276.

[63] G.L. Clark and W. Duane, *J. Opt. Soc.* **7** (1923) 455–482.

[64] L. Pauling and P.H. Emmett, *J. Am. Chem. Soc.* **47** (1925) 1026–1030.

[65] R.W. James and W.A. Wood, *Proc. Roy. Soc.*, *London*, **A109** (1925) 598–620.

[66] P. Groth, *Chemische Kristallographie*, Engelmann, Leipzig, **1** (1906) p. 306.

[67] H.A. Tasman and K.H. Boswijk, *Acta Cryst.* **8** (1955) 59–60.

[68] A.H. Armstrong, W. Duane, and R.J. Havighurst, *Proc. Nat. Acad. Sci.* **11** (1925) 218–221.

[69] F.J. Ewing and L. Pauling, *Z. Krist.* **68** (1928) 223–230.

[70] P. Scherrer and P. Stoll, *Z. Anorg. Chemie* **121** (1922) 319–320.

[71] R.G. Dickinson, *J. Am. Chem. Soc.* **44** (1922) 276–288.

[72] R.W.G. Wyckoff and E. Posnjak, *J. Am. Chem. Soc.* **43** (1921) 2292–2309.

[73] R.W.G. Wyckoff, *The Structure of Crystals*, The Chemical Catalog Co., New York (1924), p. 344.

[74] J.E. Lennard-Jones and B.M. Dent, *Phil. Mag.* **3** (1927) 1204–1227.

[75] M. Kirkpatrick and L. Pauling, *Z. Krist.* **63** (1926) 502–506.

[76] B. Gossner and O. Krauss, *Z. Krist*, **88** (1934) 298–303.

[77] L. Pauling, *Z. Krist.* **91** (1935) 367–368.

[78] D.F. Mullica, J.D. Korp, W.O. Milligan, G.W. Beall, and I. Bernal, *Acta Cryst.* **B36** (1980) 2565–2570.

[79] L. Pauling, The Structure and Entropy of Ice and of Other Crystals with Some Randomness of Atomic Arrangement, *J. Am. Chem. Soc.* **57** (1935) 2680–2684.

[80] L. Pauling, *Z. Krist.* **69** (1929) 35–41.

[81] F. Trombe and M. Foex, *Ann. d. chimie* **19** (1944) 417–445.

[82] A.F. Schuch and J.H. Sturdivant, *J. Chem. Phys.* **18** (1950) 145.

[83] A.W. Lawson and T.Y. Tang, *Phys. Rev.* **76** (1949) 301–302.

[84] P.W. Bridgman, *Z. Krist.* **67** (1928) 363–376.

[85] W.L. Bragg and J. West, *Proc. Roy. Soc.*, London **A 114** (1927) 450–473.

[86] W.H. Taylor and W.W. Jackson, *Proc. Roy. Soc.*, *London* **119A** (1928) 132–146.

[87] St. Naray-Szabo, W.H. Taylor, and W.W. Jackson, *Z. Krist.* **71** (1929) 117–130.

[88] C.W. Burnham, *Z. Krist.* **118** (1963) 337–360.

[89] J.H. Sturdivant and L. Pauling, *Z. Krist.* **69** (1929) 557–559.

[90] Beiträge zur Kenntnis des Feinbaues des Brookits und des physikalischen Verhaltens sowie der Zustandsänderungen der drei natürlichen Titandioxyde, A. Schröder, *Z. Krist.* **67** (1928) 485–542.

[91] W.H. Zachariasen, *Z. physik. Chem.* **123** (1926) 134–150.

[92] L. Pauling, *Z. Krist.* **69** (1929) 415–421.

[93] W.H. Zachariasen, *Z. Krist.* **69** (1929) 187–189.

[94] Hk. Müller-Buschbaum and H.G. von Schnering, *Z. Anorg. Allgem. Chemie* **340** (1965) 232–245.

[95] O. Greis, *Journal of Solid State Chemistry* **34** (1980) 39–44.

[96] P. Aldebert and J.P. Traverse, *Mater. Res. Bull.* **14** (1979) 303.

[97] L. Pauling and M.D. Shappell, The Crystal Structure of Bixbyite and the C-Modification of the Sesquioxides, *Z. Krist.* **75** (1930) 128–142.

[98] W.H. Zachariasen, *Z. Krist.* **67** (1928) 455–464.

[99] The Uniqueness of an X-ray Crystallographic Analysis, A.L. Patterson, *Phys. Rev. (abstract)* **55** (1939) 682. '…Langmuir and Wrinch (*Nature* **142** (1928) 581–583 have suggested that if the positions of the peaks in the F^2 series are known, the structure is uniquely determined. In the present paper it is shown that if the peaks in the F^2 series can be resolved, the structural determination can be made unique, except for the fundamental ambiguity of a center of symmetry. The proof depends on a knowledge of the atoms which compose the crystal and the demonstration that there is only one way in which a given set of atoms can produce a given F^2 series…'. Patterson's remark about the fundamental ambiguity of a center of symmetry is beside the point, we feel, and his 'suggested' is perhaps more than the Langmuir–Wrinch article would justify, because it, after an incomprehensible introductory part beginning with 'In atomic space S_1, two points A & B with intensities $\pm r$, $\mp s$, respectively, …', turns to what purports to be conclusive proof, from the existing Patterson projections for insulin, that insulin has Wrinch's C_2 cyclol structure, which it never did. At one point Langmuir and Wrinch write that their introductory argument 'also indicates the type of mathematical problem on the solution of which the interpretation of vector maps depends. It is essentially akin to the problem of finding a square root.' Altogether, L. and W. seem neither to address Patterson's problem nor make much sense.

[100] See Patterson and Pattersons, *Fifty Years of the Patterson Function*, edited by Jenny P. Glusker, Betty K. Patterson, and Miriam Rossi, International Union of Crystallography, Oxford University Press (1987), especially the articles 'The Structure of Bixyite', 36–41, by Jenny P. Glusker, and 'Patterson and Bixbyite', 42–44, by Linus Pauling, and Part III, Section C., 'Homometrics'.)

[101] H. Dachs, *Z. Krist.* **107** (1956) 370–395.

[102] L. Pauling, *Proc. Nat. Acad. Sci.* **15** (1929) 709–712.

[103] G. Bruni and A. Ferrari, *Rendiconti Acc. dei Lincei* **2** (1925) 457.

[104] G. Bruni and A. Ferrari, *Rendiconti Acc. dei Lincei* **4** (1926) 10.

[105] A. Ferrari, *Rendiconti Acc. dei Lincei* **6** (1927) 56.

[106] V.M. Goldschmidt, Geochemische Verteilungsge Setze der Elemente, VIII. Untersuchungen über Bau und Eigenschaften von Krystallen, *Skrifter utgitt av det det Norske Videnskaps-Akademi I Oslo 1. Matematisk-Videnskapslig Klause* (1926(1927)) [8] 1–156, p. 147.

[107] The Crystal Structure of Cadmium Chloride, Linus Pauling and J.L. Hoard, *Z. Krist.* **74** (1930) 546–551.

[108] A. Ferrari, A. Celari, and F. Giorgi, *Rendiconti Acc. dei Lincei* **9** (1929) 782.

[109] Adolfo Ferrari, Antonio Braibanti, and Gino Bigliardi, *Acta Cryst.* **16** (1963) 846–847.

[110] G.L. Clark and G.R. Yohe, X-ray Investigation of Optically Active Com-

pounds. I. A Proof of Molecular Asymmetry in Optically Active Phenylaminoacetic Acid, *J. Am. Chem. Soc.* **51** (1929) 2796–2807.

[111] L. Pauling and R.G. Dickinson, Objections to a Proof of Molecular Asymmetry of Optically Active Phenylaminoacetic Acid, *J. Am. Chem. Soc.* **53** (1931)
3820–3823.

[112] G.L. Clark and L.W. Pickett, *J. Am. Chem. Soc.* **53** (1931) 167–177.

[113] M.L. Huggins, *J. Am. Chem. Soc.* **53** (1931) 3823–3826.

[114] G.L. Clark, *J. Am. Chem. Soc.* **53** (1931) 3826–3831.

[115] Dickinson's thesis has two chapters. The first, two crystal structures already
published at the time, is represented in the thesis only on the title page. The
second, on sodium chlorate and then as yet unpublished, constitutes the entire
body of Dickinson's thesis. (The paper, on both $NaClO_3$ and $NaBrO_3$: R.G.
Dickinson and E.A. Goodhue, *J. Am. Chem. Soc.* **43** (1921) 2045–2055.)

[116] Dickinson asserts that the five parameters needed to locate the atoms (one each
for Na and Cl, both lying on a three-fold axis, and three for the position of one
of the O atoms) are too many to be determined from his data, but he then writes
down values for all the parameters, which, nearly ten years later, were found
(W.H. Zachariasen, *Z. Krist.* **71** (1929) 517–529) to be essentially the same as
the new values, unlike all four of those proposed, 1921–1923, by others. The
latest determinations (see Beurskens-Kerssen *et al.* [117]) confirm still further
the Dickinson parameter values for sodium chlorate and sodium bromate.

[117] Sodium chlorate grows coherently on sodium bromate seeds, as has been
known since 1856 (H. Marbach, *Annalen der Physik und Chemie* **99** (1856)
451–466), but it turns out that the overgrowths, isostructural with the seeds,
have the opposite optical rotation, and it is a delicate matter to adjust both the
structure *and* the theoretical calculation of the optical activity to obtain
agreement with the experimental facts (Gezina Beurskens-Kerssen, J. Kroon,
H.J. Endeman, J. van Laar, and J.M. Bijvoet, *Crystallography and Crystal
Perfection*, G.N. Ramachandran (Ed.), Academic Press, London and New
York (1963), pp. 225–236.

[118] On crystallization from water, each individual crystal is optically active, but the
population of crystals produced is sometimes of only one rotation, sometimes
of the other, and sometimes mixed. Dilip K. Kondepudi, Rebecca J. Kaufman,
and Nolini Singh (Chiral Symmetry Breaking in Sodium Chlorate Crystallization, *Science* **250** (1990) 975–976) reported on this, and on how sufficient
stirring greatly favors unmixed crystallization. Pretty much the same facts were
reported already by Soret in 1899, as summarized by P. Groth (*Z. Krist.* **34**
(1901) 630, and *Chemische Krystallographie* **2** (1908) 86), but this was overlooked in the *Science* article and in the special commentary that occurred in the
same issue, 'A Stirring Tale of Crystal Growth', *Science*, p. 913.

[119] J.H. Adriani, *Z. Phys. Chem.* **33** (1900) 453–476.

[120] H.A.J. Oonk and J. Kroon, *Acta Cryst.* **B32** (1976) 500–504.

[121] F. Baert, R. Fouret, H.A.J. Oonk, and J. Kroon, *Acta Cryst.* **B34** (1978)
222–226.

[122] H.A.J. Oonk, K.H. Tjoa, F.E. Brants, and J. Kroon, *Thermochimica Acta* **19**
(1977) 161–171.

[123] C.L. Burdick and J.H. Ellis, *J. Am. Chem. Soc.* **39** (1917) 2518–2525.

[124] L. Pauling and L.O. Brockway, *Z. Krist.* **82** (1932) 188–194.
[125] W.F. de Jong, *Z. Krist.* **68** (1928) 522–529.
[126] L. Pauling and R. Hultgren, *Z. Krist.* **84** (1933) 204–212.
[127] L. Pauling and J. Sherman, *Z. Krist.* **84** (1933) 213–216.
[128] W.H. Zachariasen, *Skr. Norske Vid.-Akad., Oslo, I. Mat.-Natv. Kl.* (1928) [4] 74–82.
[129] K.S. Pitzer, *Z. Krist.* **92** (1935) 131–135.
[130] Edwin McMillan and L. Pauling, *J. Am. Chem. Soc.* **49** (1927) 666–669.
[131] K. Lewkonja, *Z. anorg. Chem.* **52** (1907) 452–456.
[132] You-Chi Tang and L. Pauling, *Acta Cryst.* **5** (1952) 39–44.
[133] A. Ölander, *Z. phys. Chem. (A)* **168** (1934) 274–282.
[134] *Introduction to Quantum Mechanics*, L. Pauling and E.B. Wilson, McGraw-Hill, New York & London (1935).
[135] L. Pauling and J. Sherman, *Z. Krist.* **81** (1932) 1–29.
[136] D.R. Hartree, *Camb. Phil. Soc. Proc.* **24** (1928) 89–110, 111–132.
[137] R.W. James and G.W. Brindley, *Phil. Mag.* **12** (1931) 81–112.
[138] G. Bergman, J.L.T. Waugh, and L. Pauling, *Acta Cryst.* **10** (1957) 254–259.
[139] S. Samson, *Nature* **195** (1962) 259–262.
[140] L. Pauling and R. Marsh, *Proc. Nat. Acad. Sci.* **38** (1952) 112–118.
[141] Rather than listing all Pauling's works in this area, we shall be content with a summary reference: The Configuration of Polypeptide Chains in Proteins, L. Pauling and R.B. Corey, *Fortschr. der Chem. organische Naturstoffe* **XI** (1954) 180–239.
[142] L. Pauling and R.B. Corey, *Proc. Nat. Acad. Sci.* **39** (1953) 84–97.
[143] L. Pauling, *Proc. Nat. Acad. Sci.* **87** (1990) 244–245.
[144] The Structure of the Carboxyl Group. II. The Crystal Structure of Basic Beryllium Acetate. L. Pauling and J. Sherman, *Proc. Nat. Acad. Sci. USA* **20** (1934) 340–345.
[145] P. Selvam and K. Yvon, *Phys. Rev. V* **B39** (1989) 12329–12330.
[146] There are too many references in this area for us to cite them all; we shall be content with noting one of Pauling's nearly twenty papers on these materials since 1985 — *Proc. Nat. Acad. Sci.* **86** (1989) 8595–8599 — which is immediately followed by a paper describing the other, 'quasicrystal' side of the story and agreeing, Pauling has said with characteristic humor, 'with what I say, *well*, part of what I say' — P.A. Bancel, P.A. Heiney, P.M. Horn and P.J. Steinhardt, *Proc. Nat. Acad. Sci.* **86** (1989) 8600–8601.
[147] R.B. Corey and L. Pauling, *Rev. Sci. Inst.* **24** (1953) 621–627.
[148] First operation 1939–1940. First paper: The Use of Punched Cards in Molecular Structure Determinations, P.A. Shaffer, Jr., V. Schomaker, and L. Pauling, *J. Chem. Phys.* **14** (1946) 648–658.
[149] L. Pauling, Banquet talk, Am. Crystallographic Ass. Meeting, Pasadena (1955).

4

Some problems of nitrogen stereochemistry

J. E. Boggs,[a] L. S. Khaikin,[b] V. I. Perevozchikov,[c]
M. V. Popik,[b] N. I. Sadova,[b] S. Samdal,[d] and L. V. Vilkov[b]
[a]University of Texas, Austin, TX, USA
[b]Moscow State University, Moscow, USSR
[c]All-Union Correspondence Polytechnic Institute, Moscow, USSR
[d]Oslo College of Engineering, Oslo, Norway

INTRODUCTION

The main thrust of the scientific work of Linus Pauling has been to connect the results of research on the bulk physical properties of matter with the chemical structure of molecules. He interpreted the chemical structures as a system of chemical bonds affected by resonance. As an inductive chemical concept, the idea of the resonance effect can now be further refined by application to newly available experimental data. The past decades have been highly productive in the collection by means of electron diffraction and microwave spectroscopy of a huge structural data base for many kinds of free molecules.

THE STRUCTURE OF THE SIMPLEST AMIDES: H_2NCHO, H_2NNO_2 AND H_2NNO

The ideas of Pauling [1, 2] give a very simple picture for the additional resonance structures which are responsible for the observed physical and chemical properties of the amides:

(a)　　　　　　　　(b)　　　　　　　　(c)

A contribution from the resonance structure (a) can explain the practically planar configuration of the formamide molecule as determined by both computational and various experimental methods (see Table 1). Consideration of an analogous form (b) alone does not lead to an understanding of the pyramidal configuration of the amino group found in the nitramide molecule. However, the adjacent charge rule helps to explain this result.

The N-nitrosoamine molecule with the potential resonance structure shown as (c) has not yet been studied by any experimental techniques, but the results of *ab initio* calculations reported below shed light on the planarity of this molecule and the resulting correlation with Pauling's resonance concepts. The extent of pyramidality of the amino nitrogen bond in H_2NNO is found to be intermediate compared with the two other molecules.

The variations in amino nitrogen bond configuration correlate well with alterations of the pyramidal inversion and the internal rotation barrier heights (see Tables 2 and 3). Because of the larger pyramidality of the amino nitrogen in nitramide, the barrier to internal rotation of the $-NO_2$ group is lower than that observed for the $-CHO$ group in H_2NCHO or for the $-N{=}O$ group in $H_2NN{=}O$.

Table 1 — Amino nitrogen angle sums in some simple amides.

Molecule	Calculations		Experiment
	6-31G*	MP2/6-31G*	
H_2NCHO	360.0[a]	359.9[b]	360.0[c]
$H_2NN{=}O$	353.2[d]	355.8[d]	—
H_2NNO_2	337.8[e]	331.0[f]	334.0–340.3[g]

[a]Refs. [3–6]. [b]Ref. [7]. The calculation was performed using the standard Gaussian basis set of primitive functions (9s5p/4s) which was contracted to (4s2p/4s). [c]Refs. [8–10]. [d]Ref. [6]. [e]Refs. [6, 11, 12]. [f]Ref. [12]. [g]Refs. [13, 14].

Table 2 — Barrier heights for amino inversion in some simple amides, kcal/mol

Molecule	Calculations		Experiment
	6-31G*	MP2/6-31G*	
H_2NCHO	0.[a]	—	0.[b]
$H_2NN{=}O$	0.12[a]	0.05[a]	—
H_2NNO_2	1.9; 1.0[c]	3.1; 2.2[c]	2.7[d]

[a]Ref. [6]. [b]Ref. [10]. [c]Ref. [12]. The second value takes into consideration the energy differences for the zero vibrational levels. [d]Ref. [15].

Table 3 — Barrier heights for internal rotation in some amides, kcal/mol

Molecule	Calculations		Experiment
	6-31G*	MP2/6-31G*	
H_2NCHO	15.6[a] (syn NH_2/C=O) 18.6[a] (anti NH_2/C=O)	16.0[b] (syn NH_2/C=O)	17–21[c] (in solution)
H_2NN=O	19.2[d] (syn NH_2/N=O) 17.8[d] (anti NH_2/N=O)	—	—
H_2NNO_2	12.3[e]	10.1[e]	—

[a]Refs. [3–6]. [b]Ref. [3]. MP2/6-311G** was used. [c]Refs. [16–18]. [d]Ref. [6]. [e]Ref. [12]

It is extremely interesting to note that the relative magnitude of the rotation barrier heights in the *syn-* and *anti-* transition states of H_2NCHO and H_2NN=O (see Figs 1 and 2) are reversed. This difference may be a result of the influence of steric and dipole interactions.

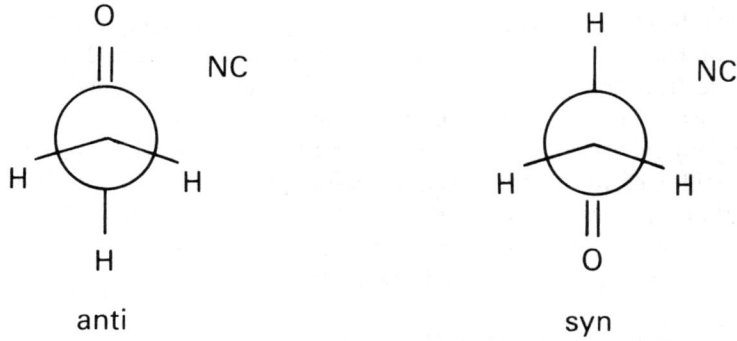

anti   syn

Fig. 1 — The two transition states for internal rotation in H_2NCHO.

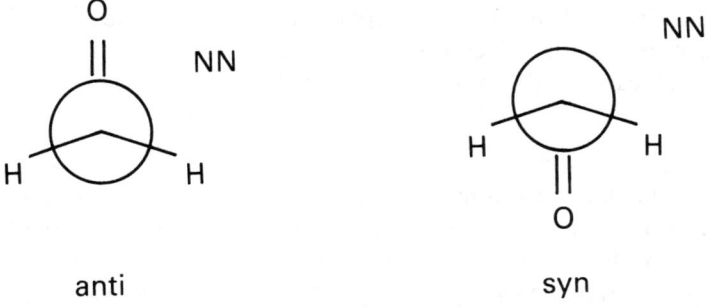

anti   syn

Fig. 2 — The two transition states for internal rotation in H_2NNO_2.

Comparison of experimental and theoretical results for H_2NCHO and H_2NNO_2

H_2NCHO

It was shown earlier by gas phase electron diffraction [8] and microwave spectroscopy [9, 10] that the central C—N bond in formamide, H_2NCHO, is much shorter than the normal single bond length of 1.454 Å found in $N(CH_3)_3$ [19], as would be expected from the resonance structure shown above.

Whether the conformation at the nitrogen atom is planar or pyramidal has been the subject of extensive dispute. The first microwave spectroscopic study by Kurland in 1955 found a very small positive inertial defect and concluded that the molecule was planar. Later, Costain and Dowling made a more complete microwave investigation and derived small torsional angles of 7° ± 5° and 12° ± 5° for the HNCO and HNCH dihedral angles. Hirota [9] in 1974 pursued the microwave investigation still further and concluded that there is only a planar single minimum. Finally, Brown *et al.* [10] in 1987 have added further support to the single minimum view.

Extensive *ab initio* calculations have been carried out for the formamide molecule. These were reviewed and extended by Boggs and Niu [20], who showed that the question of planarity at the nitrogen angle is highly sensitive to the size of the Hartree-Fock basis set used. This article also includes a discussion of certain other geometrical parameters of formamide which were disputed on the basis of experimental data alone. With the largest basis employed (6-311G** plus a second set of *d* functions on all heavy atoms) the molecule is found to be slightly pyramidal at the nitrogen atom, as was also the case in a study by Wright *et al.* [7] at the level of a double zeta plus polarization Hartree-Fock basis supplemented by second-order Moller-Plesset treatment of electron correlation (see Table 3). It may be concluded that formamide certainly has a broad, shallow minimum for the potential function for nitrogen inversion. Whether the minimum is at the strictly planar position or slightly displaced from it is still open to some reasonable question, but any deviation from planarity in the potential function involves such a small deviation in angle and energy that the molecule can be considered planar for all practical purposes. This conclusion is fully in accord with predictions that would be made on the basis of the resonance concepts of Pauling.

Calculations have shown that the two transition states for internal rotation of the aldehyde group in H_2NCHO have pyramidal configurations at the nitrogen atom (see Table 4). A slightly lower energy barrier, 15.6 kcal/mol, is found for the form in which the aldehyde oxygen atom is *syn* to the NH_2 moiety (see Fig. 1). The barrier at the *anti* transition state is 18.6 kcal/mol. It is important to note that the values found for the *syn* barriers were essentially identical in the two studies reported in Table 3 in spite of the use of significantly different calculation levels.

H_2NNO_2

The experimental data for this molecule are not highly reliable since the author [13] had to assume certain parameter values including a N=O bond length of 1.206 Å and C_{2v} symmetry of the NNO_2 group. The value derived for the N—N bond length is, however, highly correlated with the value assumed for the N=O length. For example, changing the assumed N=O distance to 1.232 Å, which is closer to that found for

Table 4 — Comparison of *ab initio* results and experimental data for H_2NCHO[a]

| Parameters | Calculations | | | | | Experiment |
| | RHF/6-31G*[b] | | | | MP2/DZ**[c] | MW[d] |
	$ES(C^1)$	$TS_t(C_{2v})$	$TS_r^a(C_s)$	$TS_r^s(C_s)$	$ES(C_1)$	
C=O	1.193	1.193	1.179	1.183	1.228	1.220
CN	1.349	1.349	1.423	1.427	1.368	1.357(2)
CH	1.091	1.091	1.094	1.088	1.102	1.098[f]
NH	0.996	0.996	1.005	1.006	1.008	1.000
NH_a	0.993	0.993	1.005	1.006	1.006	1.001
NCH	112.7	112.7	116.4	113.5	112.1	113.0
NCO	125.0	125.0	123.3	125.0	124.8	124.5(7)
OCH	122.4	122.4	120.3	121.5	123.2	122.5[f]
CNH_c	119.3	119.3	109.8	108.5	119.0	118.8(14)
CNH	121.9	121.8	109.8	108.5	121.3	121.4(45)
HNH_a	118.9	118.9	106.4	105.5	119.6	119.9
OCNH	0.16	0[f]	121.7	303.0	2.0	0.06(2)
$OCNH_a$	179.7	180[f]	238.3	57.0	177.6	179.9
HCNH	−179.9	−180[f]	−58.3	−237.0	−178.4	−179.9
$HCNH_a$	−0.3	0[f]	−301.7	−123.0	−2.8	−0.14
τ[e]	0.43	0[f]	63.4	66.0	4.4	0.20(3)
E + 168	−0.930703	−0.93070	−0.90114	−0.90569	−1.495566	

[a]Bond lengths in Å, valence and dihedral angles in degrees. ES, TS, and TS_r stand for equilibrium state and transition states for inversion and internal rotation; superior indices a (*anti*) and s (*syn*) define either TS type (see Fig. 1) or geometric parameters for ES of the molecule from viewpoint of H—N/C=O arrangement. DZ basis set contains 9s5p/4s primitive functions contracted to 4s2p/4s. E is total energy (a.u.); [b]Refs. [3–6]; [c]Ref. [7]; [d]Ref. [10]; using LAM + rotation, a model for treating wagging-inversion large amplitude motion interacting with internal rotation around the CN central bond; [e]The dihedral angle between CNH_s and CNH_a planes; [f]Fixed values.

other NO_2 derivatives, decreased the value derived for the N—N bond length by about 0.04 Å.

The main conclusion of the experimental studies was the existence of a pyramidal bond configuration at the nitrogen atom. A recent *ab initio* calculation [12] has confirmed this (see Table 5). Moreover, when H_2NNO_2 was constrained to have a planar amino group, the structure showed an appreciable shortening of the N—N bond length. This means that at this planar transition state there is resonance which is in some extent weakened in the normal pyramidal state.

Theoretical results for H_2NNO

We carried out *ab initio* calculations for H_2NNO using different approximations including the restricted Hartree-Fock (RHF) method with the basis sets 6-31G*, 6-311G*, and 6-311G**. In addition, we included electron correlation at the level of second-order Møller-Plesset perturbation theory (MP2) as shown in Table 6.

We found that the equilibrium state for H_2NNO corresponds to a pyramidal configuration of the amino nitrogen atom, but the degree of pyramidality is less than in H_2NNO_2. Inclusion of electron correlation in the calculation results in additional flattening of the pyramidal structure of the amino group.

Large changes in the geometry of H_2NNO are observed at the transition state for internal rotation. The N—N bond length, for example, is increased by about 0.1 Å above that found in the equilibrium configuration (see Table 6).

Table 5 — MP2/6-31G* computed results compared with experimental data for H_2NNO_2.[a]

Parameters	MP2/6-31G*[b]			MW, r_o^c
	$ES(C_s)$	$TS_i(C_{2v})$	$TS_r(C_s)$	
NO	1.233	1.235	1.228[f]	1.206[e]
NO′	1.233	1.235	1.235[g]	1.206[e]
NN	1.398	1.361	1.462	1.427(2)
NH	1.017	1.008	1.024	1.005(10)
NNO	116.1	116.2	115.3[f]	115.0
NNO′	116.1	116.2	117.4[g]	115.0
ONO′	127.6	127.6	127.3	130.1(3)
NNH	108.3	116.6	102.9	109.4
HNH	114.4	126.8	104.2	115.2(20)
NH/ONO′	3.7	0.[e]	0.[e]	0.[e]
NN/HNH	54.5	0.[e]	68.7	51.8(10)
ONNH	30.1	0.[e]	126.0[f]	26.5
O′NNH	−154.7	−180.[e]	−54.0g	−153.5
τ[d]	55.4	0.[e]	71.9	53.0
E + 260	−0.35192	−0.34700	−0.33583	

[a]See analogous footnote to Table 4; [b]Ref. [12]; [c]Ref. [13]; [d]Dihedral angle between two NNH planes; [e]Fixed values; [f]*anti* to NH_2; [g]*syn* to NH_2.

Table 6. — Results of *ad initio* calculations for $H_2NN=O$[a]

	RHF/6-31G*[b]					MP2/6-31G*[b]	
	$ES(C_1)$	$TS_i(C_s)$[c]	$TS_r^a(C_s)$	$TS_r^s(C_s)$	$TS_{i+r}(C_s)$	$ES(C_1)$	$TS_i(C_s)$[c]
N=O	1.183	1.186	1.160	1.164	1.163	1.236	1.237
NN	1.317	1.306	1.438	1.439	1.415	1.341	1.336
NH	1.000	0.999	1.008	1.008	0.992	1.019	1.018
NH_a	0.994	0.992	1.008	1.008	0.992	1.010	1.009
NNO	114.5	114.6	111.5	113.6	112.8	113.0	113.1
NNH	117.3	119.4	104.0	105.8	118.2	117.9	119.2
NNH_a	115.3	117.4	104.0	105.8	118.2	115.9	117.3
HNH_a	120.5	123.2	104.3	105.5	123.6	121.9	123.5
ONNH	13.5	0.[e]	125.6	304.2	90.[e]	10.5	0.[e]
$ONNH_a$	164.7	180.[e]	234.4	55.8	270.[e]	167.8	180.[e]
τ[d]	28.7	0.[e]	71.1	68.4	0.0[e]	22.7	0.[e]
E + 184	−0.826480	−0.826284	−0.798066	−0.795820	−0.778380	−1.342767	−1.342692

[a]See analogous footnote to Table 4; TS_r^a and TS_r^s are characterized by *anti* or *syn* $NH_2/N=O$ arrangements (see Fig. 1), TS_{i+r} is transition state for inversion and internal rotation which differ from TS_i by rotation of nitroso group by 90°; [b]Ref. [6]; [c]Data in Ref. [6] are in thorough agreement with results of Ref. [20]; [d]Dihedral angle between NNO_a and NNO_s planes; [e]Fixed values.

Both of the equilibrium structures as well as the structures at the transition states for configuration change by internal rotation or inversion can be rationalized on the basis of the concept of resonance introduced by Pauling.

THE EFFECT OF METHYL SUBSTITUTION IN AMIDES

The $(CH_3)_2NNO_2$ molecule

As shown by an electron diffraction study, this molecule is practically planar [22] and the central N—N bond length is equal to 1.383(3) Å (see Table 7). These results show that the amino nitrogen bond configuration is changed in going from H_2NNO_2 to $(CH_3)_2NNO_2$. The effect can be explained by the electron donor character of the methyl group which compensates for the positive charge on the amino nitrogen atom in the Pauling resonance structure which is invoked to explain flattening in H_2NNO_2.

Ab initio calculations using the 6-31G* basis set [3] confirmed some flattening of the nitrogen bond pyramidal configuration in $(CH_3)_2NNO_2$ (Table 7). However, the results show some differences with the experimental data not only in the pyramidality but also in the value of the CNC bond angle and in the N—N bond length. This problem warrants further study.

The net atomic charges from a Mulliken population analysis [3] are distributed on the heavy atoms in the following manner:

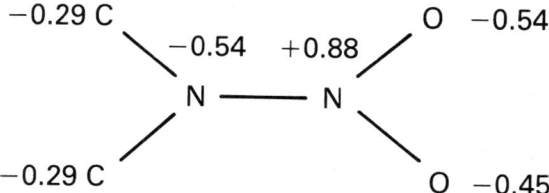

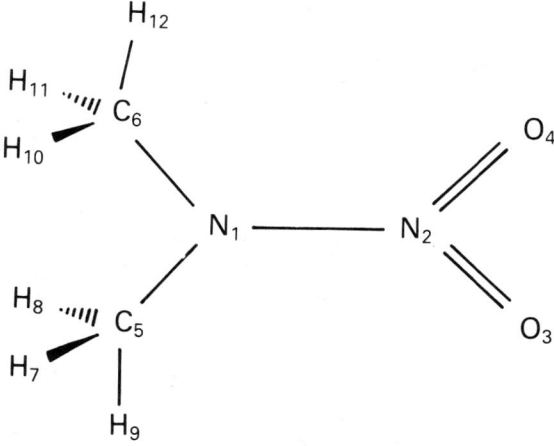

Fig. 3 — Numbering of the atoms in $(CH_3)_2NNO_2$.

Table 7 — Comparison of RHF/6-31G* computed geometry with electron diffraction data for $(CH_3)_2NNO_2$[a]

Parameters	RHF/6-31 G*[b]			ED, r_g[c]	
	$ES(C_s)$	$TS_i(C_s)$	$TS_r(C_1)$	Planar	Nonplanar
NO	1.197	1.200	1.192	1.225(1)	1.226(1)
NO'	1.197	1.200	1.181	1.225(1)	1.226(1)
NN	1.343	1.326	1.424	1.383(3)	1.402(3)
NC	1.455	1.452	1.461	1.463(3)	1.454(2)
NNO	117.4	117.3	117.9	114.8(7)	115.5(4)
NNO'	117.4	117.3	115.9	114.8(7)	115.5(4)
ONO'	125.2	125.3	126.2	130.4(13)	125.5(7)
NNC	115.8	117.4	108.7	116.2(3)	115.7(2)
CNC	120.4	125.1	113.4	127.6(6)	128.6(4)
C_5H_7	1.080			1.115(5)	1.114(5)
C_5H_8	1.077			1.115(5)	1.114(5)
C_5H_9	1.085			1.115(5)	1.114(5)
NC_5H_7	107.1			101.9(20)	101.0(16)
NC_5H_8	110.0			101.9(20)	101.0(16)
NC_5H_9	112.0			101.9(20)	101.0(16)
HCH	109.1				
NN/ONO'	1.2			0.[d]	0.[d]
NN/CNC	28.8	0.[d]	90.[d]	0.[d]	3.6
ONN/NNC	-16.1			0.[d]	-12.0
NO'/NNO	-1.1			0.[d]	
E	-337.700502				

[a]See analogous footnote to Table 4; for numbering of atoms, see Fig. 3; [b]Ref. [3]; [c]Ref. [22]; [d]Fixed values.

In view of the repulsion between the negative carbon atoms and the attraction of these carbon atoms to the positively charged nitrogen atom of the nitro group, the large CNC angle found by electron diffraction can be justified. The resonance form

$$
\begin{array}{ccc}
\text{H} & & \text{O}^- \\
\diagdown & & \diagup \\
& \text{N}^+ \!\!=\!\! \text{N}^+ & \\
\diagup & & \diagdown \\
\text{H} & & \text{O}^-
\end{array}
$$

with two adjacent atoms having the same sign of charge has a small probability. Therefore the methyl groups as electron donors in $(CH_3)_2NNO_2$ compensate the positive charge of the amino nitrogen to a greater extent than do the hydrogen atoms in H_2NNO_2.

The high barrier to internal rotation (11.35 kcal/mol) and the low barrier to inversion at the amino nitrogen (0.41 kcal/mol) obtained by *ab initio* calculations [3] confirm the utility of the resonance explanation for the $(CH_3)_2NNO_2$ molecule.

The H(CH₃)NNO and (CH₃)₂NNO molecules

Experimental studies of $(CH_3)_2NNO$ by electron diffraction [23] and by microwave spectroscopy [24, 25] showed that the heavy atoms of this molecule are coplanar and the central N—N bond length is very short, shorter than the CH_3—N bond length of 1.45 Å (see Table 8).

We carried out *ab initio* calculations for $H(CH_3)NNO$ and for $(CH_3)_2NNO$ with a 6-31G* basis set, optimizing the molecular geometries at this level [26]. The results showed that the equilibrium conformations for both molecules have all of their heavy atoms coplanar. For $H(CH_3)NNO$, the *syn* conformation of the CH_3—N and N=O bonds is energetically preferred. Again, the methyl group substituents flatten the pyramidal configuration of the amine nitrogen atom. The same conclusion could already be reached after calculations with the geometry optimized at only the 3-21G level [27]. The minimum energy conformations of the methyl groups are shown in Fig. 4.

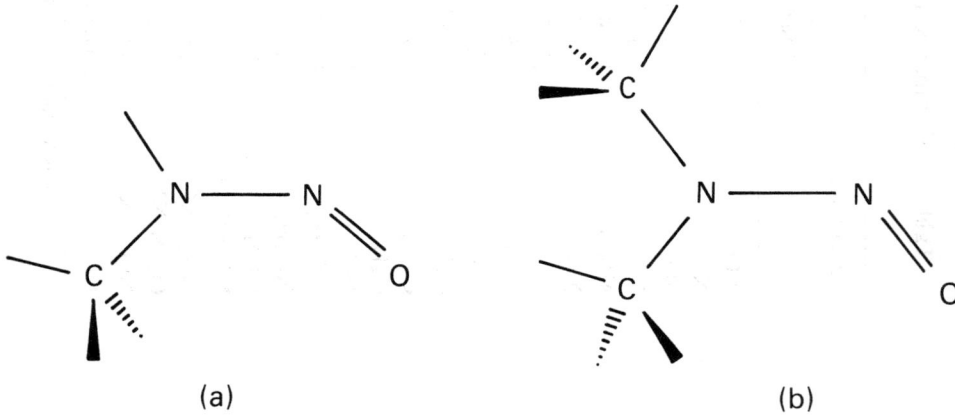

(a)   (b)

Fig. 4 — The most stable configurations of (a) $H(CH_3)NNO$ and (b) $(CH_3)_2NNO$ as obtained by computation.

Table 8 — Comparison of RHF/6-31G* computed geometries for $H_2NN{=}O$, $H(CH_3)NN{=}O$ and $(CH_3)_2NN{=}O$ with experimental data for $(CH_3)_2NN{=}O$[a]

| | RHF/6-31 G* | | | Experiment, $(CH_3)_2NN{=}O$ | | |
	$H_2NN{=}O$[b]	$CH_3NHN{=}O$[c]	$(CH_3)_2NN{=}O$[c]	ED, r_g[d]	MW, r_o[d]	MW, r_o[f]
$N{=}O$	1.183	1.191	1.194	1.235(6)	1.233(20)	1.243(33)
NN	1.317	1.303	1.300	1.344(7)	1.329(20)	1.329
NX_s	1.000	1.446	1.449	1.461(7)[h]	1.444(20)	1.442(45)
NX	0.994	0.993	1.444	1.461(7)[h]	1.452(20)	1.441(39)
NNO	114.5	115.0	115.5	113.6(6)	114.0(10)	113.5(14)
NNX_s	117.3	123.3	120.6	120.4	121.4(10)	122.6(20)
NNX_a	115.3	114.2	117.2	116.4(10)	116.1(20)	115.3(36)
X_sNX_a	120.5	122.5	122.2	123.2(7)	122.5	122.1
C_sH_p		1.081	1.081	1.129(9)[h]	1.065[h]	1.081(10)[j]
C_sH_o		1.082	1.082	1.129(9)[h]	1.065[h]	1.081(10)[j]
C_aH_p			1.079	1.129(9)[h]	1.065[h]	1.076(3)[j]
C_aH_o			1.085	1.129(9)[h]	1.065[h]	1.076(3)[j]
NC_sH_p	107.8		108.1	109.6(24)[h]	111.2[h]	109.1(4)[j]
NC_sH_o	110.7		110.6	109.6(24)[h]	111.2[h]	109.1(4)[j]
NC_aH_p			108.1	109.6(24)[h]	111.2[h]	113.2(12)[j]
NC_aH_o			110.8	109.6(24)[h]	111.2[h]	113.2(12)[j]
$ONNX_s$	13.5	0.	0.	0.	0.[i]	0.[i]
$ONNX_a$	164.7	180.	180.	180.	180.[i]	180.[i]
g	28.7	0.	0.	0.	0.[i]	0.[i]
$-E$	-184.826480	-223.860761	-262.892416			

[a] s (syn) and a (anti) indices define relative arrangement of nitroso group and X substituent on amino nitrogen (X=H or methyl C atom); methyl H atom labeled p if it is in molecular symmetry plane and o otherwise; [b]Ref. [6]; [c]Ref. [26]; [d]Ref. [23]; [e]Ref. [24]; [f]Ref. [25]; [g]Dihedral angle between NNX_a and NNX_s planes; [h]NC or CH bond length and NCH bond angle values for both methyl groups in the molecule are assumed to be equal; [i]Fixed values; [j]Each of the two nonequivalent methyl groups in the molecule preserve C_{3v} symmetry.

A Mulliken population at the RHF 6-31G* level has been reported [27] for these molecules. The electron density in the N—N bond decreases regularly with values of 0.384, 0.346, and 0.342 in the molecules H_2NNO, $H(CH_3)NNO$, and $(CH_3)_2NNO$. This is contrary to an expected increase in the resonance effect in this set of molecules. The atomic charges in $(CH_3)_2NNO$ are distributed in the following way:

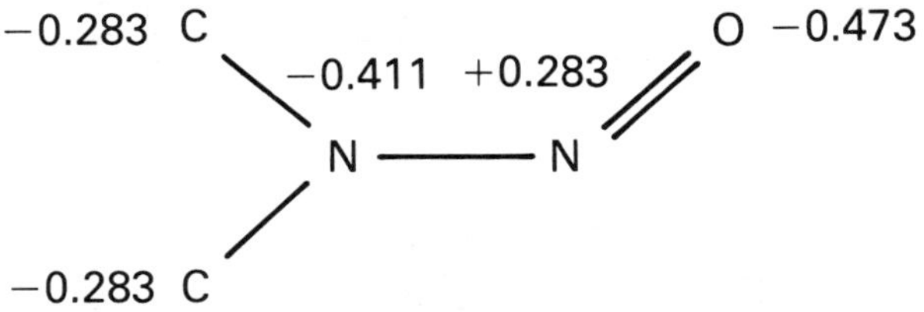

This presents a more complex picture compared with consideration of only one resonance form of the molecule. Although use of the Mulliken charge distribution is frequently criticized, using it provides one way to interpret a planar bond configuration of the amine nitrogen atom and the structure of $(CH_3)_2NNO$. Repulsion between the two negative carbon atoms and the attraction between the negative carbon atoms and the positive nitrogen atom of the nitroso group increase the CNC bond angle in $(CH_3)_2NNO$ as compared with the CNN bond angle [23].

MONOHALONITROBENZENES

Chemical observations over many years have shown a direct influence of diverse substituents on the reactivity of the benzene ring. Moreover, the mutual influence of two substituents transmitted though the ring has been well documented. Perhaps the greatest interest has focused on the effects of nitro groups and halogen atoms.

By electron diffraction, we have studied the entire series of chloro-, bromo-, and iodo- nitrobenzenes with the subsituents located in the *ortho*-, *meta*-, and *para*- positions [28–40]. Fig. 5 shows the trend in the observed carbon–halogen bond lengths in comparison with the parent halobenzenes and *ab initio* calculations for the corresponding fluoronitrobenzenes. It can be seen that the C—Cl bond lengths are shortened for the *ortho*- and *para*- derivatives. These changes are in accord with the resonance pictures:

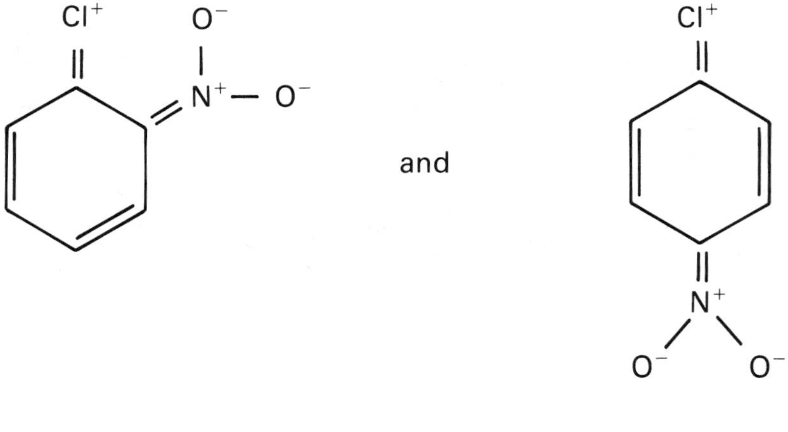

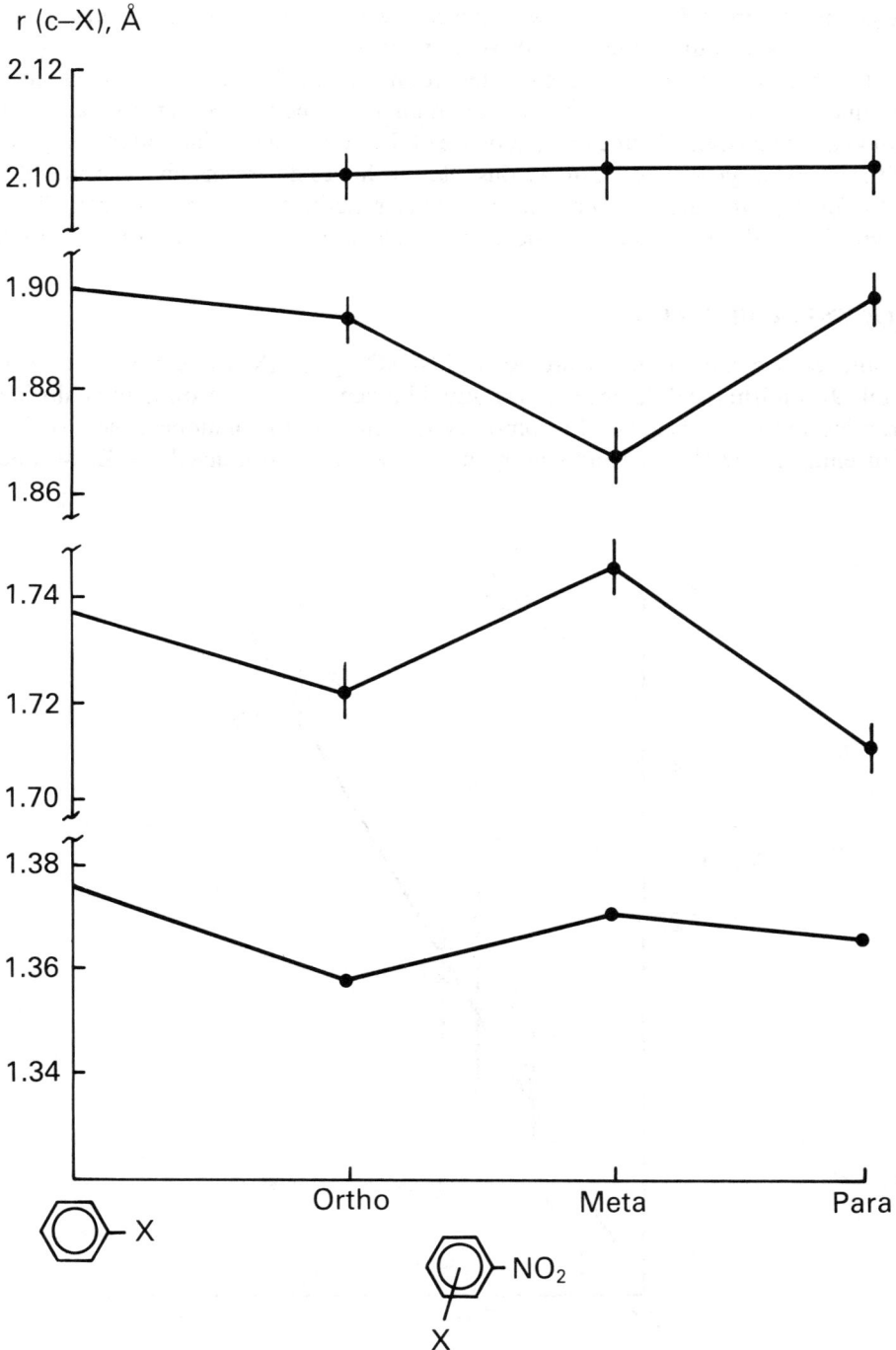

Fig. 5 — C—X bond distances for various positions of the X substituent on nitrobenzene (X = F, Cl, Br, I).

For the bromonitro- derivatives, however, we have another situation — the shortest bond length is observed for the *meta-* position. We cannot explain this behavior at present. It is interesting that we have found no significant changes in C—I bond lengths for any of the iodonitro- derivatives.

Further insight into the local interactions in the halogen derivatives can be obtained by consideration of the *ipso-* bond angles of the halobenzenes (Fig. 6) which shows an approximately linear dependence of the bond angle values on the magnitude of the electronegativity. We can use this relationship to determine an electronegativity value for the -NO_2 group, which turns out to be practically the same as for the fluorine atom. This is also in accord with the *ab initio* calculations for the fluoronitrobenzenes.

POLYNITROBENZENES

Results of electron diffraction studies of 2,6-$(NO_2)_2C_6H_3X$ where X = Cl or Br (see Table 9) confirm the difference in the mutual influence of nitro groups and chlorine or bromine atom substituents. In comparison with the monohalobenzenes, we see a shortening of the C—Cl bond length and a practically unchanged C—Br distance.

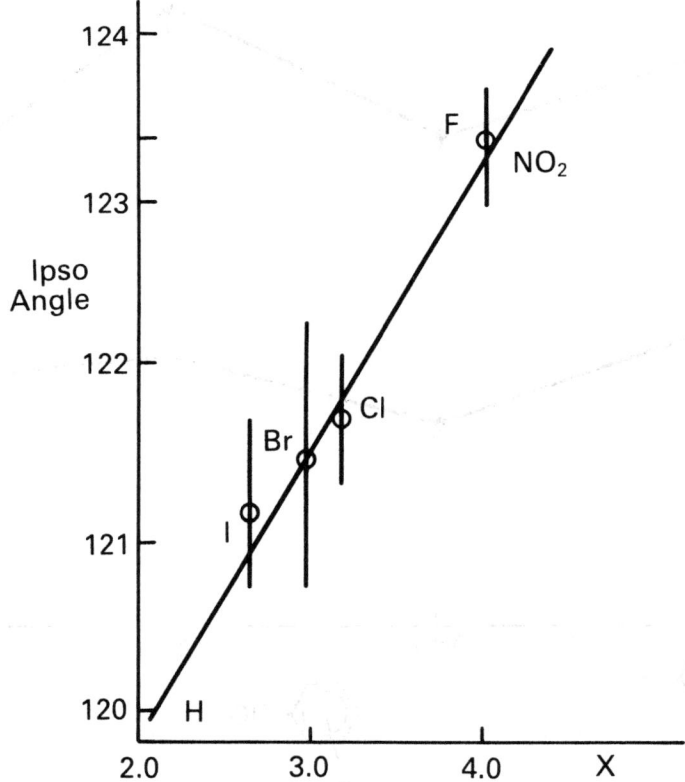

Fig. 6 — The *ipso* angle, $\angle CC_xC$, for monohalogen-substituted benzenes plotted against Pauling's electronegativity χ. Uncertainties shown by vertical bars correspond to two standard deviations. Data from Refs [38] and [41].

Table 9 — Bond lengths of nitrobenzene derivatives in Å

Molecule	C—C	C—NO$_2$	N=O	C—X (X=F,Cl,Br,I)	Ref.
o-ClC$_6$H$_4$NO$_2$	1.387(2)	1.462(12)	1.226(2)	1.721(3)	[28]
m-ClC$_6$H$_4$NO$_2$	1.383(3)	1.442(10)	1.243(3)	1.746(6)	[29]
p-ClC$_6$H$_4$NO$_2$	1.383(3)	1.469(13)	1.237(3)	1.707(4)	[30]
2,6-(NO$_2$)$_2$C$_6$H$_3$Cl	1.395(3)	1.447(4)	1.229(2)	1.712(5)	[31]
1,3,5-Cl-2,6-(NO$_2$)$_2$C$_6$H	1.400(6)	1.469(15)	1.228(7)	1.705(6)	[32]
sym-(NO$_2$)$_3$C$_6$Cl$_3$	1.399(3)	1.464(6)	1.217(2)	1.699(4)	[33]
o-BrC$_6$H$_4$NO$_2$	1.386(3)	1.494(14)	1.218(3)	1.894(6)	[34]
m-BrC$_6$H$_4$NO$_2$	1.394(3)	1.448(14)	1.238(3)	1.865(8)	[34]
p-BrC$_6$H$_4$NO$_2$	1.399(3)	1.455(12)	1.240(4)	1.895(7)	[35]
2,6-(NO$_2$)$_2$C$_6$H$_3$Br	1.393(3)	1.468(9)	1.229(3)	1.899(9)	[34]
sym-(NO$_2$)$_3$C$_6$Br$_3$	1.402(7)	1.455(16)	1.242(5)	1.886(6)	[36]
o-IC$_6$H$_4$NO$_2$	1.398(3)	1.468(18)	1.238(3)	2.101(7)	[37]
m-IC$_6$H$_4$NO$_2$	1.392(2)	1.487(11)	1.226(2)	2.100(9)	[37]
p-IC$_6$H$_4$NO$_2$	1.397(2)	1.459(15)	1.229(3)	2.102(6)	[38]
C$_6$H$_5$Cl	1.400(1)			1.737(5)	[39]
C$_6$H$_5$Br	1.399(3)			1.899(3)	[40]
C$_6$H$_5$I	1.397(6)			2.099(4)	[38]
C$_6$H$_5$F[a]	1.385			1.376	[37]
C$_6$H$_5$NO$_2$[a]	1.386	1.449	1.226		[37]
o-FC$_6$H$_4$NO$_2$[a]	1.386	1.444	1.226	1.358	[37]
m-FC$_6$H$_4$NO$_2$[a]	1.383	1.450	1.225	1.370	[37]
p-FC$_6$H$_4$NO$_2$[a]	1.383	1.446	1.226	1.366	[37]

[a]6-31G basis.

In *sym*-$(NO_2)_3C_6X_3$ (X = Cl, Br) derivatives, we have not only *ortho* but also a *para* effect of mutual interaction. Again, we find a confirmation of the difference in the interaction of chlorine or bromine atoms with the nitro groups.

DIHEDRAL ANGLES OF NO₂ GROUPS IN BENZENE DERIVATIVES

The reported electron diffraction study of nitrobenzene [41] gives an apparent small nonplanarity of the heavy atoms (Fig. 7), but the reason for this apparent nonplanarity is the time average of the torsional vibration of the nitro group. This apparent nonplanarity is also found for all non-ortho halonitrobenzenes (see Table 10), but the equilibrium structures are planar. In the ortho-halogen derivatives of nitrobenzene, we find an increase of the dihedral angles from chloro- to iodo-nitrobenzene (see Table 10). The two nitro- groups in the 2,6-$(NO_2)_2C_6H_3X$ (X = Cl, Br) derivatives have larger dihedral angles because the halogen atom cannot deviate from the bisector of the bond angle C—C(X)—C and because of steric repulsion. The largest dihedral angles are observed for the *sym*-$(NO_2)_3C_6X_3$ (X = Cl, Br) molecules. Deviations from the perpendicular positions may also be explained by the torsional vibrations.

The data obtained for many nitro derivatives show that there is only a small variation of the C—N bond lengths with the torsional angle (see Tables 9 and 10). The average C—N bond length of all non-ortho-substituted halonitrobenzenes with

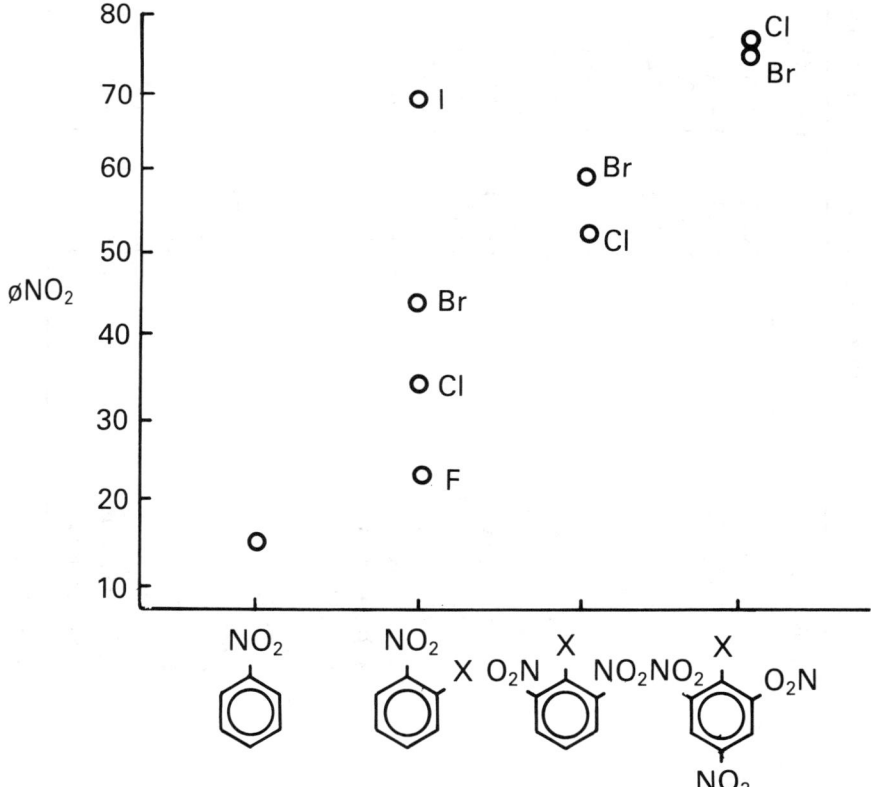

Fig. 7—Dihedral angle, ϕ_{NO_2}, for nitrobenzene and some halonitrobenzenes.

Table 10 — Dihedral angles of nitrobenzene derivatives

Molecule	∠ONO	∠CC$_{NO2}$C	∠CC$_x$C	dh∠CNO$_2$	Ref.
o-ClC$_6$H$_4$NO$_2$	123.6(1.0)	121.3(1.2)	121.6(0.7)	34(3)	[28]
m-ClC$_6$H$_4$NO$_2$	122.6(1.0)	123.0(1.5)	121.5(1.5)	13(6)	[29]
p-ClC$_6$H$_4$NO$_2$	123.0(0.9)	122.6(1.1)	120.5(0.6)	19(5)	[30]
2,6-(NO$_2$)$_2$C$_6$H$_3$Cl	123.0(0.3)	121.6(0.4)	118.9(0.6)	54(1)	[31]
1,3,5-CL—2,6-(NO$_2$)$_2$C$_6$H	129.2(3.0)	120.9(1.4)	119.7(1.5)	65(1)	[32]
sym-(NO$_2$)$_3$C$_6$Cl$_3$	131.4(2.4)	120.6(0.5)	119.4(0.5)	79(1)	[33]
o-BrC$_6$H$_4$NO$_2$	128.6(1.5)	120.4(2.4)	119.6(1.2)	43(3)	[34]
m-BrC$_6$H$_4$NO$_2$	121.8(1.4)	121.4(1.5)	121.4(1.0)	25(5)	[34]
p-BrC$_6$H$_4$NO$_2$	125.0(0.7)	121.6(0.6)	122.6(0.6)	19(3)	[35]
2,6-(NO$_2$)$_2$C$_6$H$_3$Br	125.6(0.9)	122.7(0.9)	117.1(1.2)	58(3)	[34]
sym-(NO$_2$)$_3$C$_6$Br$_3$	133.8(2.0)	121.6(0.6)	122.6(0.6)	77(1)	[36]
o-IC$_6$H$_4$NO$_2$	122.4(0.6)	122.7(1.8)	118.6(1.1)	69.5(4.3)	[37]
m-IC$_6$H$_4$NO$_2$	124.0(0.4)	124.1(0.5)	121.9(0.7)	14.0(3.8)	[37]
p-IC$_6$H$_4$NO$_2$	122.4(0.6)	122.2(0.5)	122.2(0.5)	16(1)	[38]
C$_6$H$_5$Cl			121.7(0.6)		[39]
C$_6$H$_5$Br			121.4(0.6)		[40]
C$_6$H$_5$I			121.2(0.6)		[38]
C$_6$H$_5$F			122.9		[37]
C$_6$H$_5$NO$_2$	123.4	122.2		0	[37]
o-FC$_6$H$_4$NO$_2$	123.7	119.6	120.8	23.6	[37]
m-FC$_6$H$_4$NO$_2$	123.8	122.5	122.9	0	[37]
p-FC$_6$H$_4$NO$_2$	123.6	122.1	123.2	0	[37]

[a] 6-31G basis.

presumably planar equilibrium structure is 1.461 Å, while the average C—N bond length for the ortho-substituted halonitrobenzenes with nonplanar equilibrium structures is 1.466 Å. In the many separate individual nitro derivatives studied, the accuracy is insufficient to have serious statistical significance but, in general, such dependence is small. We can consider that the *ab initio* computations for nitrobenzene by Domenicano *et al.* [41] and our own confirm this trend. For planar and orthogonal conformations, the difference in the C—N bond lengths was found to be 0.004 Å [41], and 0.009, 0.009, and 0.012 Å, respectively, for *o*-, *m*-, and *p*-fluoronitrobenzene. The experimental results are in good agreement with these findings. However, it may be concluded from these observations that the main interaction between the nitro group and the benzene ring is inductive.

CONCLUSIONS

Any empirical inductive concept has limits. The purpose of systematic studies of molecular structure is to find the borderlines of applicability of such concepts. Pauling's ideas have played an important role in the development of the central concepts of chemistry, and it is valuable for physical chemists to check them and use them where they are appropriate for the explanation of observed trends in experimental data. If some limitations are found to the validity of such concepts, they still serve to stimulate further research and further critical analysis of the ideas, the definitions, and the hypotheses.

ACKNOWLEDGMENTS

The portion of this work done at The University of Texas was supported by grants from the Texas Advanced Technology Program, The Robert A. Welch Foundation, and Cray Research, Inc.

REFERENCES

[1] L. Pauling, *The Nature of the Chemical Bond and the Structure of Molecules and Crystals; an Introduction to Modern Structural Chemisty*, Cornell University Press, Ithaca, N.Y., 1960.

[2] L. Pauling, *Nature of the Chemical Bond. The Chemical Bond, a Brief Introduction to Modern Structural Chemistry*, Cornell University Press, Ithaca, N.Y., 1967.

[3] F.R. Cordell, Thesis, The University of Texas at Austin (USA), 1987.

[4] K.B. Wilberg and K.E. Laidig, *J. Amer. Chem. Soc.* **109**, 5935 (1987).

[5] Lim Kian-Tat and M.M. Francl, *J. Phys. Chem.* **91**, 2716 (1987).

[6] L.S. Khaikin, V.I. Perevozchikov, J.E. Boggs and L.V. Vilkov, *Vestnik Moskovskogo Universiteta, Ser. Khim.* (Russian), in press, 1991.

[7] G.M. Wright, R.J. Simmonds and D.E. Parry, *J. Comput. Chem.* **9**, 600 (1988).

[8] M. Kitano and K. Kuchitsu, *Bull. Chem. Soc. Japan* **47**, 67 (1974).

[9] E. Hirota, R. Sugisaki, C.J. Nielsen and G.O. Sorensen, *J. Mol. Spectrosc.* **49**, 251 (1974).

[10] R.D. Brown, P.D. Godfrey and B. Kleibomer, *J. Mol. Spectrosc.* **124**, 34 (1987).

[11] R.P. Saxon and M. Yoshimine, *J. Chem. Phys.* **93,** 3130 (1989).
[12] J.P. Ritchie, *J. Amer. Chem. Soc.* **111,** 2517 (1989).
[13] J.K. Tyler, *J. Mol. Spectrosc.* **11,** 39 (1963).
[14] N.I. Sadova, G.E. Slepnyov, N.A. Tarasenko, A.A. Zenkin, L.V. Vilkov, I.F. Shishkov and Yu. A. Pankrushev, *Zh. Struct. Khim.* (Russian) 18, **856** (1977).
[15] D.G. Lister asnd J.K. Tyler, *Chem. Comm.,* 152 (1966).
[16] B. Sunners, L.H. Piette and W.G. Schneider, *Can. J. Chem.* **38,** 681 (1960).
[17] H. Kamei, *Bull. Chem. Soc. Japan* **41,** 2269 (1968).
[18] T. Drakenberg and S. Forsen, *J. Phys. Chem.* **74,** 1 (1970).
[19] L.V. Vilkov, V.S. Mastryukov and N.I. Sadova, *Determination of the Geometrical Structure of Free Molecules,* Mir, Moscow, 1983.
[20] J.E. Boggs and Z. Niu, *J. Comput. Chem.* **6,** 46 (1985).
[21] C.O. Reynolds and C. Thomson and *J.C.S. Faraday Trans. II* **83,** 485 (1987).
[22] R. Stølevik and P. Rademacher, *Acta Chem. Scand.* **23,** 672 (1969).
[23] P. Rademacher and R. Stølevik, *Acta Chem. Scand.* **23,** 660 (1969).
[24] A. Guarnieri, F. Rohwer and F. Scappini *Z. Naturforsch.* **30A,** 904 (1975).
[25] A. Guarnieri and R. Nicolaisen, *A. Naturforsch.* **34A,** 620 (1979).
[26] L.S. Khaikin, V.I. Perevozchikov, J.E. Boggs and L.V. Vilkov, *Vestnik Moskovskogo Universiteta, Ser. Khim.* (Russian), in press, 1991.
[27] T.K. Ha, M.T. Nguyen and P. Ruelle, *J. Mol. Struct., Theochem.* **109,** 339 (1984).
[28] O.G. Batyukhnova, N.I. Sadova, L.V. Vilkov and Yu. A. Pankrushev, *Zh. Strukt. Khim.* (Russian) **26,** 175 (1985).
[29] O.G. Batyukhnova, N.I. Sadova, L.V. Vilkov and Yu. A. Pankrushev, *J. Mol. Struct.* **97,** 153 (1983).
[30] N.P. Penionzhkevich, N.I. Sadova and L.V. Vilkov, *Zh. Strukt. Khim.* (Russian) **17,** 753 (1976).
[31] O.G. Batyukhnova, N.I. Sadova, L.V. Vilkov and Yu. A. Pankrushev, *Zh. Strukt. Khim.* (Russian) **25,** 166 (1984).
[32] E.I. Kulikova, N.I. Sadova, L.V. Vilkov and N.F. Pyatakov, *Vestnik Moskovskogo Universiteta, Ser. Khim.* (Russian) **31,** 221 (1990).
[33] E.I. Kulikova, N.I. Sadova, L.V. Vilkov and N.F. Pyatakov, *Vestnik Moskovskogo Universiteta, Ser. Khim.* (Russian) **30,** 338 (1989).
[34] O.G. Batyukhnova, N.I. Sadova, Yu. N. Syschikov and L.V. Vilkov, *Zh. Strukt. Khim.* (Russian) **29,** 53 (1988).
[35] A. Almenningen, J. Brunvoll, M.V. Popik, L.V. Vilkov and S. Samdal, *J. Mol. Struct.* **118,** 37 (1984).
[36] E.I. Kulikova, N.I. Sadova, L.V. Vilkov and N.F. Pyatakov, *Vestnik Moskovskogo Universiteta, Ser. Khim.* (Russiasn) **30,** 340 (1989).
[37] S. Samdal, private communication.
[38] J. Brunvoll, S. Samdal, H. Thomassen, L.V. Vilkov and H.V. Volden, *Acta Chem. Scand.* **44,** 23 (1990).
[39] N.P. Penionzhkevich, N.I. Sadova and L.V. Vilkov, *Zh. Strukt. Khim.* (Russian) **20,** 527 (1979).
[40] A. Almenningen, J. Brunvoll, M.V. Popik, S.V. Sokolkov, L.V. Vilkov and S. Samdal, *J. Mol. Struct.* **127,** 85 (1985).
[41] A. Domenicano, Gy. Schultz, I. Hargittai, M. Colapietro, G. Portalone, P. George and C.W. Bock, *Struct. Chem.* **1,** 107 (1990).

5

The boron–nitrogen bond: structural investigations

Roland Boese*, Norbert Niederprüm and Dieter Bläser
Universität-GH Essen, Essen, FRG

INTRODUCTION

The nature of the BN bond is disputed. This was stated in the discussion of the structures of hexagonal α-*boron nitride* and *graphite* in Gmelin's handbook [1]. For the same number of valence electrons and comparable sums of electronegativities as well as atomic radii in the BN and CC units, it seemed reasonable to suggest a replacement of CC by BN fragments and to compare their similar and different structural properties under common aspects. Owing to an always present ionic character in all bond types between boron and nitrogen, determining the bond type and the influence of substituents on structural parameters is found to be a general problem.

Extending the descriptions of classical bond types in chemistry given by Linus Pauling in *The Nature of the Chemical Bond* [2], Haaland [3] defines *dative bonds* as a new bond type on the basis of their bond-rupture behavior. In accord with the bond order–bond distance correlation of hydrocarbons, taking ethane, ethene, ethine and benzene as references, we follow the same systematic scheme. We consider the structures in the solid state of amine borane, aminoborane, iminoborane and low-substituted homologous compounds as references. For the BN single bond we suggest a strict following of the systematic of hydrocarbons and prefer to consider the bond between a four-coordinated boron and nitrogen atom only. This type of bond is the typical *dative bond*, following Haaland's suggestion. Breaking this bond type is heterolytic in contrast to the homolytic rupture of a CC single bond. Consequently for amino- and iminoboranes the three- and two- coordinated atoms will be considered. Corresponding to hydrocarbon ring compounds with three to eight members, and one or more CC units in the ring, substituted by the BN unit resulting in isoelectronic

*To whom correspondence should be addressed

systems, we also consider the aromatic and antiaromatic boron–nitrogen monocyclic ring compounds. Metal complexes are omitted, as well as seven-membered rings where no structural information has been available. The interesting feature from the stereochemical viewpoint of all these compounds is the fact that they are not only *isoelectronic* but also *isosteric*. The expression *isosteric* was introduced in 1919 by Langmuir [4] for compounds with the same number of electrons and similar conformation. The synthesis of *borazole*, now called *borazine*, by Stock and Pohland [5], and the comparison with *benzene* by E. Wiberg [6] introduced this concept to the expanding field of boron–nitrogen chemistry.

For the discussion of structural parameters we focus on the results of single-crystal X-ray diffraction, which is the most powerful tool for exact structure determination. However, the results describe the structure of molecules in the solid state influenced by intermolecular interactions. Discrepancies in comparison with structure determinations by other methods, e.g. gas-phase electron diffraction or quantum mechanical calculations, are also due to methodological differences. Mostly structural parameters of molecules determined by X-ray diffraction are compared with the respective parent compounds of low-substituted homologous compounds in order to study the influence of substituents. But most of these compounds have been determined in the gas phase or no structural information is available because many of them are liquid or gaseous at room temperature.

The crystallization of the boron–nitrogen compounds which are highly sensitive to air is a difficult business, and for those which are liquid or gaseous, special techniques must be applied [7, 8].

NON-CYCLIC BORON–NITROGEN COMPOUNDS

Amineboranes

Whereas the first amineborane $H_3N \cdot BF_3$ had already been described in 1809 [9], the parent molecule $H_3N \cdot BH_3$ (1-1) was introduced in literature one century later [5]. Owing to the partial ionic character, 1-1 is a solid, and the isosteric compound, *ethane*, is gaseous—both at ambient conditions.

Three investigations, one in the gas phase and two in the solid state of 1-1 yield bond distances of ($r_o = 167.2$ pm, $r_s = 165.7$ pm) [10], 156(5) [11] and 160(20) pm [12], respectively. In a reinvestigation by means of single-crystal X-ray structure determination, 1-1 was found in the orthorhombic space group $Pmn2_1$ (not $I4mm$ [11, 12]), with significantly deviating cell dimensions at different temperatures, indicating a strong temperature dependency [13]. A solid-state phase transition most probably occurs on cooling. The structure refinement yields a residue of 9.3%, possibly a consequence of insufficient crystal quality. The refined BN separation in the (not disordered as in [11, 12]) arrangement of molecules was found at 156.5(7) pm (Fig. 1).

In contrast to the X-ray structure of *ethane*, that had been refined in the monoclinic space group $P2_1/n$ [14], 1-1 crystallizes in an orthorhombic system with an antiparallel arrangement of the molecules (Fig. 2).

The shortest intermolecular $H_B \cdots H_N$ distance is 207 pm and the possible interaction of the dipoles could be responsible for the remarkable deviation of the BN distance from the gas-phase determination of 1-1 [10].

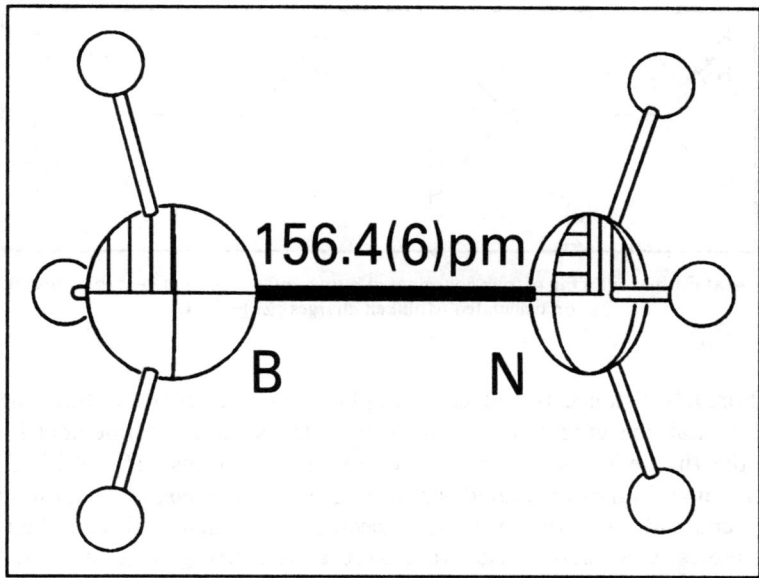

Fig. 1 — Molecular structure of **1–1**.

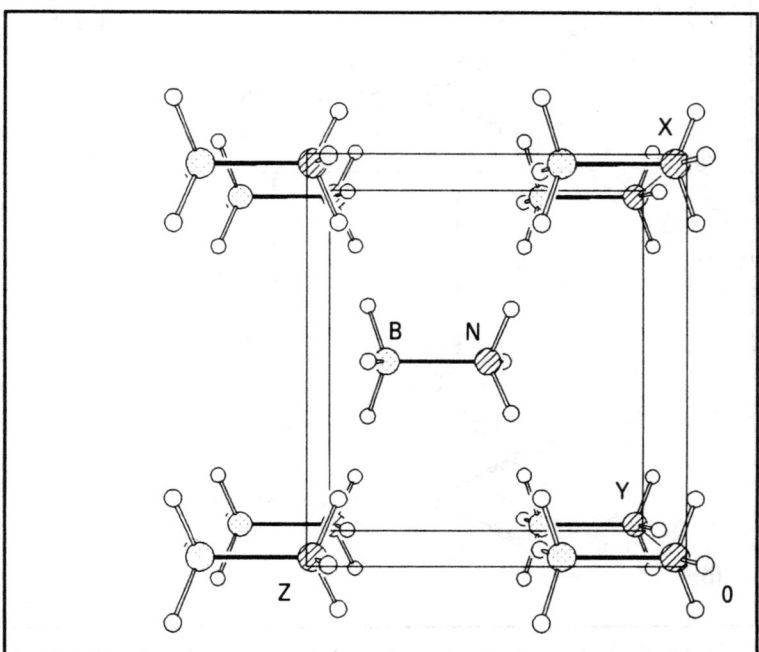

Fig. 2 — Packing plot of **1–1**, viewed along the *y*-axis.

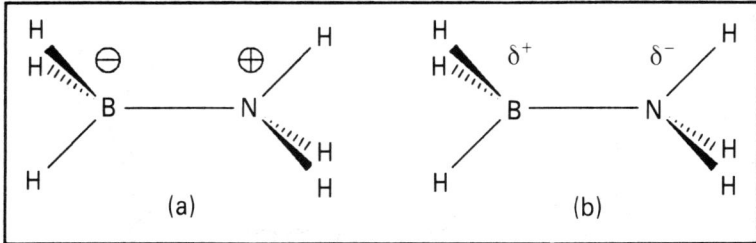

Fig. 3 — (a) Formal charging of four-coordinated boron and nitrogen in **1-1**; (b) charging based on calculated Mulliken charges [29] of **1-1**.

The short BN distance as well as the displacement factors of all atoms indicate the neglected partial charging of the atoms during the structure refinement [13], when applying the theoretical scattering power of neutral atoms [15]. With the formal description of a four-coordinated sp^3-nitrogen as an *ammonium* atom in **1-1**, a positively charged nitrogen atom is expected; the calculated Mulliken charges, however, indicates a partial negative charge at the nitrogen atom (Fig. 3). If the complete group charging of BH$_3$ and NH$_3$ is taken, the formal description can be retained.

Theoretical results taken from *ab initio* calculations (dzp hondo) [16] are displayed in Fig. 4.

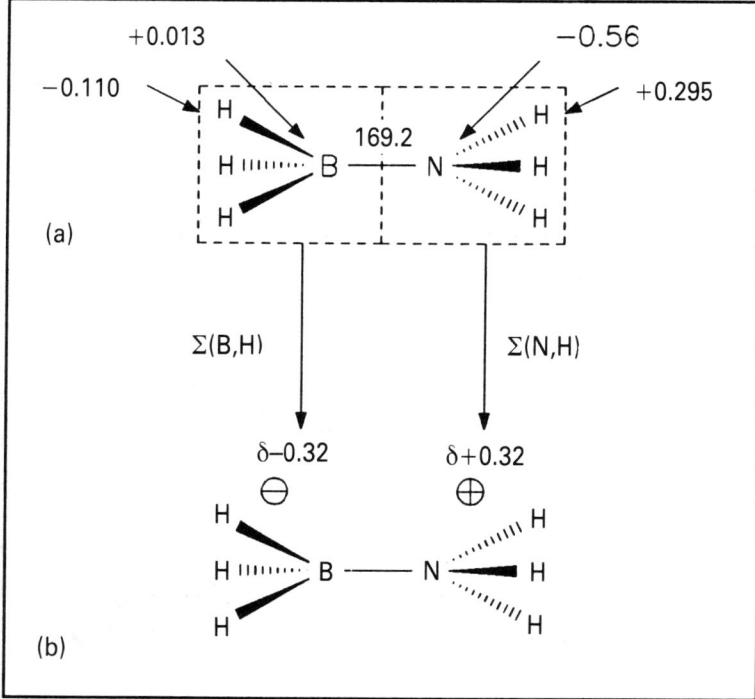

Fig. 4 — (a) Atomic and (b) group charging in **1-1** calculated by *ab initio* (dzp hondo) calculations.

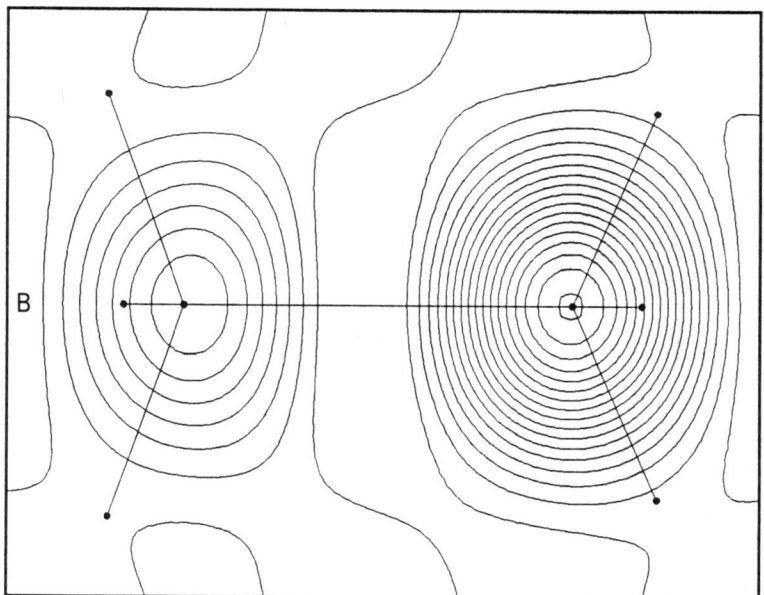

Fig. 5—Electron-density distribution in **1-1**, boron on the left, nitrogen on the right.

The calculations demonstrate the compensation of negative charge on the nitrogen atom mostly by the hydrogen atoms.

The displacement factors of the boron and the attached hydrogen atoms are inadequately higher than those for the nitrogen and its attached hydrogen atoms (see Fig. 1). Fig. 5 and Fig. 6 (perspective as in Fig. 1) illustrate this problem on the basis of electron density (ED) and difference electron density (DED) maps. The electron density found on the nitrogen atom seems unreasonably higher compared to that on the boron atom. This as well as the high electron density, which remains after subtracting the theoretical electron density of single neutral atoms at the core positions (Fig. 6), shows that the additional negative charge on the nitrogen atom was not considered satisfactoryly in the theoretical scattering power applied.

The fact that the present X-ray structure determination of **1-1** does not yield a bond length comparative with the electron diffraction and the *ab initio* calculation [16] demonstrates the difficulty in comparing structural parameters in the solid state with those for the isolated molecule. Moreover, it should be mentioned that the inadequate scattering power assumed for the atoms may additionally falsify the structural results for low-substituted molecules.

A comparable bond length for the *dative bond* was also found in a low-substituted homologous compound. In ethylenediaminebisborane (**1-2**) (Fig. 10) a BN distance of 160.0(7) pm was found [17].

In dimethylaminetrimethylborane (**1-3**) (Fig. 10) the BN distance is extended to 165.6(4) pm [18].

In order to rationalize the effect of methyl substitution on the boron and the nitrogen atom we determined the structures of trimethylamineborane (**1-4**) (Fig. 7) [19], aminetrimethylborane (**1-5**) (Fig. 8) [20] and tetramethylethylenediamine-

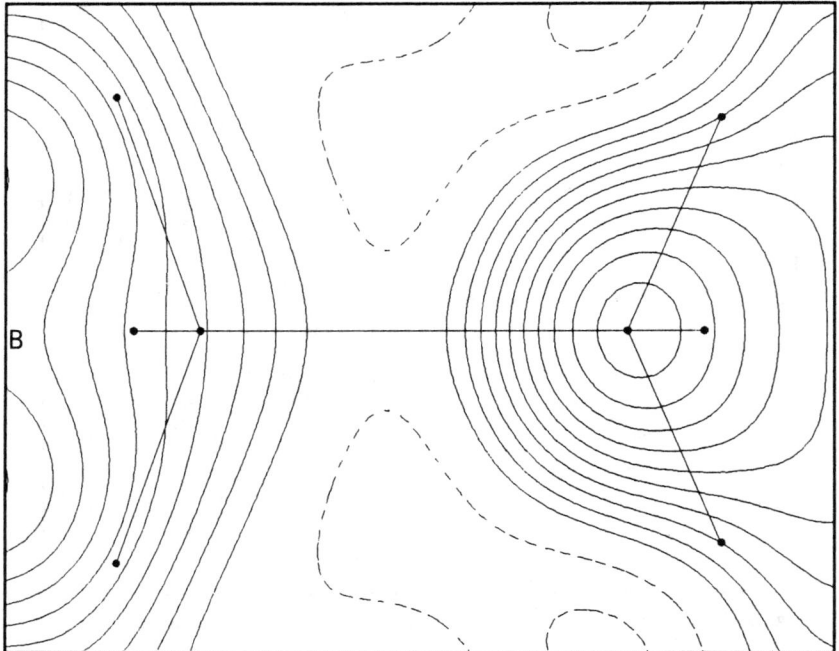

Fig. 6 — Difference electron-density distribution in **1-1**.

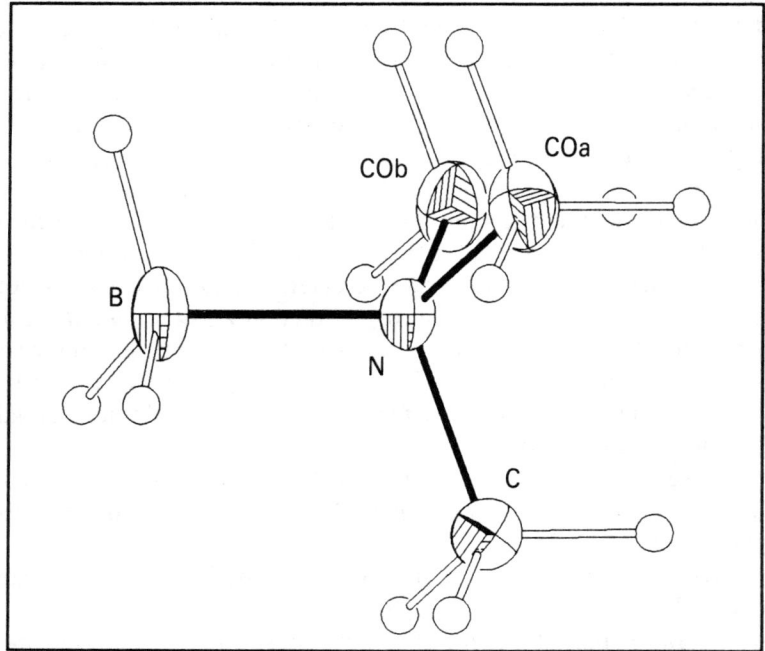

Fig. 7 — Molecular structure of **1-4**. In this and the following two figures ellipsoids are drawn at 50% probability.

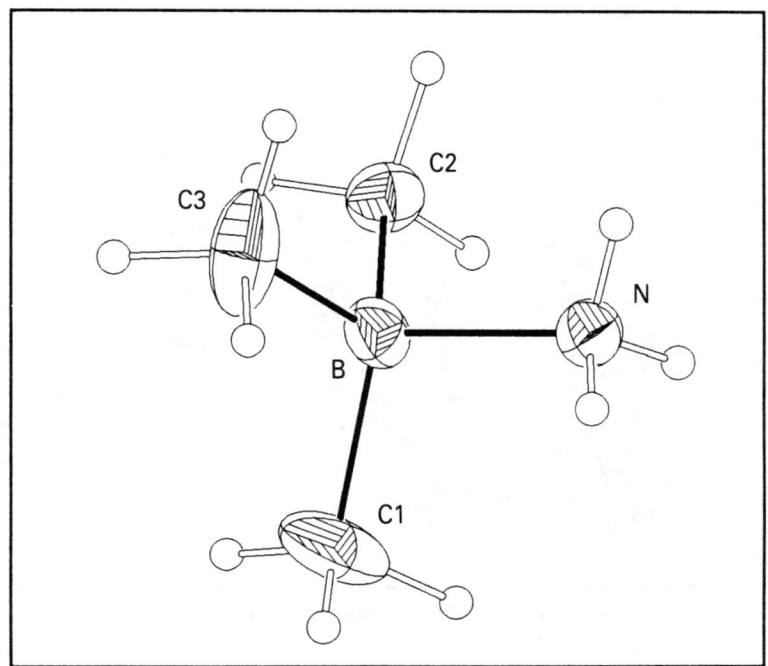

Fig. 8 — Molecular structure of **1-5**.

bis(trimethylborane) (**1-6**) (Fig. 9) [21]. For comparative reasons, Fig. 10 contains additional information on geometrical parameters.

The BN bond distances in all three compounds are elongated relative to the separation in **1-1**. The effect on the BN bond by introducing three methyl groups to the boron atom (**1-5**, BN: 164.8(1) pm) is higher than by substitution at the nitrogen atom (**1-4**, 161.6(1) pm). A dramatic increase in the BN distance occurs by substitution at the boron atom *and* at the nitrogen atom (**1-6**, BN: 173.7(4) pm). Obviously the electron-releasing influence of the methyl groups is able to reduce the electron deficiency on the boron atom, resulting in a modest incorporation of the lone pair at the nitrogen atom. Simultaneously, the BC distances increase significantly; bond lengths of 162.1(7) pm (**1-3**), 162.1(2) pm (**1-5**) and 162.6(3) pm (**1-6**) are expanded by a mean of 7.8 pm compared to the BC bond in trimethylborane (X-ray, 155.6(2) pm) [22].

Steric interaction between the substituents influencing the BN bond in **1-4** seems to be unlikely: the comparison between the CC bond lengths in neopentane (153.4(3) pm) [23] and ethane (153.2(2) pm) [14] shows no bond lengthening caused by steric interaction between the methyl groups, although the CC bonds are shorter than the BN bonds.

The NC distances in **1-4** (148.7(1) pm, CNC: 108.4(1)°) are longer than in trimethylamine (**1-7**) (NC: 144.9(1) pm, CNC: 110.8(1)°) [24] and the CNC angles are decreased. A possible explanation for these geometric changes, which are in contradiction to the VSEPR model, might be given by considering *bent bonds* in **1-7** [25]: the repulsion between the lone pair and the bonding electrons of the NC bond

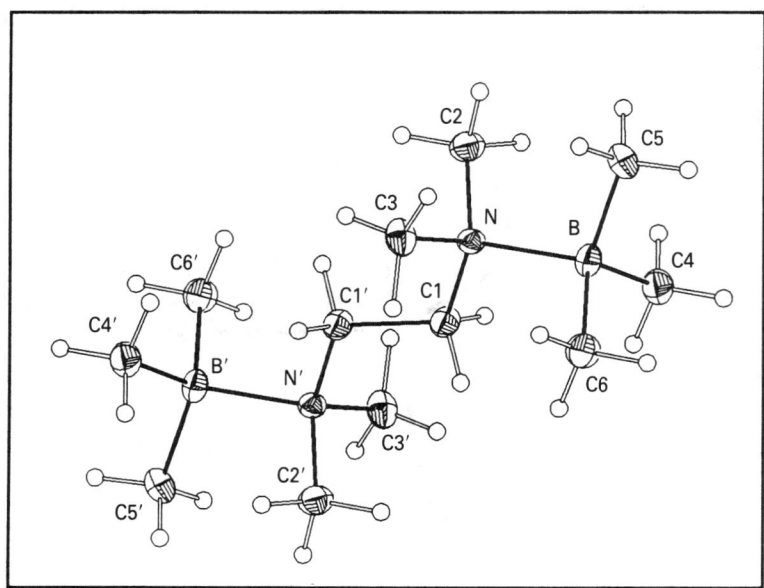

Fig. 9 — Molecular structure of **1-6**.

causes the NC bonding electrons to be shifted from the internuclear lines, resulting in a shorter NC distance. The influence of the lone pair can be relieved when forming the adduct bond to BH_3 in **1-4**, resulting in a longer NC bond distance, and simultaneously decreasing the CNC angle (Fig. 11).

The extreme BN distance of 173.7(4) pm in **1-6** is surprising. It represents the second longest BN distance in acyclic BN compounds investigated in the solid state, only exceeded by $(Me_3N—BCl_2)_2$ with 175.2 pm [26]. A long BN distance was also found in a microwave study of trimethylaminetrimethylborane (170(1) pm) **1-8** [27]. Obviously, the introduction of the third non-hydrogen substituent at the nitrogen atom in **1-6** and **1-8** has a much higher influence on the bonding situation than the substitution of two hydrogen atoms in **1-5** with two methyl groups (**1-3**) only.

While the introduced repulsive interaction in **1-3** is relieved by the bending of the nitrogen-bonded methyl groups (BNC: 114.5(3)°), the interaction in **1-6** cannot be avoided that way. Instead, additional lengthening of the NC, BC *and* BN bonds is observed (BNC$_{Me}$: 109.9(2)°, NC: 149.3(3) pm, BC: 162.6(3) pm). Recapitulating the previous results, the BN distance in **1-1** (156.5(7) pm) appears to be very short and seems to contradict those determined in a microwave study ($r_o = 167.2$, $r_s = 165.7$ pm) [10] and independent theoretical calculations (165 pm (CEPA) [28], 166.4 pm (MP3/6-31G*) [29], 166.1 pm (MP2/6-31G*) [30], 169.2 pm (dzp hondo) [16]). However, BN separations of the same magnitude can also be found in β-boron nitride (157 pm) [31], borazane (157.6(2) pm) [32] and N,N',N''-trimethylborazane (156.5(4) pm [*eee*-isomer], 157.8(5) pm [*eea*-isomer]) [33].

A reconciliation of the apparent experimental inconsistencies is presented by the results of a recent *ab initio*/IGLO study at various levels [34]: here the optimized BN separation was also found in the range of the microwave study [10]. However, fixing the distance to the value found in the X-ray experiment, the energy increases by 5.9 kJ

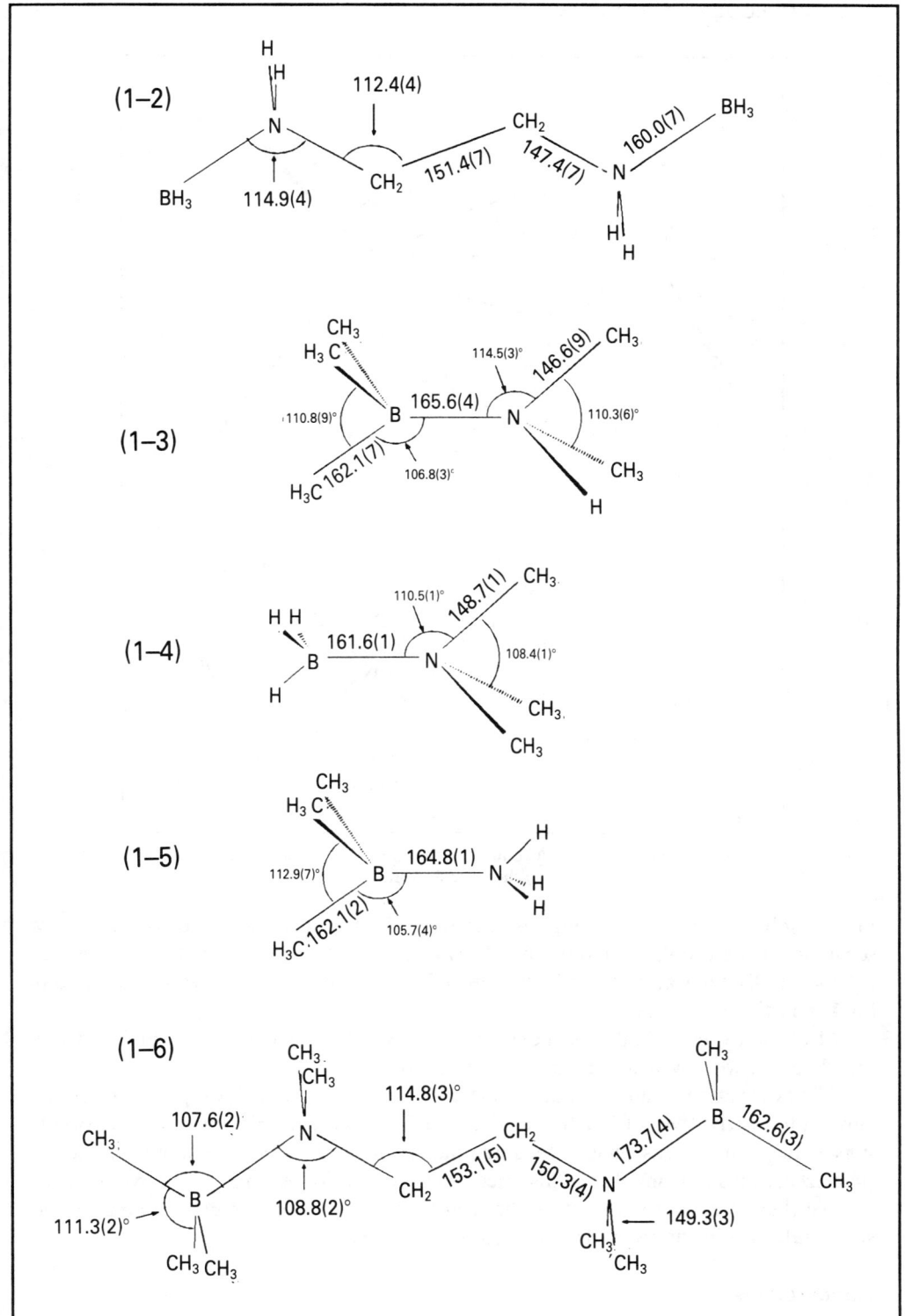

Fig. 10—Important geometrical parameters of the structures discussed. Average bond lengths are given in [pm], corresponding angles in [°].

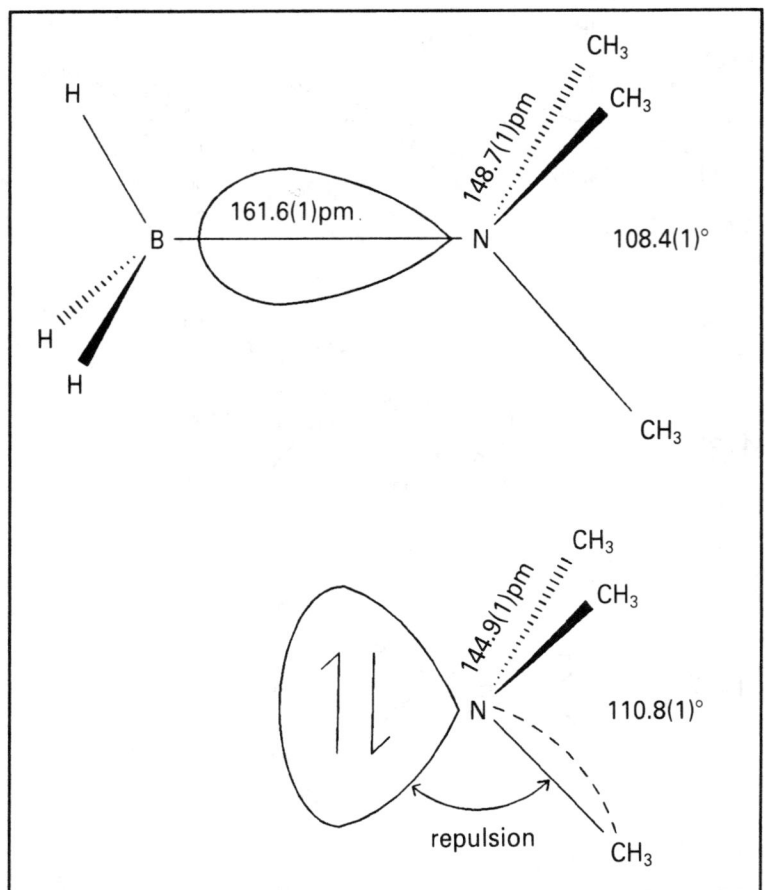

Fig. 11 — Possible explanation for NC bond elongation in **1-4** by considering *bent bonds* in **1-7** (schematic drawing).

mol^{-1} only. The potential trough is extremely shallow and the BN distance is very sensitive to the dipolar surrounding. A simulated SCRF field of water reduces the calculated BN separation to 157 pm, approximately the same value which was found for **1-1** in the crystal lattice.

Also, the experimental and theoretical ^{11}B-NMR chemical shifts, which did not match, agree well by assuming a BN distance of 150 pm.

The proceeding results indicate an extreme sensitivity of the BN separation in low-substituted derivatives of **1-1** to substituent effects and the polarity of the surroundings. Except in the gas phase, the surrounding of the molecules is always polar or polarized, and a reasonable BN distance has to be considered in a polar environment.

We therefore suggest adopting the length of the BN bond in *amineborane* in the solid state within the range 157–161 pm as a reference.

Aminoboranes

The parent compound has not been investigated by means of X-rays. Exact micro-wave data give BN distances of 139.1(2) pm for the parent aminoborane (**2-1**) [35] and an electron diffraction 139.7(2) pm for Me$_2$BNMeH (**2-2**) [36]. Among low-substituted aminoboranes with planar tricoordinated boron and nitrogen atoms, only a few structures have been determined in the solid state. Aminodimesitylborane (**2-3**), the only primary aminoborane investigated by means of X-rays, shows a BN separation of 137.5(8) pm [37]. The structure of tetramethylaminoborane (**2-4**) [38], described in 1970 (BN: 143(4) and 140(4) pm), has been frequently taken as a reference for the BN double bond in the solid state, e.g. [39] and [40].

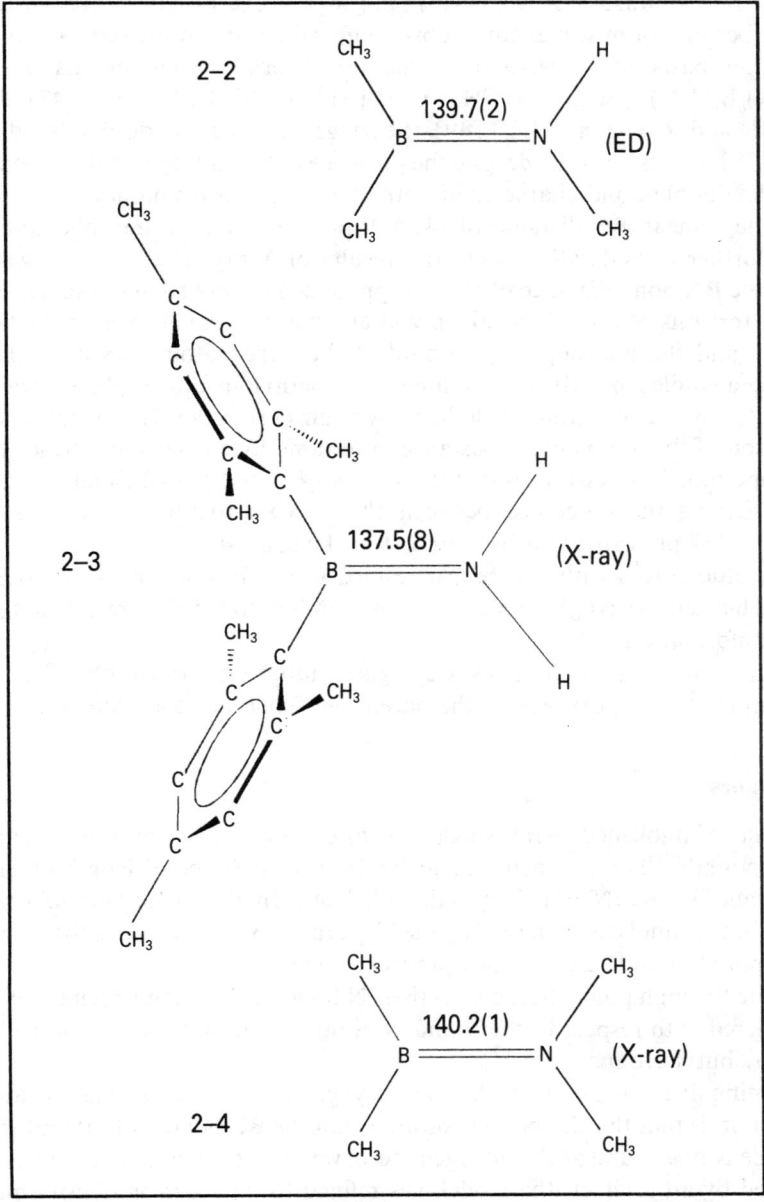

Fig. 12 — Schematic drawings of **2-2**, **2-3** and **2-4**; bond distances are given [pm].

However, we found distinctive different cell dimensions, space group and, more-over, a melting point below the measurement temperature given in [38]. In the redetermination, the BN distance was found to be 140.3(1) pm [41]. In further structural investigations of aminoboranes, BN distances shorter than 140 pm were found (dimethylaminodichloroborane, 137.9(6) pm (ED) [42]; diisopropylamino-bis(trifluoromethyl)borane 136.5(5) pm (X-ray) [43]; *tert*-butyl-(*tert*-butylamino)-tris[(*tert*-butyloxy)silylthio]borane 136.9(4) pm (X-ray) [44]; the localized BN bonds in *B*-triphenyl-*N*-tri-*tert*-butyldewarborazine 137.3(4) pm [45] and *B*-triisopropyl-*N*-tri-*tert*-butyldewarborazine 137.4(8) pm [46]). These structures are not taken into account, because of miscellaneous substituent influences on the central BN bond.

On the basis of a series of structure determinations of diaminoboranes, $(RCH_3N)_nB(CH_3)_{3-n}$ with R = CH_3, n = 1 [41], R = CH_3, H, n = 2 [47], R = CH_3, n = 3 [48] and R = H, n = 1,2,3 [49], the suggestion of a BN double-bond length of 141 pm [39] is questionable: despite the presence of two nitrogen donor atoms, which reduce the double-bond character of both BN bonds, in bis(monomethylamino)me-thylborane, a mean BN distance of 141.5(2) pm was found in the solid state [47].

For further consideration, only the results of X-ray investigations will be dis-cussed. The BN bond distance of 137.5(8) pm in **2–3** should be in the same range as in **2–1**. The torsions of the BC bonds in **2–3** allow a very small overlap of the mesityl π-systems and the unoccupied p_z-orbital at the boron atom ($\cos^2(\alpha) = 0.25$) [50]. Assuming a similar, mostly steric influence of methyl groups on the central bond of **2–4** and the isosteric tetramethylethene, we can derive the BN distance in **2–1** by comparison of the double-bond distance in tetramethylethene and ethene. The latter two compounds have distances of 134.8(1) pm [41] and 131.42(3) pm [51].

Transferring the difference between these two structures to **2–4**, an empiric distance of 137 pm can be derived for **2–1** in the solid state.

This value agrees with theoretical results, where BN distances of 137.8 pm (*ab initio*/double-zeta basis) [52] and 139.5 pm (dzp hondo) [16] were calculated for the planar conformation.

Similarly to the amineboranes, we suggest adopting a range of 137 pm $\leqslant d_{B=N}$ $\leqslant$ 139 pm for the BN distance in the parent *aminoborane* as a reference.

Iminoboranes

In the field of published X-ray single-crystal determinations of iminoboranes, only two compounds allow a conclusion to be drawn for the bond length of the parent iminoborane H—B≡N—H (**3–1**) in the solid state. In di-*tert*-butyliminoborane (**3–2**) and *tert*-butylimino[tris(trimethylsilyl)silyl]borane (**3–3**) (Fig. 13), BN distances of 125.8(4) pm [53] and 122.1(5) [54] pm were found.

Despite the high polar character of the BN bond in **3–2**, the molecule was found to be disordered with respect to the B and N atoms, obviously because of the shielding by the *tert*-butyl groups.

Assuming fixed positions for the *tert*-butyl groups in the crystal lattice and taking into account (i) that the CN bond is shorter than the BC bond and (ii) that the higher electron density is found at the nitrogen atom, we must conclude that the BN distance was biased by an artefact: the model was refined to the electron density maxima of superimposed molecules (Fig. 14).

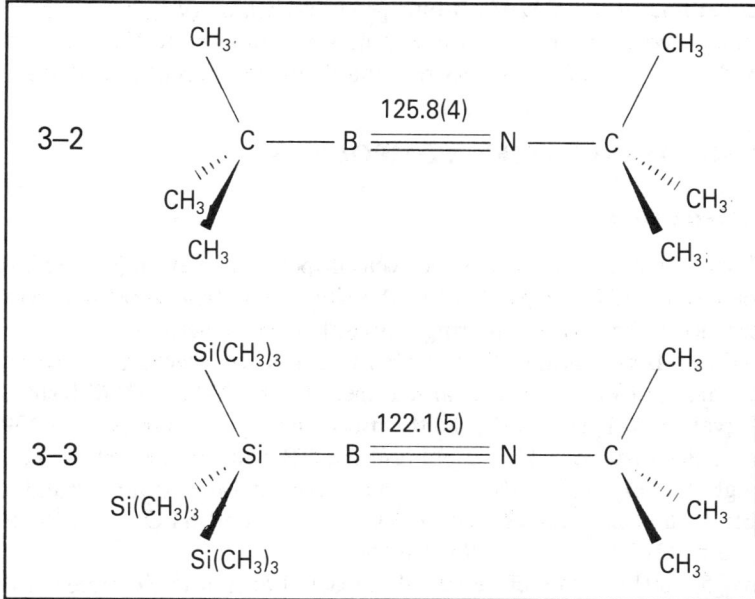

Fig. 13—Schematic drawings of **3-2** and **3-3**; BN distances are given in [pm].

We therefore believe that the BN distance was overestimated. This can be confirmed by the structure of **3-3**, which is not disordered and has a shorter BN distance. The influence of the tris(trimethylsilyl)silyl-substituent and the *tert*-butyl group should be roughly the same, providing 122 pm as the correct BN separation in **3-2**.

Similarly to the evaluation of the bond length in *aminoborane*, the BN distance in the parent *iminoborane* can be evaluated, considering the analogous CC compounds. Taking the difference between bond lengths resulting from the X-ray investigations of

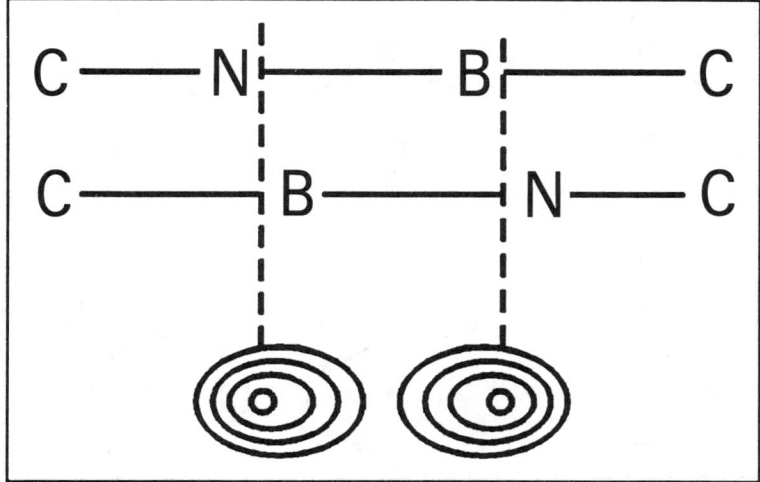

Fig. 14—Schematic drawing of the electron distribution of superimposed molecules in **3-2**.

di-*tert*-butylethine ($d_{c \equiv c}$ = 120.0(2) pm) [55] and ethine ($d_{c \equiv c}$ = 117.8(2) pm) [56] and assuming a corresponding shortening in **3-1** compared to the adopted length in **3-2**, a BN distance of 120 pm seems reasonable for the parent *iminoborane*.

CYCLIC BORON–NITROGEN COMPOUNDS

Three-membered rings

The smallest aromatic CC system is the cyclopropenylium cation (Fig. 15(a)). The CC bond distances are 137.2(1) pm [57] in the ring of the tri-*tert*-butyl derivative with lowest influence of ligands on the ring among known structures.

The cationic BN-analogon (Fig. 15(b)) is unknown, whereas the azaboriridines (Fig. 15(c)) are well known and characterized. In the *N*-butyl-*B,B'*-bis(diisopropyl-amino) derivative (**4-1**) (Fig. 16) [58] the mean endocyclic separation is 139.4(5) pm, the exocyclic BN distance 143.3(5) pm and the BB bond length 160.4(8) pm.

Although the exocyclic BN bonds have considerable double-bond character, comparable to *ab initio* calculations (STO 3–21G* and STO–31G) for the parent azadiboriridine (162.1 pm and 159.9 pm [60]).
(161.0 pm) [59, 60] and therefore reveals a partial aromatic character in the ring; a delocalization of π-electrons is possible in the ring *and* the exocyclic BN bonds. The BB bond is much shorter than in the tetraaminodiborane (169.9(9) pm) [61] and comparable to *ab initio* calculations (STO 3–21G* and STO–31G) for the parent axadiboriridine (162.1 pm and 159.9 pm [60]).

The isoelectronic diazaboracyclopropane (Fig. 15(d)) does not reveal any aromatic character; in *B*-bis(trimethlsilyl)amino-*N,N'*-bis(*tert*-butyldimethylsilyl)diaza-boracyclopropane (**4-2**) (Fig. 17) [62] the silyl groups at the ring N-atoms are twisted, resulting in an approximately tetrahedral surrounding of the N-atoms.

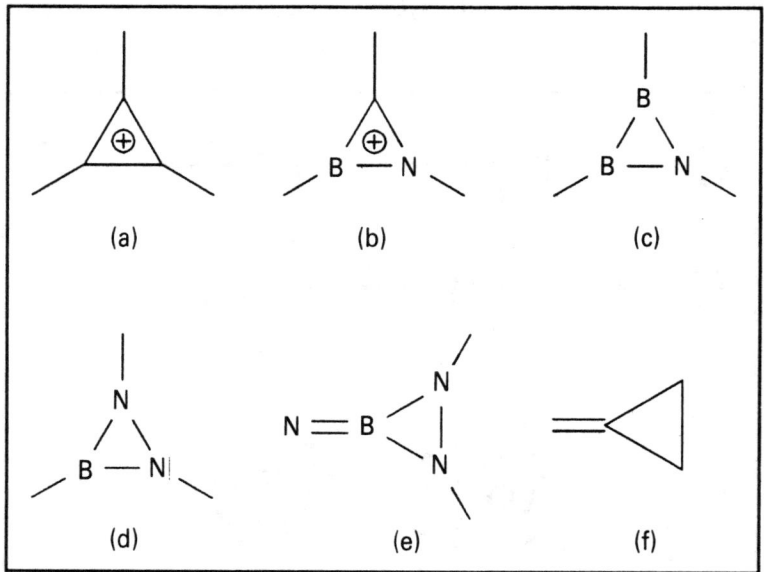

Fig. 15 — Three-membered rings mentioned in the text.

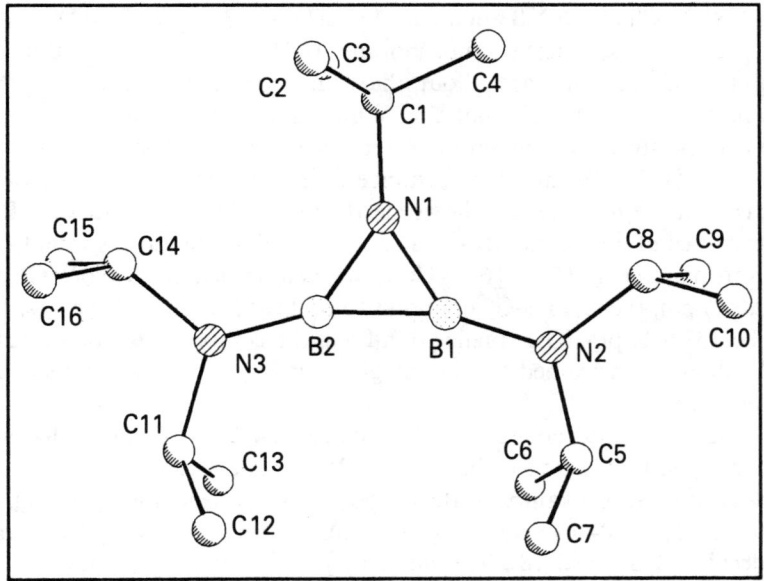

Fig. 16 — Molecular structure of **4-1**.

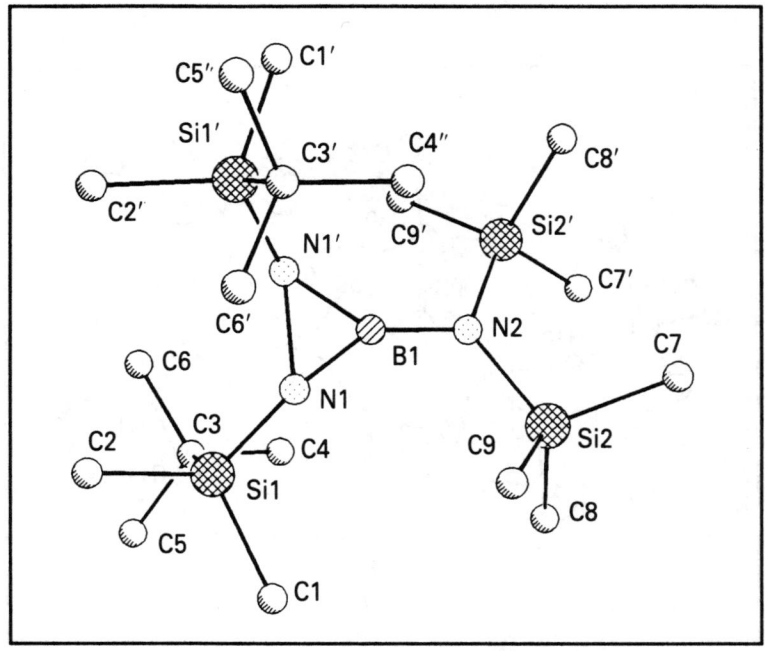

Fig. 17 — Molecular structure of **4-2**.

The BNSi, NNSi and NNB angles are 133.3(1), 117.3(1) and 54.1(1)° respectively. With respect to tris(dimethylamino)borane [48] the endocyclic BN distances (B1–N1(N1′) 142.6(2) pm) are about the same, and the exocyclic B1–N2 bond (140.6(3) pm) is shortened. The bonding situation is closer to the formula shown in Fig. 15(e). Ring strain at the boron atom (NBN 71.8(1)°) makes the NN bond relatively long (167.3(3) pm). This distance indicates additionally that there is no aromatic contribution in the ring; however, the relatively short endocyclic BN bonds provide evidence of an increased s-character. A similar situation is found in methylene-cyclopropane (Fig. 15(f)) [63]. Here, we determined the $C(sp^3)$–$C(sp^3)$ bond to be 152.6(2) pm, the $C(sp^2)$–$C(sp^3)$ bond to be 146.0(1) pm and the $C(sp^2)$–$C(sp^2)$ bond to be 131.6(1) pm. The smaller bond lengths compared to the corresponding standard values of unstrained CC bonds given in [64] are shortened by 1.8, 5 and 1.9 pm, respectively.

By adding 5 pm to the endocyclic BN distance in 4-2, 147.6 pm results. This value should give a good estimate for the sp^2-sp^3 BN bond distance.

A further interesting example is the azaboracyclopropene fragment (Fig. 15(d)) in spirocyclic 1,2,5,7,8,10-hexa-*tert*-butyl-4,9-dioxa-1,7-diaza-2,5,8,10-tetraboradispiro-[2.2.2.2]decane (4-3) (Fig. 18). The bond lengths in the ring are C1–B2: 132.9(9), C1–N1: 145.9(5) and N1–B2: 132.9(9) pm. The last mentioned is the shortest BN double-bond length that has been reported. It can be explained by the extreme ring strain accompanied by an exocyclic shift of bonding electrons.

Accepting the difference between the standard $C(sp^2)$–$C(sp^2)$ separation of 133.5 pm and the distance found in bicyclopropyl-2-enyl (128.7(1) pm) [65] as a result of ring strain, and assuming the same difference for the BN double bond in 4-3, we end

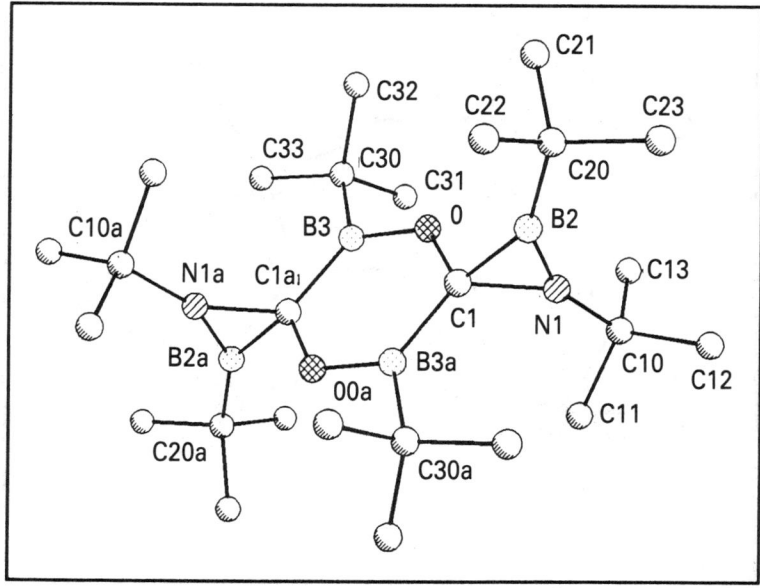

Fig. 18 — Molecular structure of 4-3.

up at a value of 137.7 pm for the unstrained BN double bond. This is close to the lower limit derived as a reference for the parent compound *aminoborane*.

Four-membered rings

The antiaromatic cyclobutadiene has been the subject of many discussions because the first X-ray investigation of the tetra-*tert*-butyl derivative [66] (CC: 146.4(3), 148.3(3) pm; twist angle of the four-membered ring: 7°) did not reveal the more pronounced bond-length alternation that was found in other derivatives [67, 68]. In a redetermination of the di-*tert*-butyl derivative at a lower temperature (125 K) [69], the alternation was more pronounced (144.1, 152.7 pm). However, the difference was still not in the expected range of 134–160 pm [70, 71]. A detailed analysis of the anisotropic displacement parameters exposed hidden disorder, which scrambled the bond distances [72]. Hence, it is not surprising that the isosteric tetra-*tert*-butyl-diazadiboretidine (**5-1**) (Fig. 19) was also found to be disordered [53]. Owing to the superimposition of the boron and nitrogen atoms (1:1) a mean BN distance of 148.6(5) pm resulted, which is exactly between those values taken as references for the BN double bond and BN single bond.

The four-membered ring was found to be much more twisted (18°) than the analogous carbon compound. This can easily be explained by the shorter BC and NC distances (148 pm, mean value) compared to the corresponding mean CC distance of 152 pm [69]. Here, intramolecular non-bonded H···H separations of 184 pm were claimed to be responsible for the torsion.

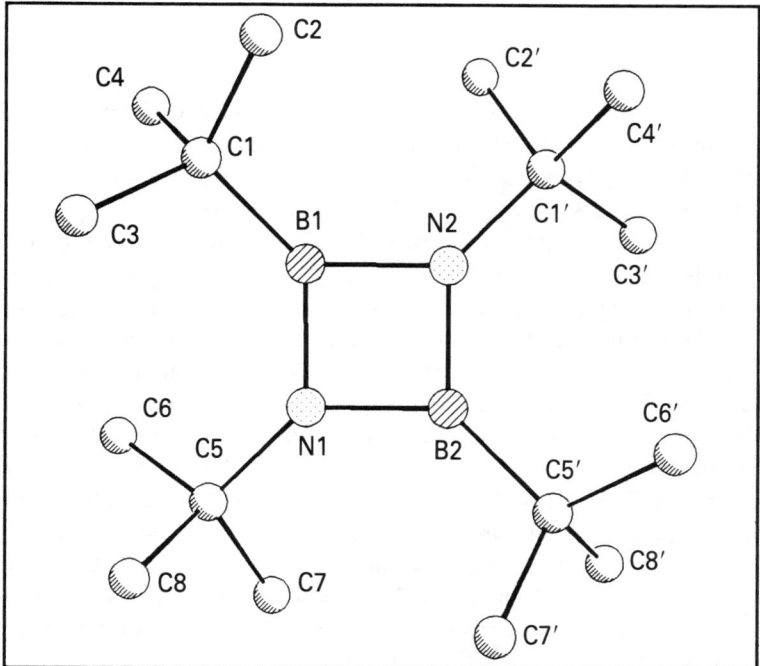

Fig. 19—Molecular structure of **5-1**, arbitrary positions for B and N atoms.

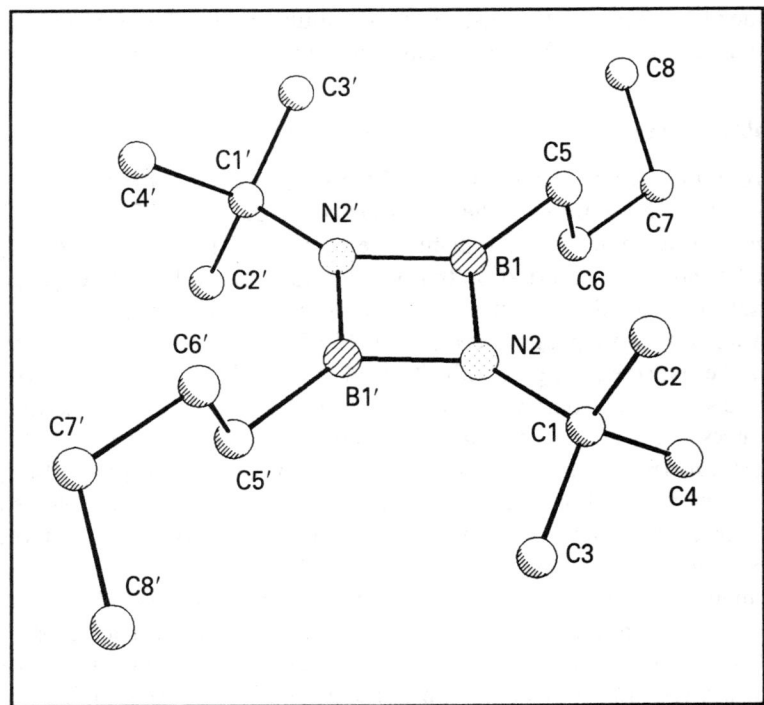

Fig. 20 — Molecular structure of **5-2**.

In the planar ring of bis(*N*-*tert*-butyl)bis(*B*-*n*-butyl)diazadiboretidine **5-2** (Fig, 20) [73], no scrambling of B and N atoms could be detected because of the different substituents. Also, no dynamic disorder was found when analyzing the anisotropic displacement parameters. Even so, the BN bond distances are equal in the ring (145.8(3), 145.9(3)), comparable with bis(*N*-di-*tert*-butyl)-bis(*B*-dipentafluorophenyl)-diazadiboretidine (143.1(7) pm) (Fig. 21) [74] and bis[*N*-bis(trimethylsilyl)]-bis[*B*-{bis(trimethylsilylamine)}]diazadiboretidine (145.4(11) pm) (Fig. 22) [75].

Although antiaromatic, the π-system must be delocalized in the planar diazadiboretidines. Obviously an essential contribution of electron density is situated in a higher localized b_g-orbital at the N atoms. A similar situation with electron-withdrawing and electron-donating substituents occurs in *trans*-bis(diethylamino)-bis(carboxyethyl-esther)cyclobutadiene (Fig. 23) [76], where the push–pull effect causes an equilibration of the endocyclic bonds (146.3 pm, mean value). Moreover, differences of 5–15° were found at the innercyclic angles, increased at the C atoms with the electron-donating substituents, and decreased at the C atoms with the electron-withdrawing substituents. This is also observed in **5-2**, owing to an intrinsic push–pull effect: innercyclic bond angles are decreased at the nitrogen atoms and increased at the boron atoms (BNB: 85.3(2)°, NBN: 94.7(2)°).

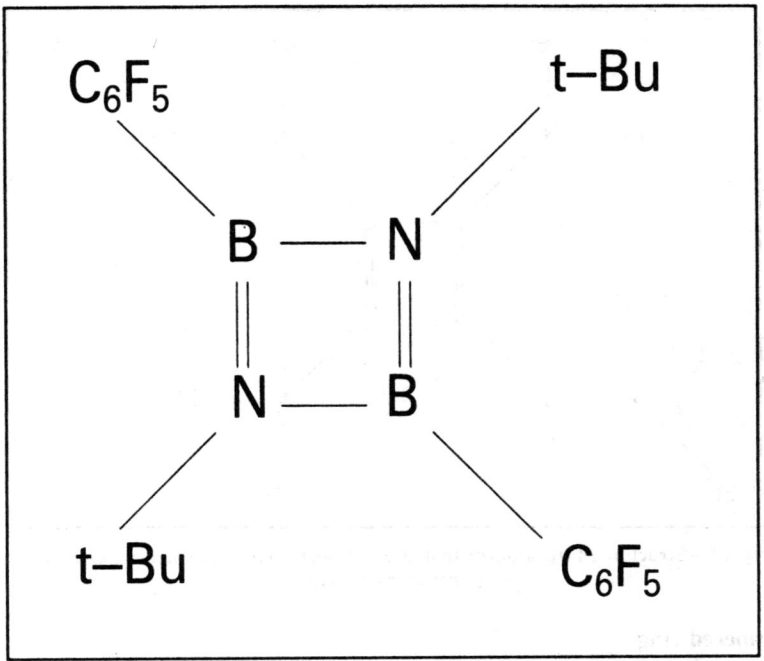

Fig. 21 — Structure of bis(*N*-di-*tert*-butyl)-bis(*B*-dipentafluorophenyl)diazadiboretidine (schematic drawing).

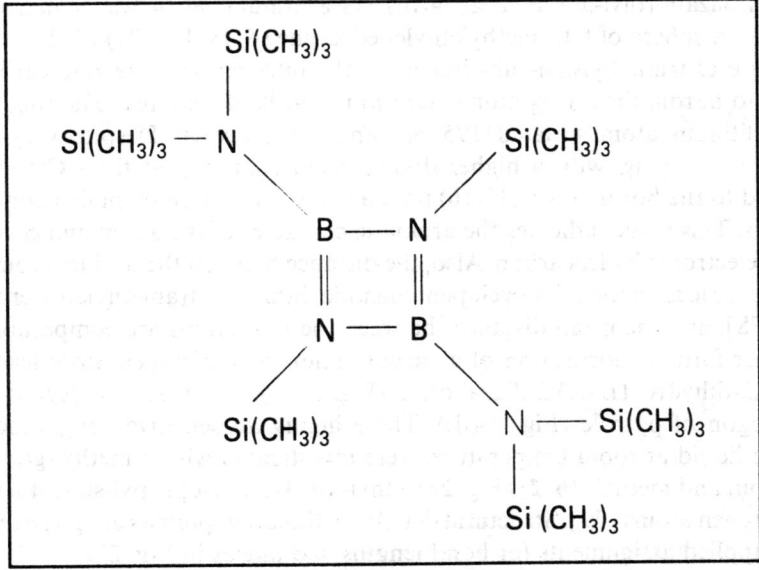

Fig. 22 — Structure of bis[*N*-bis(trimethylsilyl)]-bis[*B*-{bis(trimethylsilylamine)}]diaza diboretidine (schematic drawing).

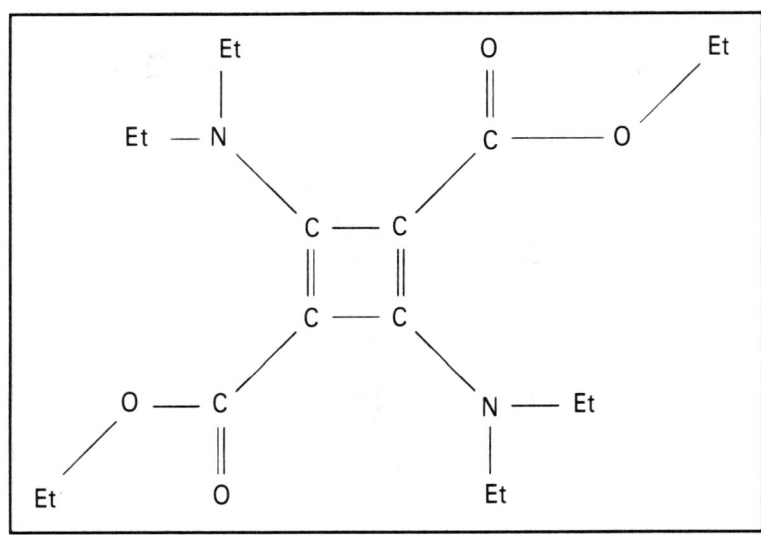

Fig. 23 — Structure of *trans*-bis(diethylamino)-bis(carboxyethylesther)cyclobutadiene
(schematic drawing).

Five-membered rings

Among the BN systems isoelectronic with the cyclopentadienide-anion, we only
consider those structures with three-coordinated ring atoms, namely the isosteric and
aromatic 6π-systems (Fig. 24). Transition metal π-coordinated systems are also
omitted.

Substitution of a CC fragment in the cyclopentadienide-anion (Fig. 24(a)) leads to
the dihydroazaborolyl-anion (Fig. 24(b)). As a lithium salt it was stabilized in the
coordination sphere of tetramethylethylenediamine (**6–1**) (Fig. 25) [77].

Because of trimethylsilyl-substituents at the nitrogen and the ring carbon atom
adjacent to boron, these ring atoms were found to be disordered. The ring is planar
and the lithium atom situated 195 pm above the center. Despite varying bond
distances in the ring, with a higher double-bond character at the CC bond *trans*-
positioned to the boron atom (136(6) pm), a mean separation of all five bonds of 145
pm results. This value indicates the aromatic character of the anion and corresponds
to the isoelectronic hydrocarbon. Also, the distance between the lithium atom and the
ring is the same as in the 1,3,4-cyclopentadienide–lithium-tetramethylethylenediamine
adduct [78], and the mean distances between the ring atoms are comparable, too.

Further formal substitution of a carbon anion by a nitrogen atom leads to the
neutral 2,3-dihydro-1H-1,3,2-diazaborole (Fig. 24(c)). It can also be regarded as the
BN-analogon of pyrrole (Fig. 24(d)). These highly air-sensitive compounds, all of
which are liquid at room temperature, were investigated with a methyl-group at the
boron atom and methyl- (**6–2**) (Fig. 26), ethyl- (**6–3**) and isopropyl-substituents (**6–4**)
at the nitrogen atoms [79]. Structural details for these compounds are given in Table 1,
and the applied assignments for bond lengths and angles in Fig. 27.

Parameters were averaged on C_2-symmetry for molecule **6–3** and **6–4**. In the
planar systems, an optimal overlap of the occupied p_z-orbitals at the nitrogen atoms
and the empty orbitals at the boron atoms is possible. Consequently we find BN

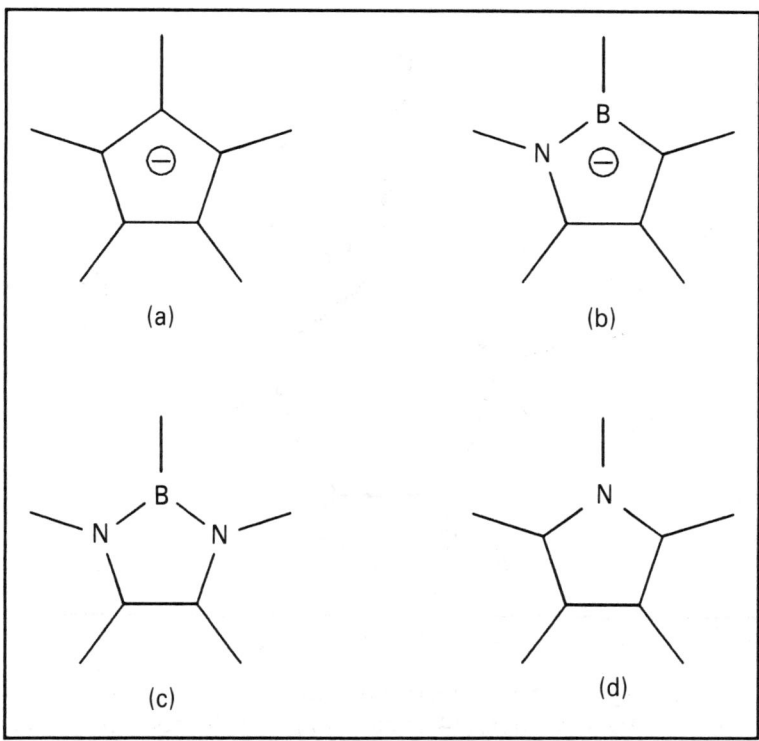

Fig. 24 — Schematic drawing of molecules mentioned in the text.

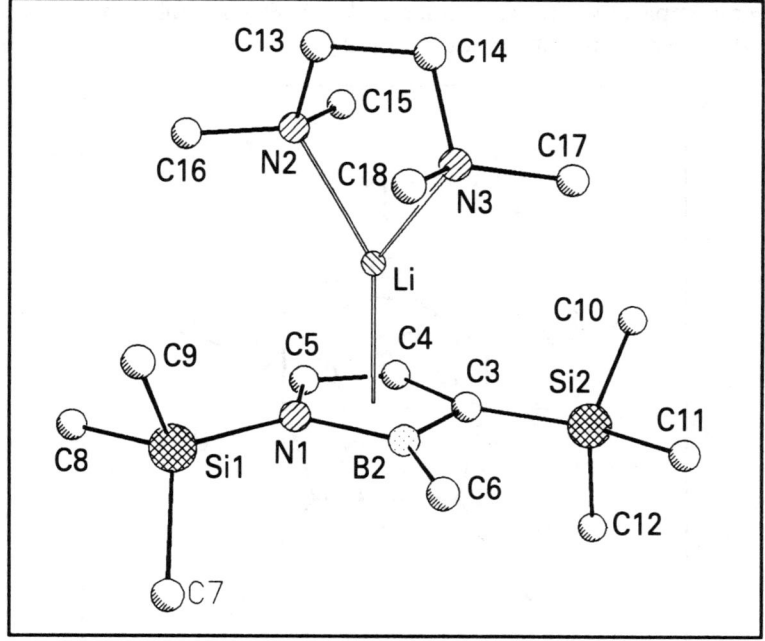

Fig. 25 — Molecular structure of **6-1**.

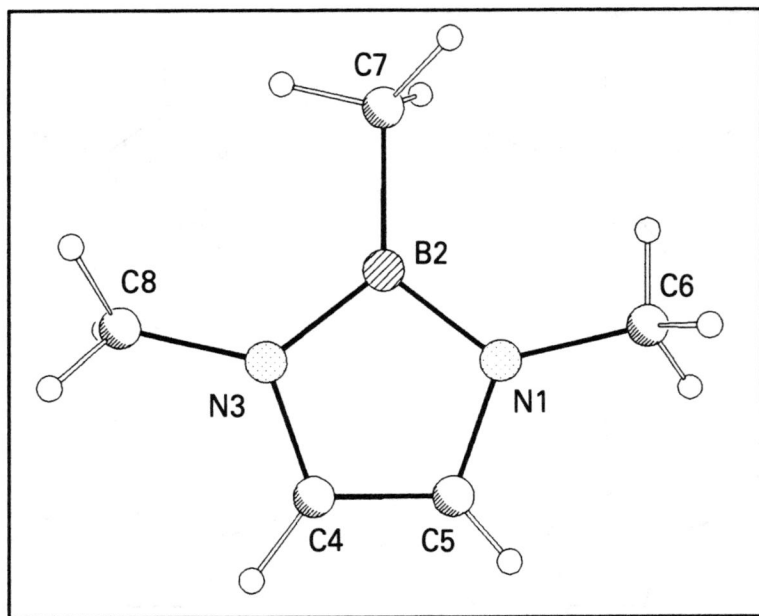

Fig. 26 — Molecular structure of **6-2**.

distances in the range of the bis(methylamino)methylborane (141.5 pm) and the bis(dimethylamino)methylborane (143.4 pm) [47]. The shortest BN distance is found in the ethyl-derivative (**6-3**) where the ethyl groups are bent in opposite directions relative to the ring system (torsion angles, $BNC_{Et}C_{Et}$: 87.1 and 97.6°, respectively).

The mean distances (*a–c*) (Table 1) indicate an aromatic character which was also proven by π-complexation to transition metals [79, 80]. Pyrrole has also been used as a 6π-ligand in transition metal complexes [81].

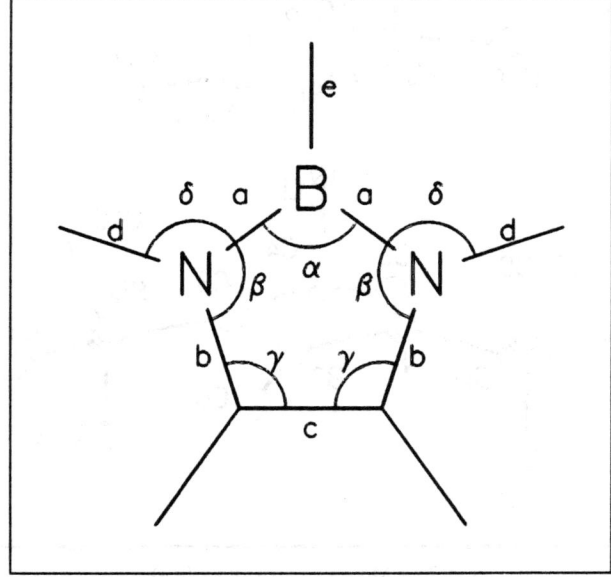

Fig. 27 — Position of bonds and angles mentioned in Table 1.

Table 1 — Bond lengths and angles of the diaza-
boroles **6-2, 6-3** and **6-4**

	6-2	**6-3**	**6-4**
R	Me	Et	iPr
a	143.3	140.8	143.0
b	139.3	141.0	140.1
c	134.9	132.7	134.2
Mean *a–c*	139.2	138.2	139.2
d	145.2	145.0	146.1
e	156.0	156.5	157.3
α	104.6	105.9	105.0
β	108.5	108.1	108.2
γ	109.2	109.0	109.3
δ	129.0	131.2	129.0
Torsion N–C–C–N	0.6	2.1	13.5
Σ angle B	360.0	359.9	360.0
Σ angle N	360.0	360.0	360.0
esd's distances	1–2	2–4	1–2
esd's angles	1	2	1

Six-membered rings

The best known of the heterocyclic aromatic ring systems in boron–nitrogen chemistry is borazine (Fig. 28(a)). Although this class of compounds has received immense theoretical interest because of its heteroaromatic character, only few X-ray investigations have been performed, which is certainly due to the liquid state of the low-substituted borazines at room temperature.

The parent *borazine* has already been characterized in 1931 [82]. In an electron diffraction study of 1969, the BN distance was found to be 143.5(2) pm [83]. Some of the structure determinations undertaken in the following years will be discussed subsequently.

Hexachloroborazine (space group *R*3) had been investigated several times because of the invariance of BN bond lengths [84–86]. Finally, the BN bond distance was determined to be 142.4(1) pm for all BN bonds. In the almost planar arrangement of atoms, the endocyclic and exocyclic BN distances in *B*-tris(dimethylamino)borazine have been found to be nearly the same (143.3(8) pm, mean value); the NC bonds were found to have a mean length of 145.8 pm [87]. This seems to be in contradiction with the electron diffraction results of *B*-monoaminoborazine, where the endocyclic and exocyclic BN bonds differ by 8 pm (BN_{endo}: 141.8, BN_{exo}: 149.8 pm) [88], but can be explained by the electron-releasing character of the nitrogen-bonded methyl groups. The donor-character of the exocyclic nitrogen atoms increases with the effect of a similar double-bond character in *all* BN bonds. Comparable to the innercyclic angles in borazine, the NBN angles (116.8°, 117.7°(borazine)) are significantly smaller than the BNB angle (123.2°, 121.1°(borazine)).

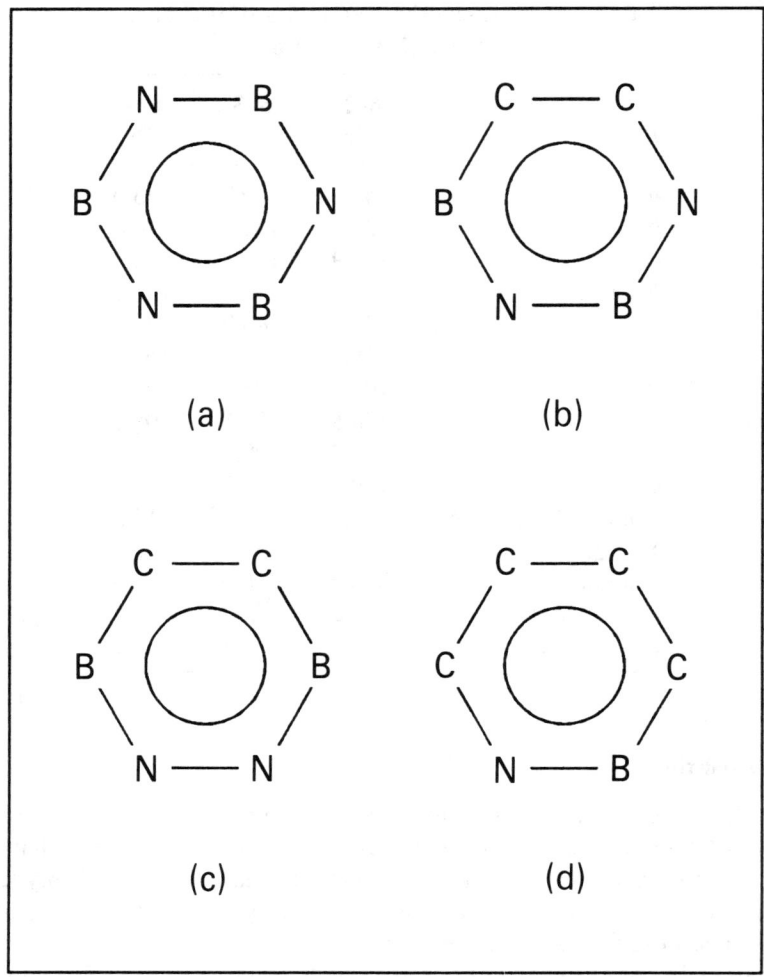

Fig. 28 — Boron-nitrogen homologs of benzene.

In *B,B′,B″*-trimethylborazine [89] and hexaethylborazine [90], a mean of 142 pm was found for the BN separation. In the latter, the ethyl groups are directed upwards and downwards in an alternating sequence.

A mean BN distance of 142.8 pm is found in *B,B′,B″*-trichloro-*N,N′,N″*-triphenyl-borazine [91], where two phenyl-groups were found to be orthogonal to the heterocyclic BN system and the third at an angle of 95°. The average BCl distance of 176.1 pm is longer than in hexachloroborazine and the NC bonds were determined to have a mean length of 145.6 pm. For hexaphenylborazine with propeller-like arrangement of the substituents (the interplanar angles lie in a range of 60.7–71.4°) a BN separation of 143.5 pm was established. With values of 153.8 and 151.2 pm for the BC and NC bonds, the difference between these two bonds is smaller than expected considering by subtracting the covalent radii. The whole structure is comparable to that of hexaphenylbenzene [92].

The derivatives of borazine already mentioned show the great variation in the geometric parameters, but in some cases do not allow the detailed comparative discussion of substituent influence on the geometric parameters because of low significance. Most of the measurements were taken at room temperature, and in some cases a distinction between boron and nitrogen atoms was not possible.

To complete the information for this class of compounds, the structures of B,B',B''-triethylborazine (**7-1**) [93] (Fig. 29), B,B',B''-triphenyl-N,N',N''-tri-n-propylborazine (**7-2**) [94] (Fig. 31), B,B',B''-triethyl-N,N',N''-tribenzylborazine (**7-3**) [95] (Fig. 32) and N,N',N''-triphenylborazine (**7-4**) [96] (Fig. 33) were determined at 125 K and below.

7-1 crystallizes with two independent molecules in the monoclinic space group $P2_1$/c. The distances in all BN bonds are equal (143.2(1) pm) within their standard deviations, the mean BC distance is 157.7 pm and the CC value is found to be 152.9 pm. With 115.7(1)° the NBN angles are smaller than the BNB angles (124.3(1)°), similar to the parent borazine (117.7, 121.1°).

The mean BCC angle of 116.1(1)° for a C(sp²)-atom is worth mentioning. On the basis of experimental and theoretical data, comparable angles were also found in other boron systems [97]. The molecules in **7-1** arrange pairwise in the crystal lattice with intermolecular B···N contacts of 377 pm (Fig. 30).

Long-distance intermolecular interactions beyond the sum of van der Waals radii were also found to be responsible for a solid-state phase transition in the isosteric triethylboroxine [98]. Further investigations with cocrystallizates of **7-1** and triethylboroxine are intended.

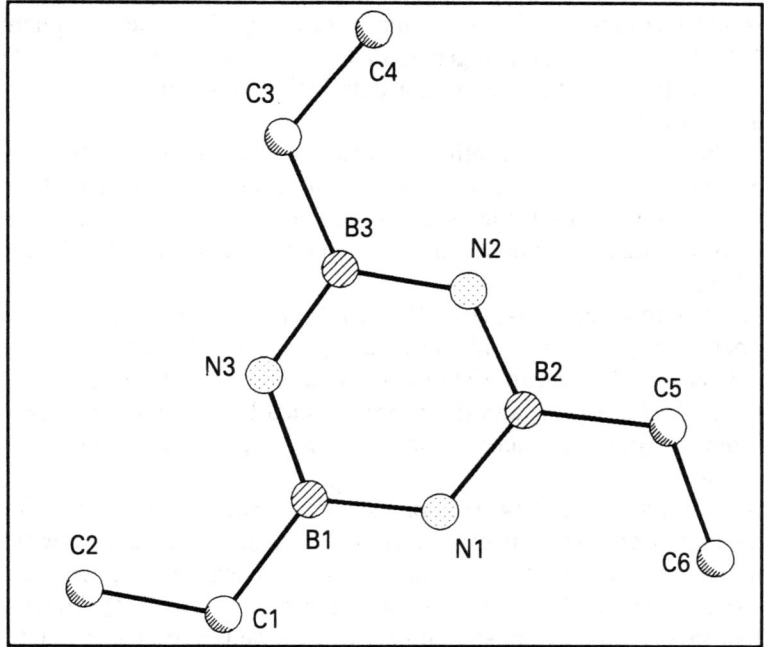

Fig. 29 — Conformation in one independent molecule of **7-1**.

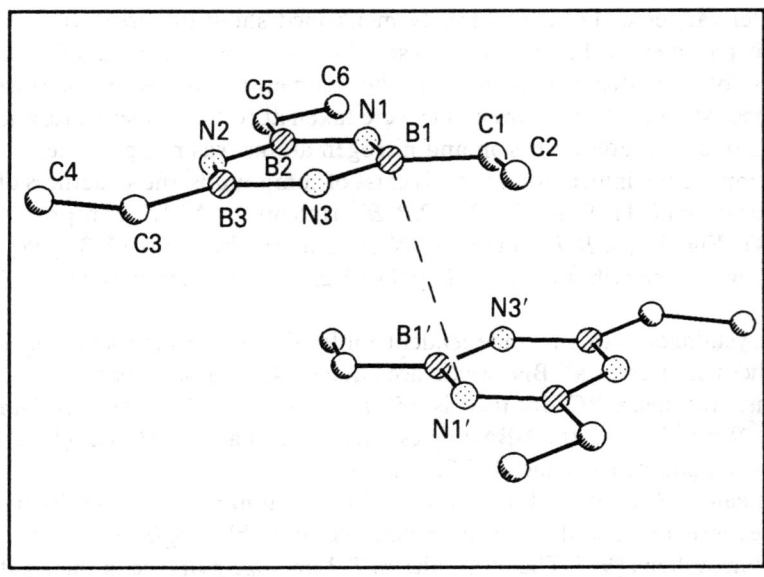

Fig. 30 — Orientation in a molecule pair of **7-1**. The broken line depicts the shortest intermolecular BN distance of 377 pm.

Inevitably two of the *n*-propyl groups in **7-2** are arranged in a *cis*-position with respect to the heterocyclus, causing an almost orthogonal arrangement of the enclosed phenyl groups (89.6°, interplanar angle) (Fig. 31). The two phenyl groups that reside between the *trans*-arranged propyl groups show interplanar angles of 98.3° and 72.3°, respectively. Within the standard deviations, the same value is found for the BN distances (143.8 pm).

In **7-3** the N-benzyl and B-ethyl groups are orientated, for steric reasons, in alternating directions (Fig. 32), the ring system remains planar and the mean BN distance is 144.8(4) pm. This is the longest BN separation for the borazines in the solid state. Also, NBN angles are smaller than the innercyclic angles at the N atoms (NBN: 116.6(5)°, BNB: 123.3(5)°).

In *N*-triphenylborazine (**7-4**) (Fig. 33) the substituents are twisted against the ring, with interplanar angles of 48.7, 42.2 and 42.0°, with two in the same and one in the opposite directions. The mean distance between the *ortho*-hydrogen atoms at the phenyl groups (CH: 108 pm) and the boron-bonded hydrogen atoms (BH: 126 pm), with 242 pm suggests interactions between the negative and positive polarized hydrogen atoms.

The mean BN distance of 143.0(1) pm is in the normal range for borazines. In the isosteric triphenylbenzene, that crystallizes in the same space group with approximately the same cell dimensions [99], the interplanar angles of the phenyl groups were found to be 39.3, 35.9 and 34.9° (in the same arrangement as **7-4**). These angles are smaller than in **7-4** which is most probably due to the larger exocyclic CC distance, allowing a configuration closer to coplanarity, which was found in triphenylboroxine [98].

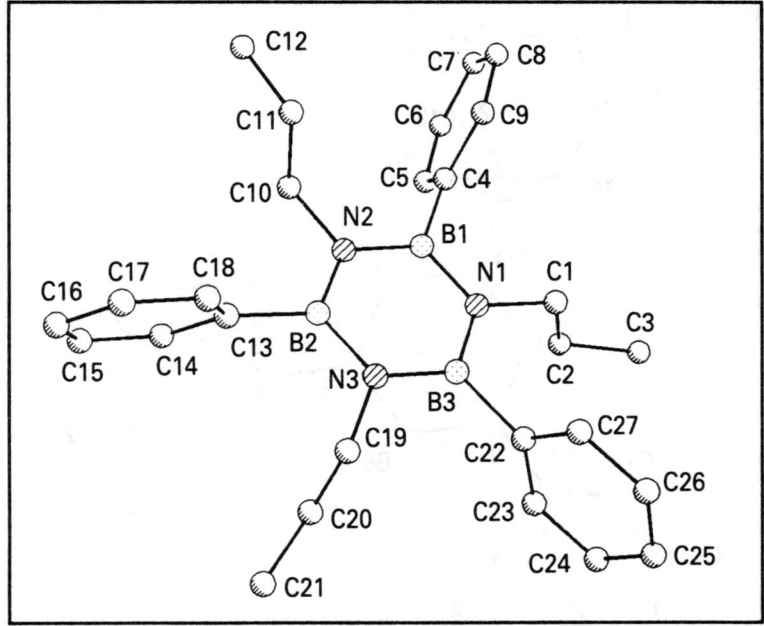

Fig. 31 — Molecular structure of **7-2**.

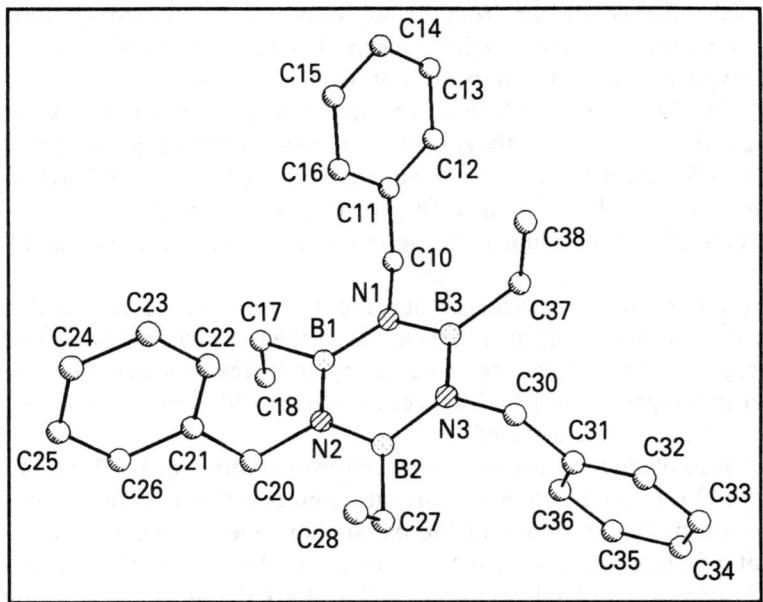

Fig. 32 — Molecular structure of **7-3**.

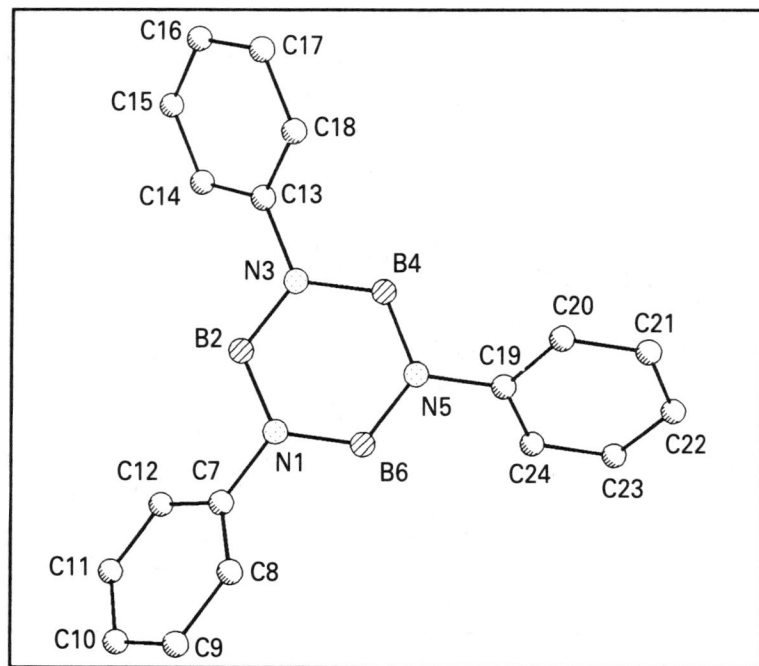

Fig. 33 — Molecular structure of **7–4**.

3,5-Di-*tert*-butyl-4,6-dimethyl-3,5-diazonia-4,6-diborata-phthalic-acid-dimethyl-ester (**7–5**) (Fig. 34) [100] corresponds to the formal description in Fig. 28(b). It represents the only known structure of this type. Owing to the interaction of the spacious substituents, the heterocyclus is found to be twisted (NBNB: $-27.3°$, NCCB: $-6.8°$, torsion angles). With a mean of 145.5 pm the BN-bond distances lie above the typical value in the borazines previously mentioned.

4,5-Diethyl-3,6-dimethyl-1,2-diaza-3,6-diborirane [101] corresponds to Fig. 28(c). With a mean distance between the ring atoms of 144.2 pm in the planar heterocyclus, a slight aromatic character can be assumed, indicated by the difference between endocyclic and exocyclic BC bonds (BC_{endo}: 156, BC_{exo}: 158 pm).

No structural information has been found for the monocyclic system of type (d) in Fig. 28.

As is common with all borazines, where the structures were determined sufficiently exactly, a smaller innercyclic angle was found at the boron atoms compared to those at the nitrogen atoms. This can be explained by an increased s-character of the boron atoms. From the preceding BN distances and those additionally listed in Table 2, a mean of 143.1 pm can be deduced.

On the basis of the two most precise structure determinations of **7–1** and **7–4**, a BN distance of 143.1(1) pm results. Since this value agrees with the corresponding result in a gas electron diffraction study of the parent borazine, neither packing-effects nor substituent influences seem to affect the geometry of the heterocyclus. Considering the equal sums of van der Waals radii in a BN and a CC fragment, the elongated BN distance demonstrates the lower aromatic character of borazine compared to benzene.

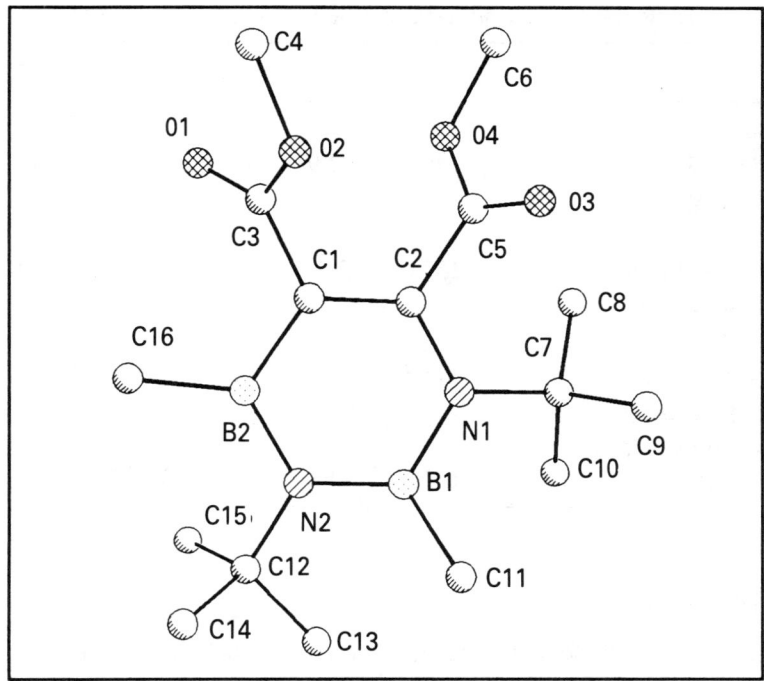

Fig. 34—Molecular structure of **7-5**.

We suggest the adoption of a BN bond length of 143 pm as a reference for the parent borazine in the solid state. This value is slighly smaller than in hexagonal boron nitride (144.6) [102], but significantly longer than in acyclic diaminoboranes with a minimum torsion of the BN bonds [47].

Table 2—Selected structural data for borazines, BN bond distances are given in [pm]

$R_{(B)}$	$R_{(N)}$	B–N	Method	Ref.
H	H	143.5(2)	ED	[83]
Cl	Cl	142.4(1)	XR	[84,85,86]
NMe$_2$	H	143.3(8)	XR	[87]
Me	H	142	ED	[89]
Et	Et	142	XR	[90]
Cl	Ph	142.8	XR	[91]
Ph	Ph	143.5	XR	[92]
Et	Benzyl	144.8(4)	XR	[95]
Et	H	143.2(1)	XR(LT)	[93]
H	Ph	143.0(1)	XR(LT)	[96]
Ph	n-Prop.	143.8(1)	XR(LT)	[94]

Table 3 — Structural data for the borazocines and COT

	a	b	α	β	γ	Lit.
X = Y = C; R,R′ = H	133.2	146.7	41.9	126.6	126.6	[103]
X = B, Y = N; R = methyl, R′ = *tert*-butyl	140.4	151.6	59.9	116.5	116.2	[104]
X = B, Y = N; R = -NCS, R′ = *tert*-butyl	140.2	145.6	57.7	115.1	121.1	[105]
X = B, Y = N; R = isopropyl, R′ = benzyl	140.5	149.7	51.8	124.1	119.0	[95]

Eight-membered rings

The antiaromatic cyclooctatetraene analogons in which all C atoms have been replaced by boron and nitrogen atoms in alternating sequence are called borazocines. Three of these species have been characterized by X-ray analysis. In all three determinations the BN-rings were found in the tub conformation with different BN distances, indicating alternating π- and σ-bonds within the ring. Table 3 shows the structural information of these heterocycles together with that of cyclooctatetraene (COT); the numbering scheme is given in Fig. 35.

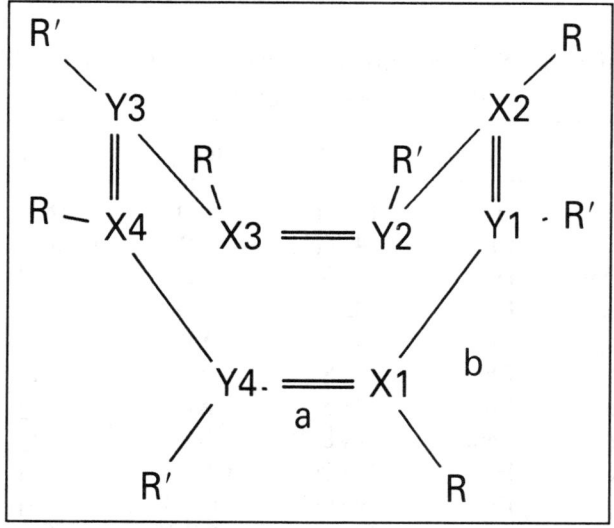

Fig. 35 — Numbering scheme for the geometric parameters given in Table 2.

COT is flatter than the borazocines with the bulky substituents; consequently the CCC angles in COT are larger. The derivative, with substituents having the lowest spatial requirements (R = isopropyl, R′ = benzyl, **8-1**), shows the smallest inter-planar angles (α). The innercyclic angles correspond to the flattening of the molecules. In **8-2**, both innercyclic angles (β and γ) are equal and differ most significantly from **8-3**. With the notable electronegativity, the -NCS substituent withdraws electron density from the boron atoms which compensate this defect by shortening the endocyclic BN single bond. The most significant BN single bond of three-valent boron and nitrogen atoms is found in **8-1** and **8-2** (BN: 151 pm). The localized cyclic double bonds (BN: 140.4 pm) are slightly longer than the upper limit assumed as a reference for the parent aminoborane; this is most probably due to the influence of the substituent in all borazocines. In COT the measured distances correspond very well with the reference values given for a $C(sp^2)$-$C(sp^2)$ single bond (146.6 pm) and a $C(sp^2)$-$C(sp^2)$ double bond 133.5 pm), respectively [64].

CONCLUSIONS

Although the nature of the BN bond is biased by ionic contributions and strongly influenced by the intramolecular and intermolecular environment, the comparison of low-substituted BN compounds with each other and with the isosteric hydrocarbons, atomic distances for the parent compounds were evaluated for the solid state. The fact should be taken into account that no completely undistorted compound can be found except in the gas phase, and that most of the structural information is derived from molecules in either the solid or the liquid state. For these, we need estimates of atomic distances in order to describe a bond length as either shortened or lengthened with respect to the ligand shell of the compound. In accordance with discussions of the carbon–carbon bond distances, the same should be possible with the data now presented for the boron–nitrogen bond. Unfortunately, no distinct values could be derived for the parent amineborane and aminoborane; the distances in these compounds are too sensitive to the dipoles of the surrounding, and therefore only ranges of estimates are given for the respective bond lengths.

The preceding estimations of distances in the BN bond for the parent compounds in the solid state should also allow us to derive a bond distance–bond order correlation comparable to the classical description by Linus Pauling [2] and in textbooks on organic chemistry [64].

With the C-C distances of a $C(sp^3)$-$C(sp^3)$ single, a $C(sp^2)$-$C(sp^2)$ double, and a $C(sp)$-$C(sp)$ triple bond and additionally the distance of benzene in the graph (Fig. 36), all taken from parent molecules in the solid state, e.g. from ethane, ethene, ethine and benzene, we find the curve slightly bent (Fig. 36). With the data suggested as references for the amine-, amino- and iminoboranes in four- three- and two-coordinated environments, we find the curves shifted to longer distances, almost linear and parallel to the curve of the CC bonds. The shift is caused by a worse overlap of contributing orbitals. The range given for the BN bond distances is in between those for the hydrocarbons and the values calculated by the *ab initio* (dzp hondo) method [16] for the isolated amine-, amino- and iminoboranes. As mentioned, the influence of the polar surrounding causes a decrease in the bond lengths through an increase in the

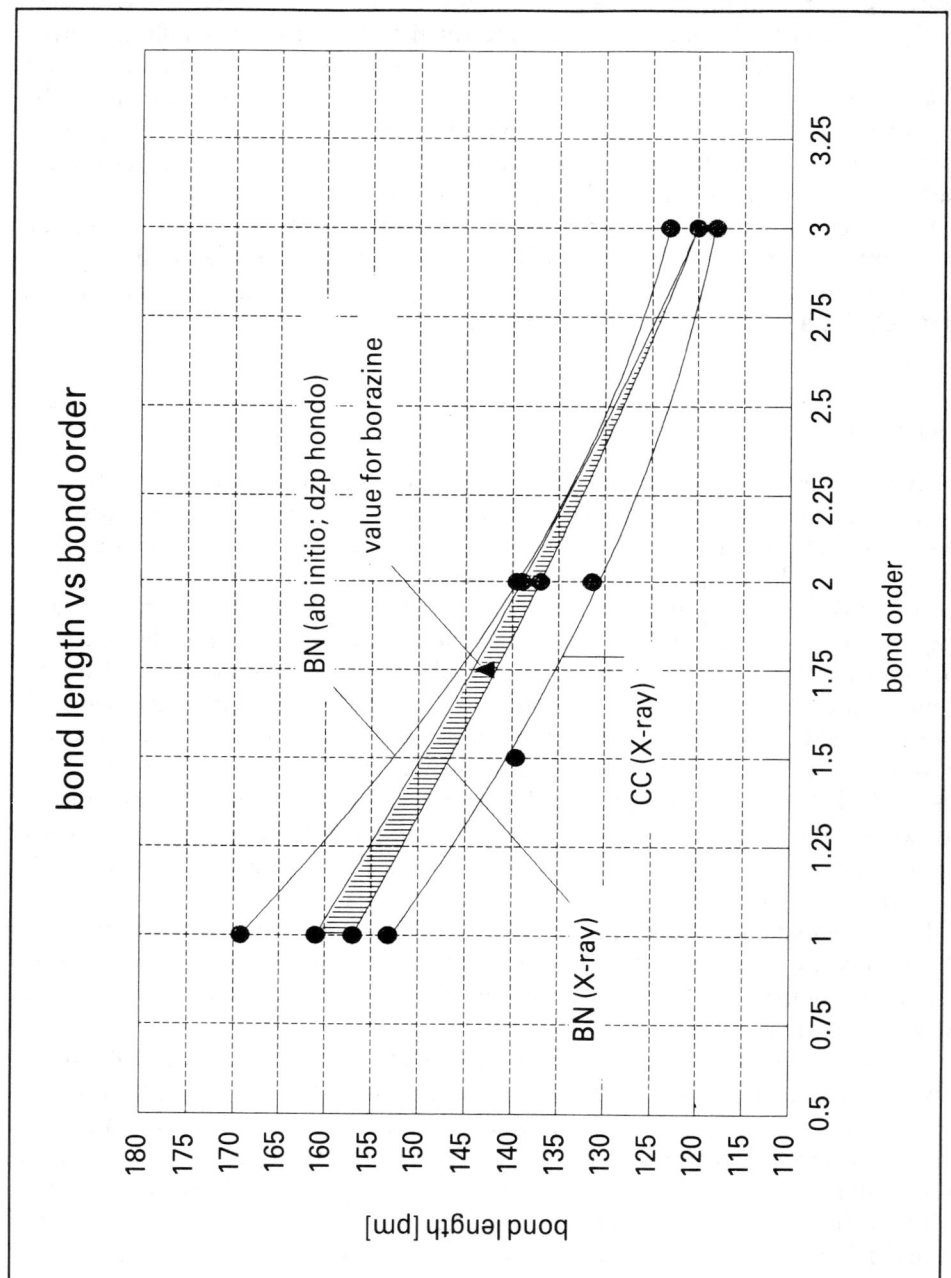

Fig. 36 — Bond order–bond-length correlation for the parent BN and CC compounds obtained by experimental and theoretical data.

polar character of the BN bonds. This is the situation as found in the evaluation of the reference values.

For borazine we have suggested a BN distance of 143 pm as a reference. Applying the range for the experimental values in Fig. 36, a bond order higher than 1.5 results for the parent borazine. This surprising consequence can only be rationalized by an additional polar contribution to the bonding situation in the BN bond, compared to the amineborane and aminoborane.

REFERENCES

[1] *Gmelins Handbuch der Anorganischen Chemie*, Ergänzungswerk zur 8. Auflage, Band 13, Borverbindungen, Teil 1, Springer-Verlag, Berlin, 1974, S. 23.

[2] L. Pauling, *The Nature of the Chemical Bond*, 3rd edn, Ithaca, N.Y., Cornell University Press, 1960.

[3] A. Haaland, *Angew, Chem.* **101** (1989) 1017; *Angew. Chem. (Int. Ed.)* **28** (1989) 992.

[4] I. Langmuir, *J. Am. Chem. Soc.* **41** (1919) 1543.

[5] A. Stock and E. Pohland, *Ber. dtsch. Chem. Ges.* **59** (1926) 2215.

[6] E. Wiberg, *Naturwissenschaften* **35** (1948) 182, 212.

[7] D. Brodalla, D. Mootz, R. Boese, and W. Oβwald, *J Appl. Cryst.* **18** (1985) 316.

[8] R. Boese and D. Bläser, *J. Appl. Cryst.* **22** (1989) 394.

[9] J.L. Guy-Lussac and J.L. Thenard, *Mem. de Phys. et de Chim. de la Soc. d'Arena'l* **2** (1809) 210.

[10] L.R. Thorne, R.D. Suenram, and F.J. Lovas, *J. Chem. Phys.* **78** (1983) 167.

[11] E.W. Hughes, *J. Am. Chem. Soc.* **78** (1956) 502.

[12] E.L. Lippert and W.N. Libscomb, *J. Am. Chem. Soc.* **78** (1956) 503.

[13] At 200K: $a = 542.1(5)$, $b = 494.5(4)$, $c = 502.3(3)$ pm.
At 125 K: $a = 552.4(6)$, $b = 473.9(4)$, $c = 502.2(4)$ pm.
The structural determination of **1–1** was done at 200 K because further cooling reduced the crystal quality.

    Orthorhombic space group *Pmn2$_1$*, $Z = 2$, 705 intensities measured, 198 unique and 191 observed ($F_0 \geqslant 4\sigma(F_0)$), 28 parameters, hydrogen positions with common isotropic displacement parameters as well as appropriate constrained angles and anisotropic displacement parameters for boron and nitrogen (B: $U_{eq.} = 0.070(2)$, H$_B$: $U_{eq.} = 0.116(4)$, N: $U_{eq.} = 0.036(1)$, H$_N$: $U_{eq.} = 0.066(4)$.

    The conventional refinement ($R = 0.093$, $R_w = 0.093$, maximum residual electron density 0.64 e Å^{-3}) gives a BN distance of 156.5(7) pm (without librational correction).

    Further details of the crystal structure investigation are available on request from the Fachinformationszentrum Energie Physik Mathematik, D-7514 Eggenstein-Leopoldshafen 2, on quoting the depository number CSD 320211, the authors' names, and the full citation of the bibliography.

[14] G.J.H. Van Nees and A. Vos, *Acta Crystallogr.* **B34** (1978) 1947.

[15] Generally increased anisotropic displacement factors are found if the scattering power applied at an atomic position during refinement is too high with respect to the real electron density and vice versa.

[16] W.F. Maier, private communications.

[17] H.-Y. Ting, W.H. Watson, and H.C. Kelly, *Inorg. Chem.* **11** (1972) 374.

[18] K. Ouzounis, H. Riffel, and H. Hess, *J. Organomet. Chem.* **332** (1987) 253.

[19] A crystal of **1–4** with the approximate dimensions of $0.42 \times 0.35 \times 0.24$ mm^3 was measured on a Nicolet R3m/V diffractometer with Mo-K$_\alpha$-radiation at 110 K. The cell dimensions, refined from the diffractometer angles of 50 centered reflections in the range $0.18 \leqslant \sin(\Theta)/\lambda \leqslant 0.3$, are $a = b = 906.37(12)$, $c = 587.72(11)$ pm, $\alpha = \beta = 90$, $\gamma = 120°$, $V = 4.1814(14).10^8$ pm^3; $Z = 3$, $d_{cal.} = 0.869$ g cm^{-3}, $\mu = 0.05$ mm^{-1}, trigonal space group *R3m*. Data collection of 4064 reflections $(\sin(\Theta)/\lambda \leqslant 0.99)$ yielded 1012 symmetry equivalent and 959 *uniquely observed* intensities $(F_0 \geqslant 4\sigma(F))$ for the structure solution with direct methods and the refinement of 22 parameters with full matrix least squares (no constraints for the hydrogen atoms and anisotropic U-values for nonhydrogen atoms) $(R = 0.035, R_w = 0.038, g = 0.0005)$.

    Futher details of the crystal structure investigation are available on request from the Fachinformationszentrum Energie Physik Mathematik, D-7514 Eggenstein-Leopoldshafen 2, on quoting the depository number CSD 320302, the authors' names, and the full citation of the bibliography.

[20] A crystal of **1–5** with the approximate dimensions of $0.32 \times 0.45 \times 0.34$ mm^3 was measured on a Nicolet R3m/V diffractometer with Mo-K$_\alpha$-radiation at 118 K. The cell dimensions, refined from the diffractometer angles of 50 centered reflections in the range $0.24 \leqslant \sin(\Theta)/\lambda \leqslant 0.3$, are $a = 915.2(2)$, $b = 647.3(2)$, $c = 963.3(2)$ pm, $\alpha = \beta = \gamma = 90°$, $V = 5.706(3) \cdot 10^8$ pm^3; $Z = 4$, $d_{cal.} = 0.849$ g cm^{-3}, $\mu = 0.04$ mm^{-1}, orthorhombic space group *Pna2$_1$*. Data collection of 10765 reflections $(\sin(\Theta)/\lambda \leqslant 0.99)$ yielded 3864 symmetry equivalent and 2495 *uniquely observed* intensities $(F_0 \geqslant 4\sigma(F))$ for the structure solution with direct methods and the refinement of 93 parameters with full matrix least squares (no constraints for the hydrogen atoms and anisotropic U-values for non-hydrogen atoms) $(R = 0.043, R_w = 0.044, g = 0.0007)$.

    Further details of the crystal structure investigation are available on request from the Fachinformationszentrum Energie Physik Mathematik, D-7514 Eggenstein-Leopoldshafen 2, on quoting the depository number CSD 320304, the authors' names, and the full citation of the bibliography.

[21] A crystal of **1–6** with the approximate dimensions of $0.22 \times 0.28 \times 0.33$ mm^3 was measured on a Nicolet R3m/V diffractometer with Mo-K$_\alpha$-radiation at 110 K. The cell dimensions, refined from the diffractometer angles of 30 centered reflections in the range $0.18 \leqslant \sin(\Theta)/\lambda \leqslant 0.3$, are $a = 1054.9(5)$, $b = 719.7(4)$, $c = 1119.0(5)$ pm, $\alpha = \gamma = 90°$, $\beta = 110.68(4)°$, $V = 7.947(6) \cdot 10^8$ pm^3; $Z = 2$, $d_{cal.} = 0.955$ g cm^{-3}, $\mu = 0.05$ mm^{-1}, monoclinic space group *P2$_1$n*. Data collection of 2323 reflections $(\sin(\Theta)/\lambda \leqslant 0.54)$ yielded 1399 symmetry equivalent and 1011 *uniquely observed* intensities $(F_0 \geqslant 4\sigma(F))$ for the structure solution with direct methods and the refinement of 76 parameters with full matrix least squares (rigid groups for hydrogen atoms, common isotropic U-values for the same groups; anisotropic U-values for nonhydrogen atoms) $(R = 0.066, R_w = 0.075, g = 0.0031)$.

    Further details of the crystal structure investigation are available on request from the Fachinformationszentrum Energie Physik Mathematik,

D-7514 Eggenstein-Leopoldshafen 2, on quoting the depository number CSD 320303, the authors' names, and the full citation of the bibliography.

[22] R. Boese, N. Niederprüm, and D. Bläser, publication in preparation.

[23] L.S. Bartell and W.F. Bradford, *J. Mol. Struct.* **37** (1977) 113.

[24] N. Niederprüm, R. Boese, and D. Bläser, publication in preparation.

[25] K.B. Wiberg and M.A. Murcko, *J. Mol. Struct. (Theochem)* **169** (1988) 355.

[26] Q. Johnson, J. Kane, and R. Schaeffer, *J. Am. Chem. Soc.* **92** (1970) 7614.

[27] P.M. Kuznesof and R.L. Kuczkowski, *Inorg. Chem.* **17** (1978) 2308.

[28] R. Ahlrichs and W. Koch *Chem. Phys. Lett.* **53** (1978) 341.

[29] J.S. Binkley and L.R. Thorne, *J. Chem. Phys.* **79** (1983) 2932.

[30] P.v.R. Schleyer, private communication.

[31] S. Geller, *J. Chem. Phys.* **32** (1960) 1569.

[32] P.W.R. Corfield and S.G. Shore, *J. Am. Chem. Soc.* **95**(5) (1973) 1480.

[33] C.K. Narula, J.F. Janik, E.N. Duesler, R.T. Paine, and R. Schaeffer, *Inorg. Chem.* **25** (1986) 3346.

[34] M. Bühl, P.v.R. Schleyer and R. Boese, *Angew Chem.*, in press.

[35] M. Sugie, H. Takeo, and C. Matsumura, *J. Mol. Spectrosc.* **123** (1987) 286.

[36] A. Almenningen, G. Gundersen, M. Mangerud, and R. Seip, *Acta Chem. Scand.* **A35** (1981) 341.

[37] R.A. Bartlett, H. Chen. H.V.R. Dias, M.M. Olmstead, and P.P. Power, *J. Am. Chem. Soc.* **110** (1988) 446.

[38] G.J. Bullen and N.H. Clark, *J. Chem. Soc. (A)* (1970) 992.

[39] P. Paetzold, *Adv. Inorg. Chem.* **31** (1987) 123.

[40] T. Franz, E. Hanecker, H. Nöth, W. Stöcker, W. Storch, and G. Winter, *Chem. Ber.* **119** (1986) 900.

[41] R. Boese, N. Niederprüm, and D. Bläser, *Structural Chemistry*, in press.

[42] F.B. Clippard Jr. and L.S. Bartell, *Inorg. Chem.* **9** (1970) 2439.

[43] D.J. Brauer, H. Bürger, F. Dörrenbach, G. Pawelke, and W. Wenter, *J. Organomet. Chem.* **378** (1989) 125.

[44] W. Wojnowski, K. Przyjemska, K. Peters, H.G. v. Schnering, T. v. Bennigsen-Mackiewicz, and P. Paetzold, *Z. Anorg. Allg. Chem.* **556** (1988) 92.

[45] P. Paetzold, J. Kiesgen, K. Krahe, H.-U. Meier, and R. Boese, *Z. Naturforsch.*, submitted for publication.

[46] P. Paetzold, C. v. Plotho, G. Schmid, and R. Boese, *Z. Naturforsch.* **39b** (1984) 1069.

[47] N. Niederprüm, R. Boese, and G. Schmid, *Z. Naturforsch.* **46b** (1991) 84.

[48] G. Schmid, R. Boese, and D. Bläser, *Z. Naturforsch.* **37b** (1982) 1230.

[49] A. Almenningen, G. Gundersen, M. Mangerud, and R. Seip, *Acta Chem. Scand.* **A35** (1981) 341.

[50] H. Nöth, *Z. Naturforsch.* **38b** (1983) 692.

[51] G.J.H. Van Nees and A. Vos, *Acta Crystallogr.* **B35** (1979) 2593.

[52] O. Gropen and H.M. Seip, *Chem. Phys. Let.* **25** (1974) 206.

[53] P. Paetzold, C. v. Plotho, G. Schmid, R. Boese, B. Schrader, D. Bougeard, R. Gleiter, and W. Schäfer, *Chem. Ber.* **117** (1984) 1089.

[54] M. Haase, U. Klingebiel, R. Boese, and M. Polk, *Chem. Ber.* **119** (1987) 1117.

[55] R. Boese and D. Bläser, unpublished results.

[56] G.J.H. Van Nees and F. Van Bolhuis, *Acta Crystallogr.* **B35** (1979) 2580.

[57] R. Boese and N. Augart, Z. Kristallogr. **182** (1988) 32.

[58] K.-H. v. Bonn, P. Schreyer, P. Paetzold, and R. Boese, Chem. Ber. **121** (1988) 1045.

[59] F. Dirschl, H. Nöth, and W. Wagner, J. Chem. Soc. Chem. Commun. (1984) 1533.

[60] F. Dirschl, E. Hanecker, H. Nöth, W. Rattay, and W. Wagner, Z. Naturforsch. **B41** (1986) 32.

[61] H. Fußstetter, J.C. Huffmann, H. Nöth, and R. Schaeffer, Z. Naturforsch. **B31** (1976) 1441.

[62] R. Boese and U. Klingebiel, J. Organomet. Chem. **306** (1986) 295.

[63] R. Boese, N. Niederprüm, and D. Bläser, unpublished results.

[64] P. Rademacher, Strukturen Organischer Moleküle, Hrsg. M. Klessinger, Weinheim, VCH-Verlag, 1987, Bd. 2.

[65] E. Billups, M.M. Haley, R. Boese, D. Bläser, M. Nussbaumer, R. Gleiter, and K.-H. Pfeifer, publication in preparation.

[66] H. Irngartinger, N. Riegler, K.D. Malsch, K.-A. Schneider, and G. Maier, Angew. Chem. **92** (1980) 214; Angew. Chem. (Int. Ed.) **19** (1980) 211.

[67] H. Irngartinger and H. Rodewald, Angew. Chem. **86** (1974) 783; Angew. Chem. (Int. Ed.) **13** (1974) 740.

[68] L.R.J. Delbaere, M.N.G. Janes, N. Nakamura, and S. Masamune, J. Am. Chem. Soc. **97** (1975) 1973.

[69] H. Irngartinger and M. Nixdorf, Angew. Chem. **95** (1983) 415; Angew. Chem. (Int. Ed.) **22** (1980) 403.

[70] O. Ermer and E. Heilbronner, Angew. Chem. **95** (1983) 414; Angew. Chem. (Int. Ed.) **22** (1983) 402.

[71] W.T. Borden and E.R. Davidson, J. Am. Chem. Soc. **1032** (1980) 7958.

[72] J.D. Dunitz, C. Krüger, H. Irngartinger, E.F. Maverick, Y. Wang, and M. Nixdorf, Angew. Chem. **100** (1988) 415; Angew. Chem. (Int. Ed.) **27** (1988) 387.

[73] P. Paetzold, E. Schröder, G. Schmid, and R. Boese, Chem. Ber. **118** (1985) 3205.

[74] P. Paetzold, A. Richter, T. Thijssen, and S. Würtenberg, Chem. Ber. **112** (1979) 3811.

[75] H. Hess, Acta Cryst. **B25** (1969) 2342.

[76] P. Paetzold, Fortschr. Chem. Forsch. **8** (1967) 437.

[77] G. Schmid, D. Zaika, J. Lehr, N. Augart, and R. Boese, Chem. Ber. **121** (1985) 1873.

[78] P. Jutzi, E. Schlüter, S. Pohl, and W. Saak, Chem. Ber. **118** (1985) 1959.

[79] G. Schmid, M. Polk, and R. Boese, Inorg. Chem. **29** (1990) 4421.

[80] (i) G. Schmid and J. Schulze, Chem. Ber. **110** (1977) 2744. (ii) G. Schmid and J. Schulze, Angew. Chem. **89** (1977) 258; Angew. Chem. (Int. Ed. Engl.) **16** (1977) 249.

[81] N. Kuhn, K. Jendral, R. Boese, and D. Bläser, Chem. Ber., in press.

[82] A. Stock and R. Wierl, Z. Anorg. Allg. Chem. **203** (1931) 228.

[83] W. Harschbarger, G.H. Lee II, R.F. Porter, and S.H. Bauer, Inorg. Chem. **8** (1969) 1683.

[84] U. Müller, Acta Crystallogr. **B27** (1971) 1997.

[85] J.G. Haasnoot, G.C. Verschoor, C. Romers, and W.L. Groeneveld, Acta Crystallogr. **B28** (1972) 2070.

[86] M.S. Gopinathan, M.A. Whitehead, C.A. Coulson, J.R. Carruthers, and J.S. Rollett, *Acta Crystallogr.* **B30** (1974) 731.

[87] H. Hess and B. Reiser, *Z. Anorg. Allg. Chem.* **381** (1971) 91.

[88] W. Harschbarger, G.H. Lee, R.F. Porter, and S.H. Bauer, *J. Am. Chem. Soc.* **91** (1969) 551.

[89] K. Anzenhofer, *J. Mol. Phys.* **11** (1966) 495.

[90] M.A. Viswamitra and S.N. Vaidya, *Z. Kristallogr.* **121** (1965) 472.

[91] W. Schwarz, D. Lux, and H. Hess, *Cryst. Struc. Commun.* **6** (1977) 431.

[92] D. Lux, W. Schwarz, and H. Hess, *Cryst. Struc. Commun.* **8** (1979) 33.

[93] R. Boese *et al.*, unpublished results.

[94] R. Boese *et al.*, unpublished results.

[95] B. Thiele, P. Schreyer, U. Englert, P. Paetzold, R. Boese, and B. Wrackmeyer, *Chem. Ber.*, submitted for publication.

[96] R. Boese *et al.*, unpublished results.

[97] R. Boese, N. Niederprüm, D. Bläser, P.v.R. Schleyer, and M. Bühl, *Angew. Chem.*, submitted for publication.

[98] R. Boese, M. Polk, and D. Bläser, *Angew. Chem.* **99** (1987) 239; *Angew. Chem. (Int. Ed. Engl.)* **26** (1987) 245.

[99] Y.C. Lin and D.E. Williams, *Acta Crystallogr.* **B31** (1975) 318.

[100] P. Schreyer, P. Paetzold, and R. Boese, *Chem. Ber.* **121** (1988) 195.

[101] W. Siebert, R. Full, H. Schmidt, J. v. Seyerl, M. Halstenberg, and G. Huttner, *J. Organomet. Chem.* **15** (1980) 191.

[102] R.S. Pease, *Acta Crystallogr.* **5** (1952) 356.

[103] K.H. Claus and C. Krüger, *Acta Crystallogr.* **C44** (1988) 1632.

[104] T. Franz, E. Hanecker, H Nöth, W. Stöcker, W. Storch, and G. Winter, *Chem. Ber.* **119** (1986) 900.

[105] P.T. Clarke and H.M. Powell, *J. Chem. Soc. (B)* (1966) 1172.

6

Calculated structures and stabilities of fibrous macromolecules

Harold A. Scheraga
Cornell University, Ithaca, NY, USA

and

George Némethy
Mount Sinai School of Medicine, New York, USA

INTRODUCTION

In the early 1930s, considerable effort was expended to try to determine the structures of fibrous proteins such as α and β keratin [1, 2]. The culmination of this activity was the ingenious proposal by Pauling and Corey [3, 4] of the α-helix and β-pleated-sheet as the fundamental structures from which these and other proteins are constituted. These proposed structures, based on an analysis of the geometries of amino acids and peptides (including standard bond lengths and bond angles, and the planarity of the peptide group) and of reflections in the fiber X-ray diagrams of α and β keratin, and making maximal use of hydrogen bonding, were later found in the X-ray structures of globular proteins, e.g. the α-helix in myoglobin [5] and hemoglobin [6] and the β-sheet in chymotrypsin [7].

This pioneering work was followed by a flurry of activity to deduce the structures of other fibrous proteins by fiber X-ray diffraction and model building, e.g. the triple-helical structure of collagen [8–10], the threefold helix of polyglycine II [11], and the β-sheet structure of form II of silk fibroin [12].

In the original proposals of Pauling and Corey, the α-helix was considered to adopt either a right- or a left-handed helical twist. While the α-helices in proteins are all right-handed, for energetic reasons that are now understood [13–16], synthetic homopolymers of non-natural amino acids have been shown to adopt either helical twist, depending on the nature of the side chains [17–19].

In the original proposals of the parallel and anti-parallel β structures, the sheets were flat, except for the pleating along lines that are perpendicular to the direction of the strands to avoid steric hindrance, leading to the terminology 'pleated sheet'. Observations of β-sheets in globular proteins [20], however, have shown that such sheets are not flat but have a right-handed twist.

With the advent of conformational energy calculations on polypeptides and proteins [21, 22], it is now possible to account for these observations in terms of interatomic interactions. This chapter, dedicated to Linus Pauling on his 90th birthday, is therefore a summary of our calculations to account for the twists of the fundamental structures (the α-helix and β-sheet) from which fibrous and globular proteins are built, and the structural arrangements in which α-helices and β-sheets are packed. Consideration will also be given to other regular-repeating structures and their packing, viz., collagen, silk and cellulose. Finally, we will also discuss the stabilities of such structures in terms of the reversible helix–coil, helix–helix, and β-sheet–coil transitions.

COMPUTATIONAL METHODOLOGY

Empirical potential energy functions are used to compute the low-energy structures of polypeptides and proteins. The algorithm currently in use in our laboratories is ECEPP/2, which stands for Empirical Conformational Energy Program for Peptides [23, 24], an updated version of the original ECEPP algorithm [25]. ECEPP is based on rigid geometry, i.e. fixed bond lengths and bond angles, carefully selected from high-resolution crystal structures of amino acids and peptides; it allows conformations to change by varying the dihedral angles for rotation about the bonds of the polypeptide backbone and its side chains. With this program, it is possible to generate a polypeptide chain in an arbitrary conformation, and calculate its energy. For many problems, the calculations are simplified by imposing a condition of regularity, in which every residue is constrained to adopt the same conformation; however, the program is sufficiently general so that this constraint need not be implemented if desired. The energy of any initial conformation can then be minimized by any of a number of algorithms; the one in current use in our laboratories is SUMSL, which stands for Secant Unconstrained Minimization Solver [26]. Provision is also made for including conformational entropy [27, 28] and solvation [29, 30] effects.

When considering interactions between α and β structures, or between any two or more macromolecular structures, account must be taken not only of the internal degrees of freedom, i.e. the dihedral angles within the chain, but also of the external degrees of freedom, i.e. the rotations and translations of one structural element relative to another. These external degrees of freedom are expressed as a coordinate transformation, written in terms of three Euler angles of rotation and three Cartesian components of a translation vector, respectively, operating on local coordinate systems that are attached to each structural element. One of the structural elements is used to define a common reference coordinate system. For each helical (or sheet) element, the orientation of the local coordinate system is related to symmetry elements of the structure, such as the helix axis or the averaged plane of the sheet [31–33]. The total energy to be minimized is a sum of intrachain and interchain energies. An analytical expression for the derivatives of potential energy with respect to the internal and external variables is used in the minimization algorithm for packing [34].

Additional variables must be considered for computations on crystal structures. In the adaptation of the WMIN algorithm [35] used in this work, a unit cell is generated from internal coordinates of the macromolecules (computed by using ECEPP), the unit cell data, and a set of translation vectors and rotation matrices that describe symmetry relations [36, 37]. The energy is minimized with respect to the unit cell parameters, relative positions and orientations of macromolecular chains in the unit cell, and intramolecular dihedral angles.

COMPUTATIONAL RESULTS

Structure

We consider first the structural properties of single chains and assemblies of chains in various conformational arrangements. Additional details about our computations can be found in a recent review [38], and in the original papers dealing with individual structural arrangements, cited throughout this section. Furthermore, numerous studies of observed characteristic features of the assemblies of regular structures in proteins have been summarized in a detailed review [39].

Handedness of α-helices

Homopolymers of amino acids can adopt either a right- or a left-handed α-helical twist. Table 1 compares the computed and observed twists of a number of α-helical structures [40]. The energy difference, in the range of 0 to 1.5 kcal/mol per residue, is sufficient to favor one form over the other [14].

Table 1 — Comparison of calculated and experimental helix sense for several polyamino acids [40]

| | Helix sense[a] | |
Polyamino acid	Calculated	Experiment
Poly-L-asp-COOH	R	R
-L-glu-COOH	R	R
Poly-β-methyl-L-asp	L	L
-ethyl-L-asp	L	R
-n-propyl-L-asp	R	R
-isopropyl-L-asp	R	R
Poly-β-benzyl-L-asp	L	L
-p-Cl-benzyl-L-asp	R	R
-p-CN-benzyl-L-asp	R	R
-p-NO$_2$-benzyl-L-asp	R	R
-p-CH$_3$-benzyl-L-asp	R	R
Poly-γ-methyl-L-glu	R	R
-γ-benzyl-L-glu	R	R
-p-Cl-benzyl-L-glu	R	—
-p-CN-benzyl-L-glu	R	—
Poly-L-phenylalanine	R	—

[a] R = right-handed; L = left-handed.

The origin of the preference for right- or left-handedness lies in a $C^\beta H_2\cdots$backbone nonbonded interaction which favors right-handedness, as in poly(L-alanine), and interactions involving atoms of the side chain beyond $C^\beta H_2$ that can favor either sense of twist [13–16, 40]. For example, in the series of poly(*ortho, meta* and *para*)chlorobenzyl aspartates, the *ortho, meta* and unsubstituted polymers adopt left-handed α-helical forms, whereas the *para* polymer adopts a right-handed α-helical form [16]. In these polymers, the side chain takes on a transverse or longitudinal aspect with respect to the backbone, bringing the chlorine atom close enough to the backbone to influence its helical twist [15, 16]. Fig. 1 illustrates the lowest-energy conformations of the *meta* polymer, showing a favorable, attractive interaction between the C–Cl dipole and the dipole of the closest peptide group in the left-handed form; the corresponding interaction in the right-handed form is repulsive [16]. Thus, this dipole–dipole interaction plays a dominant role in leading to the preference for left-handedness. These preferences have been verified experimentally for all three chloro-substituted poly(benzyl aspartates) [41].

Variants of the α-helix

The 3_{10}-helix has a similar overall appearance to the α-helix but it is tighter, containing 3.0 instead of 3.6 residues per turn, with hydrogen bonds that are formed between residues *i* and *i*-3, rather than *i* and *i*-4 as in the α-helix [42]. The 3_{10}-helix is energetically very unfavourable for the natural amino acids [13]. As a consequence, long 3_{10}-helices are rarely observed in proteins [43], but there are many examples of about one turn of a 3_{10}-helix, formed by three or, less often, by four consecutive

Left – handed   Right – handed

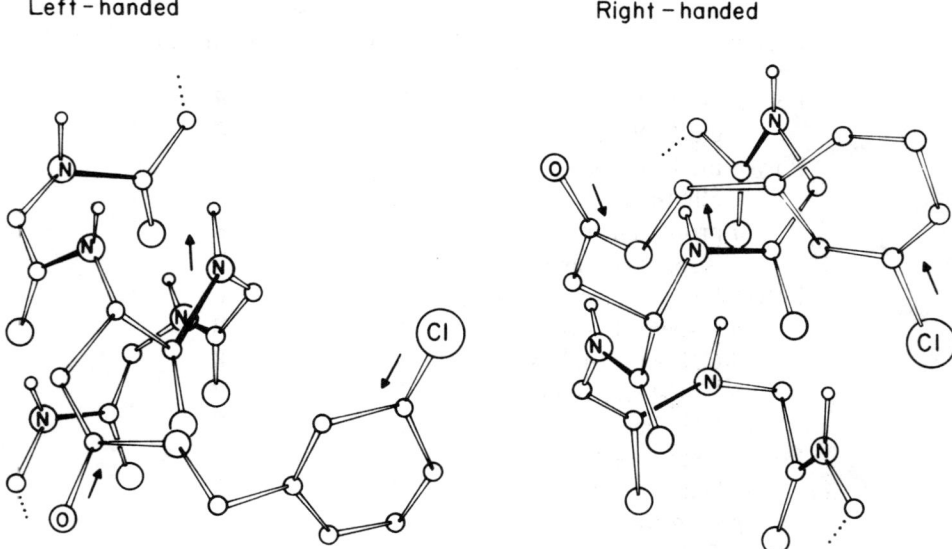

Fig. 1—Orientation of the side chains of the left- and right-handed α-helices of poly(*m*-chlorobenzyl-L-aspartate). The solid arrows represent the directions of the C–Cl, ester, and amide dipoles, respectively [16].

residues [43, 44]. A single turn of the 3_{10}-helix is a special form of the double bend; we have shown that this structure is composed of two overlapping β-bends and has a low energy [44, 45].

We have demonstrated that it is also possible to construct a second helical form with the same helical parameters (i.e. the identical height and number of residues per turn) and the same hydrogen-bonding pattern as the α-helix, but with a different set of (ϕ, ψ) dihedral angles [46]. This structure is denoted as the α_{II}-helix. Because of the change in the dihedral angles, the peptide groups are tilted so that the N—H bonds point slightly inward, toward the axis of the helix, and the C=O groups point outward. Consequently, the hydrogen bonds are lengthened and bent, and a complete α_{II}-helix is energetically less stable than the regular α-helix [46]. A single turn of an α_{II}-helix occurs frequently, however, at the C-terminus of α-helices in globular proteins [44, 46]. Our conformational analysis has demonstrated that hydrogen bonding is perturbed only marginally in this structure, because the N—H groups form bifurcated hydrogen bonds of the α- and 3_{10}-helical type, while the C=O groups do not form backbone hydrogen bonds in either form, because they are located at the end of the helix. On the other hand, an α_{II}-helical turn at the N-terminus of an α-helix is unfavourable energetically for steric reasons, and this conformation does not occur in proteins [44, 46].

The 3_{10}-helix is more stable than the α-helix for polypeptides that contain some sterically constrained amino acids, such as α-amino isobutyric acid (Aib) [47]. Many Aib-containing peptides have been shown by X-ray crystallography to form 3_{10}-helices [48–51]. A comparison of the stability of the two helical forms will be discussed in a later section on 'Stability'.

Another variant of the α-helix is the ω-helix, with the same hydrogen-bonding pattern as that of the α-helix, but with 4.0 residues per turn. Poly[β-(p-chlorobenzyl)-L-aspartate] has been shown by X-ray diffraction to exist in both the right-handed α-helical and ω-helical forms in fibers [52]. The stability of these two forms in a crystalline array has been verified by conformational energy computations [53].

Twists of β-sheets

The observed right-handed twist of β-sheets is illustrated in Fig. 2 by the computed parallel and antiparallel structures of poly(L-valine) sheets [54]. In general, side chain–backbone interactions within each strand result in a preference for a right-handed twist for L-amino acids, although there are exceptions [54, 55]. In addition, interstrand side chain–side chain interactions also make significant contributions. Thus, intrastrand interactions in an isolated extended poly(L-Ile) strand energetically favor the left-handed twist, but interstrand interactions result in the stabilization of a poly(L-Ile) β-sheet with a right-handed twist [56]. Poly(L-Ser) is exceptional, in that it is computed to favor a left-handed β-sheet [57]. This prediction is verified by the observed behaviour of Ser residues in proteins [58]. Even though Ser occurs relatively infrequently in β-sheets, it usually imparts a local deformation to the polypeptide chain that corresponds to reduced or left-handed local twisting [57, 58].

In general, increased bulk of side chains leads to a larger twist [57]. In addition, amino acid residues in β-sheets fall into two clusters in terms of their conformational preferences (Fig. 3). For one cluster, containing Ala, Ser, Thr, and aromatic residues,

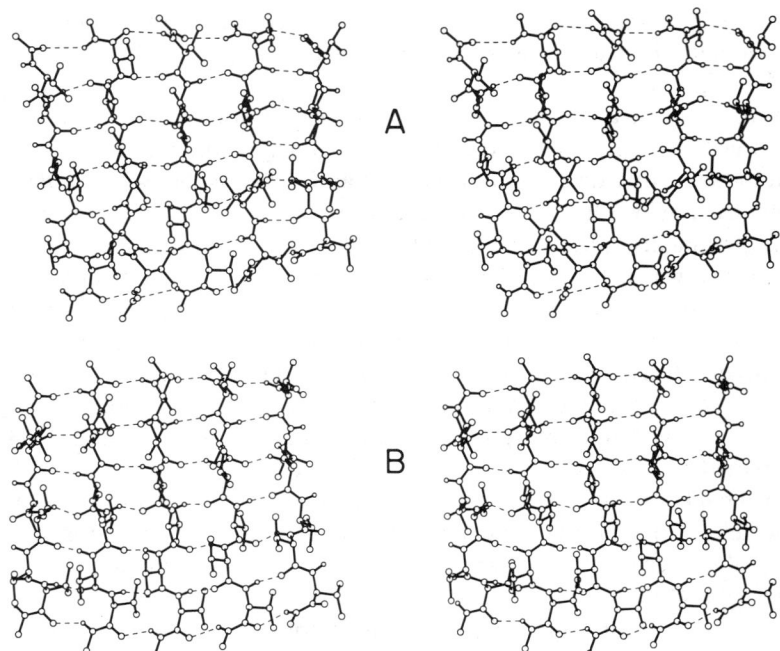

Fig. 2 — Stereo drawings of the minimum-energy β-sheets with five $CH_3CO\text{-}(L\text{-}Val)_6\text{-}NHCH_3$ chains: (A) antiparallel structure; (B) parallel structure. Hydrogen atoms of the valyl side chains and of the amino- and carboxy-terminal methyl groups have been omitted. Hydrogen bonds between neighboring chains are indicated by broken lines [54].

the β-sheets are formed by nearly fully extended chains. The chains are much less extended in β-sheets formed in the other cluster, which is composed of residues with large aliphatic side chains. The second cluster is located in a region of the (ϕ, ψ)-map that is usually considered not to belong to a β-sheet. Good hydrogen bonds are found, however, in this structure as well (Fig. 2). The presence of two clusters is related to an interesting observation. Chothia has pointed out that coiled-coil strands occur frequently in β-sheets, with an alternation of residues having dihedral angles that fall into two distinct (ϕ, ψ)-regions [59]. These regions agree closely with the location of the clusters in Fig. 3. The alternation can be correlated with the amino acid sequence in many cases, for example in the transmembrane polypeptide gramicidin A, in which L-Ala or L-Trp residues alternate with D-Val and D-Leu residues [60, 61]. The former two residues have conformations falling into the more extended region, the latter two into the less extended region [38].

The extent of twisting is a function of the amino acid sequence, but it can be enhanced strongly by interchain interactions within the β-sheet itself. We have demonstrated this for the strongly twisted two-stranded β-sheet consisting of residues 14–38 of bovine pancreatic trypsin inhibitor [62]. The twisted β-sheet is stable in the absence of the rest of the molecule. This implies that this β-sheet forms during an early stage of folding of the protein.

The computed energies have been used to predict the relative stabilities of poly(amino acid) parallel and antiparallel β-sheets [57]. The antiparallel form was

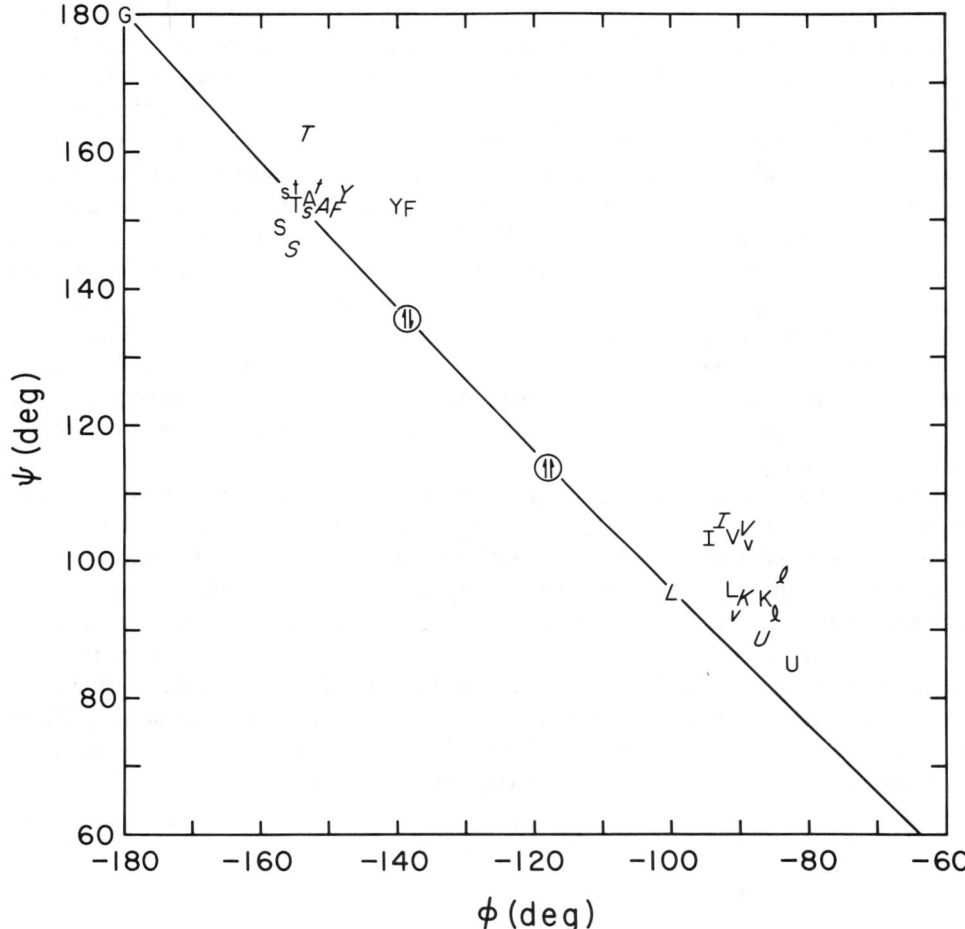

Fig. 3 — Locations of computed minimum-energy conformations of β-sheets on a (ϕ,ψ) map. The conformations of regular poly(amino acid)s are denoted by capital letters, using the one-letter abbreviations for the residues: A, Ala; F, Phe; G, Gly; I, Ile; K, Lys; L, Leu; S, Ser; T, Thr; V, Val; Y, Tyr; U, Abu. The locations of the average dihedral angles for the two kinds of residue in sequential copolymers of Leu–Val and Thr–Ser are indicated by the corresponding lower-case letters. Letters in Roman type denote parallel-chain structures; those in italic type denote antiparallel-chain structures. Symbols in circles indicate the dihedral angles of the ideal Pauling–Corey pleated sheets [4]. The diagonal line corresponds to β-sheets with no twist and $\omega = 180°$ [57].

predicted to be favored for sheets formed by residues with a small unbranched (or γ-branched) side chain (Gly, Ala, Leu), while the parallel form is favored for residues such as Val, Ile, Lys, Ser, Thr, Phe, Tyr. All of these predictions agree with experimental observations on oligopeptides, wherever data are available [63, 64].

α/α Packing

Two α-helices can pack efficiently against each other in only a limited number of ways, i.e. with a small number of relative orientations of the helix axes. Some of this restriction on the packing arrangements arises from the geometrical shape of the surfaces of the helices [38, 39]. In general, the side chains of one helix intercalate into the spaces between the side chains of the other helix. This has been described as a 'knobs into holes' arrangement [65]. In addition to this geometrical complementarity, however, interaction energies are also important in helix/helix packing. Computations on the packing of two poly(L-Ala) α-helices and of a poly(L-Ala) helix with a poly(L-Leu) helix have shown that only about ten low-energy packing arrangements can occur in each of these cases [31, 32]. The helices are nearly antiparallel in the energetically most favorable packing arrangements. The lowest-energy structure, with an orientation angle of about − 154° between the helix axes (Fig. 4), is the most frequently observed α/α packing arrangement in globular proteins [38]. Apparently, the basic patterns of packing are established by the overall geometrical and energetic features of the interacting α-helices, even though sequence-specific side chain–side chain interactions may lead to some alteration of the preferences in actual packings found in specific proteins.

The packing orientation and the inter-helix energy are influenced by both nonbonded and electrostatic interactions. The nonbonded interaction dominates the total energy of assembly of α-helices, while both interactions contribute comparably to the energy differences between various ways of packing, i.e. both are significant in choosing preferential orientations [32]. It should be noted, however, that the presence of a solvent with a high dielectric constant (such as water) weakens electrostatic interactions between α-helices, by reducing the magnitide of dipole interactions and because of the unfavorable desolvation of the helical dipoles upon association [66]. In the case of helices with nonpolar side chains, the presence of water may also provide a stabilizing effect, however, because of hydrophobic interactions.

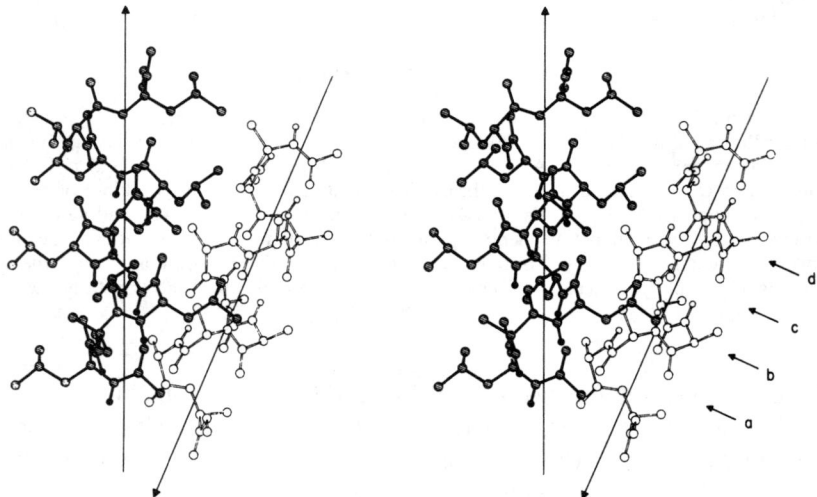

Fig. 4 — Stereoscopic pictures of an α[CH$_3$CO-(L-Leu)$_{10}$-NOCH$_3$] (shaded atoms and bonds) and an α[CH$_3$CO-(L-Ala)$_{10}$-NHCH$_3$] (open atoms and bonds) α-helix in the lowest energy packing state. The two helices are nearly antiparallel. The helix axes are indicated by arrows, with the head of the arrow pointing in the direction of the C terminus of each helix. Hydrogen atoms are omitted, except for the amide hydrogens. The arrows marked a, b, c, and d indicate regions in which the surfaces of the two α-helices are complementary [32].

Various extents of cooperativity may occur in the assembly of several α-helices [67], for example in the simultaneous interaction of the A, G, and H helices of myoglobin [68, 69], shown in Fig. 5. Our computations have indicated [67] that the association of two nearly antiparallel, tightly packed helices (G and H) is governed mainly by their pairwise interactions, without being influenced significantly by the introduction of a third helix (A) near them. On the other hand, there exist numerous ways of packing two weakly interacting and nearly perpendicular helices (A and H) by themselves, but the presence of a third helix (G) alters the relative energies of various A/H pairwise packings, and it may result in the energetic preference for a unique orientation of the A-H pair in the A-G-H complex [67]. The analysis of the packing preferences for these three α-helices suggested that the G-H pair forms earlier during folding than the interaction between helix A and the other two helices [67].

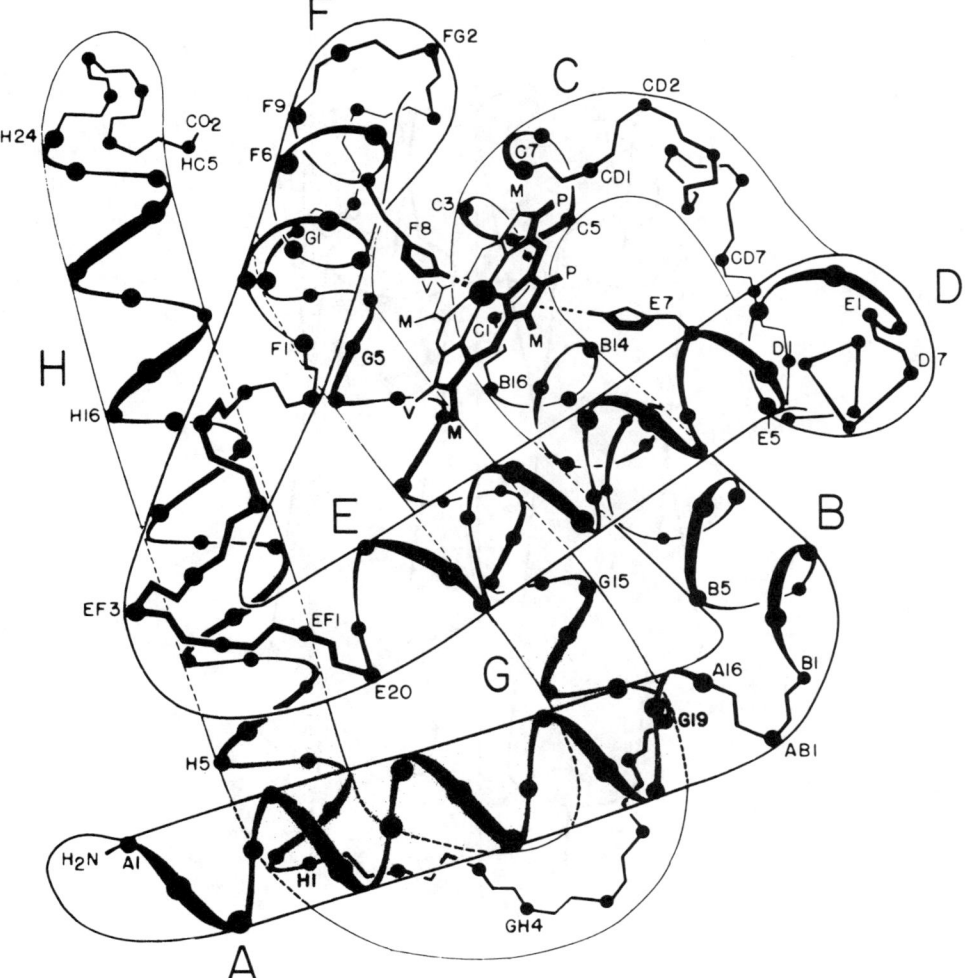

Fig. 5 — Schematic diagram of the structure of sperm whale myoglobin [68, 69] showing only C^α atoms. The α-helices A to H are labeled. The three helices (A, G and H) considered here are shaded [67].

A bundle of four α-helices is a frequently occurring structural pattern in globular proteins [44, 70]. The main structural features of this bundle are the near-antiparallel orientation of neighboring pairs of helices and a tilting of the helix axes that corresponds to a left-handed twisting of the entire bundle (Fig. 6). Both features have been explained in terms of the nonbonded and electrostatic interactions between the constituent helices [71]. In a bundle in which neighboring α-helices are sequentially

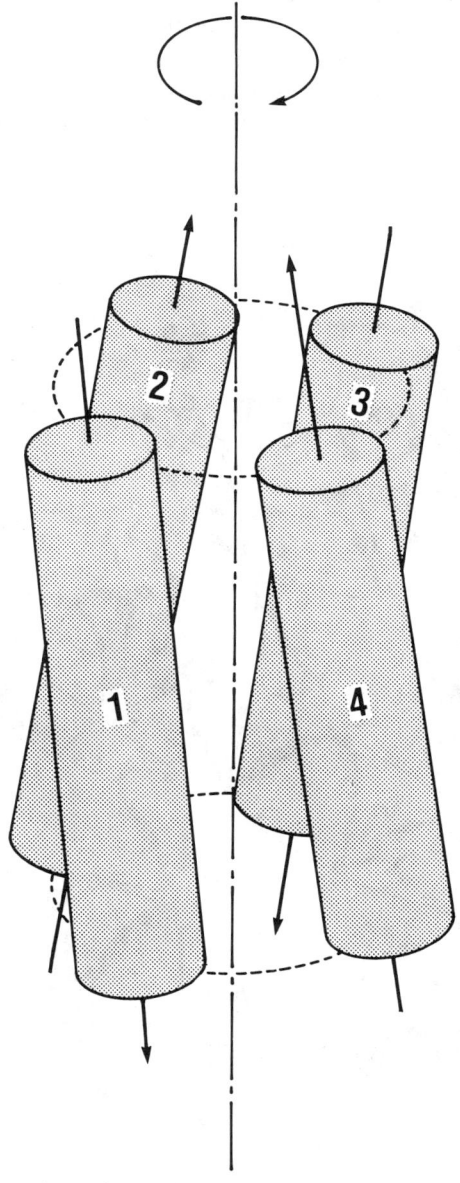

Fig. 6—Schematic illustrations of an antiparallel four-α-helix bundle. The helices are shown schematically as cylinders. Arrows indicating the helix axes point from the N- to the C-terminus [71].

connected by polypeptide links in nonhelical conformations, these links also contribute significantly to the stabilization of the bundle [72].

Packing of helices into coiled coils

Several fibrous proteins, such as α-keratin [65, 73], collagen [8–10], and tropomyosin [74], exist in the form of coiled coils that are formed by several closely interacting polypeptide chains. The coiling of two α-helices around a common axis is also the main feature of the leucine-zipper arrangement observed in some DNA-binding proteins [75, 76].

Coiled coils can be described in terms of major and minor helices. The minor helix is formed by the individual polypeptide chains. The axis of each minor helix is not straight, in contrast to the axis of simple helices, but it follows a helical path, denoted as the major helix, around the common axis of the coiled-coil structure. In simple helices, every residue along the chain must have identical backbone dihedral angles. If the residues within one repeat unit do not all have the same backbone dihedral angles, a coiled coil is formed [77]. An example of this, coiled coiling of strands in β-sheets, has been mentioned above in the section 'Twists of β-sheets'.

Crick has derived a general formula which relates the parameters of the major and minor helices for a coiled-coil polypeptide chain [78]. Geometrical relations have also been derived between the dihedral angles of polypeptide chains and the parameters that characterize the minor and major helices of coiled coils, together with the establishment of relationships between the major helix and the averaged structure of the minor helix [77]. It has been shown that severe geometrical restrictions exist for the formation of coiled-coil structures that correspond to a given backbone conformation of a polypeptide repeat unit [77].

In the triple-stranded collagen molecule [8–10], the minor helix formed by each polypeptide chain is left-handed. The axes of the minor helices follow right-handed major helices, winding around the major helical axis [79, 80]. The coiled coil may be described in terms of the translational repeat D and the azimuthal angular repeat Θ per tripeptide repeat unit along the major helical axis z [77, 79, 80], as shown in Fig. 7. Furthermore, in triple-stranded (or higher-stranded) coiled coils with screw symmetry, there exist two kinds of disposition of the equivalent repeat units in neighboring strands, viz. 'clockwise' and 'counterclockwise', depending on whether these repeat units are related to each other by right-handed or left-handed screw symmetry operations, respectively [79, 80].

α/β Packing

The association of an α-helix with a β-sheet is a frequently occurring structural motif in proteins. The packing is related to the properties of the helix and the sheet. Because of the twisting of the β-sheet, its surface is saddle-shaped, i.e. it can be described as a hyperboloid, while the α-helix is essentially a rigid rod. An energy computation carried out for a poly(L-Val) sheet interacting with a poly(L-Ala) helix predicted that there are four classes of low-energy arrangements [33]. The structures may be characterized by the angle Ω between the axes of the helix and the sheet. In the most favorable arrangements, the helix is nearly parallel or nearly perpendicular to the

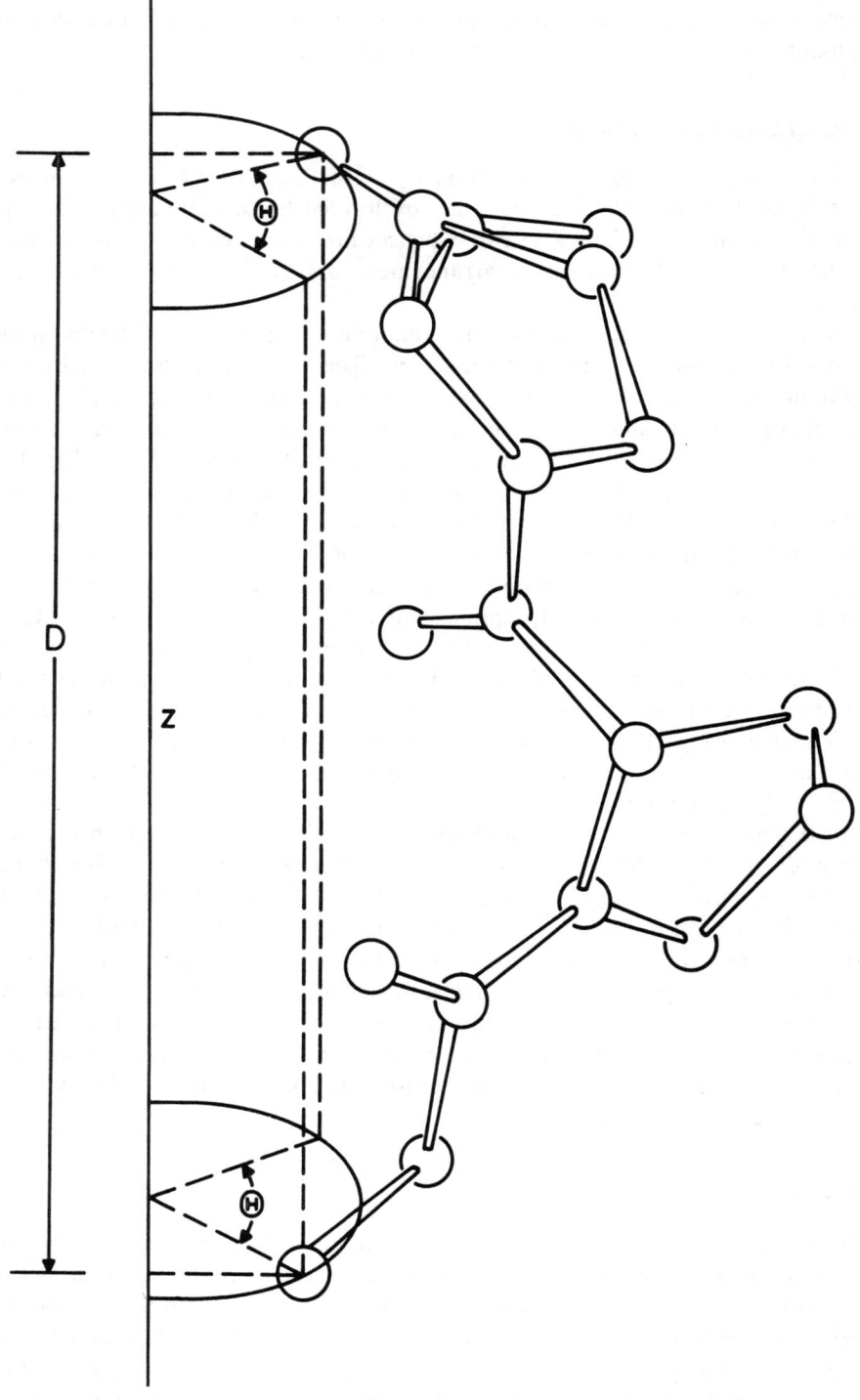

Fig. 7 — Diagram of one tripeptide unit (Gly–Pro–Pro) in a major helical coordinate system. The helix axis (z), the translational repeat per tripeptide (D), and the angular repeat per tripeptide (Θ) are indicated [79].

direction of the strands, because then the helix lies along a tangent line to the curved surface, so that it can interact with the sheet along its entire length (Fig. 8). The two remaining arrangements (not shown in Fig. 8) can be described as diagonal packings, in which only either the middle or the two ends of the helix are in contact with the sheet (with Ω near $-60°$ or $+60°$, respectively). Of these two, the first one has a low energy, while the second one is less favorable.

A histogram of observed distributions of Ω in 163 α/β packings in 39 proteins shows a large peak near $\Omega = 0°$, a broad distribution in the range of $-30°$ to $-60°$, a smaller peak near $\pm 90°$, and a few structures near $+60°$. The positions of the maxima in the distribution correspond to the preferred orientations in the computed structures [33]. As a result of sequence differences and of packing interactions with the rest of the protein molecule, the observed peaks are much broader than the computed distribution for the one computed model structure that we studied, but the grouping into well-defined classes is evident.

β/β Packing

Because of the hyperboloid shape of twisted β-sheets, as discussed above, two β-sheets can be packed efficiently in only two distinct classes of low-energy arrangements [81]. In the energetically most favored class, the strands of the two sheets are nearly parallel or antiparallel to each other, so that the two curved structures are complementary over most of their surfaces (Fig. 9). This class is seen frequently in protein crystal structures, where it has also been termed 'aligned packing' [39, 82]. In the other class, with 1–4 kcal/mol higher energies, the strands are nearly perpendicular to each other, and good packing occurs between the corner of one sheet and the interior of the other sheet [81]. In observed structures of this type, termed 'orthogonal packing', there is usually a covalent connection between a corner of the two sheets [39, 82].

βαβ Crossover packing

Two parallel strands of a β-sheet often are connected by a peptide chain that contains an α-helix. This crossover connection could, in principle, be either right- or left-handed, as shown in Fig. 10 [44, 83]. Actually, crossovers in globular proteins are always right-handed, with only two exceptions observed so far [44, 84]. This strong preference has been interpreted as a consequence of the right-handed twist of the β-sheet, resulting in a reduction of strain in the connecting chain between the strands for a right-handed crossover [84]. A comparison of the two forms of the crossover by means of conformational energy computations has established that the right-handed form is energetically much more favorable than the left-handed form [83]. Its low energy arises from favorable interactions between, the α-helix and the β-sheet, just as in α/β packing (section on 'α/β packing'), and from the absence of conformational strain in the nonhelical parts of the connecting chain.

This crossover also occurs in many proteins in the doubled form $\beta\alpha\beta\alpha\beta$, called the Rossmann fold [44, 85]. Although the two α-helices usually connect neighboring pairs of strands in the β-sheet, other connectivities may occur. A computation of the conformational energy of a structure composed of two α-helices and a three-stranded

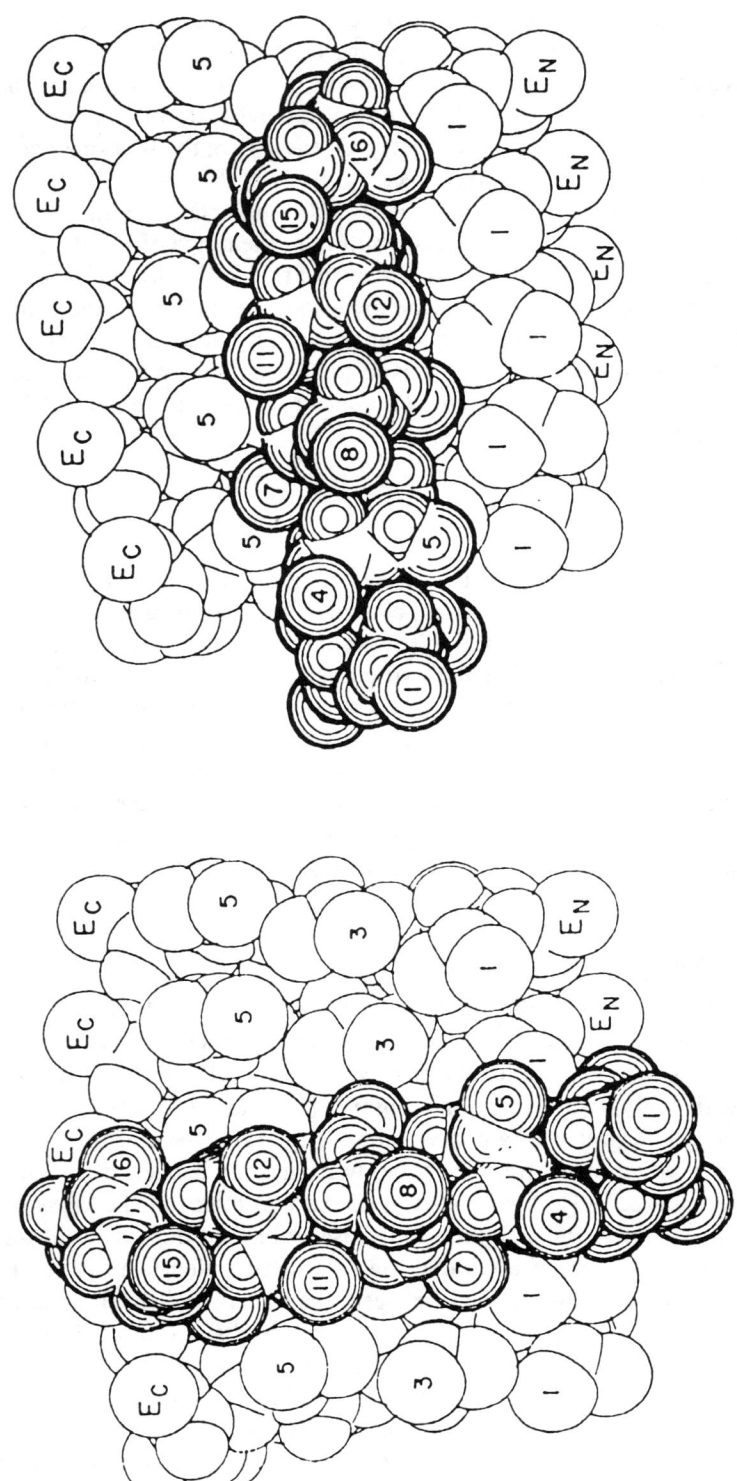

Fig. 8 — Space-filling representations of a right-handed α-helix of CH_3CO-$(L$-$Ala)_{16}$-$NHCH_3$ and a right-twisted parallel β-sheet of $[CH_3CO$-$(L$-$Val)_6$-$NHCH_3]_5$ in low-energy packing arrangements. Left: $\Omega = 4.4°$; Right: $\Omega = -89.4°$. Only heavy atoms are shown, with approximate van der Waals radii [33].

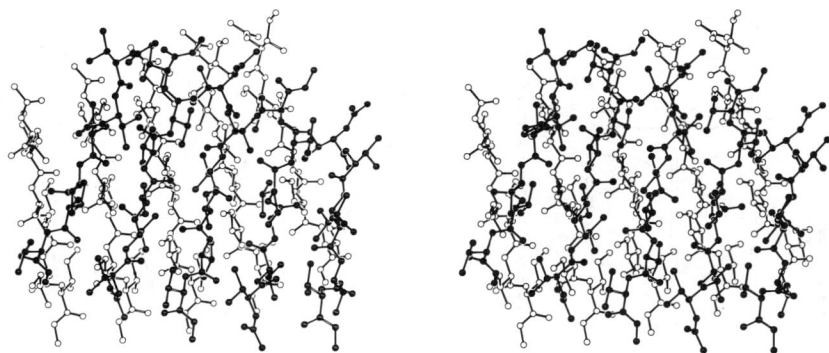

Fig. 9 — Stereoscopic picture of the low-energy near-parallel (aligned) packing of a $[CH_3CO\text{-}(\text{L-Ile})_6\text{-}NHCH_3]_5$ parallel β-sheet (open atoms) on a $[CH_3\text{-}CO\text{-}(\text{L-Val})_6\text{-}NHCH_3]_5$ antiparallel β-sheet (filled atoms). All hydrogen atoms are omitted [81].

β-sheet, with various connectivities between the five elements, has established that right-handed crossovers are also favored in $\beta\alpha\beta\alpha\beta$ structures [86].

β-Barrels

Two types of β-barrel are observed in proteins, one with parallel and one with antiparallel chains [44, 87]. The parallel-chain barrel involves α-helical segments in $\beta\alpha\beta\alpha\beta$ form, whereas the antiparallel-chain barrel generally does not involve intervening α-helical segments. β-barrels usually are not straight, with strands running approximately parallel to the axis of the barrel (as in Fig. 11(a)), but the strands are right-tilted, as shown schematically in Fig., 11(b). This tilting is in part a consequence of the intrinsic right-handed twisting of the β-sheet, but the following effect also contributes to it. Tilting improves the packing of the side chains in a β-barrel with bulky side chains, such as Val, on the outside. As a result, the energy of an eight-strand antiparallel β-barrel, in which L-Val and Gly residues alternate along the chains, is 8.6 kcal/mol lower in the right-tilted form than in the absence of tilting [88]. Conversely,

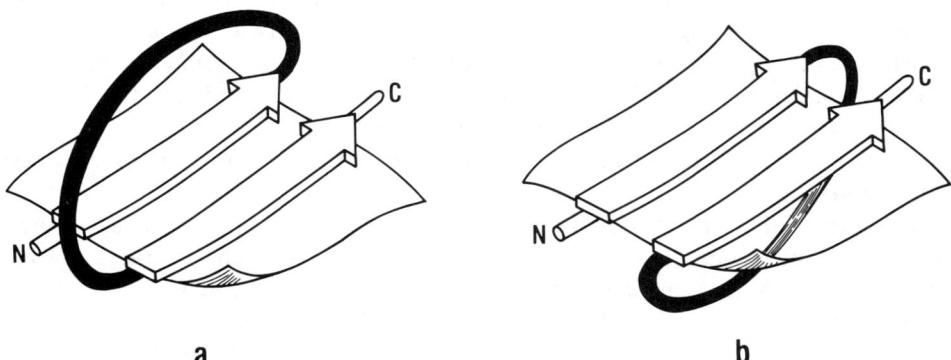

a   b

Fig. 10 — Schematic representation of (a) a right-handed and (b) a left-handed $\beta\alpha\beta$ crossover structure. [Adapted, with permission, from Richardson [44].]

(a) 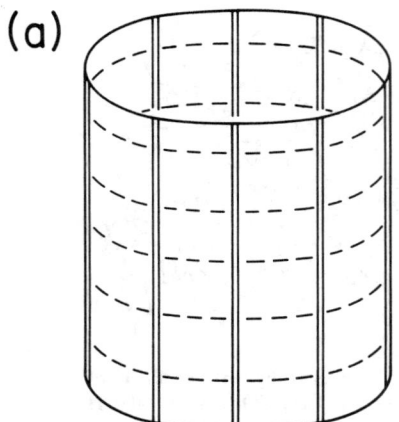   (b)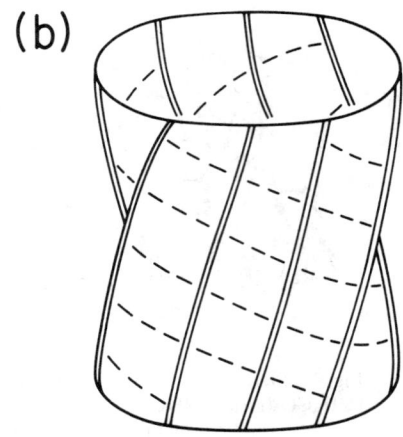

Fig. 11 — Schematic drawing of an eight-strand β barrel. (a) Non-tilted barrel with strands running parallel to the axis of the barrel. (b) Right-tilted barrel, with strands inclined with respect to the direction of the axis of the barrel [88].

left-handed tilting is energetically very unfavorable, because it would require the unfavorable left-handed twisting of the β-sheet.

The computations have also demonstrated the necessity for numerous residues with small side chains or Gly in alternating positions along the strands, because too many large residues cannot be packed into the inside of the barrel or their presence would lead to a severe distortion of the barrel [88]. On the other hand, it is favorable to have large side chains on residues in the alternating positions, i.e. pointing to the outside of a tilted barrel. This is the pattern seen frequently in antiparallel β-barrels in globular proteins [88, 89]. Tilting increases the diameter of a β-barrel slightly, so that there is more room to accommodate the internal side chains in a tilted β-barrel [90].

Silk fibroin

Silk fibroin is a block copolypeptide of crystalline domains with high Gly and Ala content and of less crystalline domains. The crystalline domains of *Bombyx Mori* silk contain the repeating amino acid sequence Gly–Ser–Gly–Ala–Gly–Ala (with some repeating Gly–Ala sequences at their ends), and they can exist in two morphologies, for which models have been proposed on the basis of fiber or powder X-ray diffraction studies [91]. The more stable form is known as silk II. A detailed structural model, consisting of packed pleated β-sheets, was first proposed for silk II by Marsh *et al.* [12] and refined by Fraser *et al.* [92]. The structure of the less stable silk I form is less well understood.

Conformational energy computations have been carried out on packed sheet model structures, composed of poly(L-Ala–Gly) chains as a simplified model [34]. Such chains can be assembled into two kinds of β-sheet, viz. those in which all Ala side chains project from the same side of the sheet (termed 'in-register sheets') and those in which the Ala side chains of adjacent strands point alternately to the two sides of the sheet (termed 'out-of-register sheets'). The computations have confirmed that the structure with the lowest energy is formed of antiparallel in-register β-sheets, packed

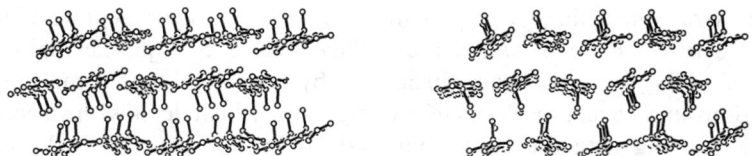

Fig. 12—Computed model structure for the crystalline domain of silk II, with 'in-register' arrangement of the sheets, viewed along the direction of the polypeptide chains. The figure shows the lowest-energy packing for three stacked antiparallel five-stranded β-sheets formed by $CH_3CO-(L-Ala-Gly)_3-NHCH_3$ chains. The sheets are perpendicular to the plane of the drawing. Hydrogen bonds within each sheet (not shown explicitly) are horizontal [34].

in such a way that the Ala-containing sides of the sheets face each other and the Gly-containing sides also face each other (Fig. 12), as in the models proposed earlier for silk II [12, 92]. The unit cell parameters of the computed structure agree closely with the observed values for silk II. A second, higher-energy computed structure has also been found, in which each strand forms a coiled coil [77], with residue conformations (for Ala and Gly, respectively) that correspond to a local right-handed and left-handed twist of a strand [55] alternating along the chains. The strands are assembled into antiparallel out-of-register hydrogen-bonded sheets. These, in turn, stack into a structure in which every pair of adjacent sheets forms the same kinds of contact (Fig. 13). The computed unit cell dimensions of the structure are consistent with observed powder X-ray reflections and the observed density of silk I [93]. Therefore, this structure is proposed as a model for the crystalline form of silk I [34]. The computations are being extended to analyze stacked sheet structures formed by the actual repeating sequence of silk, in which every third L-Ala residue has been replaced by L-Ser [94].

Collagen

The unique structure of collagen is a direct consequence of its unique amino acid sequence in which every third residue is glycine, and the intervening two residues of each triplet (denoted X and Y) are frequently proline or hydroxyproline, respectively. The triple-stranded coiled-coil helical structure of natural collagen has been derived

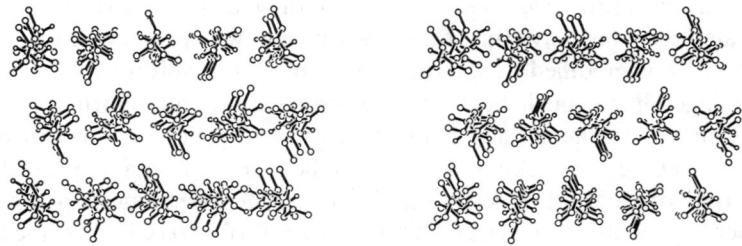

Fig. 13—Computed model structure for the crystalline domain of silk I, with 'out-of-register' arrangement of the sheets, viewed along the direction of the polypeptide chains. The figure shows the lowest-energy packing for three stacked antiparallel five-stranded sheets. Each sheet is formed of alternating $CH_3-CO-(L-Ala-Gly)_3-NHCH_3$ and $CH_3CO-(Gly-L-Ala)_3-NHCH_3$ chains. The sheets are perpendicular to the plane of the drawing. Hydrogen bonds within each sheet (not shown explicitly) are horizontal [34].

from fiber X-ray diffraction [8–10]. Synthetic Gly–X–Y poly(tripeptides) form helical fibrous structures, some of which are collagen-like [95–97]. The main structural features of collagen have been elucidated by means of conformational energy computations in a systematic series of investigations which dealt with various levels of structures, starting with the conformational analysis of the Gly–X–Y repeat unit, through the structure and stability of the triple-helical molecule, to molecular assemblies in microfibrils [97, 98].

Poly(Gly–Pro–Pro) serves as the simplest general model structure for collagen. Three equivalent poly(Gly–Pro–Pro) chains can be assembled into a three-chain structure according to various symmetry arrangements, including coiled coils with screw symmetry and having either of two dispositions of the strands (see section 'Packing of helices into coiled coils') and parallel-chain complexes formed by helices that are packed with either screw or rotational symmetry [79, 80]. The energy has been computed for every possible way of assembling three identical (Gly–Pro–Pro)$_4$ chains in regular conformations, in all of these symmetry arrangements [79]. The lowest-energy structure is a coiled-coil triple helix with screw symmetry (Fig. 14). Its helical parameters are close to those of the collagen models derived from fiber X-ray diffraction measurements [8–10]. Its atomic coordinates agree, to within an r.m.s. deviation of 0.3 Å, with the coordinates obtained subsequently for a single crystal of (Pro–Pro–Gly)$_{10}$ by means of X-ray diffraction [99]. The difference between the computed energy of the triple helix and the energies of the component polypeptide strands accounts closely for the observed enthalpy of the triple helix-to-statistical coil transition of poly(Gly–Pro–Pro), if the contribution of the free energy of hydration [29] is included in the computation [100].

The same collagen-like triple helix is the most stable structure for the assemblies formed by three poly(Gly–Pro–Hyp) [80] or three poly(Gly–Pro–Ala) [101] molecules, in agreement with X-ray powder diffraction experiments [95]. Poly(Gly–Ala–Pro) behaves differently, however [102]. Several coiled-coil triple-helical packing arrangements with low energy have been computed for this polymer, together with a parallel-chain triple-stranded complex in which the polypeptide chains take up conformations that are similar to those found in solid polyproline II [103] or polyglycine II (11). These results agree with X-ray diffraction measurements on poly(Gly–Ala–Pro) films, which can contain either collagen-like or polyproline-II-like chain assemblies, depending on the solvent used to prepare the film [104, 105]. The computations have provided an explanation of these observations because they have demonstrated that this poly(tripeptide) has several low-energy structures, the relative energy of which is modified by interactions with various solvents.

The collagen-like triple helices can be assembled into microfibrillar structures. As the first step in the theoretical analysis of the energetics of fibril formation, the geometry and energy of packing of two triple helices has been computed [106, 107]. The preferred orientation depends on the amino acid sequence. In the computed lowest-energy packing of two [CH$_3$CO-(Gly–Pro–Pro)$_5$-NHCH$_3$]$_3$ triple helices, the two molecules are arranged nearly parallel to each other, with an orientation angle of −10° between the two helix axes (Fig. 15). On the other hand, both near-parallel and near-antiparallel packings with low energies have been computed for [CH$_3$CO-(Gly–Pro–Ala)$_5$-NHCH$_3$]$_3$ triple helices. This result suggests that the observed preference for the near-parallel packing of molecules in collagen fibrils is not

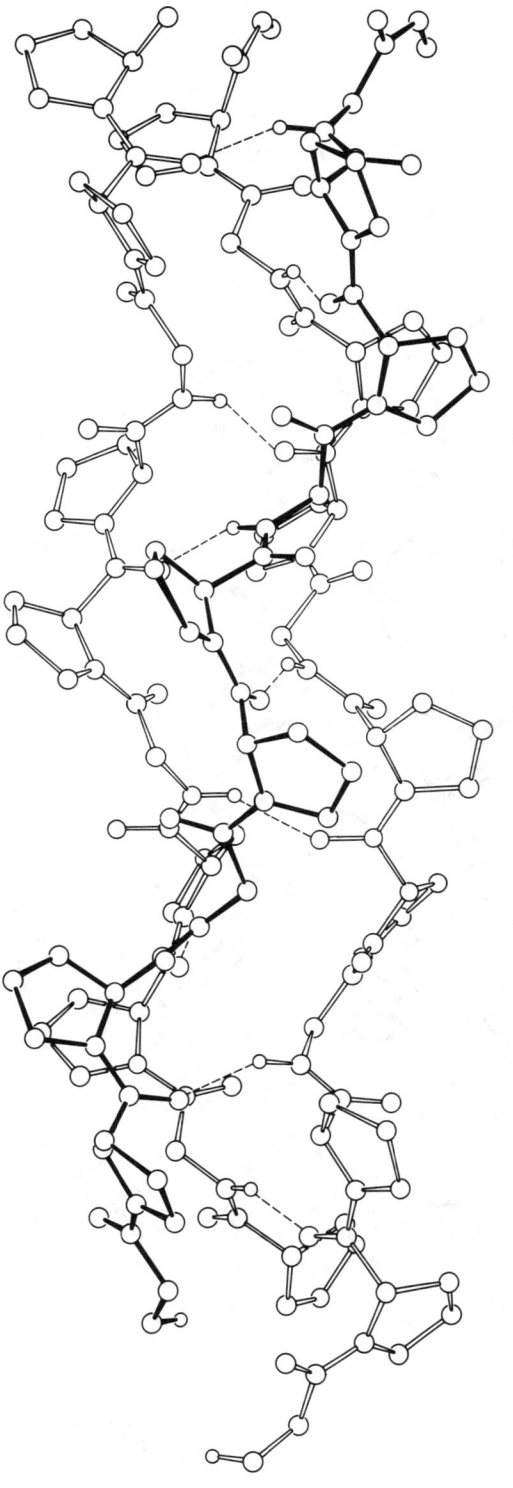

Fig. 14 — Triple-stranded coiled-coil complex of poly(Gly–Pro–Pro) of lowest potential energy [79].

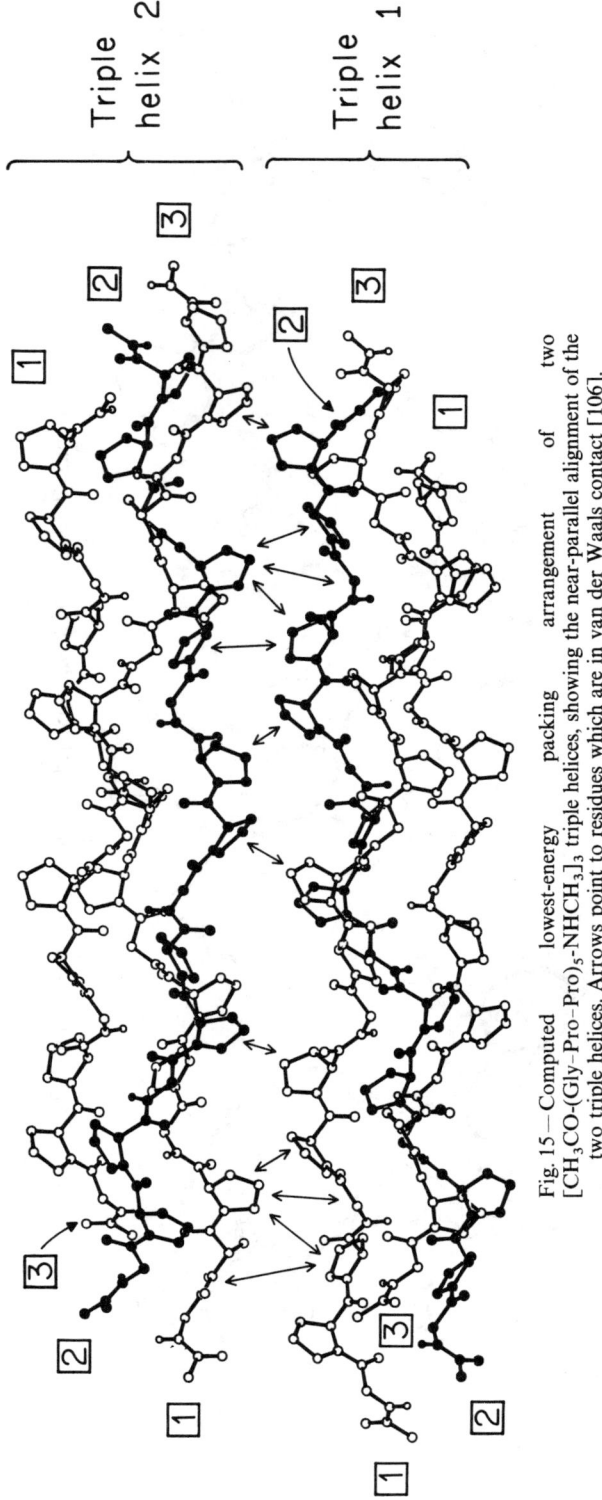

Fig. 15—Computed lowest-energy packing arrangement of two [CH₃CO-(Gly-Pro-Pro)₅-NHCH₃]₃ triple helices, showing the near-parallel alignment of the two triple helices. Arrows point to residues which are in van der Waals contact [106].

merely a nonspecific packing effect but it is an energetic consequence of specific residue–residue interactions between the triple helices and, in particular, it is due to the frequent presence of imino acids in the sequence [106].

The substitution of Hyp for Pro in position Y, where 4-Hyp is found exclusively in natural collagen as a result of post-translational hydroxylation, enhances the stability of the near-parallel packing structure [107]. Exactly the same computed packing arrangement of two triple-helical molecules is favored by the interaction energy for both $[CH_3CO\text{-}(Gly\text{-}Pro\text{-}Pro)_5\text{-}NHCH_3]_3$ and $[CH_3CO\text{-}(Gly\text{-}Pro\text{-}Hyp)_5\text{-}NHCH_3]_3$, but the stability of the packing for the latter is enhanced by 1.9 kcal/mol per Hyp residue, because of the formation of an intermolecular hydrogen bond between the side-chain hydroxyl group of Hyp in one molecule and a backbone carbonyl group in the second molecule (Fig. 16). The hydrogen-bonding OH group can be accommodated in the space between the two triple helices without any steric hindrance. This computation has provided an explanation for the observed stabilization of collagen fibrils by the presence of Hyp [108].

The triple helices in an observed collagen microfribril are not hexagonally closest-packed but there is a five-fold symmetry of packing [109, 110]. Computations on bundles of poly(Gly–Pro–Hyp) triple helices have demonstrated that optimal packing, with the maintenance of most of the favorable nonbonded and hydrogen-bonding interactions (as in Fig. 16), can be achieved easily in a bundle formed by five triple helices [111]. Assembly into smaller or larger bundles would require deviations from this optimal set of interactions. Thus, conformational energy computations have yielded an increased understanding of the structure of supramolecular assemblies of a fibrous macromolecule.

Malarial peptide helix

It is reasonable to expect that a polypeptide with multiple tandem repeats of a short amino acid sequence would exist in a helical form, just as in the case of collagen-like polypeptides. This premise has been used in conformational energy computations to predict the conformation of the immunodominant region of the circumsporozoite protein of the human malaria parasite *Plasmodium falciparum* [112, 113]. This protein contains several repeating peptide segments, each of which is at least 100 residues in length, with the $(Asn\text{-}Ala\text{-}Asn\text{-}Pro)_n$ sequence. From computations on a large number of helical and near-helical conformations, it has been concluded that two helical structures have a much lower energy than all others. One of them, a left-handed helix, is likely to exist in nonpolar environments. The other structure is stabilized by hydration, and is therefore predicted to be favored in aqueous solutions. It is a left-handed helix with 12 residues per turn, in which the Pro residues extend out from the helix (Fig. 17). Recent NMR studies have been interpreted in terms of a helix with the same two general characteristics [114].

Cellulose

Cellulose is the most ubiquitous natural polymer. It has a repeating sequence formed from glucose monomer units connected by $\beta(1\text{–}4)$ glycosidic linkages. Natural cellulose (cellulose I) consists of parallel chains, and can be converted irreversibly to

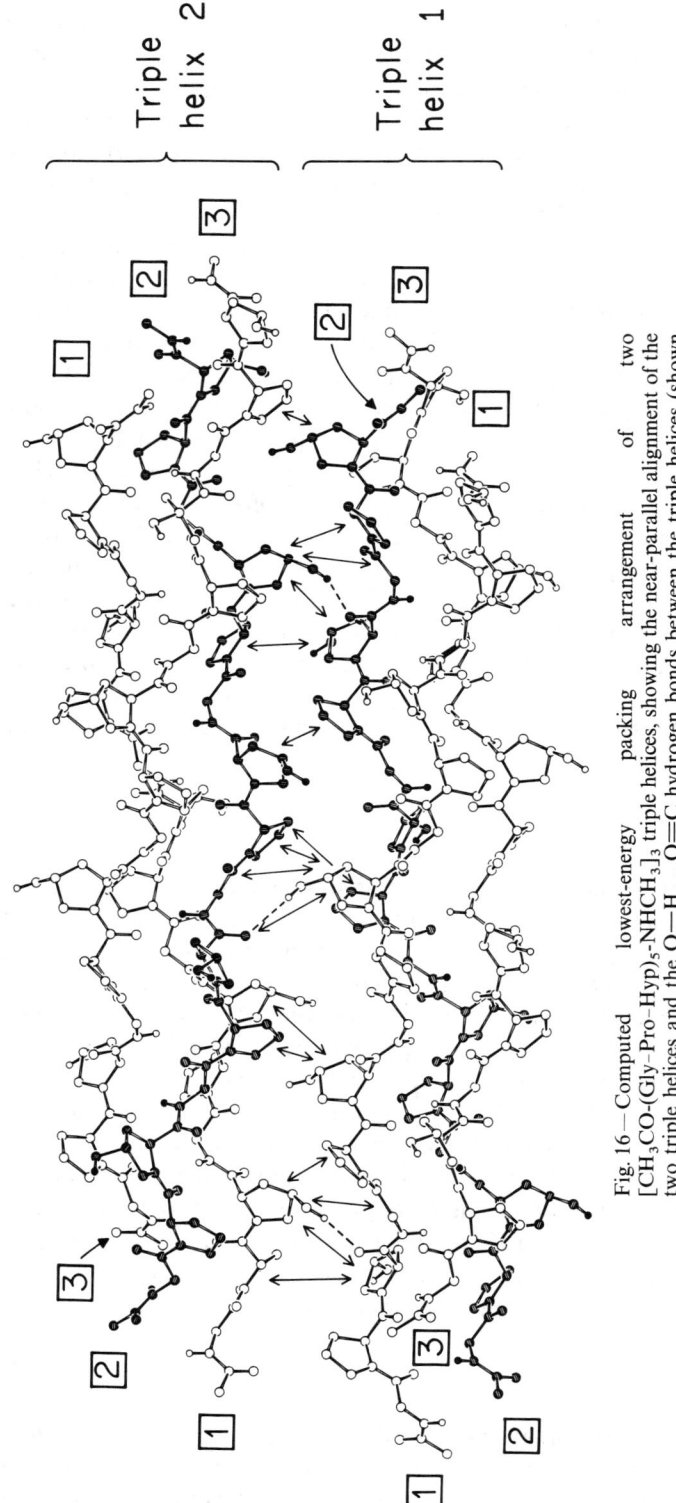

Fig. 16—Computed lowest-energy packing arrangement of two [CH₃CO-(Gly-Pro-Hyp)₅-NHCH₃]₃ triple helices, showing the near-parallel alignment of the two triple helices and the O—H . . . O=C hydrogen bonds between the triple helices (shown with dashed lines). The arrows point to residues which are in van der Waals contact [107].

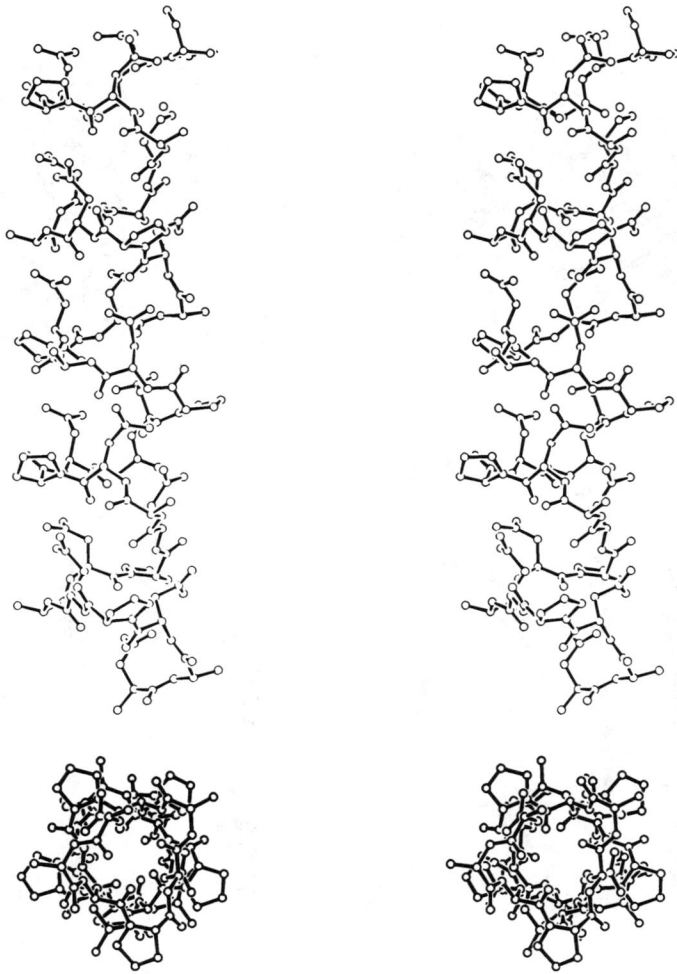

Fig. 17 — Stereo drawings of the lowest-energy conformation (computed with inclusion of the free energy of hydration) of (Asn–Ala–Asn–Pro)$_9$, viewed (top) from the side (with the C-terminus at the top) and (bottom) along the helical axis, seen from the N-terminal direction [113].

cellulose II by swelling in alkali, a process known as mercerization. It has been suggested that cellulose II has an antiparallel arrangement of chains [115].

Calculations carried out on single chains [116] and on crystalline forms [37] of cellulose I and II indicate that the structure of cellulose I is metastable with respect to that of cellulose II, in agreement with the results of the irreversible mercerization process. In the most stable computed parallel packing arrangement, the polysaccharide chains pack into sheets. Each sheet is stabilized by interchain hydrogen bonds, but no intersheet hydrogen bonds occur (Fig. 18). The computed unit cell dimensions agree closely with those determined experimentally for celluose I. On the other hand, the unit cell dimensions of the lowest-energy computed antiparallel structure are close to the experimental values for cellulose II. The latter structure also contains a specific

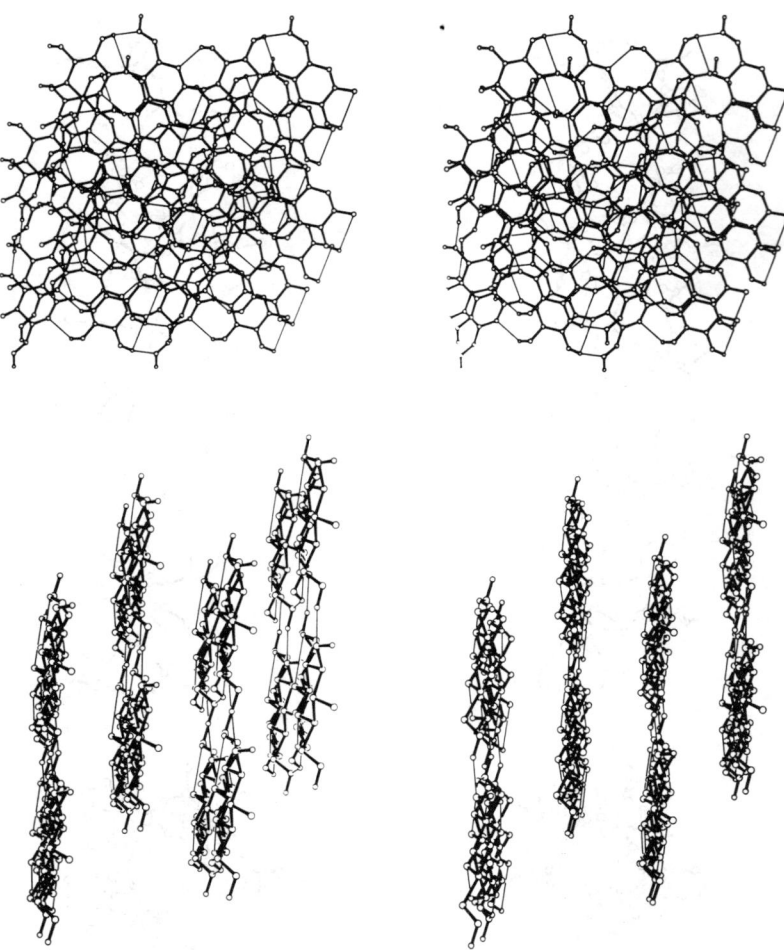

Fig. 18—Stereo views of eight two-chain unit cells of the computed lowest-energy parallel crystalline arrangement of cellulose chains, viewed (top) from above the sheets and (bottom) in a perpendicular direction. Thin lines indicate hydrogen bonds, with H...O distances not exceeding 2.8 Å. This is the structure proposed for cellulose I [37].

hydrogen bond that has been reported for cellulose II [115]. In this structure, both the chains within a sheet and between adjacent sheets are linked by a network of hydrogen bonds (Fig. 19). This arrangement has the lowest energy among all computed crystal structures.

The computations have led to a hypothesis concerning the formation and existence of the metastable cellulose I form [37]. In the biosynthetic apparatus, many cellulose chains are assumed to be produced simultaneously and in close proximity to each other, in a parallel arrangement. The formation of interchain hydrogen bonds would trap these chains in a metastable state. Conversion into a more stable arrangement would be prevented by a high activation energy. Treatment with NaOH during mercerization would disrupt the hydrogen bonds and swell the fiber, thereby enabling the chains to convert into the more stable antiparallel arrangement [115].

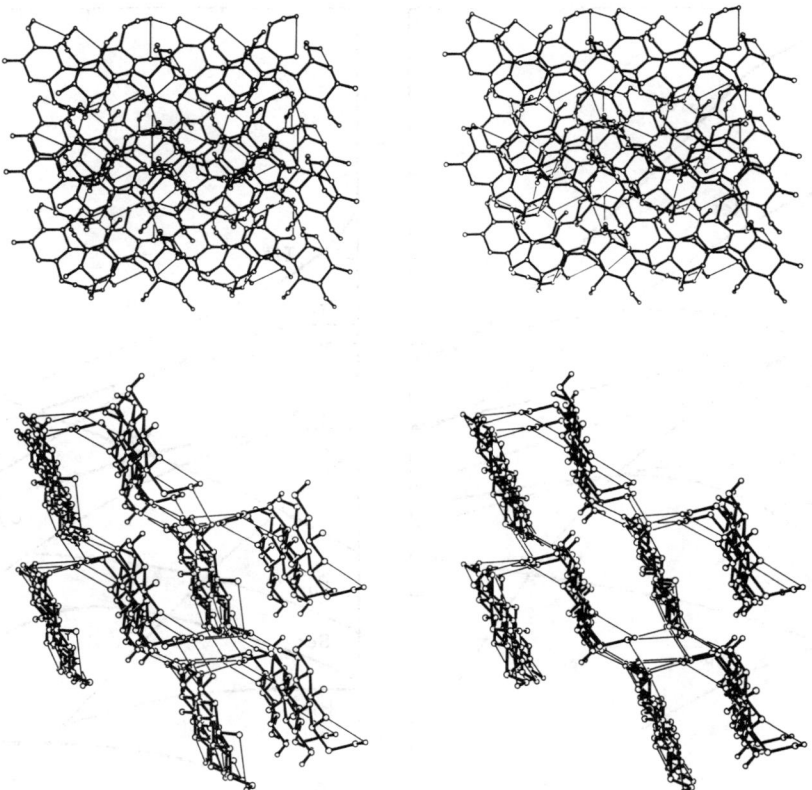

Fig. 19 — Stereo views of eight two-chain unit cells of the computed lowest-energy antiparallel crystalline arrangement of cellulose chains, viewed (top) from above the sheets and (bottom) in a perpendicular direction. Thin lines indicate hydrogen bonds, with H ...O distances not exceeding 2.8 Å. This is the structure proposed for cellulose II [37].

Stability

There is a large literature on the stability of the α-helix relative to the disordered form, and the transformation between these two forms, the so-called helix–coil transition [117]. This problem has been treated by statistical mechanics, making use of the one-dimensional Ising model, and has been addressed both for homopolymers [118, 119] and binary [120–122] and multi-component [123, 124] random copolymers of amino acids; both short- and medium-range interactions have been taken into account [125].

The relative helix–coil preferences of each of the 20 naturally occurring amino acids in water have been determined from experiments on thermally induced helix–coil transitions in host–guest binary random copolymers in which the host is a neutral water-soluble poly(amino acid) and the guest is, in turn, each of the 20 naturally occurring amino acids [126]. These results are expressed in terms of the Zimm–Bragg parameters σ and s [118]. Fig. 20 illustrates the temperature dependence of s for the various amino acids [126]. These values represent the intrinsic properties of each residue, reflecting only the interaction of a side chain and its own

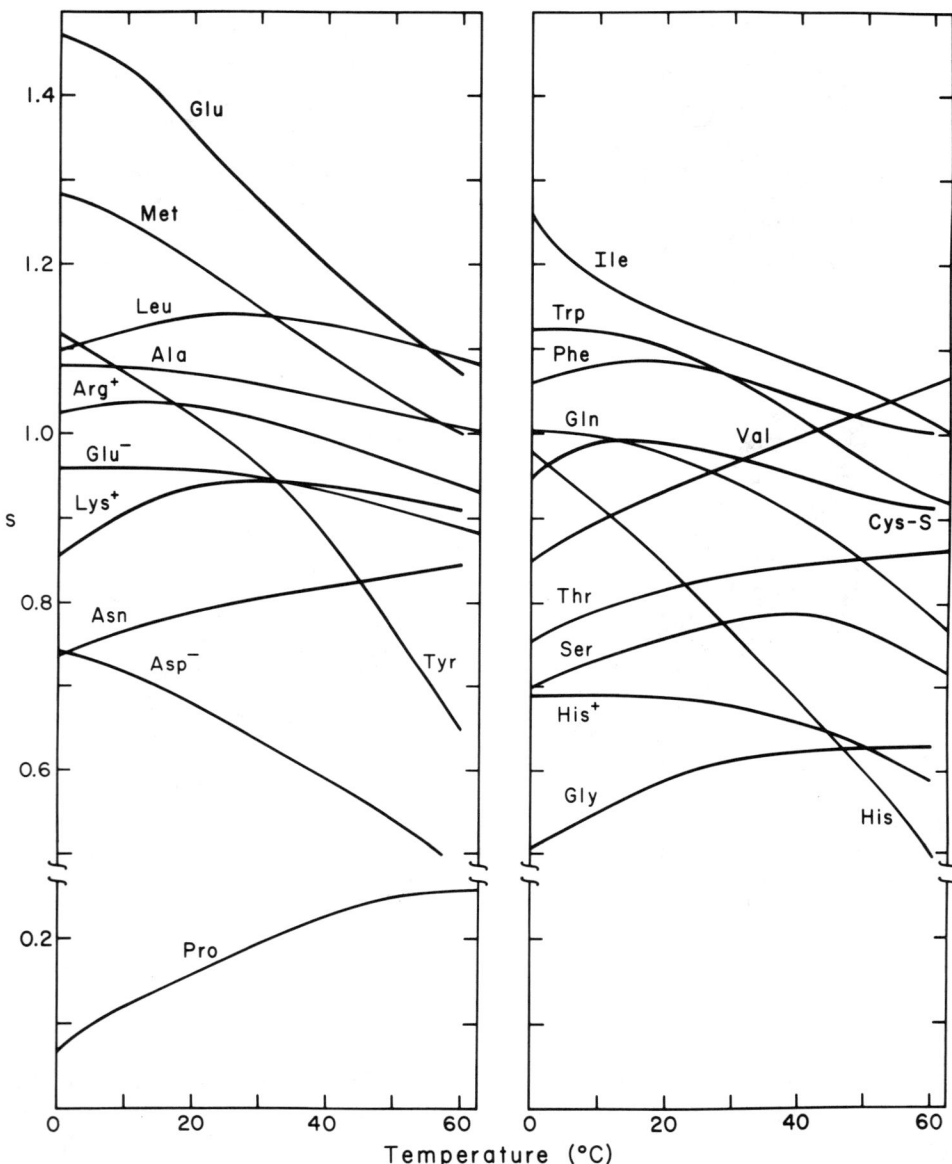

Fig. 20—Plots of *s* versus *T* for the 20 naturally occurring amino acid residues. The data were obtained with the host–guest technique [126].

backbone because long-range interactions are averaged out in the random copolymers used to obtain these values [127].

When using these values, together with the Ising model, to predict the locations of α-helices in *specific-sequence* copolymers, i.e. proteins, it is necessary to modify the values of *s* to take account of specific long-range interactions, such as salt links, charge–dipole interactions, and hydrophobic interactions [128]. These interactions are accounted for in terms of empirical parameters for each specific pair of residues that interact with each other. Numerical values of the parameters are derived from experimentally determined helix contents for specific-sequence peptides. The parameters modify the statistical weights of the respective conformational states in the Zimm–Bragg treatment [118] of the helix–coil transition.

Besides this statistical mechanical approach to the question of helix stability, the problem has also been addressed by conformational energy calculations. First of all, the helix-breaking tendencies of such residues as serine and aspartic acid can be accounted for by the tendency toward formation of side chain–backbone hydrogen bonds in *nonhelical* conformations (see Figs. 21 and 22) [129]. Secondly, the free energies of the helical and statistical coil forms in water have been calculated, and rationalize the observed behavior in terms of interatomic interactions, including those with the solvent [130–133]. For example, the *increase* of s, i.e. the stabilization of the helix, with increasing temperature for poly(L-valine) (Figs. 20 and 23) is accounted for by hydrophobic interactions between the valine side chains in the α-helix [133]. On

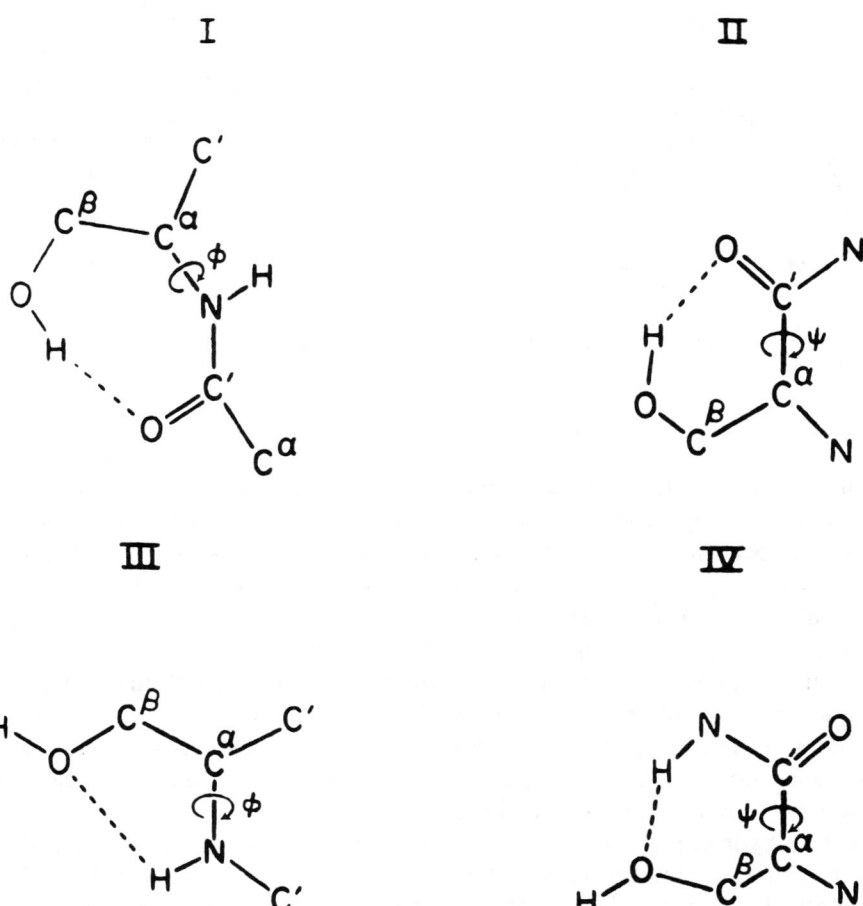

Fig. 21 — Illustration of the types of hydrogen bond between serine and threonine side chains and the backbone [129].

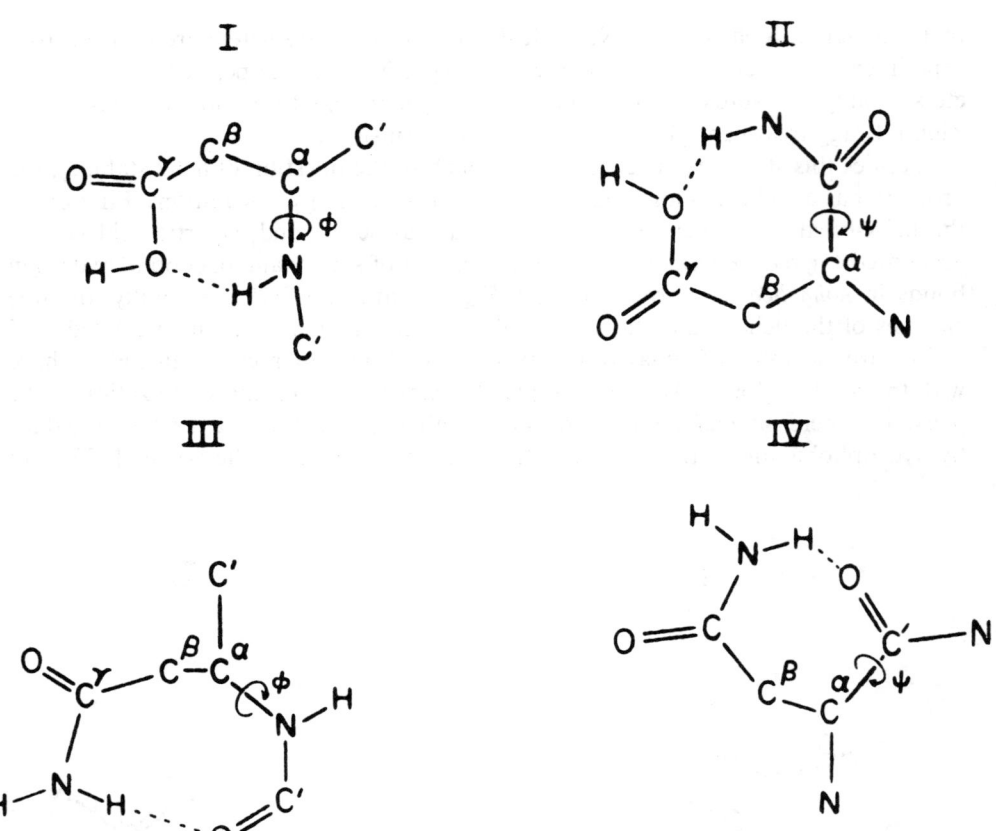

Fig. 22 — Illustration of the types of hydrogen bond between aspartic acid and asparagine side chains and the backbone [129].

the other hand, poly(isoleucine) exhibits a decrease in s with increasing temperature (Fig. 20) even though, like valine, isoleucine has a branch on the β carbon and its side chains also participate in hydrophobic interactions in the α-helix. The difference in behavior of poly(L-valine), poly(L-isoleucine) and poly(L-leucine) has been accounted for by subtle differences in the degree of hydration, i.e. hydrophobic-bond strength, in *both* the helical and coil forms of each of these polymers [134].

When homopolymers consist of ionizable groups, e.g. poly(L-lysine), the helix–coil transition is inducible by a change of pH, with the pH of the transition region depending on ionic strength. Conformational energy calculations on the helix–coil transition in aqueous salt solutions of poly(L-lysine) have accounted for the effect of pH and ionic strength on the transition [135].

If the α-hydrogen is substituted as, for example, in α-amino isobutyric acid (Aib), then the conformational energy (ϕ,ψ) map is very restricted, and the preferred form of Aib peptides is computed to be the 3_{10}- rather than the α-helical form [47]. This prediction has been verified by NMR and infrared spectroscopic measurements on solutions of oligomers of Aib [136]. The stability of the 3_{10}-helix for short poly(Aib-L-

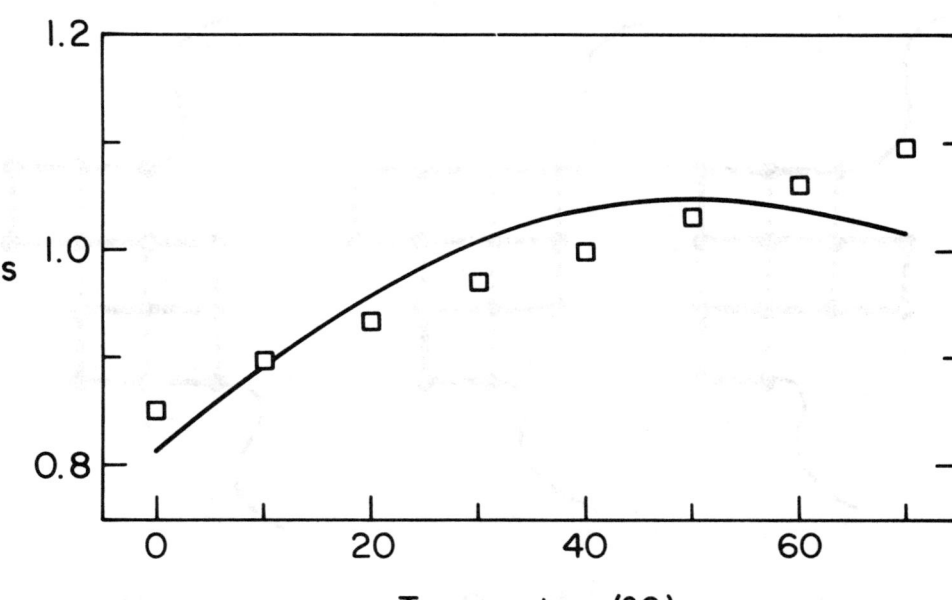

Fig. 23 — s versus T curves for poly(L-valine) in water. The squares are the experimental results (Fig. 20), and the line is the calculated one [133].

Ala) polypeptides and the increased stabilization of the α-helical form with a lengthening of the chain has been demonstrated recently [51].

The solvent-induced and thermally induced helix–coil transition has also been examined in non-aqueous solvents [137]. As a result of a change of solvent from an aqueous to a non-aqueous one, the order of the helix-stability constants, σ and s, for some amino acids is reversed [137].

In addition to the order–disorder transition, observed for α-helices, helical structures can also be induced to undergo transitions from one ordered form to another. For example, a crystalline form of poly[β-(p-chlorobenzyl)-L-aspartate] can be made to undergo a phase transition from an α-helical to an ω-helical form by heating, and rotational entropy is computed to play a role in this process [53]. Another order–order transition is the solvent induced interconversion between polyproline I (with cis peptide bonds) and polyproline II (with $trans$ peptide bonds), a process that has also been subjected to conformational energy calculations [138]. The transition has been accounted for in terms of differences in the binding of solvent components to the peptide C=O groups.

More recently, the Ising model treatment has been extended to the β-coil transition [139–144]. Fig. 24 illustrates the model used to account for the various states (turns, extended hydrogen-bonded chains, and nonbonded segments) considered by a matrix treatment. Parameters, analogous to σ and s of the helix–coil transition, were introduced to represent the statistical weights of these various states. This treatment has not yet been applied to experimental data to evaluate the relevant parameters. However, experiments have been initiated to obtain quantitative information about the tendencies of various pairs of amino acids to adopt the turn

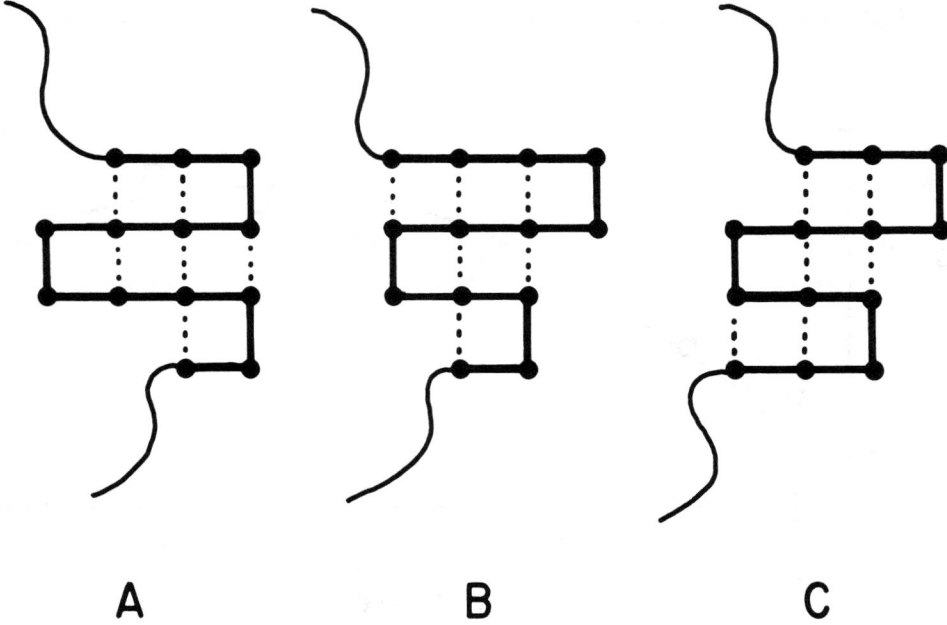

Fig. 24 — Illustrative antiparallel β-sheets [139].

conformations in hairpin-like structures [145, 146]; Table 2 illustrates the range of stabilities observed for such turns. The equilibrium constants and free energies listed in Table 2 refer to the cyclization equilibria for the closing of the disulfide bond in hexapeptides of the sequence Ac–Cys–X–Y–Z–W–Cys–NHMe. The results indicate, on the one hand, that the ease of cyclization depends strongly on the nature of the residues in the Y–Z position, where the turn is formed in the cyclic peptide (with Pro–Gly favoring turn formation) and, on the other hand, that specific residue-residue interactions (especially between positions X and W) can alter the stability of the turn.

DISCUSSION AND CONCLUSIONS

The structures of assemblies of fibrous proteins and other macromolecules can be described in many cases as a hierarchical succession of structural elements at various levels of complexity. At the scale of electron microscopic observations, fibers often consist of a series of levels of packed fibrillar and microfibrillar assemblies. The formation of long stretched-out microfibrils in a supramolecular assembly usually requires that the component molecules would be constrained to the shape of long rods, or that they be capable of taking up extended conformations which can be packed against each other. It is reasonable to expect that regular rod-, helix-, or coiled-coil-shaped macromolecules would exhibit more regularity in their primary structure than other macromolecules, such as globular proteins. Collagen provides one of the prime examples of such a hierarchical structure, but the main features described here are observed in many other fibrous systems as well, e.g. in silk, cellulose, muscle proteins, etc.

Table 2 — Equilibrium constants and standard Gibbs free energies for disulfide-exchange reactions with CVPGGC at 25° and pH 8.0 [146]

Reductant[a]	$K'_{8.0}$[b]	$(\Delta G^{\circ}_{8.0})'$ kcal·mol^{-1}
CGPGGC$_r$	6.40(0.22)	−1.10(0.02)
CKPGEC$_r$	2.50(0.08)	−0.54(0.02)
CEPGKC$_r$	2.05(0.16)	−0.43(0.05)
CGPGKC$_r$	1.42(0.08)	−0.21(0.03)
CFPGGC$_r$	1.23(0.05)	−0.12(0.02)
CKPGGC$_r$	1.21(0.11)	−0.11(0.05)
CGVVGC$_r$	0.98(0.12)	+0.01(0.07)
CVPGVC$_r$	0.78(0.03)	+0.15(0.02)
CVVVVC$_r$	≪1[c]	NA[c]
CVPGGC	(1.0)[d]	(0.00)[d]

[a] The oxidant in the equilibrium mixtures was CVPGGC.
[b] Numbers in parentheses are estimates of error.
[c] Not available.
[d] These are the values for the reference peptide.

Because of the regularities, the structures of fibrous macromolecules and of their assemblies are, to a great extent, accessible to theoretical analysis, and especially to conformational energy computations. In order to reduce computational time, it is frequently possible to take into account the regularities of individual structures and to compute the structural complexity in structural assemblies in a stepwise manner. Therefore, it has been possible to make valid theoretical predictions about complex fibrous assemblies, such as collagen fibrils, cellulose crystals, and cystalline domains of silk. The sizes of these structures exceed the sizes of globular proteins for which detailed structural computations have thus far been carried out.

Computations on the successive levels of structure, viz. polypeptide conformation, molecular structure, and supramolecular assembly, illustrate and confirm a general working principle; it is possible to account for many characteristic features of the packing patterns of polypeptides and other macromolecules in terms of local interaction energies, without having to include in all computations the long-range interactions that involve the entire structure [98, 147].

This principle can be applied to a considerable extent to globular proteins as well. Many frequently occurring and characteristic structural elements exist in globular proteins [38, 39, 44]. As shown in this chapter, their general properties can often be derived from experimental and theoretical studies of oligopeptides or fragments of proteins [38]. The formulation of the α-helix and β-sheet models by Pauling and Corey was an early, pioneering example of the recognition of the importance of characteristic substructures in proteins. The existence of common structural arrangements in which these fundamental structures are packed in proteins is another example of the working principle mentioned [98, 147]. Computations on assemblies of regularly folded polypeptide chains, reviewed here, provide explanations of many observations about the structure and stability of globular proteins in terms of interatomic interactions. In this manner, they contribute a vital step for understanding the principles of protein structure and folding [148, 149].

ACKNOWLEDGMENTS

This work was supported by research grants from the National Institute of General Medical Sciences (GM-14312) and the National Institute on Aging (AG-00322) of the National Institutes of Health, U.S. Public Health Service, and from the National Science Foundation (DMB84-01811). Part of this chapter was written while H.A.S. was a Fogarty Scholar-in-residence at the U.S. National Institutes of Health.

REFERENCES

[1] W.T. Astbury and A. Street, *Phil. Trans. Roy. Soc. Ser. A* **230** (1931) 75.

[2] W.T. Astbury and H.J. Woods, *Phil. Trans. Roy. Soc. Ser. A* **232** (1933) 333.

[3] L. Pauling, R.B. Corey and H.R. Branson, *Proc. Natl. Acad. Sci., U.S.A.* **37** (1951) 205.

[4] L. Pauling and R.B. Corey, *Proc. Natl. Acad. Sci., U.S.A.* **37** (1951) 729.

[5] J.C. Kendrew, R.E. Dickerson, B.E. Strandberg, R.G. Hart, D.R. Davies, D.C. Phillips and V.C. Shore, *Nature* **185** (1960) 422.

[6] M.F. Perutz, M.G. Rossman, A.F. Cullis, H. Muirhead, G. Will and A.C.T. North, *Nature* **185** (1960) 416.

[7] B.W. Matthews, P.B. Sigler, R. Henderson and D.M. Blow, *Nature* **214** (1967) 652.

[8] A. Rich and F.H.C. Crick, *Nature* **176** (1955) 915.

[9] G.N. Ramachandran and G. Kartha, *Nature* **176** (1955) 593.

[10] W. Traub, A. Yonath and D.M. Segal, *Nature* **221** (1969) 914.

[11] F.H.C. Crick and A. Rich, *Nature* **176** (1955) 780.

[12] R.E. Marsh, R.B. Corey and L. Pauling, *Biochim. Biophys. Acta* **16** (1955) 1.

[13] R.A. Scott and H.A. Scheraga, *J. Chem. Phys.* **45** (1966) 2091.

[14] T. Ooi, R.A. Scott, G. Vanderkooi and H.A. Scheraga, *J. Chem. Phys.* **46** (1967) 4410.

[15] J.F. Yan, G. Vanderkooi and H.A. Scheraga, *J. Chem Phys.* **49** (1968) 2713.

[16] J.F. Yan, F.A. Momany and H.A. Scheraga, *J. Am. Chem. Soc.* **92** (1970) 1109.

[17] P. Doty and J.T. Yang, *J. Am. Chem. Soc.* **78** (1956) 498.

[18] R.H. Karlson, K.S. Norland, G.D. Fasman and E.R. Blout, *J. Am. Chem. Soc.* **82** (1960) 2268.

[19] E.M. Bradbury, A.R. Downie, A. Elliott and W.E. Hanby, *Proc. Roy. Soc. (London)* **A259** (1960) 110.

[20] C. Chothia, *J. Mol. Biol.* **75** (1973) 295.

[21] G.N. Ramachandran, C. Ramakrishnan and V. Sassisekharan, *J. Mol. Biol.* **7** (1963) 95.

[22] G. Némethy and H.A. Scheraga, *Biopolymers* **3** (1965) 155.

[23] G. Némethy, M.S. Pottle and H.A. Scheraga, *J. Phys. Chem.* **87** (1983) 1883.

[24] M.J. Sippl, G. Némethy and H.A. Scheraga, *J. Phys. Chem.* **88** (1984) 6231.

[25] F.A. Momany, R.F. McGuire, A.W. Burgess and H.A. Scheraga, *J. Phys. Chem.* **79** (1975) 2361.

[26] D.M. Gay, *ACM Trans Math. Software* **9** (1983) 503.
[27] N. Gō and H.A. Scheraga, *J. Chem. Phys.* **51** (1969) 4751.
[28] N. Gō and H.A. Scheraga, *Macromolecules* **9** (1976) 535.
[29] Y.K. Kang, K.D. Gibson, G. Némethy and H.A. Scheraga, *J. Phys. Chem.* **92** (1988) 4739.
[30] J. Vila, R.L. Williams, M. Vásquez and H.A. Scheraga, *Proteins: Structure, Function, and Genetics,* **10** (1991) 199.
[31] K.-C. Chou, G. Némethy and H.A. Scheraga, *J. Phys. Chem.* **87** (1983) 2869. Erratum: *ibid.* **87** (1983) 4772.
[32] K.C. Chou, G. Némethy and H.A. Scheraga, *J. Am. Chem. Soc.* **106** (1984) 3161. Erratum; *ibid.* **107** (1985) 2199.
[33] K.-C. Chou, G. Némethy, S. Rumsey, R.W. Tuttle and H.A. Scheraga, *J. Mol. Biol.* **186** (1985) 591.
[34] S.A. Fossey, G. Némethy, K.D. Gibson and H.A. Scheraga, *Biopolymers,* submitted.
[35] W.R. Busing, ORNL-5747, Oak Ridge National Laboratory, Oak Ridge, TN (1981).
[36] L. Glasser and H.A. Scheraga, *J. Mol. Biol.* **199** (1988) 513.
[37] I. Simon, L. Glasser, H.A. Scheraga and R.S.J. Manley, *Macromolecules* **21** (1988) 990.
[38] K.C. Chou, G. Némethy and H.A. Scheraga, *Accts. Chem. Res.* **23** (1990) 134.
[39] C. Chothia, *Ann. Rev. Biochem.* **53** (1984) 537.
[40] H.A. Scheraga, *Harvey Lectures* **63** (1969) 99.
[41] E.H. Erenrich, R.H. Andreatta and H.A. Scheraga, *J. Am. Chem. Soc.* **92** (1970) 1116.
[42] J. Donohue, *Proc. Natl. Acad. Sci., U.S.A.* **39** (1953) 470.
[43] D.J. Barlow and J.M. Thornton, *J. Mol. Biol.* **201** (1988) 601.
[44] J.S. Richardson, *Adv. Protein Chem.* **34** (1981) 167.
[45] Y. Isogai, G. Némethy, S. Rackovsky, S.J. Leach and H.A. Scheraga, *Biopolymers* **19** (1980) 1183.
[46] G. Némethy, D.C. Phillips, S.J. Leach and H.A. Scheraga, *Nature* **214** (1967) 363.
[47] Y. Peterson, S.M. Rumsey, E. Benedetti, G. Némethy and H.A. Scheraga, *J. Am. Chem. Soc.* **103** (1981) 2947.
[48] B.R. Malcolm, *Biopolymers* **16** (1977) 2591.
[49] B.V.V. Prasad, N. Shamala, R. Nagaraj, R. Chandrasekaran and P. Balaram, *Biopolymers* **18** (1979) 1635.
[50] I.L. Karle, J.L. Flippen-Anderson, M. Sakumar and P. Balaram, *Int. J. Peptide Protein Res.* **31** (1988) 567.
[51] V. Pavone, E. Benedetti, B. Di Blasio, C. Pedone, A. Santini, A. Bavoso, C. Toniolo, M. Crisma and L. Sartore, *J. Biomol. Struct. Dynamics* **7** (1990) 1321.
[52] Y. Takeda, Y. Iitaka and M. Tsuboi, *J. Mol. Biol.* **51** (1970) 101.
[53] Y.C. Fu, R.F. McGuire and H.A. Scheraga, *Macromolecules* **7** (1974) 468.
[54] K.-C. Chou and H.A. Scheraga, *Proc. Natl. Acad. Sci., U.S.A.* **79** (1982) 7047.
[55] K.-C. Chou, M. Pottle, G. Némethy, Y. Ueda and H.A. Scheraga, *J. Mol. Biol.* **162** (1982) 89.
[56] K.-C. Chou, G. Némethy and H.A. Scheraga, *J. Mol. Biol.* **168** (1983) 389.

[57] K.-C. Chou, G. Némethy and H.A. Scheraga, *Biochemistry* **22** (1983) 6213.

[58] F.A. Quiocho and W.N. Lipscomb, *Adv. Protein Chem.* **25** (1971) 1.

[59] C. Chothia, *J. Mol. Biol.* **163** (1983) 107.

[60] B.A. Wallace and K. Ravikumar, *Science* **241** (1988) 182.

[61] D.A. Langs, *Science* **241** (1988) 188.

[62] K.-C. Chou, G. Némethy, M.S. Pottle and H.A. Scheraga, *Biochemistry* **24** (1985) 7948.

[63] J.S. Balcerski, E.S. Pysh, G.M. Bonora and C. Toniolo, *J. Am. Chem. Soc.* **98** (1976) 3470.

[64] C. Toniolo, G.M. Bonora, M. Palumbo and E.S. Pysh, *Pept. Proc. Eur. Pept. Symp.*, 14th (1976) 597.

[65] F.H.C. Crick, *Acta Cryst.* **6** (1953) 689.

[66] M.K. Gilson and B. Honig, *Proc. Natl. Acad. Sci., U.S.A.* **86** (1989) 1524.

[67] M. Gerritsen, K.C. Chou, G. Némethy and H.A. Scheraga, *Biopolymers* **24** (1985) 1271. Erratum; *ibid.* **24** (1985) 2177.

[68] H.C. Watson, in *Progress and Stereochemistry*, Vol. 4., B.J. Aylett and M.M. Harris, Eds, London, Butterworths, 1969, pp. 299–333.

[69] R.E. Dickerson, in *The Proteins*, Vol. II, 2nd edn, H. Neurath, Ed., New York, Academic Press, 1964, pp. 603–778.

[70] P.C. Weber and F.R. Salemme, *Nature* **287** (1980) 82.

[71] K.-C. Chou, G.M. Maggiora, G. Némethy and H.A. Scheraga, *Proc. Natl. Acad. Sci., U.S.A.* **85** (1988) 4295.

[72] L. Carlacci and K.-C. Chou, *Protein Eng.* **3** (1990) 509.

[73] R.D.B. Fraser and T.P. MacRae, *J. Mol. Biol.* **3** (1961) 640.

[74] D.A.D. Parry and J.M. Squire, *J. Mol. Biol.* **75** (1973) 33.

[75] E.K. O'Shea, R. Rutowski and P.S. Kim, *Science* **243** (1989) 538.

[76] C.R. Vinson, P.B. Sigler and S.L. McKnight, *Science* **246** (1989) 911.

[77] K. Nishikawa and H.A. Scheraga, *Macromolecules* **9** (1976) 395.

[78] F.H.C. Crick, *Acta Cryst.* **6** (1953) 685.

[79] M.H. Miller and H.A. Scheraga, *J. Polymer Sci.: Polymer Symposia* **No 54**, (1976) 171.

[80] M.H. Miller, G. Némethy and H.A. Scheraga, *Macromolecules* **13** (1980) 470.

[81] K.-C. Chou, G. Némethy, S. Rumsey, R.W. Tuttle and H.A. Scheraga, *J. Mol. Biol.* **188** (1986) 641.

[82] C. Chothia and J. Janin, *Biochemistry* **21** (1982) 3955.

[83] K.-C. Chou, G. Némethy, M. Pottle and H.A. Scheraga, *J. Mol. Biol.* **205** (1989) 241.

[84] M.J.E. Sternberg and J.M. Thornton, *J. Mol. Biol.* **105** (1976) 367.

[85] S.T. Rao and M.G. Rossmann, *J. Mol. Biol.* **76** (1973) 241.

[86] L. Carlacci and K.-C. Chou, *Biopolymers* **30**, (1990) 135.

[87] F.R. Salemme, *Prog. Biophys. Mol. Biol.* **42** (1983) 95.

[88] K.-C. Chou, A. Heckel, G. Némethy, S. Rumsey, L. Carlacci and H.A. Scheraga, *Proteins: Structure, Function, and Genetics* **8** (1990) 14.

[89] J.A. Tainer, E.D. Getzoff, K.M. Beem, J.S. Richardson and D.C. Richardson, *J. Mol. Biol.* **160** (1982) 181.

[90] K.-C. Chou, L. Carlacci and G.G. Maggiora, *J. Mol. Biol.* **213** (1990) 315.

[91] R.D.B. Fraser and T.P. MacRae, *Conformation in Fibrous Proteins and Related*

Synthetic Polypeptides, New York, Academic Press, 1973, Chapter 13.

[92] R.D.B. Fraser, T.P. MacRae, F.H.C. Stewart and E. Suzuki, *J. Mol. Biol.* **11** (1965) 706.

[93] T. Asakura, A. Kuzuhara, R. Tabeta and H. Saito, *Macromolecules* **18** (1985) 1841.

[94] S.A. Fossey, G. Némethy and H.A. Scheraga, *Macromolecules*, in preparation.

[95] A. Yonath and W. Traub, *J. Mol. Biol.* **43** (1969) 461.

[96] F.R. Brown, A. DiCorato, G.P. Lorenzi and E.R. Blout, *J. Mol. Biol.* **63** (1972) 85.

[97] G. Némethy, in *Collagen; Vol. 1: Biochemistry*, M.E. Nimni, Ed., Boca Raton, CRC Press, 1988, 79.

[98] G. Némethy and H.A. Scheraga, *Bull. Inst. Chem. Res., Kyoto Univ.* **66** (1989) 398.

[99] K. Okuyama, N. Tanaka, T. Ashida and M. Kakudo, *Bull. Chem. Soc. Japan* **49** (1976) 1805.

[100] G. Némethy and H.A. Scheraga, *Biopolymers* **28** (1989) 1573.

[101] M.H. Miller, G. Némethy and H.A. Scheraga, *Macromolecules* **13** (1980) 910.

[102] G. Némethy, M.H. Miller and H.A. Scheraga, *Macromolecules* **13** (1980) 914.

[103] P.M. Cowan and S. McGavin, *Nature* (London) **176** (1955) 501.

[104] D.M. Segal and W. Traub, *J. Mol. Biol.* **43** (1969) 487.

[105] B.B. Doyle, W. Traub, G.P. Lorenzi and E.R. Blout, *Biochemistry* **10** (1971) 3052.

[106] G. Némethy and H.A. Scheraga, *Biopolymers* **23** (1984) 2781. Erratum: *ibid.* **24** (1985) 581.

[107] G. Némethy and H.A. Scheraga, *Biochemistry* **25** (1986) 3184.

[108] T.V. Burjanadze, *Biopolymers* **21** (1982) 1489.

[109] K.A. Piez and B.L. Trus, *Biosci. Reports* **1** (1981) 801.

[110] K.A. Piez, in *The Protein Folding Problem*. D.B. Wetlaufer, Ed., Boulder, Westview Press, 1984, p. 47.

[111] J. Ballesteros, G. Némethy, M.S. Pottle and H.A. Scheraga, to be published.

[112] K.D. Gibson and H.A. Scheraga, *Proc. Natl. Acad. Sci., U.S.A.* **83** (1986) 5649.

[113] I.K. Roterman, K.D. Gibson and H.A. Scheraga, *J. Biomolec. Structure & Dynamics* **7** (1989) 391.

[114] G. Esposito, A. Pessi and A.S. Verdini, *Biopolymers* **28** (1989) 225.

[115] R.H. Marchessault and P.R. Sundararajan, in *The Polysaccharides*, G.O. Aspinall, Ed., New York, Academic Press, 1983, Vol. 2, pp. 11–95.

[116] I. Simon, H.A. Scheraga and R.S.J. Manley, *Macromolecules* **21** (1988) 983.

[117] D. Poland and H.A. Scheraga, *Theory of Helix–Coil Transitions in Biopolymers*, New York, Academic Press, 1970.

[118] B.H. Zimm and J.K. Bragg, *J. Chem. Phys.* **31**, (1959) 526.

[119] S. Lifson and A. Roig, *J. Chem. Phys.* **34** (1961) 1963.

[120] S. Lifson and G. Allegra, *Biopolymers* **2** (1964) 65.

[121] D. Poland and H.A. Scheraga, *Biopolymers* **7** (1969) 887.

[122] P.H. Von Dreele, D. Poland and H.A. Scheraga, *Macromolecues* **4** (1971) 396.

[123] A. Kidera, M. Mochizuki, R. Hasegawa, T. Hayashi, H. Sato, A. Nakajima, R.A. Fredrickson, S.P. Powers, S. Lee and H.A. Scheraga, *Macromolecules* **16** (1983) 162.

[124] J. Wojcik, A. Kidera, A.R. Leed, A. Nakajima and H.A. Scheraga, *Macromolecules* **23** (1990) 3655.

[125] H. Wako, N. Saitô and H.A. Scheraga, *J. Protein Chem.* **2** (1983) 221.

[126] J. Wojcik, K.H. Altmann, H.A. Scheraga, *Biopolymers* **30** (1990) 121.

[127] H.A. Scheraga, *Proc. Natl. Acad. Sci., U.S.A.* **82** (1985) 5585.

[128] M. Vásquez and H.A. Scheraga, *Biopolymers* **27** (1988) 41.

[129] P.N. Lewis, F.A. Momany and H.A. Scheraga, *Israel J. Chem.* **11** (1973) 121.

[130] N. Gō, M. Gō and H.A. Scheraga, *Proc. Natl. Acad. Sci., U.S.A.* **59** (1968) 1030.

[131] M. Gō, N. Gō and H.A. Scheraga, *J. Chem. Phys.* **52** (1970) 2060.

[132] M. Gō, N. Gō and H.A. Scheraga, *J. Chem. Phys.* **54** (1971) 4489.

[133] M. Gō, F.T. Hesselink, N. Gō and H.A. Scheraga, *Macromolecules* **7** (1974) 459.

[134] M. Gō and H.A. Scheraga, *Biopolymers* **23** (1984) 1961.

[135] F.T. Hesselink, T. Ooi and H.A. Scheraga, *Macromolecules* **6** (1973) 541.

[136] Y. Paterson, E.R. Stimson, D.J. Evans, S.J. Leach and H.A. Scheraga, *Int. J. Peptide and Protein Res.* **20** (1982) 468; *ibid.* **22** (1983) 128a.

[137] S. Sridhara, V.S. Ananthanarayanan, R.A. Fredrickson, B.O. Zweifel, G.T. Taylor and H.A. Scheraga, *Biopolymers* **20** (1981) 1435.

[138] S. Tanaka and H.A. Scheraga, *Macromolecules* **8** (1975) 494, 504, 516.

[139] W.L. Mattice and H.A. Scheraga. *Biopolymers* **23** (1984) 1701.

[140] W.L. Mattice and H.A. Scheraga, *Biopolymers* **23** (1984) 2879. Erratum: *ibid.* **24** (1985) 581.

[141] W.L. Mattice and H.A. Scheraga, *Macromolecules* **17** (1984) 2690.

[142] W.L. Mattice and H.A. Scheraga, in *Mathematics and Computers in Biomedical Applications*, J. Eisenfeld and C. DeLisi, Eds, Proc. 2nd IMACS International Symposium on Biomedical Systems Modelling, Elsevier, 1985, pp. 13–17.

[143] W.L. Mattice, E. Lee and H.A. Scheraga, *Can. J. Chem.* **63** (1985) 140.

[144] W.L. Mattice and H.A. Scheraga, *Biopolymers* **24** (1985) 565.

[145] P.J. Milburn, Y. Konishi, Y.C. Meinwald and H.A. Scheraga, *J. Am. Chem. Soc.* **109** (1987) 4486. Erratum: *ibid.* **109** (1987) 8123.

[146] P.J. Milburn, Y.C. Meinwald, S. Takahashi, T. Ooi and H.A. Scheraga, *Int. J. Peptide and Protein Res.* **31** (1988) 311. Erratum: *ibid.* **31** (1988) 587.

[147] H.A. Scheraga, K.-C. Chou and G. Némethy, in *Conformation in Biology*, R. Srinivasan and R.H. Sarma, Eds., Adenine Press (1983), p. 1.

[148] H.A. Scheraga, *Carlsberg Res. Commun.* **49** (1984) 1.

[149] G. Némethy and H.A. Scheraga, *FASEB J.* **4** (1990) 3189.

II. Quantum Chemistry

7

Some early (and lasting) contributions of Linus Pauling to quantum mechanics and statistical mechanics

Zelek S. Herman

Linus Pauling Institute of Science and Medicine, Palo Alto, CA, USA

INTRODUCTION

By virtually every applicable measure, Linus Pauling ranks as one of the most influential scientists of all time. As the result of over nine hundred publications during the past seventy-plus years [1-3] his contributions have profoundly influenced development of science, medicine, and social behavior. His textbook *The Nature of the Chemical Bond and the Structure of Molecules and Crystals*, in its three editions [4], has been reckoned to be the most often-cited textbook of all time [5-8]. The book, *Introduction to Quantum Mechanics, with Applications to Chemistry* [9], co-authored with E. Bright Wilson, Jr., holds the record for the book having been published for the longest period of time, without revision, by McGraw-Hill (from 1935 until 1985, when it was reissued by Dover Books). This book is, in fact, his second most often-cited work [5].

The advocacy by him and his late wife, Ava Helen Pauling, assisted by many other scientists, of the cessation of resolving national disputes through war, as summarized in the book *No More War!* [10] led in no small way to the historic Atmospheric Test Ban signed in 1963 by the United States and the Union of Soviet Socialist Republics.

Moreover, Pauling's series of textbooks on introductory chemistry have significantly influenced the education of at least two generations of scientists. These books have the titles *General Chemistry* [11], *College Chemistry* [12], and *Chemistry* [13]. In addition, together with the architect Roger Hayward, Pauling has written a picture book entitled *The Architecture of Molecules* [14], which is a veritable treasure trove of information concerning the nature of chemistry, biology, and solid-state physics.

Finally, his promotion of the improvement of human well being and the treatment and prevention of disease through the use of orthomolecular medicine (a term coined by him) has led in great measure to the intake, by a significant proportion of the populations in the developed countries, of supplemental vitamins and minerals (including vitamin C, in particular) in spite of the now waning opposition of the great majority of orthodox medical practitioners. Pauling's work in the field of orthomolecular medicine has been described in a number of books written by him [15–18], and his continuing research in this area has been summarized in the book *How to Live Longer and Feel Better* [19].

Pauling's many awards, his two unshared Nobel prizes (Chemistry, 1954; Peace, 1962), and some forty-five honorary degrees attest to the significance and impact of his contributions — no mean feat for a person who was not permitted to receive a high-school diploma until some forty-five years after his peers did [20].

Many of the specific contributions of Linus Pauling to science and medicine are well known to the current generation of people working in these fields, and specific citations to them will not be given here. For example, in the area of the nature of the chemical bond, he has done revolutionary research in valence-bond theory, the theory of hybrid bond orbitals, the theory of resonance, the concept of electronegativity, the structure of crystals by X-ray diffraction, the structure of molecules by the diffraction of electrons, the development of molecular biology, and the structure and properties of metals and intermetallic compounds. In the domain of the structure and properties of hemoglobin and other proteins, Pauling and his coworkers made significant contributions regarding the nature of the bonding of oxygen to hemoglobin, the study of the magnetic properties of hemoglobin and related compounds, the elucidation of the structure of the alpha helix, and the development of the theory of molecular evolution. In medicine, some of his important contributions involve the structure of antibodies and the nature of serological reactions, the invention of the Pauling oxygen meter, the elucidation of the cause of sickle-cell anemia, and a theory of general anesthesia. He also made important contributions concerning the structural significance of the hydrogen bond, the concept of complementary bonding in biological systems, and the fundamental mechanism of enzyme action. He has also contributed a large body of work on the structure of atomic nuclei.

However, there exist a number of early papers by Linus Pauling in the areas of quantum mechanics and statistical mechanics that seem not to be well known among younger scientists. Nevertheless, these papers are characterized by innovative ideas and rigorous mathematical development. Because of their elegance the material contained in these papers should be in the ken of scientists working today. Therefore, an attempt will be made in this chapter to describe some of these ideas, although a discussion of their significance to later work will be left, in general, to experts in the various fields.

THE ENTROPY OF SUPERCOOLED LIQUIDS AT THE ABSOLUTE ZERO

In August 1925, Linus Pauling and his colleague at the California Institute of Technology, Richard C. Tolman, published a paper [21] in which they employed the methods of statistical mechanics to investigate the ways in which the entropy of a supercooled liquid or glass is greater than that of the corresponding crystal at the absolute zero of temperature. In formulating the third law of thermodynamics, Nernst [22] and Planck [23] assumed that a given substance at the absolute zero would have the same entropy in the form of a supercooled liquid as in the crystalline form.

However, Lewis and Gibson subsequently pointed out [24] that a supercooled liquid or glass at the absolute zero might be expected in many cases to have a greater entropy than the corresponding crystal owing to the possible greater randomness in the arrangement of the molecules in the liquid, and Gibson and Giauque showed experimentally [25] that glycerol in the form of a glass does indeed appear to have a greater entropy at the absolute zero than the same amount of glycerol in the crystalline form.

Using the methods of statistical mechanics, Pauling and Tolman demonstrate that the 'excess entropy of a glass is related in a simple and definite manner to the properties of the glass when regarded as a quantum system with its degrees of freedom in the *next to the lowest quantum state*.' Moreover, they show that, contrary to an earlier proposal, 'different crystalline forms of a substance have the same entropy at the absolute zero, irrespective of the number of atoms in the unit of structure of the crystal.'

To prove these conclusions, Pauling and Tolman consider a system containing N molecules in the form of a vapor with potential energy $N\chi$ and N' molecules condensed in the form of a perfect crystal with potential energy $N'\chi'$. Then they show that the logarithm of the vapor pressure p of a perfect crystal at very low temperature is given by

$$\ln p = (\chi - \chi')/kT + (5/2)\ln kT + (3/2)\ln(2\pi M/h^2) - \ln \sigma \tag{1}$$

where M is the molecular mass, k is Boltzmann's constant, T is the absolute temperature, h is Planck's constant, and σ is the symmetry factor for the molecule.

In order to obtain a similar expression for the vapor pressure of a supercooled liquid, they first realize that in a perfect crystal having a given size, shape, and orientation, each molecule occupies a definite equilibrium position about which it will oscillate when the temperature is raised. The total number of exemplars of the crystals possible will be found by permutations of the atoms which do not involve any change in the equilibrium positions of the molecules. However, for the case of a supercooled liquid, especially one composed of complicated molecules, there may exist a number of different ways of fitting the molecules together and still obtaining a glass occupying the same space. Thus, for a supercooled liquid, one must consider rearrangements which involve a shifting of the actual positions occupied by molecules as well as permutations of the atoms without any change in the equilibrium positions of the molecules. These notions then lead to the following expression for the logarithm of the vapor pressure p' of the glass at very low T:

$$\ln p' = (\chi - \chi')/kT + (5/2)\ln kT + (3/2)\ln(2\pi M/h^2) - \ln \sigma$$
$$- d \ln \phi(N')/dN' \tag{2}$$

where $\phi(N')$ denotes the number of possible rearrangements of the molecules of the glass. Nevertheless, the last term in equation (2) must be a constant, say $\ln a$, since they consider a large enough sample of the glass so that the vapor pressure is independent of the size of the sample. Therefore, equation (2) becomes

$$\ln p' = (\chi - \chi')/kT + (5/2)\ln kT + (3/2)\ln (2\pi M/h^2) - \ln \sigma - \ln a \tag{3}$$

Pauling and Tolman then show that, in this expression, the quantity a may be regarded as the average number of ways in which a single molecule can be arranged in the liquid.

To obtain the difference in entropy between the crystal and the glass, they next allow the crystal to form one mole of vapor by reversible evaporation at the vapor pressure and then change the pressure on the gas to the vapor pressure of the glass and carry out a reversible condensation.

The entropy change ΔS is consequently

$$\Delta S = S_{\text{crystal}} - S_{\text{glass}}$$
$$= N_o k(\ln p - \ln p') = R \ln a \tag{4}$$

where N_o is Avogadro's number and R the gas constant. Therefore, Pauling and Tolman conclude that 'the entropy of supercooled liquids at the absolute zero may be expected to vary from zero for simple molecules to a few calories per degree per mole for complicated molecules.'

THE SCREENING CONSTANT OF RELATIVISTIC OR MAGNETIC X-RAY DOUBLETS

On 27 October 1926 the *Zeitschrift für Physik* received from Linus Pauling, who was then residing in Munich as a Fellow of the John Simon Guggenheim Foundation at the Institute for Theoretical Physics, the manuscript bearing the title 'Die Abschirmungskonstanten der relativischen oder magnetischen Röntgenstrahlendubletts (The Screening Constant of Relativistic or Magnetic X-Ray Doublets).' This paper was published [26] in 1926, and it represents the first time that quantum mechanics was applied in an effective way to atoms with more than one electron. Wentzel [27] had previously made a theoretical attempt to treat this problem by assuming that the nuclear charge is reduced by the shielding of the inner electrons, the amount being called the screening constant. However, Wentzel found a large discrepancy from experiment for the values of the screening constant.

In this paper, Pauling demonstrates 'daß die Diskrepanz, die Wentzel fand, nicht reell ist, sondern auf einem Irrtum in seinen Gleichungen beruht (that the discrepancy which Wentzel found is not real but depends on an error in his calculations).' He begins his investigation, as did Wentzel, by summarizing the treatment using the old quantum theory. In this treatment the Hamiltonian contains a potential energy term that is expressed as a series in the effective charges provided by each orbit divided by the radius of the orbit. Then the Wilson–Sommerfeld quantization rules [28, 29] are imposed. This reduction leads to values of the radial coordinate r as a function of the quantum number k of the old quantum theory, enabling one to determine the relativistic or magnetic doublet energy separation:

$$\Delta v = \text{constant}(Z - S)^4/k^2\,n^3 \tag{5}$$

In this expression, Z is the atomic number, S the screening constant, and k and n the azimuthal and principal quantum numbers, respectively, of the old quantum theory. However, he points out that Wentzel made a critical error in his derivation of the screening constant by assuming that the effective angular momentum quantum number is the same in the different regions of space. Pauling does not make this assumption in his derivation of an expression for the value of the screening constant, and he thus arrives at the proper expression of the screening constant [30].

Pauling then proceeds, as Wentzel did, to the quantum mechanical derivation of the value of $\langle r \rangle$, which is given by the integral $\iiint r \, \Psi_{nlm}^2 \, dV$, wherein Ψ_{nlm} is the hydrogenic eigenfunction resulting from the solution of the Schrödinger wave equation [31]. He points out that Waller [32] had provided a method for evaluating this type of intergral, leading to the expression:

$$\langle r \rangle = (a_o n^2/Z) \left\{ 1 + [1 - l(l + 1)/n^2] \right\} \tag{6}$$

where a_o is the Bohr radius ($= 0.5282$ Å), and n and l are the principal and orbital angular momentum quantum numbers, respectively. Pauling remarks that this expression is in complete agreement with the classical expression for the average value of r, except that k^2 has been replaced with $l(l + 1)$. Similar expressions result for the various electron shells, so that he is able to present a table with the values of the theoretically computed screening constant, as well as the values deduced from experiment. He concludes that the agreement between the theoretical and experimental values, while not perfect, is 'recht befriedigend (truly satisfactory)' and the differences must be ascribed to the approximate validity of the derivation of the equation for the screening constant in terms of classical perturbation theory and especially the gross idealization of an electron shell as a surface charge on a sphere [33].

THE INFLUENCE OF A MAGNETIC FIELD ON THE DIELECTRIC CONSTANT OF A DIATOMIC DIPOLE GAS

'The Influence of a Magnetic Field on the Dielectric Constant of a Diatomic Dipole Gas' is the title of a paper published [34] by Linus Pauling in the January 1927 issue of *The Physical Review*. In this paper he shows that according to the old quantum theory, for a diatomic dipole molecule in crossed magnetic and electric fields, there is predicted to be spatial quantization with respect to the direction of the magnetic field for experimentally realizable values of the field strengths, so that the application of a strong magnetic field to a gas such as hydrogen chloride should produce a very large change in the dielectric constant of the gas. However, he finds by applying quantum mechanics to the problem that the dielectric constant does not 'depend on the direction characterizing the spatial quantization, so that no effect of a magnetic field would be predicted.' This is in accord with what is found experimentally. Therefore, the lack of this effect 'provides an instance of an apparently unescapable and yet definitely incorrect prediction of the old quantum theory.'

Pauling begins his investigation with the treatment of the problem by the old quantum theory. In the previous year he had published two papers [35, 36] using this approach and found that though there is a discrepancy between experiment and the results of the old quantum theory, 'it is not easy to reject with the straight-forward and well-grounded quantum theory calculations' [35]. Nevertheless, Pauling presents treatments by both the old quantum theory and the new quantum mechanics 'because the difference in the results is very interesting and furnishes additional evidence for the new mechanics' [34].

According to the old quantum theory, a diatomic molecule without electronic angular momentum and considered as a rigid assembly of charged mass points can be characterized by its moment of inertia I, electric moment μ, and quadrupole moment κ. The diatomic molecule undergoes precession in crossed electric and magnetic fields, with strengths E and H, respectively, according to the equations

$$\omega = 3\mu^2 \, E^2 I/2 \, p^3 \tag{7}$$

$$\omega_M = \kappa H/2Ic \tag{8}$$

where $-\omega_M$ is the angular precessional frequency of the total angular momentum vector about H in the presence of the magnetic field alone; $-\omega \cos \Theta$, in which Θ is the angle between the total angular momentum vector and E in the presence of the electric field alone; c is the velocity of light; $p = jh/2\pi$ is the magnitude of the total angular momentum vector and has a component in the direction of H of magnitude $mh/2\pi$; j and m have half-integral values (*vide infra*); and h is Planck's constant. Furthermore, under the influence of the electric field E the gas becomes electrically polarized in the field direction with the value of the polarization per unit volume given by

$$P = (3E/4\pi)[(\varepsilon - 1)/(\varepsilon + 2)] = N\langle\mu\rangle + N\alpha E \tag{9}$$

where ε is the dielectric constant of the gas, N the number of molecules per unit volume, α the coefficient of induced polarization of the gas, and $\langle\mu\rangle$ the average value for all molecules of μ which is the time-average of $\mu \cos \theta$, θ being the polar angle. The problem in the old quantum theory is to obtain an expression for $\langle\mu\rangle$. It turns out that

$$\langle\mu\rangle = \mu^2 \, EC/kT \tag{10}$$

where

$$C = \frac{\displaystyle\sum_j \sum_m (1/j^2)[(3 \, m^2/j^2) - 1]\exp(-\sigma j^2)}{4\sigma \displaystyle\sum_j \sum_m \exp(-\sigma j^2)} \tag{11}$$

and $\sigma = h^2/8\pi^2 IkT$, with k being Boltzmann's constant and T the absolute temperature. This result had been obtained by Pauling in his 1926 papers [35, 36] and earlier by Pauli [37]. Pauling, however, now evaluates the sum in equation (11) using integral values for j and m, and assumes σ to be very small; in his 1926 papers, Pauling had evaluated C as a function of σ assuming half-integral values for j and m, i.e. that the sum in equation (11) has $j = 1/2, 3/2, 5/2, \ldots, \infty$ and $m = -j, -j + 1, \ldots, 0, \ldots, j - 1, j$, and he gave a table of values of C [35, 36] (Parenthetically, it might be noted that this choice of quantum numbers resulted from the need of the old quantum theory to assign half-integral values of m in order to explain the fine structure in the rotation–oscillation [38] and pure rotation spectrum [39] of hydrogen chloride.)

Next, Pauling employs the old quantum theory to evaluate the effect of a strong magnetic field in addition to the electric field. In this case he finds that

$$\langle\mu_\psi\rangle = [(3 \cos^2 \psi - 1)/2](\mu^2 EC/kT) \tag{12}$$

where ψ is the angle the strong magnetic field makes with the electric field. Thus, he states that '*in the presence of a strong magnetic field making an angle ψ with the electric* *field the polarization due to permanent dipoles will according to the old quantum theory be $(3 \cos^2 \psi - 1)/2$ times its value in the absence of the magnetic field.'* This turns out not to be what is found experimentally.

At this juncture, Pauling proceeds to the treatment of the problem employing the methods of quantum mechanics. Mensing and Pauli [40] and Van Vleck [41] had recently shown that the polarization excited in a molecule by a static external electric field could be calculated according to quantum mechanics. Using the methods of quantum mechanics, Pauling finds that

$$\langle \mu \rangle = (\mu^2 E/3kT) \left\{ \sigma \sum_{j=0}^{\infty} (2j+1)\exp[-\sigma j(j+1)] \right\}^{-1} \tag{13}$$

where in quantum mechanics $j = 0, 1, 2, \ldots, \infty$. He remarks that this expression reduces to the classical equation of Debye for small values of σ, i.e. for high temperatures.

Pauling then considers the influence of a strong magnetic field in addition to the electric field. In this situation he finds that

$$\mu_\psi(j, m) = [(3 \cos^2 \psi - 1)/2] \, \mu(j, m), \qquad j \neq 0 \tag{14}$$

in agreement with the results of the classical theory. In this expression

$$\mu(j, m) = (8 \pi^2 I \, \mu^2 E/h^2)[2j-1)(2j+3)]^{-1}$$
$$\times \{[3 m^2/j (j+1)] - 1\}, \qquad j \neq 0 \tag{15}$$

For $j = 0$,

$$\mu_\psi(0, 0) = \mu(0, 0) = (8 \pi^2 I \, \mu^2 E/3h^2) \tag{16}$$

When summing over j, m in equation (14) to derive $\langle \mu_\psi \rangle$, it is found, as for the case of an electric field alone, that only those molecules in the lowest state contribute to the polarization, which is still given by equation (13) and 'is consequently independent of the direction of quantization.' Thus, *'on the basis of the quantum mechanics a magnetic field should not influence the dielectric constant of a gas such as hydrogen chloride'* [42].

Finally, Pauling remarks that, based on these considerations, an experiment was conducted by Mott-Smith and Daily [43] at the California Institute of Technology to gage the effect of a magnetic field upon the value of the dielectric constant of hydrogen chloride and of nitric oxide. Mott-Smith and Daily found that *'in no case was any change in the dielectric constant detected,* within a limit of error of about one part in 100,000 in ε' [44].

THE THEORETICAL PREDICTION OF THE PHYSICAL PROPERTIES OF MANY-ELECTRON SYSTEMS

On 1 January 1927, Arnold Sommerfeld communicated to the *Proceedings of the Royal Society* in London a paper by Linus Pauling with the title 'The Theoretical Prediction of the Physical Properties of Many-electron Atoms and Ions. Mole Refraction, Diamagnetic Susceptibility, and Extension in Space' [45]. One of the salient features of this paper is that it presented in print for the first time, as far as can be determined, pictures of the radial dependence of the eigenfunctions for hydrogen-like atoms and pictures of the electron radial distribution functions for these eigenfunctions as well.

Pauling begins his investigation by discussing the notion of a screening constant that reduces the effective nuclear charge felt by an electron owing to the screening of the nuclear charge by the other electrons. He notes that Sommerfeld had employed this notion to derive from relativistic considerations with the old quantum theory an equation for the relativistic or magnetic doublet separation in many-electron systems. Furthermore, a similar equation had just been derived by Heisenberg and Jordan [46] from quantum mechanics and the idea of the electron spin, and Pauling describes how it is possible to evaluate theoretically values for the screening constant.

After stating that 'the important problem of the theoretical evaluation of the properties of many-electron atoms and ions has so far received little attention, compared with that devoted to spectral term values,' Pauling proceeds to a derivation from the Schrödinger wave equation for hydrogen-like atoms of the eigenfunctions in terms of Ferrers' associated Legendre functions and Laguerre polynomials. Having obtained this result, he discusses the significance of the conjugate square of the wave function as the electron density and $4\pi r^2 \Psi^*\Psi$ as the electron radial distribution function. He then presents pictures of the dependence of the eigenfunctions on the distance from the nucleus for hydrogen-like atoms and the corresponding electron radial distribution functions. He then discusses various ways for the idealization in many-electron systems of an electron shell as a uniform distribution of electricity on the surface of a sphere. The average value of this shell is given by $\langle r \rangle = \iiint r \, \Psi^*\Psi \, dV$, and the evaluation of this integral is permitted by the work of Waller [32], whose equations lead to the result

$$\langle r \rangle = (a_o n^2/Z)\{1 + [1 - l(l + 1)/n^2]/2\} \tag{17}$$

where $a_o = h^2/4\pi^2 \, \mu e^2$, in which h is Planck's constant, μ is the reduced mass, and e is the charge on the electron, Z is the atomic number, and n and l are the principal and angular momentum quantum numbers, respectively. Pauling points out here that in the old quantum theory, the corresponding expression for $\langle r \rangle$ contained the old quantum theory azimuthal quantum number k squared in place of $l(l + 1)$.

At this stage he considers the quantization of electron orbits, that is, an electron orbit penetrating a number of electron shells. He employs classical mechanics and quantizes with rules of the old quantum theory, the 'values of the azimuthal quantum number k chosen in such a way as to cause our formulas to approximate as closely as possible to the quantum mechanics.' In the resulting formulas he derives he notes that 'in order to approximate as closely as possible to the quantum mechanics, we shall use throughout for k^2 the quantity $l(l + 1)$; for often in the quantum mechanics $l(l + 1)$ occupies the place formerly given to k^2, as we have seen in the case of $\langle r \rangle$.'

With this in hand, Pauling then treats the theoretical determination of the mole refraction R, defined by the equation

$$R = V(v^2 - 1)/(v^2 + 2) = (4\pi N_o/3) \, \alpha \tag{18}$$

where V is the molar volume, v the refractive index, N_o Avogadro's number, and α the polarizability. The polarizability relates the magnitude of the electric field strength $\mathbf{E}$ to the second-order Stark effect energy $(-\alpha E^2/2)$. The problem is then to compute the polarizability. For hydrogen-like atoms, Epstein [47], Wentzel [48], and Waller [32] were able to derive an expression for the polarizability using the Schrödinger wave mechanics, leading to

$$R = [N_o h^6/12(2\pi)^5 \, m^3 \, e^6 \, Z^4] \, n^4 \, (17n^2 - 3m^2 - 9n_3^2 + 19)$$
$$= (0.0470/Z^4)n^4(17n^2 - 3m^2 - 9n_3^2 + 19) \tag{19}$$

where m and n_3 are subsidiary quantum numbers, for the mole refraction. To generalize this result to many-electron systems, Pauling next introduces an averaging procedure for m and n_3^2 and the notion of a mole refraction screening constant S_R. This results in

$$R = \text{constant } n^6(Z - S_R)^{-4} \tag{20}$$

By using his results concerning the quantization of penetrating orbits, he is further able to derive an expression for the value of S_R. Pauling then presents a table wherein the values of the mole refraction screening constant are compared with those derived from the experimental values of R for a number of atoms and ions. it is seen from this comparison that 'for elements with only a few electrons the agreement is complete, and that it becomes less satisfactory as the electron number of the structure increases.' He attributes the disagreement for ions with many electrons to the assumption that the value of the screening constant is independent of Z. He also presents a table in which the theoretical values of R (which apply only to free ions in the gaseous states) for various univalent ions are compared with the experimental values of R for dilute solutions of the alkali halides and for alkali halide crystals, and he concludes that 'our predicted values for gaseous ions show that ions in solution are indeed more similar to gaseous ions than are ions in crystals, as far as the mole refraction is concerned.' Moreover, because of the possible dependence of S_R on Z, Pauling next assumes that S_R is a linear function of Z, and he uses the solution values for the bromide and iodide ions to estimate the dependence of S_R on Z. Having done this, he presents a table of predicted values of the mole refraction for many atoms and univalent ions; these values are desirable because 'they may be compared with core polarizabilities deduced from the energy levels of non-penetrating alkali-like electron states in order to test the spectral theory used in the deduction.'

At this point, Pauling proceeds to a consideration of the diamagnetic susceptibility. This quantity depends on $\langle r^2 \rangle$. Using the same concepts he employed for the mole refraction, he obtains an expression for $\langle r^2 \rangle$ in terms of a diamagnetism screening constant and wherein the old quantum theory k^2 is replaced by $l(l + 1)$. This permits him to present a table comparing the theoretical values of the diamagnetism screening constants with those derived from experiment. Again, the agreement is excellent for elements with only a few electrons, and an increasing error is found for increasing electron number. This prompts him to correct empirically for the dependence of the values of the screening constant on atomic number assuming the same constancy of the ratios of the screening constants as for the mole refraction screening constants and then to provide values for the diamagnetic susceptibilities of atoms and univalent ions. The agreement with the experimental values of the diamagnetic susceptibility, where available, allows him to conclude that 'our theoretical values of the diamagnetic susceptibilities of atoms and ions are not incompatible with the experimental data.'

Pauling next explores the effect of the notion of the screening constant on the electron distribution in atoms and ions, i.e. 'atomic sizes'. This involves the integral

$\iiint r \, \Psi^* \Psi \, dV$, and his previous considerations permit him to derive values of the size screening constant, which again can be corrected for dependence on Z. Using these values he is able to compute the electron radial distribution function for atoms and univalent ions, and he presents pictures of the electron radial distribution function for Na^+ and for Cl^- along with those obtained from X-ray crystallographic studies on sodium chloride crystals. He concludes that 'we may accordingly say that an atom is composed of a nucleus embedded in a ball of electricity (the two K electrons with small contributions from other shells), which in turn is surrounded by more or less distinctly demarcated thick concentric shells, containing essentially the L, M, N, etc., electrons.' Furthermore, 'the high maximum of the electron density at the nucleus given by our calculations provides considerable justification for the method of determining crystal structures with the aid of the relative intensities of Laue spots produced by crystal planes with complicated indices.'

Finally, Pauling treats the subject of interatomic distances in terms of the forces between atoms. In this case he is concerned with the integral for the electron repulsion force. He points out that the derivative with respect to internuclear separation of the potential energy is zero at the equilibrium separation, and by considering the hydrogen halides and using hydrogenic eigenfunctions, he solves the derivative equation to obtain values for the equilibrium internuclear distances in certain hydrogen halides under the assumption that the halogen ion is not deformed by the hydrogen ion. He thus obtains a value in agreement with experiment for HF and values somewhat larger than the experimental ones for HCl and HBr, indicating 'that the deforming effect of the hydrogen ion on the halide ions is of greater relative importance for these ions than for the fluoride ion.'

Pauling concludes this paper with a note added on 10 February 1927 stating that in December 1926, Van Vleck 'has discussed the mole refraction and the diamagnetic susceptibility of hydrogen-like atoms with the use of the wave mechanics, obtaining results identical with our equations' for such atoms [49].

THE QUANTUM MECHANICAL TREATMENT OF THE HYDROGEN MOLECULE AND THE HYDROGEN MOLECULE-ION

In the June 1928 issue of *Chemical Reviews*, Linus Pauling published a paper with the title 'The Application of the Quantum Mechanics to the Structure of the Hydrogen Molecule and Hydrogen Molecule-ion and to Related Problems' [50]. One of the interesting aspects of this paper is that Pauling introduced, for the first time, the notion that the Pauli exclusion principle can be satisfied by constructing a determinant (now called the 'Slater determinant') of spin-orbit functions. Slater, in fact, pointed out in his scientific autobiography [51] that Pauling was the first person to show this.

Pauling begins this paper by relating that many attempts had been made to describe in terms of the old quantum theory the structures of the hydrogen molecule, the hydrogen molecule-ion, and the helium atom. All of these attempts were unsuccessful in that they led to results that were definitely incompatible with the observed properties of these substances, and this fact was one of those that led to the rejection of the old quantum theory and the introduction of the new quantum mechanics, which, 'in contradiction to the old quantum theory leads to results in agreement with experiment within the limit of error of the calculation.'

Before attacking the problem of applying quantum mechanics to the hydrogen molecule and hydrogen molecule-ion, Pauling gives a summary of such observed properties of these systems as ionization potentials, heats of dissociation, frequencies of nuclear oscillation, and moments of inertia. Then he presents the now well-known quantum mechanical treatment of the hydrogen atom, leading to eigenfunction solutions of the Schrödinger equation in terms of the Laguerre polynomials and Ferrers' associated Legendre functions, and he presents an illustration showing the eigenfunction, electron density, and electron radial distribution function as functions of the distance from the nucleus for the ground state of the hydrogen atom. He emphasizes that this treatment does not account for the fine-structure of the hydrogen spectrum unless one introduces the notion that the electron has a spin, whose vector can take either one of two possible orientations in space, as deduced by Uhlenbeck and Goudsmit [52] from the empirical study of line spectra, and he states that 'this result is of particular importance for the problems of chemistry.'

Pauling then proceeds to the treatment of two-electron systems by means of Schrödinger perturbation theory [53] and points out that the energy of such systems is lowered by an amount, called the *resonance energy*, due to the interchange of electrons in the wave function, over the energy given by eigenfunctions in which there is no interchange of electrons. The resonance phenomenon was discovered in 1926 by Heisenberg [54] and by Dirac [55], and Pauling emphasizes that '*the interchange energy of electrons is in general the energy of the non-polar or shared-electron chemical bond.*'

Next, Pauling attacks the problem of the hydrogen molecule-ion using the Born–Oppenheimer approximation [56] of fixed nuclei. By transforming the Schrödinger equation into elliptical coordinates, he shows that it becomes separable into three differential equations, wherein the energy must be determined as a function of the internuclear separation. He states that 'many efforts have been made to solve these equations analytically, but so far they have all been unsuccessful, and little has been published regarding them ... It is probable, in view of the vigor with which it is being attacked, that the problem will be solved before very long.' Nevertheless, Pauling remarks that the problem has already been solved numerically by Burrau [57] for the normal hydrogen molecule-ion, and he gives a detailed account of these calculations 'since the journal in which they were published is often not available.' He then discusses Burrau's results in terms of the energy of H_2^+ as a function of internuclear separation and its relationship to the heat of dissociation of H_2 and H_2^+. He also presents illustrations showing Burrau's results for the electron density contours and the average electron density as a function of position and comments that 'it will be seen that the electron is most of the time in the region between the nuclei, and can be considered as belonging to them both, and forming a bond between them.'

At this point, Pauling goes on to the treatment of the hydrogen molecule-ion by means of perturbation theory. This is a treatment that has not been heretofore published. He finds that the 'resonance energy leads molecule formation only if the eigenfunction is symmetric in the two nuclei.' He also remarks on the inadequacies of a number of calculations made by other researchers.

It is at this juncture that Pauling proceeds to the quantum mechanical treatment of the hydrogen molecule. This treatment was to form the basis of the valence-bond, or Heitler–London–Slater–Pauling, method. The attack is that of Heitler and London [58] in terms of the first-order perturbation treatment of the interaction of two hydrogen atoms. He begins with two hydrogen atoms, and the two electrons are interchanged within the framework of the Born–Oppenheimer approximation. The Schrödinger equation leads to a sum of integral equations, the most difficult of which was solved analytically by Sigiura [59]. The first-order perturbation treatment yields results only in approximate agreement with experiment, as to be expected from the similar treatment of the hydrogen molecule-ion. Nevertheless, from an analysis of the results, Pauling finds that the 'conclusion can be drawn that in the hydrogen molecule the interchange energy of the two electrons is the principal cause of the forces leading to molecule formation.' He then discusses the excited states of the hydrogen molecule, and he notes that one of the excited states has a very large equilibrium distance and small oscillational frequency, suggesting that 'the molecule is here not non-polar, but is a polar compound of H^+ and H^-.'

Finally, Pauling discusses the application of quantum mechanics to systems containing more than two electrons. It is for such systems that one must explicitly take into account the spin of the electron in order to satisfy the Pauli exclusion principle [60], and he states: '*Only eigenfunctions which are antisymmetric in the electrons; that is, change sign when any two electrons are interchanged, correspond to existant* [sic] *states of the system.*' He mentions that for the two-electron systems already considered, the Pauli exclusion principle is satisfied fortuitously. However, 'if the system contains more than two electrons explicit consideration must be given the spins.' As an example, he gives the problem of the interaction of two helium atoms. In this case, there are the four spin eigenfunctions $\psi\alpha$, $\psi\beta$, $\phi\alpha$, and $\phi\beta$ to be occupied by the four electrons. According to him, 'the only eigenfunction allowed by Pauli's principle for the system is

$$\Psi(He_2) = a \begin{vmatrix} \psi(1)\alpha(1) & \psi(1)\beta(1) & \phi(1)\alpha(1) & \phi(1)\beta(1) \\ \psi(2)\alpha(2) & \psi(2)\beta(2) & \phi(2)\alpha(2) & \phi(2)\beta(2) \\ \psi(3)\alpha(3) & \psi(3)\beta(3) & \phi(3)\alpha(3) & \phi(3)\beta(3) \\ \psi(4)\alpha(4) & \psi(4)\beta(4) & \phi(4)\alpha(4) & \phi(4)\beta(4) \end{vmatrix}$$

(a is a factor of such value as to make the eigenfunction normalized). It will be seen that this is antisymmetric, for interchanging any two electrons is equivalent to interchanging two rows of the determinant, and hence changing its sign.' This is the first mention in the literature of the so-called 'Slater determinant'. Using perturbation theory Pauling shows that in He_2, 'there are no forces tending to molecule formation, but instead repulsion at all distances.' He concludes the paper with the notions that the interaction of two alkali metal atoms should be similar to that of two hydrogen atoms, owing to the completed shells of the ions, and that atoms of the second column of the periodic table, such as mercury, should interact in a way similar to two helium atoms, so that 'the attractive forces would be at most very small.'

THE MOMENTUM DISTRIBUTION IN HYDROGEN-LIKE ATOMS

In the 1 July 1928 issue of *The Physical Review*, Boris Podolsky, later to gain fame for the Einstein–Podolsky–Rosen paradox, and Linus Pauling published a paper entitled 'The Momentum Distribution in Hydrogen-like Atoms' [61]. In this paper, Podolsky and Pauling derived for the first time the momentum eigenfunctions for hydrogen-like atoms, and they demonstrated that the root mean square of the total momentum of the electron is equal to the momentum of the electron in a circular Bohr orbit with the same quantum number.

Podolsky and Pauling begin their investigation by observing that the normal eigenfunction $\Psi_{nlm}(r, \theta, \phi)$ is the transformation function in the Dirac transformation theory from the Cartesian coordinates of the electron to the quantum numbers, and it may be denoted by $(x, y, z/n, l, m)$. Similarly, the transformation function from the momenta p_x, p_y, p_z can be denoted by $(p_x, p_y, p_z/n, l, m)$ and can be used to derive a distribution function in momentum space, namely the momentum distribution function. If the momentum distribution function is known for a set of values of n, l, and m, then it is an easy matter, according to them, to calculate the probability that the electron has a total momentum lying within a given range.

They then note that the momentum transformation function $(p_x, p_y, p_z/n, l, m)$ can be obtained from the Cartesian transformation function $(x, y, z/n, l, m)$ according to the work of Jordan [62]:

$$(p_x, p_y, p_z/n, l, m) = h^{-3/2} \int_{-\infty}^{\infty} \int_{-\infty}^{\infty} \int_{-\infty}^{\infty} \exp[(-2\pi i/h)(xp_x + yp_y + zp_z)]$$
$$\times (x, y, z/n, l, m)\, dx\, dy\, dz \tag{21}$$

where h is Planck's constant. Consequently, by transforming to spherical coordinates, the momentum eigenfunction $\Upsilon_{nlm}(P, \Theta, \Phi)$, where P is the total momentum vector and the angles Θ and Φ denote the orientation of the momentum vector relative to the Cartesian coordinates, becomes:

$$\Upsilon_{nlm}(P, \Theta, \Phi) = h^{-3/2} \int_{-\infty}^{\infty} \int_0^{\pi} \int_0^{2\pi} \exp\{(-2\pi i/h)[\sin\theta \sin\Theta \cos(\Phi - \phi)$$
$$+ \cos\theta \cos\Theta]rP\}\Psi_{nlm}(r, \theta, \phi)\, r^2 \sin\theta\, dr\, d\theta\, d\phi \tag{22}$$

wherein $\Psi_{nlm}(r, \theta, \phi)$ is the normal hydrogenic eigenfunction in terms of Ferrers' associated Legendre polynomials of degree l and order m and the associated Laguerre polynomial depending on l [63].

Next, Podolsky and Pauling rewrite the momentum eigenfuncton as

$$\Upsilon_{nlm}(P, \Theta, \Phi) = \frac{h^{-3/2}}{(2\pi)^{1/2}} \left(\frac{(2l+1)(l-m)!}{2(l+m)!}\right)^{-1/2} \left(\frac{(2\gamma)^{l+1}}{(n+1)!}\right)$$
$$\times \left(\frac{\gamma(n-l-1)!}{n(n+l)!}\right)^{-1/2}$$
$$\times \int_0^{\infty} I_2 \exp(-\gamma r)r^{l+2}L_{n+l}^{2l+1}(2\gamma r)\, dr \tag{23}$$

where

$$I_2 = \int_0^{\pi} \{\exp[(-2\pi i/h)]rP\}\cos\theta \cos\Theta\, I_1 P_l^m(\cos\theta)\sin\theta\, d\theta \tag{24}$$

$$I_1 = \int_0^{2\pi} \exp[\pm im\phi - (2\pi i/h)rP \sin\theta \sin\Theta \cos(\Phi - \phi)]d\phi \tag{25}$$

$$\gamma = 4\pi^2 \mu e^2 Z/nh^2 = Z/na_o \tag{26}$$

$P_l^m(\cos\theta)$ and $L_{n+l}^{2l+1}(2\gamma r)$ are the Ferrers' associated Legendre and Laguerre polynomials, respectively, μ is the reduced mass, Ze the nuclear charge, and a_o the Bohr radius 0.5282 Å.

Following a series of abstruse mathematical transformations, Podolsky and Pauling arrive at the following expression for the momentum eigenfunctions of hydrogen-like atoms:

$$\Upsilon_{nlm}(P,\Theta,\Phi) = \frac{\exp(\pm im\Phi)}{(2\pi)^{1/2}}\left(\frac{(2l+1)(l-m)!}{2(l+m)!}\right)^{1/2} P_l^m(\cos\theta)$$

$$\times \left(\frac{-(-i)^l \pi 2^{2l+4} l!}{(\gamma h)^{3/2}}\right)\left(\frac{n(n-l-1)!}{(n+l)!}\right)^{1/2}\left(\frac{\zeta^l}{(\zeta+1)^{l+2}}\right)$$

$$\times C_{n-l-1}^{l+1}[(\zeta^2-1)/(\zeta^2+1)] \tag{27}$$

in which

$$\zeta = 2\pi P/\gamma h = nP/Zp_o = P/p_n \tag{28}$$

$C_{n-l-1}^{l+1}[(\zeta^2-1)/(\zeta^2+1)]$ are the Gegenbauer C functions [64], p_o $(=2\pi\mu e^2/h)$ is the momentum of the electron in a circular Bohr orbit with $n=1$ and $Z=1$, corresponding to a hydrogen atom in the normal state, and p_n $(=Zp_o/n)$ is the momentum of the electron in a circular Bohr orbit characterized by the total quantum number n and the nuclear charge Ze. Podolsky and Pauling discuss a number of ways of obtaining the Gegenbauer $C_k^v(x)$ functions and provide a table containing specific values of them for particular values of n and l.

Then they proceed to evaluate the probability that the electrons have a momentum lying in the range between P and dP. This is given by

$$\Xi_{nl}(P) = \int_0^\pi \int_0^{2\pi} |\Upsilon_{n\chi}(\Psi,\Theta,\Phi)|^2 P^2 \sin\Theta \, d\Theta \, d\Phi \tag{29}$$

which becomes upon integration

$$\Xi_{nl}(P) = (a_o/Zh)\left(\frac{2^{4l+6}n^2(l!)^2(n-l-1)!}{(n+l)!}\right)\left(\frac{\zeta^{2l+2}}{(\zeta^2+1)^{2l+4}}\right)$$

$$\times \{C_{n-l-1}^{l+1}[(\zeta^2-1)/(\zeta^2+1)]\}^2 \tag{30}$$

The diagonal elements of the P^2 matrix are equal, according to Podolsky and Pauling, to the average values of the square of the momentum in the various quantum states:

$$\langle P_n^2 \rangle = \int_0^\infty P^2 \Xi_{nl}(P) \, dP = p_n^2 = (2\pi\mu e^2 Z/nh)^2 \tag{31}$$

They therefore conclude that 'p_n^2 is just the average value of the square of the momentum of the electron in a Bohr orbit with total quantum number n; so that the root mean square momentum for a hydrogen-like atom is the same in the quantum mechanics as in the old quantum theory, in each case depending only on the principal quantum number n.'

THE ROTATIONAL MOTION OF MOLECULES IN CRYSTALS

On 1 August 1930, Linus Pauling published a paper [65] entitled 'The Rotational Motion of Molecules in Crystals' in *The Physical Review*, in which he applied the methods of quantum mechanics to deduce the entropy of molecules in crystals. The impetus for this investigation was a calculation by Giauque and Johnston [66] on the difference in entropy of gaseous molecular hydrogen and crystalline hydrogen. This calculation showed a deviation from value zero expected for crystalline hydrogen at absolute zero. Pauling found that the introduction of quantum mechanics to the statistical mechanics of molecules in crystals yields that the allowed states of the system can approximate either of two extremes, rotation and oscillation of the atoms comprising the molecule, or 'can lie between these two extremes, approximating neither more closely than the other.' Unlike the situation resulting from classical mechanics, the quantum mechanical description shows that 'the transition from one extreme to the other is unbroken.'

Pauling begins his investigation by considering a crystal composed of diatomic molecules arranged in their equilibrium orientation. If the crystal is characterized by having two equilibrium orientations about the polar axis, measured by the polar angle θ, possible for the molecules (as in orthorhombic crystalline iodine), the average potential interaction of a given molecule with the surrounding molecules can be expressed by the periodic function

$$V = V_0(1 - \cos 2\theta) \tag{32}$$

in which V_0 is a constant. This leads to the Schrödinger wave equation

$$(1/\sin^2 \theta)\,(\partial^2\psi/\partial\phi^2) + (1/\sin \theta)\{\partial[\sin \theta(\partial\psi/\partial\theta)]/\partial\theta\}$$
$$+ (8\pi^2 I/h^2)[E - V_0(1 - \cos 2\theta)] = 0 \tag{33}$$

where I is the moment of inertia of the molecule, h is Planck's constant, and ϕ the azimuthal angle. Pauling then points out that while this eigenvalue equation had not been solved, except for limiting cases, in the situation for a molecule with motion restricted to a plane, equation (33) becomes Mathieu's equation:

$$\partial^2\psi/\partial\theta^2 + (4\alpha + 16q \cos 2\theta)\psi = 0 \tag{34}$$

wherein

$$\alpha = (2\pi^2 I/h^2)\,(E - V_0) \tag{35}$$

and

$$q = (\pi^2 I V_0/2h^2) \tag{36}$$

Pauling observes that equation (34), coupled with the requirement that the wavefunction be periodic in θ, yields solutions known as the Mathieu functions [67]. For $q = 0$, equation (34) reduces to the equation for the plane rotator, with energy levels

$$E_n = (n^2 h^2/8\pi I) \tag{37}$$

For q large, on the other hand, the eigenfunction ψ is approximately a combination of Hermite orthonormal functions, namely the eigenfunctions for the harmonic oscillator, with energy levels given by

$$E_n = (n + 1/2)hv_0 \tag{38}$$

in which

$$v_0 = h(2q)^{1/2}/\pi I \tag{39}$$

Pauling then points out that the investigation in terms of the orientation is not essential and that, 'We can define the motion of the molecule in a given state as oscillational in case the eigenfunctions for that state can be closely approximated by a combination of Hermite functions and the energy of the state is given approximately as $(n + 1/2)hv_0$. For rotational motion the eigenfunction and energy level should approximate those for a free rotator.'

Next, Pauling evaluates the conditions under which these contingencies prevail. He finds

$$n + 1 < 2\pi(IV_0)^{1/2}/h \leftrightarrow \text{oscillational motion}$$

$$n + 1 > 2\pi(IV_0)^{1/2}/h \leftrightarrow \text{rotational motion}$$

and estimates the uncertain quantity V_0 in these expressions from the heat capacity for the solid (using the Einstein model) for various molecules. This, coupled with the moment of inertia obtained from band spectral data, allows him to present a table in which the values of $n_0 + 1$, at which the transition from oscillational to rotational motion occurs, are listed for various molecules. The value of $n_0 + 1$ for H_2 in this table 'shows that *even in the lowest state* the molecules are rotating freely, the intermolecular forces producing only small perturbations from uniform rotation.'

On the other hand, the other extreme occurs for I_2 in that the transition from oscillation to rotation takes places at about $n = 300$. Pauling asserts that at the melting point of I_2, the molecules are in states of n between 10 and 15, 'so that there are no rotating molecules in this crystal. This agrees with the fact that equilibrium positions for the atoms have been found by X-ray methods.'

Pauling then proceeds to the calculation of the entropy of ordinary crystalline hydrogen, in which there are three molecules with rotational quantum number $j = 1$ for every one with $j = 0$. He observes that the associated eigenfunctions approximate those of the free molecules, so that the rotating molecules interact with each other as though they were nearly spherically symmetrical, implying that the crystal should have a close-packed structure, in agreement with the known cubic symmetry of ordinary crystalline hydrogen. The entropy of mixing of ordinary crystalline hydrogen thus has the value $-(1/4)R \ln(1/4) - (3/4)R \ln(3/4)$. The antisymmetrical molecules have a quantum weight 1 in the normal rotational state ($j = 0$), and the symmetrical molecules have a quantum weight 9, corresponding to the three rotational eigenfunctions with $j = 1$, each of which is associated with any of the three spin functions [68]. Therefore, he finds that the entropy of solid hydrogen at temperatures just below the melting point is

$$S = -(1/4)R \ln(1/4) - (3/4)R \ln(3/4) + (3/4)R \ln 9 + S_{tr}$$
$$= 4.39 \text{ e.u.} + S_{tr} \tag{40}$$

in which S_{tr} is the translational entropy. This value agrees well with the experimental value $4.3 \pm 0.1 + S_{tr}$ of Giauque and Johnston [66].

At very low temperatures ($<5K$) the solid solution of H_2 becomes unstable relative to phases of definite composition, and the entropy falls to

$$S = n_A R \ln 3 + S_{tr} \tag{41}$$

where n_A is the mole-fraction of symmetric molecules, since the entropy of mixing and of the quantum weight 3 for $j = 1$ are lost at the same time. Pauling notes that only at temperatures less than about 0.001 K will the spin-quantum-weight entropy be lost.

A consideration of his calculated values of $n_0 + 1$ for CH_4, N_2, O_2, and the hydrogen halides allows Pauling to further predict that these molecules 'oscillate at low temperatures but go over mainly to rotational states before the melting point is reached.' Thus he predicts that 'a methane crystal between 20 and 90.6 K (the melting point) should consist of rotating molecules in cubic close-packing; below 20 K the tetrahedral molecules oscillate about equilibrium positions.' Pauling concludes that, 'In general it is to be expected that rotational motion of molecules and complex ions of sufficiently low moment of inertia will set in below the melting point of the crystals.' He points out that such a condition is met by molecules containing hydrogen atoms and one heavy atom.

THE ENTROPY OF ICE

In December 1935, Linus Pauling published an article in the *Journal of the American Chemical Society* entitled 'The Structure and Entropy of Ice and of Other Crystals with Some Randomness of Atomic Arrangement' [69]. This article, in which statistical methods that had much bearing on his later work on the nature of metals and intermetallic compounds are employed, provides an explanation for the observed residual entropy of ice at low temperatures in terms of structural randomness.

Pauling begins his investigation by stating that 'the arrangement of oxygen atoms (but not of hydrogen atoms) in crystals of ice is known'. However, he observes, there is a question concerning whether a given hydrogen atom is located midway between the two oxygen atoms it connects or is closer to one than to the other. He answers that it is the latter situation that prevails, i.e. in ice each hydrogen atom is located about 0.95 Å from one oxygen atom and about 1.81 Å from another.

He then makes the following assumptions concerning the structure of ice. (1) In ice, each oxygen atom has two hydrogen atoms bonded to it at distances of about 0.95 Å, forming a water molecule with the HOH angle being about 105°, as in the gas molecule. (2) The orientation of each water molecule in ice is such that its two hydrogen atoms are directed approximately toward two of the four oxygen atoms surrounding it tetrahedrally, forming hydrogen bonds. (3) Adjacent water molecules are oriented such that only one hydrogen atom lies approximately along each oxygen–oxygen axis. And (4) Under normal circumstances the interaction between non-adjacent molecules is not sufficient to appreciably stabilize any one of the many configurations satisfying the previous conditions over the other possible configurations.

These conditions imply that an ice crystal can exist in any one of a large number of configurations, and the crystal can change from one configuration to another either by rotation of some of the molecules or by the motion of the hydrogen nuclei, with each moving a distance of about 0.86 Å away from the potential minimum situated at

0.95 Å from one oxygen atom to another one 0.95 Å from an adjacent oxygen atom. Thus, on cooling to low temperatures, the ice crystal freezes into one of the possible configurations, but in a reasonable period of time it does not become a perfect crystal with no randomness of molecular orientation. The number of possible configurations available to a crystal of ice containing N water molecules can be simply calculated. According to the second assumption a particular water molecule can orient itself in any of six ways. Nevertheless, he notes that the probability that the nearest-neighbor molecules will permit a given orientation is 1/4 since each adjacent molecule has four tetrahedral directions, two of which are occupied by hydrogen atoms and two of which are vacant, yielding a probability of 1/2 that a given direction is available for each hydrogen of the original molecule and a probability of 1/4 that both hydrogens can be situated in the given orientation. Therefore, the total number of configurations W possible for the N molecules comprising the crystal is $(6/4)^N = (3/2)^N$, and the resulting entropy S is

$$S = k \ln W = k \ln(3/2)^N = R \ln(3/2) \tag{42}$$

where R is the gas constant. Pauling also offers an alternative derivation of equation (42) using assumption (1).

The residual entropy of ice due to this lack of regularity is thus calculated to be $R \ln (3/2) = 0.805$ e.u. while the observed entropy discrepancy of ice at low temperatures, based on the work of Giauque and Ashley [70], is 0.87 e.u. Pauling concludes that the agreement in the theoretical and experimental entropy values provides strong support for the postulated structure of ice [71].

He further predicts that, owing to the existence of hydrogen bonds that permit some randomness of atomic arrangement, many other crystals will be found to have residual entropy at very low temperatures. As examples he cites diaspore ($AlHO_2$) and staurolite ($Al_5FeHO_{13}Si_2$), both of which contain OHO groups with hydrogen bonds, and he predicts a residual entropy of $R \ln 2$ for these crystals, using arguments similar to those for ice. For formic acid and other monocarboxylic acids forming dimers, he predicts a residual entropy of $(1/2) R \ln 2$ since the interaction within a carboxyl group is expected to be strong enough to allow only the two configurations in which one hydrogen nucleus is attached to each carboxyl group to be stable at low temperatures.

However, Pauling points out that 'hydrogen bonds may not always lead to residual entropy.' In the crystal lepidocrocite, $FeHO_2$, for example, there exist infinite strings of oxygen atoms joined by hydrogen bonds, and there are only two accessible configurations available per string — a lack of definiteness that 'does not lead to an appreciable residual entropy.' Finally, he is not able to 'predict with confidence' whether 'the hydrogen nucleus between two fluorine atoms connected by a hydrogen bond has the choice of two positions (as for oxygen atoms) or not,' leading to a potential function for the hydrogen nucleus which has two minima rather than one, but he does state that 'hydrogen bonds between unlike atoms, as in NH_4F, would not lead to residual entropy.'

CONCLUSIONS

The scientific contributions discussed in this chapter are of such important and lasting quality that were he to have done nothing else in science, Linus Pauling would be worthy of renown in the fields of theoretical chemistry and physics. These contributions reveal his innovative thought processes, scientific acumen, and careful consideration of mathematical detail.

ACKNOWLEDGMENTS

This chapter is dedicated to Professor Linus Pauling on the occasion of his ninetieth birthday. The author thanks him for a critical reading of the manuscript and for many stimulating conversations regarding the subject matter discussed herein.

The date of Professor Pauling's ninetieth birthday, 28 February 1991, very nearly coincides with the tenth anniversary of the author's association with Professor Pauling as his collaborator and assistant in the fields of theoretical chemistry and physics and biostatistics. It has been a fruitful collaboration with one of history's greatest scientists and a great human being as well. The debt of gratitude owed to Professor Pauling by the author for his mentorship can never be repaid.

The author thanks Professor Zvonimir B. Maksić for his kind hospitality in the past and for inviting the author to write this chapter. *Hvala!*

REFERENCES AND NOTES

[1] G. Albrecht, 'Scientific Publications of Linus Pauling', in *Structural Chemistry and Biology: A Volume Dedicated to Linus Pauling by his Students, Colleagues, and Friends*, A. Rich and N. Davidson (Eds), W.H. Freeman, San Francisco, 1968, pp. 888–907.

[2] Z.S. Herman and D.B. Munro, *Croatica Chemica Acta* **61**, C9–C34 (1988).

[3] Z.S. Herman and D.B. Munro, 'The Publications of Professor Linus Pauling from 1920 to 1990', in *The Pauling Catalogue: Ava Helen and Linus Pauling Papers at Oregon State University*, C.S. Mead, J. Wallace, Z.S. Herman, and D.B. Munro (Eds), Kerr Library Special Collections, Oregon State University, Corvallis, Oregon, 1991, pp. 207–305.

[4] L. Pauling, *The Nature of the Chemical Bond and the Structure of Molecules and Crystals: An Introduction to Modern Structural Chemistry*, Cornell University Press, Ithaca, New York, 1st edition, 1939, 419 pp; 2nd edition, 1940, 450 pp; 3rd edition, 1960, 644 pp. Translated into French, German, Japanese, Russian, and Spanish.

[5] Z.S. Herman, 'The Twenty-five Most Cited Publications of Linus Pauling', in *The Roots of Molecular Medicine: A Tribute to Linus Pauling*, R.P. Huemer (Ed.), W.H. Freeman, New York, 1986, pp. 254–259.

[6] *The Scientist* **3(2)**, 10 (23 Jan. 1989); *Current Contents (Physical Sciences)* **29(34)**, 11 (21 Aug. 1989).

[7] E. Garfield, *Current Contents (Physical Sciences)* **29(34)**, 3 (21 Aug. 1989).

[8] S.G. Brush, *Current Contents (Physical Sciences)* **30(20)**, 7 (14 May 1990).

[9] L. Pauling and E. Bright Wilson, Jr., *Introduction to Quantum Mechanics, with Applications to Chemistry*, McGraw-Hill, New York, 1935, 468 pp; Dover Books, New York, 1985, 468 pp. Translated into Italian and Japanese.

[10] L. Pauling, with illustrations by R. Hayward, *No More War!*, Dodd, Mead & Co., New York, 1958, 254 pp; Enlarged Apollo Edition: Dodd, Mead & Co.,

New York, 1962, 262 pp; 25th Anniversary Edition: Dodd, Mead & Co., New York, 1983, 304 pp. Translated into German, Japanese, Russian, Slovakian, and Swedish.

[11] L. Pauling, *General Chemistry: An Introduction to Descriptive Chemistry and Modern Chemical Theory*, W.H. Freeman, San Francisco, 1st edition, 1947, 618 pp; 2nd edition, 1953, 710 pp; 3rd edition, 1970, 959 pp; Dover edition: Dover Publications, Inc., New York, 1988, 959 pp. Translated into French, German, Gujurati, Hebrew, Italian, Japanese, Portuguese, Rumanian, Russian, Sinhala, Spanish, and Swedish.

[12] L. Pauling, with illustrations by R. Hayward, *College Chemistry: An Introductory Textbook of General Chemistry*, W.H. Freeman, San Francisco, 1st edition, 1950, 705 pp; 2nd edition, 1955, 685 pp; 3rd edition, 1964, 832 pp. Translated into Hindi and Japanese.

[13] L. Pauling and Peter Pauling, *Chemistry*, W.H. Freeman, San Francisco, 1975, 767 pp. Translated into Polish and Russian.

[14] L. Pauling and R. Hayward, *The Architecture of Molecules*, W.H. Freeman, San Francisco, 1964, 107 pp. Translated into German and Japanese.

[15] L. Pauling, *Vitamin C and the Common Cold*, W.H. Freeman, San Francisco, 1970, 122 pp; Bantum Books, New York, 1971, 112 pp; Ballantine Books, London, 1972, 132 pp. Translated into Danish, Dutch, French, German, Japanese, Norwegian, Portuguese, and Swedish.

[16] D. Hawkins and L. Pauling (Eds), *Orthomolecular Psychiatry: Treatment of Schizophrenia*, W.H. Freeman, San Francisco, 1973, 697 pp.

[17] L. Pauling, *Vitamin C, the Common Cold, and the Flu*, W.H. Freeman, San Francisco, 1976, 230 pp. Translated into Japanese and Portuguese.

[18] E. Cameron and L. Pauling, *Cancer and Vitamin C*, Linus Pauling Institute of Science and Medicine, Palo Alto, CA, 1979, 238 pp; Warner Books, New York, 1981, 231 pp. Translated into French and Japanese.

[19] L. Pauling, *How to Live Longer and Feel Better*, W.H. Freeman, New York, 1986, 322 pp; Avon Books, New York, 1987, 413 pp. Translated into Croatian, French, German, Japanese, and Portuguese.

[20] Pauling was not allowed to fulfill the requirements for high school graduation of having taken two terms of American history. He wanted to take these courses concurrently instead of sequentially, as stipulated by school authorities. However, he was awarded an honorary high school diploma from Washington High School by the Portland Board of Education in 1962. According to his official biographer, Professor Robert J. Paradowski, Pauling is the only person ever to have been awarded an honorary diploma by Washington High School.

[21] L. Pauling and R.C. Tolman, *J. Am. Chem. Soc.* **47**, 2148 (1925).

[22] W. Nernst, *Nachr. Kgl. Ges. Wiss. Göttingen, Math.-physik. Klasse* **1906**, 1.

[23] M. Planck, *Thermodynamik*, Walter de Gruyter and Co., Berlin, 6th edition, 1921, p. 273.

[24] G.N. Lewis and G.E. Gibson, *J. Am. Chem. Soc.* **42**, 1529 (1920).

[25] G.E. Gibson and W.F. Giauque, *J. Am. Chem. Soc.* **45**, 93 (1923).

[26] L. Pauling, *Z. Physik* **40**, 344 (1926).

[27] G. Wentzel, *Z. Physik* **37**, 911 (1926).

[28] W. Wilson, *Phil. Mag.* **29**, 795 (1915).

[29] A. Sommerfeld, *Annal. Physik* **51**, 1 (1916).
[30] Pauling has a footnote here stating that 'Herr Wentzel stimmt nach freundlicher Rücksprache mit dem Verfasser hierin überein (Following amiable consultation Mr. Wentzel agrees with the author on this).'
[31] E. Schrödinger, *Annal. Physik* **79**, 36 (1926).
[32] I. Waller, *Z. Physik* **38**, 635 (1926).
[33] Professor Pauling has told this author that his satisfaction with the results obtained in his paper provided him with a strong incentive for his further work in quantum mechanics.
[34] L. Pauling, *Phys. Rev.* **29**, 145 (1927).
[35] L. Pauling, *Proc. Natl. Acad. Sci. USA* **12**, 32 (1926).
[36] L. Pauling, *Phys. Rev.* **27**, 568 (1926).
[37] W. Pauli, Jr., *Z. Physik* **6**, 319 (1921).
[38] W.F. Colby, *Astrophysical J.* **58**, 303 (1923).
[39] M. Czerny, *Z. Physik* **34**, 227 (1925).
[40] L. Mensing and W. Pauli, Jr., *Physik. Z.* **27**, 509 (1926).
[41] J.H. Van Vleck, *Nature* **118**, 226 (1926).
[42] Pauling notes here that after his paper had been submitted for publication, a note in which a similar conclusion is stated by Kronig (*Proc. Natl. Acad. Sci. USA* **12**, 488, 608 (1926)) appeared.
[43] L.M. Mott-Smith and C.R. Daily, *Phys. Rev.* **28**, 978 (1926).
[44] Professor Pauling recently related to this author that he should have already recognized in 1926 that the value of the total angular momentum is given by $[j(j + 1)]^{1/2}h/2\pi$, not $jh/2\pi$ (as in the old quantum theory), and he would have saved himself a lot of trouble. He further mentioned that in spite of the fact that it is emphasized in L. Pauling and S. Goudsmit, *The Structure of Line Spectra*, McGraw-Hill, New York, 1930, that the value of the total angular momentum is given by $[j(j + 1)]^{1/2}h/2\pi$, a number of physicists to this day continue to use the incorrect value of $jh/2\pi$.
[45] L. Pauling, *Proc. Roy. Soc. (London)* **A114**, 181 (1927).
[46] W. Heisenberg and P. Jordan, *Z. Physik* **37**, 263 (1926).
[47] P.S. Epstein, *Nature* **118**, 444 (1926); *Phys. Rev.* **28**, 695 (1926).
[48] G. Wentzel, *Z. Physik* **38**, 518 (1926).
[49] J.H. Van Vleck, *Proc. Natl. Acad. Sci. USA* **12**, 662 (1926).
[50] L. Pauling, *Chem. Rev.* **5**, 173 (1928).
[51] J.C. Slater, *Solid-state and Molecular Theory: a Scientific Autobiography*, John Wiley and Sons, New York, 1975, pp. 61–62. According to Slater, 'It was then an obvious step to replace Dirac's determinantal function ... by an identical expression in which the orbitals ... are replaced by spin orbitals. This is antisymmetric in interchange of the four coordinates, *x*, *y*, *z*, and spin, of any two electrons. And it obeys the exclusion principle, in that it automatically vanishes if any two spin orbitals are identical in all four quantum numbers. This method, though not familiar, was not really new; it had been used by Pauling in 1928, in discussing the repulsion of two helium atoms. It is the only step which had to be taken to incorporate the spin properly into the problem; no further use of the group theory was needed.' There is another interesting observation by Slater on p. 64 of his book concerning the efforts in 1929 to hire prominent people in the

physics and chemistry departments at Harvard. He writes, 'Moves were also under way in the chemistry department. By then I had been long enough at Harvard so that I knew the faculty pretty well, and James B. Conant, who was still a chemistry professor, was trying very hard to get Linus Pauling to join the department. I had known Pauling only by reputation, but I helped Conant with the entertainment of Pauling when he came to visit, and tried my best to induce him to come to Harvard. He didn't, but I got to know him, and was much impressed.'

[52] G.E. Uhlenbeck and S. Goudsmit, *Naturwissenschaften* **13**, 953 (1925); *Nature* **107**, 264 (1926).

[53] E. Schrödinger, *Annal. Physik* **81**, 109 (1926); *Phys. Rev.* **28**, 1049 (1926).

[54] W. Heisenberg, *Z. Physik* **38**, 411 (1926); **39**, 499 (1926); **41**, 239 (1927).

[55] P.A.M. Dirac, *Proc. Roy. Soc. (London)* **A112**, 661 (1926).

[56] M. Born and J.R. Oppenheimer, *Annal. Physik* **84**, 457 (1927).

[57] O. Burrau, *Kgl. Danske Videnskabernes Selskab. Math.-fys. Meddelelser* **7**, 14 (1927).

[58] W. Heitler and F. London, *Z. Physik* **44**, 455 (1927).

[59] Y. Sigiura, *Z. Physik* **45**, 484 (1927).

[60] W. Pauli, Jr., *Z. Physik* **31**, 765 (1925).

[61] B. Podolsky and L. Pauling, *Phys. Rev.* **34**, 109 (1929).

[62] P. Jordan, *Z. Physik* **40**, 809 (1927).

[63] The hydrogen-like eigenfunctions are derived in, for example, Ref. [9], pp. 112*ff.*

[64] L. Gegenbauer, *Wiener Sitzungsber.* **70**, 6 (1870). An explicit expression for the Gegenbauer C functions is given in E.T. Whittaker and G.N. Watson, *A Course in Modern Analysis*, Cambridge University Press, London, 3rd edition, 1920, p. 329.

[65] L. Pauling, *Phys. Rev.* **36**, 430 (1930).

[66] W.F. Giauque and H.L. Johnston, *J. Am. Chem. Soc.* **50**, 3221 (1928).

[67] E. Mathieu, *Liouville's J.* **13**, 137 (1868). Pauling notes that E.U. Condon, *Phys. Rev.* **31**, 891 (1928), pointed out that the Mathieu functions of even order are the eigenfunctions for the plane pendulum.

[68] Under normal circumstances, hydrogen can be considered as two distinct molecular species, one, *para hydrogen*, having the nuclear spins opposed and existing only in even nuclear rotational states (for the normal electronic state), and the other, *ortho hydrogen*, having the nuclear spins parallel and existing only in the odd nuclear rotational states. For a discussion of the quantum mechanics of *ortho* hydrogen and *para* hydrogen, see Ref. [9], pp. 355–358.

[69] L. Pauling, *J. Am. Chem. Soc.* **57**, 2680 (1935).

[70] W.F. Giauque and M.F. Ashley, *Phys. Rev.* **43**, 81 (1933).

[71] Pauling later pointed out (Ref. [11], 3rd edition, pp. 433–434) that a more careful counting of the possible arrangements leads to a value of the residual entropy of $R \ln 1.5068 = 0.815$ e.u., in almost perfect agreement with the revised experimental value of 0.813 e.u.

8

Electronegativity, configuration energy, and the periodic table

Leland C. Allen

Princeton University, Princeton, NJ, USA

IN HONOR OF LINUS PAULING'S CONTRIBUTIONS TO CHEMICAL BONDING

Pauling's contributions to the theory of the chemical bond commenced in 1928, immediately following Heitler and London's invention of the quantum mechanical valence bond method in their application to H_2 and G.N. Lewis' development of his classical dot structures from 1916 to 1923. During the next decade, Pauling developed the valence-bond-like models which have largely dominated chemists' understanding of representative element chemistry for 60 years. A principal reason has been their accessibility to chemists: Pauling has been the leading creator of novel, useful, and mathematically simple models for the complex electronic structure that underlies all of chemistry. Pauling's resonance theory accounted for the delocalization of π orbitals and atomic charges, hybrid orbitals explained molecular shape, and maximum overlap determined internuclear separation and rationalized atomic radius. His extended concern with the cataloguing of atomic radii long ago made him chemistry's authority on this basic question, and his 1947 list of metallic and covalent radii remains the standard reference. Electronegativity provided the key parameter needed to describe polar covalence. His 1932 *JACS* paper introduced atomic electronegativity and stated that its difference between two single bonded atoms measured the ionic character of this bond — the same fundamental concept we still believe to be its central feature. Relative values for 10 atoms (H, P, I, S, C, Br, Cl, N, O, and F) were given, and 60 years later none deviates by more than 5% from today's numbers, with only a single-order reversal due to a 2% difference in values! In this 1932 article he found regularities among differences in molecular heats of combustion by positioning atoms along a scale 'in the way that genes are mapped in a chromosome from crossover data', thereby making practical use of his attendance at Professor Thomas Hunt

Morgan's famed Cal Tech lectures on genetics and displaying the multidisciplinary approach to science that has been a hallmark of Pauling's contributions.

The extreme simplicity of Pauling's electrostatic valence rule along with its amazing range of applicability — even today the dominant organizing scheme for the structure of mineral silicates — is another example of his versatility and style. His bond order–bond distance equation is yet another. This scheme has been used recently with great success to obtain the geometry of chemical reaction pathways from the X-ray study of many closely related crystal structures. Similarly, Pauling's enunciation of the coplanarity of the peptide bond provided a fundamental building block in understanding protein structure.

Pauling was the first to address the nature of the hydrogen bond successfully, and his simple ionic model correctly recognized its largely electrostatic origin, explained its directionality, its internuclear separation and its bond strength in terms of the electronegativity of the two atoms bonded to H. Application and elucidation of hydrogen bonding in diverse systems followed, e.g. NH_4HF_2, a crystal structure entirely determined by hydrogen bonds, the structure and entropy of ice, the flickering clusters of ice I and ice II in liquid water and, of course, the consummate insight of his 1951 paper, 'The structure of proteins: two hydrogen-bonded helical configurations of the polypeptide chain'.

Pauling's approach has been characterized by quick insights and innovative hypotheses based on deep and wide ranging factual knowledge of nearly all areas of science. His model of mutually complementary surface regions on hemoglobin as a molecular disease and the origin of sickle-cell anemia, and his molecular theory of general anesthesia have both led to extensive and ongoing medical research programs. During the long period of his career, many of the problems he has addressed have been, of course, subjected to sustained and sophisticated analysis beyond the scope of his simple models. However, the truly remarkable and unique feature of Pauling's contributions to chemical bonding has been that most of the qualitative physical pictures his models desribe remain at the core of our thinking and talking about the phenomena in question.

CONFIGURATION ENERGIES AND COMPLETION OF THE PERIODIC TABLE

The research reported here was initiated as an attempt to define electronegativity more precisely. In addition to this goal, it has led to a new definition of the Periodic Table. For pedagogical reasons, it seems easiest to first present the case for a new Periodic Table and then to show how a more precise definition of electronegativity appears as one of its properties.

By 1923 the Periodic Table was fully established as the central method for organizing chemical phenomena. In addition to Mendeleev's original table, the discoveries of noble gas atoms and the atomic number, Z, along with Lewis' octet rule, had set the stage for use of the semi-classical Bohr–Sommerfeld atomic model to rationalize the observed chemical patterns in terms of the basic laws of physics. Thus in his 1922 paper [1], Bohr used the two quantum numbers, n and l, to assign electronic configurations to the atoms. In spite of this triumphant joining of chemistry to physics, chemists and physicists have continued to regard the Periodic Table from different viewpoints. Chemists see the Periodic Table as a sequence of groups whose atoms have common physical and chemical properties, with each period terminated by an inert atom. The two-dimensional array of atoms is associated with the two quantum numbers, n and l; thus the chemical similarity of groups arises from a common number and shape (l value) of valence electrons and the atomic size increases with increases in n as a group is descended. Physicists see it as a collection of valence electron configurations that depict the properties of atomic shell structure. Across a row (period), configuration occupancy increases as successive l designated subshells are filled. Columns (groups) specify configuration size (shell quantum number), n, while holding the l value and occupancy constant.

The chemist's atom-based Periodic Table fully represents the periodic law of chemical behavior, and appears complete. All the atoms are displayed in a two-dimensional array: a third dimension would have to repeat those already shown. The physicists' Periodic Table is configuration-based and incomplete: configuration occupancy (and shape), the horizontal dimension; configuration size, the vertical dimension. The missing dimension is configuration *energy* because the full information content of the quantum numbers n and l includes specification of the energy levels as well as the size and shape of the atomic orbitals. That an energy component has been long absent from the Periodic Table has come about for two reasons [2]. First is that chemists have been by far the largest users of the Periodic Table and, in addition to its appearance of completeness as an atom-based array, chemists have employed electron affinities, various ionization potentials, and electronegativity scales, to augment it in their explanation of chemical phenomena. The second reason is that Niels Bohr's singular invention of electronic configuration, which assured universal acceptance of the Periodic Table, was made four years before Schrödinger's energy eigenvalue equation identified energy as the central parameter needed to explain the structure of matter.

The explicit form of the expression needed to define configuration energy (for the representative elements) is immediately obvious:

$$CE = (a\varepsilon_s + b\varepsilon_p)/(a + b)$$

where a, b are the number of s and p electrons, ε_s, ε_p their spherically averaged ionization potentials [3]. This is just the ionization potential of an average valence electron, and Table 1 lists configuration energies (in electron-volts) for a large collection of atoms [4]. These numbers were obtained from the U.S. National Bureau of Standards high-resolution energy level tabulation [5] using the multiplet averaging technique given by Slater [3]. Values for the transition elements were estimated by an approximate procedure described elsewhere [6]. As noted, chemists have long augmented their use of the Periodic Table with various electronegativities' scales, thus *de facto* making electronegativity a part of the Periodic Table. It therefore follows from our reasoning above that configuration energy should be a logical definition of electronegativity.

At the beginning of this section we stated that the research reported here was begun in an attempt to more precisely define electronegativity, and in this paragraph we briefly summarize those early efforts because they provide an independent way of showing that the one-electron energy, $(a\varepsilon_s + b\varepsilon_p)/(a + b)$, is the simplest and most appropriate definition of electronegativity. In arriving at this conclusion, three factors

Table 1 — Configuration energies (electron–volts)

Atom	Config. energy[a]	Atom	Config. energy[a]	Atom	Config. energy[b]
H	13.61	K	4.34	Sc	6.8
He	24.59	Ca	6.11	Ti	7.4
		Ga	10.39	V	8.1
Li	5.39	Ge	11.80	Cr	8.6
Be	9.32	As	13.08	Mn	9.2
B	12.13	Se	14.34	Fe	9.9
C	15.05	Br	15.88	Co	10.4
N	18.13	Kr	17.54	Ni	11.0
O	21.36			Cu	10.8
F	24.80	Rb	4.18		
Ne	28.31	Sr	5.70		
		In	9.79	Y	5.9
Na	5.14	Sn	10.79	Zr	6.6
Mg	7.65	Sb	11.74	Nb	7.4
Al	9.54	Te	12.76	Mo	8.2
Si	11.33	I	13.95	Tc	9.0
P	13.33	Xe	15.27	Ru	9.8
S	15.31			Rh	10.6
Cl	16.97	Cs	3.89	Pd	11.3
Ar	19.17	Ba	5.21	Ag	11.7
		Zn	9.39		
		Cd	8.99		
		Hg	10.44		

[a] Energy level data from National Bureau of Standards tables and from A.A. Radzig and B.M. Smirnov, *Reference Data on Atoms, Molecules and Ions*, Springer-Verlag, 1985. See Ref. 1 of Ref. 4 for detailed listing of sources, method for using data, and conversion to other units.
[b] Rough estimates based on fractional occupancy interpolation of relativistically corrected Hartree–Fock solutions for integral occupancy s^1d^{n+1} and s^2d^n configurations carried out by Dr Joseph B. Mann of the Los Alamos National Laboratory, Los Alamos, New Mexico.

were taken into consideration. (1) Reviews of electronegativity strongly indicated that it was an energy quantity closely related to ionization potentials [7]. (2) All of the many different previous definitions involve atoms or atomic orbitals that are slightly perturbed, promoted or otherwise modified, therefore it made sense to at least investigate the simplest possible *unmodified* system: the orbital energies in neutral, ground state, free atoms. (3) An extensive search of journal articles, reviews and textbooks dating back to 1932 showed that the values from only two scales, those of Pauling [8] and Allred and Rochow [9], have been used extensively by practicing chemists to rationalize molecular charge distributions and characterize bonding. Given these three factors, many different algebraic combinations of ε_s and ε_s were empirically explored: weighted sums and differences, multiplicative forms, expressions with different power laws, etc., to see what combination would best reproduce the Pauling and the Allred and Rochow scales. From this experimentation we found that $(a\varepsilon_s + b\varepsilon_p)/(a + b)$ was clearly superior. Fig. 1 gives a comparison of the three scales.

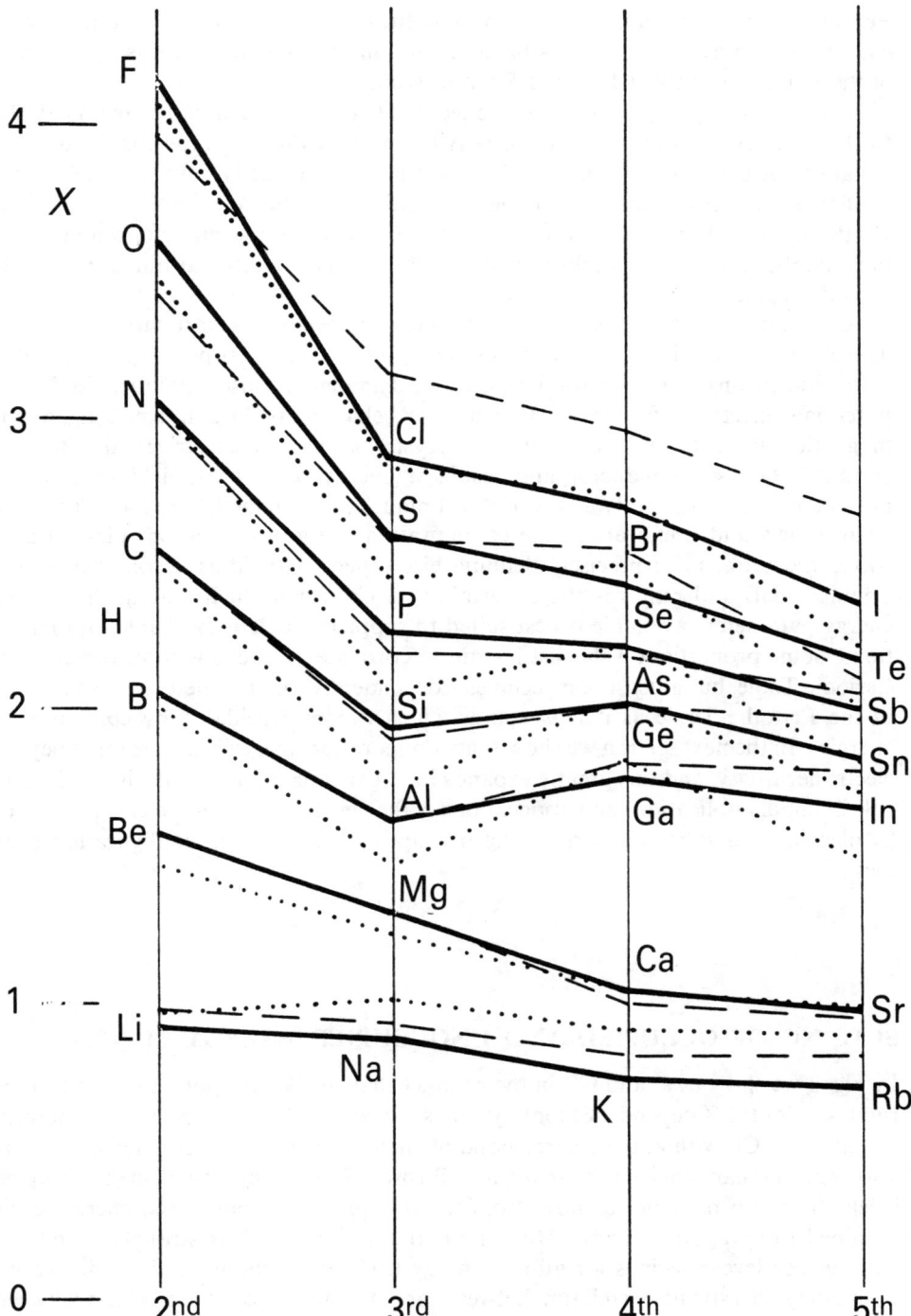

Fig. 1. — Comparison of three electronegativity scales (Pauling units). $\chi_{spec} = (a\varepsilon_s + b\varepsilon_p)/(a + b)$, solid lines. Pauling scale, dashed lines. Allred and Rochow scale, dotted lines.

Because of its high accuracy (achieved by virtue of the precision inherent in the NBS energy level data) and its physics-based definition, the new scale acts as a refinement of the Pauling and the Allred and Rochow scales.

From the two paragraphs above we see that two independent approaches have led to the same expression for electronegativity and that this expression is the defining equation for configuration energy. Since the Periodic Table is simply a collection of configurations, configuration energies explicitly characterize its energy dimension. Furthermore, Schrödinger's equation and the two configuration-determining quantum numbers, *n* and *l*, establish energy as a *necessary* additional dimension of the Periodic Table.

Contemporary inorganic and general chemistry textbooks start with the Periodic Table's two-dimensional atomic array and present many properties as correlative third dimensions, e.g. polarizabilities, radii, diamagnetic susceptibilities, ionization potentials, electron affinities [10], acid–base behavior, boiling and melting points, magnetic moments, electronegativities, crystal structures, electrical and thermal conductivities, semiconductor and insulator energy band gaps, field gradients at nuclei etc., This is certainly a powerful technique, and it immediately forces us to ask what is new and useful about the configuration energy formulation. First of all it solves the problem of precisely defining electronegativity (thus accomplishing our original goal) and removes the uncertainty in choosing which among the various energy parameters available is best suited to supplement Periodic Table rationalizations. Some properties on this list remain as correlates of the new three-dimensional Periodic Table, but at least one becomes a dependent variable. There are also features of the Periodic Table as it now stands which can be elucidated by configuration energies. In the next section we show that configuration energy has a meaning beyond electronegativity, and this greatly expands its organizing ability. With this additional role in mind, implications and applications of the new formulation are given in section 'Molecular and solid-state organizing principles obtained from configuration energies'.

SPACING OF OCCUPIED AND UNOCCUPIED ENERGY LEVELS

In Fig. 2 we show ε_s, ε_p and CE in their funnel-shaped effective potentials, and to the right we plot the Z dependence for a typical set of atoms. Not only is there an increase (negative) in CE with Z, but a corresponding increase in energy level spacing is clearly apparent. Similar plots can be made for all rows of representative element occupied levels. It is also possible to show, from the occupied and unoccupied energy levels obtained from Thomas–Fermi–Dirac potentials [6], that CE is strongly correlated with energy level spacings for atomic energy levels throughout the Periodic Table. Ultimately, this strong correlation between large CE and large spacings, and between small CE and small energy level spacings, is simply a consequence of the funnel-shaped nature of atomic effective potentials. For metals and alloys the energy levels are often so close together that the appropriate quantity is the density of levels rather than their spacings. We have noted in the previous section that chemists and

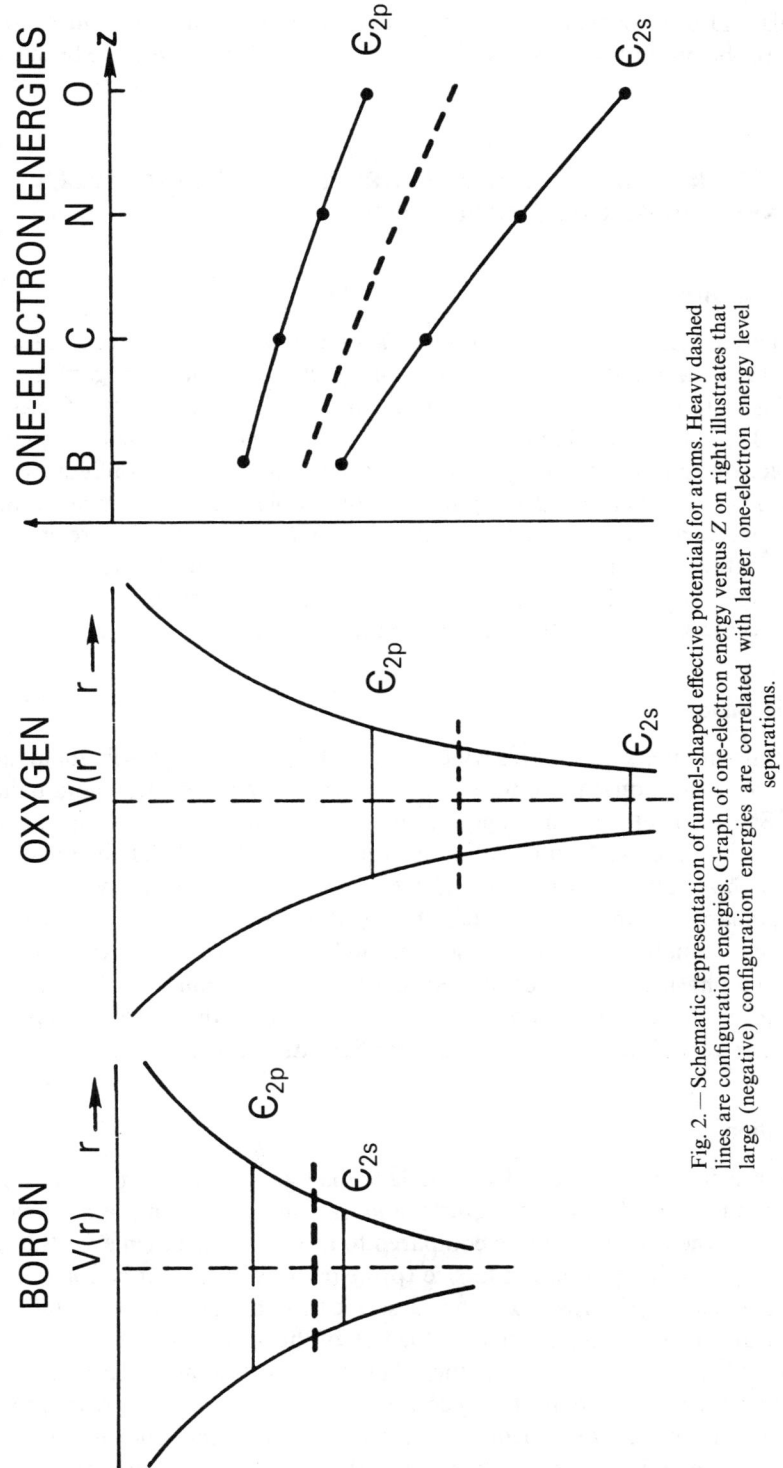

Fig. 2. — Schematic representation of funnel-shaped effective potentials for atoms. Heavy dashed lines are configuration energies. Graph of one-electron energy versus Z on right illustrates that large (negative) configuration energies are correlated with larger one-electron energy level separations.

chemistry textbooks have frequently employed electron affinity, various ionization potentials and electronegativities to supplement the Periodic Table in rationalizing bonding problems, but density of states and energy level spacings have been much less often cited.

MOLECULAR AND SOLID-STATE ORGANIZING PRINCIPLES OBTAINED FROM CONFIGURATION ENERGIES

Electronegativity

In the first edition of *The Nature of the Chemical Bond*, written seven years after his seminal 1932 paper, Pauling introduced his section on electronegativity with the qualitative definition, 'the power of an atom in a molecule to attract electrons to itself', and it is now possible to make his statement more specific and quantitative. Thus, electronegativity is one property of configuration energy; configuration energy is embedded in the Periodic Table; the Periodic Table is chemistry's most important *molecule*-organizing scheme, therefore the phrase, *in a molecule*, has reference to the Periodic Table itself [11]. This seems especially appropriate because configuration energies have properties beyond those attributed to electronegativity, and their addition to the Periodic Table greatly enhances its power.

Metalloid Band

There are no numbers or symbols which specify the position of the diagonal line in the current two-dimensional Periodic Table and its accompanying metalloid band (B, Si, Ge, As, Sb, and Te) nor any suggestion as to why a line separating metals from nonmetals should exist. Configuration energies fulfill this role. Their magnitudes for the atoms above this line are large and their energy levels are widely separated — exactly the characterizing physical features of the nonmetals. All atoms below the diagonal have smaller average energies and higher density of states than those above the line: the physical properties of metals. The metalloid band is a flat region in the configuration energy dimension, the only such region with approximately constant values (their variation is only 7% of the configuration energy range).

Metallization

Students are frequently puzzled by the fact that groups in the p-block appear to possess significantly different structures and chemistry for compounds made from elements near the top of a column compared to those further down even though their configurations are only differing in size (principal quantum number, n), e.g. carbon compared to lead, s^2p^2. Most textbooks describe this phenomenon as metallization or dehybridization, and recognize that as the size of the atoms increases, the s, p, and d energy levels are getting closer together. This leads to the increased mixing of these orbitals and because different mixing combinations occur for levels with only a small energy separation between them directionality is lost and coordination number increases. But there is nothing in the current Periodic Table to anticipate or quantify this behavior: again, configuration energies provide the missing information which is necessary for its completion. As Z increases across a row, CE increases almost linearly (because subshells are being filled and electrons in the same shell have a nuclear screening ability only 40% of that of inner shell electrons). The combination of an approximately linear increase across a row and a CE increase up a group approximately proportional to the reciprocal of the radius produces a minimum change along the metalloid band and defines the position of the diagonal separation line. It should be noted that quantitative explanation of the metal–nonmetal diagonal line and metallization requires both the average energy level and density of levels properties of configuration energy.

Covalent, metallic, ionic bonding

Nearly everyone instinctively uses the Periodic Table to help classify compounds and individual bonds as covalent, metallic or ionic. Looking at the Periodic Table one easily finds rules for the total number of bonds (valence) that an atom may possess by virtue of sharing or transferring electrons or filling holes in valence shells and subshells, but no measure of covalent, metallic or ionic bonding is present. The addition of configuration energies to the Periodic Table is therefore simply articulating and making more quantitative an intuition that many practicing chemists have had for many years. Assignment of CE values permits quantitative specification of bonding type from metallic to covalent for elemental compounds left to right along rows. For non-elemental compounds and for bonding gradations involving ionic bonds, two forms of map that supplement the Periodic Table have proved especially useful. The first of these are van Arkel-Ketelaar triangles, and an example is shown as Fig. 3. Quantitative, almost uniform gradations, horizontally as well as along the metallic–ionic and covalent–ionic legs, are shown by CE and ΔCE, respectively, and the change in values along each leg is approximately half the full range of CE (for simplicity and generality, stoichiometries have been suppressed). Because of these features, quantitative van Arkel-Ketelaar triangles can prove extremely useful in the design of engineered materials. Structural classification diagrams, a second form of binary compound mapping useful for supplementing the Periodic Table, have been recently discussed by Puddephatt and Monaghan [12], and an application to representative element hydrides is given as Table 2. In this diagram, ΔCE clearly separates their structures into three diagonal regions: ionic, polymeric and covalent. In the saline (ionic) hydride region, on the left side of the ionic–polymeric dividing line, $\langle \Delta CE \rangle = 7.40$ eV and along the right side of the polymeric–covalent dividing line, $\langle \Delta CE \rangle = 2.10$ eV, while in the interesting polymeric region, ΔCE is very nearly constant, with $\langle \Delta CE \rangle = 3.92$ eV. The three traditional covalent subdivisions: electron deficient, electron sufficient, and electron excess, follow group categorization IIIA; IVA; VA, VIA, and VIIA, respectively. In addition, the covalent region can be further separated into the two diagonal sections indicated by the dotted line: to the left of this line, CE_H is greater than the element to which it is bound, but to the right, ΔCE changes sign. The change in polarity across this line results in significantly different chemical reactions.

   Because the vast majority of chemical phenomena continue to be explicable in terms of ionization potentials and one-electron orbitals, it has been long recognized

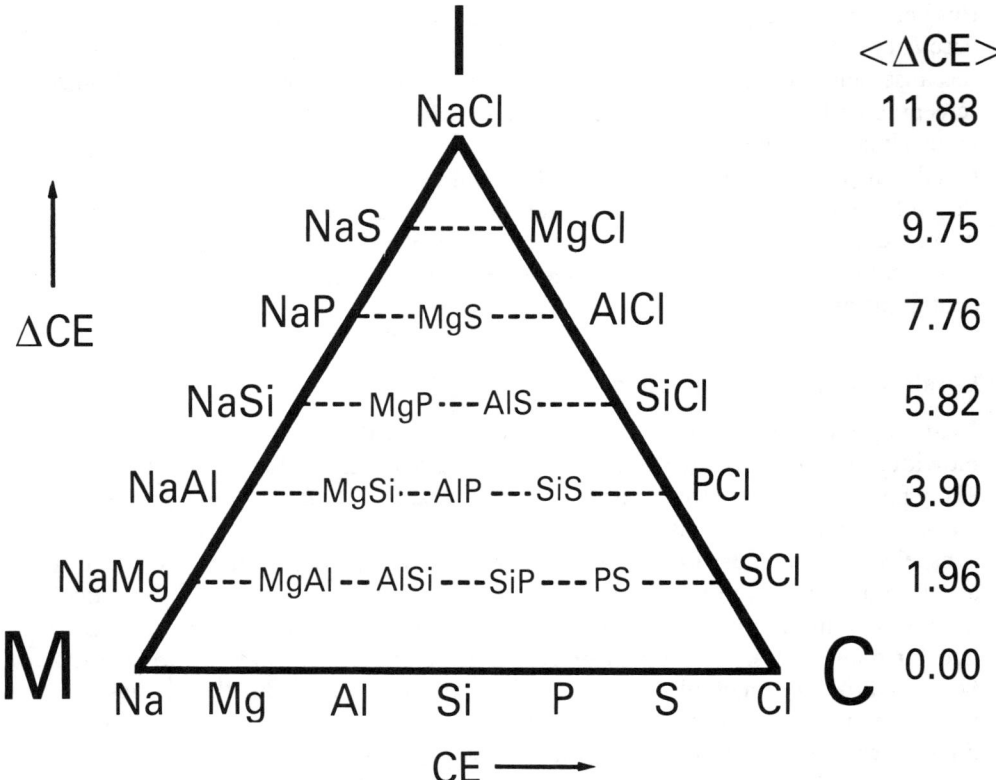

Fig. 3.—A van Arkel-Ketelaar triangle illustrating the continuous gradation between bonding types and its detailed quantification by configuration energy changes. (The smaller lettering used inside the triangle was required for graphical presentation reasons and has no chemical significance).

that ionic, covalent and metallic bonding questions can be answered qualitatively within the framework of the Hartree–Fock approximation. Therefore wavefunctions and ε_is satisfying the atomic Hartree–Fock equations, $\mathcal{H} \phi_i(1) = \varepsilon_i \phi_i(1)$, are appropriate for defining the ground state, free atoms of the Periodic Table and their

Table 2 — Structural classification for hydrides of the representative elements

IA	IIA	IIB	IIIA	IVA	VA	VIA	VIIA
LiH	BeH_2		BH_3	CH_4	NH_3	H_2O	HF
NaH	MgH_2		AlH_3	SiH_4	PH_3	H_2S	HCl
KH	CaH_2	ZnH_2	GaH_3	GeH_4	AsH_3	H_2Se	HBr
RbH	SrH_2		InH_3	SnH_4	SbH_3	H_2Te	HI
	Ionic		Polymeric			Covalent	

configuration energies. Since the Hartree–Fock equations are mathematically and physically well defined, this identification adds significantly to the claim that configuration energies are necessary for Periodic Table completion. A great deal is known about the limitations as well as the advantages of Hartree–Fock solutions, and their most important shortcoming is an inability to accurately predict binding energies. Extension of the usual restricted Hartree–Fock method to the Hyper Hartree–Fock or Unrestricted Hartree–Fock schemes often corrects this shortcoming, but, in general, configuration energies can only yield qualitative binding energy trends.

The hydrogen bond, A—H···B, is another bonding category that figures prominently in chemistry and biochemistry. This bond is successfully described by Hartree–Fock wavefunctions, and its principal features (dimerization energy, internuclear separation, force constants, IR intensity enhancement, charge transfer, and cooperativity) can be characterized by configuration energies associated with the proton donor (A—H) and the electron donor (B) [6, 13].

Atomic radii

In addition to the empirical radii that chemists usually employ and which derive from study of many molecules and solids (and which Pauling has played such an important role in developing), there are the radii identified with the outer radial maxima of free atom atomic orbitals. The close association of the two definitions has long been known [14], and because configuration energies are themselves free atom quantities it is logical to adopt the latter definition. We can then demonstrate, Fig. 4, that radius is a function of configuration energy. In the past, radius has been most frequently regarded as an independent atomic property. A simple relationship between radii and CE is expected because most of the valence-energy-determining electronic charge is at the outer radial maxima. Thus for representative elements:

$$\langle r \rangle = k(CE) = k(CE)^{-b}$$

where k is a constant and b a positive number between 1 and 2, both of which depend solely upon the group of the Periodic Table which is specified. Since configuration energy is an occupation weighted average over the valence shell electrons, $\langle r \rangle$ is similarly constructed. The Fig. 4 insert gives the variation of b versus group number. For each group there is a characteristic b value that specifies the particular flair of its atomic funnel-potential, and b varies continuously in a p-block or s-block as one moves horizontally across the table.

The Mott metal to nonmetal transition

For four decades, solid-state physicists have been exploring this transition in doped semiconductors, solutions of metals in ammonia, expanded fluid alkali metals, etc., and it obeys an equation of the form [15],

$$n_c^{1/3} a = 0.26$$

where n_c is the critical electron concentration of the solid for the onset of electronic metallic character, and a is the effective radius of the localized electron (atomic) center

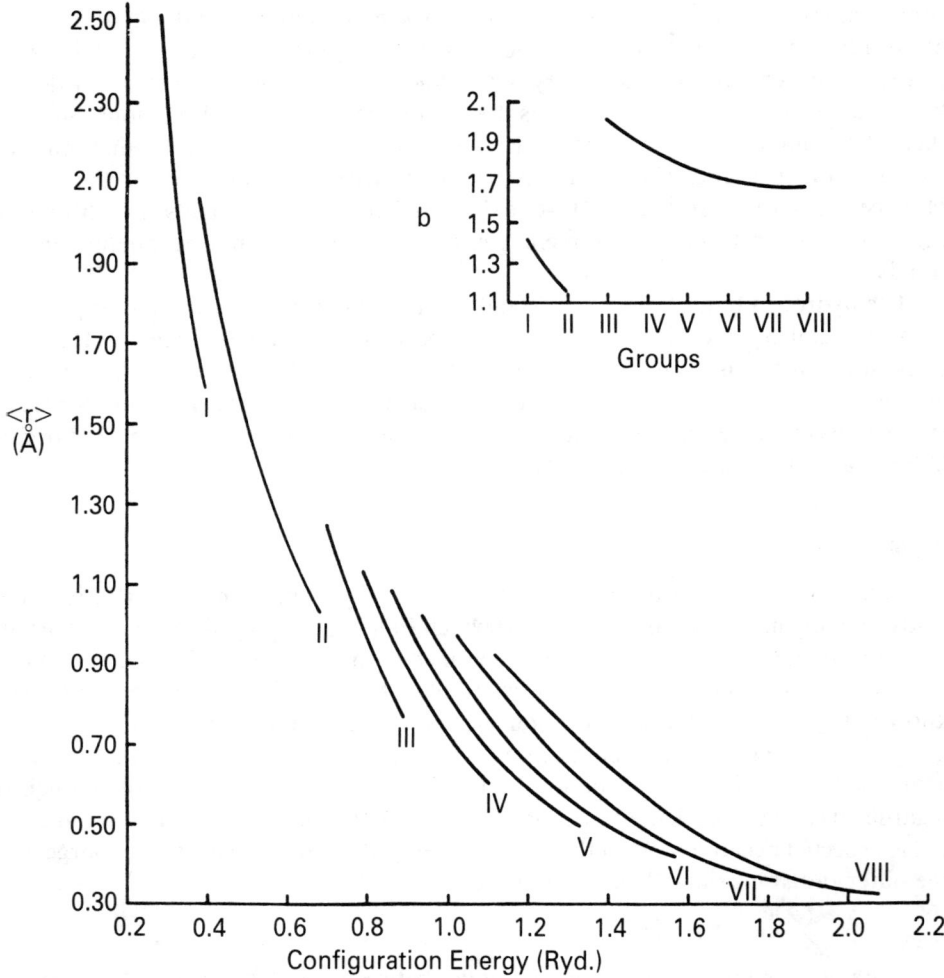

Fig. 4.—Atomic radii (determined by occupation weighted outer radial maxima of p and s orbitals) for representative elements versus configuration energies (Rydbergs) for the groups in the Periodic Table. Functional form of relationship is $\langle r \rangle$ = constant (configuration energy)$^{-b}$. A plot of b versus groups is given as inset.

in the material studied. Many of the systems of interest involve a more specialized type of transition than that characterized by the diagonal line in the Periodic Table, but the expanded fluid alkali metals are akin to 'metal-doped vacuum', and as pure metals they can be associated with a configuration energy and corresponding radius. Thus it is gratifying to find that the a values found experimentally are accurately represented by the Group I line in Fig. 4 [15].

Oxidation state limitations in the representative elements

Maximum covalent ligation equal to the number of valence shell electrons is the rule for representative elements (e.g. fluorides in their group oxidation states) with the well known exceptions N, O, F, Cl, Br, He, Ne, Ar, and Kr. This bonding pattern is entirely controlled by configuration-energy magnitudes: these atoms possess the highest nine values in the Periodic Table and they hold their electrons sufficiently tightly so that even with fluorine ligands it is not possible to engage all of their electrons in covalent bonds. On the other hand, in largely ionic solids (e.g. oxides), coordination numbers higher than that possible for covalent bonding are allowed because Coulomb interactions are dominant and there is no saturation of valence. Noble-gas chemistry is a dramatic example of the need for configuration energies in explaining the structure of matter, e.g. He and Be are both s^2 configurations, but one is an inert free atom while the other is a metal with high electrical conductivity.

HOMO–LUMO interactions

In much of chemical structure and reactivity, energy level spacings (or density of states), particularly of the unoccupied levels, play a dominant part in observed phenomena, and these are ordered by configuration energies. The opportunity of using d orbital for octet rule expansion in second row atoms, but not in the first row, is a familiar example. The need to use p orbitals on Li and Be to explain the chemistry of these atoms is another example. More generally, the 'frontier orbitals', especially the energy gaps between HOMOs and the first few unoccupied orbitals above them, have become an important method for organizing trends in chemical reactivity.

Multiple bonding between representative elements

This occurs most prominently among the first row atoms C, N, and O, and is much less prevalent for successively high rows. This phenomenon is again traceable to configuration energy. CE decreases down a group, and the ratio of valence radius to inner shell radius also decreases rapidly down a group. For first row atoms, high CE and small core allow the atoms to penetrate deeply into each other, thereby realizing large π orbital overlap, but for higher rows this possibility decreases very rapidly.

Transition elements

The formula for transition metal CE is the same as that for the representative atoms given in section 'Configuration energies and completion of the Periodic Table, of this chapter except that the one-electron energies are those for the ns and $(n-1)d$ orbitals. In contrast to the pure core $(n-1)s$ and $(n-1)p$ orbitals, the radial tails of the $(n-1)d$ orbitals partly extend into the ns valence region, and their one-electron energies are correspondingly closer to those of ns than to the $(n-1)s$ and $(n-1)p$ values. The unusual shape and energy of the $(n-1)d$ orbitals gives rise to $(n-1)d^p ns^2$ and $(n-1)d^{p+1} ns$ configurations which have unusually close-lying total energies, producing the well known non-periodic behavior of the lowest energy configurations for the three transition element series. Because these two configurations have nearly the same energy, a configuration interaction wavefunction is required, thereby greatly increasing the difficulty in interpreting the results. Fortunately, a method which solves this problem, the Hyper Hartree–Fock method (HHF), was devised some time ago by Slater, Mann, Wilson, and Wood [16]. HHF is an extension of the usual Hartree–Fock method which yields total energies only slightly less accurate than the configuration interaction solution and provides the desired $\varepsilon_{(n-1)d}$ and ε_{ns} values. The price to be paid for HHF is use of fractional occupation numbers, $(n-1)d^{p+x}ns^{2-x}$, where $0 \le x \le 1$, but this does not really present an interpretation problem more serious than the fractional atomic charges we have accepted for many years. Another unusual $(n-1)d$ orbital feature is their change from valence participation, with orbital coefficients approximately half that of the ns in left-side elements (Sc, Y, La) to pure core with no valence region participation at Zn, Cd, and Hg. Their contribution to the number of bonding valence electrons is measured by the square of the ratio of their radial amplitude to that of the ns radial amplitude at the position of the ns outer radial maxima. Interpolation of recent calculations by Mann [17] on relativistically corrected integral occupation number configurations of the transition elements by approximate second-order perturbation theory yields rough estimates for transition metal configuration energies as given in Table 1.

The change in $(n-1)d$ participation from sizable on the left of the Periodic Table to zero as the ds disappear into the core at the right end is responsible for the successive decrease in maximum oxidation state available to the late transition elements. It should be noted that the origin of this oxidation state limitation is fundamentally different than the origin of representative element maximum oxidation limitations.

BEYOND THE PERIODIC TABLE

Perhaps the single most important attribute of configuration energies as the third dimension of the Periodic Table is their ability to make direct connection with molecular orbital and energy band theory, the conceptual and computational methods that dominate contemporary approaches to understanding the structure and reactivities of molecules and solids. These methods focus on the one-electron energy levels required to describe detailed properties of particular molecules and solids, and the obviously appropriate atomic reference is to an average atomic energy level and to the relative spacing of its levels.

Likewise, the continued success of the Periodic Table, and especially the enhancement of its organizing power with the addition of configuration energies, suggests that it would be worthwhile trying to identify an atomic configuration energy in a molecule or solid. Just as one recognizes that 'molecules are made out of atoms', a molecular wavefunction (to a first approximation) is made out of a linear combination of atomic orbitals. Using this wavefunction it is possible to project out a one-electron energy weighted average atomic energy that is an *in situ* atomic configuration energy, and this has been called the energy index for atom A in a molecule, EI_A [18]. Quite recently we have found that rotational barrier heights, accurate to within a few percent, can be obtained for formamide, ethane, methylamine, and methanol from the change in EI as the barrier is traversed, $\Delta(EI_A + EI_B)$, where A and B are the atoms which define the bond around which rotation takes place. If equal success is achieved with further examples, this will be a very exciting development because the localization of our model to only the two atoms defining the bond makes it the simplest possible representation of this phenomenon. Similar encouraging results are obtained for the inversion barriers in NH_3 and PH_3. A closely related quantity, the bond polarity index, BPI_{AB}, estimates the effective bond dipole in a molecule, and this leads to a useful definition of group electronegativity. The considerable benefit to be realized by these indices is their ability to unify, simplify, and quantify many of the electronic properties used by organic and inorganic chemists in their everyday research activities.

SUMMARY

It was found that Pauling's and Allred and Rochow's electronegativity values, by far the most widely employed scales, could be reproduced (for representative elements) by a free atom average energy, $(a\varepsilon_s + b\varepsilon_p)/(a + b)$. In parallel with this empirical finding it was recognized that the assignment of Periodic Table configurations using the quantum numbers n and l points logically to a configuration *energy* definition, $(a\varepsilon_s + b\varepsilon_p)/(a + b)$. A strong correlation between configuration energy and energy level spacing was also found. Combining these three observations leads to completion of the Periodic Table by addition of an energy dimension and thereby an explanation for its diagonal line separating metals from nonmetals and of the metallization which occurs on descending a group. Configuration energy provides a quantitative measure which differentiates covalent, metallic and ionic bonding, and electronegativity, a manifestation of the bond polarity aspect of configuration energy, is likewise an intrinsic property of the Periodic Table.

As chemical science progresses, there is a vast increase in the number of compounds discovered, and the Periodic Table becomes ever more important as the single most effective method for organizing their properties. The configuration energy dimension significantly increases its capability and at the same time imparts an elegant simplicity.

ACKNOWLEDGMENTS

The author wishes to thank graduate student, Eugene Knight, and Professor Joel Liebman for many helpful discussions and useful suggestions.

REFERENCES

[1] N. Bohr, *Nature* (1923) **112**, 29.
[2] There is another form of the Periodic Table that also includes energy level information: a graph of n versus l, with each nl point labeled by its generic one-electron energy level. Accompanying the nl point is a horizontal list of atoms (ordered according to Z) whose orbitals are being filled. This table emphasizes orbital filling according to relative energy, i.e. the 'aufbauprinzip' as a basis for classification of the elements (H.C. Longuet-Higgins, *J. Chem. Ed.* (1957) **34**, 30; W.R. Walker and G.C. Curthoys, *ibid* (1956) **33**, 69). If the horizontal lists of atoms are omitted, this array becomes the usual mnemonic for the $n + 1$ rule of orbital filling. (J.E. Huheey, *Inorganic Chemistry*, 3rd edn, Harper & Row, 1983, p. 28).

[3] J.C. Slater, *Phys. Rev.* (1955) **98**, 1039. This reference defines and rationalizes the concept of one-electron atomic energies, and gives a table of experimentally derived values.

[4] At this point a critic might argue that in the detailed description of molecular binding there is no 'average valence electron' nor spherical atom, but the Periodic Table must remain a first-order approximation which is made up of 'average atoms' suitable for binding into any molecular or solid-state environment.

[5] L.C. Allen *J. Am. Chem. Soc.* (1989) **111**, 9003. This reference extends and updates the table of one-electron energies given by J.C. Slater (ref. [3]), discusses the concept of electronegativity and makes a detailed comparison of the average one-electron energies, $(a\varepsilon_s + b\varepsilon_p)/(a + b)$, with Pauling's and Allred and Rochow's electronegativity values.

[6] L.C. Allen, Configuration Energies of the Elements and Their Use in Explaining the Structure of Matter, *J. Am. Chem. Soc.* (in press).

[7] H.O. Prichard and H.A. Skinner, *Chem. Rev.* (1955) **55**, 745. J. Mullay, in *Structure and Bonding*, Eds K.D. Sen and C.K. Jørgensen, Springer-Verlag, New York, 1987, p. 1.

[8] L. Pauling, *J. Am. Chem. Soc.* (1932) **54**, 3570. An update of Pauling's values has been given by A.L. Allred *J. Inorg. Nucl. Chem.* (1961) **17**, 215.

[9] A.L. Allred and E.G. Rochow, *J. Inorg. Nucl. Chem.* (1958) **5**, 264.

[10] Electron affinities are important quantities, but they have often been falsely claimed to be periodic. A graph of the extensive list of values, (H. Hotop and W.C. Lineberger, *J. Phys. Chem. Ref. Data* (1985), **14** 731) clearly shows this not to be true.

[11] Previous definitions of electronegativity have employed *molecular* descriptors either explicitly (dipole moment, force constant, bond energy, covalent radius) or implicitly (atomic excited states, atomic electron affinities), but in section 'Spacing of occupied and unoccupied energy levels' we showed that these complications are unnecessary for matching the numerical values that have been used successfully by practicing chemists for many years. In fact, the periodic properties of the molecular descriptors are simply a few of the many correlations with molecular and solid-state properties made by the Periodic Table and, in particular, by its atomic configuration energies.

[12] R.J. Puddephatt and P.K. Monaghan, *The Periodic Table of the Elements*, 2nd edn, Clarendon Press, Oxford, 1986.

[13] L.C. Allen *J. Am. Chem. Soc.* (1975) **97**, 6921.

[14] J.C. Slater *Quantum Theory of Molecules and Solids*, Vol. 2, McGraw-Hill, 1965, Chapter 4.

[15] N.F. Mott, *Metal–Insulator Transitions*, Taylor and Francis, London, 1974; P.P. Edwards and M.J. Sienko, *J. Chem. Educ.* (1983) **60**, 691; *Chem. in Britain* (1983) **19**, 39; *Acc. Chem. Res.* (1982) **15**, 87.

[16] J.C. Slater, J.B. Mann, T.M. Wilson and J.H. Wood, *Phys. Rev.* (1969) **184**, 672.

[17] J.B. Mann, private communication.

[18] L.C. Allen, D.A. Egolf, E.T. Knight and C. Liang *J. Phys. Chem.* (1990) **94**, 5602.

9

Electronegativity scales

Ralph. G. Pearson
University of California, Santa Barbara, CA, USA

INTRODUCTION

The concept of electronegativity (EN) is almost as old as chemistry itself. Berzelius classified elements as electronegative or electropositive, depending on whether they appeared at the anode or cathode upon electrolysis of their salts. By the turn of the century it was understood that these terms referred to the electron-attracting and electron-holding power of the atoms of the element. During the twenties, the founders of physical-organic chemistry extended the use of the term to include radicals, or groups of atoms. There was a rough ordering of the EN of the various atoms and groups known.

In 1932 Linus Pauling made a landmark contribution [1]. He created an empirical scale of the ENs of the elements based on heats of formation, or, essentially, bond energies. The basic idea was that bonds between atoms of different ENs were stronger than some average value of bonds between the same atoms.

The empirical equation that he used to extract numerical values of EN is of very dubious validity, subject to many deviations [2]. It is a tribute to Pauling's genius that, in spite of this, he put the common elements in an order of EN that is universally accepted today. He also gave, in 1939, a definition of the word electronegativity: 'the power of an atom in a molecule to attract electrons to itself.' [3] Many scientists would accept this as the final definition.

Over the years a number of other scales of EN have appeared [4]. Most of these have been based on some experimental or theoretical property of the free atom. Examples are the well-known Allred–Rochow scale, the spectroscopic scale of Allen [5], and the stability ratio scale of Sanderson [6]. These scales are always judged by whether or not they agree with the original Pauling scale. Failure to agree would be considered a very serious deficiency. The fact that so many diverse properties do give the same ordering to the elements is powerful support for Pauling's interpretation.

Mulliken presented his scale in 1934 [7], $EN = (I + A)/2$. As might be expected, he was anxious to match Pauling's scale, and not to supplant it. As a result he did not use ground state values for the ionization potentials and electron affinities of the atoms. Instead he suggested values which would pertain to suitable excited, valence states. For carbon, for example, I, and A would be those of a hybrid sp^3 orbital, or other appropriate hybrid.

There was good reason to focus on hybrid orbitals and atoms in molecules. The quantum theory of the chemical bond was being actively developed, along both valence bond and MO lines. A subject of great importance was the polarity of a bond. In other words, how were the bonding electrons distributed between the bonded atoms, and what was the composition of the bonding orbital? This is a question still of great interest today.

A closely related question has to do with the net charge that may be assigned to each atom in the molecule. This question is often answered using an EN scale and the EN equalization principle of Sanderson [8]. If two atoms, or groups, have different ENs, electrons will flow from that of lesser EN to that of higher EN, until the two are equalized. This principle has no proof, but is intuitively very appealing.

An important new viewpoint on EN was contained in a paper by R.G. Parr, and his coworkers, published in 1978 [9]. In this work, electronegativity was explored by the use of density functional theory [10]. The main conclusions of this paper, and those that followed, will be presented next.

ABSOLUTE ELECTRONEGATIVITY

Consider a chemical system consisting of several nuclei and N electrons. The nuclei generate a potential v. Holding the nuclei fixed in position, the ground state electron density functional, ρ, is that which satisfies the variational equation

$$\delta[E[\rho]-\mu N[\rho]] = 0 \tag{1}$$

The quantity μ is the Lagrange multiplier, which ensures that the integral of ρ over the volume is equal to N. It is called the electronic chemical potential. From the calculus of variations, it follows that $\mu = (\partial E/\partial N)_v$.

The name comes from the thermodynamic equation

$$T\,dS = dE + P\,dV - \mu\,dN \tag{2}$$

At zero pressure and temperature, we also have $(\partial E/\partial N) = \mu$. For the ordinary chemical potential of thermodynamics, N is the number of molecules in the system. The electronic chemical potential in a single molecule plays somewhat the same role that the ordinary chemical potential plays in macroscopic thermodynamics. If the system is not in equilibrium, μ will not be constant: electron density will transfer from regions of high μ to those of lower μ, changing ρ. Finally, μ will be constant, ρ will be the correct electron density for the ground state, and E will have its minimum value, for the given set of nuclear positions.

Suppose we now have two chemical systems, originally isolated from each other. In general they will have different chemical potentials μ_C^0 and μ_D^0. If they are now allowed to interact, there will again be a flow of electron density from high μ to low μ, until finally a single value of μ_{CD} prevails. It is often convenient to think of the new chemical potential of each subsystem. Then we have, at equilibrium,

$$\mu_C = \mu_D = \mu_{CD} \tag{3}$$

The various values of μ can be found from quantum mechanical calculations [11]. However, it is much more useful to have a method using experimental results. If we draw a plot of E vs. N for any system, then μ is simply the instantaneous slope of such a curve. Experimentally we only know points on the curve for integral values of N, from data such as ionization potentials and electron affinities.

We do not know the instantaneous slope of the curve. However, if we pick the neutral species (or any other) as our starting point, we do know the mean slope for the change from $(N - 1)$ to N electrons. It is simply equal to $-I$. Also the mean slope from N to $(N + 1)$ electrons is simply $-A$. Using the method of finite differences, we can approximate the slope at N as $-(I + A)/2$.

$$-\mu = -\left(\frac{\partial E}{\partial N}\right)_v \simeq \frac{I + A}{2} = \chi \qquad (4)$$

The quantity $\chi = -\mu$ is called the absolute electronegativity.

The name comes in part from its similarity to the Mulliken EN. But it is not the same. Since density functional theory deals only with equilibrium systems, I and A now refer to ground state values, not valence state ones. The absolute part of the name comes from the relationship to the electronic chemical potential, a fundamental property of the system. Actually, Iczkowski and Margrave had earlier proposed that $-(\partial E/\partial N)_v$ be called the electronegativity [12].

There is a second reason to call χ the absolute EN. As already mentioned, if two systems are brought together, there is a flow of electron density which equalizes the electronic chemical potentials. But, in view of (4), the values of χ are also equalized, that is $\chi_C = \chi_D = \chi_{CD}$. Thus we have a rigorous proof of electronegativity equalization. Further progress can be made, if we know how the chemical potential changes as the number of electrons in the system (or subsystem) changes. For reasons that will become clear, we define the absolute chemical hardness, η, as follows [13].

$$2\eta = \left(\frac{\partial \mu}{\partial N}\right)_v = \left(\frac{\partial^2 E}{\partial N^2}\right)_v \qquad (5)$$

Thus hardness is the resistance of the chemical potential to a change in the number of electrons. Hardness is also the curvature of the plot of E vs. N.

From the method of finite differences again, we obtain an operational definition of η:

$$\eta \simeq \frac{(I - A)}{2} \qquad (6)$$

The units of η are the same as those of χ, eV. For any ion, atom, radical or molecule, we can calculate η, if I and A are known. The chemical softness, σ, of a system is also defined as the inverse of the hardness, $\sigma = 1/\eta$.

For isolated reactants, C and D, we can write

$$\mu_C = \mu_C^0 + 2\eta_C \Delta N \qquad (7)$$

$$\mu_D = \mu_D^0 - 2\eta_D \Delta N \qquad (8)$$

where ΔN is the fractional number of electrons transferred from D to C. Applying electronegativity equalization, or $\mu_C = \mu_D$, we find,

$$\Delta N = \frac{(\chi_C^0 - \chi_D^0)}{2(\eta_C + \eta_D)} \qquad (9)$$

Thus the difference in electronegativity drives the process, and the sum of the hardness parameters acts as a resistance.

While (9) seems very reasonable, it is only valid for molecules still quite far from each other. As the molecules approach, there will be additional effects on μ_C and μ_D due to changes in the external potentials, v_C and v_D. For an isolated molecule these potentials are due only to the nuclei, but upon closer approach, v_C must include the potential due to the nuclei and electrons of D, and vice versa. These will change as R, the intermolecular distance, changes. Also there will be significant changes in ρ_C and ρ_D, not only because of electron transfer, but because of reorganization corresponding to covalent bonding.

Nevertheless, we can assume that a large value of ΔN in equation (9) means a strong interaction between C and D. In particular, consider C and D as neutral Lewis acids and bases:

$$C + :D \rightarrow C - D \tag{10}$$

This requires that $\chi_C^0 > \chi_D^0$. The bond between C and D is a coordinate covalent bond, with some ionic character. Both contributions to the bond strength will increase with ΔN.

We can also rate a series of C molecules reacting with a single D molecule, or the reverse case, by calculating the several values of ΔN. Many interesting applications of (9) have been found in this way [14]. In some cases, bond energies can be correlated. In other cases, the heights of the energy barriers to reaction can be correlated.

The initial lowering of the energy due to the electron transfer can also be found [13].

$$\Delta E = \frac{-(\mu_C^0 - \mu_D^0)^2}{4(\eta_C + \eta_D)} \tag{11}$$

This equation also has some useful applications. For example, it can be used to improve the energy calculated from an approximate wave function [15].

While equation (9), in principle, can be used to calculate bond polarities, it is not as reliable as methods using the Pauling scale. This is a result of the secondary changes mentioned above. However, the absolute scale of EN can be used in a unique way to probe bond polarity. For example, consider a molecule X–Y, consisting of two atoms or radicals held together by a bond. The polarity of the bond could determine whether the molecule behaves as X^+, Y^- or X^-, Y^+. The same question can be asked by looking at the reactions

$$X^-(g) + Y^+(g) = X - Y(g) = X^+(g) + Y^-(g) \tag{12}$$

The difference in energy between the products on the right and those on the left is easily found.

$$\Delta E = (I_X - A_Y) - (I_Y - A_X) = 2(\chi_X - \chi_Y) \tag{13}$$

If X has a greater absolute electronegativity than Y, ΔE is positive. This means that X–Y acts as X^-, Y^+. The answer is a thermodynamic one, and involves no assumption. While the example is in the gas phase, the same result would usually be found in solution. The sum of the solvation energies of X^- and Y^+ would be very nearly equal to the sum for X^+ and Y^-, except when X or Y is the hydrogen atom.

THE TWO SCALES

To summarize, at this point we have two different scales, both called EN scales. One is the Pauling scale, including all those that attempt to match the originals. The other is the so-called absolute scale. They are defined quite differently, and are used in quite different ways. The Pauling scale is used quite successfully to estimate bond polarities and net charges on atoms in molecules. The absolute scale is essentially a measure of the chemical reactivity of a free atom, molecule, radical or ion.

Since the Pauling scale has no meaning with respect to molecules, or even ions, we can only compare the two for atoms and radicals. Table 1 shows the EN values for the first few elements in the two scales, using the spectroscopic values, converting them to Pauling units [16] for the Pauling scale.

The two scales are roughly parallel, but there are definite deviations. For example, hydrogen is much more electronegative on the absolute scale. Also B and Al are less EN than Be and Mg. This disagrees with our expectation that EN increases smoothly as one goes from left to right in any row.

But this by no means is proof that the Pauling scale is more correct than the absolute scale. The former scale tells us that bonds between non-metallic elements and Be or Mg are more ionic than bonds to B and Al. The latter scale tells us that it is easier to remove a 2p or 3p electron from B or Al, than 2s or 3s electron from Be or Mg. Both scales are equally correct, if used in their own domain.

Hydrogen is a somewhat special case. The high value of χ_A is consistent with the reactivity of H atoms with electron donors [14b]. The relatively low value of χ_P is consistent with the evidence that hydrogen is protonic in most of its bonds to non-metallic elements. The proton, lacking lone pairs, can burrow deep into the electron cloud of another atom. The hydride ion would be kept at a distance. This effect favors the protonic form in a bond.

Table 2 lists the Pauling and absolute ENs for several radicals. The two orders of EN are now inverted, with the χ_P values changing very little. Also included are the

Table 1 — Electronegativities of some elements in Pauling units and in absolute units

	χ_P [a]	χ_A [b]		χ_P	χ_A
H	2.30	7.18	—	—	—
Li	0.91	3.01	Na	0.87	2.85
Be	1.58	4.90	Mg	1.29	3.75
B	2.05	4.29	Al	1.61	3.23
C	2.54	6.27	Si	1.92	4.77
N	3.07	7.30	P	2.25	5.62
O	3.61	7.54	S	2.59	6.22
F	4.19	10.41	Cl	2.87	8.30

[a] From reference [5].
[b] In eV.

Table 2—Electronegativities of some radicals in the two scales compared to reactivity towards oxygen

	$\chi_P{}^a$	χ_A	$k \times 10^{12\,b}$
$t\text{-}C_4H_9$	2.82	3.31	23.4
$i\text{-}C_3H_7$	2.78	3.55	14.1
C_2H_5	2.77	4.06	4.4
C_2H_3	2.77	4.85	6.7
CH_3	2.76	4.96	2.0
SH	2.61	6.40	$<10^{-5}$

[a] L.C. Allen, D.A. Egolf, E.T. Knight and C. Liang, *J. Phys. Chem.* **94** (1990) 5602.
[b] Rate constant at 298 K, cm^3 molecule^{-1} s^{-1}, for reaction with O_2.

second-order rate constants for the reaction:

$$R\cdot + O_2 \overset{k}{\to} RO_2 \tag{14}$$

Since O_2 has $\chi_A = 6.40$ eV, it is the electron acceptor towards all the radicals. The smallest values of χ_A denote the best electron donors. The SH radical has the same χ_A as O_2 and is unreactive since no driving force exists. The rate constants agree very well with χ_A, but not at all with χ_P.

These examples illustrate the different meanings and uses of the two EN scales. Unfortunately, the fact that the same label, electronegativity, is used for both creates ample opportunity for confusion and misunderstanding. Since the Pauling scale has the advantage of seniority and long established usage, one solution may be to find another name for absolute EN.

This presents some difficulties. If the essential meaning of electronegativity is the ability to attract and hold electrons, there is no compelling reason to restrict it to combined atoms. The extension to molecules and ions seems to be a natural and useful step. Donor–acceptor interactions are at the very heart of chemical bonding.

An alternative, of course, is to use the term 'electronic chemical potential'. But this also conflicts with an existing usage, that of the thermodynamic chemical potential. The relationship between the total energy of a chemical system and the μ of equation (1) is given by [9, 17]:

$$E = N\mu - (K - 1)T - V_{ee} + V_{nn} \tag{15}$$

V_{ee} and V_{nn} are the total electron–electron and nuclear–nuclear repulsion energies. T is the total kinetic energy and K is a parameter equal to unity for a one- or two-electron system, and about 1.57 for polyelectron systems.

If we apply (15) to the process of forming a molecule from its atoms, we obtain

$$\Delta E = E_{el} = \Delta N\mu - \Delta(K - 1)T - \Delta V_{ee} + \Delta V_{nn} \tag{16}$$

Table 3—Changes in total energy for forming some molecules from atoms[a]

Molecule	$-\Delta E^b$	$-\Delta N\mu$	$\Delta(K-1)T$	ΔV_{ee}	ΔV_{nn}
H_2	0.134	0.171	0	0.677	0.714
LiF	0.147	−0.068	−0.086	5.712	4.118
CO_2	0.419	−0.705	0.244	63.512	62.622
CO	0.289	0.110	0.979	21.720	22.514
C_2H_6	0.721	0.456	0.401	41.873	41.932
BeO	0.072	−1.560	1.080	13.723	13.127

[a] All numbers are from near Hartree-Fock calculations, and are in atomic units.
[b] ΔE is also the electronic energy of statistical thermodynamics.

For references, see W.G. Richards *et al.*, *A Bibliography of ab initio Molecular Wave Functions*, Clarendon Press, Oxford, 1971.

This ΔE is the electronic energy of statistical thermodynamics, which is the largest part by far of the ordinary chemical potential.

Taking the results of Hartree–Fock calculations for some atoms and molecules, we find the numbers given in Table 3. While $\Delta N\mu$ is larger than ΔE, it is small compared to other terms such as ΔV_{ee} and ΔV_{nn}. Furthermore, $\Delta N\mu$ can be either positive or negative, whereas ΔE is always negative. At $T = 0$, $P = 0$, E_{el} and the ordinary chemical potential must always decrease for a spontaneous process, such as forming a stable molecule from its atoms [18].

In summary, while the electronic chemical potential of density functional theory is part of the thermodynamic chemical potential, it cannot be used as a substitute. Knowing $N\mu$ for reactants and possible products does not give information on changes in bond energies or equilibrium constants.

A more useful quantity is the grand potential, defined as $\Omega = E - N\mu$. This quantity does have a minimum value for open systems. Important conclusions can be drawn from it [10, 19].

Unless a better name can be found, it appears that the term 'absolute EN' will be around for some time. This means that all scientists must be aware of the difference between χ_p and χ_A, and make it clear which scale they have in mind.

REFERENCES

[1] L. Pauling, *J. Am. Chem. Soc.* **54** (1932) 3570.
[2] R.G. Pearson, *J. Am. Chem. Soc.* **110** (1988) 7684.
[3] L. Pauling, *The Nature of the Chemical Bond*, Cornell University Press, Ithaca, 1939.
[4] For a review, see J. Mullay, *Electronegativity, Structure and Bonding*, Springer Verlag, New York, 1987, Vol. 66, p. 1.
[5] L.C. Allen, *J. Am. Chem. Soc.* **11** (1989) 9003.
[6] R.T. Sanderson, *Polar Covalence*, Academic Press, New York, 1983.
[7] R.S. Mulliken, *J. Chem. Phys.* **2** (1934) 782.

[8]   R.T. Sanderson, *Science* **121** (1955) 207.

[9]   R.G. Parr, R.A. Connelly, M. Levy and W.E. Palke, *J. Chem. Phys.* **68** (1978) 3801.

[10] For an introduction, see R.G. Parr and W. Yang, *Density Functional Theory for Atoms and Molecules*, Oxford Press, New York, 1989.

[11] See reference [4] for several examples. Also, J. Robles and L.J. Bartolotti, *J. Am. Chem. Soc.* **106** (1984) 3723.

[12] R.P. Iczkowski and J.L. Margrave, *J. Am. Chem. Soc.* **83** (1961) 3547.

[13] R.G. Parr and R.G. Pearson, *J. Am. Chem. Soc.* **105** (1983) 7512.

[14] (a) R.G. Pearson, *Inorg. Chem.* **27** (1988) 734; (b) *J. Org. Chem.* **54** (1989) 1523.

[15] R.G. Pearson and W.E. Palke, *Int. J. Quant. Chem.* **37** (1990) 103.

[16] The units of the Pauling scale are rarely used, but they are $(eV)^{1/2}$.

[17] R.G. Pearson, unpublished results.

[18] The zero-point energy must also be included, but this is small.

[19] P.K. Chataraj, H. Lee and R.G. Parr, *J. Am. Chem. Soc.*, **113** (1991) 1855.

10

BeO: the strongest neutral Lewis acid

Wolfram Koch

IBM Heidelberg Scientific Center, Heidelberg, FRG

Gernot Frenking

Philipps-Universität Marburg, Marburg, FRG

INTRODUCTION

In 1923, the same year J.N. Brønsted formulated his proton-based acid–base theory [1], G.N. Lewis introduced a different and more general definition of an acid and a base [2]. The underlying concept of his acid–base theory is the formation of an adduct between a molecule (or atom) which acts as a *two-electron (or electron-pair) donor* (the Lewis base) and an *acceptor* molecule (or atom) which has a vacant orbital of the correct symmetry to accept the electron-pair (the Lewis acid). While the definition of a base as an electron-pair donor in the Lewis theory is equivalent to the one used in Brønsted's acid–base theory (a base is a proton acceptor), Lewis' definition of an acid as an electron-pair acceptor is much broader than the definition given by Brønsted (an acid is a proton donor). For example, in all Brønsted acid–base reactions $AH + B \rightarrow A^- + BH^+$, the acid is the proton donor AH. In the Lewis picture, however, the proton itself and not the proton donor AH is the acid, since it is the vacant 1s atomic orbital of H^+ which accepts the pair of electrons from the base B. One problem pertinent to the Lewis acid–base concept is that it is not possible to determine the absolute strength of a particular Lewis acid or base. Unlike the Brønsted approach, where the strength of acids (and the respective conjugate bases) is defined with respect to H^+ and can easily be classified according to their pK_a values, the Lewis acidity depends on the nature of the base. This is due to the fact that the character and strength of the donor–acceptor bond formed between a Lewis acid and a particular base may be completely different from the bond that the same acid forms with another base. The possible types of bond formed cover the whole spectrum from mainly electrostatic bonds based on coulombic interactions (the ionic bond in LiF, for example, can be understood as the donor–acceptor complex formed from the Lewis acid Li^+ and the base F^-) to mostly covalent bonds which are governed by orbital interactions (as for example, in H_3NBH_3). This led to the qualitative classification of a Lewis acid or base as being *hard* (i.e. those which prefer electrostatic-type interactions) or *soft* (i.e. those which prefer orbital-controlled interactions) and to the empirical rule that hard acids prefer to bind to hard bases, and soft acids prefer to bind to soft bases (the HSAB principle) [3].

Despite the difficulties to give a quantitative stability ranking of Lewis acids or bases one can describe the extremes, i.e. what should be the strongest or weakest acid or base, respectively. Since a Lewis base participates either through direct donation of an electron-pair or through electrostatic interactions (such as charge/charge or charge/dipole interactions) in an acid–base reaction to form an adduct, those neutral molecules which hold their valence electrons very tightly and exhibit no dipole moment should be at the very low end of the Lewis basicity scale. Acids, which are able to form a bond to even this class of molecules or atoms, will consequently be considered as very strong Lewis acids. The weakest neutral electron donor one can think of is beyond doubt the helium atom. Its first ionization energy, which to a first approximation is a measure of how easily the electrons can be employed in a chemical reaction, is by far the highest known (24.6 eV [4]), while the atomic dipole polarizability ($0.205 Å^3$ [5]), which determines the interaction energy between a neutral atom and a molecule with a permanent dipole moment, is the lowest of all elements. However, very recently we were able to show that even He can act as a Lewis base once a strong enough acid is used [6–9]. For example, using doubly positively charged electron acceptors, the binding energies of the dative bond between He and the acceptor can be well above 80 kcal/mol [6]. Also for singly positively charged electron acceptors, He bonds with dissociation energies in excess of 20 kcal/mol were found [6]. Finally, after a systematic investigation the search for a *neutral* Lewis base capable of binding He was also successful: using beryllium oxide as the electron-accepting Lewis acid, a neutral molecule HeBeO is formed with a theoretically predicted ground state He—Be bond dissociation energy of *ca.* 3 kcal/mol [7]. In this chapter we will demonstrate that *BeO should be considered the strongest neutral Lewis acid*. Besides recapitulating noble gas BeO compounds, we will present quantum chemical investigations of BeO adducts with some other very weak Lewis bases, such as N_2, H_2, CO, ethylene, and acetylene.

The chapter is organized as follows. After a very brief description of the quantum chemical methods used, we will give an overview on how we came up with BeO as the most promising candidate as a bonding partner for helium, which then triggered all our further investigations on noble gas and BeO compounds. This will be followed by an analysis of the electronic structure of BeO in order to explain its extraordinary bonding capabilities. Finally, we will present our computational results on Lewis acid–base adducts of BeO with some neutral closed-shell molecules.

COMPUTATIONAL METHODS

Our theoretical results have been obtained using standard quantum chemical *ab initio* techniques. Molecular geometries were obtained at the Hartree–Fock and correlated MP2 levels using standard basis sets of high to very high quality. Using the optimized geometries, energy calculations have been carried out using a larger basis set and full fourth-order Møller–Plesset perturbation theory (MP4). For some systems, complete active space SCF(CASSCF) calculations were carried out as well. All stationary points were confirmed to be minima by having only positive eigenvalues in the force constant matrix. Differences in zero-point vibrational energies are accounted for in the calculation of relative energies. The exact specification of the theoretical level used will be given in the text. For a detailed description of the methodology, basis sets and electron correlation treatment, the reader is referred to the monograph by Hehre *et al.* [10].

FROM CF^{2+} TO HeBeO

Our interest in donor–acceptor adducts which finally led to the discovery of HeBeO was triggered more than six years ago by a computational investigation of CF^{n+} cations [11]. In that study we found that with increasing positive charge (up to $n = 2$) the carbon–fluorine bond becomes significantly stronger and shorter. Both CF^+ and CF^{2+} are theoretically predicted as very stable cations with surprisingly strong and short C—F bonds. The C—F bond lengths computed at MP2/6-31G(d) are 1.173 and 1.146 Å for CF^{2+} and CF^{2+}, respectively. The latter exhibits one of the shortest C—F bond lengths ever reported. Even the triply charged cation, CF^{3+}, has a rather short carbon–fluorine bond. The C—F bond in the neutral CF radical, on the other hand, is considerably longer, i.e. 1.291 Å [11]. The significant strengthening of the C—F bond with electron removal, in spite of the coulomb repulsion introduced, was attributed to very strong π-electron donation of fluorine into the vacant p_x and p_y atomic orbitals on C. Thus a formal C—F multiple bond is formed. In a subsequent CASSCF study on the corresponding neon diatomics, CNe^+ and CNe^{2+}, the same trend was seen [12]. Singly charged CNe^+ is essentially unbound in its ground state, while taking out a second electron leads to a significant stabilization. The $X^1\Sigma^+$ ground state of CNe^{2+} shows a deep minimum at an internuclear distance of 1.561 Å. A different interpretation of these unusual, strongly bound diatomics is given by employing a donor–acceptor model, based on L. Pauling's discussion of the He_2^{2+} dication, published in 1933 [13]. Although the most favorable dissociation limit for $X^1\Sigma^+ CNe^{2+}$ corresponds to $C^+(^2P) + Ne^+(^2P)$, in the bonding region the wavefunction is dominated by 'charge-polarization' terms (i.e. C^{2+}-Ne). Thus, around r_e, CNe^{2+} can be viewed as the result of a Lewis acid–base reaction involving the neon atom as the electron donor and the doubly positively charged $C^{2+}(^1S)$ as the acceptor. Further, it was shown by systematic studies on trends in binding energies of XNg^{n+} (X = Li—F, Ng = He, Ne, Ar, $n = 1, 2$) that the model of donor–acceptor interaction is better suited for an understanding of the binding in these molecules than isoelectronic reasoning [8, 14, 15].

Encouraged by the result that neon could be incorporated as an electron donor into a molecule, we commenced a series of investigations on helium containing positively charged systems. Many examples were calculated which showed that helium is indeed capable of forming rather strong and short bonds with carbon in dications and cations [6]. However, it became evident that it is not only the positive charge, but even more so the electronic state of the binding partner which is of crucial importance for the bond strength and bond length of the helium bond. For example, He_2C^{2+} has a fairly long (1.605 Å) He—C atomic distance in its $^1A_1$ ground state, but a much shorter bond (1.170 Å) is found in the $^3B_1$ excited state. Very short carbon–helium bonds (1.08–1.10 Å) are found for the $^1\Sigma^+$ (4π) states of $HeCC^{2+}$, $HeCCHe^{2+}$, and $HeCC^+$. The bond dissociation energies of the He—C bond in these electronic states (which do not necessarily represent the electronic ground states) yielding neutral He and a cationic or dicationic C_2 fragment are predicted to be as high as 89.9 kcal/mol for $HeCC^{2+}$ [6].

Again, these helium compounds are best understood as donor–acceptor molecules consisting of He as the electron donor (or Lewis base) and a suitable acceptor (or Lewis acid) fragment, like CC^{2+}. Strong helium bonds are formed when the Lewis acid provides low-lying, empty σ-orbitals accepting the He electrons and a high positive charge which adds strong, stabilizing charge–induced dipole electrostatic interactions. Using this simple model, the strong influence of the electronic state of the acceptor can easily be rationalized (Scheme 1). In the case of, for example, He_2C^{2+}, the donor–acceptor interaction of the occupied He 1s atomic orbital (AO) with the lowest unoccupied AO of ground state $C^{2+}(^1S)$, i.e. the 2p AO, yields $He_2C^{2+}(^1A_1)$. This orbital interaction is, however, much weaker than the interaction of the half-filled 2s and 2p AOs in excited $C^{2+}(^3P)$ with the $He(1s)^2AO$. Hence, the $He_2C^{2+}(^3B_1)$ molecule, resulting from the latter interaction, features significantly shorter bonds. However, the stronger binding interactions do not compensate for the $C^{2+}(^1S) \rightarrow (^3P)$ excitation energy. Thus, although more weakly bound, He_2C^{2+} assumes a $^1A_1$ electronic ground state [6]. The strong binding capabilities of the $^1\Sigma_g^+(4\pi)$ electronic state of CC^{2+} can be rationalized along similar lines. As shown in Scheme 2 this state has a very low-lying empty $2\sigma_u$-orbital available for donor–acceptor interaction with He. In contrast, the ground state of CC^{2+} ($^1\Sigma_g^+(0\pi)$) has no π electrons, but the $2\sigma_u$ and $3\sigma_g$ MOs are occupied (Scheme 2). Hence, there are no low-lying, vacant σ-MOs and we were unable to locate a minimum structure containing a He—C bond which correlates to the $^1\Sigma_g^+(0\pi)$ state of CC^{2+}.

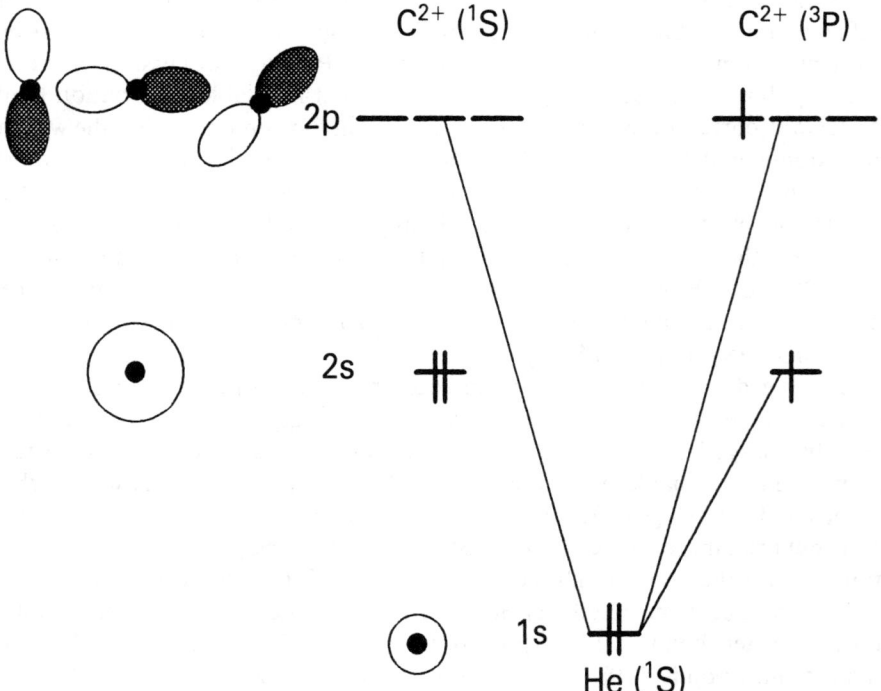

Scheme 1 — Donor-acceptor orbital interaction between $He(^1S)$ and $C^{2+}(^1S)$ and (^3P).

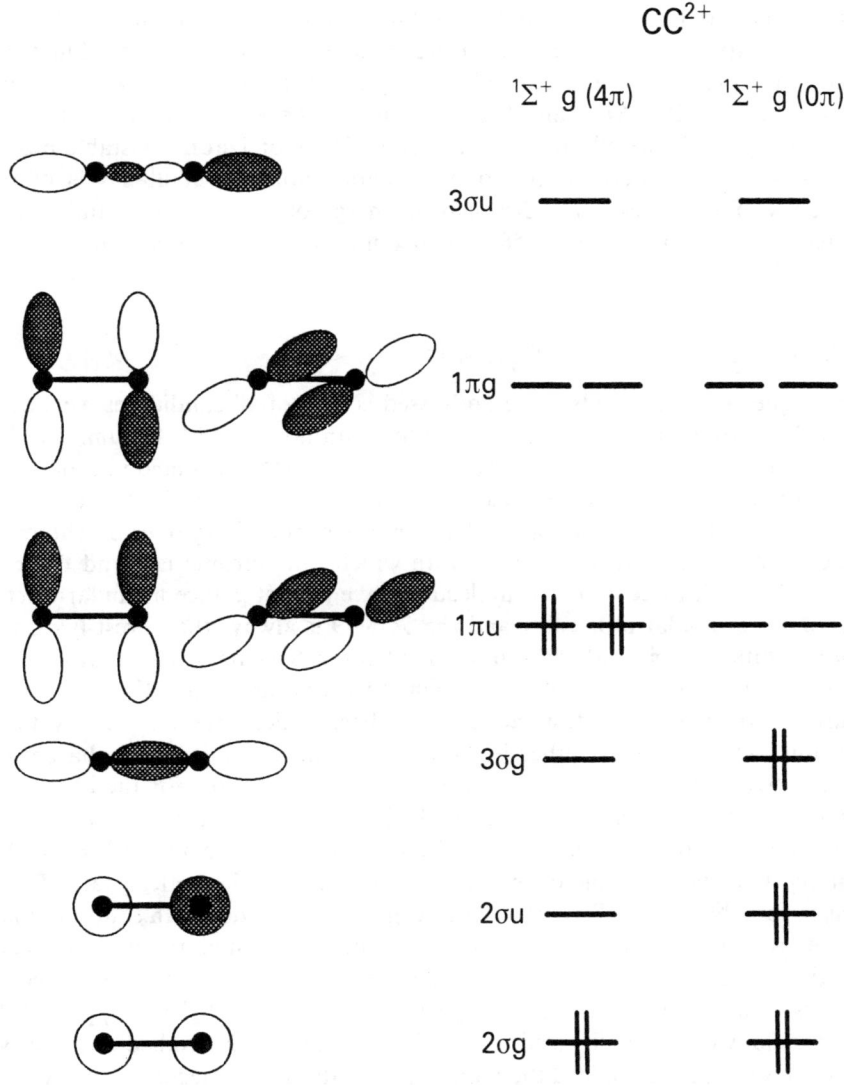

Scheme 2 — Orbital diagram for $CC^{2+}({}^1\Sigma_g^+(4\pi))$ and $({}^1\Sigma_g^+(0\pi))$.

The discovery that He binds that strongly to carbon in $HeCCHe^{2+}$ and $HeCCH^+$, and that the electronic state, not the positive charge, is decisive for the strength of the Lewis acid–base complex, prompted us to search for a *neutral* Lewis acid which might be strong enough to bind to He. Thus we had to find a neutral electron acceptor which has (i) a strong dipole moment to introduce dipole–induced dipole interactions between He and the acceptor, (ii) a high positive partial charge at the acceptor site and (iii) low-lying vacant σ-orbitals. Since CC^{2+} and CCH^+ proved to be such excellent electron acceptors [6], the first candidates we looked at were the isoelectronic neutral compounds B_2, BCH, and HCB. However, neither HeBBHe, nor HeBCH nor HeCBH were computed as being stable [7]. How should the electronic structure of the acceptor be changed to make it more suitable? While B_2, BCH and HCB do fulfil the requirement of having low-lying, vacant σ-orbitals, it is apparently the lack of the ability to polarize the He atom which prevents formation of an He adduct. So we turned to acceptors with an increased polarity and performed calculations using the isoelectronic series BN, BeO, and LiF (in their $^1\Sigma^+(4\pi)$ electronic states) as possible Lewis acids [16]. While all attempts to locate HeBN or HeLiF as stable molecules failed, HeBeO was indeed found to be a stable molecule at all levels of theory employed with a He—Be bond dissociation energy of *ca.* 3 kcal/mol and a He—Be bond length between 1.52 and 1.58 Å, depending on the theoretical approach used [6, 7, 16].

WHAT MAKES BeO SO PECULIAR?

The next question that needs to be addressed is why of all candidates we studied is only BeO a strong enough Lewis acid to form an adduct with helium. As already stated above, there are certain requirements for a neutral Lewis acid in order to form acid–base adducts with very weak electron donors: (i) a strong dipole moment, (ii) a positive partial charge, and (iii) a low-lying empty orbital. Why does beryllium oxide seem to offer the optimal compromise with which isoelectronic LiF and BN cannot compete [16]? All three acceptor molecules have at first glance a similar electronic structure. All are polar molecules, and the 5σ MO is always the lowest-lying vacant acceptor orbital, as schematically shown for BeO in Scheme 3. The relative electron affinities (EA) of the three molecules might to a first approximation be used as a measure for their affinities towards a Lewis base which interacts mostly through orbital interactions. On the other hand, dipole moments and partial charge on the electropositive end of the acceptor are the determining factors for the ability of the Lewis acids to interact with a base through electrostatic interactions. Table 1 shows our theoretically predicted electron affinities and dipole moments of LiF, BeO, and BN, along with the available experimental and previous theoretical data. The data computed for BN at the MP2 and MP3 level must be regarded with caution, since the single-determinant-based methods used are not well suited to give an accurate description of the BN $^1\Sigma^+(4\pi)$ electronic state. Therefore we also computed these properties using a multireference configuration interaction (MRCI) approach [17], which should yield reliable results for all three molecules studied. At the MP2/6-31G(d) level, BeO has indeed the highest EA of all three molecules. However, if the MP3/6-31G(d) level is used to compute the EA of BN, a value much larger than the one for BeO results. The same qualitative picture can be seen at the MRCI level. Thus, although BeO does have a fairly large EA, this cannot be the only factor responsible for its extraordinary bonding abilities. The same applies to the dipole moments μ. Depending on the level of calculation, LiF is predicted to either have a slightly larger (MP2 [16]) or slightly smaller μ than BeO (MRCI [17]). Thus, if only dipole–induced dipole forces were responsible for the attractive interactions between He and the Lewis acid, HeLiF should have been found to be a stable species, with a stability comparable to that of HeBeO.

The key to the peculiar binding ability of BeO can be found when considering more factors than only electron affinity and dipole moment. The electrostatic interaction between an approaching Lewis base and a polar acceptor is determined not only by the dipole moment of the acceptor but also by its positive partial charge.

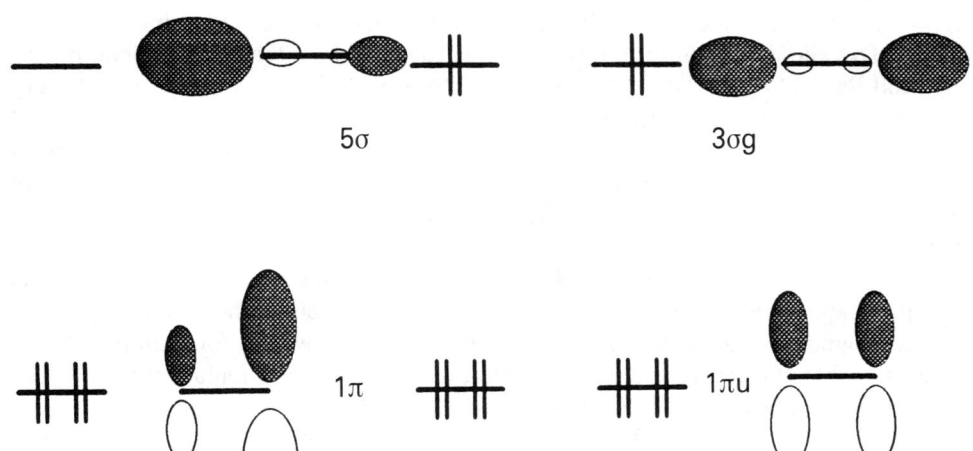

Scheme 3 — Valence molecular orbitals of BeO, N_2, and CO.

Table 1 — Electron affinities EA (in eV) and dipole moments μ (in Debye) of LiF, BeO, and BN

Species	MP2/6–31G(d,p)[a]		MP3/6–31G(d,p)		MRCI[b]		Exptl. Calc.	
	EA	μ	EA	μ	EA[c]	μ	EA	μ
LiF	0.33	5.85			0.10	6.32	0.31[d]	6.33[e]
BeO	1.98	5.40			2.06	6.43	2.15[d]	7.1[f]
BN	1.85	1.34	3.33	2.81	3.14	1.96		2.03[g]

[a] Electron affinities have been obtained using the 6–31 + G(d,p) basis set.
[b] ref. [17].
[c] MRCI energies are corrected for higher than double excitations using the multireference analog of the Davidson correction.
[d] ΔSCF value taken from Y. Yoshioka and K.D. Jordan, *J. Chem. Phys.* **73**, 5899 (1980), Y. Yoshioka and K.D. Jordan, *Chem. Phys.* **56**, 303 (1981).
[e] Exptl. value taken from D.E. Stogryn and A.P. Stogryn, *Mol. Phys.* **11**, 371 (1966).
[f] Exptl. value taken from M. Yoshimine, *J. Phys. Soc. Jpn.* **25**, 1100 (1968).
[g] Calculated (CI) value from S.P. Karna and F. Grein, *Chem. Phys.* **98**, 207 (1985).

The larger this charge, the larger the attractive charge-(induced) dipole interaction. Of course, this is the most prominent reason why cations are such efficient Lewis acids. If we turn to our three systems, it is clear that LiF, although it is a very polar molecule with a large dipole moment, cannot have a positive partial charge on Li which exceeds $+1$. BN, owing to the comparatively small difference in the electronegativities between B and N, is considerably less polar and does not exhibit a significant positive partial charge on the boron atom. In BeO, however, there is a strong electron transfer from Be to O. The bond has significant ionic character and the positive partial charge on beryllium may well be larger than $+1$ [18]. Thus, for an approaching electron acceptor, BeO, although a neutral molecule, appears like a cation. To validate this interpretation we investigated the interaction of helium with Be^+ and Be^{2+} [14]. At the MP4/6–311G(2df,2pd) level of theory, $HeBe^+$ is only very weakly bound. The dissociation energy D_0 for the $X^2\Sigma^+$ ground state is merely 0.2 kcal/mol. In contrast, $HeBe^{2+}$ is significantly stronger bound ($D_0 = 18.9$ kcal/mol) [14]. Obviously, He–Be interactions in HeBeO are considerably stronger than in $HeBe^+$, but much weaker than in $HeBe^{2+}$, in accordance with the simple electrostatic picture of BeO as having a partial charge larger than $+1$ (but necessarily smaller than $+2$) on Be. If the partial charge on Be is indeed the crucial factor responsible for the acceptor capability in BeO, it should be possible to replace the oxygen by other similar fragments. That is indeed the case. Calculations show that, for example, BeNH is also capable of forming adducts with helium. All attempts to replace the beryllium atom with, for example, Mg, Al, or B, however, failed [19].

In conclusion, the unusual Lewis acid strength of BeO has its origin in an optimal combination of a rather low-lying empty σ-orbital, a high dipole moment and, most importantly, an extraordinarily high positive partial charge on Be, giving BeO a 'pseudo-cationic' character, which exceeds a charge of $+1$ at Be [16].

LEWIS ACID–BASE ADDUCTS WITH BeO

In the preceding section we described the Lewis acid–base adduct of He with BeO. However, owing to its strong acceptor ability, BeO also binds to other, usually inert compounds. In an extension of our previous work we studied the higher homologs of HeBeO, i.e. NeBeO and ArBeO [16]. Both adducts were identified as stable species. At our highest level of theory (MP4/6–311G(2df,2pd)//MP2/6–31G(d,p) for NeBeO and MP4/6–311G(d,p)//MP2/6–31G(d,p) for ArBeO) the dissociation energies are 2.2 and 7.0 kcal/mol for the Ne and Ar compound, respectively [16]. The relatively low binding energy of NeBeO is at first glance a bit surprising. Using the simple donor–acceptor model, one would expect that going from He to Ne and Ar the attractive interaction between the noble gas atom and BeO will become stronger, since the polarizability and hence Lewis basicity increases with increasing size of the atom. However, unlike helium, neon and argon possess filled p-orbitals in their valence shells, which introduce repulsive p–π interactions involving the occupied 1π molecular orbitals in BeO.

The acceptor abilities of BeO are of course not limited to light noble gas atoms. Other candidates are the diatomics N_2, CO, and H_2, and the polyatomic molecules acetylene (C_2H_2) and ethylene (C_2H_4), each a poor closed-shell electron donor [20]. Our calculations showed that all these molecules do indeed form stable Lewis acid–base adducts with BeO [21]. Table 2 contains the computed dissociation energies of the adducts into $BeO(X^1\Sigma^+)$ and the ground state donor fragments computed at the MP4/6–311G(2df,2pd) level of approximation, while the geometries optimized at MP2/6–31G(d,p) are displayed in Scheme 4.

Table 2 — Calculated (MP4/6-311G(2df,2pd)//MP2/6-31G(d,p)) dissociation energies D_e and D_0 (in kcal/mol)

Reaction		D_e	D_0
$NNBeO(^1\Sigma^+)$ (1)	$\longrightarrow$ $N_2(^1\Sigma_g^+) + BeO(^1\Sigma^+)$	32.3	30.0
$NNBeO(^1A_1)$ (2)	$\longrightarrow$ $N_2(^1\Sigma_g^+) + BeO(^1\Sigma^+)$	11.2	10.3
$OCBeO(^1\Sigma^+)$ (3)	$\longrightarrow$ $CO(^1\Sigma^+) + BeO(^1\Sigma^+)$	43.3	40.8
$COBeO(^1\Sigma^+)$ (4)	$\longrightarrow$ $CO(^1\Sigma^+) + BeO(^1\Sigma^+)$	20.4	18.4
$HHBeO(^1A_1)$ (5)	$\longrightarrow$ $H_2(^1\Sigma_g^+) + BeO(^1\Sigma^+)$	18.5	14.9
$C_2H_2BeO(^1A_1)$ (6)	$\longrightarrow$ $C_2H_2(^1\Sigma_g^+) + BeO(^1\Sigma^+)$	41.3	39.6
$C_4H_2BeO(^1A_1)$ (7)	$\longrightarrow$ $C_2H_4(^1A_1) + BeO(^1\Sigma^+)$	45.2	42.6

The nitrogen molecule forms two different donor–acceptor adducts with BeO. A linear molecule (**1**) is found when N_2 acts as a σ-donor, while a cyclic, C_{2v} symmetric structure (**2**) results from donor–acceptor interaction between BeO and the π-system of the N_2 triple bond. Also for carbon monoxide, two different stable Lewis acid–base complexes have been located, depending on whether BeO interacts with the carbon or the oxygen side of CO as electron donor (**3** and **4**, respectively). Both adducts have a linear structure. Cyclic compounds, as in the case of N_2, were not localized. Molecular hydrogen, on the other hand, gives rise only to a cyclic adduct (**5**). The electron-donating part in acetylene and ethylene is the π-bond, and only cyclic adducts are formed with beryllium oxide (**6** and **7**, respectively). As can be seen from the data in Table 2, the computed dissociation energies D_0 of the donor–acceptor adducts into BeO and the donor are fairly large, ranging from 10.3 kcal/mol for the cyclic N_2 complex up to 42.6 kcal/mol for C_2H_4BeO. All these systems should therefore be observable under suitable experimental conditions.

The linear N_2BeO compound **1** formed as an end-on complex with N_2 as σ-donor is computed to have a N—Be bond dissociation energy of 30.0 kcal/mol. **1** is 18.6 kcal/mol more stable than the cyclic π-complex **2**, reflecting a higher σ- than π-donor ability of N_2. The better donor ability of the $3\sigma_g$ orbital can be attributed to interaction with the $2\sigma_g$ orbital, which makes the $3\sigma_g$ orbital essentially non-bonding. It acts mostly as a lone pair orbital, sticking out on both sides of N_2, as shown in the schematic representation of the relevant valence orbitals of N_2, CO, and BeO in Scheme 3. Thus, there is enhanced overlap between this MO and the 5σ acceptor MO on BeO. In addition, while electron donation from the $1\pi_u$ MO causes a weakening of the N—N triple bond, donation from $3\sigma_g$ leaves the N—N bond almost unchanged. This is also reflected in the geometrical changes of N_2 upon interaction with BeO in an end-on or side-on manner, respectively. In **1**, the N—N distance is virtually the same as in free molecular nitrogen. The formation of the side-on adduct **2**, however, is accompanied by a significantly lengthening of 0.015 Å of the N—N bond. Qualitatively very similar results are obtained when the interaction of N_2 with the prototype of a cationic Lewis acid, i.e. H^+, is studied. Linear N_2H^+ is significantly stronger bound than the cyclic alternative, which is computed to be 46.5 kcal/mol less stable (MP4/6-311G(2df,2pd)//MP2/6-31G(d,p)) than the linear form [19]. Unlike **2**, cyclic N_2H^+ is not even a minimum, but a saddle point on the N_2H^+ potential energy

Scheme 4

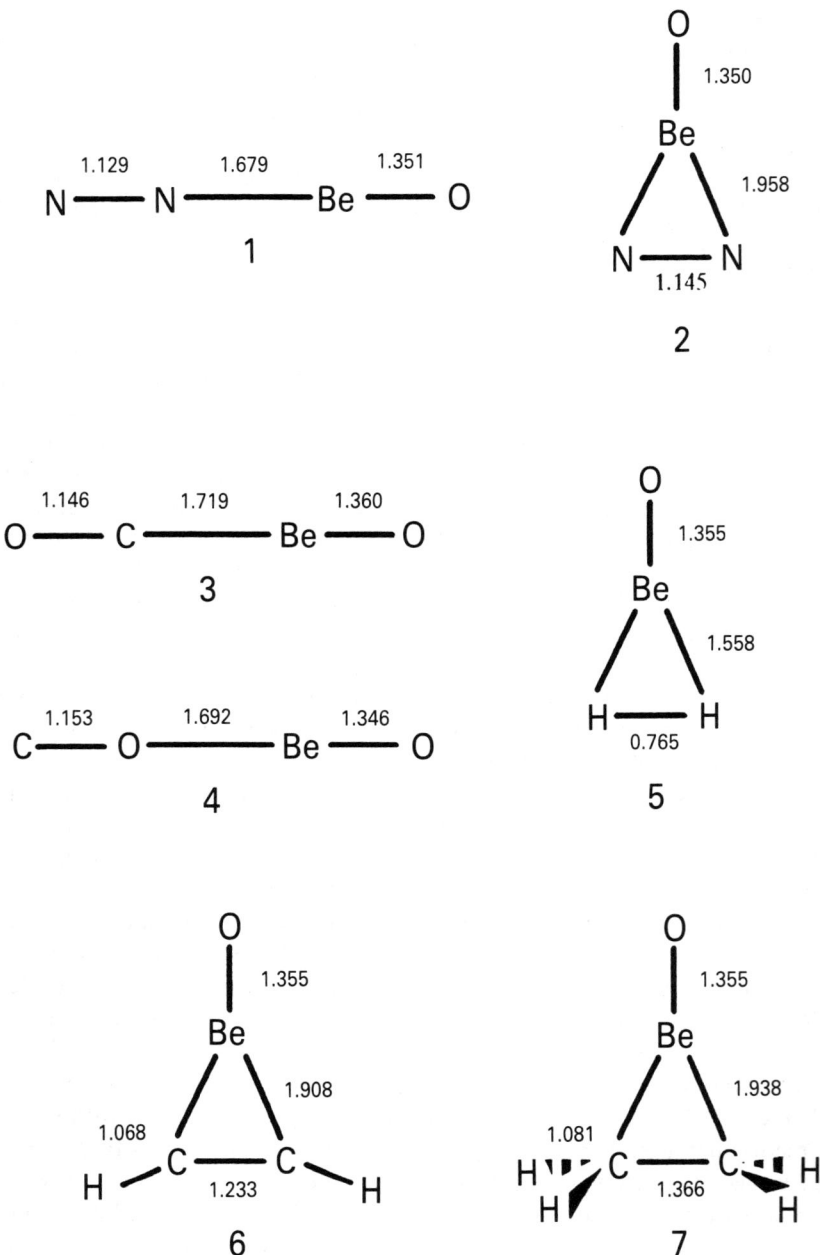

Scheme 4, cont'd

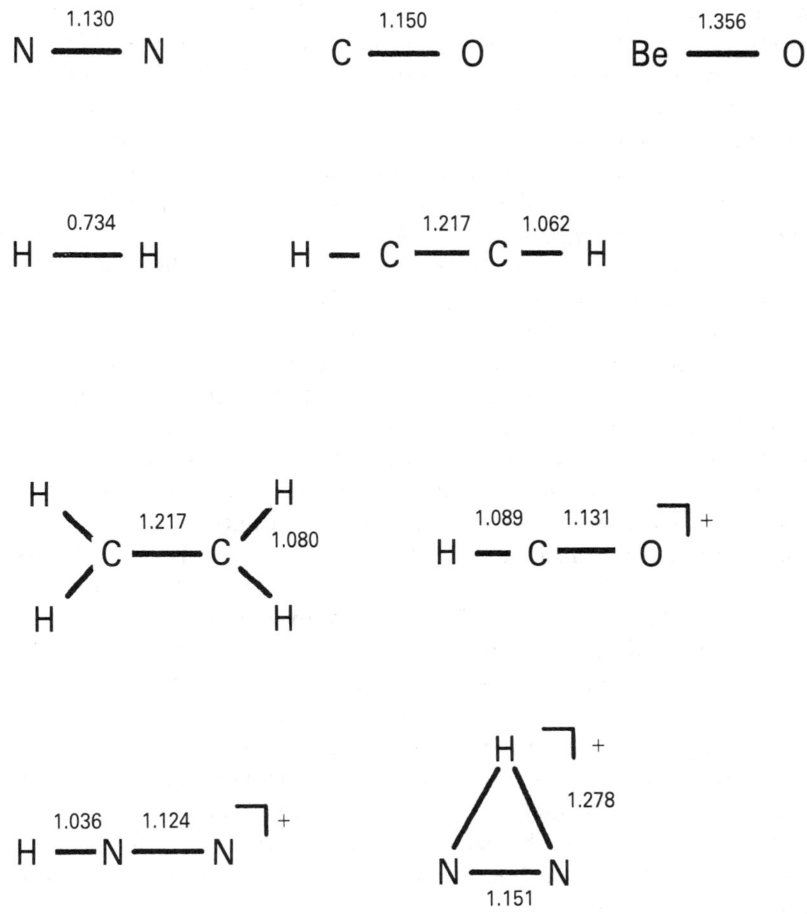

Scheme 4 — Optimized geometries (at MP2/6-31G(d,p), in Å and degree).

surface. The N—N bond in linear N_2H^+ is 0.006 Å shorter than in N_2. This is probably a consequence of the polarization of the N_2 unit due to protonation, which introduces attractive electrostatic interaction between the nitrogen atoms. In the C_{2v} isomer, however, the N—N bond is lengthened by 0.021 Å.

OCBeO (**3**) is 22.4 kcal/mol more stable than the second linear isomer, COBeO (**4**). This parallels the results found for protonation of CO. Carbon-protonated CO is computed to be *ca.* 40 kcal/mol more stable than the oxygen-protonated isomer, COH^+ [22]. The HOMO of carbon monoxide is the 5σ orbital, which is mainly located on C, giving rise to a stronger donor–acceptor interaction on this side of the molecule (see Scheme 3). The asymmetry of the 5σ orbital also explains why CO (with carbon as the active atom) is a considerably better electron donor than N_2, as demonstrated by the much higher dissociation energy of the Be—C bond of 40.8 kcal/mol (as compared to 30.0 kcal/mol for the N—Be bond in **1**) and the higher proton affinity of CO as compared to N_2 (143.2 and 118.2 kcal/mol, respectively [22]). If, on the other side, the oxygen of CO is the electron-donating center, CO is inferior to N_2 as a Lewis acid.

Protonated H_2 assumes an equilateral triangular (D_{3h}) equilibrium geometry [23]. The proton forms a 2-electron–3-center bond, employing the only available occupied $1\sigma_g$ MO on H_2. Since this interaction leads to a significant weakening of the H—H bond, the proton affinity of H_2 is much lower (101.3 kcal/mol [24]) than the PA of N_2 and CO. Using BeO instead of the proton as the electron acceptor does not qualitatively change this picture. Also H_2BeO (5) is theoretically predicted to be cyclic [25]. The electron donation from the $H_2\sigma$-bond into the BeO acceptor orbital causes an increase of the H—H distance from 0.734 Å in H_2 to 0.765 Å in 5. The dissociation energy is computed as 14.9 kcal/mol, almost 26 kcal/mol lower than for OCBeO.

The similarities described above for interactions between H^+ and N_2, CO, H_2, and the corresponding BeO chemistry also apply for the hydrocarbon Lewis bases, acetylene and ethylene. Interaction of a cation or the positive end of a dipole or an induced dipole with the π-electrons of the C—C multiple bond is known to be the initial step in the electrophilic addition to carbon–carbon double and triple bonds [26]. Symmetrically bridged, positively charged intermediates, such as, for example, the well known bromonium ion, $C_2H_4Br^+$, are formed [26]. The same applies to protonation, where it is now well established that the resulting cations are in both cases symmetrically H-bridged species [27]. BeO behaves similarly. The adducts with acetylene and ethylene are both C_{2v} symmetric molecules. Since this donor–acceptor interaction involves electron donation from the occupied π-orbitals of the multiple bond into the vacant 5σ acceptor orbital on BeO, a weakening of the parent π-system is the consequence, very similar to the situation described previously for N_2. This manifests itself by a lengthening of the CC bond of 0.016 and 0.032 Å for acetylene and ethylene, respectively. The dissociation energy of the compounds is relatively high. At MP4/6-311G(d,p) we compute 39.6 kcal/mol for C_2H_2BeO (6) and 42.6 kcal/mol for C_2H_4BeO (7). The latter is the highest dissociation energy of all BeO adducts studied in this investigation. The slightly higher dissociation energy for 7 reflects the known lower reactivity of a carbon–carbon triple bond towards electrophilic additions compared to a double bond.

SUMMARY

The examples presented in this chapter give ample evidence that BeO is an extremely powerful Lewis acid, whose strength is only surpassed by cationic electron acceptors. Its electron-accepting abilities are strong enough to form stable donor–acceptor adducts with even the least reactive molecular or atomic Lewis bases, such as the light noble gas atoms or the nitrogen molecule. The origin of the peculiar properties of beryllium oxide is the combination of a large dipole moment, a low-lying vacant acceptor molecular orbital, and an unusually high positive partial charge on the Be atom. Although neutral, BeO has great similarities to a positively charged electron acceptor, like, for example, the proton. Many of the reactions or resulting structures known from the interaction of a cation with an electron donor have their counterparts when BeO is used as Lewis acid.

ACKNOWLEDGMENT

We thank C. Thümmler for carefully reading the manuscript.

REFERENCES AND NOTES

[1] J.N. Brønsted, *Rec. Trav. Chim.* **42**, 718 (1923).
[2] For a recent monograph on Lewis acid–base theory, see: W.B. Jensen, *The Lewis Acid–Base Concept*, Wiley, New York (1980).
[3] See, for example, T. Ho, *Hard and Soft Acids and Bases Principle in Organic Chemistry*, Academic Press, New York (1977).
[4] *Handbook of Chemistry and Physics*, 65th edn, CRC Press, Boca Raton (1984).
[5] T.M. Miller and B. Bederson, *Adv. At. Mol. Phys.* **13**, 1 (1977).
[6] W. Koch, G. Frenking, J. Gauss, D. Cremer and J.R. Collins, *J. Am. Chem. Soc.* **109**, 5917 (1987).
[7] W. Koch, J. Collins and G. Frenking, *Chem. Phys. Lett.* **132**, 330 (1986).
[8] G. Frenking, W. Koch and J.F. Liebman in *From Atoms to Polymers. Isoelectronic Analogies*, J.F. Liebman, A. Greenberg (eds), VCH, New York (1989).
[9] G. Frenking and D. Cremer, *Structure and Bonding* **73**, 18 (1990) and further references cited therein.
[10] W.J. Hehre, L. Radom, P.v.R. Schleyer and J.A. Pople, *Ab initio Molecular Orbital Theory*, Wiley, New York (1986).
[11] W. Koch and G. Frenking, *Chem. Phys. Lett.* **114**, 178 (1985). See also: J. Senekowitsch, S.V. ONeil, H.-J. Weruer and P.J. Knowles *J. Chem. Phys.* **93** 562 (1990).
[12] W. Koch and G. Frenking, *J. Chem. Phys.* **86**, 5617 (1987).
[13] L. Pauling, *J. Chem. Phys.* **1**, 56 (1933). Although Pauling did not explicitly use the donor–acceptor picture adopted here, his VB-based interpretation of the bond in He_2^{2+} employs similar qualitative arguments.
[14] G. Frenking, W. Koch, J. Gauss, D. Cremer and J.F. Liebman, *J. Phys. Chem.* **93**, 3397 (1989).
[15] G. Frenking, W. Koch, J. Gauss, D. Cremer and J.F. Liebman, *J. Phys. Chem.* **93**, 3410 (1989).
[16] G. Frenking, W. Koch, J. Gauss and D. Cremer, *J. Am. Chem. Soc.* **110**, 8007 (1988).
[17] MRCI calculations of experimental re with orbitals determined in a full-valence CAS calculation. The complete CAS configuration list was used as the MRCI reference space. An atomic natural orbital (ANO) type [14s9p4d3f] → (6s5p3d2f) basis set (P.-O. Widmark, P.-Å. Malmqvist and B.O. Roos, *Theor. Chim. Acta*, in press) was employed. W. Koch, unpublished.
[18] We do not make use of computed partial atomic charges (e.g. using Mulliken polulation analysis) in our discussion. Atomic partial charges are not physical observables and there is no unambiguous way of computing them.
[19] W. Koch and G. Frenking, unpublished.
[20] Part of this has been reported in a communication: G. Frenking, W. Koch and J.R. Collins, *J. Chem. Soc., Chem. Commun.*, 1147 (1988).
[21] W. Koch, J. Collins and G. Frenking, to be published.

[22] D.J. DeFrees and A.D. McLean, *J. Comput. Chem.* **7**, 321 (1986).

[23] R.E. Christoffersen, S. Hagstrom and F.P. Prosser, *J. Chem. Phys.* **40**, 236 (1964).

[24] S.G. Lias, J.F. Liebman and R.D. Levin, *J. Phys. Chem. Ref. Data* **13**, 695 (1984).

[25] H_2BeO was very recently also studied by Nicolaides and Valtazanos. Their results are similar to ours: C.A. Nicolaides and P. Valtazanos, *Chem. Phys. Lett.* **176**, 239 (1991).

[26] J. March, *Advanced Organic Chemistry*, Wiley, New York (1985).

[27] P. Vogel, *Carbocation Chemistry*, Elsevier, Amsterdam (1985).

11

SCF–multiple scattering analysis of the electronic structure and Lewis acidity of the aluminium trihalides

E. M. Berksoy
McGill University, Montreal, Canada

and

M. A. Whitehead
Oxford University, Oxford, UK

Dedicated to Linus Pauling, scientific genius and unique maverick who inspires, on his 90th birthday.

INTRODUCTION

Aluminium trihalides AlX_3 (X = F, Cl, Br, I) have attracted much interest [1–7], because the AlX_3 are strong Lewis acids and have interesting chemical bonding. Despite the similar chemistry of the AlX_3 and BX_3, the AlX_3 dimerize, whereas the BX_3 do not. Previous theoretical studies included the prediction of the geometry of $AlCl_3$ by an *ab initio* method [1], geometry optimization calculations of AlF_3^2 and $AlCl_3^3$ with an *ab initio* method and extended Hückel calculations on all the AlX_3 series by Slanina *et al.*[4]. The electronic structure of $AlCl_3$ was calculated by the extended Hückel method [5], and the ionization potentials (IP's of $AlCl_3$ were calculated with an *ab initio* method [6]. Experimental IP's of the aluminium trihalides except AlF_3 were reported by Lappert *et al* [7]. IP's of AlF_3 were found experimentally and compared to SCF results by Dyke *et al* [8]. However, there is no single rigorous quantum-mechanical calculation of the electronic structures and properties of all the monomers. Furthermore the correlation between the theory and the observed chemistry is for only one of the compounds.

To relate the observed chemistry of the aluminium trihalides to their electronic structures, SCF–MS calculations [9] have been applied to the aluminium trihalides and their dimers. The electronic structure of the aluminium trihalide dimers and the nuclear quadrupole resonance (n.q.r.) coupling constants of the ^{27}Al and the halogens calculated by the SCF–multiple scattering method have been published [10]; the n.q.r. related experimental studies of the aluminium trihalides are given in references therein. The bonding behaviour of the halogens in the aluminium trihalides and the Lewis acidity order of the aluminium trihalides are elucidated using the SCF–MS method. The SCF–MS method does not give accurate total energies because of the muffin-tin approximation, nevertheless the molecular properties from the MS wavefunctions are accurate, and the SCF–MS is a suitable method for analyzing the relative bonding trends for elements across the row as well as down the columns of the periodic table [9c]

METHOD OF CALCULATION

The SCF–MS calculations were performed with an updated version of the SCF–MS code [1], which did not contain the correlation correction to the exchange potential, and was therefore modified [2] to include the Vosko–Wilk–Nusair (VWN) [3] exchange-correlation potential. The computational details of the aluminium trihalide dimers have been given in Ref. [2]. The computational conditions of the monomers were the same as for the dimers. Calculations were performed at the experimental geometries [4] with $X\alpha$ exchange and VWN [3] exchange-correlation potentials. For the $X\alpha$ calculations, Schwarz's α values [5] were used for the atomic regions, the inner and outer sphere α values being taken as the weighted averages of the atomic α values. The partial waves l were taken as 2 for the aluminium and halogens, and 4 for the outer sphere as in the dimers. The calculations used two different schemes for choosing the sphere sizes (Table 1). The first was based on the Pauling ionic radius definition [6] (Scheme I) in which the effective nuclear charges of the anion and cation are inversely proportional to their radii [7]. Keeping this ratio constant in both the monomer and dimer, the spheres were overlapped; the overlap which gave the experimental dimerization energy with the $X\alpha$ potential was chosen as the preferred overlap for all aluminium trihalides (Table 1). In Scheme II the sphere radii were chosen according to the Norman criterion [8]. As in the dimers, calculations were performed using both the $X\alpha$ and the VWN potentials in Scheme I. The VWN one-electron energies are about 0.03 Ry more negative than the $X\alpha$ one-electron energies, and the percentage compositions of the molecular orbitals differ by $\pm 0.2\%$. In

Table 1 — The sphere sizes used in the calculations (a.u.)

Molecule	Scheme	R_{out}	R_{Al}	R_X
AlF_3	I (25%)	5.59560	1.33167	2.51537
	II	4.77749	2.08434	1.89725
$AlCl_3$	I (20%)	6.73872	1.80799	2.84591
	II	6.31237	2.34731	2.61955
$AlBr_3$	I (22%)	7.18390	2.34945	2.89425
	II	7.00798	2.49153	2.91832
AlI_3	I (15%)	7.53411	2.37296	2.92321
	II	7.66615	2.54021	3.25524

Scheme II, the calculations were performed only with the preferred VWN exchange-correlation potentials. The quasi-relativistic wavefunctions [9] were used for the bromine and iodine in Schemes I and II with the VWN potential.

RESULTS AND DISCUSSION

Electronic structure

The aluminium trihalide monomers, AlX_3, have D_{3h} symmetrty [4, 10] and eight occupied valence orbitals, four of which span the doubly degenerate e irreducible representations of D_{3h} symmetry (Table 2). These molecular valence orbitals in all the aluminium trihalides are divided into two major groups: lower lying bonding and higher lying non-bonding orbitals. The one-electron energies, and percentage composition of $AlCl_3$, are given in Table 2 to exemplify the aluminium trihalide results. The lowest occupied valence orbital just above the core orbitals is $1a'_1$, the bonding orbital between Al(s) and Cl(s). The next is the doubly degenerate $1e'$ bonding between Al(p) and Cl(s) which is mainly localized on the halogen(s). The next bonding orbital is $2a'_1$, which has the highest Al(s) contribution to the Al(s) and Cl(p) bonding among the valence bonding orbitals; as the electronegativity of the halogen decreases, the percentage contribution of Al(s) in this orbital smoothly increases, starting from 2.5(18.9) in AlF_3 to 16.1(27.6) in $AlCl_3$ to 33.8(34.5) in $AlBr_3$ and reaching the highest percentage 41.9(35.9) in AlI_3. The figures are the Scheme I results while Scheme II results from the table are given in parentheses in the text. The next orbitals $2e'$ and $1a''_2$ are basically the bonding between Al(p) and halogen (p) orbitals. For $AlCl_3$ (Table 2), the $2e'$ bonding orbital is made of 6.6(16.0)% of Al(p) and 30.6(25.8)% of Cl(p). Similarly, the $1a'_2$ bonding orbital is made of Al(p) and Cl(p) but has less Al(p) 1.6(6.1)% and more Cl(p), 32.7(31.2)% compared to the $2e'$. The $3e'$ orbital is mainly a non-bonding halogen (p) orbital with very small percentage contribution of Al(p) varying between 0.1 and 1.0% in the aluminium trihalides. The next orbitals $1e''$ and $1a'_2$ are the non-bonding halogen (p) orbitals. As is seen for $AlCl_3$ in Table 2, basically each Cl contributes 33% of a p orbital to either the $1e''$ and $1a'_2$ orbitals. The $1a'_2$ orbital is important, being the highest occupied molecular orbital (HOMO) in these compounds. The lowest unoccupied molecular orbital (LUMO) is $3a'_1$, which is an anti-bonding orbital. The wavefunction contour maps of the $2a'_1$ and $3e'$ orbitals of $AlCl_3$ are given in Fig. 1. The $2a'_1$ is a bonding orbital and $3e'$ is a non-bonding orbital, illustrative of the wavefunction contours for all these aluminium trihalides. The contour maps show the qualitative difference between the orbitals. In the bonding orbital, Fig. 1(a), the electron cloud is shared between Al and each Cl, whereas in the non-bonding orbital, Fig. 1(b), no electron cloud occurs between Al and Cls.

The total charges on each atom in Table 3 show that the charge on Al increases from the more electronegative fluorine to less electronegative iodine in the trihalides. This increase is more profound in Scheme I since Scheme I reflects the real nature of the monomer much better than Scheme II: Scheme I makes the ionicity larger for the ionic monomers, AlF_3 and $AlCl_3$, because of less charge on Al than in Scheme II; for the most covalent monomers, $AlBr_3$ and AlI_3 the sphere sizes of Scheme I get closer to those of Scheme II; consequently the results of Table 3 approach each other.

Table 2—The one-electron energies, ε, (Ry) and percentage compositions of the molecular orbitals (MOs) of $AlCl_3$

| MO | ε | Al | | | Cl | | |
		s	p	d	s	p	d
			$X\alpha(I)$				
$3a'_1$†	−0.1512	27.87	—	2.55	15.77	3.25	4.17
$1a'_2$‡	−0.6882	—	—	—	—	33.29	0.04
$1e''$	−0.7084	—	—	0.50	—	33.16	0.01
$3e'$	−0.7141	—	0.62	1.10	0.00	32.80	0.00
$1a''_2$	−0.7280	—	1.63	—	—	32.72	0.07
$2e'$	−0.7641	—	6.62	0.55	0.09	30.56	0.30
$2a'_1$	−0.8695	16.20	—	0.60	0.80	25.99	0.95
$1e'$	−1.5724	—	1.17	0.36	32.76	0.04	0.03
$1a'_1$	−1.5922	2.66	—	0.27	32.08	0.20	0.08
			$VWN(I)$				
$3a'_1$†	−0.2287	27.64	—	2.67	15.83	3.28	4.12
$1a'_2$‡	−0.6137	—	—	—	—	33.29	0.04
$1e''$	−0.6337	—	—	0.50	—	33.16	0.06
$3e'$	−0.6454	—	0.76	1.69	0.00	32.44	0.07
$1a''_2$	−0.6572	—	1.62	—	—	32.72	0.07
$2e'$	−0.7433	—	6.60	0.55	0.09	30.56	0.30
$2a'_1$	−0.8924	16.07	—	0.61	0.80	26.03	0.94
$1e'$	−1.5526	—	1.18	0.36	32.76	0.04	0.03
$1a'_1$	−1.5691	2.69	—	0.27	32.06	0.20	0.08
			$VWN(II)$				
$3a'_1$†	−0.1578	28.29	—	16.84	9.31	4.07	4.92
$1a'_2$‡	−0.6467	—	—	—	—	33.31	0.02
$1e''$	−0.6728	—	—	2.22	—	32.57	0.02
$3e'$	−0.6901	—	0.03	4.49	0.00	31.69	0.13
$1a''_2$	−0.6996	—	6.08	—	—	31.16	0.15
$2e'$	−0.8064	—	16.01	2.92	0.45	25.84	0.73
$2a'_1$	−0.9245	27.63	—	1.45	1.76	20.73	1.15
$1e'$	−1.5724	—	4.51	1.72	30.83	0.30	1.13
$1a'_1$	−1.5978	8.15	—	1.20	29.44	0.58	0.20

$X\alpha(I)$ represents Scheme I with the $X\alpha$ potential, VWN(I) and VWN(II) Schemes I and II with Vosko–Wilk–Nusair (VWN) exchange–correlation potential.
† LUMO.
‡ HOMO.

Ionization potential

In the SCF–MS method, the ionization potentials (IPs) are calculated using Slater's transition state method, in which half an electron is removed from an orbital and the SCF–MS calculation is iterated to self-consistency. A study was made of all the transition states for all the orbitals in $AlCl_3$. Table 4 gives the IPs of $AlCl_3$ calculated

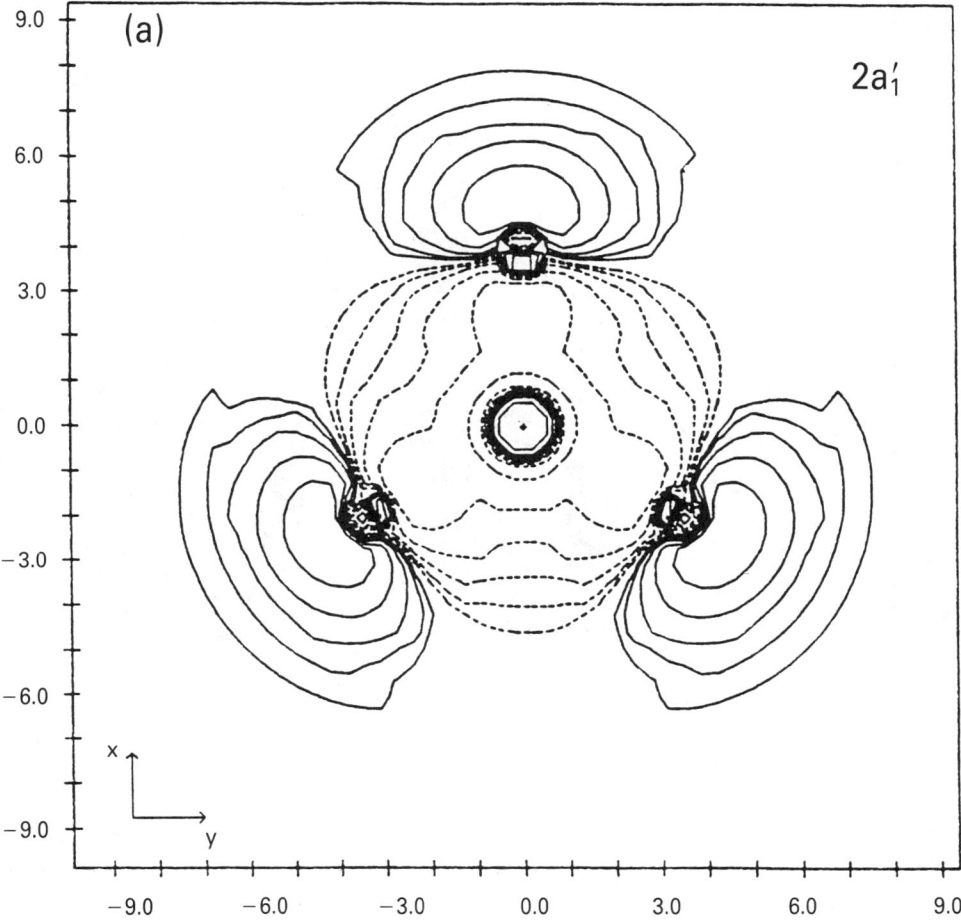

Fig. 1.—Wavefunction contour diagrams of selected molecular orbitals of $D_{3h}AlCl_3$ molecule (Scheme I with Xα exchange potential). (a) $2a_1'$ on the xy plane, (b) doubly degenerate $3e'$ on the xy plane. The contour values are in (electrons/a_0^3)$^{1/2}$ where a_0 is the Bohr radius, starting from the outermost ± 0.005, ± 0.01, ± 0.02, ± 0.04, ± 0.08, and ± 0.16. Positive wavefunction contours are indicated by solid lines, and negative by dashed lines.

using the Slater transition state in which half an electron is taken away from the orbital and reiterated until self-consistency is achieved. The electrons are taken from the orbitals across the table and ionization of the orbitals of $AlCl_3$ calculated for the orbitals down the table. The IPs are almost the same when transition state calculations are applied to each orbital in turn (underlined numbers). The IPs are almost identical for the highest occupied orbitals ($1a_2'$ to $1a_2''$). There is no change when half an electron is taken from any of the $1a_2'$, $1e''$, $3e'$ and $1a_2''$ orbitals; IPs of lowest occupied orbitals, $2e'$, $2a_1'$, $1e'$ and $1a_1'$, are approximately 0.10 eV higher, but remain constant through the series. Thus it does not matter whether the transition state calculations are done for all orbitals. Similar results were also found for IPs of BX_3 given by Preston et al. [11]. Wrinn and Whitehead [12] found that if the half-electron is taken from the highest lying totally symmetric orbital of planar

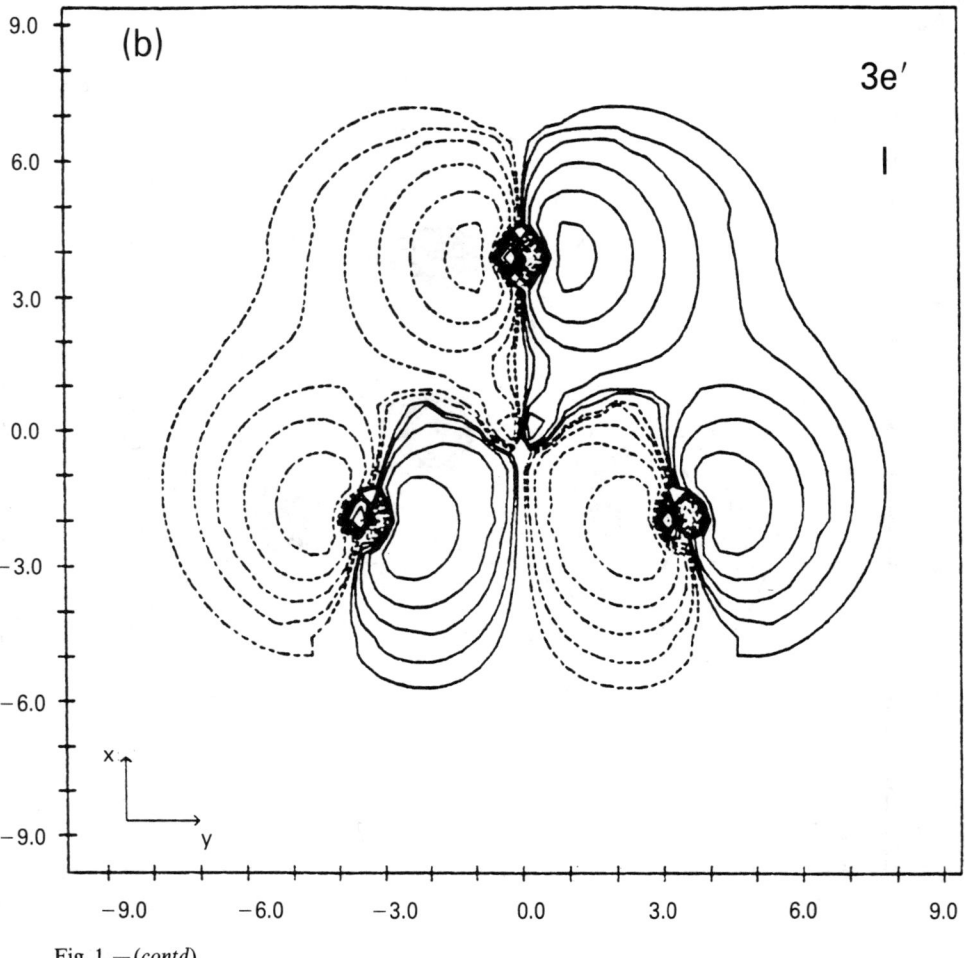

Fig. 1. — (*contd*)

molecules, the comparison with experiment is better than if half an electron is taken from any other orbital. The theoretical calculations are usually compared with the highest occupied orbitals from experiment. The results for aluminium trihalide are given along with the available experimental and calculated *ab initio* IP values in Table 5. Experimental values and *ab initio* calculations of the IP of $AlCl_3$ were given by Lappert *et al.* [13, 14]. The SCF–MS results agree with experiment, and are better than the *ab initio* calculation in which Koopmans' theorem was used. The IPs of AlF_3 are in better agreement with experiment for the highest occupied levels than *ab initio* results [15]. In all the aluminium trihalides, better agreement with experiment is achieved with Scheme I than Scheme II, showing that Scheme I, using the Pauling ionic radius definition, describes the nature of the molecule better than Scheme II. The calculated IPs of $AlBr_3$ are about 1 to 2 eV lower than those of $AlCl_3$, while the IPs of AlI_3 are only 0.5 to 1 eV lower than those of $AlBr_3$, as is expected on substitution of chlorine by the less electronegative bromine and iodine atoms. The more electronegative F makes the IPs of AlF_3 about 2 to 3 eV higher than the IPs of $AlCl_3$.

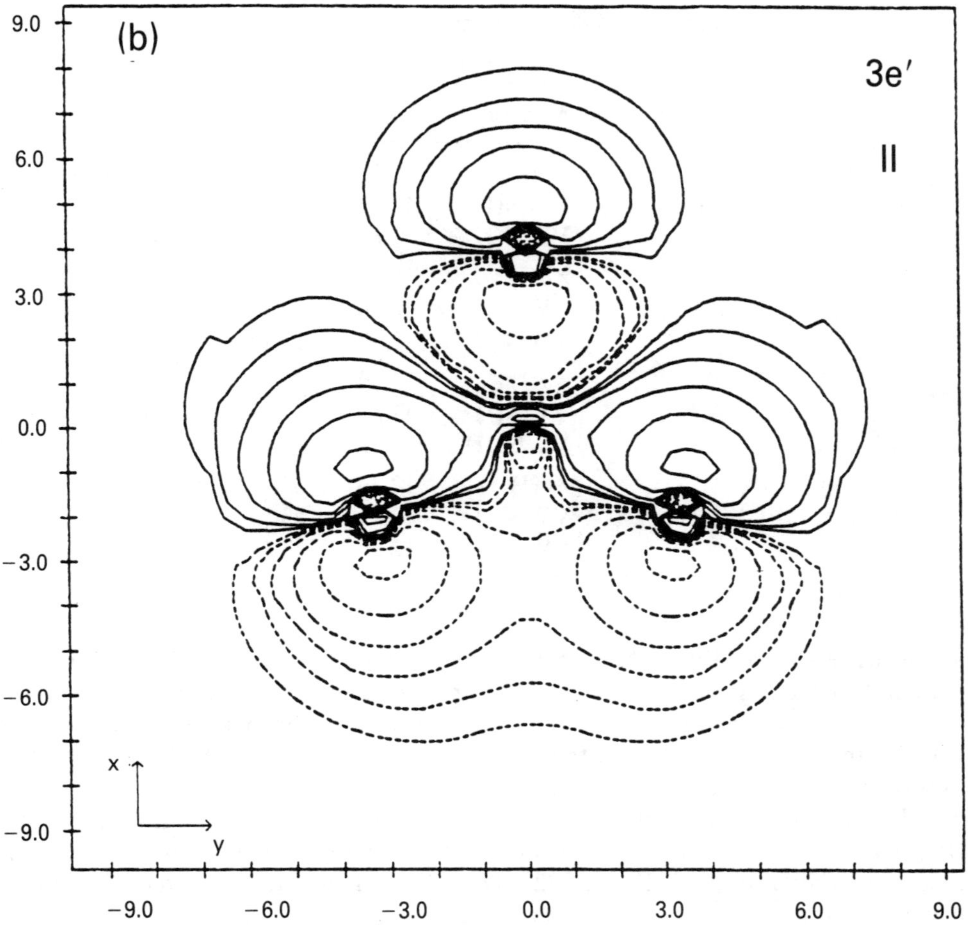

(b)

3e′

||

X

y

Fig. 1. — (contd)

Another calculation was performed on $AlCl_3$ to compare Slater's and Williams *et al.*'s [16] transition state definition. Williams *et al.*'s definition gives a more accurate value for the IPs by summing 1/4 of the ground state eigenvalue and 3/4 of the transition state energy calculated by taking away 2/3 of an electron. Table 6 compares the Slater transition state calculations for all the orbitals of $AlCl_3$ with the transition state calculations of Williams *et al.* Comparison with the experiment is also given in Table 6. Table 6 shows that Williams *et al.*'s definition gives about 0.02 eV higher IPs than Slater's, and gives almost the same agreement with experiment. However, it is even less theoretically reasonable than the Slater transition state, which accounts for the self-interaction problem in density functional theory.

Lewis acidity

The relative Lewis acidity of the aluminium trihalides and similar compounds, such as boron trihalides, has been studied by many workers [17–22]; usually the information

Table 3 — The net atomic charge distribution in the AlX_3

Molecule	Method	Al	X
AlF_3	$X\alpha(I)$	10.18	9.94
	VWN(I)	10.18	9.94
	VWN(II)	11.65	9.30
$AlCl_3$	$X\alpha(I)$	10.86	17.64
	VWN(I)	10.86	17.64
	VWN(II)	12.07	17.18
$AlBr_3$	$X\alpha(I)$	11.89	35.28
	VWN(I)	11.86	35.29
	VWN(II)	12.12	35.18
AlI_3	$X\alpha(I)$	12.30	53.09
	VWN(I)	12.21	53.13
	VWN(II)	12.26	53.14

about the relative acidity of these compounds is extracted from their thermodynamic data and sometimes combined with semi-empirical studies. A theoretical insight into the relative Lewis acidity, the power to accept an electron pair, and the relation between this relative acidity and the electronic structure from the SCF–MS calculations is given for the aluminium trihalides.

The relative Lewis acidity of the similar compounds (BX_3) has the order $BF_3 \ll BCl_3 < BBr_3$, which is contrary to the expected order from the relative electronegativities of the halogens [22]. The relative Lewis acidity of the aluminium trihalides parallels the boron trihalides. Several workers [18–21] suggested that this acidity trend was in the reverse order of the π-bonding between the metal and the halogen. The π-bonding in F compounds is greatest, and the π-bonding decreases along the series Cl, Br to I. Increased π-bonding increases the reorganization energy,

Table 4 — Ionization potentials of $AlCl_3$ (eV) using VWN potential with Scheme II from which the 1/2 e is taken

MOs	$1a_2'$	$1e''$	$3e'$	$1a_2''$	$2e'$	$2a_1'$	$1e'$	$1a_1'$
$1a_2'$	_11.73_	11.72	11.72	11.70	11.80	11.81	11.85	11.85
$1e''$	12.07	_12.06_	12.06	12.04	12.14	12.15	12.19	12.19
$3e'$	12.31	12.29	_12.29_	12.27	12.37	12.39	12.42	12.42
$1a_2''$	12.41	12.40	12.40	_12.38_	12.48	12.50	12.52	12.52
$2e'$	14.01	14.00	14.00	13.98	_14.12_	14.16	14.14	14.15
$2a_1'$	15.60	15.59	15.59	15.58	15.74	_15.81_	15.74	15.75
$1e'$	24.46	24.45	24.44	24.42	24.53	24.58	_24.59_	24.59
$1a_1'$	24.79	24.78	24.76	24.89	24.80	24.90	24.93	_24.93_

Table 5 — The ionization potential (IP) of the aluminium trihalides (eV)

Molecule	MO	SCF–MS calculation			ab initio[a]	Exp.[a]
		Xα(I)	VWN(I)	VWN(II)		
	$1a_2'$	15.09	15.44	14.65	16.72	15.45
	$1e''$	15.34	15.70	15.03	17.21	
	$3e'$	15.11	15.45	15.29	17.49	16.10
AlF_3	$1a_2''$	15.51	15.88	15.28	17.73	
	$2e'$	15.60	15.97	16.59	18.38	17.07
	$2a_1'$	15.77	16.13	17.55	19.55	20.09
	$1e'$	33.49	33.72	33.85	41.64	—
	$1a_1'$	33.64	33.87	34.01	42.00	—

					ab initio[b]	Exp.[b]
	$1a_2'$	12.22	12.82	11.73	13.54	12.01
	$1e''$	12.48	13.08	12.07	14.12	12.47
	$3e'$	12.57	13.17	12.31	14.36	12.73
$AlCl_3$	$1a_2''$	12.72	13.33	12.41	14.79	13.33
	$2e'$	13.26	13.86	14.01	15.54	14.04
	$2a_1'$	14.70	15.29	15.60	17.36	15.97
	$1e'$	24.34	24.89	24.46	—	—
	$1a_1'$	24.60	25.16	24.79	—	—

						Exp.[c]
	$1a_2'$	10.18	10.95	10.86	—	10.91
	$1e''$	10.44	11.21	11.13	—	11.53
	$3e'$	10.61	11.36	11.32	—	11.74
$AlBr_3$	$1a_2''$	10.75	11.50	11.44	—	12.37, 13.01
	$2e'$	12.02	12.77	13.02	—	13.32
	$2a_1'$	14.04	14.80	14.95	—	15.23
	$1e'$	22.31	23.84	23.92	—	—
	$1a_1'$	22.58	24.05	24.13	—	—

						Exp.[c]
	$1a_2'$	9.54	10.37	9.81	—	9.66
	$1e''$	9.82	10.63	10.10	—	10.15, 10.18
	$3e'$	9.95	10.76	10.27	—	10.46, 10.56
AlI_3	$1a_2''$	10.20	10.98	10.44	—	11.24
	$2e'$	11.24	12.03	11.99	—	11.77, 12.77
	$2a_1'$	13.55	14.41	14.15	—	14.10
	$1e'$	19.30	21.49	20.94	—	—
	$1a_1'$	19.75	21.80	21.28	—	—

[a] Ref. [15]; [b] Ref. [13]; [c] Ref. [14].

the energy required to transform the acid into an electronic state in which it can accept a pair of electrons from a donor. An increased reorganization energy makes it difficult for an acid to react with a base, and consequently reduces the acidity. The electronic structure analysis in the previous section and the calculated ionization potentials of these compounds show that the π-bonding, as well as a σ-bonding, is strong in AlF_3 and decreases through $AlCl_3$, $AlBr_3$ and AlI_3.

Beside the π-bonding of the halogen and the metal, it is also important to investigate other molecular properties which can give information about the relative acidity. The LUMO and the HOMO of the compounds are good examples of such

Table 6 — Comparison of IPs of AlCl$_3$ transition state calculations used Slater's and Williams *et al.*'s definition with experiment (eV)

MOs	Slater	Williams	Exp.[a]
1a$_2'$	11.73	11.75	12.01
1e''	12.06	12.08	12.47
3e'	12.29	12.31	12.73
1a$_2''$	12.38	12.40	13.33
2e'	14.12	14.13	14.04
2a$_1'$	15.81	15.83	15.97
1e'	24.59	24.61	—
1a$_1'$	24.93	24.94	—

[a] Ref. [13].

properties, because when a Lewis acid accepts an electron pair from a donor, the LUMO of the acid is the most likely to accommodate these electrons. The relation of the LUMO of similar compounds to their acidity was discussed by Kato *et al* [17]. They suggested that the acidity of these monomers and dimers related to the 'almost' unfilled pπ orbitals of the metal; however, their suggestion was based on a semi-empirical estimation of the LUMO, which may not be correct. One of the merits of the SCF–MS method is the correct estimation of the anti-bonding orbitals [23]. The LUMOs of the aluminium trihalides and the percentage composition of these empty

Table 7 — The percentage composition of the LUMO of AlX$_3$

Molecule	Method	Al s	Al p	Al d	X s	X p	X d
AlF$_3$	Xα(I)	2.62	—	0.19	30.92	0.57	0.89
	VWN(I)	2.55	—	0.21	30.84	0.61	0.96
	VWN(II)	33.84	—	25.77	11.67	1.50	0.29
AlCl$_3$	Xα(I)	27.87	—	2.55	15.77	3.25	4.17
	VWN(I)	27.64	—	2.67	15.83	3.28	4.12
	VWN(II)	28.29	—	16.84	9.31	4.07	4.92
AlBr$_3$	Xα(I)	33.53	—	5.24	4.20	9.67	6.54
	VWN(I)	33.81	—	4.84	4.35	9.84	6.26
	VWN(II)	31.84	—	6.42	4.80	9.43	6.34
AlI$_3$	Xα(I)	32.16	—	5.21	1.25	13.84	5.79
	VWN(I)	32.69	—	5.24	1.45	14.02	5.22
	VWN(II)	29.54	—	3.59	2.84	11.09	8.36

anti-bonding orbitals are given in Table 7. The $3a'_1$ orbital is the LUMO in the aluminium trihalides, and symmetry ensures that there is no metal p contribution to this LUMO $3a'_1$ orbital. During acid–base interaction the donor will coordinate to form an adduct with the metal of the acceptor. From Table 7 it can be seen that in aluminium, only the s orbitals are available to form such an adduct with the donor. This will lead to σ-bonding between an acceptor and a donor. This supports the idea of Cotton and Leto [18] that an acceptor (BX_3 or AlX_3) is ready to form an adduct when a σ-bond can be formed between the acceptor and donor.

To find the order of the relative acidity of these aluminium trihalides the percentage composition of the LUMO of each aluminium trihalide will be compared. Scheme I, in which the sphere sizes really reflect the natural characteristics of the molecule, ionic or covalent, shows that AlF_3 has the smallest Al(s) orbital contribution compared to the other aluminium trihalides, and that the Al(s) contribution increases smoothly, through the Cl, Br and I compounds. Thus, while AlF_3 is unable to accept an electron pair from a donor, the inability decreases along the halogens from Cl, Br to I. It is also important to consider the energy gap between the HOMO and the LUMO [24, 25], because when the difference is very large it is difficult for the LUMO to accept electrons. AlF_3 has the largest (0.7 Ry) energy difference, and the difference smoothly decreases from F to I (Table 8). Therefore from these comparisons, the relative Lewis acidity order of the aluminium trihalides is as follows:

$$AlF_3 \ll AlCl_3 < AlBr_3 < AlI_3$$

The percentage compositions of the LUMO of each aluminium trihalide, especially the ionic trihalides such as AlF_3 and $AlCl_3$, with Scheme I are different than with Scheme II. However, both Schemes I and II give essentially the same percentage compositions of the LUMO for the covalent trihalides ($AlBr_3$ and AlI_3). Thus, when

Table 8 — The energy difference, Δ, between the HOMO and LUMO of AlX_3 (Ry)

Molecule	Method	$\Delta\|\varepsilon(HOMO) - \varepsilon(LUMO)\|$
AF_3	Xα(I)	0.7051
	VWN(I)	0.6963
	VWN(II)	0.6256
$AlCl_3$	Xα(I)	0.5370
	VWN(I)	0.5416
	VWN(II)	0.4889
$AlBr_3$	Xα(I)	0.3826
	VWN(I)	0.3850
	VWN(II)	0.3907
AlI_3	Xα(I)	0.3145
	VWN(I)	0.3150
	VWN(II)	0.3389

the molecules are covalent, the choice of sphere sizes, Scheme I or Scheme II, makes no difference. However, when the molecules are ionic, it is better to choose the sphere sizes that reflect the real nature of the molecule, Scheme I. Similar results were obtained for the dimers of these aluminium trihalides [2].

To see whether the relative acidity of the aluminium trihalide monomers changes when they dimerize, the percentage compositions of the LUMOs of the dimers are given in Table 9. The percentage composition of the Al(s) orbital shows a similar increase from Al_2F_6 to Al_2I_6 molecule in Scheme I as in the monomers. For example, again concentrating on Scheme I results, in the VWN(I) calculations, the Al(s) orbital percentage composition is 1.60 for Al_2F_6 and increases to 9.61 for Al_2Cl_6 and 12.69 for Al_2Br_6, and reaching 16.09 for Al_2I_6. In the aluminium dimers, the percentage contributions of Al(p) orbital is very small. The energy gap, Δ, between HOMO and LUMO of the dimers has the same decrease as the monomers going from Al_2F_6 to Al_2I_6 (Table 10). These results show that aluminium trihalide monomers keep their acidity order when they dimerize.

CONCLUSION

The electronic structure and bonding of the aluminium trihalides were analysed by the SCF–MS method. A population analysis of the molecular orbital and the ionization potentials of the aluminium trihalides suggests that the π- and the σ-bonding is strongest in AlF_3 and decreases along the Cl-, Br-, and I-containing trihalides. Along with the analysis of HOMO and LUMO of these aluminium trihalides, the relative Lewis acidity order is found as $AlF_3 \ll AlCl_3 < AlBr_3 < AlI_3$, which supports the previous findings that state that the relative acidity order is in the reverse order of the relative electronegativities of the halogens in the aluminium trihalides. The IPs of the aluminium trihalides agree with experiment, and are much better than *ab initio* for $AlCl_3$ and an SCF result for the highest occupied orbitals of AlF_3.

Table 9 — The percentage compositions of the LUMO of Al_2X_6

Molecule	Method	Al			X_B			X_T		
		s	p	d	s	p	d	s	p	d
Al_2F_6	$X\alpha$(I)	1.67	0.70	0.10	34.89	2.58	2.68	2.43	1.10	0.17
	VWN(I)	1.60	0.68	0.10	35.13	2.36	2.64	2.46	1.12	0.17
	VWN(II)	16.00	0.50	3.30	14.52	0.19	0.35	6.83	0.67	0.07
Al_2Cl_6	$X\alpha$(I)	9.80	2.70	0.23	—	5.47	24.93	0.31	2.09	1.04
	VWN(I)	9.61	2.73	0.23	—	5.35	25.35	0.30	2.05	1.02
	VWN(II)	14.91	0.03	3.96	6.83	0.59	5.32	3.71	2.45	3.02
Al_2Br_6	$X\alpha$(I)	12.43	3.67	0.18	—	7.73	18.90	0.21	2.29	1.06
	VWN(I)	12.69	3.58	0.18	—	7.78	18.57	0.22	2.35	1.03
	VWN(II)	15.37	2.77	0.10	—	6.57	13.50	0.48	2.86	2.50
Al_2I_6	$X\alpha$(I)	15.78	1.48	1.92	1.53	2.23	7.48	0.35	8.35	1.09
	VWN(I)	16.09	1.47	1.80	1.70	2.41	6.82	0.36	8.46	1.03
	VWN(II)	16.24	1.10	1.85	1.42	2.03	5.63	0.72	8.06	2.09

Table 10 — The energy difference, Δ, between the HOMO and LUMO of Al_2X_6 (Ry)

Molecule	Method	$\Delta|\varepsilon(\text{HOMO}) - \varepsilon(\text{LUMO})|$
Al_2F_6	Xα(I)	0.6033
	VWN(I)	0.5938
	VWN(II)	0.6501
Al_2Cl_6	Xα(I)	0.4275
	VWN(I)	0.4270
	VWN(II)	0.5013
Al_2Br_6	Xα(I)	0.3645
	VWN(I)	0.3686
	VWN(II)	0.4075
Al_2I_6	Xα(I)	0.1702
	VWN(I)	0.1715
	VWN(II)	0.2021

ACKNOWLEDGEMENTS

This research was partly supported by the N.S.E.R.C. (Canada). The McGill Computing Centre is thanked for the computing time and facilities on the Amdahl 5850.

REFERENCES

[1] S.P. So and W.G. Richards, *Chem. Phys. Letters* **32**, 231 (1975).

[2] J.B. Collins, P.V.R. Schleyer, J.S. Binkley and J.A. Pople, *J. Chem. Phys.* **64**, 5142 (1976).

[3] L.A. Curtis, *Int. J. Quantum Chem.* **14**, 709 (1978).

[4] Z. Slanina and S. Beran, *J. Mol. Struct. (Theochem.)*, **92**, 1 (1983).

[5] H. Kato, K. Yamaguchi, T. Yonezawa and K. Fukui, *Bull. Chem. Soc Jpn.* **38**, 2144 (1965).

[6] M.F. Lappert, J.B. Pedley, G.J. Sharp and M.F. Guest, *J. Chem. Soc. Faraday Trans. 2* **72**, 539 (1975).

[7] (a) M.L. Lappert, J.B. Pedley, G.J. Sharp and N.P.C. Westwood, *J. Electron Spect. and Relat. Phenom.* **3**, 232 (1974); (b) G.K. Barker, M.F. Lappert, J.B. Pedley, G.J. Sharp and N.P.C. Westwood, *J. Chem. Soc. Dalton Trans.* 1765 (1975).

[8] J.M. Dyke, C. Kirby, A. Morris, B.W.J. Gravenor, R. Klein and P. Rosmus, *Chem. Phys.* **88**, 289 (1984).

[9] For reviews on the multiple scattering method, see (a) K.H. Johnson, *Adv. Quantum Chem.* 7, 143 (1973); (b) J.C. Slater, *Quantum Theory of Molecules and Solids*, (McGraw-Hill, New York, 1974) Vol. 4; (c) D.A. Case, *Ann. Rev. Phys. Chem* **33**, 151 (1982).

[10] E.M. Berksoy and M.A. Whitehead, *J. Chem. Soc., Faraday Trans 2* **84**, 1707 (1988).

[11] M. Cook and D.A. Case, Program XASW, NRCC Catalog Vol. 1, *Quantum Chem. Program Exchange Bulletin* **1**, 98 (1981).

[12] (a) S.H. Vosko, L. Wilk and M. Nusair, *Can. J. Phys.* **58**, 1200 (1980); L. Wilk and S.H. Vosko, *J. Phys. C: Solid State Phys.* **15**, 2139 (1982).

[13] JANAF Thermochemical Tables, edited by D.R. Stull and H. Prophet, NSRDS–NBS37, Washington, D.C. (1971).

[14] K. Schwarz, *Phys. Rev. B* **5**, 2466 (1972).

[15] L. Pauling, *The Nature of Chemical Bonding* (Cornell University Press, Ithaca, NY, 3rd edn., 1960).

[16] J.E. Huheey, *Inorganic Chemistry: Principles of Structure and Reactivity* (Harper and Row, New York, 3rd edn. 1983).

[17] (a) J.G. Norman, *J. Chem. Phys.* **61**, 4630 (1974); (b) *Mol. Phys.* **31**, 1191 (1976).

[18] J.H. Wood and A.M. Boring, *Phys. Rev. B* **18**, 2701 (1978); D.D. Koelling and B.N. Harmon, *J. Phys. C* **10**, 3107 (1977).

[19] J.T. Preston, J.J. Kaufman, J. Keller, J.B. Danese and J.W.D. Connolly, *Chem. Phys. Lett.* **37**, 55 (1976).

[20] M.C. Wrinn and M.A. Whitehead, *J. Chem. Soc. Faraday Trans.* **86**, 889 (1990); E.M. Berksoy, *Multiple Scattering Calculations of Large Inorganic Systems*, Ph.D. Thesis, McGill University, 1990.

[21] A.R. Williams and E. de Groot, *J. Chem. Phys.* **63**, 628 (1975).

[22] F.A. Cotton and J.R. Leto, *J. Chem. Phys.* **30**, 993 (1959).

[23] H.C. Brown and R.R. Holmes, *J. Am. Chem. Soc.* **78**, 2173 (1956).

[24] D.G. Brown, R.S. Drago and T.F. Bolles, *J. Am. Chem. Soc.* **90**, 5706 (1968).

[25] B. Swanson, D.F. Shriver and J.A. Ibers, *Inorg. Chem.* **8**, 2182 (1969).

[26] D.B. Beach and W.L. Jolly, *J. Phys. Chem.* **88**, 4647 (1984).

[27] R.G. Pearson, *J. Chem. Educ.* **64**, 561 (1987).

[28] Z. Zhou and R. Parr, *J. Am. Chem. Soc.* 111, 7371 (1989).

12

Valence bond mixing: The 'LEGO' way. From resonating bonds to resonating transition states

Sason Shaik
Ben Gurion University, Beer-Sheva, Israel

INTRODUCTION

On page 570 of his monumental book, *The Nature of the Chemical Bond* [1a], Linus Pauling points out the work of 'Eyring and Polanyi and their collaborators' in the area of chemical reactivity [2]. He then writes: 'It is to be hoped that the quantitative treatments (of Eyring and Polanyi, Evans and their collaborators [1b]) can be made more precise and more reliable; but before this can be done effectively, an extensive development of *the qualitative theory* [1b] of chemical reactions must take place, probably in terms of resonance'.

When I was kindly asked by Zvonomir Maksic to participate in this homage to Linus Pauling on the occasion of his 90th birthday, my immediate answer was of course positive. Later on I became all the more eager to participate when I encountered the above statement in *The Nature of the Chemical Bond*, and found there those roots of continuity which bridged my work to Linus Pauling's across a time gap of a few decades. Indeed, since 1981 [3] I have been engaged in developing a qualitative theory of reactivity in terms of what I call 'valence bond (VB) mixing' and what Linus Pauling refers to as 'resonance'. Alongside this qualitative theory I have also been seeking ways to test the qualitative ideas isomorphically by accurate quantitative means. Together with my collaborators, Addy Pross, Philippe Hiberty and the Orsay Group, we have moved quite a way [4] in this direction, in the spirit of that statement on page 570 in *The Nature of the Chemical Bond*.

The term 'resonance' has become somewhat detached from its original meaning and is associated in organic chemistry mainly with molecules such as benzene. Consequently 'resonance' conveys to many the electronic delocalization in the sense portrayed by benzene and its likes. I therefore prefer the term 'VB mixing' to describe the process of construction of a state wavefunction from VB building blocks. This is the 'LEGO' way [4d] of understanding the whole by piecing it up from constituents.

Like Linus Pauling (e.g. on page 563 of Reference [1a]), I too feel that this principle underlies much of our scientific thinking. In chemistry in particular this philosophy has rendered great service, starting from the classical structural theory which views a molecule as a collection of 'bonds' assembled in a certain architecture, and moving on to the concept of transferable functional groups, and to the construction of molecular orbitals from transferable group orbitals. The way of VB mixing is just that; it is a way of allowing chemical trends to be patterned and new ones to be predicted by constructing the whole from component building blocks.

What I would like to do in this chapter is to present some applications and insights of VB mixing from the simple species, *the bond*, to the more complex one, *the transition state* of a chemical reaction. In this manner I feel that I can project some sense of the continuity of qualitative valence bond ideas from the days of *The Nature of the Chemical Bond* [1a] to our own work in the present days on 'curve crossing diagrams' [3,4].

ELEMENTS OF VB THEORY

Much as in MO theory, in VB mixing too, the qualitative tools are the energy gap, which separates the configurations, and the matrix element, which determines their coupling. According to well known perturbation theoretic arguments, the mixing coefficient, λ, and energy stabilization, due to VB mixing, will be proportional to the matrix element that couples the VB configurations and inversely proportional to the energy gap that separates them. The type of matrix elements have been reviewed [5], and for qualitative purposes they are also simplified and couched in terms of the resonance integrals, β, used in MO language.

Another important factor in VB theory is the repulsion which arises from the overlap of two orbitals which contain electrons with identical spins. This overlap repulsion occurs for triplet electron pairs, and for cases with three and four electrons in two orbitals. We will make use of these matrix elements and overlap repulsion terms, but not derive them again [5]. All these elements of VB theory are applicable whether the VB theory uses pure AOs, hybrids, or fragment orbitals which take advantage of the role of orbital symmetry [3,5].

RESONATING BONDS

We define resonating bonds as those which owe their existence to resonance bonding.

Charge-shift resonance in odd-electron bonding

All odd-electron bonds are resonating [1a]. Thus, the one-electron bond H_2^+ owes $\geqslant 80\%$ of its bonding ($D = 61$ kcal/mol) [6] to the resonance of the two forms, shown in **1**. This is a *charge-shift resonance* which couples the two forms by a Hückel-type β resonance integral of the two AOs which participate in the electron shift [5]; in this

$$\text{H}\boldsymbol{\cdot} \quad \text{H}^+ \longleftrightarrow \text{H}^+ \quad \boldsymbol{\cdot}\text{H}$$

1

case these AOs are the two 1s orbitals of the H centers. The charge-shift resonance is common to all the situations where the VB configurations differ from one another by *single-electron shifts*, between two centers (or fragment orbitals). Thus, three-electron bonds are also resonating, and the most celebrated example is the O_2 molecule whose π bonding in the ground state, $^3\Sigma_g$, is a resonating three-electron bond, as shown in **2** [7].

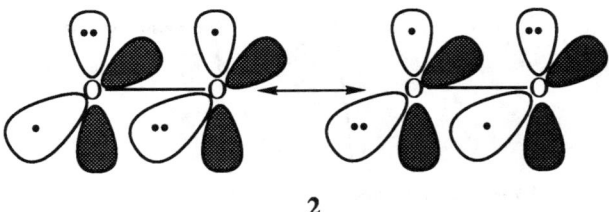

2

Resonating odd-electron bonding is ubiquitous in contemporary chemistry [8], and many novel species which involve three-electron bonds have been created and are now well known in organic chemistry, e.g. $(RS\!\therefore\!SR)^+$ and $(R_2N\!\therefore\!NR_2)^+$, and so on, where the dot-over-line symbol denotes the three electrons (line = an electron pair, and a dot = one electron). These bonds owe their entire existence to the charge-shift resonance; as such they are charge-shift resonating bonds.

Charge-shift resonance in electron-pair bonding

Pauling considers [1a] all covalent bonds to be resonance-bound. The resonance bonding originates in two conceptually different resonance types. The first one is the spin-pairing energy which arises from the *exchange-resonance* due to the mixing of the two parts of the wavefunction which differ by spin interchange. For example, for an A—B bond the exchange-resonance is shown pictorially in the mixing diagram [5] in **3**, and the corresponding resonance energy is given by $\sim 2\beta S$, where S is the overlap

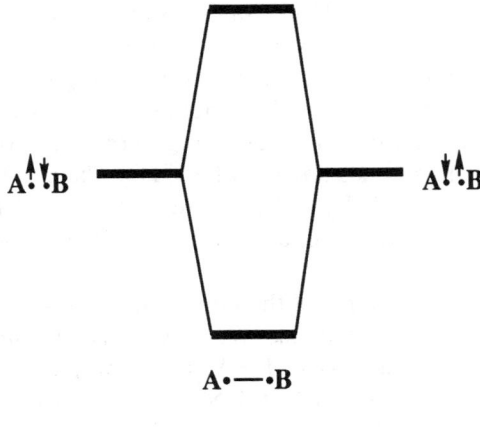

3

integral of the two orbitals which are occupied by the two electrons in **3**. The resulting covalent form of the bond is depicted by two electrons connected by a line, and is also called the Heitler–London (HL), structure. The antibonding combination in **3** is the triplet state of the bond [5] and will not concern us in this chapter.

The other source of resonance bonding arises from the charge-shift resonance between the HL form and the lowest lying polar form as shown in **4**. (In homonuclear

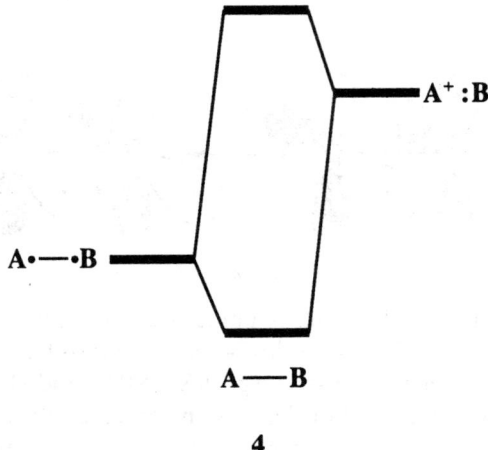

4

A—A cases, the polar form is the charge-resonance form $A^+A^- \leftrightarrow A^-A^+$.) The antibonding combination in **4** corresponds to the $^1\sigma\sigma^*$ excited state of the bond [5]. The two forms in the diagram are coupled by the charge-shift β resonance, but since they are separated by a large energy gap, the charge-shift resonance is only partly manifested into bonding. In perturbation theoretic arguments the resulting resonance energy will be given by:

$$RE(AB/A^+B^-) = \beta^2/[E_{AB} - E_{A^+B^-}] \tag{1}$$

Recently in collaboration with the Orsay group [9], we have shown that the exchange-resonance, **3**, is unimportant in some bonds, e.g. O—O, N—N, F—F, C—F, etc., and the *raison d'etre* of these bonds lies in the charge-shift resonance. What is more important is that this distinction was essential for understanding two observations about these bonds: firstly, the atomic origins of the lone-pair bond-weakening effect, which operates in these bonds, as discussed by Sanderson [10]; secondly, the connection of this bond weakening to the negative deformation density which is observed for these bonds and their like [11]. The bond-weakening effect was found to be due to the cost of hybridization which counterbalances the exchange-resonance in the covalent HL form, and made this form unbound or only slightly so [9].

Indeed, whenever the HL form of the bond is destabilized, any bonding must be sustained by the charge-shift resonance. We may envision sources of destabilization other than costly hybridization, as in F—F. For example, there exist electron-pair bonded species such as $(He—He)^{2+}$, $(H_3N—NH_3)^{2+}$, which are metastable and separated by a barrier from the fragments, e.g. $2He^+$, which lie more than 100 kcal/mol lower in energy than the bound species [12]. In these $(A—A)^{2+}$ systems, exchange-resonance is counterbalanced by the repulsion between the two positively charged centers. The main source of bonding that is left for the species is the exchange resonance due to the interaction of the HL form with the doubly charged form, A^{2+}: $A \leftrightarrow A: A^{2+}$. This was Pauling's conclusion in 1933 [13]. Gill and Radom [12c] have investigated a variety of such species by MO-based computations and interpreted the origins of the metastability of A_2^{2+} to a curve crossing between the HL and the doubly charged form, $A^{2+}: A \leftrightarrow A: A^{2+}$. A modern VB study will be required to choose between the two explanations [14]. It appears though that the generally very large gap between the HL and doubly charged configurations makes it more likely that the experimentally observed He_2^{2+} and $N_2H_6^{2+}$ species are *metastable resonating bonds* that owe their existence to the charge-shift resonance.

Why do some bonds get stronger by protonation, while other similar bonds become weaker?

Most R—X bonds (R is alkyl) become weaker by heteroatom protonation, with the exception of CH_3—NR'_2 bonds which become much stronger upon protonation on the nitrogen atom [15]. For example, in CH_3—NH_2 the C—N bond energy is ~ 80 kcal/mol, but becomes ~ 105 kcal/mol in CH_3—NH_3^+. This behavior is very different from the CH_3—OH which undergoes 25 kcal/mol C—O bond weakening upon protonation, or for that matter $(CH_3)_3C$—OH and $(CH_3)_3C$—NH_2, which are weakened by 40–72 kcal/mol.

The VB mixing diagrams in Fig. 1 provide a simple and lucid explanation for the unusual bond strengthening of CH_3—NH_2 and the more expected bond weakening in other cases. Thus, the unprotonated CH_3—NH_2 bond owes most of its bonding to the spin-pairing in the HL covalent form. Using Pauling's averaging formulae [1a], the covalent bond energy contribution can be estimated as 71 kcal/mol. The charge-shift resonance contributes therefore only ~ 9 kcal/mol to the overall bonding. This is due to the fact that the VB mixing gap between the configurations, Fig. 1(a), is governed by $IP(CH_3) - EA(NH_2)$, where the two terms are the ionization potential and electron affinity respectively. The electron affinity of NH_2 is very poor, and consequently the polar form $CH_3^+NH_2^-$ is very high lying and only a small part of the charge-shift resonance is manifested into bonding.

Two things occur upon protonation, as indicated in Fig. 1(b). Firstly, the HL form is destabilized and loses most of its bonding. The arithmetic averaging formula leads to ~ 16 kcal/mol of covalent bonding [16], while some additional stabilization is due to the polarization interaction of the positive charge on NH_3^+ with CH_3. Some of this HL destabilization is due to the fact that the protonated HL form has to be generated by a costly electron shift from the second configuration which itself becomes carbocationic upon protonation. The energy gap between the configurations is governed now by $IP(NH_3) - IP(CH_3)$, which is very small. All the effects together bring the two configurations almost into degeneracy, and the charge-shift resonance is now fully manifested into bonding. Thus, the CH_3—NH_3^+ bond owes its unusual strength and its *raison d'etre* to the charge-shift resonance.

The bond weakening experience by other bonds, e.g. CH_3—OH, $(CH_3)_3C$—OH, $(CH_3)_3C$—NH_2, upon protonation has an equally simple rationale by just considering the gap factor in the protonated R—XH^+ bond, i.e. the quantity $IP(XH) - IP(R)$. In all these weakened bonds the gap is large and the charge-shift resonance is not fully manifested into bonding. In fact a large body of data on bond strengths of protonated

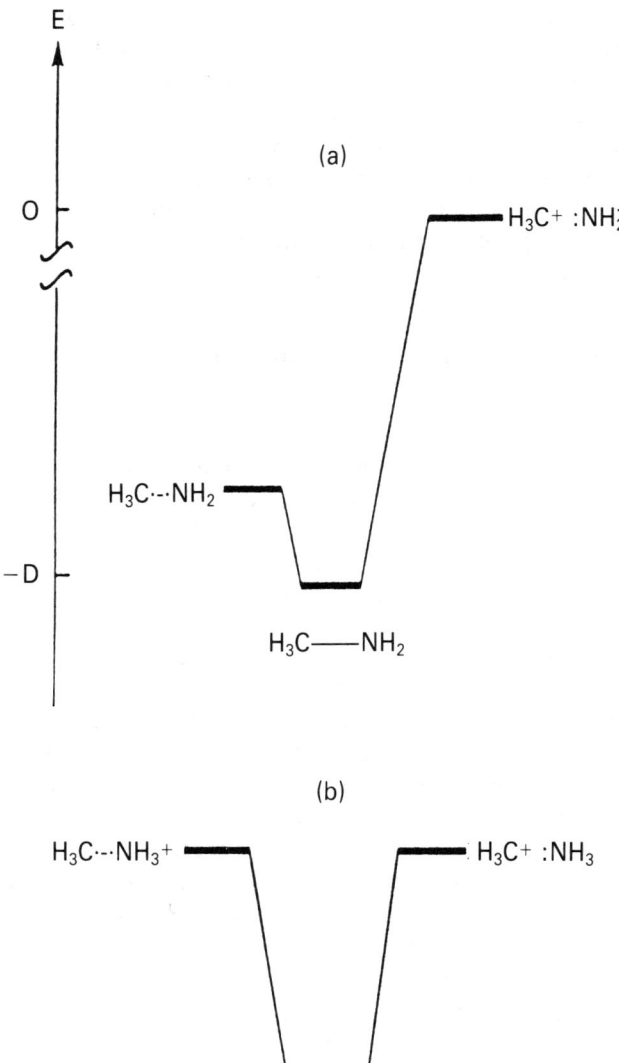

Fig. 1. — VB mixing diagrams [5], showing (a) the formation of a H_3C-NH_2 bond, and (b) the formation of the protonated $H_3C-NH_3^+$ bond.

and unprotonated species can be understood and new data can be predicted in terms of the simple idea of VB mixing. Some of the bonds are expected to be charge-shift resonating, like $CH_3-NH_3^+$, and stronger than their non-protonated analogs, while other bonds will be weakened by protonation and not enjoy full charge-shift resonance. These bonding changes are certainly expected to manifest themselves in the chemistry of the corresponding species.

Resonating Si—X bonds?

The peculiar reluctance of the 'primarily ionic' Si—X bonds to undergo heterolytic cleavage in solution [17] is probably another manifestation of the charge-shift resonance. Indeed, bonds like S—F and Si—O possess large charge-shift resonance energy, as may be estimated by use of Pauling's averaging formulae or by modern *abinitio* VB computations such as those reported in Ref. [9]. These large resonance energies mean that the bonds possess covalency despite their 'ionic' appearance in the sense of charge distribution [9]. In fact, even Ph_3SiClO_4 appears to be covalent in the solid state, as found recently by Prakash *et al* [18].

In solution there will be a competition between two opposing tendencies. On the one hand, the charge-shift resonance will tend to create strong affinity and covalency between the ions: R_3Si^+ and any negative ion (or an electron-pair donor). On the other hand, the solvation energy will tend to separate the ions. Since the charge-shift resonance is apparently large, it may well be that the Si—X covalency that binds the ions will persist even in heterolytic media and will prevent thereby occurrence of S_N1-type chemistry. Clearly the nature of the Si—X bond cannot be understood by either one of the traditional motifs: ionic and covalent. The concept of charge-shift resonance offers an attractive alternative to understand the nature of this bond.

Resonating hypercoordination

There appears to be a consensus these days that hypercoordinated molecules of main elements are not hypervalent, in the sense that $sp^3d^n(n = 1,2)$ hybridizations are inappropriate to describe these hypercoordinated species [19]. Hypercoordinated bonding must be conceptualized then in terms of the s and p AOs alone. We propose a resonating hypercoordination that originates in the charge-shift resonance. For example, XeF_2 should owe its bonding to the charge-shift resonance between a no-bond form and the ionic structures (mainly $F:^-$ $^+Xe \cdot \cdot F$ and its mirror-image form). Thus once again, while bonding in hypercoordinated molecules involves considerable 'ionic character' by criterion of charge distribution [19], the *raison d'etre* of this bonding is still the charge-shift resonance, and not so much the ionicity *per se*.

Clearly, the propensity of an atom to form strong charge-shift resonating bonds is likely to endow such an atom with an ability to form stable hypercoordinated molecules, e.g. XeF_2 yes but $XeCl_2$ less likely, and so on. When is an atom expected to be a resonating charge-shift binder? This we believe is a question worthy of future exploration [20].

A comment on resonance bonding in π-systems

A few years ago we called into question the role of resonance in π-systems, notably benzene [21]. Nowadays, we are not the only ones asking these questions [22]. Nevertheless because the questioning has generated some misunderstanding [23] and so much resistance, we wish to put this question in perspective with the importance we attribute to resonance bonding.

There is no argument that in the hexagonal geometry of benzene there is significant resonance stabilization due to the mixing of the two Kekule forms. Our own estimate of the resonance energy — or what we call the quantum mechanical resonance energy (QMRE) — is ~ 85 kcal/mol [21]. In fact, square cyclobutadiene also possesses significant resonance energy: ~ 30 kcal/mol in our studies and some 21–22 kcal/mol in the RGVB study of Voter and Goddard [24]. It is by definition that any mixing should lower the energy — this is a rule of quantum mechanics which we did not question.

Our own inquiry focused on an entirely different question: whether π-delocalization is by itself an independent driving force or rather is it a byproduct of a structural constraint? Our answer was that the π-component of benzene is a 'transition state' that wishes to localize to three short π (only) bonds, but is prohibited from doing so by the σ-frame which strongly prefers a *regular* hexagon. Being trapped in this regular geometry, the π-'transition state' undergoes full delocalization by resonance and thereby lowers its energy (relative to a single Kekule form at the regular hexagonal geometry). It follows therefore that π-delocalization is a byproduct of a structural choice made by the σ-frame and is not the root cause of this structural choice, as we were accustomed to think and teach.

This view of benzene frees the concept of resonance and VB mixing from the constraints of global stability and brings us to resonating transition states.

RESONATING TRANSITION STATES

Looking at reactivity with VB theory is not much different than looking at bonding, and this is already apparent in the seminal work of Bell, Evans and Polanyi [2b,c], who located the transition state as the lowest point on the seam of crossing of the VB curves which emanate from reactants and products.

To apply the VB mixing principle to chemical reactivity it is necessary to repeat the process along an entire reaction coordinate [3–5]. At each point we identify the contributing VB structures, and by mixing them we generate states. There are quite a few ways of carrying out the VB mixing; but we are interested in the way that will lead to the utmost lucidity and qualitative insight. As such, we found [3–5] that a separate consideration of how each structure varies along the reaction coordinate leads to the most lucid insight about chemical reactivity. In so doing, VB curves are traced along the reaction coordinate, and each one describes a particular bonding situation or ionicity. In all the reactions where bonds are broken and replaced by new ones, an intersection of two VB curves occurs; one describing the bonding motif of reactants and the other describing the bonding motif of the products. The ensuing VB mixing of these two curves and others results in resonance bonding, which is more usually known by the term 'avoided crossing' [25], and which we believe to be *a general mechanism of barrier formation in chemical reactions* [3].

To highlight the key features of the model it has been named 'the curve-crossing diagram' or alternatively 'the avoided-crossing diagram' (earlier names are VBCM and SCD [4a–c]). The many insights which are provided by the model have been reviewed numerous times [4], and what we wish to focus upon in the remainder of this section is the mechanism of activation and the particular role of the *transition state as a resonating species.*

The mechanism of activation

Consider a *single-step reaction* which involves bond breaking and making, and is generalized by $R \rightarrow P$ (reactants → products). The two curves that intersect along the reaction coordinate are made up of those VB structures which describe, on the one hand, the reactants' bonds and, on the other hand, the products' bonds, in terms of the usual mixing between HL and polar structures with some additional structures which can mix into each bonding motif [3,4h]. These curves, also called the Lewis curves, are shown in Fig. 2, and each one of them is seen to start as a ground state at one extreme and to end as an excited state (pseudo state) at the other extreme.

At the crossing point we have two degenerate wavefunctions, describing largely reactant and product bonding. Unless the two wavefunctions differ by their total symmetry (which will be a case of real crossing, as discussed on page 197 of Ref. [5]) they will mix and generate bonding and antibonding combinations, with Ψ_c the bonding combination being the transition state (TS). The TS is then a unique species along the reaction coordinate, and serves as a switching point which allows transit from the bonding motif of the reactant to that of the product, by virtue of *being both at the same time and locus.*

The barrier to the reaction can be expressed most simply as:

$$\Delta E_f^{\ddagger} = \Delta E_c - B \tag{2}$$

Equation (2) states that the barrier to the reaction is given by the height of the crossing point minus the resonance energy, B, due to avoided crossing. Since the two bonding motifs which avoid the crossing are initially separated by an excitation gap, G_R, then equation (2) is actually revealing a mechanism of barrier formation. Thus, in order to transit from R to P we must close the G_R gap by raising the energy of the reactants' bonds (by virtue of bond distortion and overlap repulsive interactions) until

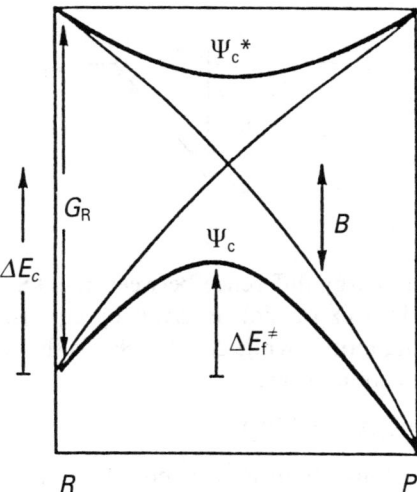

Fig. 2. — An avoided-crossing diagram [3–5] for a single-step reaction symbolized by $R \rightarrow P$. The excitation gap G_R is the energy required to excite the reactants from their bonding motif to the products' bonding motif, in a frozen geometry corresponding to that of R.

such point where the reactant bonding motif attains equal energy to the bonding motif of the product which descends in energy toward the crossing point. Then the energy is lowered by resonance mixing, which delocalizes the electrons over the old and new bonds. Since the height of the crossing point is a fraction of the gap, G_R, then with some assumptions on the size of this fraction and on the value of B it is possible to discuss the variation of barriers for many reaction types such as S_N2, radical addition, nucleophilic attacks on double bonds, atom abstractions, and so on [4].

The resonance energy of the transition state

We have recently [5,26] begun to investigate the transition state resonance energy with an aim to quantify it and eventually relate it to fundamental properties of the TS, such as its geometry, charge distribution and electron count. *Ab initio* VB computations and Morokuma-type MO analyses provide some quantitative values. Thus, for example, for an S_N2 transition state like CH_5^-, the resonance energy is ∼16 kcal/mol [4h], while for the H_3^- analog a value of 28 kcal/mol is obtained [26]. For the H_3 transition state we found a resonance energy of 43 kcal/mol [4j], while for radical addition to olefins [4g] the resonance energies are of the order of 20 kcal/mol. The largest resonance energy that we have computed [21] is 119 kcal/mol for the H_6 hexagonal transition state in the bond exchange reaction of $3H_2$. We are also trying to understand the physical meaning of our computational findings, and the attempts are so far quite encouraging.

Our qualitative reasonings are based on diagram **5**, which shows the avoided-crossing situation in terms of a VB mixing diagram. The resonance energy, B, can be

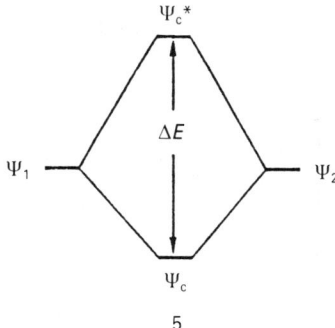

5

expressed in terms of the energy difference between the TS, Ψ_c, and its antibonding counterpart state, Ψ_c^*. This expression is given by equation (3), where S_{12} is the overlap of the two bonding motifs, while $\Delta E(\Psi_c^*, \Psi_c)$ is the electronic excitation from the TS to its excited state counterpart.

$$B = [(1 - S_{12})/2]\Delta E(\Psi_c^*, \Psi_c) \qquad (3)$$

Equation (3) allows us to think about B in effective and lucid terms. Thus, much as in ground state molecules, where excitation energies are related to bond strengths, in TSs too the stronger the bonds, the larger the $\Delta E(\Psi_c^*, \Psi_c)$, and the larger in turn is the reasonance energy of the TS. Indeed if we calibrate a simple Hückel-type method to reproduce bond energies, we can also emulate the *ab initio* values of B in a satisfactory manner [26].

The relation between B and the bond strengths in the TS will of course depend on the nature of atoms which constitute the TS and on its geometry. TSs of weak binders will possess small resonance energies, while TSs of strong binders will possess large resonance energies. Following a similar logic, very stretched bonds of a TS will endow it with a small resonance energy, while compact bonds will endow the TS with large resonance energy. This is indeed what we and others find by a quantitative exploration of B [26,27].

Another important trend is the dependence of $\Delta E(\Psi_c^*,\Psi_c)$ on the number of electrons which are delocalized over the reactive bonds. Thus, much as in ground state molecules in transition states too, certain electron counts (the Hückel $4n$ formula) will possess orbital degeneracy in a cyclic structure, and therefore $\Delta E(\Psi_c^*,\Psi_c)$ will be small and accordingly the resonance energy will also be small. These are the 'antiaromatic' TSs for the Woodward–Hoffmann forbidden reactions [28]. Other electron counts ($4n + 2$) will not have orbital degeneracies, and hence are expected to possess large $\Delta E(\Psi_c^*,\Psi_c)$ and accordingly high resonance energy. These are the 'aromatic' TSs for the Woodward–Hoffmann allowed reactions. Equation (3) thus recovers the orbital symmetry rules and makes a connection to MO theory. These symmetry properties of the TS resonance energy have been derived independently by Evans and Warhurst [29], and recently by Bernardi, Robb and collaborators [30], using a projection of MCSCF wavefunctions into VB configurations.

Finally, the S_{12} term in equation (3) is also important. This term reveals that, when everything else is constant, the more similar are the two bonding motifs, the smaller the resonance energy of the TS. This effect was observed [26] for TSs with 4 electrons delocalized over 3 centers as in the S_N2 reaction. In this case there exists one configuration, called the triple-ion, $Y:^-R^+:X^-$, which is common to both the reactant, $Y:^-$ (R—X), and the product, (Y—R)$:X^-$, wavefunctions. The larger the weight of this configuration, the smaller the B. Indeed, for a completely ionic TS there is one VB configuration and no resonance energy. This trend, which is indicated by equation (3), is fully supported by detailed *ab initio* VB computations [26,31].

The resonance energy of the TS emerges therefore as a probe of transition state properties; the geometry of the TS, its bonds' strength, its electron count and orbital nature, and finally in some cases the resonance energy of the TS are also an indicator of its charge distribution. This is quite an important property then, and there is an incentive to learn how to quantify the resonance energy for transition states of reactions under real conditions. This may not sound all that practical, but this is how we usually start on a new track.

CONCLUDING REMARKS

I have presented here applications of VB mixing to a range of problems, from the bond to the transition state of a chemical reaction [3–5]. These applications highlight the idea of resonance bonding and show the unity with which VB theory looks at chemistry. Linus Pauling has led the VB way in *The Nature of the Chemical Bond* [1a]. Some 52 years later this way is still useful and enjoying a renaissance of conceptual [3–5,30,32] and methodological developments [33].

ACKNOWLEDGMENT

This research was supported during 1989–1991 by the Basic Research Foundation administered by the Israel Academy of Sciences and Humanities. I am indebted to Mr Michael Dorfman whose assistance made it possible to complete the manuscript and meet the deadline despite the delays caused by the war in the Persian Gulf.

REFERENCES

[1] (a) L. Pauling, *The Nature of the Chemical Bond*, 3rd edn., Cornell University Press, Ithaca, New York, 1960. (b) My own emphasis and additions.

[2] (a) H. Eyring. *J. Chem. Phys.* **3** (1935) 107. (b) M.G. Evans and M. Polanyi, *Trans. Faraday Soc.* **31** (1935) 875; M.G. Evans and M. Polanyi, *Trans. Faraday Soc.* **34** (1938) 11. (c) R.P. Bell, *Proc. Roy. Soc. London* **154A**, (1936) 414.

[3] S.S. Shaik, *J. Am. Chem. Soc.* **103** (1981) 3692.

[4] (a) A. Pross and S.S. Shaik, *Acc. Chem. Res.* **16** (1983) 363. (b) S.S. Shaik, *Prog. Phys. Org. Chem.* **15** (1985) 197. (c) A. Pross, *Adv. Phys. Org. Chem.* **21** (1985) 99. (d) S.S. Shaik, *Pure Appl. Chem.* **63** (1991) 195. (e) A. Pross, *Acc. Chem. Res.* **18** (1985) 212. (f) T.H. Lowry and K.S. Richardson, *Mechanism and Theory in Organic Chemistry*, Harper and Row, New York, 1987, pp. 604–608; 359–360. (g) S.S. Shaik and P.C. Hiberty, in: *Theoretical Models for Chemical Bonding*, Ed. Z.B. Maksic, Springer Verlag, Heidelberg, 1991. (h) G. Sini, S.S. Shaik, J.M. Lefour, G. Ohanessian and P.C. Hiberty, *J. Phys. Chem.* **93** (1989) 5661. (i) G. Sini, P.C. Hiberty and S.S. Shaik, *J. Chem. Soc. Chem. Commun.* (1989) 722 (j) P. Maitre, P.C. Hiberty, G. Ohanessian and S.S. Shaik, *J. Phys. Chem.* **94** (1990) 4089.

[5] S.S. Shaik, in: *New Theoretical Concepts for Understanding Organic Reactions*, Kluwer, Dordrecht, 1989, NATO ASI Series **C267**. J. Bertran and I.G. Ciszmadia, Eds.

[6] (a) B.N. Dickinson, *J. Chem. Phys.* **1** (1933) 317. (b) E.A. Hylleraas, *Z. Physik* **71** (1931) 739.

[7] W.H. Goddard, III, T.H. Dunning, Jr, W.J. Hunt and P.J. Hay, *Acc. Chem. Res.* **6** (1973) 368.

[8] (a) K.-D. Asmus, *Acc. Chem. Res.* **12** (1979) 463. (b) W.K. Musker, *Acc. Chem. Res.* **13** (1980) 200. (c) R.W. Alder, *Acc. Chem. Res.* **16** (1983) 321. (d) T. Clark, *J. Am. Chem. Soc.* **110** (1988) 1672. (e) P.M.W. Gill and L. Radom, *J. Am. Chem. Soc.* **110** (1988) 4931. (f) M.C.R. Symons and S.P. Mishra, *J. Chem. Res. Synop* (1981) 214.

[9] G. Sini, P. Maitre, P.C. Hiberty and S.S. Shaik, *J. Mol. Struct. (THEOCHEM.)*, **229** (1991) 163.

[10] R.T. Sanderson, *Polar Covalence*, Academic press, New York, 1983.

[11] J.D. Dunitz and P. Seiler, *J. Am. Chem. Soc.* **105** (1983) 7056.

[12] (a) M. Guilhaus, A.G. Brenton, J. H. Beynon, M. Rabrenovic and P. v. R. Schleyer, *J. Phys. B* **17** (1984) L605. (b) B. Frlec, D. Gantar, L. Golic, and I. Leban, *Acta Crystallogr.* **37** (1981) 666. (c) P.M.W. Gill and L. Radom, *J. Am. Chem. Soc.* **111** (1989) 4613.

[13] L. Pauling, *J. Chem. Phys.* **1** (1933) 56.

[14] P. Maitre, Ph.D. Thesis, Université de Paris-Sud, Orsay, France, 1990. The computations in the thesis appear to favor resonating-bonding.

[15] (a) M. Meot-Ner, Z. Karpas and C.A. Deakyne *J. Am. Chem. Soc.* **108** (1986) 3913. (b) K. Hiraoka and P. Kebarle, *J. Am. Chem. Soc.* **99** (1977) 360.

[16] Since the bond energy $D(H_3N-NH_3^{2+})$ is negative (see Ref. [12c] above), only the arithmetic mean can be used:

$$D_{HL}(H_3C-NH_3^+) = 0.5[D(H_3C-CH_3) + D(H_3N-NH_3^{2+})]$$

This formula leads to 16 kcal/mol as the HL bond energy of $H_3C-NH_3^+$.

[17] Y. Apeloig, in: *Heteroatom Chemistry*, Chapter 2, Ed. E. Block, VCH, New York, 1909.

[18] G.K.S. Prakash, S. Keyaniyan, R. Aniszfeld, L. Heigler, G.A. Olah, H.K. Choi and R. Bau, *J. Am. Chem. Soc.* **109** (1987) 5123.

[19] (a) A.E. Reed and P. v. R. Schleyer, *J. Am. Chem. Soc.* **112** (1990) 1434. (b) E. Magnusson, *J. Am. Chem. Soc.* **112** (1990) 7940.

[20] S. Shaik, P.C. Hiberty and P. Maitre, in preparation.

[21] S.S. Shaik, P.C. Hiberty, G. Ohanessian and J.-M. Lefour, *J. Phys. Chem.* **92** (1988) 5086.

[22] For earlier and more recent treatments, see: (a) R. Bar and S.S. Shaik, *New J. Chem.* **8** (1984) 411. (b) N.D. Epiotis, *New J. Chem.* **8** (1984) 11. (c) J.P. Malrieu, *New J. Chem.* **10** (1986) 61. (d) L. Salem, in: *Molecular Orbital Theory of Conjugated Systems*, Benjamin; Reading, MA, 1972, pp 103–106; 494–495. (e) R.S. Berry, *J. Chem. Phys.* **35** (1961) 29; 2253. (f) E. Heilbronner, *J. Chem. Educ.* **66** (1989) 471. (g) K. Jug and A.M. Koster, *J. Am. Chem. Soc.* **112** (1990) 6772.

[23] N.C. Baird, *J. Org. Chem.* **51** (1986) 3907; P.C. Hiberty, S.S. Shaik, G. Ohanessian, J.-M. Lefour, *J. Org. Chem.* **51** (1986) 3908.

[24] A.F. Voter and W.A. Goddard, III, *J. Am. Chem. Soc.* **108** (1986) 2830.

[25] L. Salem, *Science (Washington D.C.)* **191** (1976) 822.

[26] S.S. Shaik, E. Duzzy and E. Bartuv, *J. Phys. Chem.* **94** (1990) 6574.

[27] O.K. Kabbaj, F. Volatron and J.P. Malrieu, *Chem. Phys. Lett.* **147** (1988) 393.

[28] R.B. Woodward and R. Hoffmann, *The Conservation of Orbital Symmetry*, Academic Press, New York, 1970.

[29] G. Evans and E. Warhurst, *Trans. Faraday Soc.* **34** (1938) 614.

[30] F. Bernardi, M. Olivucci and M.A. Robb, *Acc. Chem. Res.* **23** (1990) 405.

[31] G. Sini, Ph.D. Thesis, Université de Paris-Sud, Orsay, France, 1990.

[32] See, for example, recent monographs and papers: (a) N.D. Epiotis, J.R. Larson and H.L. Eaton, *Lect. Notes Chem.* **29** (1982) 1–305; N.D. Epiotis, *Lect. Notes Chem.* **34** (1983) 1–585. (b) R.D. Harcourt, *Lect. Notes Chem.* **30** (1982) 1–260. (c) M.H. McAdon and W.A. Goddard, III, *Phys. Rev. Lett.* **55** (1985) 2563; *J. Phys. Chem.* **91** (1987) 2607. (d) R. McWeeny, *Theor. Chim. Acta* **73** (1988) 115. (e) P.C. Hiberty and C. Leforestier, *J. Am. Chem. Soc.* **100** (1978) 2012. (f) M.B. Lepetit, B. Ouijia, J.P. Malrieu and D. Maynau, *Phys. Rev. A* **39** (1989) 3274; 3289; J.P. Malrieu and D. Maynau, *J. Am. Chem. Soc.* **104** (1982) 3021; 3029. A. Clotet, J.P. Daudey, J.P. Malrieu, J. Rubio and F. Spiegelmann, *Chem. Phys.* **147** (1990) 293. (g) A. Warshel, *Acc. Chem. Res.* **14** (1981) 284; J.K. Hwang, G. King, S. Creighton and A. Warshel, *J. Am. Chem. Soc.* **110** (1988) 5297. (h) D.G. Klein, *Pure. Appl. Chem.* **55** (1983) 299. (i) P.W. Anderson, *Science* **235** (1987) 1196. (j)

S.S. Shaik, *J. Am. Chem. Soc.* **104** (1982) 5328. (k) Z.G. Soos, *Isr. J. Chem.* **23** (1983) 337.

[33] (a) F.B. Bobrowicz and W.A. Goddard, III, in: *Methods of Electronic Structure Theory*, Ed. H.F. Schaefer, Plenum Press, New York, 1977. (b) D.L. Cooper, J. Gerratt and M. Raimondi, *Adv. Chem. Phys.* **59** (1987) 319. (c) J. Verbeek, Ph.D. Thesis, University of Utrecht, Holland, 1990. (d) J.H. van Lenthe and J.H. Balint-Kurti, *J. Chem. Phys.* **78** (1983) 5699. (e) P. Maitre, J.-M. Lefour, G. Ohanessian and P.C. Hiberty, *J. Phys. Chem.* **94** (1990) 4082.

13

A comparison of *ab initio* valence bond methods in terms of compactness and of qualitative description of the chemical bond

P.C. Hiberty, E. Noizet, P. Maître and G. Ohanessian
Université de Paris-Sud, France

INTRODUCTION

One important facet of quantum chemistry is its ability to provide chemists with models on which they can rely for qualitative purposes. The first successful model was that of Lewis [1], further supported by the valence bond (VB) theory of Slater and Pauling [2]. This way of thinking was later eclipsed by molecular orbital (MO) theory, in which the concept of a local two-electron bond is put into the background. Yet a chemical reaction is nothing more than a phenomenon of breaking of and/or forming of local bonds, and in that respect the qualitative VB concepts remain an irreplaceable tool in understanding chemical reactivity.

Such concepts regained some importance for a decade, particularly with the theory of the diabatic curve-crossing diagrams of Shaik and Pross [3]. In this context, *ab initio* VB methods are welcome, as they provide wavefunctions that can be interpreted in terms of the familiar Lewis bonding schemes, and may help the development of qualitative theories [4].

On the computational side, the accuracy of the Hartree–Fock method is found more and more wanting in meeting modern standards, and one rediscovers that an elegant way of taking a good deal of valence electron correlation into account is to go back to VB theory. *Ab initio* VB methods are now able to yield some wavefunctions that are considerably better than the Hartree–Fock ones, while not being much less compact, so that the most realistic view of a molecule turns out to correspond to a set of local two-electron bonds (or lone pairs), just as in the good old days. Moreover, such VB descriptions are excellent starting points for limited CI expansions, because the localization of the orbitals allows us to concentrate our computational effort on separate small-dimensional fragments.

Thus VB theory provides some compact wavefunctions which are expressed in a language familiar to chemists. But there are several brands of *ab initio* VB theories, corresponding to different definitions of the atomic orbitals (AOs), thus leading to different formal descriptions of the chemical bond. This addresses the fundamental question of the nature of the two-electron bond, which can be considered as a 100% covalent bond between 'distorted' AOs or as a mixed covalent-ionic bond between purely local AOs. This question will be examined in this chapter, the F_2 molecule serving as a computational test case to probe which description is most faithful to the true nature of the bond.

Ab initio VB methods also differ by their compactness. This is another important aspect, as it regards the effective feasibility of VB computations, and it will be discussed with a view to demarcate the different domains of various VB theories, with a computational application to C—H bond breaking in methane.

THE DIFFERENT BRANDS OF *AB INITIO* VB THEORIES

All VB theories share the common feature that the orbitals they deal with are more or less localized on single centres, rather than being fully delocalized as in MO theories. They differ on the ways these orbitals are defined, and one may highlight three categories. (i) The orbitals are orthogonal to each other, but remain as much as possible localized on a single centre. They are generally obtained through Lowdin's procedure [5], but may also arise from unitary transformation of a set of MOs by some standard localization procedure. Both techniques have been used by Malrieu *et al.* [6] in performing orthogonalized valence bond (OVB) readings of CASSCF wavefunctions. The computational advantages of dealing with orthogonal Slater determinants are obvious, with, however, the inconvenience that the significant orthogonalization tails of the so-defined orbitals may render the wavefunction difficult to interpret [7]. (ii) The wavefunction is composed of a series of formally covalent bonds and/or doubly occupied core and lone-pair orbitals. The orbitals involved in a bond pair are singly occupied and non-orthogonal to each other. Optionally, the orbitals involved in different bond pairs may be constrained to be orthogonal to each other (strong orthogonality restriction). This idea, first proposed by Hurley, Lennard-Jones and Pople (HLJP) [8], has been further developed by Goddard [9] with his generalized valence bond (GVB) method. This latter method is generally, but not necessarily, used in the perfect-pairing (PP) approximation, which means that the wavefunction displays a single VB structure, in which the singly occupied orbitals are coupled two by two in a unique way. In the spin-coupled valence bond (SCVB) method later developed by Gerratt *et al.* [10], the strong orthogonality restriction is systematically dropped, and all the spin-couplings between single occupied orbitals are generally allowed. In all these methods, which we will refer to in this chapter as HLJP methods, the orbitals are optimized for self-consistency and are allowed to be delocalized on more than one centre. As a consequence, the orbitals involved in a bond pair are generally mostly localized on a single centre but also bear some delocalization tails, especially on the second centre to which the first one is bonded. However, the delocalization generally also spreads to the other neighbouring centres, more or less importantly according to the specific case, and this can lead to difficulties in interpreting the wavefunction in terms of chemical bonding schemes. Despite this reservation, the HLJP procedure is still an elegant way of gathering the covalent and ionic parts of a bond into a unique configuration state function (see below), which has the advantage of compactness. (iii) The third type of method consists of fully preserving the atomic or fragment character of the orbitals that are used by restricting them to be strictly localized on a single centre or at least a single fragment. This of course does not prevent the orbitals from being variationally optimized, but the optimization is performed without delocalization. Then a bond pair, even if the bond is homonuclear, is described as being composed of a major covalent part (say 80%) [11] and a minor zwitterionic part, which both appear explicitly. This method, sometimes called the 'classical' valence bond (and will be so named in this chapter), somewhat sacrifices the compactness of the wavefunction but has the advantage of eliminating any pollution of the AOs by delocalization tails, thus dealing with VB structures as close as possible to the qualitative concept of the chemical bonding scheme.

IONIC COMPONENTS OF A HOMOLYTIC BOND: PHYSICAL OR ARTEFACTUAL?

The comparison between classical VB and HLJP descriptions of the single bond of H_2 was reported long ago [12], and it is well known that the formally covalent HLJP function implicitly contains some ionic classical VB structures, as illustrated below with the example of H_2. Dropping the normalization factors, and calling a and b the atomic orbitals of each hydrogen in a minimal basis set, the HLJP wavefunction of H_2 is:

$$\Psi = |a'\bar{b}'| + |b'\bar{a}'| \tag{1}$$

where a' and b' are mainly, but not entirely, localized on a single centre:

$$a' = \lambda a + \mu b; \lambda > \mu \tag{2a}$$

$$b' = \lambda b + \mu a \tag{2b}$$

It is well known that putting (2) into (1) would lead to the equivalent expression:

$$\Psi = (\lambda^2 + \mu^2)(|a\bar{b}| + |b\bar{a}|) + 2\lambda\mu(|a\bar{a}| + |b\bar{b}|) \tag{3}$$

which now has the form of a classical VB wavefunction with a major covalent component and a minor ionic one.

From the above equations, HLJP and classical VB methods may be considered to differ only by a computational aspect, and to yield the same classical picture of the two-electron bond. However, some supporters of HLJP theory went one step further, and, for a few years, argued for a novel view of the chemical bond as a qualitative concept [10,13]. In this conception, the mathematical form of the HLJP wavefunction reflects the true nature of the bond, which is 100% covalent in the case of a homolytic bond, and links together two delocalized, or distorted, orbitals, and the ionic terms of classical VB methods are considered as non-physical artefacts which just compensate for the lack of generality of strictly localized orbitals. Polar bonds are also described as formally fully covalent, but the polarity is retrieved through the unsymmetrical distortion of the orbitals. This view strongly contradicts the classical conception, in which the ionic terms of the classical VB description correspond to a physical reality, that of a charge fluctuation accompanying any bond, even a homolytic one, and are considered as present, though hidden, in the HLJP description.

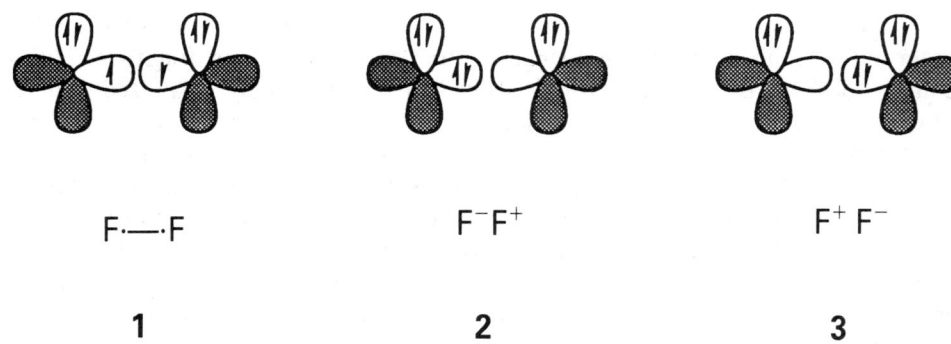

Fig. 1. — The three classical VB structures involved in the single bond of F_2. Only one lone pair is shown out of three on each fluorine atom.

This debate on the nature of the bonds appears largely as a matter of semantics if one only considers the simple example of H_2. Indeed, equations (1) and (3) are strictly equivalent in minimal basis sets and they remain quasi-equivalent when larger basis sets are used. However, the situation is different when a larger molecule is considered, because now the bonding electrons have an environment, and this environment has no reason to be the same in the covalent and ionic components of the bond. This can be illustrated with the example of the F_2 single bond, classically described as a major covalent structure **1**, in resonance with two minor ionic structures **2** and **3** (Fig. 1). As regards the orbitals surrounding the single bond, it is clear that one lone pair of the left fluorine atom does not experience the same electric field as would be the case if the bonding electrons were shared by both atoms as in **1**, both located on the left fluorine as in **2** or on the right fluorine as in **3**. The HLJP conception of the bond does not take this effect into account, as the formally 100% covalent bond is surrounded by an averaged environment, while the classical VB conception is quite compatible with a separate optimization of the inactive bonds for each VB structure, thus allowing the charge fluctuation to be accompanied by an instantaneous relaxation of the environment. If this effects is important, it means that the ionic terms are physical and that their proper treatment is essential to an accurate theoretical description of the chemical bond. If not, then the homolytic single bond may as well be considered as a 100% covalent coupling between two distorted (or delocalized) orbitals, without loss of physical meaning. To probe the importance of this effect, we have studied the homolytic single bond of a diatomic molecule, the F_2 dimer, by both HLJP and optimized classical VB methods.

THEORETICAL METHODS

Two different methods, both dealing with strictly localized AOs, will be used in this chapter to perform classical VB calculations. The first method uses a minimal number of VB structures with AOs optimized for the molecule, while in the second method the AOs are only optimized for the fragments and a selection of VB structures is performed so as to obtain wavefunctions similar, to first order, to those of the first method. Both are aimed at being applied to the study of chemical reactions, and are thus required to properly describe bond-breaking and bond-forming. The dissociation of a diatomic molecule, the F_2 dimer, has been chosen as an illustrative example of the first method, while the dissociation of methane into methyl and hydrogen fragments has been studied with the second method.

Direct optimization of the strictly localized AOs

This is the most conceptually simple way of performing a classical VB calculation, and the one that leads to the most compact wavefunction. The molecule is first considered as composed of interacting fragments, and strictly localized fragment orbitals are defined. Here, the fragments are the two fluorine atoms. The valence AOs are then divided into two sets. The first set is composed of the 'inactive' orbitals, which are not directly involved in the bonding change occurring in the reaction. In the current example, these orbitals are those of the six lone pairs of F_2. The second set includes the 'active' orbitals, in this case the two axial orbitals that are singlet-coupled in the σ bond. We then make the approximation that the intra-pair correlation energy of the inactive electrons does not change much throughout the reaction process, and that its neglect merely uniformly shifts our calculated potential surface. On the other hand, the electrons populating the active orbitals are correlated, and to that aim all the valence bond structures relevant to the correct description of the active bonds are included in the wavefunction. As an example, the correct description of the σ bond in F_2 only requires the three VB structures **1–3**, one which is purely covalent and two ionic ones, displayed in Fig. 1.

Then the coefficients of these VB structures and all the orbitals, active or inactive, are simultaneously optimized, *each VB structure having its specific set of AOs, each being different from the others*. This means that, for example, one π lone pair of the left fluorine in **1** is different from the corresponding lone pair in **2** or **3**. It is also important to note that the orbitals of one VB structure are not optimized so as to minimize the energy of that particular structure, but so as to lower the whole wavefunction involving all VB structures [14]. Throughout the optimization, the orbitals rearrange their size, their hybridization, and may distort if the basis set involves polarization functions, but remain localized on a single fragment.

The method for optimizing orbitals consists of generating *local* Brillouin states corresponding to intrafragment monoexcitations, to perform a CI in the space so-defined, and to use the CI coefficients to transform the orbitals in the following way [15]. If one calls $\Psi_{(i \to j)}$ such a Brillouin state, generated from the starting function Ψ by a promotion of one electron from an orbital ϕ_i to an orbital ϕ_j located on the same fragment, the result of the CI is the new function Ψ_{CI} (dropping normalization):

$$\Psi_{CI} = \Psi + \sum_{j \neq i} C_i^j \Psi_{(i \to j)} \tag{4}$$

Then a new set of orbitals $\{\phi_i'\}$ can be formed as:

$$\phi_i' = \phi_i + \sum_{j \neq i} C_i^j \phi_j \tag{5}$$

The procedure is repeated iteratively until all Brillouin states have a negligible coefficient in the CI. This method, which we will call in what follows the 'optimized classical VB', will be used below in the computational study of the F_2 bond [16].

CI among a set of selected VB structures, with Hartree-Fock optimized fragment AOs

This second method has been described in detail in earlier publications [17]. Briefly, it obeys the same principle as the first one, except that the process is truncated to the first iteration of the process, and that the VB wavefunction is simply taken as that arising from the CI among Brillouin states. The role of these monoexcitations is then equivalent to an optimization of the orbitals, and the final result is independent, to first order, of the quality of the starting orbitals. The method is of course valid only if the orbitals that are used are not too far from being optimal, and this can be controlled by making sure that the coefficients of the Brillouin states, in the CI, are small compared with those of the parent VB structures. Reasonably good orbitals can be defined from Hartree-Fock calculations performed on the separate fragments.

Technical details

The standard 6-31G* basis set of AOs has been used for the F_2 molecule, while a Huzinaga-Dunning basis set [18] complemented by a set of p orbitals on the leaving hydrogen has been chosen for the C-H dissociation of methane, to make possible a comparison with an earlier calculation by Carter and Goddard [19]. The starting Hartree-Fock fragment orbitals have been computed with the Monstergauss program [20], and the VB calculations for both reactions have been performed with the VB program written by Lefour and Flament [21].

THE DISSOCIATION OF F_2

The dissociation of the F_2 dimer is a classical test case for computational methods, and one knows that the Hartree-Fock method is particularly inadequate since it yields an unbound dimer [22], lying some 33 kcal/mol, in its experimental geometry, above the separated atoms, while the experimental potential well is about 38 kcal/mol [23]. On the other hand, it has been shown that quite large basis sets are necessary to reproduce this latter value by computation [24], so that our results would be best compared with some full CI calculations performed in the same DZ + P basis set. Such a study is not available in the literature, but Laidig, Saxe and Bartlett [25] have performed a series of accurate MO-CI calculations on the F_2 dimer, using a modified Dunning-Huzinaga (9s5pld/4s3pld) basis set of about the same type as ours, and estimated the full CI limit of their calculated dissociation energies as being 29.0 ± 1.5 kcal/mol. As for their calculated equilibrium distances, they cluster around the values 1.43-1.44 Å in their best calculations, vs 1.412 Å experimentally, so that we have chosen the value 1.43 Å as a good compromise for our calculations.

Generalized valence bond calculation

The active space in this simple reaction is limited to two electrons and two orbitals, those forming the bond pair along the F-F axis. The inactive space is made up of the 2s and 2p lone pairs of both fluorine atoms. In the basis set that we have used, the GVB wavefunction is already much better than the Hartree-Fock one, since it yields a potential well with the right sign. However, the dissociation energy is rather small, 15.7 kcal/mol, and this value is only improved by one kcal/mol if the equilibrium distance is optimized, leading, however, to the unrealistic value of 1.499 Å. To our knowledge, no SCVB has been performed on the F_2 molecule, but previous experience shows that GVB and SCVB wavefunctions, if both of them exhibit a unique coupling scheme, are not very different [26]. We have further analysed the content of the GVB wavefunction by projecting it onto a basis set of classical VB structures of the types **1**, **2** and **3**, built with purely local AOs. This projection involves no approximations and does not change the nature of the wavefunction [27], leaving its energy, in particular, unchanged, but allows one to calculate the weights [28] of the ionic components that are implicitly contained in the GVB formalism.

The results of this analysis, displayed in the last entry of Table 1, show that the F_2 bond is largely dominated by its covalent component, just as in any typical homopolar bond, e.g. H_2, and allow one to generalize the conclusions that we will draw below on this particular example.

Optimized classical valence bond calculations

The active space is the same as in the previous calculation, and the wavefunction is calculated by a simple 3X3 CI among the three VB structures **1-3**, each having its own specific set of AOs. The starting orbitals arise from a Hartree–Fock calculation on the F atom, and the calculation using these orbitals corresponds to iteration 0 in Table 1. The corresponding energy is quite poor, showing that a naive VB calculation using the AOs of the free atoms just as they are, without further optimization or without generating complementary VB structures, would be quite inaccurate. However, as soon as the AOs are optimized *and allowed to be different from each other* in different VB structures, the energy drops considerably (iteration 1 in Table 1), and converges rapidly, within five iterations.

It is interesting to look at the coefficients and the weights [28] of **1**, **2** and **3** as the wavefunction converges to its final form. The classical VB calculation at iteration zero (using the AOs of the free atom) and the GVB calculation both overestimate

Table 1 — Results of an optimized classical VB calculation of F_2, at an equilibrium distance of 1.43 Å

| Iteration | Energy (au) | D_e (kcal/mol) | Coefficients (weights) | |
			Covalent **1**	Ionic **2** or **3**
0	−198.71314	−4.6	0.840 (0.813)	0.194 (0.094)
1	−198.75972	24.6	0.772 (0.731)	0.249 (0.134)
2	−198.76494	27.9	0.754 (0.712)	0.258 (0.144)
3	−198.76572	28.4	0.751 (0.709)	0.260 (0.146)
4	−198.76600	28.5	0.752 (0.710)	0.259 (0.145)
5	−198.76608	28.6	0.750 (0.707)	0.261 (0.146)
Projected GVB	−198.74554	15.7	(0.768)	(0.116)

covalency, while the ionic weights and coefficients gradually increase as the classical VB calculation converges, indicating that the relief of the mean-field environment constraint favours the ionic structures more than the covalent one. Indeed, in both iteration zero and the GVB calculation, the inactive orbitals are optimized for an average neutral atomic population and are consequently well adapted to the covalent structure **1**. On the other hand, the ionic structures **2** and **3** are not well accommodated in such surroundings. As a consequence, they lie too high in energy and cannot contribute to the wavefunction as much as they should, thus leading to underestimated ionic coefficients and to a poor binding energy.

The energetic improvement brought by an adequate treatment of the ionic components is quite significant, as can be judged from the difference between the GVB energy and that of iteration 5: 12.9 kcal/mol, which brings the final dissociation energy of F_2, in the optimized classical VB description, amazingly close to the estimated full CI limit in a similar basis set. We believe that this excellent agreement is not fortuitous, for two reasons: (i) the approximation that is made — double occupancy of the inactive orbitals and its consequent neglect of most of the intra-atomic correlation energy throughout the reaction — is well defined and physically grounded; as least when the bond is not polar, and (ii) all the classical VB calculations that we have performed to date have yielded very good dissociation energies and reaction barriers [4,29], as compared with the best MO-CI calculations using the same basis sets. Lastly, it should be noted that our optimized classical VB wavefunction is not the most general one of its type, and that its energy could still be lowered by replacing the doubly occupied active AO of each ionic structure by a pair of singlet-coupled orbitals of different sizes, so as to take into account the radial correlation of active electrons. This improvement does not alter the compactness of the wavefunction nor its interpretation in terms of **1–3**. Before such calculations are effectively done and appear in a forthcoming paper, it remains that, on the basis of calculated bond energies, a simple description of the F_2 bond in terms of **1–3** captures the essence of the bond while a 100% covalent description does not, even if full generality is given to the orbitals.

COMPACTNESS OF VARIOUS VB APPROACHES

No VB method can beat the HLJP approach, at least in the perfect-pairing approximation, on the ground of compactness. With the exception of delocalized systems, the description of a multi-bond molecule only necessitates one (formally all-covalent) VB structure. As an example, the perfect-pairing SCVB energy of methane, in the DZ + P basis set, is already quite close (only 1.8 kcal/mol higher) to the full SCVB energy calculated with all possible spin couplings between the eight valence orbitals [26].

The classical VB method is by nature less compact, since three VB structures (four determinants) are needed to describe a single bond, vs only one VB structure (two determinants) in the HLJP method. As a consequence, the whole VB description of a multi-bond molecule yields a complicated wavefunction when the molecule gets large. Indeed, the number of possible bonding schemes is 3^n, n being the number of bonds in the molecule, if all the combinations between covalent and ionic bonds are considered, in the perfect-pairing approximation. Even worse, this latter approximation is no more valid in resonating systems such as conjugated molecules and one now could generate, in all rigour, 175 bonding schemes for the π system of benzene, 1764 for that of cyclooctatetraene, etc. Of course, many of these bonding schemes might be *a priori* discarded as being physically unrealistic, such as the multi-ionic structures or those displaying funny orbital couplings. However, such a truncation would be somewhat arbitrary in that the loss of accuracy it would lead to would remain uncertain. It is thus our opinion that the classical VB approach is not well suited to the all-VB treatment of large or conjugated molecules.

Such a treatment, however, is not at all necessary in a vast domain of quantum chemistry: the study of chemical reactions. Indeed, wide classes of reactions proceed through elementary steps only involving a few active orbitals and electrons (typically three or four), and the remaining orbitals can be treated at the uncorrelated level, just as we did for the F_2 dissociation reaction and a number of other ones [4, 29]. Then the inactive electrons are frozen into doubly occupied, e.g. Boys-localized, orbitals. The number of bonding schemes is thus greatly reduced, amounting to six for all the family of S_N2 reactions [30], eight for radical exchanges [4d], substitutions, etc.

Now a distinction must be made between the numbers of bonding schemes, which matters for the simplicity of interpretation, and the dimension of the CI space, which matters for the practical feasibility of the calculation. In the optimized classical VB method, these two numbers are identical, as each bonding scheme is described by a single, optimized, VB configuration. On the other hand, the ordinary (not optimized) classical VB method (see section 'CI among a set of selected VB structures, with Hartree–Fock optimised fragment AOs') obviously deals with larger CI spaces, because one bonding scheme is described by a linear combination of several VB configurations. Can this latter method be applied successfully to the study of a reaction, while keeping a tractable CI space? In a recent SCVB study of reactions (6) and (7), Sironi *et al.* have raised strong doubts about the compactness of the classical VB approach as applied to the same reactions, and have put forward the number of

$$CH_4 \rightarrow CH_2 + H_2 \tag{6}$$

$$CH_4 \rightarrow CH_3 + H \tag{7}$$

485649472 VB structures, which actually correspond to the dimension of the full CI [31]! As a reply to this statement, we have carried out the study of reaction (7) using the method described in section 'CI among a set of skeletal VB structures, with Hartree–Fock optimised fragment AOs', at different levels of compactness and accuracy, in order to give an order of magnitude of the CI dimensions actually required in classical VB studies of chemical reactions.

THE DISSOCIATION OF ONE CH BOND IN METHANE

We have considered reaction (7) as an interaction between two fragments, H and CH_3. As in the study of the same reaction by Carter and Goddard [19], CH_3 and CH_4 have been calculated in their experimental geometry [32]. The occupied space of orbitals is partitioned into core, inactive and active sets: $(1a_1)$, $(2a_1, 1e)$ and $(3a_1)$ for the methyl fragment, respectively. For the hydrogen fragment, the 1s orbital is active. The fragment orbitals have been obtained by separate Restricted Hartree–Fock open-shell calculations [33] on the neutral fragments, which yielded some sets of purely localized SCF-occupied and virtual orbitals. These SCF-occupied orbitals are used to construct the three VB structures which form the elementary CI space (VB3). These VB structures, from which the final CI space is built, are those corresponding to the covalent and ionic elementary VB structures of the bond being broken, i.e.

$$H_3C \cdot — \cdot H \qquad H_3C^+H^- \qquad H_3C^-H^+$$
$$\quad \textbf{4} \qquad\qquad \textbf{5} \qquad\qquad \textbf{6}$$

For each of them, all intrafragment monoexcitations can be generated (VB3*(S_{act} + S_{in})),S_{act} and S_{in} corresponding to all monoexcitations among active and inactive orbitals, respectively. This leads to 79 non-orthogonal VB configurations. Total energies for CH_4 and CH_3 + H are reported in the first row of Table 2, along with the corresponding dissociation energy of 101.7 kcal/mol. Note that it is important to optimize the inactive orbitals for each elementary structure separately (this is the role of the S_{in} monoexcitations). Neglect of this set of configurations would lead to a loss of 5.4 kcal/mol in binding.

The VB3*(S_{act} + S_{in}) set of VB configurations can be considered as the minimal CI space that must be generated to fall into line with the method [17] mentioned in section 'CI among a set of selected VB structures, with Hartree–Fock optimized fragment AOs'. At this level, the agreement between the calculated dissociation energy and the experimental value (110.3 kcal/mol) is not unreasonable, in consideration of our relatively modest basis set of orbitals. However, our method allows for an extension of the CI space to fully correlate the active electrons, provided the inactive ones remain uncorrelated, and this can be done by adding all the double excitations among the orbitals of the active set. This leads to the VB3*(SD_{act} + S_{in}) CI space of dimension 149, which yields an improved dissociation energy of 108.7 kcal/mol.

The effect of the D_{act} excitations is mainly to ensure the correlation of the active electrons in the ionic structures **5** and **6**. Indeed, if the D_{act} excitations generated from **1** are omitted (entry 3 of Table 2), the calculated dissociation energy is hardly altered (108.3 kcal/mol), while the dimension of the CI drops to 129.

It is interesting to compare our selection of VB configurations with the CCCI method of Carter and Goddard [19], based on GVB orbitals. This latter method

Table 2—Total energies (in atomic units) for CH_4 and CH_3 + H and the corresponding bond dissociation energies (in kcal/mol). (BDE exp: 110.3 kcal/mol.) S_{act} (D_{act}) corresponds to all local monoexcitations (diexcitations) from the active orbitals and S_{in} from the inactive ones. All CI dimensions are given in C_{3v} symmetry. Entry 5 is taken from Ref. [19]

Entry	Method	Dimension	CH_4	CH_3 + H	D_e
1	VB3*(S_{act} + S_{in})	79	−40.229412	−40.067305	101.7
2	VB3*(S_{act} + S_{in} + D_{act}(ionic only))	129	−40.239968	−40.067305	108.4
3	VB3*(SD_{act} + S_{in})	149	−40.240478	−40.067305	108.7
4	CCCI-like	165	−40.244042	−40.067305	110.9
5	CCCI		−40.244928	−40.067305	111.5

starts from a 3-configuration CI function in the bond pair being broken, which is formally equivalent to the result of a CI among the VB3 space above. It then adds all double excitations from the bond pair (SD_{act}) and all singles from other occupied valence orbitals (S_{in}) to all virtuals, leading to 111.5 kcal/mol for the dissociation energy. The same configuration list can be obtained from the VB3*($S_{act} + S_{in}$) set by adding the doubles from the active set and all *interfragment* singles. In the specific framework of our classical VB method, dealing with strictly localized orbitals, it can be expected that the latter set of excitations is of minor importance since it corresponds, in general, to interfragment charge transfers among inactive orbitals. Indeed the resulting wavefunction, including 165 spin eigenfunctions in C_{3v} symmetry (row 4 in Table 2), yields a dissociation energy only slightly lower (110.9 kcal) than the previous ones. Interestingly, this latter value is almost similar to the CCCI dissociation energy obtained by Carter and Goddard, although these authors use GVB-(1/2) optimized orbitals while our CCCI-like calculation uses the Hartree–Fock orbitals of the free fragments. This indicates that the CCCI method generates a coherent CI space whose results are rather stable under variations in the set of orbitals. In the classical VB method, the CI spaces that we use (entries 2 and 3 of Table 2) should have the same property, since they differ from the CCCI space only by a set of weakly stabilizing VB configurations.

Lastly, as far as compactness is concerned, the small dimensions of the CI spaces (129–149) that we have dealt with in this study speak for themselves and strongly contradict the above-mentioned criticism [31].

CONCLUDING REMARKS

It follows from our study that the HLJP and classical VB methods have disjoint domains of application. Thanks to their compactness, GVB and SCVB methods are well suited to the full-VB description of molecules. On the quantitative side, they are close to CASSCF quality and offer a nice compromise between simplicity and accuracy; moreover, they constitute excellent reference functions for truncated CI expansions if greater accuracy is required [19]. On the interpretative side, their application to standard molecules such as methane, the π systems of benzene, naphthalene, etc. have provided some bonding pictures in agreement with chemical expectations, and there is no reason why they should not be successful in more exotic systems, provided they are constituted of bonds of major covalent type. We would, however, express some reservations for systems in which ionic structures might be predominant. Indeed, the HLJP description of such systems is still formally covalent, and the only way ionic structures manifest themselves is through a strong delocalization of the orbitals. As a consequence, the interpretation of the wavefunction may be difficult or misleading, e.g. in some S_N2 transition states ($[NRX]^-$), especially near the S_N1 borderline where the ionic ($N^- R^+ X^-$) VB structure may be overwhelmingly important, or in 1,3 dipoles [34]. The dilemma arises from the *a priori* assumption of covalency, and it cannot be solved by adding ionic structures because this would cause some redundancy with the full freedom of the orbitals to delocalize throughout their optimization. Apart from these well characterized special cases, the physical picture offered by the HLJP model appears to be correct if one keeps in mind that the formally covalent bonds have to be considered as involving a small implicit ionic component which is in fact present, though hidden, in the wavefunction. Indeed, the study of F_2 (but the reasoning is general) shows that charge fluctuation is a real feature of the bonds and not a computational artefact, as the ionic terms demonstrate their physical existence by their need to be properly treated, with an instantaneous relaxation of the valence space following the charge fluctuations. It is to be noted that the GVB method, for example, has long been known to disfavour ionic terms, and that various methods, ranging from empirical corrections [35] to selected CI [19], have been devised to remedy for this.

The classical VB description is by nature more detailed, and consequently less compact, than the HLJP one. However, this inconvenience disappears if the VB information is reduced to the only interesting part of a molecular system, for example the set of active bonds throughout a chemical reaction, the remaining part being treated at the doubly occupied MO level. This makes classical or optimized classical methods well suited to the study of reactivity, and particularly to the quantitative calculation of diabatic potential surfaces [4, 29], rather than to the all-VB description of large multi-bond molecules. On the qualitative side, these methods yield a unified and balanced picture of homopolar or heteropolar bonds, and delocalized systems whatever their ionic character, since no special *a priori* assumption is made on the nature of the bonds. Lastly, their excellent ability to quantitatively describe the phenomenon of bonding, no further CI being necessary, shows that the physical picture they offer, albeit not the simplest one, is close to the true nature of the bond.

REFERENCES AND NOTES

[1] G.N. Lewis, *J. Amer. Chem. Soc.* **38** (1916) 762.

[2] J.C. Slater, *Phys. Rev.* **38** (1931) 1190.

[3] (a) S.S. Shaik, *J. Amer. Chem. Soc.* **103** (1981) 3692; (b) A. Pross and S.S. Shaik, *Acc. Chem. Res.* **16** (1983) 363; (c) A. Pross, *Adv. Phys. Org. Chem.* **21** (1985) 99; (d) A. Pross, *Acc. Chem. Res.* **18** (1985) 212; (e) S.S. Shaik, in *New Concepts for Understanding Organic Reactions*, J. Bertran and I.G. Csizmadia (Eds), Kluwer, Dordrecht, 1989: NATO ASI Series Vol. C267; (f) S.S. Shaik and P.C. Hiberty, in *Theoretical Models of the Chemical Bonding*, Part 4, Z.B. Maksic (Ed.), Springer, Heidelberg, 1991.

[4] (a) G. Sini, P.C. Hiberty and S.S. Shaik, *J. Chem. Soc. Chem. Comm.* (1989) 772; (b) G. Sini, S.S. Shaik, J.M. Lefour, G. Ohanessian and P.C. Hiberty, *J. Phys. Chem.* **93** (1989) 5661; (c) G. Sini, G. Ohanessian, P.C. Hiberty and S.S. Shaik, *J. Amer. Chem. Soc.* **112** (1990) 1407; (d) P. Maître, P.C. Hiberty, G. Ohanessian and S.S. Shaik, *J. Phys. Chem.* **94** (1990) 4089.

[5] P.O. Löwdin, *J. Chem. Phys.* **18** (1950) 365.

[6] (a) G. Trinquier and J.P. Malrieu, *J. Phys. Chem.* **94** (1990) 6184; (b) A. Clotet, J.P. Daudey, J.P. Malrieu, J. Rubio and F. Spiegelman, *Chem. Phys.* **147** (1990) 293. See also: K. Ruedenberg, M.W. Schmidt, M.M. Gilbert and S.T. Elbert, *Chem. Phys.* **71** (1982) 41, 51, 65.

[7] G. Ohanessian and P.C. Hiberty, *Chem. Phys. Lett.* **137** (1987) 437.

[8] A.C. Hurley, J. Lennard-Jones and J.A. Pople, *Proc. Roy. Soc.* **A220** (1953) 446.

[9] (a) W.J. Hunt, P.J. Hay and W.A. Goddard, *J. Chem. Phys.* **57** (1972) 738; (b) W.A. Goddard and L.B. Harding, *Ann. Rev. Phys. Chem.* **29** (1978) 363; (c) F.B.

Bobrowicz and W.A. Goddard, in *Methods of Electronic Structure Theory*, H.F. Schaefer (Ed.), Plenum Press, New York, 1977, pp. 79-127.

[10] (a) D.L. Cooper, J. Gerratt and M. Raimondi, *Adv. Chem. Phys.* **69** (1987) 319; (b) D.L. Cooper, J. Gerratt and M. Raimondi, *Int. Rev. Phys. Chem.* **7** (1988) 59; (c) D.L. Cooper, J. Gerratt and M. Raimondi, in *Valence Bond Theory and Chemical Structure*, D.J. Klein and N. Trinajstic (Eds), Elsevier, 1990, p. 287; (d) D.L. Cooper, J. Gerratt and M. Raimondi, in *Advances in the Theory of Benzenoid Hydrocarbons*, I. Gutman and S.J. Cyvin (Eds), *Top. Current Chem.* **153** (1990) 41.

[11] P.C. Hiberty and G. Ohanessian, *Int. J. Quant. Chem.* **27** (1985) 259.

[12] J.C. Slater, *J. Chem. Phys.* **43** (1965) S11.

[13] (a) D.L. Cooper, J. Gerratt, M. Raimondi and S.C. Wright, *Chem. Phys. Lett.* **138** (1987) 296; (b) D.L. Cooper, J. Gerratt and N. Raimondi, *Nature (London)* **323** (1986) 699; (c) F. Penotti, J. Gerratt, D.L. Cooper and M. Raimondi, *J. Mol. Struc. (Theochem)* **169** (1988) 421.

[14] R. Broer, CECAM Workshop for Valence Bond Theory, Orsay, France, 1990. An orbital optimization in **1**, **2** and **3** separately leads to a very disappointing dissociation energy.

[15] For a general use of Brillouin states in MCSCF methods, see F. Grein and T.C. Chang, *Chem. Phys. Lett.* **12** (1971) 44.

[16] A preliminary report of this calculation has been given at the CECAM Workshop for Valence Bond Theory, Orsay, France, 1990.

[17] (a) P.C. Hiberty, *16ème Congrés des Chimistes Théoriciens d'Expression Latine*, Lyon–Villeurbanne, France, 1986. (b) P.C. Hiberty and J.M. Lefour, *J. Chem. Phys.* **84** (1987) 607; (c) P. Maître, J.M. Lefour, G. Ohanessian and P.C. Hiberty, *J. Phys. Chem.* **94** (1990) 4082.

[18] T.H. Dunning, Jr., *J. Chem. Phys.* **53** (1970) 2823.

[19] E.A. Carter and W.A. Goddard, *J. Chem. Phys.* **88** (1988) 3132.

[20] M. Peterson and R. Poirier, MONSTERGAUSS, Department of Chemistry, University of Toronto, Canada, 1981.

[21] J.M. Lefour and J.P. Flament, DCMR, Ecole Polytechnique, 91128 Palaiseau Cedex, France.

[22] (a) K. Hijikata, *Rev. Mod. Phys.* **32** (1960) 445; (b) P.J. Hay, W.R. Wadt and L.R. Kahn, *J. Chem. Phys.* **68** (1978) 3059.

[23] E.A. Colbourn, M. Dagenais, A.E. Douglas and J.W. Raymonds, *Can. J. Chem.* **54** (1976) 1343.

[24] M.R.A. Blomberg and P.E.M. Siegbahn, *Chem. Phys. Lett.* **81** (1981) 4.

[25] W.D. Laidig, P. Saxe and R.J. Bartlett, *J. Chem. Phys.* **86** (1987) 887.

[26] P.C. Hiberty and D.L. Cooper, *J. Mol. Struc. (Theochem)* **169** (1988) 437.

[27] (a) P.C. Hiberty and C. Leforestier, *J. Amer. Chem. Soc.* **100** (1978) 2012; (b) P.C. Hiberty and G. Ohanessian, *J. Amer. Chem. Soc.* **104** (1982) 66.

[28] The weight W_m of a valence bond structure V_m whose coefficient in the total wavefunction is C_m is calculated by Chirgwin and Coulson's formula: $W_m = C_m^2 + \Sigma_{n \neq m} C_m C_n \langle V_m | V_n \rangle$ See: B.H. Chirgwin and C.A. Coulson, *Proc. Roy. Soc. London* **A201** (1950) 196.

[29] (a) O.K. Kabbaj, M.B. Lepetit, G. Sini, P.C. Hiberty and J.P. Malrieu, *J. Amer. Chem. Soc.*, in press; (b) G. Sini, P. Maître, P.C. Hiberty and S.S. Shaik, *J. Mol.*

Struc. (*Theochem*), in press; (c) O.K. Kabbaj, Thèse, Laboratoire de Chimie Théorique, Faculté des Sciences, Rabat, Maroc (1990); (d) G. Sini, Thèse, Laboratoire de Chimie Théorique, Université de Paris-Sud, 91405 Orsay Cedex, France (1990).

[30] (a) S.S. Shaik, *Prog. Phys. Org. Chem.* **15** (1985) 197; (b) S.S. Shaik, H.B. Schlegel and S. Wolf, *Theoretical Aspects of Physical Organic Chemistry: Application to the S_N2 Transition State*, Wiley, New York, in press.

[31] '*According to the Weyl formula, there are 485 649 472 classical VB structures for a system of 8 electrons and 34 orbitals (DZP basis with a frozen core), and this gives some idea of the computational complexity that has to be overcome.*' from: M. Sironi, D.L. Cooper, J. Gerratt and M. Raimondi, *J. Amer. Chem. Soc.* **112** (1990) 5054.

[32] CH_3 is planar. The C—H bond lengths are 1.094 Å in CH_4 and 1.079 Å in CH_3.

[33] E.R. Davidson, *Chem. Phys. Lett.* **21** (1973) 565.

[34] P.C. Hiberty, in *Valence Bond Theory and Chemical Structure*, D.J. Klein and N. Trinajstic (Eds), Elsevier, 1990, p. 221.

[35] M. Goodgame and W.A. Goddard, *Phys. Rev. Lett.* **54** (1985) 7.

14

Doctrine conflicts dividing the valence bond tribe

Jean-Paul Malrieu
Université Paul Sabatier, Toulouse, France

INTRODUCTION AND GENERAL REMARKS

While the competition between valence bond (VB) and molecular orbital (MO) approaches once appeared open, the success of the latter over the former seemed nearly complete a few years ago, and even if the last ten years have seen some revival of the VB approach [1], the VB community remains a small tribe among the quantum chemistry people. Despite such an uncomfortable situation (and partly because of it), this VB community is deeply divided by doctrinal conflicts (as frequently occurs among minorities). The main purpose of this chapter is to clarify the content of these conflicts. This conflict largely concerns the choice of the elementary building blocks, namely the one-electron functions, to be used in the Slater determinants or configurations. It will be shown that one may be led to opposite choices according to the relative importance attributed to computation and interpretation.

Reminder 1: the two objectives of quantum chemistry

The objective of quantum chemistry is in principle twofold. It should:

(i) produce reliable numerical results regarding the energetic (and spectroscopic, but this is again energetic) properties of molecules, and many quantum chemists like to speak of their activity as of a 'theoretical or numerical spectrometer' confirming or competing with physical experiments. This is the dominant trend today in our discipline.

(ii) produce models and interpretations, as grounded as possible, as rigorous as possible, but sufficiently easy to follow in a deductive or pictorial way. These models should help to rationalize the types of chemical structure, and trends in physical or chemical properties. They should help to stimulate the imagination of the chemist in proposing new combinations of atoms designed to exhibit new properties. The rational combinatory activity of chemistry, the way chemistry builds its own objects, is the real singularity of chemistry among sciences (a science of artefacts, in the literal meaning of the word (product of art)) and theoretical chemists will never be chemists if they only consider themselves as human protheses of their computational spectrometers. In my opinion the modelling activity is as important as the technological efficiency and numerical accuracy of our *ab initio* tools. But the production of rigorous pictures and models in quantum chemistry receives much less effort than does the technological improvements.

It will be shown here that conflicts in the VB community largely originate from different weightings of the two above-mentioned objectives.

Reminder 2: the technological disadvantages of traditional VB methods

The technological disadvantages of traditional VB methods are:

(i) the non-orthogonality of the monoelectronic functions, resulting in the non-orthogonality of the Slater determinants, ϕ_i, and the computational cost of the matrix elements $\langle \phi_I | H | \phi_J \rangle$, in contrast with the MO–CI problem.

(ii) the lack of pre-eminence of a peculiar single determinant, and the lack of any clear hierarchy of determinants, a situation which is much worse than for the MO–CI problem where an SCF step defines a good zeroth-order description and where one may gather (singly plus) doubly excited determinants as a first class of perturbers, the next one involving triples and quadruples, and so on, as evidenced by perturbative arguments.

Reminder 3: advantages and disadvantages of VB theory for interpretative purposes

(i) If the N monoelectronic functions are local, atomic orbitals for instance, associated with a certain region of space, each VB determinant represents a given distribution of the n electrons of the system in the molecular space. The VB language should provide the most physical picture of the electronic population and especially of its *fluctuation*; electronic delocalization and correlation, atomic polarization and/or promotion should preferably (although not necessarily) be traced in that language, which permits a local scrutiny of the physical phenomena. Notice, however, that local information may in principle be obtained from an MO–CI wave function. For instance, one might perfectly answer the question: what are the relative probabilities to find the carbon atom of CH_4 in the ^3P (s^2p^2) configuration, the $^5S(sxyz)$ configuration or any other spin state associated with these space parts? Karafiloglou has, for instance, developed such local VB-type glances at an MO–CI wave function [2]. The challenge is important since some basic questions, supposedly sufficiently established to be taught at elementary levels, such as hybridization, are not yet clearly demonstrated from accurate (i.e. using extended basis sets and intensive treatment of electronic correlation) calculations. Some *excellent* quantum chemists think that hydridization does not occur, and that CH_4 may be built from a C ^3P s^2p^2 ground state of fluctuating orientation, and we have no firm proof to disagree with this view.

The fascinating fact, for someone looking from some distance at the quantum chemistry trends, is that such basic questions, formulated half a century ago by the pioneering work of L. Pauling, were not answered and finally forgotten or considered with contempt by the leading family of quantum chemists, as an obsolete, irrelevant or unanswerable question (three *different* attitudes in principle, having in common a unique sensitivity to the accurate computation of observables). Such questions are actually meaningful and may receive well-defined accurate answers.

Another area of superiority of VB approaches is that they provide some kind of direct or intrinsic diabatization since the functions generated from the asymptotic state of the atoms may be followed *before* their interactions. Such analyses are possible from MO–CI calculations by unitary transformations of the adiabatic eigenfunctions, but the VB approach seems a more direct way. And actually this idea has received very interesting qualitative and quantitative applications [3].

(ii) from an interpretative point of view, one defect of the traditional VB language is again in the non-orthogonality of one- and n-electronic functions since it becomes impossible to define a set of disjoint events and speak in terms of probabilities as the chemist would like. This remark has led some of us to work in local orthogonal basis sets of monoelectronic wave functions, defining orthogonal valence bond (OVB) expansions and thus making it possible to speak in terms of probabilities. There is a counterpart to that strategy, namely the existence of orthogonalization tails, which might in principle destroy the locality of the elementary functions and weaken the interpretation. This question will be discussed here in some length.

Methodological remark: realism or consistency?

VB analyses of the wave function frequently face the following comment from referees: 'either you practise a non-orthogonal analysis, and due to the non-orthogonality of the neutral and ionic VB component, the numbers are meaningless, or you use an orthogonal basis set and the AOs having tails lose their local character and are meaningless'. In order words, any VB decomposition is a waste of time. The amazing feature is that such remarks are frequently formulated by authors who use Mulliken's population analysis, handling the overlap dilemma through a much more uncontrolled device when large basis sets are used. But one must listen to their arguments, and try to think about them. One may say that

(i) the overlap problem is a real one since it avoids one thinking in terms of probabilities. However, given the monoelectronic functions, the VB expansion of the wave function is unique and one may analyse the *coefficients* if not the probabilities. Comparing analogous systems or comparing the VB expansion of the SCF wave function with the exact one may provide very useful information and give a clear picture of the local effects of electronic correlation (for instance, see ref [4]).

(ii) if orthogonal basis sets are used, they retain some *approximate* atomic character, and if clear general trends appear in the wave function, they may be used as interpretative tools, leading to very significant and pictorial descriptions of the correlated wave function. And now since the events are orthogonal, all the probability theory instruments of analysis are available (correlation, entropy, etc. . . . .) [5].

Our assumption is that one must sacrifice a naive realism (for instance the use of strictly atomic orbitals) and try to find a *consistent grid* of interpretation. Examples will be given below, showing that general statements regarding the local electronic effect of electronic correlation may be formulated using either non-orthogonal or orthogonal basis sets.

THE FOUR TYPES OF MONOELECTRONIC FUNCTION AND THEIR RESPECTIVE ADVANTAGES

There are four different types of monoelectronic function to build VB determinants or configurations. One may either use functions defined from the free atom, or use molecularly variational monoelectronic functions. This is the primary choice; a secondary choice concerns the orthogonal or non-orthogonal character of these functions. The overall partition is schematized in Table 1 together with the corresponding strategies.

Atomic definitions of the monoelectronic functions: the three strategies

The VB approach was initially formulated in minimal basis sets and used, for instance, the self-consistent field (SCF) AOs for the ground state (or eventually of the relevant excited state) of the free atom. This seems a natural choice. It is well known, however, that the exact solution of H in such a minimal basis set (exact CI $\equiv$ exact VB in the same basis set) is far from the exact one and is unable to provide satisfactory binding energies. Before analysing the possible strategies to go besides the minimal basis set, let us comment on the orthogonality problem. For H_2, the neutral non-orthogonal VB component

$$\phi_N = \frac{a^0 \bar{b}^0 + b^0 \bar{a}^0}{\sqrt{2(1 + S^2)}} \quad \text{(where } a^0 \text{ and } b^0 \text{ are the } 1s \text{ AOs of the H atoms and } S = \langle a^0 | b^0 \rangle)$$

is known to give an important part of the binding energy. A supplementary stabilization comes from the interaction with the ionic state

$$\phi_I = \frac{a^0 \bar{a}^0 + b^0 \bar{b}^0}{\sqrt{2(1 + S^2)}}$$

Table 1 — The four types of monoelectronic wave function used in VB-type calculations

	Non-orthogonal	orthogonal
Atomic definition	Traditional VB	OVB
Variational		CASSCF
definition	GVB	VB* OVB formulation

VB*: a non-orthogonal VB expansion of the CASSCF function is possible (see text).

If one performs the $S^{-1/2}$ transformation between a^0 and b^0,

$$\{a', b'\} = S^{-1/2}\{a^0, b^0\} \quad a' = a^0(1 + \tfrac{3}{8}S^2) - \frac{S}{2}b^0$$

one may well generate a 'neutral' OVB state

$$\phi'_N = \frac{a'\bar{b}' + b'\bar{a}'}{\sqrt{2}}$$

This state is now *unbound*. It is actually of higher energy than the triplet state, which may indifferently be written

$$\phi^T = \frac{a^0\bar{b}^0 - b^0\bar{a}^0}{\sqrt{2(1 + S^2)}} = \frac{a'\bar{b}' - b'\bar{a}'}{\sqrt{2}}$$

in terms of non-orthogonal or orthogonal AOs, since it is unique in the minimal basis set. It is clear that

$$\langle \phi'_N | H | \phi'_N \rangle - \langle \phi_T | H | \phi_T \rangle = 2K_{a'b'}$$

which is necessarily positive. This neutral state is now strongly repulsive, and the existence of the bond comes from the interaction with the orthogonal ionic state

$$\frac{a'\bar{a}' + b'\bar{b}'}{\sqrt{2}} = \phi'_I$$

In such a case, the neutral–ionic mixing will appear to be very strong.

As a *third way*, one might have kept ϕ_N and have defined $\phi''_I = \lambda(\phi_I - \langle \phi_N | \phi_I \rangle \phi_N)$ according to a Schmidt orthogonalization, where $\langle \phi_N | \phi_I \rangle = 2S/(1 + S^2)$ and λ is a normalization factor. Then ϕ''_I may well appear with a very weak component in the wave function. This would give some kind of *orthogonal VB* description *using non-orthogonal one-electron functions*.

Table 2 — VB strategies starting from atomic orbitals

	Non-orthogonal VB	Orthogonal VB	
		Symmetrical orthogonal of valence AOs	Hierarchized orthog. of configurat.
			Neutral Ionic
			$S^{-1/2}$ project in ⊥subspace $+S^{-1/2}$
Advantages	Retains the free-atom character	Strong orthogonality —comput. = MO–CI —probabilities = OK	Neutral situations remain neutral —probabilities = OK weak orthog.
Disadvantages	Non-orthogonality —computation diffic. —no probability	Deviates from the free-atom meaning	—comput. diffic.

For a more complex problem, where one has several neutral VB functions, one might orthogonalize the neutral determinants (or configurations) among themselves through an $S^{-1/2}$ transform, and then project the ionic determinants (or configurations) in the complementary subspace and orthogonalize them among themselves in a symmetrical manner (cf Table 2). This would provide a general orthogonal VB basis set *without* orthogonalization of the AOs. The low energy of the neutral configurations would be kept, since they remain linear combinations of non-orthogonal neutral configurations. This procedure has no special advantage from a practical viewpoint, since the matrix elements between the orthogonal VB determinants would take a long time to calculate, but it would provide a set of disjoint events to think of in terms of probabilities. This third way, which one may call *hierarchized orthogonalization* of VB configurations (or determinants), would deserve some practical exploration. It might be refined by defining more classes, for instance by considering the singly ionic VB configurations before the doubly ionic ones, and so on. Notice that for the two-centre problem, the symmetrical orthogonalization of all the VB configurations gives the same result as the symmetrical orthogonalization of the AOs.

Variational monoelectronic functions: GVB and valence CASSCF

The elementary problem

From the early work of Coulson and Fisher [6], a second family of VB descriptions has been developed, under the name of generalized valence bond (GVB) formalism [7]. For a two-electron (single-bond) problem, the wave function is written as

$$\psi(1,2) = (a\bar{b} + b\bar{a})/\sqrt{2(1 + S'^2)}, \quad \text{with } S' = \langle a|b \rangle$$

where a and b are *optimized*; they are centred on one atom but have tails on the neighbouring one. This wave function, in that precise case (see below), is identical to the complete active space self-consistent field (CASSCF) wave function [8].

$$\psi_{CAS} = \lambda\sigma_g^2 - \mu\sigma_u^2$$

where σ_g and σ_u are variationally optimized, as well as λ and μ. Actually, one may always write σ_g and σ_u in terms of yet unspecified orbitals a and b:

$$\sigma_g = \frac{a + b}{\sqrt{2(1 + S')}} \quad \sigma_u = \frac{a - b}{\sqrt{2(1 - S')}}$$

and in ψ_{CAS} one may cancel the weight of the 'ionic' components $a\bar{a}$ and $b\bar{b}$ by putting

$$\frac{\lambda}{2(1 + S')} - \frac{\mu}{2(1 - S')} = 0$$

which defines S:

$$S' = \frac{\lambda - \mu}{\lambda + \mu}$$

and in turn, a and b from σ_g and σ_u:

$$a = \sqrt{\frac{1 + S'}{2}}\, \sigma_g + \sqrt{\frac{1 - S'}{2}}\, \sigma_u$$

$$b = \sqrt{\frac{1 + S'}{2}}\, \sigma_g - \sqrt{\frac{1 - S'}{2}}\, \sigma_u$$

Thus $\{a,b\}$ or $\{\sigma_g, \sigma_u\}$ define the *same valence subspace* and the same MOs. A third basis is given by $\{a', b'\}$:

$$a' = (\sigma_g + \sigma_u)/\sqrt{2} \quad \text{and} \quad b' = (\sigma_g - \sigma_u)/\sqrt{2}$$

which is orthonormal. One may work *indifferently* with one of these basis sets, and it is somewhat surprising to learn that 'the non-orthogonal orbitals a and b are suited for the singlet state while orthogonal functions a' and b' are only relevant for the triplet state' [9].

The advantage of using non-orthogonal orbitals is that the optimal valence wave function (i.e. the CASSCF one) reduces here to a *unique* space part. If one works in a minimal basis set a^0, b^0:

$$a = \frac{\cos \varphi a^0 + \sin \varphi b^0}{\sqrt{1 + \sin 2\varphi S}} \qquad b = \frac{\sin \varphi a^0 + \cos \varphi b^0}{\sqrt{1 + \sin 2\varphi S}}$$

$\sin \varphi$ brings the delocalization tail ($\pi > \varphi > 0$), and the tail is *in phase* while the orthogonalization tail was *out of phase*. Then

$$a\bar{b} + b\bar{a} = \frac{a^0\bar{b}^0 + b^0\bar{a}^0 + \sin 2\varphi(a^0\bar{a}^0 + b^0\bar{b}^0)}{1 + \sin 2\varphi S}$$

which is formally neutral, contains the traditional ionic VB component.

Notice that

$$S' = \langle \cos \varphi a^0 + \sin \varphi b^0 | \sin \varphi a^0 + \cos \varphi b^0 \rangle / (1 + S \sin 2\varphi)$$

$$= \frac{S + \sin 2\varphi}{1 + \sin 2\varphi S}$$

which is larger than S since

$$\frac{S + \sin 2\varphi}{1 + \sin 2\varphi S} > S$$

is satisfied if $\sin 2\varphi > \sin 2\varphi S$, i.e. if $S < 1$ (since $\sin 2\varphi > 0$) which is always true. Then it is easy to visualize (cf. Fig 1) the relative orientation of the set of monoelectronic functions and their overlap.

Generalization

The idea of an optimal unique space part with a combination of various spin parts of appropriate multiplicity is exploited in the GVB formalism. Then for a singlet ne^-

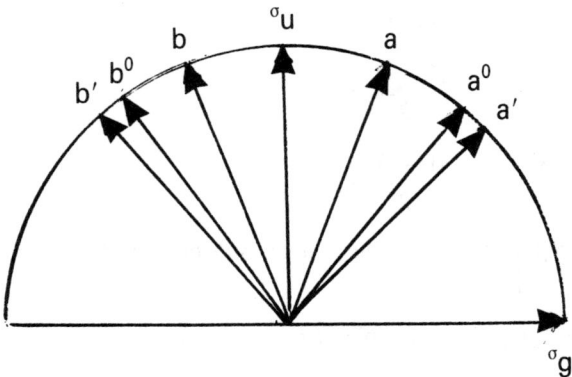

Fig. 1 — Relative orientation of: purely atomic orbitals a^0; orthogonal atomic orbitals a'; GVB orbitals a, in a two-electron two-centre problem

problem,

$$\psi^{\text{GVB}} = (ab...n)\left(\sum_I C_I S_I\right)$$

where $a, b,...n$ are a unique set of optimized one-electron functions and the C_Is are optimized parameters of the various singlet spin functions S_I (in general, products of two-electron singlets). Such a very limited function proved to be quite efficient in many cases.

The valence CASSCF function consists in practising a full valence CI within a set of valence MOs (the active MOs) which are variationally optimized simultaneously [8]. This special multiconfigurational SCF function dissociates properly into the HF solutions of the free atom when the bonds are broken and incorporates in the best way all the left–right (or more generally internal or non-dynamical) parts of the electronic correlation of the valence electrons. This becomes a new standard zeroth-order description. There are cases where the CASSCF function refuses to keep a valence character (see, for instance, refs [10, 11]), namely for strongly ionic bonds like Li^+F^- where the charge fluctuation is negligible (and where the GVB approach would certainly fall close to the closed-shell description). But, even in that case, state average CASSCF treating both $\text{Li}\cdot\text{F}\cdot$ and Li^+F^- makes the computation of an optimal valence set of MOs possible.

Ruedenberg *et al.* [12] have discussed in some length the fact that the valence CASSCF procedure actually defines an optimized molecularly adapted valence space, as does the GVB approach. These spaces are strictly identical for the $2e^-/2$ orbital problem only; elsewhere, the resulting valence spaces produced by the two approaches are slightly different. Ruedenberg *et al.* [12] also suggested that even if obtained in terms of symmetry-adapted bonding and antibonding MOs, this valence space may be expressed in terms of atomic-like functions. To obtain them one may for instance

(i) localize the active MOs through a localization procedure such as the Boys' [13] criterion. While this criterion, when applied to the SCF-occupied MOs, provides bond and lone pair doubly occupied MOs, when acting on the full valence space, it defines a set of orthogonal nearly atomic molecular orbitals, which may be considered as orthogonal optimal (molecularly adapted) valence atomic orbitals [14].

(ii) project the orbitals a^0 of the free atom in the space of the active MOs:

$$a'' = \sum_{\substack{i \\ \text{active valence MOs}}} |i\rangle\langle i|a^0\rangle$$

One so obtains a non-orthogonal basis of the valence space on which one might develop the CASSCF function ψ in a VB expansion. This would be interesting since the orbitals a'' are the functions belonging to the molecularly adapted valence space which have the largest overlap with the free atom orbitals. This would furnish a VB expansion calculated in a large basis AO basis set with the *same length* as the VB expansion of the minimal basis set.

Another solution to obtain a non-orthogonal atomic-like basis set might consist, if one wants to define n_A valence atomic orbitals on atom A, in searching the n_A lowest eigenvectors of the matrix defining the distance of the valence electrons to the atom A (position r_A),

$$\langle i|\bar{d}_A^2|j\rangle = \langle i|(r - r_A)^2|j\rangle \qquad i, j \text{ active}$$
$$\bar{d}_A^2|a_k\rangle = l_k|a_k\rangle \qquad k = 1, n_A$$

This procedure would define an orthogonal subset of n_A vectors belonging to the valence space and as close as possible to the nucleus of atom A.

Notice that these procedures may be applied on the set of GVB orbitals as well, thus offering a passage from the compact GVB form to a more traditional VB expansion.

(iii) one may symmetrically orthogonalize the orbitals a'', so obtaining orbitals of type a' [15].

From orthogonal nearly atomic molecular orbitals provided by procedures (i) and (iii), the CASSCF wave function ψ^{CAS} may be re-expressed in an OVB expansion:

$$\psi^{CAS} = \sum_{\substack{I \\ \text{OVB}}} C_I \phi_I$$

Since the ϕ_Is are orthogonal, the quantities C_I^2 may be considered as probabilities. Several papers [14–16] have exploited this strategy (in its variants (i) and (iii)) and proved their consistency and interest.

One might argue that the use of a variational CASSCF procedure in a large basis set, followed by a localization or a projection + orthogonalization procedure, will destroy the atomic character of the NAMOs and break any connexion between the coefficients of an OVB expansion and those of a traditional VB expansion. So far we may say that, expressed in terms of determinants, *the hierarchy of the OVB–CASSCF coefficients is identical to the hierarchy of the VB ones* obtained in the minimal basis set.

As an example, Table 3 gives a comparison of the largest VB and CASSCF–OVB coefficients for the N_2 molecule. The VB coefficients are from a minimal basis set [

Table 3—The hierarchy of neutral VB and OVB determinants in N_2 ($N_A \equiv N_B$)

A		B	VB[a]	OVB[b]
xyz	·	$\overline{xyz}$	0.063	0.183
$xy\bar{z}$	·	$\overline{xy}z$	0.053	0.165
$\bar{x}yz$	·	$x\overline{yz}$	0.050	0.158
x^2y	·	$\bar{y}z^2$	0.037	0.1344
x^2z	·	$y^2\bar{z}$	0.040	0.1325

[a] Minimal basis set VB calculation [5].
[b] CASSCF calculation in a double zeta + polarization basis set [6].

and the CASSCF–OVB ones from a double zeta + polarization basis set [6]. The non-orthogonality problem prevents a direct comparison, but it is satisfying to see that the hierarchy is maintained; this shows that the physical effects of the valence electronic correlation (reduction of the charge fluctuation per atom, reduction of the charge fluctuation per orbital and increase of the atomic spin momentum fluctuation) are meaningful, as is apparent in both expansions.

Local character of the various monoelectronic functions

The strict AOs a^0, b^0 of the free atom are usually considered as meaningful and a correct grounding in building a VB analysis, while orthogonal MOs such as $\{a', b'\}$ (either coming from the orthogonalization of $\{a^0, b^0\}$ or from the CASSCF solution) are sometimes considered as being polluted by their tails to such an extent that they would become meaningless. On the other hand GVB supporters consider their orbitals as being the most relevant ones. It is interesting to compare them, especially from their more or less local character. The comparison may be performed in the minimal basis set first, and it is clear that we must compare

$$a' = a_0(1 + \tfrac{3}{8}S^2) - \frac{S}{2}b_0$$

$$a^0$$

$$a = \cos \varphi a^0 + \sin \varphi b^0 \qquad \frac{\pi}{4} > \varphi > 0$$

If one examines the fluctuation of the position of an electron with respect to its centroid

$$\langle \tilde{a}|\Delta_z^2|\tilde{a}\rangle = \langle \tilde{a}|(z - z_a^0)^2|\tilde{a}\rangle$$

with

$$z_z^0 = \langle \tilde{a}|z|\tilde{a}\rangle$$

(centroid of the orbital $\tilde{a} = a'$, a^0 or a), it has been well established that the fluctuation of position increases when one goes from a' to a^0 and from a^0 to a. In that sense the orthogonal functions are more local.

One may look at the same problem differently and ask the following question: when do we obtain a more ionic character of the ionic determinants $|\tilde{a}\tilde{a}|$? One may wonder what is the heirarchy of ionicity of the three types of determinant $|a'\bar{a}'|$, $|a^0\bar{a}^0|$ or $|a\bar{a}|$. Notice that, taking the origin in the centre of the bond, and working in the symmetry-adapted MO basis,

$$\langle \sigma_g | z | \sigma_u \rangle = \mu \quad \text{while} \quad \langle \sigma_g | z | \sigma_g \rangle = \langle \sigma_u | z | \sigma_u \rangle = 0$$

Then using the equations

$$\tilde{a} = (\sqrt{1 + S}\, \sigma_g + \sqrt{1 - S}\, \sigma_u)/\sqrt{2}$$

$$\langle \tilde{a} | z | \tilde{a} \rangle = \mu\sqrt{1 - S^2}$$

The dipole moment of the ionic configuration $\tilde{a}\tilde{a}$

$$\mu(\tilde{a}\tilde{a}) = 2\mu\sqrt{1 - S^2}$$

is at a maximum when $S = 0$, i.e. for the orthogonal basis. The *optimal ionic configuration* (i.e. of maximum dipole) is the one *described with orthogonal atomic orbitals*. Remembering that the overlap between the GVB orbitals a and b, $S' = \langle a | b \rangle$, is larger then the overlap S between the free AOs $S = \langle a^0 | b^0 \rangle$, one sees that

$$\mu(a'\bar{a}') > \mu(a^0\bar{a}^0) > \mu(a\bar{a})$$
$$\mu(\text{OVB}) > \mu(\text{VB}) > \mu(\text{GVB})$$

The same conclusions are true for the OVB functions obtained from a CASSCF calculation. Table 4 gives the dipole moment μ of the ionic single determinant

$$\phi_{A^- B^+} = |a'\bar{a}'|$$

of H_2, where a' is obtained as the $(\sigma_g + \sigma_u)/\sqrt{2}$ combination of the two valence active MOs σ_g and σ_u resulting from a CASSCF calculation in a double zeta + polarization ($2s$, $1p$) basis set. One sees that the ratio μ/r (where r is the bond length) tends to 1, by *upper* values, which confirms that the orthogonal ionic forms have a larger dipole moment than the VB ionic form $|a^0\bar{a}^0|$, and again the higher locality of the orthogonal functions.

Table 4 — Dipole moment of an OVB–CASSCF ionic form of H_2

r_{AB}	a.u.	1.4	2.4	5
μ	a.u.	1.82	2.61	5.01
$\dfrac{\mu}{r}$		1.300	1.087	1.002

GVB VERSUS VALENCE CASSCF: THE MISSING EFFECT

The success of the GVB formalism is that it is able to give a rather accurate description of the wave function, correctly including the electronic delocalization and the major internal correlation effects, using non-minimal basis sets, with a *unique space part*. Regarding the compactness of the wave function, the SCF approach, which permitted so elegantly to include mean repolarization, hybridization and relaxation, is now defeated, and one understands the evident pride of the promoters of this technique.

One should add as a major success that it led to a very new picture of metallic clusters in terms of interstitial MOs [17] which was not accessible from the classical HF description (but which were perfectly available from unrestricted HF calculations [18]).

However, this advantage is only partial, since

(i) the computation remains quite difficult,
(ii) the compactness itself prevents any direct analysis of the charge fluctuation. In that sense the GVB approach is not a tool for analysing and understanding the wave function. Competing on the numerical side with the MO–CI approach it sacrifices the analytic goal of the VB tradition which is intrinsically contradictory with compactness,
(iii) the identity of GVB and valence CASSCF functions is restricted to the special case of $2e^-$ in two orbitals. In general the valence CASSCF solution offers more degrees of freedom; besides the definition of an optimal basis set, the number of degrees of freedom is equal to the number of proper spin eigenfunctions for an ne^- in n orbital problem with *one e^- per orbital* plus the $n(n-1)/2$ overlaps of the orbitals in the GVB method, while the CASSCF function leaves the overlap degrees of freedom but also the underlined restrictive condition, which is much more severe. This will be illustrated below for the $4e^-$-, 4-orbital problem already. So that it is clear that for the same dimension of the active space (and except for $2e^-$ in two orbitals):

$$E_{CASSCF} < E_{GVB}$$

(iv) oppositely, the lone pairs introduce two MOs in GVB calculations (incorporating radial or angular atomic corrrelation) and only one in valence CASSCF descriptions.

It is interesting to see what is the main difference between the GVB functions and the valence CASSCF function. This study will be performed on a double-bond problem, such as the C=C bond of ethylene, but the conclusion is general. On that problem the CASSCF active space is defined from four valence MOs:

$$\{\sigma_g, \sigma_u^*, \pi_u, \pi_g^*\}$$

while the GVB function maintaining the σ/π separation defines four orbitals $\{\sigma_a, \sigma_b, \pi_a, \pi_b\}$ located on A and B respectively. We shall assume that the two spaces of four monoelectronic functions are identical, i.e.

$$\sigma_g = \frac{\sigma_a + \sigma_b}{\sqrt{2(1 + S_\sigma)}} \qquad \sigma_a = \sqrt{\frac{1 + S_\sigma}{2}} \sigma_g + \sqrt{\frac{1 - S_\sigma}{2}} \sigma_u \qquad \text{etc.} \ldots$$

One may also define, by one of the previously discussed procedures, orbitals σ_a^0, σ_b^0, π_a^0, π_b^0 belonging to the same valence space and as localized as possible on atoms A and B respectively:

$$\sigma_a = \lambda\sigma_a^0 + \mu\sigma_b^0 \qquad \pi_a' = \lambda'\pi_a^0 + \mu'\pi_b^0 \qquad \text{etc.} \ldots$$

Let us examine first the physical content of the GVB function. The GVB function may be written

$$\psi^{\text{GVB}} = |\sigma_a\sigma_b\pi_a\pi_b|[C_1(\alpha\beta - \alpha\beta)(\alpha\beta - \beta\alpha)/2$$
$$+ C_2((\alpha\beta + \beta\alpha)(\alpha\beta + \beta\alpha)/2 - \alpha\alpha\beta\beta - \beta\beta\alpha\alpha)]$$
$$= C_1\psi^{\text{GVB-PP}} + C_2\psi^1(T_\sigma T_\pi)$$

where $\psi^{\text{GVB-PP}}$ is the perfect pairing coupling description with singlet pairings in both the σ and the π subspaces, while the second component $\psi^1(T_\sigma T_\pi)$ involves triplet couplings in both the σ and the π subspaces. The usual expressions only imply products of 'bond' singlets, but this way of writing is correct, identical to the traditional one, since one can only write two linearly independent singlet states. Notice that in our formulation the second spin function is orthogonal to the first.

Now it is interesting to decompose both $\psi^{\text{GVB-PP}}$ and $\psi^1(T_\sigma T_\pi)$ in their local components:

$$\psi^{\text{GVB-PP}} = \tfrac{1}{2}|[(\lambda\sigma_a^0 + \mu\sigma_b^0)(\lambda\bar{\sigma}_b^0 + \mu\bar{\sigma}_a^0) + (\lambda\sigma_b^0 + \mu\sigma_a^0)(\lambda\bar{\sigma}_a^0 + \mu\bar{\sigma}_b^0)]$$
$$* [(\lambda'\pi_a^0 + \mu'\sigma_b^0)(\lambda'\bar{\pi}_b^0 + \mu'\bar{\pi}_a^0) + (\lambda'\pi_b^0 + \mu'\pi_a^0)(\lambda'\bar{\pi}_a^0 + \mu'\bar{\pi}_b^0)]|$$

In this function the largest coefficient $(\lambda^2\lambda'^2 + \mu^2\mu'^2)$ concerns the neutral determinants of the types

$$\phi_N = |\sigma_a^0\bar{\sigma}_b^0\ \pi_a^0\bar{\pi}_b^0| \quad \text{and} \quad \phi_{N'} = |\sigma_a^0\bar{\sigma}_b^0\ \pi_b^0\bar{\pi}_a^0| \quad \text{(or symmetrical)}$$

The second class of determinants with coefficients $\lambda^2\lambda'\mu'$ or $\lambda\mu\lambda'^2$ respectively are singly ionic VB determinants with one charge transfer in either the π or the σ system:

$$\phi_{\text{SI}_\pi} = |\sigma_a^0\bar{\sigma}_b^0\ \pi_a^0\bar{\pi}_a^0| \qquad \phi_{\text{SI}_\sigma} = |\sigma_a^0\bar{\sigma}_a^0\ \pi_a^0\bar{\pi}_b^0|$$

The third class of determinants involves two charge transfers, one of the σ and one of the π subsystem, but they lead to two types of determinant of deeply different character:

$$\phi_{\text{DI}} = |\sigma_a^0\bar{\sigma}_a^0\ \pi_a^0\bar{\pi}_a^0|$$

which is doubly ionic, and

$$\phi_{N''} = |\sigma_a^0\bar{\sigma}_a^0\ \pi_b^0\bar{\pi}_b^0|$$

which is neutral, although with double occupancies in both the σ and the π subsystems. They appear with the same coefficient $\lambda\lambda'\mu\mu'$. Now the constraints and defects of the GVB function appear very clearly:

(i) the coefficient of the doubly ionic structure is determined by the coefficients of the singly ionic ones, which is not a major defect.
(ii) the coefficients of the neutral determinants ϕ_N and $\phi_{N'}$ are identical, while the former determinants satisfy the atomic Hund's rule, having two parallel spins on

A and B, while ϕ_N violates it, i.e. is of higher energy and should have a smaller coefficient.

(iii) $\phi_{N''}$, although neutral, has the same coefficient as ϕ_{DI} which is doubly ionic, and this is really bad.

Now, what is repaired by the inclusion of the second component in the GVB function? It is easy to demonstrate that

$$\sigma_a \sigma_b (\alpha\beta + \beta\alpha) = k(\sigma_a^0 \bar\sigma_b^0 - \sigma_b^0 \bar\sigma_a^0)$$
$$|\sigma_a \sigma_b| = k|\sigma_a^0 \sigma_b^0|$$
$$|\bar\sigma_a \bar\sigma_b| = k|\bar\sigma_a \bar\sigma_b|$$

i.e. the triplet combinations written from σ_a and σ_b are purely neutral in the classical VB sense. The same is true for the triplet combinations written from π_a and π_b; hence the configuration $\psi^1(T_\sigma T_\pi)$ is purely neutral and has

$$\psi^1(T_\sigma T_\pi) = |\sigma_a^0 \bar\sigma_b^0 \pi_a^0 \bar\pi_b^0| + \text{sym.}$$
$$- |\sigma_a^0 \bar\sigma_b^0 \pi_b^0 \bar\pi_a^0| + \text{sym.}$$
$$+ |\sigma_a^0 \sigma_b^0 \pi_b^0 \bar\pi_a^0| + \text{sym.}$$
$$+ |\bar\sigma_a^0 \bar\sigma_b^0 \pi_b^0 \pi_a^0| + \text{sym.}$$

where sym. holds for the result of the A $\leftrightarrow$ B permutation. Hence the component $C_2 \psi(T_\sigma T_\pi)$ in ψ^{GVB} will

(i) introduce components on situations with two α spins in the σ system and two β spins in the π system (or reverse situations),

(ii) increase the weight of the neutral component ϕ_N which satisfies the atomic Hund's rule and decrease the weight of $\phi_{N'}$ which violates it.

This is the unique benefit, in that case, of going from GVB–PP to the full (spin-coupled) GVB. This benefit is rather weak, as will be shown later. The major defect of the GVB–PP was the equality of the coefficients of ϕ_{DI} and $\phi_{N''}$, and this is not corrected. One sees that the GVB formalism correctly dictates the balance between electronic delocalization and left–right correlation within each bond, but does *not* introduce *any charge correlation between the bonds*, at least when keeping the σ/π separation.

The exact wave function (and already the CASSCF one) has a larger coefficient on $\phi_{N''}$ than on ϕ_{DI}, and this effect would be obtained by including another VB configuration, namely

$$\psi' = |1/2|(\sigma_a \bar\sigma_a - \sigma_b \bar\sigma_b)(\pi_a \bar\pi_a - \pi_b \bar\pi_b)$$

which may be shown to be purely ionic. Actually,

$$\sigma_a \bar\sigma_a - \sigma_b \bar\sigma_b = \lambda^2 \sigma_a^0 \bar\sigma_a^0 + \mu^2 \sigma_b^0 \bar\sigma_b^0 + \lambda\mu(\sigma_a^0 \bar\sigma_b^0 + \sigma_b^0 \bar\sigma_a^0)$$
$$- (\lambda^2 \sigma_b^0 \bar\sigma_b^0 + \mu^2 \sigma_a^0 \bar\sigma_a^0 + \lambda\mu(\sigma_a^0 \bar\sigma_b^0 + \sigma_b^0 \bar\sigma_a^0))$$
$$= (\lambda^2 - \mu^2)(\sigma_a^0 \bar\sigma_a^0 - \sigma_b^0 \bar\sigma_b^0)$$

and the same is true for the π part. Hence

$$\psi' = k|\sigma_a^0 \bar\sigma_a^0 \pi_a^0 \bar\pi_a^0 + \text{sym.} - \sigma_a^0 \bar\sigma_a^0 \pi_b^0 \bar\pi_b^0 + \text{sym.}|$$
$$= k|\phi_{DI} - \phi_{N''} + (\text{sym.})|$$

Superposing ψ' to ψ^{GVB} will lead to the expected increase in the component $\phi_{N''}$ and decrease in the component ϕ_{DI}.

ψ' introduces the interaction between the transition dipoles in the two bonds, i.e. the coupling between the left–right movement of the electrons in the two bonds (leftward in σ when rightward in π), or the *interbond* dynamic electric interaction.

Then it is very easy to make a correspondence between the GVB function and the MO–CI expansion within the same valence space. If ϕ_0 is the $|\sigma_g^2\pi_u^2|$ ground state single determinant;

$$\psi^{GVB-PP} = C_0\phi_0 + C_1\phi(\sigma \to \sigma^*)^2 + C_2\phi(\pi \to \pi^*)^2$$
$$+ C_1C_2\phi(\sigma \to \sigma^*)^2.(\pi \to \pi^*)^2$$

Numerical calculations (G. Trinquier, private communication) show that the σ/π interbond correlation is twice smaller than the intrabond correlation of the π bond and twice larger than the intrabond correlation of the σ bond (which is weakly correlated).

So far, the derivation has concerned a GVB treatment keeping the σ/π separation. It is clear that in that case there is no overlap between the σ and π one-electron functions, and some degree of freedom is lost. If one leaves this separation, one is led to a banana-bond description, i.e.

$$\psi = (aba'b')\left(\sum_I C_I S_I\right)$$

where the four functions a, b, a', b' are equivalent for symmetry reasons. Then one has one more degree of freedom (three overlaps instead of two) and the energy will be somewhat lower. Some authors have claimed that double bonds were actually bent banana bonds; this statement is rather naïve since the GVB function may be considered as an approximation of a CASSCF wave function, or part of the CASCI wave function built from the GVB active MOs (sometimes called GVB–CI), and these wave functions are necessarily invariant under a unitary transformation of the active MOs from banana to $\sigma + \pi$ representation. The superiority of the banana-bond description is simply linked to the GVB (single space part) constraint.

One may, however, wonder whether the extra degree of freedom of the bent-bond description correctly manages the coupling between the transition dipoles of the two bonds. It is rather difficult to evaluate the numerical effect of the σ/π separation release from ref. [19] since it compares a GVB–PP with orthogonality between pairs with a true GVB calculation, and the 6.5 kcal/mol energy gain cannot be entirely attributed to the σ/π into banana representation change. One should recall that the $\sigma-\pi$ correlation originating from $^1(\sigma \to \sigma^*)\cdot^1(\pi \to \pi^*)$ excitations (i.e. from ionic components) contribute 11 kcal/mol to the correlation energy (G. Trinquier, private communication). From an analytical point of view, it is possible to show that again the difference between the coefficients of the neutral $|a\bar{a}b'\bar{b}'|$ and doubly ionic $|a\bar{a}a'\bar{a}'|$ configurations essentially comes from the mixing with the ionic configuration not included in the GVB function.

AN EMBARRASSING INHERITANCE: THE TRADITIONAL COUPLING OF TWO ELECTRON SINGLETS

In order to diminish the number of functions, most VB calculations use linearly independent products of two-electron singlet functions $ab(\alpha\beta-\beta\alpha)$. This tradition is supported by the mathematical literature, including Rumer's work, but it may be considered as both inefficient from a computational point of view and obscuring the interpretation.

From a computational point of view, the time consumed is not in the research of the lowest eigenstate of a matrix but in the generation of the matrix. Direct CI algorithms have been greatly improved when the spin-symmetry heavy algebra has been abandoned and when one has gone back to single determinants (or pairs of single determinants) [20]. This remark is of no help in the GVB approach, where there is only one space part, but one may notice that for empirical Hamiltonians such as the Hubbard [21] or the P.P.P. [22] or CNDO [23] Hamiltonians, where the atomic orbitals are considered as orthogonal, the use of products of singlet pair functions is really inefficient since one goes from an orthogonal basis of VB single determinants to a smaller but non-orthogonal basis, while the matrix elements between single determinants are very easy to calculate:

$$\langle \phi_I | H | \phi_J \rangle = \beta_{pq} \qquad \text{if } \phi_I = a_p^+ a_q \phi_J$$
$$= 0 \qquad \text{otherwise}$$

the matrix elements between non-orthogonal spin-singlets take much longer time. We have written [24] a small direct VB–CI program for semi-empirical Hamiltonians, which finds the solution of a non-symmetrical problem involving up to 1.10^6 determinants with a computation time of 25' on a microvax 2. This is far better than the usual specific diagrammatic-VB algorithms [25].

The other paradox is on the conceptual side. As we have shown in the preceding example, going from GVB–PP to GVB essentially introduces spin-fluctuations. The space part is fixed and considering the various singlet spin-couplings consists in playing with the spin on these various functions, i.e. in giving different weights to the various spin-distributions of the n_α electrons in the $2n_\alpha$ orbitals (when $n_\alpha = n_\beta$). This means that the GVB determination of the monoelectronic space functions $a, b,...2n$ determines both an optimal valence space of one-electron functions and a non-orthogonal basis within this space, i.e. optimal delocalization tails around each atom. But the mixing of the various spin-couplings (singlet functions) essentially plays with spin-ordering. This reminds one of the Heisenberg problem, and this actually happened in the interstitial picture of alkali clusters in a convincing manner [17, 18] as it occurred for the π system of conjugated molecules [26].

THE LIMIT OF MEAN-FIELD DESCRIPTIONS

Both the GVB and the CASSCF descriptions work within a mean-field approxima-tion: the valence space is optimized and is unique; it gives the best set of valence orbitals consistent with the best weight given to the neutral (O)VB structures A·B· or ionic of opposite polarities (A^+B^- and A^-B^+). This valence space is neither optimal for A·B· nor for A^+B^- nor for A^-B^+. In that sense the valence orbitals of these procedures are a *compromise* and are adapted to a mean field. To make this evident, compare, for instance, the GVB interstitial orbitals for Li clusters [7] with those obtained from a UHF calculation [18]: these orbitals, which are very far from atomic orbitals, are practically undistinguishable.

The variational optimization of the valence space is definitely a great improvement, but the use of a *unique* valence space is a major restriction; it necessarily misses important physical effects. Actually, the CASSCF limit of the $2e^-$, 2 OM description of the single-bond Li_2 provides only about 60% of the dissociation energy [16]. The same ratio is obtained for the $6e^-$, 6 OM CASSCF description of N_2 [16]. Quantum chemistry now has to go far beyond these limits, and must incorporate the dynamical (or external) correlation effects.

These effects may be viewed as going through single and double excitations from each of the valence configurations. A multireference single + double CI acting on all the valence configurations will give the same result whatever the choice of the MO basis set within the valence space; they may be symmetry-adapted, orthogonal or non-orthogonal atom-like functions. If one uses atom-like (or more generally local) orbitals, i.e. a (O)VB expansion of the valence space of Ne^- determinants. Then one may distinguish between

— single excitations, which will permit a relaxation of each of the VB functions. This relaxation incorporates two effects:

(i) the orbitals on atom A would prefer to be more concentrated in A^+ and more diffuse in A^- (this may be seen as an instantaneous orbital breathing of each of the VB forms)
(ii) the instantaneous electric field in each of these VB forms (especially the ionic ones) will polarize the valence orbitals; for instance, in a F^+F^- ionic VB situation of F_2, the π orbitals are distorted leftwards by mixing with d orbitals by the instantaneous field of the σ dipole, while they are distorted rightwards in the other ionic structure and undistorted (or symmetrically distorted by the repulsion) in F·F·. This effect is an instantaneous repolarization.

— double excitations will correlate the pairs of valence electrons which are not the same in the neutral and the ionic VB structures. In H_2 for instance, in the neutral VB forms ab, double excitations will bring dispersion energies, while in ionic forms the two electrons are closer, in the same AO, and will avoid each other through $(1s_a^2 \to 2S_A^2)$ excitation, giving radial correlation, and through $(1s_a^2 \to 2p_A^2)$ excitations, giving rise to the so-called angular correlation. These correlation effects are larger than those acting on the neutral forms, and they should contribute to the bond energy.

Hiberty *et al.* [27] have systematically performed classical VB calculations in which they introduce all the single excitations from each of the valence VB configurations, i.e. incorporating the instantaneous relaxation of each of these VB forms and they in general have obtained excellent dissociation energies in problems where GVB is insufficient (such as F_2). This would tend to indicate that the instantaneous relaxation of the VB forms is the dominant contribution of dynamical correlation to the binding energy.

Clotet *et al.* [16] have included both single and double excitations from all OVB–CASSCF determinants in Li_2 and N_2, and have shown the crucial impact of these dynamical correlation effects on the dissociation energy, but they have not split the relative role of single and double excitations, and work is in progress from both sides, i.e. incorporating the effect of double excitations in VB calculations and separating the effect of singles and doubles in the dressed-CASSCF calculations.

The alternative is not between GVB or an enormous full VB. Previous work and a physical analysis show that singles and doubles on the minimal VB would overcome the defects of traditional minimal basis set VB and incorporate the leading physical effects, namely:

(i) the correlated fluctuation of positions of the electrons in the valence shell
(ii) the static (a) and dynamic (b) relaxation of the valence space (single excitations on the VB determinants)
(iii) the local dynamical correlation effects (double excitations).

The number of configurations would remain limited in many problems, especially if atomic natural orbitals were used to define the AO basis. The only problem is the size consistency question, if one truncates at the single + double excitation level.

In the preceding partition of physical effects,

— GVB incorporates most of the effects (i) and (iia) (but misses some important non-dynamical correlation effects, as shown in section 'GVB versus valence CASSCF: the missing effect'
— CASSCF incorporates effects (i) and (iia) perfectly.

Effect (iib) is not included in these mean-field descriptions, but is incorporated in the work of Hiberty et al. [25]. One may actually refer to the generalized Brillouin's theorem: the optimization of the monoelectronic space warrants that

$$\langle \psi_{MC} | H | a_{l*}^+ a_m \psi_{MC} \rangle = 0$$

where m is an active orbital and $l*$ an inactive orbital, but of course

$$\langle \psi_{MC} | H | a_{l*}^+ a_m \phi_l \rangle \neq 0$$

despite the fact that ϕ_l is a component of the ψ^{MC} wave function. Of course one may raise the objection that the VB expansion including the valence ($\simeq$ minimal basis set) configurations plus all singles and doubles on these mother configurations will become too long and lead to an uninterpretable wave function. One should, however, be reminded that the theory of effective Hamiltonians is available to combine the requirements of numerical precision on the one hand and compactness (or intelligibility) on the other hand [28]. The valence configurations may define a model space (with possible subdivision into a main model space spanned by neutral VB configurations and an intermediate model space spanned by the other valence configurations, if one uses the intermediate effective Hamiltonian formulation [29] to avoid intruder-states problems, and one may build on this model space (either perturbatively, according to one version of the quasi degenerate perturbation theory, or variationally), an effective VB Hamiltonian which would be chosen by the dynamical correlation effects. This has been done with an orthogonal basis of the model space, i.e. on an OVB approach, using either orthogonalized valence orbitals (for the Li_2 problem [30]) or nearly atomic MOs from a valence CASSCF calculation (for the Li_2 and N_2 problems [15]). But this might be done as well in non-orthogonal formulations (or re-expressing the dressed CASCI Hamiltonian in terms of the non-orthogonal VB determinants spanning the same valence space (as defined in subsection 'Generalization'. Research into such dressed -VB Hamiltonians would really be an exciting challenge.

REFERENCES AND NOTES

[1] For recent reviews on VB theories, see the contributions to the book *Valence Bond Theory and Chemical Structure*, edited by D.J. Klein and N. Trinajstic, Elsevier, Amsterdam, 1990, and especially those of R. McWeeny (p. 13) and of M. Raimondi and M. Sironi (p. 111). See, also, D.L. Cooper, J. Geratt and M. Raimondi, *Adv. Chem. Phys.* **67**, 319 (1987).

[2] P. Karafiloglou, *Chem. Phys.* **128**, 373 (1988).

[3] S.S. Shaik, *Nouv. J. Chim.* **6**, 159 (1982); S.S. Shaik and H. Pross, *J. Amer. Chem. Soc.* **104**, 2708 (1982); S.S. Shaik, *Progress Phys. Org. Chem.* **15**, 197 (1985); G. Sini, G. Ohanessian, P.C. Hiberty and S.S. Shaik, *J. Amer. Chem. Soc.* **112**, 1407 (1990); F. Bernardi, *J. Amer. Chem. Soc.* **109**, 544 (1987).

[4] P. Karafiloglou and J.P. Malrieu, *Chem. Phys.* **104**, 383 (1986); P.C. Hiberty, in ref. [1], p. 221 and references herein.

[5] For large molecules and semi-empirical systems, see: P.C. Hiberty and C. Leforestier, *J. Amer. Chem. Soc.* **100**, 2012 (1978); P.C. Hiberty and G. Ohanessian, *J. Amer. Chem. Soc.* **106**, 6963 (1984); P.C. Hiberty and G. Ohanessian, *Int. J. Quant. Chem.* **27**, 245 and 259 (1985); M.B. Lepetit, J.P. Malrieu, D. Maynau and B. Oujia, *Phys. Rev. A* **39**, 3274 and 3289 (1989). The last two references include OVB analysis on Li clusters *ab initio* wave functions.

[6] C.A. Coulson and I. Fisher, *Phil. Mag.* **40**, 386 (1949).

[7] This terminology was first introduced by Goddard and coworkers: W.J. Hunt, P.J. Hay and W.A. Goddard, *J. Chem. Phys.* **57**, 738 (1972), but in many works of this group, some constraints are introduced, for instance perfect pairing of the electrons (one spin part only) and/or orthogonality between the one-electron functions belonging to different pairs. We accept the term GVB for the fully free version of the theory, which is frequently called 'spin-coupled VB' by a second group of authors.
D.L. Cooper, J. Geratt and M. Raimondi, *Adv. Chem. Phys.* **67**, 319 (1987). See also from the name authors ref. [1], p. 287. In the following, the restrictions of GVB to perfect-pairing restriction will be called GVB-PP.

[8] This method was originally proposed under the name 'Fermi–Sea multiconfigurational Hartree–Fock by Nicolaïdes and Beck.
D.R. Beck and C.A. Nicolaïdes, *Int. J. Quant. Chem.* **58**, 17 (1974); C.A. Nicolaïdes and D.R. Beck, *Chem. Phys. Lett.* **36**, 79 (1975).
The term CASSCF is found in: B. Roos, P.R. Taylor and P.E.M. Siegbahn, *Chem. Phys.* **48**, 157 (1980). See also B. Roos, *Adv. Chem. Phys.* **LXIX**, 339 (1987).

[9] Anonymous referee of *J. Chem. Phys.* (1990).

[10] C.W. Bauschlicher and S.R. Langhoff, *J. Phys. Chem.* **89**, 4246 (1988).

[11] A. Sanchez de Meras, M.B. Lepetit and J.P. Malrieu, *Chem. Phys. Letters*, **172**, 163 (1990).

[12] K. Ruedenberg, M.W. Schmidt, M.M. Gilbert and S.T. Elbert, *Chem. Phys.* **71**,

41, 51 and 65 (1982). B. Lam, M.W. Schmidt and K. Ruedenberg, *J. Phys. Chem.* **89**, 222 (1985).

[13] S.F. Boys, in *Quantum Theory of Atoms, Molecules and Solid State*, P.O. Löwdin (Ed.), New York, Academic Press, 1966, p. 253.

[14] G. Trinquier, *J. Amer. Chem. Soc.* **113**, 144 (1991); G. Trinquier and J.P. Malrieu, *J. Amer. Chem. Soc.* in press, work concerning the double bridge of diborane.

[15] G. Trinquier and J.P. Malrieu, *J. Phys. Chem.* **94**, 6184 (1990).

[16] A. Clotet, J.P. Daudey, J.P. Malrieu, J. Rubio and F. Spiegelmann, *Chem. Phys.* **147**, 293 (1990).

[17] M.H. McAdon and W.A. Goddard, *Phys. Rev. Lett.* **55**, 2563 (1985); M.H. McAdon and W.A. Goddard, *J. Phys. Chem.* **91**, 2607 (1987).

[18] M.B. Lepetit, J.P. Malrieu and F. Spiegelmann, *Phys. Rev. B.* **41**, 8093 (1990).

[19] W.E. Palke, *J. Amer. Chem. Soc.* **108**, 6543 (1986).

[20] P.J. Knowles and N.C. Handy, *Chem. Phys. Lett.* **111**, 315 (1984).

[21] J. Hubbard, *Proc. Roy. Soc. London A* **276**, 238 (1963).

[22] R. Pariser and R.G. Parr, *J. Chem Phys.* **21**, 466 and 767 (1953); J.A. Pople, *Trans. Far. Soc.* **49**, 1375 (1953).

[23] J.A. Pople, D.P. Santry and G.A. Segal, *J. Chem. Phys.* **43**, S129 (1965).

[24] B. Oujia and J.P. Malrieu, to be published.

[25] Z.G. Soos and S. Ramasesha, ref. [1], p. 81 (and previous references herein).

[26] D. Maynau, M. Said and J.P. Malrieu, *J. Amer. Chem. Soc.* **105**, 5224 (1983); D. Maynau, Ph. Durand, J.P. Daudey and J.P. Malrieu, *Phys. Rev. A.* **28**, 3193 (1983); M. Said, D. Maynau, J.P. Malrieu and M.A. Garcia Bach, *J. Amer. Chem. Soc.* **106**, 571 (1984); J.P. Malrieu, in ref. [1], p. 135; V. Mujica, N. Loncia and O. Gocsinski, *Phys. Rev. B.* **132**, 4178 (1985).

[27] P. Maître, J.M. Lefour, G. Ohanessian and P.C. Hiberty, *J. Phys. Chem.* **94**, 4082 (1990).

[28] Ph. Durand and J.P. Malrieu, *Adv. Chem. Phys.*, K.P. Lawley (Ed.), *Ab initio Methods in Quantum Chemistry*, Wiley, 1987, p. 31.

[29] J.P. Malrieu, Ph. Durand and J.P. Daudey, *J. Phys. A. Math. Gen.* **18**, 809 (1985).

[30] S. Evangelisti, J.P. Daudey and J.P. Malrieu, *Phys. Rev. A* **35**, 4930 (1987).

15

The covalent radius in density-functional theory

Manoj K. Harbola and Robert G. Parr
University of North Carolina, Chapel Hill, NC, USA

INTRODUCTION

In recent years, density-functional theory [1], in addition to serving as a convenient computational tool [2] to calculate the electronic structure of large molecules, has provided rigorous definitions to concepts such as electronegativity [3, 4], hardness [5, 6], and covalent radius [3], that have existed in chemistry for a long time. In the following, we survey how these concepts are given their precise physical meaning employing density-functional theory. In particular we describe how covalent radius can be uniquely defined within this framework.

Density-functional theory is based on a theorem by Hohenberg and Kohn [7], which states that the ground-state density $\rho(\mathbf{r})$ of an atomic or molecular electronic system determines uniquely (up to a constant) the external (nuclear) potential $v(\mathbf{r})$ which the electrons are moving in. Since the density also gives the number N of electrons, we can accordingly express the ground-state energy, which is determined by N and $v(\mathbf{r})$ via the Schrödinger equation, a functional of the ground-state density. Thus the ground-state energy can be written as

$$E[\rho] = \int v(\mathbf{r})\rho(\mathbf{r})\, d\mathbf{r} + F[\rho] \tag{1}$$

where $F[\rho]$ is a universal functional of the density. Further, by the variational principle for the energy, the energy functional of equation (1) assumes its lowest value for the ground state. Thus for the ground state, the following Euler equation is satisfied:

$$\frac{\delta E}{\delta \rho} = v(\mathbf{r}) + \frac{\delta F[\rho]}{\delta \rho} = \mu = \text{constant} \tag{2}$$

where μ is the Lagrange multiplier that ensures that the total number of electrons be N.

Equation (2) can be written in the following form if we allow the number of electrons to change [8]:

$$\int \frac{\delta E}{\delta \rho} \delta\rho(\mathbf{r}) \, d\mathbf{r} = \int \mu \delta\rho(\mathbf{r}) \, d\mathbf{r} \tag{3}$$

which gives

$$\mu = \left(\frac{\partial E}{\partial N}\right)_{v(\mathbf{r})} \tag{4}$$

Equation (4) gives μ its physical meaning: it is the chemical potential or escaping tendency of the electrons in the system.

ELECTRONEGATIVITY AND HARDNESS

The chemical potential of equation (4) may be recognized as essentially the negative of the Mulliken absolute electronegativity of the system. By considering a system with fractional electron numbers [9] as a statistical mixture in a grand-canonical ensemble, the resultant continuous E vs N curve is found to be a series of straight-line segments. Thus, the three-point finite difference approximation gives [9]

$$\mu = -\frac{I + A}{2} \tag{5}$$

where I and A are the ionization potential and electron affinity, respectively. But the electronegativity definition of Mulliken is [4]

$$\chi_M = \frac{I + A}{2} \tag{6}$$

Hence

$$\mu = -\chi_M \tag{7}$$

With this definition of electronegativity, the Sanderson Principle [10, 11] of electronegativity equalization follows. When atoms combine together to form a molecule, their chemical potentials become that of the molecule.

Another chemical concept that has been given its precise definition via density-functional theory is that of hardness [5, 6]. The absolute hardness η measures [12] how the chemical potential of a species changes as its number of electrons is changed. Thus

$$\eta = \frac{1}{2}\left(\frac{\partial \mu}{\partial N}\right)_v = \frac{1}{2}\left(\frac{\partial^2 E}{\partial N^2}\right)_v \tag{8}$$

where the factor of $\frac{1}{2}$ is arbitrary. Again, three-point finite approximation is appropriate; it gives

$$\eta = \frac{I - A}{2} \tag{9}$$

The inverse of hardness is softness [13].

$$S = \frac{1}{2\eta} = \left(\frac{\partial N}{\partial \mu}\right)_v \tag{10}$$

With these definitions of hardness and softness, the HSAB principle [14] and the principle of maximum hardness [15] have both been rationalized [12, 16, 17].

COVALENT RADIUS

Using the concepts discussed above, we shall now demonstrate how precise meaning can be given to the covalent radius of an atom. In the next section we shall also see how the concepts of covalent radius, electronegativity and hardness are integrated when we define the covalent radii of atoms using density-functional theory.

We first recall how the covalent radius of an atom is defined. The covalent radius of an atom A is one half of the equilibrium distance of the diatomic A–A bond [3]. The concept of covalent radius is meaningful because it is observed that when two different atoms A and B form a molecule, the equilibrium distance between A and B is close to the sum of their respective covalent radii.

The justification for the covalent radius as an atomic property can be stated as follows. We have empirically (to a certain accuracy) $2R(AB) = R(AA) + R(BB)$ and $2R(AC) = R(AA) + R(CC)$, with $R(AA)$ the same in the two cases, and so on. $R(AA)$ consequently is independent of B and C, hence a property of the atom A only; call it $2R_A$, where R_A is the covalent radius of A.

The covalent radius of the ground state of an atom, like any property of it, by density-functional theory must be determined by the electron density. Or, equivalently, it can be determined by the classical electrostatic potential (from which the density can be found using Poisson's equation). The only question is: how?

In fact, the covalent radius of an atom may be naturally defined [18] as the radial distance, at which the total electrostatic potential (nuclear + electronic) is equal to the chemical potential of that atom. Thus at the covalent radius r_c

$$\mu = -\frac{I + A}{2} = v^*(r_c) = -\frac{Z}{r_c} + \int \frac{\rho(\mathbf{r}')}{|\mathbf{r}_c - \mathbf{r}'|} d\mathbf{r}' \tag{11}$$

where Z is the nuclear charge and $\rho(\mathbf{r})$ is the electronic density. Equivalently, if $F[\rho]$ of equation (1) is written as $F[\rho] = J[\rho] + G[\rho]$, where $J[\rho]$ is the Coulomb energy of electrons and $G[\rho]$ is another universal functional, then

$$\mu = -\frac{Z}{r_c} + \int \frac{\rho(\mathbf{r}')}{|\mathbf{r}_c - \mathbf{r}'|} d\mathbf{r} + \frac{\delta G[\rho]}{\delta \rho} \tag{12}$$

Comparison with equation (11) gives

$$\frac{\delta G[\rho]}{\delta \rho} = 0 \text{ at } r = r_c \tag{13}$$

Thus an equivalent definition of the covalent radius is that it is that radial distance at which $[\delta G/\delta \rho] = 0$.

Fig. 1 shows the distances thus determined in comparison with the experimentally determined covalent radii [19]. The agreement between the two is excellent.

It is incumbent upon us to note that a very similar proposal to relate electronegativity, covalent radius, and electrostatic potential was made a long time ago by Walter Gordy [20]. Gordy argued qualitatively that the electronegativity of an atom could

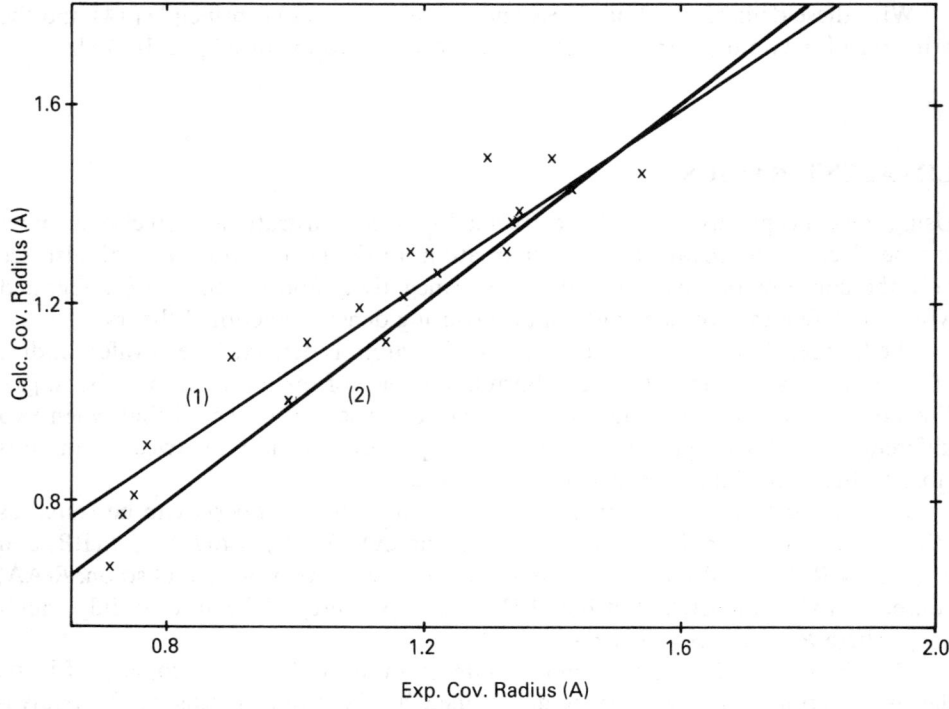

Fig. 1. — Line (1) is a plot of r_c as determined by equation (11) vs the single-bond covalent radius [19] of atoms. Line (2) would correspond to an exact equality between the two quantities.

be equated to the electrostatic potential felt by one of its valence electrons. Our results forcefully vindicate Gordy's ideas.

Balbas, Alonso and Vega [21] have examined some of these ideas in detail, using a density-functional pseudo potential method. They find confirmation.

HARDNESS FROM ELECTROSTATIC POTENTIALS

We now show how the equality between the chemical potential (negative electronegativity) and the electrostatic potential at the covalent radius of an atom can be exploited to relate hardness η and the change in electrostatic potential as the number of electrons in an atom is changed [22].

For atoms from equation (11)

$$\mu = v^*(r_c) \tag{14}$$

so that applying a three-point difference formula we obtain

$$\eta = \frac{1}{2}\left(\frac{\partial \mu}{\partial N}\right)_v = \frac{[v^{*-}(r_c) - v^{*+}(r_c)]}{4} \tag{15}$$

where $v^{*-}(r_c)$ and $v^{*+}(r_c)$ are the electrostatic potentials for the negative and positive ions respectively, evaluated at the covalent radius r_c of the neutral atom. Constant v in the formula for η does not mandate the argument r_c for the ions; in taking this argument we are making an assumption.

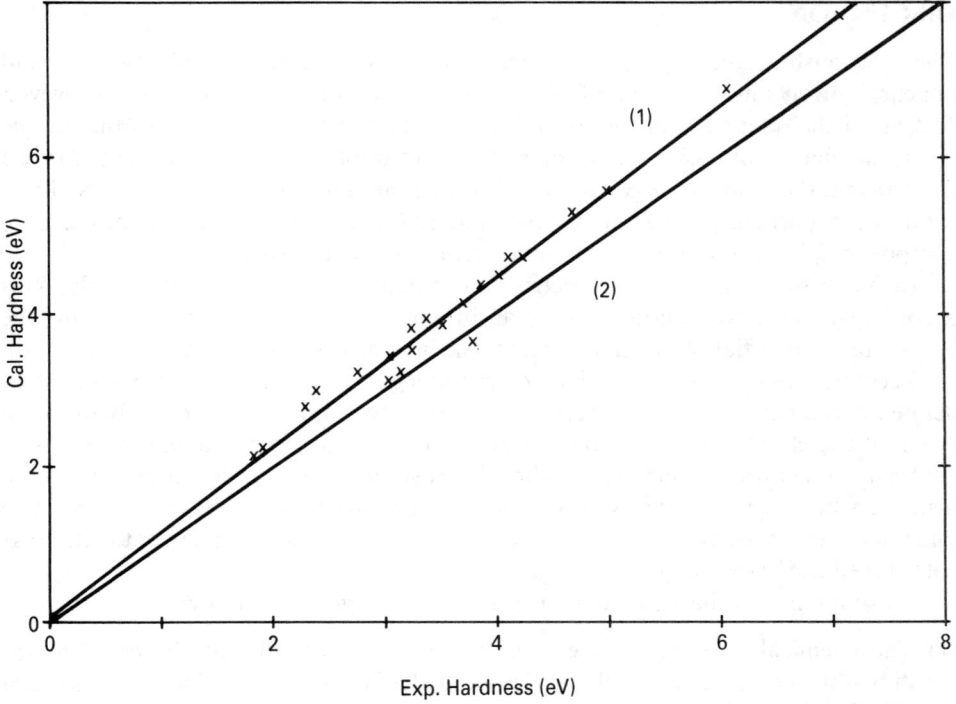

Fig. 2. — Line (1) is a plot of hardness determined from the electrostatic potential at the covalent radius of an atom (equation (15)) vs the experimental hardness [12]. Line (2) would correspond to an exact equality between the two quantities.

In Fig. 2 we compare hardnesses calculated from equation (15) with the experimental hardnesses [22]. It is clear that equation (15) gives a good measure of hardness.

To obtain additional understanding, let us rewrite equation (15) as

$$v^{*-}(r_c) - v^{*+}(r_c) = -\frac{Z}{r_c} + \int \frac{\rho_{N+1}(\mathbf{r})}{|\mathbf{r}_c - \mathbf{r}|}\, d\mathbf{r} + \frac{Z}{r_c} - \int \frac{\rho_{N-1}(\mathbf{r})}{|\mathbf{r}_c - \mathbf{r}|}\, d\mathbf{r}$$

$$= \int \frac{\rho_{N+1}(\mathbf{r}) - \rho_{N-1}(\mathbf{r})}{|\mathbf{r}_c - \mathbf{r}|}\, d\mathbf{r} \qquad (16)$$

which in finite different approximation gives

$$v^{*-}(r_c) - v^{*+}(r_c) = 2 \int \frac{f(\mathbf{r})}{|\mathbf{r}_c - \mathbf{r}|}\, d\mathbf{r} \qquad (17)$$

where $f(\mathbf{r})$ is the so-called Fukui function [23], and

$$\eta = \frac{1}{2} \int \frac{f(\mathbf{r})}{|\mathbf{r}_c - \mathbf{r}|}\, d\mathbf{r} \qquad (18)$$

Thus chemical hardness is one half of the electrostatic potential at the covalent radius due to the Fukui function.

DISCUSSION

The relationship, among various chemical quantities demonstrated above, would appear to nicely unify and quantify these various concepts. It must be noted, however, that all of the quantities in the formulas are expected to be sensitive to the unique particular details of each case of interest. For example, when a molecule is formed from atoms, the atoms change. In a way it is like the classical thermodynamics of real solutions, for which properties are not predictable from the properties of the pure components, but for which a systematic theory exists, nevertheless.

In our discussion here, the electrostatic potential plays a significant role. This accords with the well-known fact that in the valence regions of an atom, the electrostatic potential plays an important role in controlling chemical reactivity [24].

Recently, electrostatic potentials for mononegative ions have been employed to define the radii of negative ions [25]. The natural definition follows directly from the fact that the electrostatic potential for negative ions exhibits a maximum (this is in contrast to neutral atoms where the electrostatic potential is monotonic)—the distance where the potential is maximum is defined to be the radius of the ion. The ionic radii determined in this manner for halide ions are in good agreement with those determined [26] crystallographically.

To summarize, it has been demonstrated that, to good accuracy:

(1) The chemical potential is the total electrostatic potential at the covalent radius — the potential due to the nucleus and all of the electrons. This behaves much like the potential due to the nucleus and the electrons within r_c, incidentally [18] (or the potential due to a nucleus with an effective charge), so that one may think of it that way if one wants.

(2) The hardness is the electrostatic potential at the covalent radius due to the Fukui function. The latter being known to control site selectivity in chemical reactions [23], this leads one to expect this formula for hardness to be of use in modeling chemical bond formation.

The clear implication is that 'the radial distance, where the chemical potential is equal to the electrostatic potential in the free atom, is of special significance for the bonding properties of atoms' [21]. Nevertheless, the fundamental problem remains: to show why, in fact, twice the covalent radius defined in this way should be the equilibrium distance in the homonuclear diatomic molecule.

Acknowledgment

Aided by a research grant to the University of North Carolina from the National Science Foundation.

REFERENCES

[1]   R.G. Parr and W. Yang, *Density-Functional Theory of Atoms and Molecules*, Oxford University Press, New York, 1989.

[2]   J.K. Labanowski and J.W. Andzelm, *Density-Functional Methods in Chemistry*, Springer-Verlag, New York, 1991.

[3]   L. Pauling, *The Nature of the Chemical Bond*, 3rd edn, Cornell University Press, Ithaca, N.Y., 1938.

[4]   R.S. Mulliken, *J. Chem. Phys.* **2** (1934) 782.

[5]   R.G. Pearson, *J. Am. Chem. Soc.* **85** (1963) 3533.

[6]   R.G. Pearson (Ed.), *Hard and Soft Acids and Bases*, Dowden, Hutchinson and Ross, Stroudsburg, PA, 1973.

[7]   P. Hohenberg and W. Kohn, *Phys. Rev.* **136** (1964) B864.

[8]   R.G. Parr, R.A. Donnelly, M. Levy and W.E. Palke, *J. Chem. Phys.* **68** (1978) 3801.

[9]   J.P. Perdew, R.G. Parr, M. Levy and J.L. Balduz, *Phys. Rev. Lett.* **49** (1982) 1691.

[10] R.T. Sanderson, *Science* **114** (1951) 670.

[11] R.T. Sanderson (Ed.), *Chemical Bonds and Bond Energy*, Academic Press, New York, 1976.

[12] R.G. Parr and R.G. Pearson, *J. Am. Chem. Soc.* **105** (1983) 7512.

[13] W. Yang and R.G. Parr, *Proc. Natl. Acad. Sci. USA* **82** (1985) 6723.

[14] R.G. Pearson, *J. Am. Chem. Soc.* **85** (1963) 3533.

[15] R.G. Pearson, *J. Am. Chem. Ed.* **64** (1987) 561.

[16] P.K. Chattaraj and R.G. Parr, *J. Am. Chem. Soc.* **113** (1991) 1854.

[17] P.K. Chattaraj, H. Lee and R.G. Parr, *J. Am. Chem. Soc.* **113** (1991) 1855.

[18] P. Politzer, R.G. Parr and D.R. Murphy, *J. Chem. Phys.* **79** (1985) 3859; see also P. Politzer, R.G. Parr and D.R. Murphy, *Phys. Rev. A* **31** (1985) 2806.

[19] J.E. Huheey, *Inorganic Chemistry*, Harper and Row, New York, 1972.

[20] W. Gordy, *Phys. Rev.* **69** (1946) 604.

[21] L.C. Balbas, J.A. Alonso and L.A. Vega, *Z. Phys. D* **1** (1976) 215.

[22] M.K. Harbola, R.G. Parr and C. Lee, *J. Chem. Phys.*, **99** (1991) 6055.

[23] R.G. Parr and W. Yang, *J. Am. Chem. Soc.* **106** (1984) 4049.

[24] P. Politzer and D. Truhlar (Eds.), *Chemical Applications of Atomic and Molecular Electrostatic Potentials*, Plenum, New York, 1981.

[25] K.D. Sen and P. Politzer, *J. Chem. Phys.* **89** (1989) 4370.

[26] R.D. Shanon, *Acta Crystallogr. A* **32** (1976) 751.

16

Bent multiple bonds in normal and hypervalent molecules

Peter A. Schultz and Richard P. Messmer
General Electric Corporate Research and Development, Schenectady, NY, USA

INTRODUCTION

Much of our qualitative understanding of electronic structure is couched in the language of local orbitals and bonds, within the conceptual framework of the valence bond (VB) picture. The wide variety of tools this interpretational scheme offers to analyze and predict electronic structure has been given its most powerful articulation in Pauling's classic *The Nature of the Chemical Bond* [1]. One of the longest standing unresolved questions within the valence bond picture has been: what is the best qualitative description of the bonding in molecules with multiple bonds? In 1931, Pauling [2] and Slater [3] independently noted that a quantum mechanically consistent valence bond theory allowed at least two very different descriptions of multiple bonds: one as a set of σ and π bonds, and another as a set of symmetrically equivalent bent bonds. The two alternatives are illustrated in Fig. 1 for ethylene.

Pauling has long been a proponent, in fact the principle proponent, of the latter view [4], arguing that the bent-bond model has distinct theoretical advantages over the σ,π-bond model in empirically describing molecular properties. However, computational limitations precluded a direct first-principles test of the bent-bond–σ,π-bond-model question, and in the intervening decades the *computational* advantages of operating in a symmetry-restricted basis has slanted the thinking about the bonding to the point that the σ,π-bonding model is widely accepted as theoretically 'correct' and the bent-bond picture is considered an obsolete unsubstantiated model.

Today, the computational environment, both hardware and software, has matured to the point where, for the first time, the calculations implicit in the comparison of the two bonding models are possible so that the question of the better bonding model can be directly addressed. In this chapter we describe the results of such calculations which directly compare the bent-bond and σ,π-bond models and find the bent-bond model *energetically* the superior representation of the bonding. We find this to be the case not only in the standard molecules for which this has long been an issue, but also in molecules for which this outcome is somewhat of a surprise. Thus, the ideas of Pauling and Slater have greater validity than has generally been accorded them, and may also

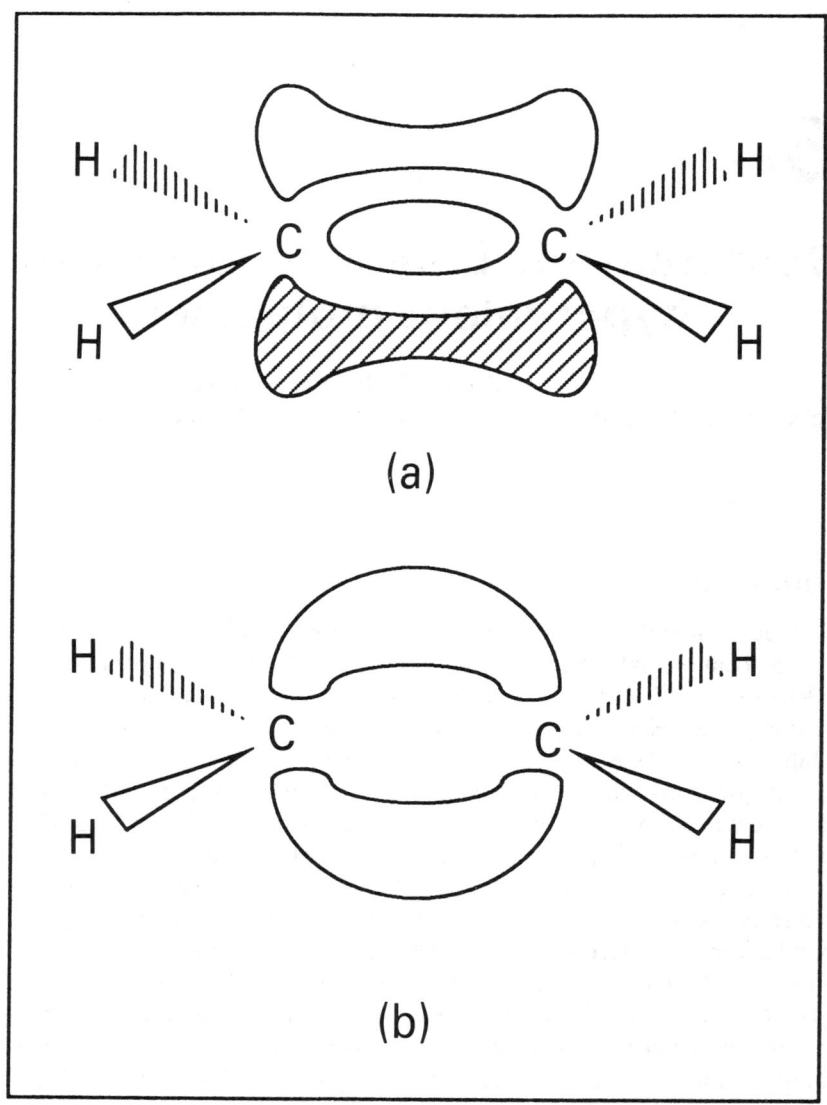

Fig. 1 — Bonding model alternatives for ethylene: (a) σ,π representation; (b) bent-bond representation.

have more general applicability than either anticipated. Using a bonding interpretation of a computational valence bond theory, we use an energy criterion to weigh the relative merits of the two bonding models. Comparing the variationally determined lowest energy solution for each, the superiority of the bent-bond model is quantitatively demonstrated for representative molecules with double bonds (CO_2 [5]), triple bonds (C_2F_2 [6, 7]), a mixed-bonding description (CO), conjugated bonding (benzene [8 ,9]), and for a hypervalent species (SO_2 [10]).

BACKGROUND

Before quantum mechanics, experimental chemists had devised a very successful structural chemistry based upon interlocking tetrahedra. Looking at such a representation of the CO_2 molecule, in Fig. 2(a), one can readily visualize the double bent bonds between the central carbon atom and the terminal oxygen atoms that this simplistic structural chemistry implies. Hence, there were bent bonds even before there were electrons to comprise them! With the advent of quantum mechanics, it was quickly realized that a quantum mechanical bonding theory could support two alternative bonding models: a modernized bent-bond model generalized from the old structural chemical ideas, and a new model in terms of σ and π bonds. The quantum mechanical σ,π description benefitted from computational efficiencies derived from the segregation of orbitals by symmetry, and the fact that these symmetry orbitals can be used to characterize the electronic excitations of the system. Yet Pauling argues effectively that the bent model made intrinsically more physical sense *for the ground state*, based upon its greater empirical usefulness in describing molecular structure [1].

One powerful application of the bent-bond/tetrahedral-bonding model predicts the carbon–carbon double-bond and triple-bond lengths in ethylene and acetylene to within 0.02 Å using the single-bond length and representing the bonds' multiple bonds as circular arcs beginning at the atom at tetrahedral angles. Extending these arguments, Pauling also very successfully rationalized bond angles in a wide selection of molecules with multiple bonds. Given these semi-empirical successes, and the fact that analogous analysis did not exist within the σ,π model, Pauling preferred the bent-bond model, even venturing so bold a statement as to say that the σ,π-bond description was 'a passing fad' [4].

Unfortunately, the calculations needed to test these conjectures were not possible at the time, and, in fact, the intrinsic algebraic and computational complexities inherent in the quantum mechanical valence bond theory made calculations for any molecule beyond H_2 impractical [11]. Thus, while the language of Pauling's valence

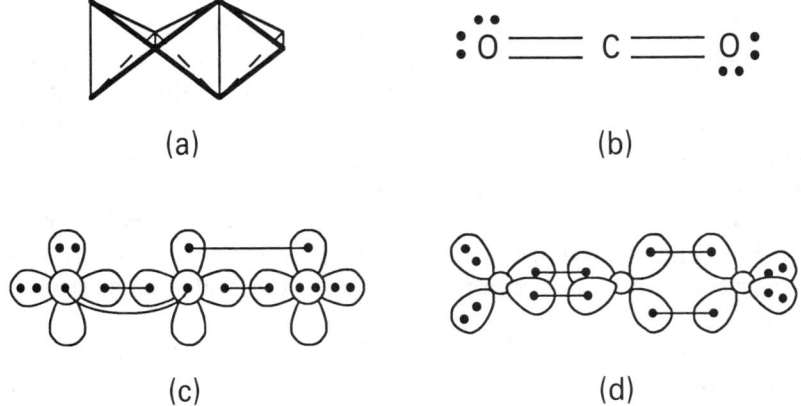

Fig. 2 — Schematic representations of bonding in CO_2: (a) classic structural model for CO_2; (b) Lewis structure for the bonding in CO_2; (c) σ,π-bond orbital model with double bond as $\sigma + \pi$ bond; (d) Ω-bond orbital model with double bond as two equivalent bent bonds.

bond interpretational scheme flourished, the computational theory corresponding to it did not.

If the lack of computational expression limited the development of the valence bond theories, the same was not true of the molecular orbital theory of Mulliken and the Chicago school [12]. The simplest wave function consistent with the basic tenets of quantum mechanics is a single Slater determinant of doubly occupied mutually orthogonal orbitals:

$$\Psi = \mathscr{A}[\phi_1^2\alpha\beta\phi_2^2\alpha\beta\dots\phi_n^2\alpha\beta] \tag{1}$$

The self-consistent Hartree-Fock wave function based on this molecular orbital (MO) wave function proved much simpler computationally, and MO theory quickly became the dominant computational paradigm for electronic structure.

The general character of the canonical molecular orbitals (CMOs) arrived at by diverse computational techniques (from *ab initio* to semi-empirical) is very similar, and via Koopmans' theorem (for Hartree-Fock wave functions) directly relate the electronic ground state and excited and ion states in a very simple fashion. The spatial forms of the delocalized CMOs, however, bear little relation to the useful concept of two-electron–two-centre bonds that form the foundation of structural chemistry, and fly in the face of the chemical fact that local bonds have properties largely independent of the global environment. The key to reconcile the computational delocalized CMOs and the existing chemical lore is the invariance of the total MO wave function of equation (1) to unitary transformations among the doubly occupied CMOs ϕ_i [13], including those transformations which generate localized bond orbitals [14]. It is in this context that the question of the relative worthiness of the σ,π-bond model vs. bent-bond model was first computationally examined.

The first localized MO study for multiple bonds was that of Foster and Boys [15] on the formaldehyde molecule, using their principle of exclusive orbitals. Their result was in accord with the bent-bond ideas, the C=O double bond in H_2CO being described by two equivalent bent bonds after the localization. Using the method of Edmiston and Ruedenberg [16], Kaldor found bent bonds to be the best 'energy localization calculations were computationally intensive, negating the computational advantages of the MO approach. Second, the same orbital invariance which made localized MO procedures possible also rendered them somewhat arbitrary. As

While localized MO approaches enjoyed many successes, and lent support to Pauling's bent-bond ideas, they suffered from many difficulties as well. First, the localization calculations were computationally intensive, negating the computational advantages of the MO approach. Second, the same orbital invariance which made localized MO procedures possible also rendered them somewhat arbitrary. As graphically demonstrated by von Niessen [19], different localization prescriptions would yield qualitatively different results, one producing bent bonds, another σ,π bonds, and a third, orbitals not compatible with any classical bonding structure. With no variational measure, the 'better' description was a matter of taste; all descriptions were theoretically equally valid. Third, the MO wave function is so primitive as to not include the basic electronic correlations necessary to dissociate the H_2 molecule, and this deficiency makes conclusions about the bonding based on this wave function tentative anyway.

Only limited calculations incorporating electronic correlation effects were directed toward divining the nature of the multiple bond. Klessinger performed calculations for CO and N_2 [20], and ethylene and acetylene [21]; he concluded that a σ,π basis was superior to a bent-bond basis for describing the mutiple bonds. In 1972, self-consistent valence bond (see below) calculations [22] yielded σ,π bonds as the lowest variational solutions for ethylene and acetylene (though the bent-bond alternative were *not* calculated for comparison). Thus the debate stood until recently. Localized MO results usually favored bent bonds, but were intrinsically inconclusive. What few correlated calculations addressing this issue, which could (in principle) energetically distinguish different bond models, favored the σ,π-bond model.

Recent calculations, however, have reopened this question, and indicate that Pauling's conjectures about multiple bonds may be correct. We have recently begun to systematically examine the question of the better bonding model, and using correlated wave functions directly interpretable within the valence bond picture, show that, indeed, 'bent bonds are better'. First, we digress to give a brief description of the theoretical methods used.

HIERARCHY OF INDEPENDENT PARTICLE INTERPRETABLE WAVE FUNCTIONS

The question of the 'better' bonding model is only well-posed in an independent pasrticle (IP) picture. Theoretically, the wave functions used must be representable as:

$$\Psi_{IP} = \mathscr{A}[\varphi_1(\mathbf{r}_1)\, \varphi_2(\mathbf{r}_2)\dots \varphi_n(\mathbf{r}_n)\, \Theta\{\mathbf{n}\}] \tag{2}$$

where the φ_i are spatial orbitals occupied by exactly one electron and $\Theta\{\mathbf{n}\}$ is an n-electron spin eigenfunction. The classical valence bond theory is an IP model where the φ_i correspond to atomic hybrids. A wave function that cannot be written in this form *cannot* be rigorously interpreted within an IP picture, hence cannot be associated with a simple bonding picture and cannot meaningfully compare different bonding models.

The most general approach within the IP context of equation (2) is the generalized valence bond (GVB) wave function of Ladner and Goddard [23]. In the GVB approach, the spatial forms of each of the singly occupied orbitals φ_i, and the spin couplings among them, are self-consistently determined without any overlap or spin coupling restrictions. This is essentially a full generalization of the classical valence bond theory. This method, however, compounds the complexities of the VB approach and two simplifications are made to yield a more tractable method. First, pairs of orbitals are singlet coupled, postulating a particular form:

$$\Theta\{\mathbf{n}\} = (\alpha(1)\beta(2) - \beta(1)\alpha(2))\, (\alpha(3)\beta(4) - \beta(3)\alpha(4))\dots \tag{3}$$

for the spin function (α denoting spin up and β spin down). This constitutes the perfect pairing (PP) approximation, leading to

$$\Psi = \mathscr{A}[\varphi_{1\mu}\, \varphi_{1\nu}\, (\alpha\beta - \beta\alpha)\, \varphi_{2\mu}\, \varphi_{2\nu}\, (\alpha\beta - \beta\alpha)\dots \varphi_{m\mu}\, \varphi_{m\nu}\, (\alpha\beta - \beta\alpha)] \tag{4}$$

A second constraint further simplifies the equations. While the orbitals $\varphi_{i\mu}$, $\varphi_{i\nu}$ within each pair continue to be permitted a variationally determined overlap, orbitals in different pairs are constrained to be orthogonal:

$$\langle \varphi_i | \varphi_j \rangle = 0,\, i \neq j \tag{5}$$

the so-called strong orthogonality (SO) condition. The two constraints defined by equations (3) and (5), first suggested in 1953 [24], and later implemented by Goddard and coworkers in the early 1970s [25, 26], lead to a practical computational scheme denoted here as SOPP–GVB [27]. Observe that one final constraint, forcing the orbitals within a pair to be the same:

$$\langle \varphi_{i\mu} | \varphi_{i\nu} \rangle = 1 \tag{6}$$

yields the molecular orbital wave function of equation (1). Hence MO wave functions are IP wave functions, the lowest in the IP hierarchy, also. Yet while the MO wave functions are invariant to changes in the orbitals related by unitary transformations, the same is not true of the more general IP wave functions. Changing the orbitals in any way, in principle, changes the energy. As a consequence, wave functions associated with different bonding models will have variationally distinct energies. These general IP wave functions can, therefore, meaningfully test the more appropriate bonding model.

The calculations which follow involve wave functions throughout this hierarchy of IP interpretable wave functions. Hartree–Fock and SOPP–GVB calculations employ the 'GVB–PP' [27] method. The full GVB calculations employ a self-consistent IP interpretable constrained configuration interaction method [28, 29]. Calculations incorporating chemical resonance of different VB structure wave functions, or 'structure interaction' (SI)—by analogy to the configuration interaction (CI) in molecular orbital theory—use our implementation [30] of the resonating-GVB method of Voter and Goddard [31]. All calculations are done in the experimental geometries, and standard double zeta basis sets [32] are used for all atoms, augmented by d-Gaussian polarization functions ($\zeta_C = 0.75$, $\zeta_O = 0.85$, $\zeta_F = 1.34$, and $\zeta_S = 0.53$).

'BENT BONDS ARE BETTER'

Double bonds: CO_2

The CO_2 molecule is linear, usually described as a central carbon atom with double bonds to each of the terminal oxygen atoms, i.e. using the Lewis structure of Fig. 2(b). There are sixteen valence electrons, eight involved in the C—O bonds the other eight in oxygen lone pairs. The classical bonding description within the conventional σ,π-bond model is schematically depicted in Fig. 2(c). A σ bond and a π bond make up each double bond, the two π bonds in planes perpendicular to each other, and there is a σ and a π lone pair on each oxygen. An SOPP–GVB calculation (defining eight SOPP pairs among the sixteen valence electrons), imposing σ,π symmetry restrictions on the orbitals, variationally yields a wave function that can be associated with this idealized bonding picture. The calculated energy, as shown in Table 1, is 2.46 eV lower than the Hartree–Fock result. A surprising thing occurs, however, when the artificially imposed symmetry restrictions are removed. The resulting variational SOPP–GVB wave function is 0.31 eV lower in energy, and the IP interpretable orbitals φ_i change, taking spatial forms corresponding to the valence bond model of Fig. 2(d) instead. Representative orbital contour plots of these orbitals are presented in Fig. 3. There are two sets of equivalent double bent bonds between the carbon and each of the oxygen

Table 1 — Computed total energies for the CO_2 molecule, in
Hartree (1 Hartree = 27.21161 eV).

Wave function	σ,π constrained	Symmetry relaxed
Hartree–Fock	− 187.67447	− 187.67447[a]
8 SOPP–GVB pairs	− 187.76505	− 187.77632
+ SI	− 187.76510	− 187.81397

[a]The H–F wave function is identical for both.

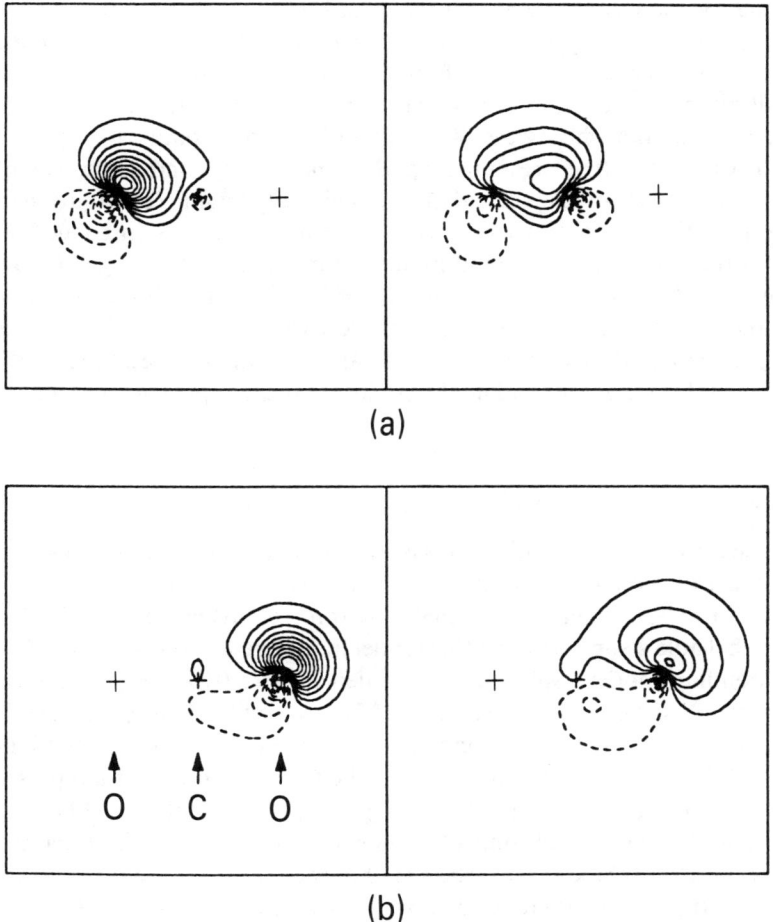

(a)

(b)

Fig. 3 — Representative orbital amplitude contour plots from the variational Ω-bond eight-pair SOPP–GVB calculation for CO_2. Solid lines represent positive contours and dashed lines, negative contours. (a) One of the four sets of singlet-coupled C—O bond pair orbitals. (b) One of the four sets of singlet-coupled bent oxygen lone pair orbitals.

atoms. Each consists of an orbital mostly on the carbon paired with an orbital mostly on the oxygen. The planes of the two sets of double bonds are perpendicular to one another. In addition, there are two radially correlated bent lone pairs on each oxygen atom, a radially compact orbital being paired with a more radially diffuse orbital. The C—O bond formed by the overlapping bent-bond hybrids has a shape reminiscent of the Greek letter Ω, hence they are referred to as Ω bonds.

The astute observer will note that the structures defined in Fig. 2 have D_{2d} symmetry, while experimentally the molecule has $D_{\infty h}$. To remedy this classically, chemical resonance is invoked to obtain a description with the proper symmetry. One bonding structure, denoted as I, when taken together with the symmetrically related structure I', obtained by a 90° rotation about the molecular axis, has the correct symmetry: $I + I'$. The corresponding quantum calculation involves an SI calculation of the SOPP–GVB wave functions Ψ_I and $\Psi_{I'}$: $\Psi_{SI} = \Psi_I + \Psi_{I'}$. The difference between the σ,π model and the Ω-bond model increases to 1.33 eV at this more general level of theory, increasing the margin for bent bonds by over a volt.

These calculations [5] marked the first quantitative demonstration that the bent-bond model is the energetically preferred description of the bonding in a molecule with multiple bonds. Independently, Palke published results [33] of approximately variational calculations lifting the SO restrictions in the orbitals of ethylene that also produced bent bonds (the presumable higher-energy σ,π-bond result was not calculated) as the result of the calculations. This important contribution explicitly demonstrated that restrictions on the GVB wave functions introduce significant biases into the σ,π-bond vs. Ω-bond-model comparison, as the original SOPP–GVB calculation mentioned above [22] found a σ,π-bond solution for ethylene.

It is uncanny how closely the results of these calculations parallel the description of the bonding offered by Pauling long ago. After several decades, the first theoretical vindication of this viewpoint had finally come. Further support was quickly forthcoming.

Triple bonds: C_2F_2

To examine the bonding in triple bonds is a natural next step, and Fig. 4 schematically illustrates the two bonding alternatives for a carbon–carbon triple bond. One can have either a σ bond and two π bonds, or three equivalent Ω bonds. The results presented in Table 2 support the latter model in preference to the former. Comparing the two bonding models, splitting three valence pairs (the C—C bond pairs) into SOPP pairs while treating the rest of the electrons with doubly occupied Hartree--Fock orbitals, we find the Ω-bond model favored over the σ,π-bond model by 0.26 eV. Orbital amplitude contour plots of the C—C σ bond and a representative π bond are shown in Figs 5(a) and 5(b). In Fig. 5(c), a representative Ω bond from this calculation is also shown. Splitting all 22 valence electrons into SOPP pairs results in a wave function with three C—C Ω bonds, three bent lone pairs on each F, and C—F σ bonds, a picture very aptly described by classical interlocking tetrahedra. Imposing σ,π orbital symmetry restrictions results in a σ,π-bond-model alternative 0.48 eV higher in energy than the optimal SOPP–GVB result with Ω bonds.

As was the case for ethylene, the SO constraints will bias the bonding comparison, likely against the Ω bond model, suggesting the margin favoring Ω bonds should be

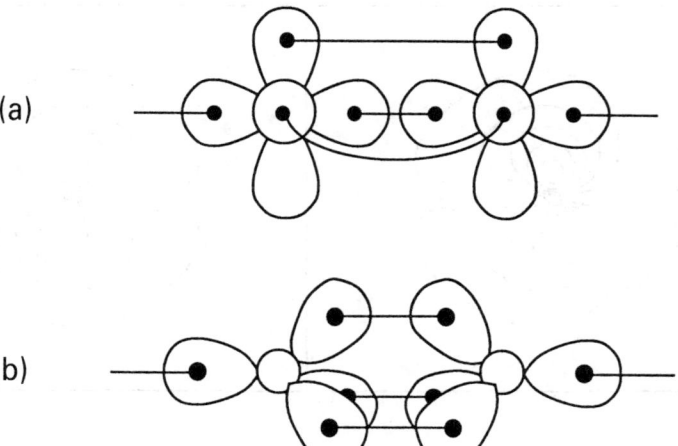

Fig. 4 — Two bonding alternatives for C—C triple bond: (a) σ,π-bond model with one σ bond and two π bonds, (b) Ω-bond model with three equivalent bent bonds.

even larger. In principle, the PP constraints can also be significant. On the basis of CI calculations intended to test their impact on the description of multiple bonds, Bauschlicher and Taylor (BT) [34] and Carter and Goddard [85] argued that the spin restrictions biased the bonding comparison in the opposite direction. In particular, BT claimed this bias was large enough to reverse the energetic order given for the two bonding models in the SOPP calculations for C_2F_2 [6], and that, therefore, the Ω-bond model was an artifact of the limitations in the SOPP approach. The BT analysis was flawed [7], however, as the particular CI expansions employed in the calculations were incompatible with an IP interpretation and hence could not be associated wtih any classical bonding description, much less compare two different ones. This unfortunate (and common) confusion regarding the difference between IP interpretational orbitals, the φ_i uniquely defined in the IP hierarchy in the previous section, and the computational basis orbitals, which can be arbitrarily chosen, notwithstanding, the basic question of the effect of the SOPP restrictions on the GVB

Table 2 — Computed total energies for the C_2F_2 molecule, in Hartree.

Wave function	σ,π-bond model	Ω-bond model
Hartree–Fock	−274.52520	−274.52520[a]
11 SOPP–GVB pairs	−274.65831	−274.67608
3 SOPP–GVB pairs	−274.57524	−274.58467
3 PP–GVB pairs	−274.57524[b]	−274.59053
3 GVB pairs	−274.58858	−274.59566

[a]The H–F wave function is identical for both.
[b]Same as the SOPP–GVB result, as SO is not a constraint on the σ,π wave function.

(a)

(b)

(c)

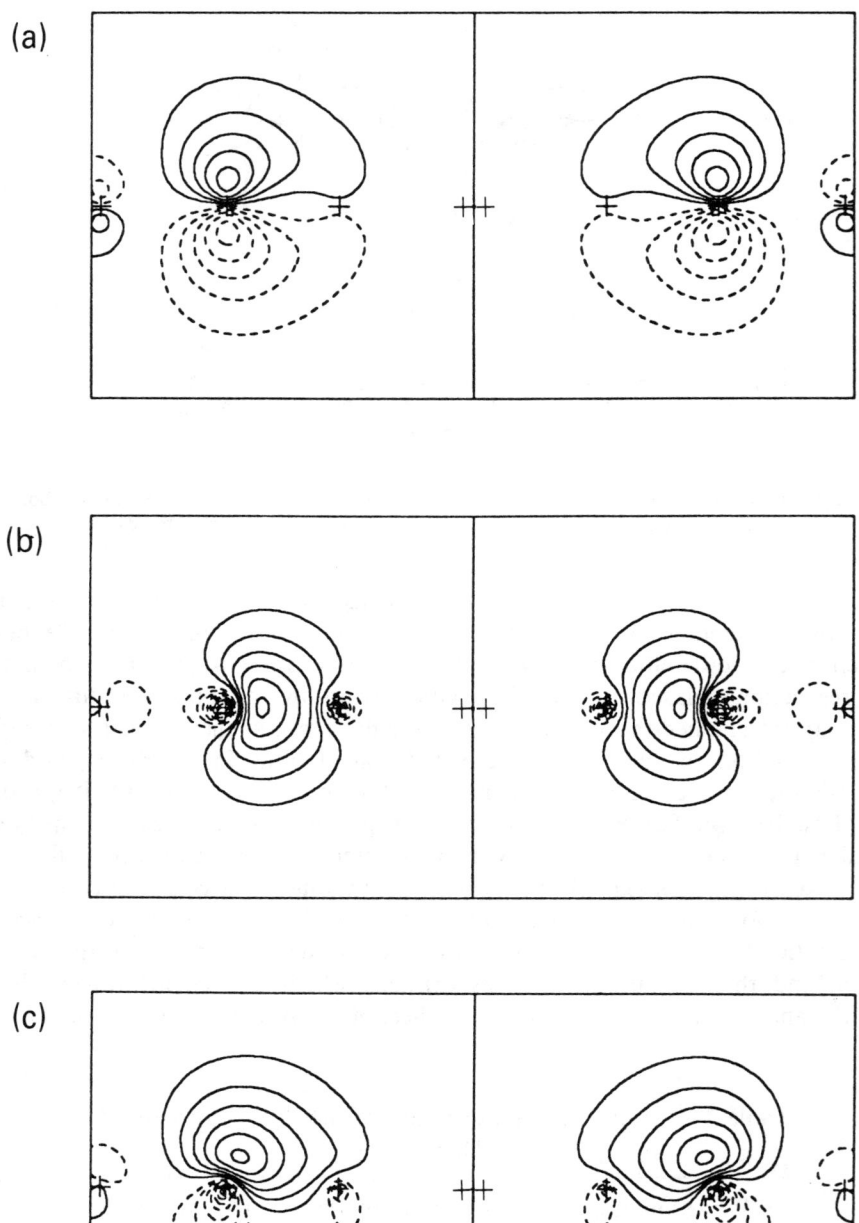

Fig. Fig. 5.—Orbital amplitude contour plots of IP orbitals from three pair SOPP–GVB calculations for C_2F_2: (a) C—C σ-bond pair orbitals from σ,π-bond-model result; (b) C—C π-bond pair orbitals from σ,π-bond-model result; (c) C—C bent-bond pair orbitals from Ω-bond-model result.

wave function are valid ones. To answer this question properly requires that the full unrestricted GVB calculations be done.

A method was developed [28] to perform the full GVB calculations, and for the first time to apply this approach to study multiple bonds. Indeed, these GVB calculations are among the few calculations of their kind in the literature. The original GVB approach was applied profitably to a few chemical systems [23, 36], but ultimately abandoned as computationally cumbersome. The approach has been rediscovered recently [37], and given a general implementation, but based on published results for benzene, may have numerical difficulties [9].

In the GVB calculations for C_2F_2, all but the C—C bonding electrons are treated at the Hartree–Fock level, six GVB orbitals being used for the six electrons of the triple bond. Narrowing the correlated treatment focuses attention on the bond and yields a tractable problem. The results listed in Table 2 indicate that at this level of IP theory, the Ω-bond model is favored over the σ,π-bond model by 0.19 eV. As the GVB wave function is the most general that supports an orbital interpretation of the bonding, the conclusion that the Ω-bond model is superior is theoretically definitive for this molecule.

Mixed Bonding: CO

In addition to the canonical triple-bond structure C≡O, Pauling argued that two symmetrically related double-bond C=O structures also should contribute significantly to the overall description of the molecule. The Ω-bond representations of these structures, denoted *I, II*, and *II'*, are depicted in Fig. 6. In *I*, there are three C—O Ω-bond pairs and radially split lone pairs on both the carbon and oxygen atoms, with a formal charge transfer from the oxygen to the carbon, In *II* and *II'*, there are two C—O Ω-bond pairs, two radially split bent lone pairs on the oxygen atom, and an

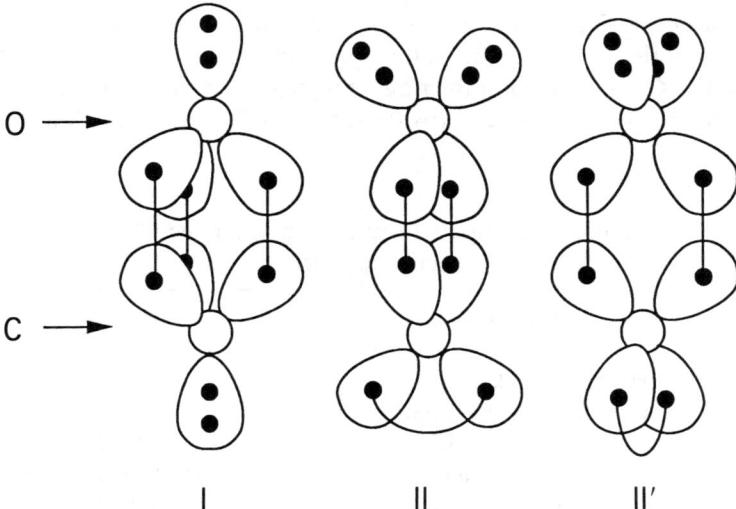

Fig. 6—Schematic representations of CO classical valence bond alternatives in the Ω-bond model.

angularly split lone pair on the carbon atom. The quantum mechanical implementation of this resonating valence bond description involves an SI wave function of the form

$$\Psi_{SI} = c_I \Psi_I + c_{II} (\Psi_{II} + \Psi_{II'})$$ (7)

where Ψ_I, Ψ_{II}, and $\Psi_{II'}$ are many-electron IP wave functions associated with the corresponding bonding structures of Fig. 6. The calculated SI coefficients are those that give the lowest energy for Ψ_{SI}. In the calculations presented here, SOPP–GVB wave functions are used to describe the individual structures. All ten valence electrons are treated in five SOPP pairs.

In the single-structure calculations, the most stable wave function takes the form of triple σ,π bonds. The triple Ω-bond SOPP–GVB wave function is 0.12 eV higher. We also obtained the double-bond structures and, as listed in Table 3, the double Ω-bond structures are higher still in energy. However, as Pauling predicted, these structures do contribute significantly to the total SI wave function. The total energy stabilization afforded by the resonance interaction yields an energy half an electron volt lower than the calculated energy in the σ,π-bond context. This calculation lends plausibility to the semi-empirical arguments Pauling used for describing the bonding in this molecule.

Conjugated molecules: benzene

Benzene represents the prototypical π-electron system. The extraordinary success of π-electron theories in describing this molecule has discouraged speculation about alternative bonding schemes, even by Pauling, the foremost proponent of the bent-bond concept [4]. In light of the results discussed above for simple molecules with multiple bonds, suggesting that bent bonds might also be the better description of the bonding in benzene is a reasonable hypothesis. Recent calculations directed toward examining this question [8, 9, 28] support this conjecture.

For benzene, a single classical bonding structure is not adequate to describe the ground state of the molecule. The 'natural' description would have alternating double and single C—C bonds around the ring, implying a threefold symmetry for benzene. Kekulé's 'oscillating' structures provide the basis for understanding the observed

Table 3 — Computed total energies for the CO molecule, in Hartree.

Wave function	Calculated energy
Hartree–Fock	− 112.75852
Five SOPP–GVB pairs	
I—σ,π-bond model	− 112.83032
I—Ω-bond model	− 112.82587
II—Ω-bond model	− 112.82089
$I + II + II'$—Ω-bond model	− 112.84794

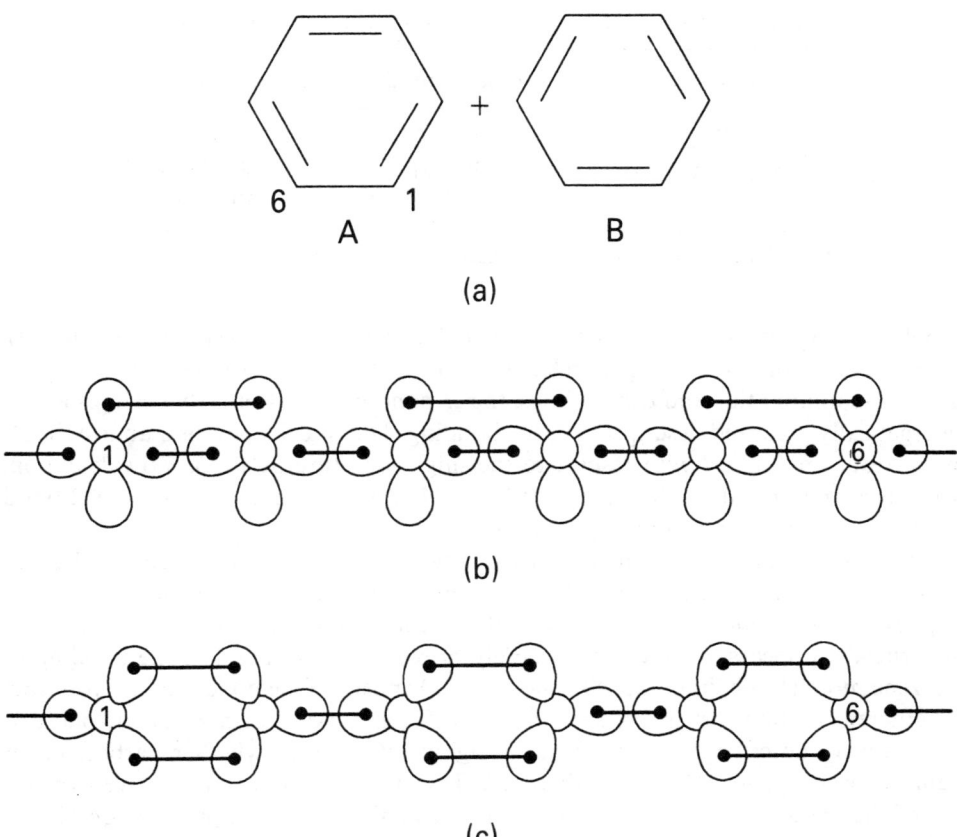

(a)

(b)

(c)

Fig. 7 — Valence bond description of benzene: (a) the two Kekule′ structures A and B, necessary to describe the molecule, each consisting of single and double bonds about the ring; (b) with the ring unraveled (1 to 6) and σ,π symmetry constraints on the orbitals enforced, the double bonds would consist of a σ and a π bond; (c) in the Ω-bond model, single σ bonds would alternate with double bent bonds about the ring.

sixfold symmetry of the molecule via the resonance of two such structures A and B as in Fig. 7(a). The issue, then, is whether the individual structures A and B are better represented in the σ,π-bond model, with double bonds consisting of a σ bond and a π bond as in Fig. 7(b), or in the Ω-bond model, with double bonds in the form of equivalent bent bonds as in Fig. 7(c).

Eighteen electrons, those associated with the C—C bonds, are treated as SOPP pairs. To simplify the calculations, the C cores and C—H bond electrons are treated as doubly occupied Hartree–Fock orbitals. To further simplify the calculations, d-Gaussian polarization functions are not included in the carbon atom basis set. Two separate SOPP-GVB wave functions are calculated: $\Psi_A^{\sigma\pi}$ — the optimal result with σ,π bonds, and Ψ_A^{Ω} — the optimal SOPP-GVB result with bent bonds. The total energies of the σ,π model and Ω-bond-model wave functions are compared using the SI wave functions $\Psi_{SI} = c(\Psi_A + \Psi_B)$. The results of the calculations are tabulated in Table 4.

The energies calculated for the single-structure SOPP-GVB wave functions are nearly the same, with the σ,π-bond model 0.06 eV lower than the Ω-bond model (with

Table 4 — Computed total energies for benzene, in Hartree.

Wave function	σ,π-bond model	Ω-bond model
Hartree–Fock	− 230.64037	− 230.64037[a]
9 SOPP–GVB pairs	− 230.75046	− 230.74834
+ SI	− 230.76236	− 230.77752

[a]The Hartree–Fock wave function is identical for both (see Ref. [18]).

d polarization functions this order is reversed, but the difference is still not significant). As previously, one would expect relaxation of SO restrictions to favor the Ω bonds. Even discounting this and other biases, the Ω-bond model ultimately emerges as the energetically preferred description of the bonding. The SI calculations stabilize the Ω-bond result by 0.79 eV with respect to the single-structure SOPP–GVB calculation, while the σ,π-bond SI result only nets 0.32 eV. Hence, in the final analysis, the Ω-bond model is convincingly favored by 0.41 eV over the σ,π-bond model.

This surprising conclusion radically diverges from the existing theoretical paradigm for this molecule. However, the valence bond calculations unambiguously support this new bonding description. Yet this conclusion is not quite so radical as it superficially appears, for the model merely brings the description of the bonding in benzene more closely into line with the rest of chemistry. One no longer has to resort to different carbon atoms, *sp*, *sp²*, *sp³*, for different environments; rather, one can qualitatively describe the bonding as roughly tetrahedral, whether carbon be in benzene or acetylene, ethylene or diamond. This universality is an attractive byproduct of the bent-bond ideas. The next case we examine, however, offers a completely different outlook, and shows how the bent-bond concept can be used to develop a novel understanding of a completely new system.

Hypervalent molecules: SO_2

The qualitative description of the bonding in sulfur oxides is a fascinating theoretical question that has posed a difficult dilemma for classical bonding models. Take SO_2 as a concrete example [38]. If the oxygen and sulfur atoms are described as the tetrahedral species they are usually considered to be, several resonance structures of formally ionic species are needed to describe the bonding [1, 39], as Fig. 8(a) illustrates. If, instead, the usual bonding behaviour is imposed without transfer of charge (a description for which Purser constructs an excellent empirical case [38]), as in Fig. 8(b), the sulfur atoms requires more than the four orbitals permitted by its *sp* valence. This fact has led SO_2 and related molecules to be described as hypervalent molecules.

Sulfur dioxide has been the subject of much theoretical work recently (see Ref. [10] and references therein) to better understand its electronic structure. Here, we discuss the results of one particular study [10] which adopted an IP approach, using SOPP–GVB calculations to divine the nature of bonding. The results of this reveal a startling new description of these molecules and a more general hybridization capability for the sulfur.

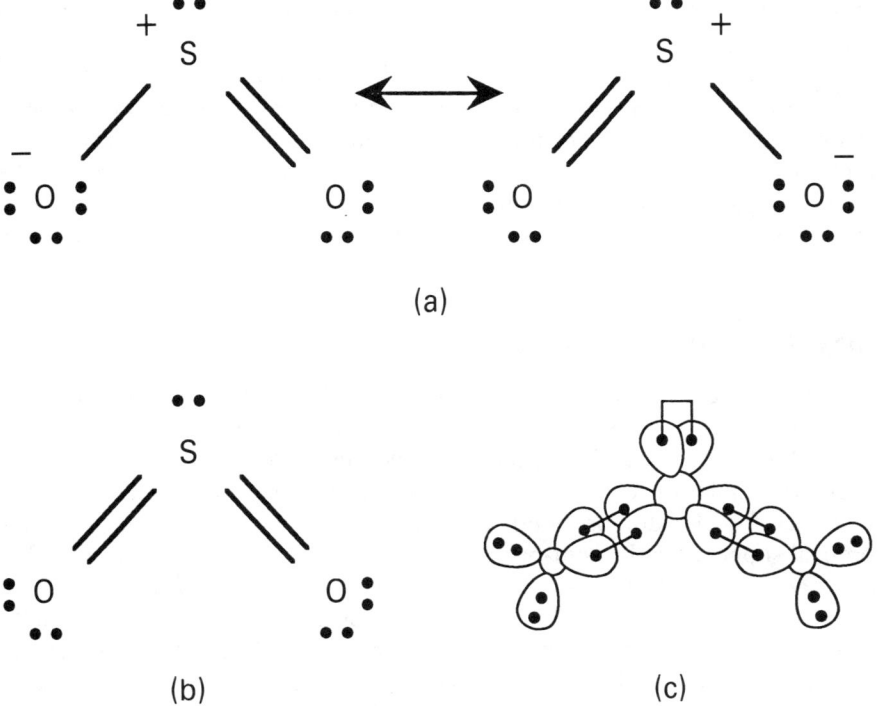

Fig. 8—Bonding descriptions for SO_2: (a) classic description involving resonance structures with formal transfer of charge; (b) hypervalent representation; (c) schematic representation of variational SOPP–GVB calculation.

The results of the SOPP–GVB calculations for SO_2, splitting all valence pairs, are schematically depicted in Fig. 8(c). The optimal SOPP–GVB result yields double Ω bonds between the sulfur atom and each oxygen atom, along with two radially split lone pairs on each oxygen atom and an angularly split lone pair on the sulfur atom. This results has several interesting consequences.

First, note that the bonding is described by a single classical VB structure. Furthermore, SOPP–GVB geometry optimization calculations with this wave function yield an equilibrium structure in excellent agreement with experiment. This testifies to how well this wave function, and hence the bonding model associated with it, describes the bonding in the molecule.

Second, the sulfur is 'hypervalent', i.e., *six* local orbitals arranged in as trigonal prism describe its bonding rather than the four permitted by the *sp* valence configuration. That this is *not* due to promotion into *d* orbitals is demonstrated by calculations excluding *d* orbitals from the sulfur basis set. The *sp*-basis calculation obtains the same qualitatively hypervalent result. While quantitatively important in the calculation, *d* orbitals are qualitatively irrelevant in the description of the bonding.

Third, the Lewis–Langmuir octet rule is clearly violated about the sulfur atom. A ten-count of electrons in five pairs (four S—O bonds and the lone pair) surround the sulfur atom. The key to making this possible is the electron-withdrawing power of the oxygen atoms. The electronegative oxygen atoms polarize the S—O bonds strongly enough away from the sulfur atom that five electron pairs can access the large (relative to the first row atoms) sulfur core.

Lastly, note that this bonding description has no σ,π analog (without invoking d orbitals). This would involve a σ and π bond from both oxygen atoms to the sulfur atom and there is only one π orbital on the sulfur to accommodate the two π bonds. Without the Ω-bond model, this simple description would not be possible. Yet, it leads to a very compact and simple picture that provides a very accurate description of the molecular properties, and it is this goal which is the principal objective of theory.

SUMMARY AND CONCLUSIONS

The systems discussed in this chapter constitute only single representative molecules from many different classes of molecules for which bent bonds have been theoretically discovered to be the better description of the bonding than the artificially constrained σ,π-bond model. However, evidence points toward this being generally true. Since the first studies for the double bonds in CO_2 [5] and C_2H_4 [33] gave substantive theoretical support for the bent-bond model, numerous studies have returned to the subject and confirmed this result in a variety of systems [5–10, 28, 29, 33, 40, 41]. A systematic survey [28, 29] of this issue found several molecules prefer bent bonds using the SOPP–GVB approximation. A majority of SOPP results, however, yielded the usually accepted σ,π-bond model. With removal of the SO and PP constraints, these results were reversed, and the Ω-bond model was found to be the better model in the full GVB calculations.

Given the difficulty of these calculations it is little surprise that it has taken so long to provide quantitative justification for the empirically successful bent-bond model. As the GVB wave functions used in this work constitute the pinnacle of the IP hierarchy, this judgement appears definitive. Results for ethylene and acetylene, among others, support the bent-bond picture, so that, along with the benzene discussed above, the principal members of the π-electron canon have been shown to be better described using bent bonds.

The results for benzene and for SO_2 are surprises that extended the validity and applicability of the model to systems previously unforeseen, though entirely sensible in hindsight. The flexibility provided by the bent-bond model leads to new under-standing and novel concepts with which to understand the bonding in these systems. One conceptual advantage of the MO approach has been the compact and simple description of ion and excited states of molecules. Recent work [42–44] has shown how only slight added complexity in a bent-bond-based VB model can also describe the ion states, while simultaneously taking into account electronic correlation effects occasionally crucial in describing their properties [44].

An interesting new development worth noting here takes advantage of the bent-bond representation of the bonding to design model force fields that include classical effective particles to explicitly take account of some electronic degrees of freedom [45]. Locating these new pseudo-particles at the centers of orbitals and treating their interactions with atomic cores and with each other via model functional potentials, one can design a viable scheme for molecular simulations. With a bent-bond picture, these particles have natural positions they would not have with a σ,π-bond picture, and therefore one might expect that the potentials and simulations so constructed might have a greater range of validity and transferability than is currently available. Only the future will tell how successful this new avenue will be. And it all started 60 years in the past with the ideas of Pauling and Slater.

REFERENCES

[1] L. Pauling, *The Nature of the Chemical Bond*, 3rd edn., Cornell University Press, Ithaca, NY, 1960.

[2] L. Pauling, *J. Am. Chem. Soc.* **53,** 1367 (1931).

[3] J.C. Slater, *Phys. Rev.* **37,** 481 (1931).

[4] L. Pauling, Kekulé Lecture, in *Theoretical Organic Chemistry*, Butterworth, London, 1959.

[5] R.P. Messmer, P.A. Schultz, R.C. Tatar, and H.-J. Freund, *Chem. Phys. Lett.* **126,** 176 (1986).

[6] R.P. Messmer and P.A. Schultz, *Phys. Rev. Lett.* **57,** 2653 (1986).

[7] R.P. Messmer and P.A. Schultz, *Phys. Rev. Lett.* **60,** 860 (1988).

[8] P.A. Schultz and R.P. Messmer, *Phys. Rev. Lett.* **58,** 2416 (1987).

[9] R.P. Messmer and P.A. Schultz, *Nature* **329,** 492 (1987).

[10] C.H. Patterson and R.P. Messmer, *J. Am. Chem. Soc.* **112,** 4138 (1990).

[11] J.C. Slater, *Quantum Theory of Molecules and Solids*, Vol. 1, McGraw-Hill, NY, 1963.

[12] R.S. Mulliken, *Rev. Mod. Phys.* **4,** 1 (1932).

[13] V. Fock, *Z. Physik* **61,** 126 (1930).

[14] C.A. Coulson, *Trans. Farad. Soc.* **38,** 433 (1942); J.E. Lennard-Jones, *Proc. Roy. Soc. (London)* **A198,** 1,14 (1949).

[15] S.F. Boys, *Rev. Mod. Phys.* **32,** 296 (1960); J.M. Foster and S.F. Boys, *Rev. Mod. Phys.* **32,** 300 (1960).

[16] J.E. Lennard-Jones and J.A. Pople, *Proc. Roy. Soc. (London)* **A202,** 166 (1950); C. Edmiston and K. Ruedenberg, *Rev. Mod. Phys.* **35,** 457 (1963).

[17] U. Kaldor, *J. Chem. Phys.* **46,** 1981 (1967).

[18] M.D. Newton, E. Switkes, and W.N. Lipscomb, *J. Chem. Phys.* **53,** 2645 (1970).

[19] W. von Niessen, *Theoret. chim. Acta. (Berlin)* **27,** 9 (1972); *ibid.* **29,** 29 (1973).

[20] M. Klessinger, *J. Chem. Phys.* **46,** 3261 (1967).

[21] M. Klessinger, *Int. J. Quant. Chem.* **4,** 191 (1970).

[22] P.J. Hay, W.J. Hunt, and W.A. Goddard III, *J. Am. Chem. Soc.* **94,** 8293 (1972).

[23] R.C. Ladner and W.A. Goddard III, *J. Chem. Phys.* **51,** 1073 (1969).

[24] A.C. Hurley, J.E. Lennard-Jones, and J.A. Pople, *Proc. Roy. Soc. (London)* **A220,** 446 (1953).

[25] R.A. Bair, W.A. Goddard III, A.F. Voter, A.K. Rappé, L.G. Yaffe, F.W. Bobrowicz, W.R. Wadt, P.J. Hay, and W.J. Hunt, GVB2P5 Program (unpublished); R.A. Bair, Ph.D. thesis, Caltech (1980); F.W. Bobrowicz and W.A. Goddard III, in *Methods in Electronic Structure Theory*, Vol. 3 of *Modern Theoretical Chemistry*, H.F. Schaefer III (Ed.), Plenum, NY, 1977.

[26] W.J. Hunt, P.J. Hay, and W.A. Goddard III, *J. Chem. Phys.* **57,** 738 (1972).

[27] The authors of Refs [25, 26] used the unfortunate nomenclature GVB–PP to describe the method, neglecting to make explicit the SO constraint also present

in the method. For our current purposes, this is an important distinction and we will use more specific designations.

[28] P.A. Schultz, Ph.D. thesis, University of Pennsylvania (1988).

[29] P.A. Schultz and R.P. Messmer, to be published.

[30] C.M. Kao and P.A. Schultz, RGVB Program (unpublished).

[31] A.F. Voter and W.A. Goddard III, *Chem Phys.* **57,** 253 (1981); A.F. Voter, Ph.D. thesis, Caltech (1982).

[32] T.H. Dunning, Jr. and P.J. Hay, in *Methods of Electronic Structure*, Vol. 3 of *Modern Theoretical Chemisty*, H.F. Schaefer III (Ed.), Plenum, NY, 1977.

[33] W.E. Palke, *J. Am. Chem. Soc.* **108,** 6543 (1986).

[34] C.W. Bauschlicher and P.R. Taylor, *Phys. Rev. Lett.* **60,** 859 (1988).

[35] E.A. Carter and W.A. Goddard III, *J. Am. Chem. Soc.* **110,** 4077 (1988).

[36] W.A. Goddard III and R.C. Ladner, *Int. J. Quantum Chem.* **IIIS,** 63 (1969); W.A. Goddard III, *ibid.* 593 (1970); R.J. Blint, W.A. Goddard III, R.C. Ladner, and W.E. Palke, *Chem. Phys. Lett.* **5,** 302 (1970); W.A. Goddard III and R.C. Ladner, *J. Am. Chem. Soc.* **93,** 6750 (1971); W.A. Goddard III and R.J. Blint, *Chem. Phys. Lett.* **14,** 616 (1972); C.W. Wilson and W.A. Goddard III, *J. Chem. Phys.* **56,** 5913 (1972).

[37] D.L. Cooper, J. Gerratt, and M. Raimondi, *Nature* **323,** 699 (1986).

[38] G.H. Purser, *J. Chem. Educ.* **66,** 710 (1989).

[39] F.A. Cotton and G. Wilkinson, *Advanced Inorganic Chemistry*, 5th edn, Wiley, NY, 1988.

[40] H.-J. Freund and R.P. Messmer, *Surf. Sci.* **172,** 1 (1986).

[41] W.A. Goddard III as cited in Ref. [33].

[42] R.P. Messmer, P.A. Schultz, S.H. Lamson, C.H. Patterson, and H. Wang, in *The Challenge of d and f Electrons*, D.R. Salahub and M.C. Zerner (Eds), ACS, Washington DC, 1989.

[43] R.P. Messmer, *J. Mol. Struct. (THEOCHEM)* **169,** 137 (1988).

[44] C.M. Kao and R.P. Messmer, *Chem. Phys. Lett.* **106,** 183 (1984).

[45] R.P. Messmer, W.-X. Tang, and H.-X. Wang, *Phys. Rev. B* **42,** 9241 (1990).

17

Linus Pauling, phosphorus–phosphorus bond distances, and the strain energy of P_4

Benjamin M. Gimarc and D. Scott Warren
University of South Carolina, Columbia, SC USA

Linus Pauling is surely the greatest chemist of the twentieth century. Although the citation for his 1954 Nobel Prize for chemistry mentions his experimental work on the conformational structure of proteins, it is his conceptual contributions to chemistry that most of us use every day and with which we frame new questions for research. What is truly remarkable about his conceptual contributions is the amount of insight he had to have to produce them. It is as though Pauling has inside information that is not available to the rest of us, a special relationship with nature that allows him to glimpse intimate relationships in the physical world, often before the data are available to confirm them.

Recently, we became interested in homoatomic clusters of the representative elements. Phosphorus is a good example. White phosphorus exists as regular tetrahedral P_4 units. The ion P_5^- is known as a planar pentagonal ring, P_6^{4-} is a planar hexagon, P_7^{3-} is a cage-shaped cluster, and so on to large clusters such as P_{10}^{6-}, P_{11}^{3-}, P_{16}^{2-}. This ability to form homoatomic clusters containing more than two or three atoms is quite common among the lower period main-group elements, but, with the notable exception of carbon, absent among the first row elements N, O, and F on whose properties many of our ideas of chemical valence are based.

We needed standard values of lengths of phosphorus–phosphorus single, double, and triple bonds with which to compare bond distances in the phosphorus clusters, both experimental and calculated. Naturally, the first place we turned to look for them was Pauling's book *The Nature of the Chemical Bond* [1]. The first edition was published in 1939, but the copy on our shelf is the 1948 second edition. Sure enough, it contains a table showing single-, double-, and triple-bond covalent radii for phosphorus. A footnote to the table mentions that the values were based on a paper by Pauling published in 1932 [2]. The single-bond radius came from half the P—P distance in black phosphorus, determined by X-ray studies of the solid, and in P_4, from gas-phase electron diffraction experiments [3, 4]. The triple-bond radius is half the P≡P bond distance in P_2 as determined by Herzberg in 1932 [5]. However, the first compounds containing P=P double bonds were not prepared and characterized until the 1980s. Pauling got his result by multiplying the single-bond value by 0.9.

Today, a wide variety of compounds as well as elemental phosphorus contain bonds between pairs of phosphorus atoms that have been described as single, double, and triple bonds [6-21]. Table 1 contains some representative distances for the three kinds of bond as determined in recent years by X-ray diffraction, gas-phase electron diffraction, and rotational and vibrational spectroscopy. P—P single bonds are around 2.22 Å as in P_4, H_2P—PH_2, and S_3P—PS_3^{4-}. P=P double bonds in compounds of the type R—P=P—R range from 2.01 to 2.03 Å. Herzberg has updated his P≡P triple-bond distance in P_2. The best value is 1.894 Å, a little longer than the one from 1932. These distances are in remarkable agreement with those Pauling proposed years ago: 2.20 Å, 2.00 Å, and 1.86 Å, respectively.

With the development of *ab initio* SCF MO techniques for calculation of molecular electronic structures, the molecule P_4 has received considerable theoretical attention [22-27]. Several authors have commented on the apparently highly strained regular tetrahedral structure of P_4 in which twelve P—P—P angles of 60° deviate considerably from what might be expected as the normal valence angle of 93° around phosphorus in unstrained PH_3 [22-26]. Almost four decades ago, Pauling and Simonetta [28] estimated the strain energy of P_4 as about 23 kcal/mol and noted that this value was comparable to that accepted at the time for cyclopropane, C_2H_6. They concluded that the strain energy of P_4 was much smaller than might have been anticipated on the basis of standard valence models. The authors of the only recent paper attempting to determine a quantitative measure of strain in P_4 concluded that the strain energy of this molecule is small or possibly even zero [27]. Sufficient thermochemical data are available with which to estimate the strain energy in tetrahedral P_4. The result shows that P_4 has only a modest strain energy, roughly comparable to that in cyclopropane as Pauling and Simonetta found.

Table   1 — Representative   phosphorus–phosphorus bond distances

Bond type	Distance (Å)
P—P	2.223 $(P_4)^a$
	2.2191, 2.218 $(H_2P—PH_2)^{b,\,c}$
	2.2182 $((CF_3)_2P—P(CF_3)_2)^d$
	2.19–2.267 $(P_2S_6^{4-})^{e-j}$
	2.20–2.24 $(P_2Se_6^{4-})^{h,\,i,\,k}$
P=P	2.004–2.034 $(R—P=P—R)^{l,\,m,\,n}$
P≡P	1.894 $(P_2)^o$

[a] Ref. [6]; [b] Ref. [7]; [c] Ref. [8]; [d] Ref. [9]; [e] Ref. [10];
[f] Ref. [11]; [g] Ref. [12]; [h] Ref. [13]; [i] Ref. [14];
[j] Ref. [15];
[k] Ref. [16]; [l] Ref. [17]; [m] Ref. [18]; [n] Ref. [19];
[o] Ref. [20, 21].

Table 2 — Standard heats of formation for some phosphorus compounds[a]

Substance	$P_4(g)$	$PH_2(g)$	$PH_3(g)$	$P_2H_4(g)$
ΔH_f°(kcal/mol)	14	30	5.5	5

[a] Refs [29, 30].

Strain energy is usually the difference between an experimental thermochemical quantity and a comparable value calculated on the basis of a model that does not include strain. The standard heats of formation, ΔH_f°, available from well-known tabular sources [29, 30], are shown in Table 2. Standard bond energies, $D(X—Y)$, from Sanderson [31] appear in Table 3.

The origins of tabular values of average bond energies are usually anonymous, as might seem appropriate for numbers to be used in varied or unspecified applications, but Sanderson shows explicitly that his $D(P—P)$ is that for a bond in tetrahedral P_4 and, hence, it already contains the effects of strain. Starting with the heats of formation for $PH_2(g)$ and $P_2H_4(g)$ we can get an unstrained $D(P—P)$ which must be the enthalpy change for equation (1):

$$H_2P—PH_2(g) \rightarrow 2PH_2(g) \tag{1}$$

$$\Delta H = 2\Delta H_f^\circ[PH_2(g)] - \Delta H_f^\circ[P_2H_4(g)] = 2 \times 30 - 5$$
$$= 55 \text{ kcal/mol} = D(P—P)(\text{unstrained})$$

It is satisfying that the unstrained P—P bond energy corresponds to a stronger bond than does the strained quantity (51 kcal/mol) obtained by Sanderson. The strain energy of an individual P—P bond must therefore be 55-51 = 4 kcal/mol and since there are six P—P bonds in tetrahedral P_4, the strain energy in P_4 should be $6 \times 4 = 24$ kcal/mol, almost exactly the value found by Pauling and Simonetta and a little less than the currently recognized value of the strain energy of cyclopropane (28 kcal/mol) that contains only three 60° valence angles [32].

Another estimate of the strain energy in P_4 can be obtained by comparing experimental and calculated values of enthalpy change for equation (2):

$$P_4(s) + 6H_2(g) \rightarrow 4PH_3(g) \tag{2}$$

The heat of reaction for equation (2) is four times the heat of formation for $PH_3(g)$ or:

$$\Delta H_2^{exp} = 4 \times \Delta H_f^\circ[PH_3(g)] = 22 \text{ kcal/mol}$$

Table 3 — Standard bond energies according to Sanderson[a]

Bond X—Y	P—P	P—H	H—H
$D(X—Y)$ (kcal/mol)	51	76	104

[a] Ref. [31].

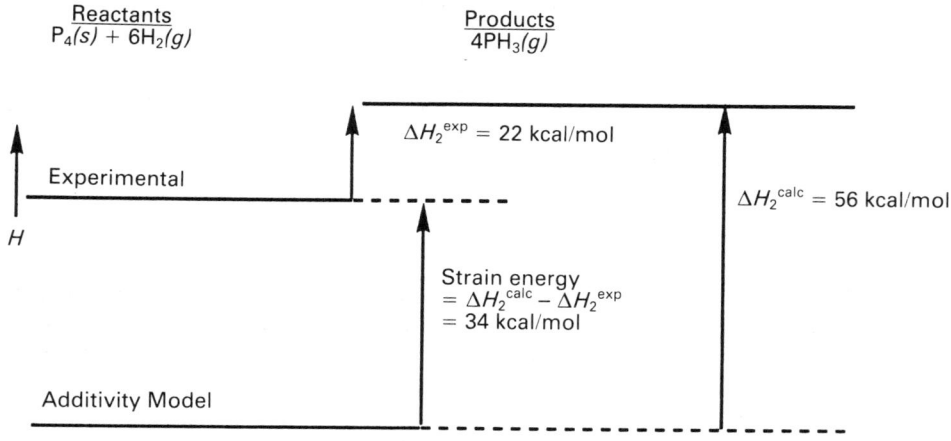

Fig. 1. — Relative enthalpies of products, and strained and unstrained model of reactants.

Compare this experimental value with one calculated for equation (2) using the bond additivity model:

$$\Delta H_2^{calc} = \Delta H_f^{\circ}[P_4(g)] + 6D(P\text{—}P) + 6D(H\text{—}H) - 12D(P\text{—}H)$$
$$= 14 + 6 \times 55 + 6 \times 104 - 12 \times 76 = 56 \text{ kcal/mol}$$

The smaller value of ΔH_2^{exp} compared to that for ΔH_2^{calc} indicates that starting materials must be at higher enthalpy than expected on the basis of the additivity model. That extra enthalpy must be due to the strain energy of P_4 not included in the bond additivity model. The difference, $\Delta H_2^{calc} - H_2^{exp} = 56 - 22 = 34$ kcal/mol, is the strain energy of P_4. The various quantities can be interpreted using an enthalpy diagram as shown in Fig. 1. This value of the strain energy is comparable to that of bullvalene, $C_{10}H_{10}$ (35 kcal/mol), a molecule containing a cyclopropane ring [33].

These strain energy estimates must be taken with reservation because of uncertainties in the experimental heats of formation and the standard value of $D(P\text{—}H)$ which has a multiplier of 12 in the work above. But clearly, the strain energy of P_4 is rather modest, perhaps in the range of 24 to 34 kcal/mol.

The moral of these two tales is that Linus Pauling, working with crude data or no data at all, was able to arrive at estimates of interesting physical quantities that have aged remarkably well.

This work has been supported by grant number CHE-9012216 from the National Science Foundation to the University of South Carolina.

REFERENCES

[1] L. Pauling, *The Nature of the Chemical Bond*, 2nd edn., Cornell University Press, Ithaca, 1948, p. 164.
[2] L. Pauling, *Proc. Natl. Acad. Sci. (USA)* **18** (1932) 293.
[3] R. Hultgren and B.E. Warren, *Phys. Rev.* **47** (1935) 808.
[4] L.R. Maxwell, V.M. Mosley, and S.B. Hendricks, *J. Chem. Phys.* **3** (1935) 698.
[5] G. Herzberg, *Ann. Physik* **15** (1932) 677.

[6]   N.J. Brassington, H.G. Edwards, and D.A. Long, *J. Raman Spectrosc.* **11** (1981) 346.

[7]   J.R. Durig, L.A. Carreira, and J.D. Odom, *J. Am. Chem. Soc.* **96** (1974) 2688.

[8]   B. Beagley, A.R. Conrad, J.M. Freeman, J.J. Monaghan, B.G. Norton, and G.C. Holywell, *J. Mol. Structure* **11** (1972) 371.

[9]   H.L. Hodges, L.S. Su, and L.S. Bartell, *Inorg. Chem.* **14** (1975) 599.

[10] G.D. Dittmar and H. Schäfer, *Z. Naturforsch.* **29B** (1974) 312.

[11] M.Z. Jandali, G. Eulenberger and H. Hahn, *Z. Anor. Allg. Chem.* **470** (1980) 39.

[12] M. Bouchetiere, P. Toffoli, P. Khodadad, and N. Rodier, *Acta Cryst.* **B34** (1978) 384.

[13] W. Klinger, G. Eulenberger, and H. Hahn, *Z. Anorg. Allg. Chem.* **401** (1973) 97.

[14] M.Z. Jandali, G. Eulenberger, and H. Hahn, *Z. Anorg. Allg. Chem.* **447** (1978) 105.

[15] P. Toffoli, P. Khodadad, and E. Rodier, *Acta Cryst.* **C39** (1983) 1485.

[16] H. Yun and J.A. Ibers, *Acta Cryst.* **C43** (1987) 2002.

[17] M. Yoshifuji, I. Shima, and N. Inamoto, *J. Am. Chem. Soc.* **103** (1981) 4587.

[18] A.H. Cowley, J.E. Kilduff, J.G. Lasch, S.K. Mehrotra, N.C. Norman, M. Pakulski, B.R. Whittlesey, J.L. Atwood and W.E. Hunter, *Inorg. Chem.* **23** (1984) 2582.

[19] T. Busch, W.W. Schoeller, E. Niecke, M. Nieger and H. Westermann, *Inorg. Chem.* **28** (1989) 4334.

[20] G. Herzberg, *Molecular Spectra and Molecular Structure. I. Spectra of Diatomic Molecules*, Van Nostrand, Princeton, 1959, p. 561.

[21] K.P. Huber and G. Herzberg, *Constants of Diatomic Molecules*, Van Nostrand, New York, 1979.

[22] G. Trinquier, J.-P. Malrieu, and J.-P. Daudey, *Chem. Phys. Lett.* **80** (1981) 552.

[23] G. Trinquier, J.-P. Daudey, and N. Komiha, *J. Am. Chem. Soc.* **107** (1985) 7210.

[24] R. Ahlrichs, S. Brode, and C. Ehrhardt, *J. Am. Chem. Soc.* **107** (1985) 7260.

[25] M.W. Schmidt and M.S. Gordon, *Inorg. Chem.* **24** (1985) 4503.

[26] K. Raghavachari, R.C. Haddon, and J.S. Binkley, *Chem. Phys. Lett.* **122** (1985) 219.

[27] W.S. Schoeller, V. Staemmler, P. Rademacher, and E. Niecke, *Inorg. Chem.* **25** (1986) 4382.

[28] L. Pauling and M. Simonetta, *J. Chem. Phys.* **20** (1952) 29.

[29] M.W. Chase, Jr., C.A. Davies, J.R. Downey, Jr., D.J. Frurip, R.A. McDonald, and A.N. Syverad, *JANAF Thermochemical Tables*, 3rd edn, Part II.

[30] M.K. Karapet'yants and M.L. Karapet'yants, *Thermodynamic Constants of Inorganic and Organic Compounds*, Ann Arbor-Humprey Sci. Pub., Ann Arbor, 1970.

[31] R.T. Sanderson, *Chemical Bonds and Bond Energy*, 2nd edn., Academic Press, N.Y., 1976.

[32] A. Greenberg and J.F. Liebman, *Strained Organic Molecules*, Academic Press, N.Y., 1978, p. 66.

[33] M. Månsson and S. Sunner, *J. Chem. Thermodyn.* **13** (1981) 671.

18

Theoretical studies of the Mills–Nixon effect

Z. B. Maksić,[a,b] M. Eckert-Maksić,[a] M. Hodošček,[c] W. Koch[d] and D. Kovaček[a]
[a]Rudjer Bošković Institute, Zagreb, Croatia, Yugoslavia
[b]University of Zagreb, Zagreb, Croatia, Yugoslavia
[c]Boris Kidrič Institute, Ljubljana, Slovenia, Yugoslavia
[d]IBM Heidelberg Scientific Center, Heidelberg, FRG

INTRODUCTION

The Mills–Nixon effect [1] has been a subject matter of long-standing research interest, debates and controversies. This is not surprising because fused benzenoid systems involving small rings combine two exciting topics which have fascinated chemists for a long time, namely aromaticity and angular strain. The former is a stabilizing principle, whereas the latter increases destabilization and chemical reactivity. For interesting reviews concerning the chemistry of fused MN-systems the reader is advised to consult the work of Billups, Thummel, Vollhardt and others [2-5]. Originally, the hypothesis was put forward by Mills and Nixon in 1930 who studied electrophilic substitution reactions in benzocycloalkenes ($n = 3, 4$) depicted in Fig. 1. It appeared that the substitution was favoured in the beta position. They interpreted this finding by a double bond localization in indan which prefers the Kekulé structure shown in Fig. 1(b). The opposite was a case in tetralin (Fig. 1(c)). Although the experimental data which led to the Mills–Nixon hypothesis were not unambiguous as

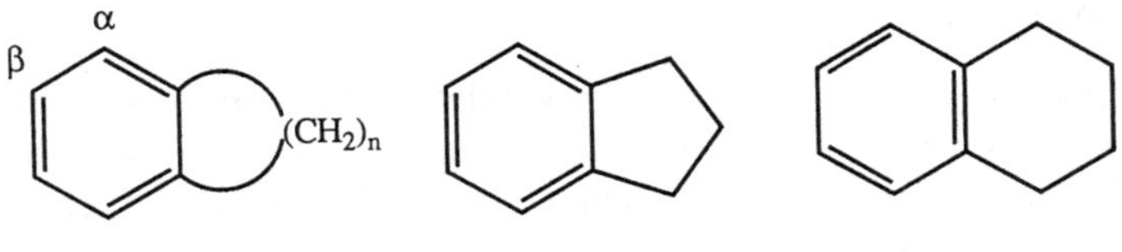

Fig. 1 — (a) Schematic representation of benzocycloalkenes. (b) Predominating Kekulé structure in indan according to the Mills–Nixon hypothesis. (c) Predominating Kekulé structure in tetralin according to the Mills–Nixon idea.

it turned out later on [6], subsequent valence bond studies of Sutton and Pauling [7] gave a distribution of bond distances which was in accordance with the original postulate.

On the other hand, the early calculations of Longuet-Higgins and Coulson who employed the modified HMO method indicated an anti-MN effect in the original molecular systems [8]. These two studies made by leading researchers of that time illustrate rather nicely the difference in opinion on the MN phenomena. It is a consequence of the elusiveness of the MN systems. They are large and highly strained. Their size frequently precludes a careful theoretical scrutiny. On the other hand high reactivity prevents syntheses of some crown-case molecules.

Direct experimental evidence provided by the X-ray technique is ambiguous. Some data are in favour of the MN bond fixation [9–13] whilst other data do not exhibit any significant bond alternation [14–16]. Serious doubts about the existence of the MN effect are raised by recent NMR measurements of chemical shifts [17] and $^4J(H—C \doteq \doteq \doteq C—Me)$ proton–proton spin–spin coupling constants [18, 19]. Theoretical results are not unequivocal either. They are divided in two pro [7, 20–23] and contra [8, 24] sets, respectively. We note in passing that a conclusive theoretical evidence in favour of the MN effect in some prototype systems will be presented in this chapter.

Interpretation of the MN phenomena is of general interest. It was realized at the outset by Mills–Nixon [1] and adopted later by Sutton and Pauling [7] that deviation of the bond angles around the carbon junction atoms are the driving force for the bond alternation within the benzene moiety. The rehybridization argument in discussing properties (predominantly acidity) of fused benzenes was employed by Fraenkel et al. [25] followed by Finnegan [26] and Streitwieser et al. [27]. The effect of rehybridization on spin densities in radical anions of fused aromatics was studied in a number of papers by Rieke et al. [28–30] and in the references cited therein. The first to realize the structural implications of the rehybridization effect which were directly related to the MN postulate seem to be Randić and Maksić [31]. Results presented here lend support to this conjecture.

2. THEORETICAL APPROACH

2.1 Selection of the theoretical model

There is a sort of relation

$$A * S \cong \text{const.} \tag{1}$$

which resembles Heisenberg's uncertainty relationship. Here S is a size of the examined molecule which involves both a number of atoms and a number of electrons. A stands for accuracy of the calculations. If the size is larger, accuracy and reliability of results are smaller. In theoretical treatments of large systems we seek an approximate method which is the best compromise between feasibility and accuracy.

Our target molecules here are relatively large aromatic molecules fused to small rings(s). In developing a suitable approximate method for their treatment we shall replace studied molecules with smaller model systems. They have to be small enough to be amenable to sophisticated calculations and yet they have to possess immanently the most characteristic features of the fused compounds. For example, benzo[1, 2:3, 4:5, 6]tricyclobutene is successfully modelled by an angularly deformed benzene (Fig. 2). The H—C—C angle is squeezed to the value of 94° in a symmetrical D_{3h} way. This distortion has a profound influence on the CC distances in benzene (Table 1) where results of calculations of methods widely differing in degree of sophistication are given. Their range is extended from highly semi-empirical schemes (MINDO/3, MNDO, AM1) to the MP2/6-311G *ab initio* procedure. Perusal of the data presented in Table 1 shows that the difference in interatomic distances between C_1—C_1 and C_1—$C_{1'}$ denoted carbon atoms is highly pronounced by all theoretical methods. It appears that the former is invariably longer, and the latter shorter, than the CC

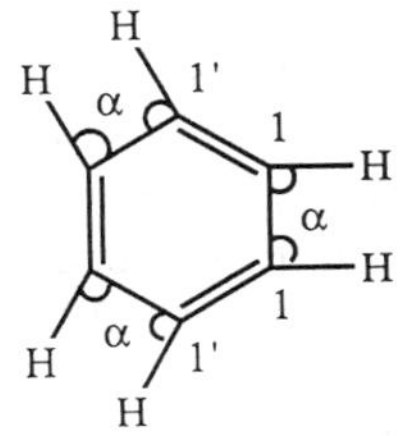

Fig. 2 — Hypothetical model system describing benzo[1,2;3,4;5,6]tricyclobutene. The CCH angle has a value taken from the target molecule.

Table 1 — CC bond distances in benzene and its hypothetical angularly distorted form as obtained by various theoretical methods (in Å)

α	Distance	MINDO/3	MNDO	AM1	3-21G	6-31G
120°	C_1—C_1	1.407	1.407	1.395	1.385	1.388
	C_1—$C_{1'}$	1.407	1.407	1.395	1.385	1.388
94°	C_1—C_1	1.496	1.524	1.501	1.461	1.493
	C_1—$C_{1'}$	1.350	1.339	1.332	1.365	1.342
Δ		−0.146	−0.185	−0.169	−0.096	−0.151

α	Distance	6-31G*	6-31G**	6-311G	MP2/6-311G	(3-21G)$_{corr.}$
120°	C_1—C_1	1.386	1.386	1.388	1.408	1.399
	C_1—$C_{1'}$	1.386	1.386	1.388	1.408	1.399
94°	C_1—C_1	1.489	1.484	1.495	1.508	1.467
	C_1—$C_{1'}$	1.339	1.341	1.340	1.368	1.381
Δ		−0.150	−0.143	−0.155	−0.140	−0.086

distance in a real benzene molecule. This is in line with the MN hypothesis. The method of choice should be economical, applicable to large systems and reliable enough. Since some of the studied fused systems involve atoms possessing lone pairs, semi-empirical methods can be immediately dismissed since they are utterly unsuccessful in treating atoms with couples of electrons with internally saturated spins. On the other hand the MP2/6-311G *ab initio* approach is certainly more reliable, but it is not feasible in large molecules, particularly since geometric optimization is required. Hence, a desired method lies somewhere in between these two extremes. Survey of the calculated distances shows that inclusion of polarization functions diminishes the absolute value of the difference Δ as evidenced by the 6-31G and 6-31G** results. Further, explicit, albeit approximate, treatment of the electron correlation tends also to decrease this difference as comparison of data obtained by 6-311G and MP2/6-311G procedures shows. In this respect it is useful to note that the CC bond distance in the ground state of benzene is 1.386 Å offered by the 6-311G** basis set which is considered to be close to the Hartree–Fock limit. The influence of the electron correlation at the MP4(SDQ)/6-311G** level of theory is estimated to be + 0.012 Å yielding $d(CC) = 1.3979$ Å in benzene [32]. It appears that the 3-21G benzene CC distance is in excellent agreement with the Hartree–Fock limit and, more importantly, inclusion of higher order corrections in Møller–Plesset perturbation method, together with saturation of the 6-311G basis set by adding polarization functions, would jointly decrease a value of $\Delta = d(C_1-C_1) - d(C_1-C_{1'})$ implying that an estimate of 0.096 Å obtained by a small 3-21G basis set is better than expected at the first sight. This is gratifying since 3-21G calculations can be applied to large molecules. It is also possible to increase performance of the 3-21G SCF method by some empirical adjustments. A scaling procedure involves selection of the small characteristic molecules with available experimental bond distances, execution of 3-21G calculations with the geometry optimization and least-squares fitting of the measured distances. The linear relationships read [33]:

$$d(C—C)_{SC} = 0.845\ d(C—C)_{3-21G} + 0.223\ \text{Å} \qquad (2)$$

$$d(C=C)_{SC} = 0.896\ d(C=C)_{3-21G} + 0.158\ \text{Å} \qquad (3)$$

The average absolute errors are 0.005 Å and 0.006 Å, respectively, implying that achieved accuracy is very good. It should be pointed out that empirical adjustments embodied in the scaled 3-21G procedure involve the effect of electron correlation introduced in an empirical way. In this respect it is noteworthy that the $(3-21G)_{SC}$ CC distance in benzene is 1.399 Å (Table 1) and is in almost perfect agreement with the MP4(SDQ)/6-311G** value of 1.3979 Å. Hence, our theoretical model of choice will be the scaled 3-21G SCF scheme. More sophisticated basis sets will be utilized too whenever possible.

2.2 Interpretation of the phenomenon

After identifying the studied phenomenon in the model system and selecting a suitable theoretical tool for its examination, one should try to understand the underlying mechanism leading to the MN-type of bond alternation in benzene ring.

Since a deformation takes place in the plane of the molecule, which is at the same time the nodal plane for π-electrons, the angularly distorted benzene represents a clear cut case where the changes in σ-skeleton can be — to good approximation — separated from π-electrons. Further, it is reasonable to assume that CC bond alternation in deformed benzene arises mainly due to the rehybridization of carbon atoms since perturbation is confined to the plane of the molecule. One can easily find out that an effect of squeezing of $H—C_1—C_1$ angle from ideal 120° value to only 94° is opening of the interhybrid angle of hybrids describing C—H and $C_1—C_{1'}$ bonds. Consequently, their s-characters increase leading to a simultaneous decrease in the 'annelated' $C_1—C_1$ bond. This intuitive result is borne out by the MNDO calculation performed on the MP2/6-311G geometries of the ground-state benzene and its hypothetical deformed structure (Table 2). The hybridization s-characters extracted from the MNDO wave function show that the initial sp^2-canonical hybridization is significantly changed by C—H bending deformation. The 'annelated' $C_1—C_1$ bond assumes sp^3 hybridization which is highly unusual for a planar system formed by the carbon atoms of the coordination 3. Hence, these bonds are considerably longer. On the other hand the 'exo' bond $C_1—C_{1'}$ has a high s-content of 37.5% causing its appreciable shortening. We note in passing that it is in full accordance with the analogous hybridization picture in small rings possessing exo-double bond(s) [34].

Analysis of the maximum overlap hybrid orbitals shows that the bending angles are: $\delta_{11} = -19.1°$ and $\delta_{11'} = 17.1°$ for $C_1—C_1$ and $C_1—C_{1'}$, bonds, respectively. A negative sign implies that the 'annelated' bond is bent inside the ring whereas the oposite is the case in adjacent $C_1—C_{1'}$ bonds. Hence, the angularly distorted benzene exhibits alternation of bond distances and 'in–out' bending of the electron density. This pattern is confirmed by the X-ray measurements in benzocyclobutene [9].

Finally, it is very important to note that a descent in symmetry and redistribution of the σ-electron density in the plane of benzene leads to significant change in π-electron bond orders. They are substantially lower and higher in 'annelated' and 'exo' bonds, respectively. Hence, π-electron distribution also contributes to the bond alternation in the same way as σ-electrons. We shall see later on that this is a condition for a strong MN-effect (*vide infra*). It is important to note that the

Table 2 — MNDO hybridization indices and π-bond orders calculated for the MP2/6-311G geometries of benzene and its hypothetical angularly distorted form[a]

α	Bond	s-character	π-bond-orders
120°	$C_1—C_1$	32.7–32.7	0.67
	$C_1—C_{1'}$	32.7–32.7	0.67
94°	$C_1—C_1$	24.4–24.4	0.42
	$C_1—C_{1'}$	37.5–37.5	0.86
Δ'		26.2	0.44

[a]Δ' denotes either a difference in s-character or in π-bond order per single CC bond.

underlying picture is rehybridization which yields a first order effect in line with the physical basis of the iterative maximum overlap model [35].

3. APPLICATIONS. RESULTS AND DISCUSSION

3.1 Appraisal of the model

Since the high quality X-ray data are available for benzocyclobutene and benzo-[1,2:4,5]dicyclobutene, these molecules will serve as test cases for the adopted theoretical model. Results given in Table 3 show a very good agreement with measured values as evidenced by low average absolute error of $\simeq 0.003$ Å. It is particularly gratifying that differences between annelated and adjacent bonds are well reproduced and that both experiment and theory predict a weak MN bond fixation in these systems.

A stronger bond localization is found in biphenylene [37]. This result is rationalized by a cooperative action of σ and π electrons. For this purpose it is useful to recall covalent Kekulé structures of this molecule (Fig. 3). Spin-pairing schemes d and e involving unfavourable cyclobutadiene moiety can be immediately abandoned. Kekulé structures b and c lead to a uniform shortening of all CC bonds increasing their double bond character. The most stable scheme, a, contributes to the asymmetry

Table 3 — Comparison of $(3\text{-}21G)_{SC}$ bond distances with available experimental data for some benzocyclobutenes and biphenylene (in Å)

Molecule	Bond	$(3\text{-}21G)_{SC}$	Δ	Exp.	Δ
	$C_1\!-\!C_1$	1.400	0	1.391[a]	0
	$C_1\!-\!C_2$	1.386	-0.014	1.385	-0.006
	$C_2\!-\!C_3$	1.410	0.010	1.400	0.009
	$C_3\!-\!C_3$	1.401	0.001	1.399	0.008
	$C_1\!-\!C_4$	1.523	—	1.518	—
	$C_4\!-\!C_4$	1.574	—	1.576	—
	$C_1\!-\!C_1$	1.403	0	1.399[a]	0
	$C_1\!-\!C_2$	1.396	-0.007	1.394	-0.005
	$C_1\!-\!C_3$	1.523	—	1.521	—
	$C_3\!-\!C_3$	1.572	—	1.575	—
	$C_1\!-\!C_1$	1.429	0	1.426[b]	0
	$C_1\!-\!C_2$	1.370	-0.059	1.372	-0.054
	$C_2\!-\!C_3$	1.428	-0.001	1.423	-0.003
	$C_3\!-\!C_3$	1.386	-0.043	1.385	-0.041
	$C_1\!-\!C_{1'}$	1.512	—	1.514	—

[a]Reference [9].
[b]Reference [36].

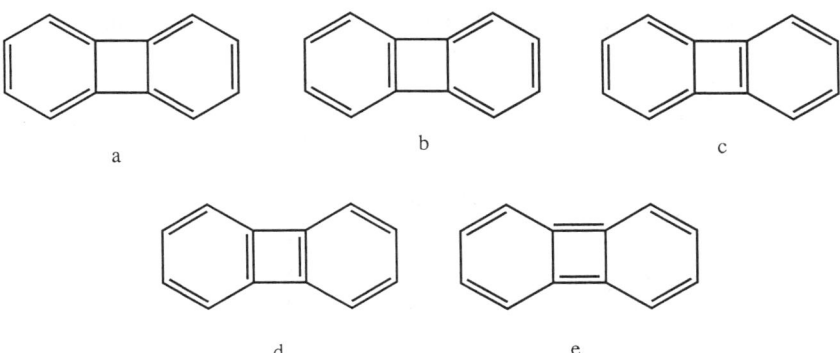

Fig. 3 — Covalent Kekulé structures of biphenylene.

of fused and adjacent bonds in the same sense as rehybridized local σ orbitals at the carbon junction atoms. Hence a relatively strong MN effect is operative in biphenylene. It is useful to compare adjacent and distal C_1—C_2 and C_3—C_3 bonds, respectively, relative to the benzene. The changes in bond lengths are -0.029 Å and -0.013 Å respectively. There is a considerable difference in average s-contents in C_1—C_2 and C_3—C_3 bonds. They are 40.0–34.6 and 34.8–34.8 as obtained by the MNDO method (as percentages). The reference hybridization in benzene is 32.7–32.7 (Table 2). The HMO π-bond orders for adjacent and distal bonds are 0.68 and 0.69 as compared to the benzene value of 0.667, which is determined by symmetry. It appears that a shortening of the distal bond is caused mainly through the π electron density whereas the opposite is true for the adjacent C_1—C_2 bonds where rehybridization is very effective. It is also clear that hybridization effect prevails.

We conclude that the $(3\text{-}21G)_{SC}$ SCF model is adequately designed for studying MN effect and that a cooperative action of σ and π electrons leads to a more pronounced bond fixation in the benzene fragment.

3.2 The size effect of the fused cycloalkene

It is of interest to examine archetype molecules indan and tetralin (Table 4). Calculations show that there is a weak MN effect in the former molecule. The situation is less clear in tetralin, but loosely speaking one can say that a tiny anti-MN effect takes place in it. Both results are in full accordance with the expectations of Mills and Nixon [1] and with the first theoretical results of Sutton and Pauling [7].

If the rehybridization picture holds, then it is easy to anticipate that a stronger MN effect takes place when a smaller ring is annelated. Further, since the effect should be roughly additive a bond simultaneously adjacent to two fused small rings should exhibit much stronger localization. This is corroborated by the results obtained for benzocyclobutadienes [39] and benzocyclopropenes [41].

Structural parameters of benzocyclobutadienes are compared with the corresponding values calculated in benzocyclobutenes in Table 5. The C_1—C_2—X angle in the former family of molecules is smaller than 90° indicating an even higher rehybridization effect. Additionally, a more pronounced π-electron localization is

Table 4 — Predicted bond distances in indan and tetralin (in Å)

Molecule	Bond	$(3\text{-}21G)_{SC}$	Δ
	C_1-C_1	1.402	0
	C_1-C_2	1.394	−0.008
	C_2-C_3	1.402	0
	C_3-C_3	1.399	−0.003
	C_1-C_4	1.508	—
	C_4-C_5	1.540	—
	C_1-C_1	1.403	0
	C_1-C_2	1.404	0.001
	C_2-C_3	1.394	−0.009
	C_3-C_3	1.400	−0.003
	C_1-C_4	1.511	—
	C_4-C_5	1.522	—
	C_5-C_5	1.522	—

expected in view of the presence of distal π-bonds which perturb the aromatic sextet. This intuitive conjecture is supported by the data produced by the $(3\text{-}21G)_{SC}$ SCF model. The Mills–Nixon bond fixation is highly pronounced in benzocyclobutadienes. The shortest adjacent exo-bond is found in benzo[1,2;3,4;5,6]tricyclobutadiene. Its fragments closely match the structure of 1,2-dimethylencyclobutene. We note with satisfaction that our results are in good agreement with X-ray data of the (heavily) substituted benzocyclobutadiene which also indicate a strong MN alternation of bond distances [40].

Hybridization and π-bond orders provide a simple transparent interpretation of the bonding and structural features of these systems (Table 6). In the first place, longer annelated bonds in benzocyclobutadienes are a consequence of lower s-character and π-bond orders. The opposite holds for the adjacent exo-bonds. It would be useful to get some idea about the relative contributions of σ- and π-electrons. In order to delineate their effects a model of distorted benzene is utilized again. The C_1-C_2-H angle is put equal to the corresponding values calculated in the corresponding benzocyclobutadienes (they read 88.4°, 87.4° and 86.8°, respectively). The CC bond distances are taken from the optimized $(3\text{-}21G)_{SC}$ structures of benzocyclobutadienes too. These frozen structures avoid any bias which might be introduced by optimizing bond distances of the deformed model benzenes. The latter reflect only σ-effect caused by fusion since distal π-bonds are missing. In spite of that, the calculated π-bonds orders resemble very closely the actual bond orders in the benzocyclobutadienes studied. Inclusion of the peripheral π-bond in the actual molecules enhances π-localization, but this effect is of lesser importance. The main part of the π-electron localization is induced by the rehybridization and subsequent descent in symmetry of the benzene nucleus. This example illustrates rather nicely that rehybridization of the

Table 5—Characteristic bond distances in benzocyclobutenes and benzocyclo-butadienes as estimated by the scaled 3-21G basis set *ab initio* calculations (in Å)

Molecule	Bond	X=CH$_2$	Δ	X=CH	Δ
	C$_1$–C$_2$	1.400	0	1.435(1.401)[a]	0
	C$_2$–C$_3$	1.386	−0.014	1.360(1.367)	−0.075
	C$_3$–C$_4$	1.410	0.010	1.444(1.445)	0.009
	C$_4$–C$_5$	1.401	0	1.376(1.370)	−0.059
	C$_1$–X	1.523		1.529(1.529)	
	X–X	1.574		1.356(1.357)	
	C$_1$–C$_2$–X	93.3°		88.5°(89.1°)	
	C$_1$–C$_2$	1.410	0	1.488	0
	C$_2$–C$_3$	1.387	−0.023	1.346	−0.142
	C$_3$–C$_4$	1.420	0.010	1.490	0.002
	C$_1$–C$_6$	1.377	−0.033	1.334	−0.154
	C$_1$–X	1.523		1.507	
	X–X	1.574		1.362	
	C$_1$–C$_2$	1.420	0	1.522	0
	C$_2$–C$_3$	1.377	−0.043	1.331	−0.191
	C$_1$–X	1.523		1.493	
	X–X	1.573		1.365	
	C$_1$–C$_2$	1.486		1.513	
	C$_2$–C$_3$	1.334		1.332	
	C$_1$–X	1.525		1.495	
	X–X	1.556		1.359	
	C$_1$–C$_2$	1.463		—	
	C$_1$–X	1.341		—	
X=X	X–X	1.336		—	

[a]X-ray data of the substituted benzocyclobutadiene [40], where C(CH$_3$)$_3$ groups are placed at the site X whereas other positions are substituted by methyls.

carbon junction atom represents the most important mechanism in determining the MN effect.

Fusion of the three-memebered ring to benzene leads to a strong MN localization too. This is evidenced by the 6-31G* results (Table 7). They are somewhat confusing at the first sight since both types of bonds — annelated and exo-adjacent bonds — are shorter than in the free benzene. However, it is well known that small rings dictate the

Table 6 — s-Characters and π-bond orders in benzocyclobutenes, benzocyclobuta-dienes and deformed benzenes mimicking the latter family of compounds as estimated by the MNDO method. Distances are obtained by the $(3\text{-}21\text{G})_{SC}$ procedure

Molecule	Bond	s-Characters (%)			π-Bond orders		
		X=CH$_2$	X=CH	X=H	X=CH$_2$	X=CH	X=H
(benzocyclobutene, positions 1–6)	C$_1$—C$_2$	29.7–29.7	28.2–28.2	26.7–26.7	0.61	0.47	0.53
	C$_2$—C$_3$	38.2–33.1	40.4–34.2	39.9–34.3	0.70	0.82	0.79
	C$_3$—C$_4$	32.9–32.8	31.6–31.8	32.0–31.8	0.63	0.48	0.54
	C$_4$—C$_5$	33.7–33.7	34.6–34.6	34.9–34.9	0.70	0.82	0.78
	C$_1$—X	31.5–22.5	32.0–25.9		0.14	0.12	
	X—X	21.9–21.9	35.6–35.6		0.08	0.97	
(positions 1–6)	C$_1$—C$_2$	28.6–29.1	25.4–26.1	24.2–25.1	0.58	0.30	0.43
	C$_2$—C$_3$	38.9–33.5	41.5–35.4	40.8–35.3	0.73	0.90	0.85
	C$_3$—C$_4$	32.6–32.6	30.8–30.8	31.2–31.2	0.59	0.34	0.44
	C$_1$—C$_6$	39.0–39.0	41.4–41.4	41.1–41.1	0.72	0.90	0.86
	C$_1$—X	33.0–20.6	34.7–27.2		0.14	0.18	
	X—X	22.0–22.0	34.9–34.9		0.08	0.95	
(positions 1–3)	C$_1$—C$_2$	29.4–29.4	24.8–24.8	23.8–23.8	0.54	0.21	0.36
	C$_2$—C$_3$	38.5–38.5	41.8–41.8	41.5–41.5	0.75	0.91	0.90
	C$_1$—X	32.3–22.4	34.6–27.2		0.14	0.21	
	X—X	22.0–22.0	34.7–34.7		0.08	0.94	
(positions 1–3)	C$_1$—C$_2$	29.7–29.7	28.7–28.7	35.3–35.3	0.23	0.18	0.43
	C$_2$—C3	40.9–36.0	41.7–35.9	35.3–30.1	0.96	0.96	0.90
	C$_1$—X	22.6–22.8	29.7–27.5		0.13	0.21	
	X—X	22.3–22.3	34.5–34.5		0.09	0.94	
(positions 2)	C$_1$—C$_2$	31.6–31.6			0.25		
	C$_1$—X	36.8–36.2			0.97		
X=X	X—X		36.4–36.4			1.00	

behaviour of larger (fused) systems and in particular it is a deformation caused by the three-membered ring which determines rehybridization at junction atoms. Taking into account that rehybridization is a predominant effect, it follows that the annelated bond in benzocyclopropene should be compared to a double bond in a free cyclopropene. If this bond distance is taken as a reference (1.279 Å by the 6-31G* basis set) then fusion to benzene causes its lengthening of 0.053 Å! The adjacent C$_2$=C$_3$ bond is compressed relative to benzene by − 0.016 Å. This interpretation is

Table 7 — Mills–Nixon effect in benzocyclopropenes as estimated by the 6-31G* procedure[a] (in Å)

Molecule	Bond	6-31G*	Δ
	C_1-C_2	1.332	0.053
	C_2-C_3	1.370	−0.016
	C_1-C_5	1.494	
	C_1-C_1	1.359	0.066
	$C_1-C_{1'}$	1.355	−0.031
	C_1-C_5	1.497	
	C_1-C_1	1.279	0
	C_1-C_2	1.494	

[a]The change Δ is determined relative to benzene CC bond distance (which reads 1.386 Å for the 6-31G* basis set) for the exo-bond. On the other hand annelated bond is gauged against the C=C distance in free cyclopropene.

corroborated by actual rehybridization [41, 42]. The MN effect is magnified by multilateral fusion of three-membered rings as expected (Table 7).

3.3 The effect of substituents

The judicious choice of substituents at particular positions may amplify or diminish the MN effect. We shall illustrate this point with a couple of examples. Let us consider benzo[1,2;3,4]dicyclobutenes (Table 8). Substitution of fluorines at the benzene sites increases asymmetry in the π-electron ring. This is easily understood by the Walsh–Bent rule which states that the electronegative substituent prefers a *vis-à-vis* hybrid orbital placed at the bonded electropositive atom possessing increased p-content. Concomitantly, the carbon hybrid orbitals at the site 2 forming CC bonds of the benzene ring exhibit increased s-character. Hence, C_1-C_2 bond distances become shorter. In contrast, the electropositive Li atom leads to a weak anti-MN effect because C_2—Li carbon hybrid AO has more s-character than in the C—H counterpart. Increased p-character in C_1-C_2 bonds leads to their lengthening and consequently to the inverse MN localization.

Heteroanalogues of benzocyclopropenes provide another nice example of the influence of electronegative substituents. Results offered by the 6-31G basis set are presented in Table 9. One observers that C_1-C_2 and C_2-C_3 bond distances

Table 8—Amplification and diminution of the MN effect by substituents in benzocyclobutenes by the $(3\text{-}21G)_{SC}$ procedure

Molecule	Bond	$(3\text{-}21G)_{SC}$	Δ
	$C_1\!-\!C_1$	1.403	0
	$C_1\!-\!C_2$	1.396	-0.007
	$C_1\!-\!C_3$	1.523	
	$C_3\!-\!C_3$	1.572	
	$C_1\!-\!C_1$	1.400	0
	$C_1\!-\!C_2$	1.388	-0.012
	$C_1\!-\!C_3$	1.517	
	$C_3\!-\!C_3$	1.573	
	$C_1\!-\!C_1$	1.409	0
	$C_1\!-\!C_2$	1.413	0.004
	$C_1\!-\!C_3$	1.529	
	$C_3\!-\!C_3$	1.572	

decrease along the series as electronegativity of the substituent increases. This is in line with the Walsh–Bent rule since s-character is transferred from $C_1\!-\!X$ bonds to neighbouring $C_1\text{-}C_2$ and $C_2\!-\!C_3$ bonds. Further, additional substitution of electronegative fluorine atoms at positions 3 and 6 will produce even more pronounced localization (Table 10). Since rehybridization is essentially a local effect the influence of substituents at sites 3 and X will be roughly additive. This is indeed the case (Tables 9 and 10). The calculated s-characters and π-bond orders by the MNDO method (Table 11) give support to the intuitive interpretation described above. It is important to notice that π-bond orders of the $C_1\!-\!C_2$ bonds in benzocyclopropene heteroanalogues decrease, thus contributing to the increase in its length. However, calculations show that this bond actually decreases, illustrating once again that rehybridization prevails.

4. CONCLUDING REMARKS

Summarizing our calculations we can say that MN effect usually takes place in benzenes fused to smaller ring(s). A similar phenomenon is observed in annelated

Table 9 — Relevant structural parameters in benzene and hetero-analogues of cyclopropabenzene as obtained by 6-31G calculations (in Å)

Molecule	X	1–2	2–3	3–4	4–5	1–X
	—		1.388	1.388	1.388	1.388	—
	CH_2	1.340	1.368	1.408	1.395	1.512
	NH	1.327	1.352	1.427	1.384	1.493
	O	1.310	1.347	1.434	1.383	1.499

naphtalenes [44]. Electronegative substituents yield pronounced MN bond fixation when placed at the critical sites in the fused molecules. On the other hand, electropositive substituents at these positions can reduce the extent of the MN effect, cancel it, or even lead to anti-MN type of localization. The origin of the MN effect is detected in rehybridization at the junction atoms as the most important phenomenon. It is followed by the redistribution of π-electrons which monitor and follow changes in the σ-skeleton upon ring fusion, Interaction between π-electrons themselves is less important but still significant in the final shaping of the annelated aromatics. Concomitantly, if π-electrons counteract σ-rehybridization, then a mild or weak MN effect takes place. On the other hand, the synergistic action of σ- and π-electrons leads to strong MN bond alternation. The importance of σ-redistribution casts doubts on π-electron only methods or topological treatments of fused aromatic systems. The same holds for experimental NMR techniques which indirectly 'measure' π-bond orders by, for example $^4J(H—C \doteq \doteq \doteq C—CH_3)$ spin–spin coupling constants [8, 9].

Table 10 — Relevant structural parameters in α,α-difluoroderivatives of heteroanalogues of cyclopropabenzene as estimated by 6-31G procedure (in Å)

Molecule	X	1–2	2–3	3–4	4–5	1–X
	—	—	1.387	1.378	1.378	1.387	—
	CH$_2$	1.348	1.352	1.399	1.388	1.506
	NH	1.333	1.341	1.415	1.379	1.485
	O	1.314	1.339	1.420	1.379	1.488

Finally, we would like to suggest an improved definition of the MN effect by stating that it is essentially a variation of bond distances within the aromatic fragment when annelated to smaller ring(s). The annelated bond is longer than in a free cycloalkene or its heteroanalogue, whereas adjacent exo-bonds become shorter than in a parent aromatic molecule.

Table 11—Hybridization parameters (in percentage) and bond orders in hetero-analogues of cyclopropabenzene and their α,α-difluoro derivatives as obtained by MNDO wave functions calculated by employing 6-31G geometries

X	Bond	s-Characters	π-Bond orders	s-Characters	π-Bond orders
	1—2	25.6–25.6	0.63	25.4–25.4	0.61
	2—3	42.1–32.2	0.69	41.7–39.0	0.69
CH_2	3—4	32.6–32.8	0.64	39.5–31.9	0.60
	4—5	33.3–33.3	0.69	33.7–33.7	0.72
	C—X	30.2–17.8	0.12	31.2–17.6	0.12
	3—H	33.5	—	—	—
	3—F	—	—	22.4–14.5	—
	1—2	26.7–26.7	0.58	26.7–26.7	0.57
	2—3	46.3–32.0	0.74	45.7–38.7	0.72
NH	3—4	32.3–32.2	0.58	39.5–31.3	0.55
	4—5	33.6–33.6	0.74	34.2–34.2	0.76
	C—X	24.2–6.8	0.12	26.0–6.8	0.12
	3—H	34.2	—	—	—
	3—F	—	—	23.2–15.4	—
	1—2	29.0–29.0	0.56	28.8–28.8	0.57
	2—3	48.4–31.5	0.76	47.9–38.2	0.74
	3—4	32.3–31.9	0.55	39.9–31.1	0.54
O	4—5	34.1–34.1	0.77	34.9–34.9	0.77
	C—X	20.8–3.8	0.09	21.7–3.8	0.09
	3—H	34.7	—	—	—
	3—F	—	—	23.3–14.6	—

ACKNOWLEDGEMENT

The financial support of the Scientific Research Council of Croatia and the International Office of Kernforschungsanlage Jülich Germany (M.E.M. and Z.B.M.) is gratefully acknowledged.

We also would like to thank our co-workers A. Lesar, D. Margetić, D. Mitić and K. Poljanec for performing some of reported calculations. The computer center at IBN HDSC is acknowledged for providing significant computing resources and excellent service.

REFERENCES

[1] W.H. Mills and I.G. Nixon, *J. Chem. Soc.* (1930) 2510.

[2] W.E. Billups, *Acc. Chem. Res.* **11** (1978) 245; W.E. Billups, W.A. Rodin and M.M. Haley, *Tetrahedron* **44** (1988) 1305.

[3] R.P. Thummel, *Isr. J. Chem.* **22** (1982) 11.

[4] B. Halton, *Chem. Rev.* **89** (1989) 1161.

[5] A.J. Barkovitch, E.S. Strauss and K.P.C. Vollhardt, *Isr. J. Chem.* **20** (1980) 225.

[6] G.M. Badger, *Quant. Rev. Chem. Soc.* **5** (1951) 147.

[7] L.E. Sutton and L. Pauling, *Trans. Farad. Soc.* **31** (1935) 939.

[8] H.C. Longuet-Higgins and C.A. Coulson, *Trans. Farad. Soc.* **42** (1946) 756.

[9] R. Boese and D. Bläser, *Angew. Chem.* **100** (1980) 293.

[10] R. Boese, D. Bläser, K. Gomman and U.H. Brinker, *J. Am. Chem. Soc.* **111** (1989) 1501.

[11] R. Neidlein, D. Christen, V. Poignée, R. Boese, D. Bläser, A. Gieren, C. Ruiz-Pérez and T. Hübner, *Angew. Chem. Int. Ed. Engl.* **27** (1988) 294.

[12] W.E. Billups, M.Y. Chow, K.H. Leavell, E.S. Lewis, J.L. Morgrave, R.L. Sass, J.J. Shieh, P.G. Werness and J.L. Wood, *J. Am. Chem. Soc.* **95** (1973) 7878.

[13] J.L. Crawford and R.E. Marsh, *Acta Cryst.* **B29** (1973) 1238.

[14] R.E. Cobbledick and F.W.B. Einstein, *Acta Cryst.* **B32** (1976) 1908.

[15] R.P. Thummel, J.D. Korp, I. Bernal, R.L. Harlow and R.L. Soulen, *J. Am. Chem. Soc.* **99** (1977) 6916.

[16] J.D. Korp, R.P. Thummel and I. Bernal, *Tetrahedron* **33** (1977) 3069.

[17] R. H. Mitchell, P. D. Slowey, T. Kamada, R. V. Williams and P.J. Garratt, *J. Am. Chem. Soc.* **106** (1984) 2431.

[18] M.J. Collins, J.E. Gready, S. Sternhell and C.W. Tansey, *Austr. J. Chem.* **43** (1990) 1547.

[19] M. Barfield, M.J. Collins, J.E. Gready, P.M. Hatton, S. Sternhell and C.W. Tansey, *Pure & Appl. Chem.* **62** (1990) 463.

[20] C.S. Cheung, M.A. Cooper and S.L. Manatt, *Tetrahedron* **27** (1971) 689, 701.

[21] B. Halton and M.P. Halton, *Tetrahedron* **29** (1973) 1717.

[22] P.C. Hiberty, G. Ohanessian and F. Delbecq, *J. Am. Chem. Soc.* **107** (1985) 3095.

[23] A. Stanger and K.P.C. Vollhardt, *J. Org. Chem.* **53** (1988) 4889.

[24] Y. Apeloig and D. Arad, *J. Am. Chem. Soc.* **108** (1986) 3241; Y. Apeloig, M. Korni and D. Arad, in Strain and its Implications in Organic Chemistry, A.de Meijere and S. Beechert, Eds., Klumer Academic Publishers, 1989.

[25] G. Fraenkel, Y. Ashai, M.J. Mitchell and M.P. Cava, *Tetrahedron* **20** (1964) 1179.

[26] R.A. Finnegan, *J. Org. Chem.* **30** (1065) 1333.

[27] A. Streitwieser, Jr., G.R. Ziegler, P.C. Mowery, A. Lewis and R.G. Lawler, *J. Am. Chem. Soc.* **90** (1968) 1357.

[28] R.D. Rieke, *J. Org. Chem.* **36** (1971) 227.

[29] S.E. Bales and R.D. Rieke, *J. Org. Chem.* **37** (1972) 3866.

[30] R.D. Rieke, S.E. Bales, C.F. Meares, L.I. Rieke and C.M. Milliren, *J. Org. Chem.* **39** (1974) 2276.

[31] M. Randić and Z.B. Maksić, *J. Am. Chem. Soc.* **93** (1971) 64.

[32] J.E. Boggs (private communication). We thank Professor Boggs for this information prior to publication.

[33] Z.B. Maksić, M. Eckert-Maksić, D. Kovaček, M. Hodošček, K. Poljanec and J. Kudnig, *M. J. Mol. Structure (THEOCHEM)*, in press.

[34] M. Eckert-Maksić, Z.B. Maksić, A. Skancke and P.N. Skancke, in Z.B. Maksić (Ed.), *Modelling of Structure and Properties of Molecules*, Ellis Horwood, Chichester, 1987, p. 67.

[35] Z.B. Maksić and A. Rubčić, *J. Am. Chem. Soc.* **99** (1977) 4233.

[36] J.K. Fawcett and J. Trotter, *Acta Cryst.* **20** (1966) 87.

[37] M. Eckert-Maksić, M. Hodošček, D. Kovaček, Z.B. Maksîĉ and K. Poljanec, *Chem. Phys. Lett.* **177** (1990) 49.

[38] J.E. Bloor, M. Eckert-Maksić, K. Poljanec and Z.B. Maksić, to be published.

[39] Z.B. Maksić, M. Eckert-Maksić, D. Kovaček and D. Margetić, to be published.

[40] W. Winter and H. Straub, *Angew. Chem.* **90** (1978) 142.

[41] W. Koch, M. Eckert-Maksić and Z.B. Maksić, to be published.

[42] M. Eckert-Maksić, Z.B. Maksić, M. Hodošček and K. Poljanec, *Int. J. Quant. Chem*, in press.

[43] A.D. Walsh, *Disc. Farad. Soc.* **2** (1949); H.A. Bent, *Chem. Rev.* **61** (1961) 275.

[44] M. Hodošček, D. Kovaček and Z.B. Maksić, to be published.

19

A new way of calculating bond dissociation energies

Elfi Kraka, Dieter Cremer and Sture Nordholm
University of Göteborg, Göteborg, Sweden

INTRODUCTION

Knowledge of accurate values of homolytic bond dissociation energies (BDEs) is one of the prerequisites for the understanding of free radical reactions and reaction mechanism in general. BDEs lead to atomization and bond energies, which are used to characterize and to rationalize the nature of the chemical bond and molecular structure [1]. Therefore, considerable efforts have been devoted to the determination of BDEs for many molecules. However, even nowadays, with improved and refined techniques, experimentally determined BDEs rarely have an accuracy of ± 1 kcal/mol or better. Since BDEs are mostly based on experimental heats of formation and since these are extremely difficult to measure for dissociation products which are short-lived free radicals, errors as large as 5 kcal/mol are to be expected. This is reflected by the existing compilations of heats of formation and BDE values [2–13].

Theory does not have to cope with the difficulties of experiment, and, therefore, it has become a primary source for accurate molecular data. As a matter of fact, highly accurate BDE values have been obtained from *ab initio* calculations [14, 15]. But also in the case of theoretical BDEs, one has to pay a considerable price to get this accuracy. As is well known, simple restricted Hartree–Fock (RHF) theory fails to describe homolytic bond dissociation correctly. Unrestricted HF (UHF) theory describes homolytic bond dissociation qualitatively correctly but does not lead to reasonable BDE values. Therefore, one has to go beyond the HF level in order to get a better calculational description of bond dissociation. Highly accurate BDE values are obtained when multi-configuration SCF (MCSCF) is combined with CI in order to describe static and dynamic correlation in reactant and products adequately [15–20]. This, of course, requires considerable computational effort, and sufficiently accurate BDE values can only be obtained for relatively small molecules. Therefore, simpler methods have been looked for, which may lead to reasonable BDE values. For example, single-determinant-based correlation methods such as Møller–Plesset perturbation theory [21] or configuration interaction (CI) have been tested [22]. Although calculated BDE values are better than HF results, differences as large as 50 kcal/mol from experimental values are not very encouraging. Recently, these investigations have been improved by including bond functions in the basis set used [23–26]. An additional improvement has been found by using coupled cluster (CC) theory rather than MP or CI methods [27].

An alternative way has been used by Pople and co-workers [28–30]. These authors considered combinations of bond dissociation reactions that lead to formal (isogyric) reactions with a constant number of unpaired electron spins. By using exact experimental data together with the computed BDE values, they were able to calculate BDEs for AH_n molecules with an accuracy of 1 kcal/mol or better.

While these more recent approaches for calculating BDEs are quite promising, it remains to be seen whether they can be applied in a routine way. For example, the use of bond functions leads to excellent results in the case of AH_n molecules [24–26], but extension of the calculations to heavy atom bond dissociation seems to increase the rms error in calculated BDE values to 10 and more kcal/mol [23]. Also, the use of isogyric reactions strongly depends on the availability of accurate heats of reaction for an appropriate reference reaction. This is the case for AH bond dissociation (reference reaction: $H_2 \rightarrow 2H$), but not necessarily for homolytic dissociation of an arbitrary bond AB.

In order to calculate BDE values for a variety of homolytic bond dissociation reactions including both AH and heavy atom bonds AA and AB at relatively low costs, it will be necessary to stick to methods that are similar in their cost requirements to HF and that can be run with basis sets of moderate size, for example split valence basis sets such as 6-31G(d) or 6-31G(d,p) [31]. Such a method, of course, should start from a wavefunction that describes dissociation correctly. This implies that the method in question has to be based on a small MCSCF approach which covers the most important static (non-dynamic) correlation effects needed for proper description of dissociation at the *ab initio* level. We have chosen GVB for this purpose since GVB can be considered as a simple systematic MCSCF description.

As for the assessment of dynamic correlation effects, we have used local spin density (LSD) functional theory [32]. LSD functionals offer a way of predicting correlation effects at a cost level which is essentially that of HF. It has been shown that LSD functionals can lead to useful predictions of molecular properties [32, 33]. In this chapter we will investigate what level of accuracy is achieved if BDE values of a variety of molecules are determined by a method that combines the calculational advantages of both GVB and LSD. A clear assessment of this new GVB–LSD method can only be made if results are compared with BDE values from HF, HF–LSD and GVB calculations obtained with the same basis set at the same geometry. Thus, we will present here for the first time a thorough comparative investigation of BDE values obtained at the four levels of theory described above.

Our investigation has been stimulated by earlier research carried out along similar lines and with similar intentions. In 1974, Lie and Clementi investigated local density functionals (LDF) designed to yield proper dissociation potential curves for simple diatomic molecules [34, 35]. These authors stressed the necessity of using density functionals in connection with a proper reference function that describes homolytic dissociation correctly. For this purpose, they added to the HF function a few configurations that guaranteed proper dissociation on at least a qualitative basis. This approach may be termed a MCSCF–LDF description of homolytic dissociation. Unfortunately, a routine implementation of the MCSCF–LDF approach requires for each molecule considered a decision on how many and which configurations have to be included. For example, for Li_2, just one additional configuration $(1\sigma_g^2, 1\sigma_u^2, 2\sigma_u^2)$ turned out to be necessary, while for N_2 nine additional singlet- or triplet-coupled configurations were needed [35]. Also, the density functional used by Lie and Clementi was a preliminary one and subject to further improvements. Because of this and the limited number of test examples, a positive assessment of the utility of the MCSCF–LDF approach could not be made.

USE OF LOCAL SPIN DENSITY FUNCTIONAL THEORY

In this work, LSD is used to calculate the correlation energy E_c as a corrective term for the HF energy, as was suggested by Stoll and co-workers [36–38]. Within this LSDC (C for correlation) approach the total energy is given by

$$E = E(\text{HF}) + E_c(\text{LSDC}) \tag{1}$$

with

$$E_c(\text{LSDC}) = \int d\mathbf{r}\rho(\mathbf{r})\varepsilon_c[\rho_+(\mathbf{r}),\rho_-(\mathbf{r})] - \int d\mathbf{r}\rho_+(\mathbf{r})\varepsilon_c[\rho_+(\mathbf{r}),0]$$
$$- \int d\mathbf{r}\rho_-(\mathbf{r})\varepsilon_c[0,\rho_-(\mathbf{r})] \tag{2}$$

where $\rho(\mathbf{r})$, $\rho_+(\mathbf{r})$, and $\rho_-(\mathbf{r})$ denote total electron density as well as α- and β-spin density distribution:

$$\rho(\mathbf{r}) = \rho_+(\mathbf{r}) + \rho_-(\mathbf{r}) \tag{3}$$

We use for ε_c the Vosko–Wilk–Nusair (VWN) functional [39] that is based on accurate Monte Carlo data for the homogeneous electron gas calculated by Ceperley and Alder [40]. Kemister and Nordholm [41, 42] have extended the VWN para-metrization to arbitrary polarization according to the method of von Barth and Hedin [43]. In addition, Kemister and Nordholm have used Gaussian basis functions to evaluate ε_c, which significantly facilitates the inclusion of the LSDC algorithm into standard *ab initio* packages. We have taken the LSDC programs from the Ph.D. thesis of Kemister [44], rewritten them for routine use in multipurpose *ab initio* programs and adapted them to our program package COLOGNE90 [45]. The modified LSDC programs can be used at the restricted and the unrestricted HF level in an iterative and non-iterative way (see below). They allow the use of electron density distributions resulting from HF, MP, Cl, CC, QCI, GVB, MCSCF and CASSCF calculations [46].

At the HF level of theory, there are two ways of including the LSDC function-al [41]. For example, the LSD correlation energy E_c can be calculated at the end of the SCF iterations by using the converged density distributions $\rho_+(\mathbf{r})$ and $\rho_-(\mathbf{r})$. Alternatively, the LSDC functional can be included directly in the SCF iterations in order to minimize the sum $E(\text{HF}) + E_c$ rather than the HF energy alone. This can be done by extending the Fock operator according to

$$F = F_{HF} + F_c \tag{4}$$

with

$$F_c^+ = \varepsilon_c[\rho_+(\mathbf{r}),\rho_-(\mathbf{r})] - \varepsilon_c[\rho_+(\mathbf{r}),0] + \rho_+(\mathbf{r})\varepsilon_c'[\rho_+(\mathbf{r}),\rho_-(\mathbf{r})]$$
$$- \rho_+(\mathbf{r})\varepsilon_c'[\rho_+(\mathbf{r}),0] \tag{5}$$

where

$$\varepsilon_c' = \delta\varepsilon_c'/\delta\rho_+ \tag{6}$$

and similar equations for β spin electrons.

Both the fixed and the iterative HF–LSDC procedures have been applied at the HF level of theory throughout this chapter. As was found in earlier work [41], the correlation energies obtained by the two procedures are very similar, differing in most cases by less than 3 mHartree. Since the calculation of BDE values leads to partial cancellation of these differences, final BDEs differed by less than 1 kcal/mol in all cases considered. Therefore, only those LSDC correlation energies that are calculated after the SCF iteration (fixed HF–LSDC energies) will be discussed in the following.

The fact that the two ways of calculating HF–LSDC energies lead to similar values suggests that the VWN functional depends only little on changes in the HF electron density distribution. We have checked the dependence of ε_c on $\rho(\mathbf{r})$ further by feeding into the VWN functional various types of correlation corrected density distribution, resulting from MP, CI, CC, QCI and GVB calculations [46]. In all cases, the dependence of the calculated LSDC energy on corrections in $\rho(\mathbf{r})$ turned out to be small. These observations led us to couple the LSDC functional to GVB in the simplest way possible: First the GVB calculation is carried out, and upon reaching convergence, $\rho(\mathbf{r})$, $\rho_+(\mathbf{r})$ and $\rho_-(\mathbf{r})$ are calculated at the GVB level and used to calculate the LSDC energy. This leads to the GVB–LSDC energy according to

$$E(GVB-LSDC) = E(GVB) + E_c(LSDC) \tag{7}$$

The energies obtained in this way were used to determine the BDEs.

CALCULATION OF BOND DISSOCIATION ENERGIES D_e

The BDE of a molecule AB is defined as the difference of the energy of AB and those of the dissociation products A and B, where AB and the dissociation products are all at their equilibrium geometry.

AB → A + B

One has to distinguish between D_e, D_o and D_T values which refer to energy differences taken at the bottom of the potential well, at the zero vibrational level or some averaged vibrational level corresponding to temperature T K. Theory leads to D_e values, which with the help of calculated or experimental frequencies can be transformed to D_o or D_T values. Alternatively, experimental D_o values can be transformed to D_e values for reasons of comparison. We have used the latter approach using experimental frequencies in order to eliminate error sources that stem from inaccurate *ab initio* frequencies.

While in all cases investigated in this work AB is a closed-shell system, the dissociation products A and B are open-shell molecules or atoms. Accordingly, we have calculated AB at the restricted HF (RHF) level, while A and B have been calculated at the unrestricted HF (UHF) level. This is an economic way of calculating BDEs, but it is not without ambiguity. Comparison of RHF and UHF energies is problematic owing to the increased flexibility of the UHF wavefunction. A consistent description of the dissociation process can only be achieved by calculating the potential energy surface of the system AB along the dissociation coordinate by one and the same method. Since RHF leads to heterolytic dissociation, with BDE values being far too high, a reasonable description can only be obtained by using the UHF approach for the calculation of the dissociation energy. In many cases, UHF and RHF possess the same solution at or close to the equilibrium of AB. For these cases, our approach of calculating AB at RHF and the dissociation products at UHF is justified. There are, however, other molecules for which a unique UHF solution exists throughout the dissociation process. Then, BDE values are obtained which are larger than those reported here. But even the UHF values are considerably smaller than the true BDE values.

Since we are primarily concerned with correlation-corrected BDE values and since the differences between RHF and UHF energies are largely annihilated by correlation corrections, we are justified in comparing RHF- and UHF-based energy values in the following.

USE OF GVB FOR THE CALCULATION OF DISSOCIATION ENERGIES

The generalized valence bond (GVB) method is a variational version of the valence bond (VB) approach [47, 48]. The method allows one to correlate individual electron pairs that are associated with specific bonds via localized MOs. In the case of homolytic bond dissociation, only the orbitals describing the breaking bond form GVB pairs.

$$[pair] = [\phi_a(1)\phi_b(2) + \phi_b(1)\phi_a(2)][\alpha(1)\beta(2) - \beta(1)\alpha(2)] \qquad (8)$$

while all other occupied orbitals are considered to form an HF core.

$$[core] = [\phi_{c1}(1)\phi_{c1}(2)\alpha\beta\phi_{c2}(1)\phi_{c2}(2)\alpha\beta \dots] \qquad (9)$$

Pair orbitals ϕ_a and ϕ_b are variationally optimized (as are core orbitals ϕ_c) overlapping one-electron GVB orbitals. While HF orbitals are orthogonal, GVB orbitals in general are not. But for computational reasons the GVB pair can be rewritten in terms of natural orbitals χ.

$$[pair]^{NO} = [c_g\chi_g^2 + c_u\chi_u^2][\alpha(1)\beta(2) - \beta(1)\alpha(2)] \qquad (10)$$

with

$$c_g^2 + c_u^2 = 1 \qquad (11)$$

where χ_g and χ_u correspond to the orthogonal bonding and antibonding natural orbitals of the GVB pair. In the perfect pairing (PP) version of GVB, electrons are always singlet coupled and the GVB–PP wavefunction for a closed-shell molecule with several pairs takes the form:

$$\Psi^{GVB-PP} = A[[core][pair(1,2)][pair(3,4)]\ldots].$$

The GVB–PP approach can be considered as a simple MCSCF method [48]. It recovers static correlation effects and gives the correct functional form for proper bond dissociation.

In this work, the GVB–PP approach has been used to account for static correlation effects influencing BDEs. Thus if the bond A—B is broken in the dissociation, only the electron pair associated with the corresponding localized AB bond orbital has been correlated. None of the other electron pairs that are responsible for bonding within A or B has been considered in the dissociation process. For example, static correlation effects in the dissociation of the CC bond in H_3CCH_3, H_2CCH_2 and HCCH have been described by correlating 1, 2, and 3 CC bond electron pairs, respectively, but none of the CH bond electron pairs.

While this approach seems to be straightforward at first sight, it leads to unreasonable results if hetero atoms with electron lone pairs are involved in the dissociation process. For example, there is just an OO single bond in hydrogen peroxide, H_2O_2, but from experimentally as well as theoretically determined properties of H_2O_2 it becomes clear that the lone pairs at the two O atoms participate to some extent in bonding [49, 50]. In the equilibrium geometry of H_2O_2, one lone pair at each O atom is collinear with the neightbouring OH bond (see Schemes 1, 2b). It delocalizes into the $\sigma^*(OH)$ orbital (anomeric effect) thus leading to partial π-character of the OO bond and thereby increasing its stability. At the same time, lone pair–lone pair repulsion leads to some weakening of the OO bond, but the net effect of the O electron lone pairs is still stabilizing [49, 50].

It is clear that electron lone pairs have to be considered in the dissociation process. But using the same reasoning one has also to consider interactions of vicinal bonds, e.g. the CH bonds in H_3CCH_3, H_2CCH_2 and HCCH, in the dissociation process. Repulsive interactions will decrease, and attractive interactions will increase the BDE. We have tested this and have found that contrary to the effects of electron lone pairs, the effects of vicinal CH bond electron pairs on the static correlation energy of the dissociation process are negligibly small. Therefore, only the electron lone pairs of hetero atoms have been considered in the GVB calculations.

Even though it is not difficult to identify the impact of a lone pair on the BDE, a consistent treatment of lone pair effects at the GVB level is very difficult. In the example given above, the participation of two electron lone pairs in the OO bond is obvious. What is not obvious is the role of the two remaining lone pairs with regard to OO bonding. Inclusion of all four electron lone pairs in the GVB calculation leads to static correlation effects and dissociation energies which are far too large. This indicates that not all electron lone pairs participate in OO bonding. According to the anomeric interaction of O electron lone pairs, there are just two of the four lone pairs that participate in OO bonding while the two other lone pairs may not affect in any way bonding in H_2O_2. This, however, is difficult to assess on a quantitative basis, in particular if molecules such as CH_3OH, CH_3NH_2, etc. are considered.

Since the question of lone pair participation in bonding cannot be satisfactorily answered in any case, we have adopted the following approach. For molecules AH_n we have not considered those electron lone pairs which are directed away from the AH bonds, since their influence on AH bonding is negligible. This applies to the σ lone pair of FH, H_2O and NH_3 (see Schemes 1, 1a, 2a, 3a). However, the impact of the $p\pi$ electron lone pairs at F and O (Scheme 1) cannot be neglected. Calculations show that the distribution of the $p\pi$ lone pairs directly affects the electronic charge along the AH bond and at H. For example, augmentation of the H basis set by polarization functions leads to AH bond strengthening owing to pseudo-π character of the AH bond introduced by the polarization functions. This, of course, does not mean that the AH bond gets any double-bond character, but the distribution of the $p\pi$ lone pairs has to be taken into account for an accurate *ab initio* description of AH bonding. In this chapter we have used polarization functions only for the heavy atoms. However, in view of a correct and consistent treatment of lone pair effects we include the $p\pi$ lone pairs of F and O in the GVB treatment of AH bonding and indicate this by lone pair participation numbers q of 2 and 1, respectively.

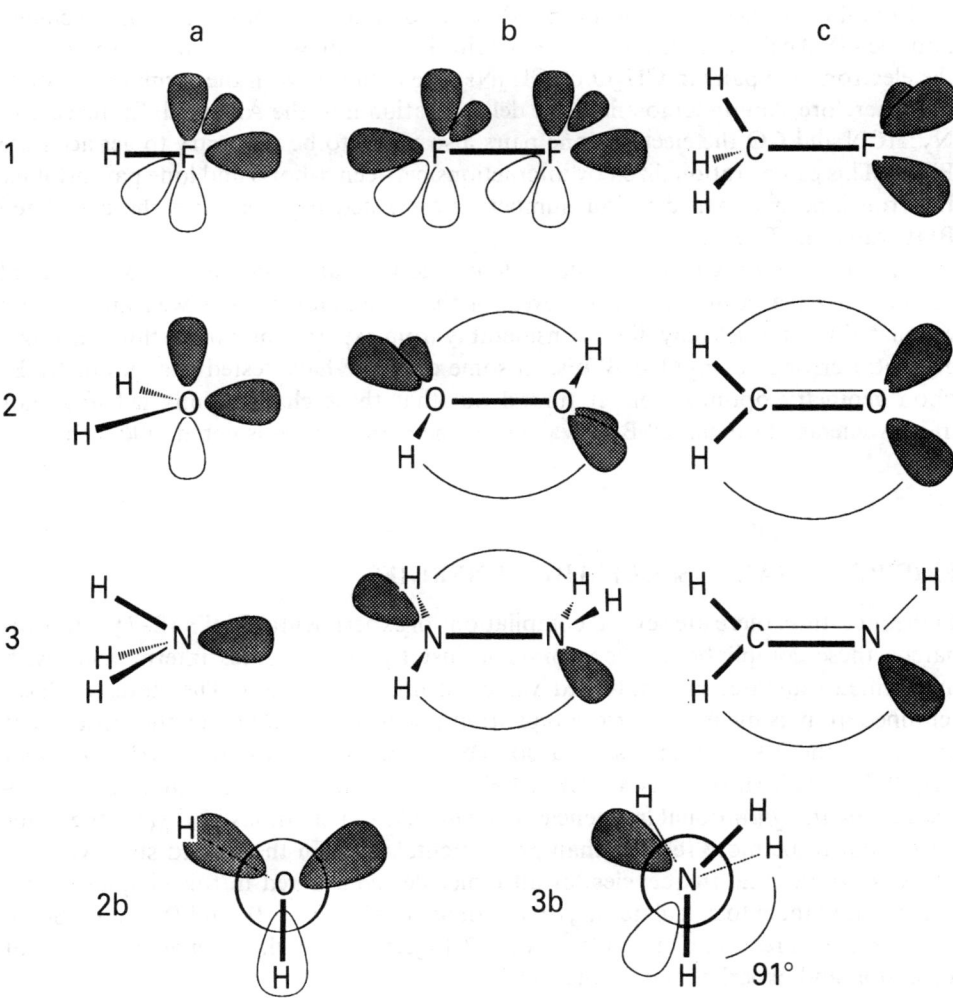

Once the lone pair participation numbers q of molecules AH_n have been fixed, these numbers are used to determine q for bonds AB in molecules H_aABH_b. Thus, $q(OO)$ for H_2O_2 is calculated to be $4 - 2\ q(OH) = 2$ in accordance with the description of the anomeric effect for H_2O_2 given above (see also Schemes 1, 2b). Nevertheless, there are several cases that need special consideration. In the equilibrium conformation of N_2H_4 (Schemes 1, 3b) the electron lone pairs are almost collinear with the vicinal NH bonds although this also implies some NH eclipsing. This indicates anomeric participation of the electron lone pairs in NN bonding. Accordingly, we have set $q(N_2H_4)$ to 2 as in the case of H_2O_2.

For F_2 (Schemes 1, 1b), ClF and Cl_2, $q = 4$ leads to better BDEs than $q = 6$. This adds support to exclude those lone pairs that are directed away from the AA or AB bond. However, for CH_3F it is no longer possible to distinguish between σ and π electron lone pairs at F (Schemes 1, 1c) since the three lone pairs are equivalent owing to symmetry. Their spatial distribution is best described by sp^3 hybrid orbitals. Therefore, we have included all three electron lone pairs in the GVB calculation ($q = 3$).

For all molecules with multiple bonds, electron lone pairs have been fully included into the GVB calculation. The reason for this is given in Scheme 1, which shows that the electron lone pairs in CH_2O or CH_2NH are collinear with the vicinal CH bonds and, therefore, can undergo anomeric delocalization into the AB bond. In the case of N_2, HCN and CO the electron lone pairs also prove to be necessary to get accurate BDEs. This can be rationalized by interactions between σ-bond and lone pair orbitals. Electron lone pair participation numbers q are listed together with the calculated BDE values in Table 2.

In order to obtain a consistent description of all compounds, experimental geometries have been used at all levels of theory employed. It is well known that calculated geometries may differ considerably from experimental ones, thus causing a geometry error in computed BDEs. In some cases we have tested changes in BDEs upon geometry optimization. It turned out that these changes only led to minor improvements of calculated BDE values for the molecules considered in this chapter.

EXPERIMENTAL DISSOCIATION ENERGIES

In the literature, there are several compilations of experimental BDEs [2–7]. Unfortunately, these compilations often comprise just D_o or D_T values intermingled with unspecified estimates and outdated values of uncertain origin. Therefore, we have refrained from using any of these compilations. Instead we have taken the most recent update of the JANAF tables as a source for heats of formation $\Delta H_f^0(298)$ and $\Delta H_f^0(0)$ [11, 12]. In two cases (BeH_2, BH_3), we had to take ΔH_f^0 values as well as geometries and vibrational frequencies from ab initio calculations [28] since the latter turned out to be more reliable than experimental data. In the second step, we have collected vibrational frequencies for all molecules considered in this chapter [9, 52, 53] and used them to calculate $\Delta H_f^0(0)$ (if not available), D_{298}, D_o and D_e values. Some of these data are given in Tables 1 and 2 together with the sources for heats of formation and experimental frequencies.

SELECTION OF TEST MOLECULES

Previous investigations of BDEs using density functional theory have been constrained to simple diatomic molecules [34, 35, 41]. Even though results on these molecules may provide first indication on the usefulness of a particular calculational approach, they do not allow one to draw any conclusion with regard to routine use of the method in the case of polyatomic molecules. Therefore, we have looked for a larger set of representative homolytic dissociation reactions that can be made up from molecules containing just first row atoms. We have settled on three groups of molecules and three groups of dissociation reactions: first, molecules AH_n that can lead to AH bond dissociation reactions; secondly, molecules A_2H_{2a} ($a = 0,1,\dots$) and AA bond dissociation reactions; finally, molecules AH_aBH_b and BDEs AB, with A a carbon atom, that are representative for bond dissociation in organic compounds.

Within this set of text examples some molecules had to be excluded. These are Be_2, B_2 and O_2. Be_2 is a van der Waals compound with a binding energy of less than 2 kcal/mol [54] and, therefore, does not fall into the group of covalently bonded molecules investigated in this chapter. The O_2 molecule has been excluded since it is well known that GVB with perfect pairing cannot describe homolytic dissociation of the $^3\Sigma_g^+$ ground state of O_2 correctly [55]. The value of the BDE of B_2 was actually calculated in this work (D_e(GVB–LSDC) = 61.3; D_e(exp) = 70 kcal/mol), but after dropping molecular oxygen, B_2 would have been the only open-shell molecule within our set of examples and, therefore, we will include it in a forthcoming investigation of BDE values of open-shell molecules.

As discussed above, the inclusion of electron lone pairs into the GVB calculations turned out to be a stumbling block in the beginning. This was particularly true with regard to F_2. Comparative calculations for halogen compounds such as HCl, Cl_2 and ClF indicated that reasonable results are obtained when two rather than three electron lone pairs per halogen atom are included in the GVB calculation. Even though the additional calculations involved molecules with a second period atom, they are included in the present work which is thus based on a sample set of 27 molecules and 27 different dissociation reactions.

RESULTS AND DISCUSSION

In Table 1, heats of formation ΔH_f^0 at 298 and 0 K are given for those atoms and molecules that occur in the dissociation reactions investigated. Also listed are sources for vibrational frequencies needed for the calculation of D_e values from experimental D_o values and those for the experimental geometries used in the calculations. Table 2 gives calculated and experimental D_e values for the 27 dissociation reactions investigated. HF, HF–LSDC, GVB and GVB–LSDC values are plotted against experimental D_e values in Figs 1, 2, 3 and 4, respectively. A more detailed analysis of the calculated BDEs is given in Table 3 as well as Figs 5, 6 and 7.

BDEs calculated at the HF level of theory are always too small. Discrepancies are as large as 100 kcal/mol and more. Fig. 1 shows that computed HF BDEs do not correlate with experimental BDE values. It is noteworthy that the BDEs of both F_2 and H_2O_2 are negative (-35.2 and -1.5 kcal/mol, respectively) which means that these molecules are not stable at the HF level of theory. This has been reported several

Table 1—Heats of formation ΔH_f^0 and zero-point energies (ZPE) for all molecules[a].

Molecule	Sym	$\Delta H_f^0(298)$	$\Delta H_f^0(0)$	ZPE	ΔH_f^0 References	Freq	Geom
LiH	$C_{\infty v}$	33.61	33.65	1.94	12	52	51
BeH$_2$	$D_{\infty h}$	39.50	39.86	10.16	28	28	28
BH$_3$	D_{3h}	74.80	74.71	10.86	28	28	28
CH$_4$	T_d	-19.895	-15.992	27.11	12	52	51
NH$_3$	C_{3v}	-10.97	-9.30	11.46	12	52	51
H$_2$O	C_{2v}	-57.80	-57.11	12.88	12	52	51
FH	$C_{\infty v}$	-65.14	-65.13	5.66	12	52	51
HCl	$C_{\infty v}$	-22.06	-22.02	4.12	12	9	51
Li$_2$	$D_{\infty h}$	51.60	51.50	0.49	12	52	51
C$_2$H$_6$	D_{3d}	-20.04	-16.27	45.46	12	52	51
C$_2$H$_4$	D_{2h}	12.54	14.58	30.87	12	52	51
C$_2$H$_2$	$D_{\infty h}$	54.19	56.37	16.19	12	52	51
C$_2$	$D_{\infty h}$	200.22	198.20	2.61	12	9	51
N$_2$H$_4$	C_2	22.79	26.22	31.82	12	52	51
N$_2$H$_2$	C_{2h}	50.90	52.61	17.18	12	52	51
N$_2$	$D_{\infty h}$	0	0	3.33	12	52	51
H$_2$O$_2$	C_2	-32.53	-31.03	15.90	12	52	51
F$_2$	$D_{\infty h}$	0	0	1.27	12	52	51
Cl$_2$	$D_{\infty h}$	0	0	0.79	12	9	51
CH$_3$NH$_2$	C_s	-5.49	-1.92	39.17	12	52	51
CH$_2$NH	C_s	27.60	29.49	24.22	59	52	51
HCN	$C_{\infty v}$	32.30	32.39	9.77	12	52	51
CH$_3$OH	C_s	-47.96	-45.33	31.15	12	52	51
CH$_2$O	C_{2v}	-27.70	-26.78	16.14	12	52	51
CO	$C_{\infty v}$	-26.42	-27.20	3.06	12	52	51
CH$_3$F	C_{3v}	-56.00	-54.08	23.95	12	52	51
ClF	$C_{\infty v}$	-12.02	-12.00	1.11	12	9	51
Dissociation products							
H(^{2}S)		52.10	51.63		12		
Li(^{2}S)		38.07	37.70		12		
BeH	$C_{\infty v}$	81.70	81.10	2.84	12	9	51
BH$_2$	C_{2v}	74.80	74.71	10.86	12	12	51
CH$_3$	D_{3h}	34.80	35.62	18.25	12	53	51
CH$_2$(^{3}B$_1$)	$C_{\infty v}$	92.35	92.23	10.45	12	b	51
CH	$C_{\infty v}$	142.00	141.18	3.91	12	9	51
C(^{3}P)		171.29	169.98		12		
NH$_2$	C_{2v}	45.50	46.19	11.46	12	53	51
NH	$C_{\infty v}$	81.4	81.2	4.47	12	9	51
N(^{4}S)		112.97	112.53		12		
OH	$C_{\infty v}$	9.32	9.17	5.10	12	9	51
O(^{3}P)		59.55	58.98		12		
F(^{2}P)		18.97	18.47		12		
Cl(^{2}P)		28.99	28.59		12		

[a] All values in kcal/mol.
[b] *Chem. Phys. Lett.* **123**, 187 (1988).

Table 2—Calculated and experimental bond dissociation energies[a].

Molecule	Bond	q^b	D_e(HF)	D_e(HF–LSDC)	D_e(GVB)	D_e(GVB–LSDC)	D_e(Exp)
LiH	LiH		31.11	49.54	41.33	59.76	57.62
BeH$_2$	BeH		74.46	100.09	82.08	107.71	100.19
BH$_3$	BH		88.70	112.72	96.88	120.90	112.81
CH$_4$	CH		85.84	109.75	95.51	119.43	112.10
NH$_3$	NH	0	79.93	103.80	91.24	115.11	116.28
H$_2$O	OH	1	80.99	105.60	100.82	125.43	125.69
HF	FH	2	86.99	111.22	115.37	139.60	140.89
HCl	ClH	2	70.71	92.97	82.17	104.43	106.37
Li$_2$	LiLi		1.72	14.00	8.90	21.18	24.39
C$_2$H$_6$	CC		69.34	86.80	78.81	96.27	96.50
C$_2$H$_4$	CC		118.50	156.87	145.13	183.50	179.67
C$_2$H$_2$	CC		180.34	199.33	215.01	234.00	234.38
C$_2$	CC		9.26	43.30	99.80	133.84	144.37
N$_2$H$_4$	NN	2	30.18	44.29	56.37	70.34	75.11
N$_2$H$_2$	NN	2	46.23	80.04	92.58	126.39	118.03
N$_2$	NN	2	107.36	167.14	163.25	223.03	228.39
H$_2$O$_2$	OO	2	−1.50	11.96	37.34	50.90	55.07
F$_2$	FF	4	−35.16	−23.57	35.55	47.14	38.21
Cl$_2$	ClCl	4	10.60	22.55	39.56	51.51	57.97
CH$_3$NH$_2$	CN	0	58.17	74.70	75.82	92.34	93.18
CH$_2$NH	CN	1	92.16	128.06	126.85	162.75	153.25
HCN	CN	1	154.12	193.43	186.87	226.18	227.27
CH$_3$OH	CO	1	58.70	75.16	78.53	94.99	97.97
CH$_2$O	CO	2	99.69	135.89	144.83	180.83	183.69
CO	CO	3	169.84	205.40	226.37	261.93	259.22
CH$_3$F	CF	3	68.86	84.64	97.98	113.76	113.87
ClF	ClF	4	4.04	17.87	47.28	61.11	60.16

[a] All values in kcal/mol.
[b] Lone pair participation number; for explanation, see text.

times for F_2 [56], but it is less known that the same is true for H_2O_2. Both Cl_2 and ClF possess positive D_e values at the HF level, but their values are 56 and 47 kcal/mol smaller than the experimental BDEs. Although the failure of HF theory seems to be particularly dramatic in the case of F_2, even larger discrepancies are calculated in the case of homolytic dissociation of multiple bonds. For example, the HF error for C_2 dissociation is 135 kcal/mol, for N_2 dissociation 121 kcal/mol and for homolytic CO dissociation 89 kcal/mol.

The inclusion of LSDC corrections leads to an improvement of homolytic dissociation energies, as has been noted before by Kemister and Nordholm [41]. Errors in calculated AH BDE values are decreased by *ca.* 20 kcal/mol, yielding for Be—H, B—H and C—H dissociation almost exactly the experimental D_e values. This is due to a 10–60 kcal/mol enlargement of HF–LSDC BDE values. On the other hand, the F_2 dissociation energy is still negative at the HF–LSDC level (−23.6 kcal/mol) and the errors for C_2, N_2 and CO are 101, 61 and 53 kcal/mol, respectively. Figs 2 and 6 clearly reveal that at the HF–LSDC level only a small, but not very significant, improvement of calculated BDEs has been achieved. A somewhat more positive assessment of the capability of LSDC to predict dissociation energies given in the literature [41] was due to the limited test examples considered. We conclude that HF–LSDC is not able to predict reasonable D_e values for homolytic dissociation reactions.

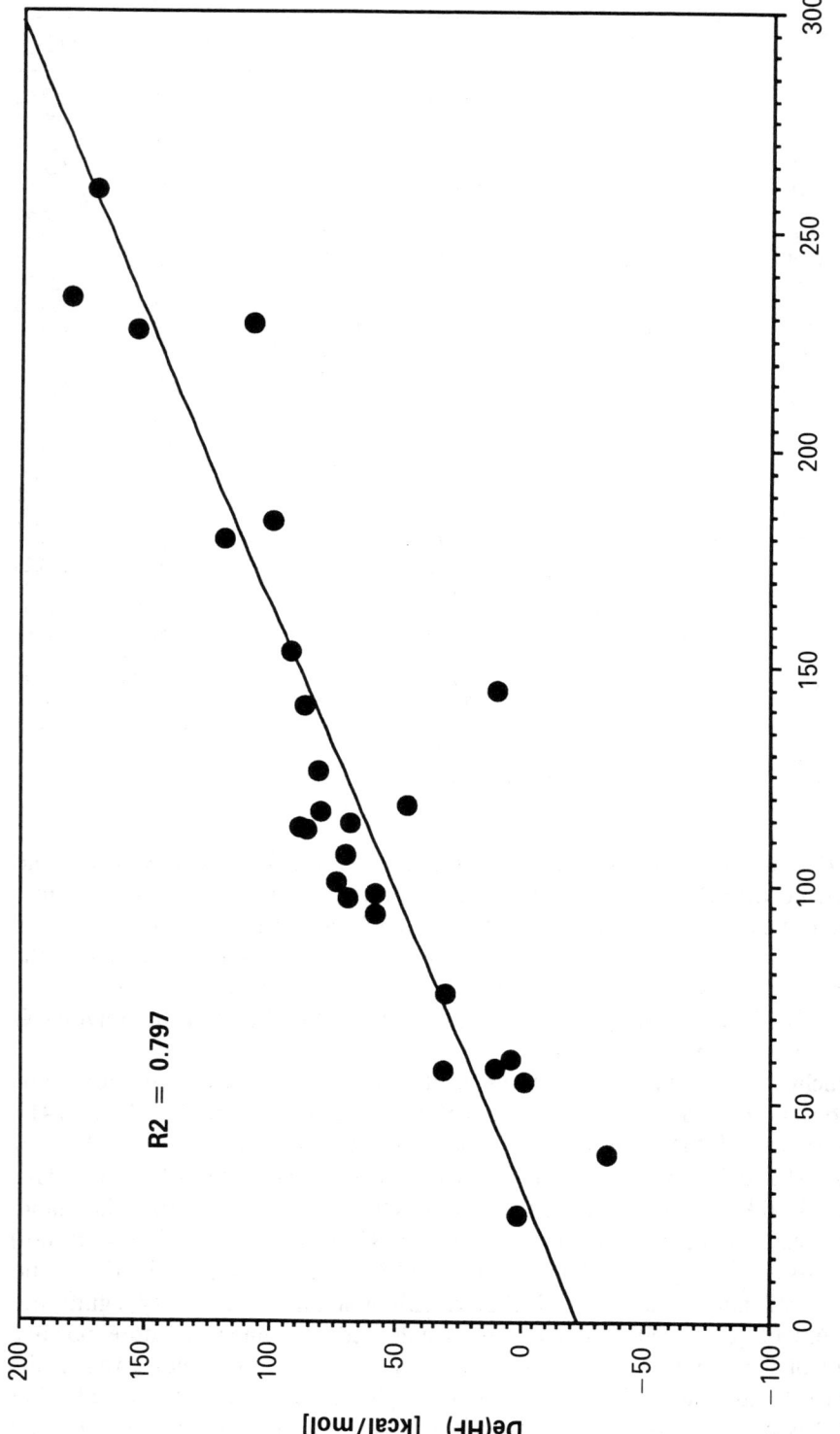

Fig. 1.—Correlation of HF bond dissociation energies D_e(HF) with experimental dissociation energies D_e(Exp). R^2 is the correlation coefficient.

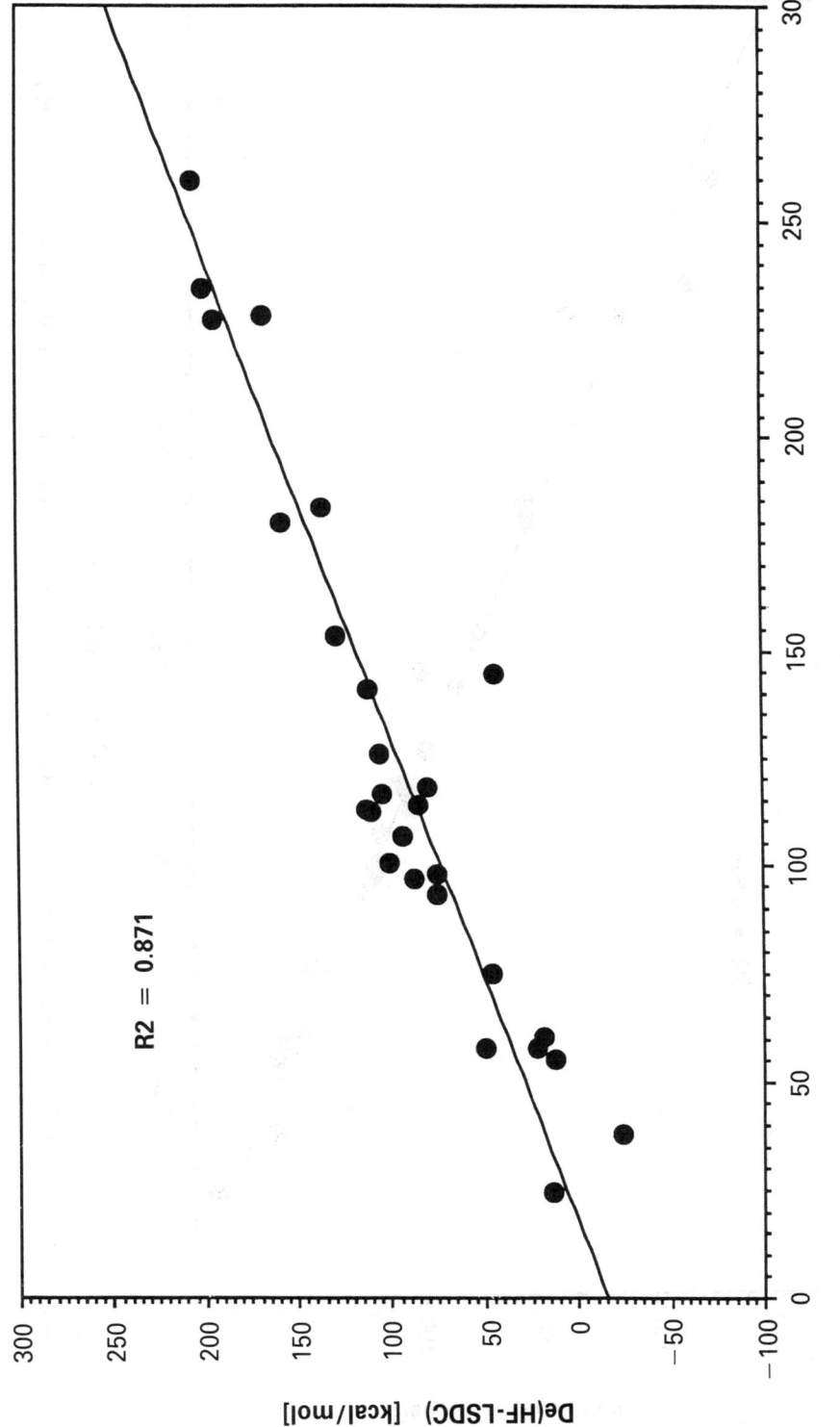

Fig. 2. — Correlation of HF–LSDC bond dissociation energies D_e(HF–LSDC) with experimental dissociation energies D_e(Exp). R^2 is the correlation coefficient.

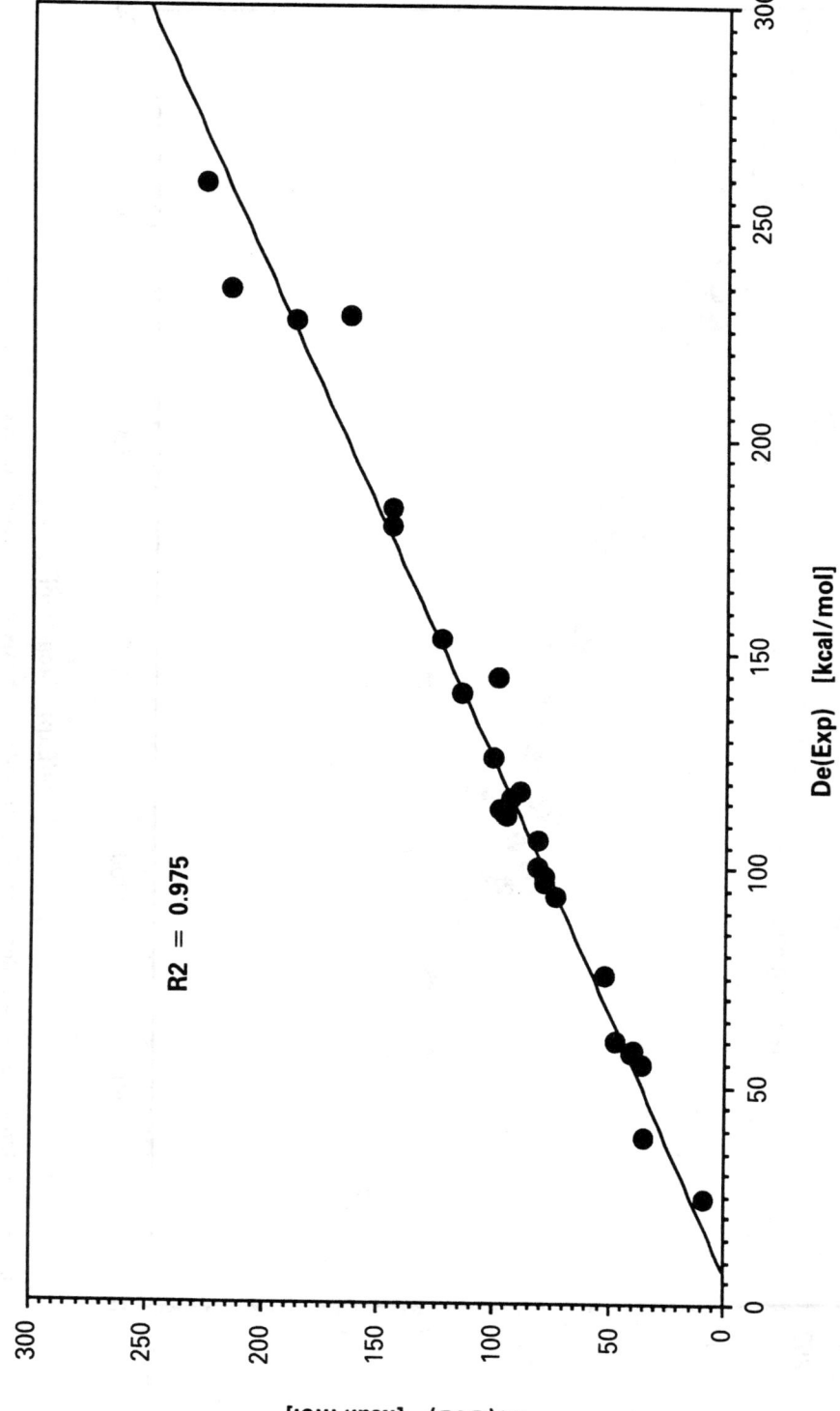

Fig. 3. — Correlation of GVB bond dissociation energies D_e(GVB) with experimental dissociation energies D_e(Exp). R^2 is the correlation coefficient.

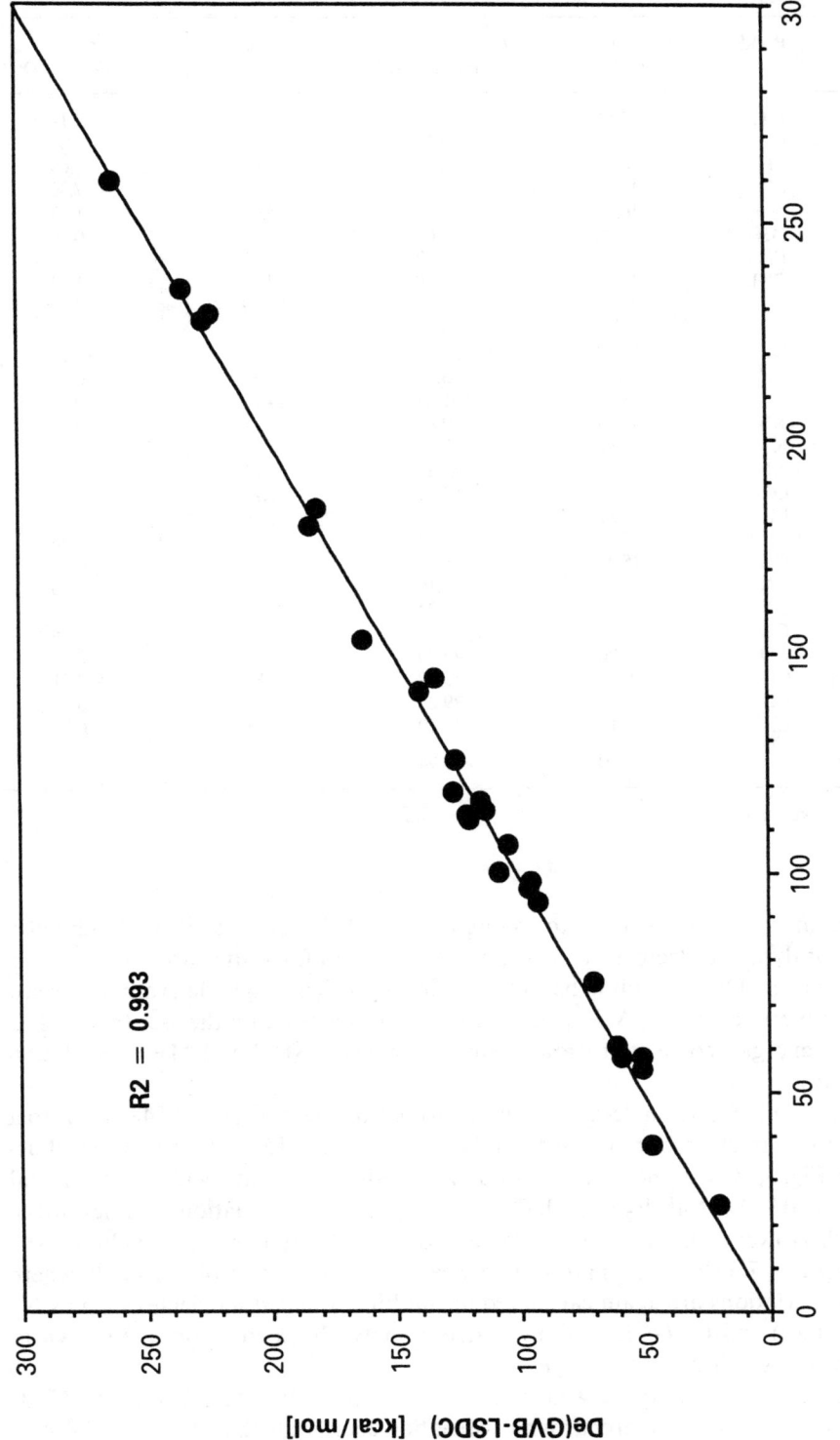

Fig. 4. — Correlation of GVB–LSDC bond dissociation energies D_e(GVB–LSDC) with experimental dissociation energies D_e(Exp). R_2 is the correlation coefficient.

Table 3—Comparison of experimental and calculated bond dissocitation energies[a].

Molecule	Bond	D_e(Exp.) − D_e(HF)	D_e(Exp.) − D_e(HF–LSDC)	D_e(Exp.) − D_e(GVB)	D_e(Exp.) − D_e(GVB–LSDC)
LiH	LiH	26.51	8.08	16.29	−2.14
BeH_2	BeH	25.73	0.10	18.11	−7.52
BH_3	BH	24.11	0.09	15.93	−8.09
CH_4	CH	26.26	2.35	16.59	−7.33
NH_3	NH	36.35	12.48	25.04	1.17
H_2O	OH	44.70	20.09	24.87	0.26
FH	FH	53.90	29.67	25.52	1.29
HCl	ClH	35.66	13.40	24.20	1.94
Li_2	LiLi	22.67	10.39	15.49	3.21
C_2H_6	CC	27.16	9.70	17.69	0.23
C_2H_4	CC	61.17	22.80	34.54	−3.83
C_2H_2	CC	54.04	35.05	19.37	0.38
C_2	CC	135.11	101.07	44.57	10.53
N_2H_4	NN	44.93	30.82	18.74	4.77
N_2H_2	NN	71.80	37.99	25.45	−8.36
N_2	NN	121.03	61.25	65.14	5.36
H_2O_2	OO	56.57	43.11	17.73	4.17
F_2	FF	73.37	61.78	2.66	−8.93
Cl_2	ClCl	47.37	35.42	18.41	6.46
CH_3NH_2	CN	35.01	18.48	17.38	0.84
CH_2NH	CN	61.09	25.19	26.40	−9.50
HCN	CN	73.15	33.84	40.40	1.09
CH_3OH	CO	39.27	22.81	19.44	2.98
CH_2O	CO	84.00	47.80	38.86	2.86
CO	CO	89.38	53.82	32.85	−2.71
CH_3F	CF	45.01	29.23	15.89	0.11
ClF	ClF	56.12	42.29	12.88	−0.95
Average error		54.50	28.94	24.09	3.96

[a] All values in kcal/mol.

The failure of LSDC is not astonishing in view of the fact that it predominantly accounts for dynamic electron correlation effects but not for static correlation effects. For this reason, LSDC cannot be better than any other single-determinant-based correlation method such as MP or CI. This is also suggested by the fact that LSDC correlation energies correlate to some extent with MP2, MP3 and MP4 correlation energies [46].

Contrary to HF and single-determinant correlation methods, GVB [48] leads to a qualitatively correct description of homolytic dissociation [57]. This is nicely illustrated by Fig. 3, which shows that GVB BDE values correlate with experimental BDEs ($R^2 = 0.975$). Both F_2 and H_2O_2 possess positive dissociation energies at the GVB level. However, GVB does not lead to quantitatively correct D_e values, as is reflected by Fig. 7. GVB values are still between 10 and 65 kcal/mol too small, where the largest deviations are again calculated for multiple-bonded molecules such as N_2 (− 65 kcal/mol) and C_2 (−45 kcal/mol). Accordingly, the correlation line shown in Fig. 3 does not go through the origin.

Our GVB results compare well with those obtained by Hay, Hunt and Goddard [57] who computed hydrocarbon dissociation energies with a DZ and a DZ + P

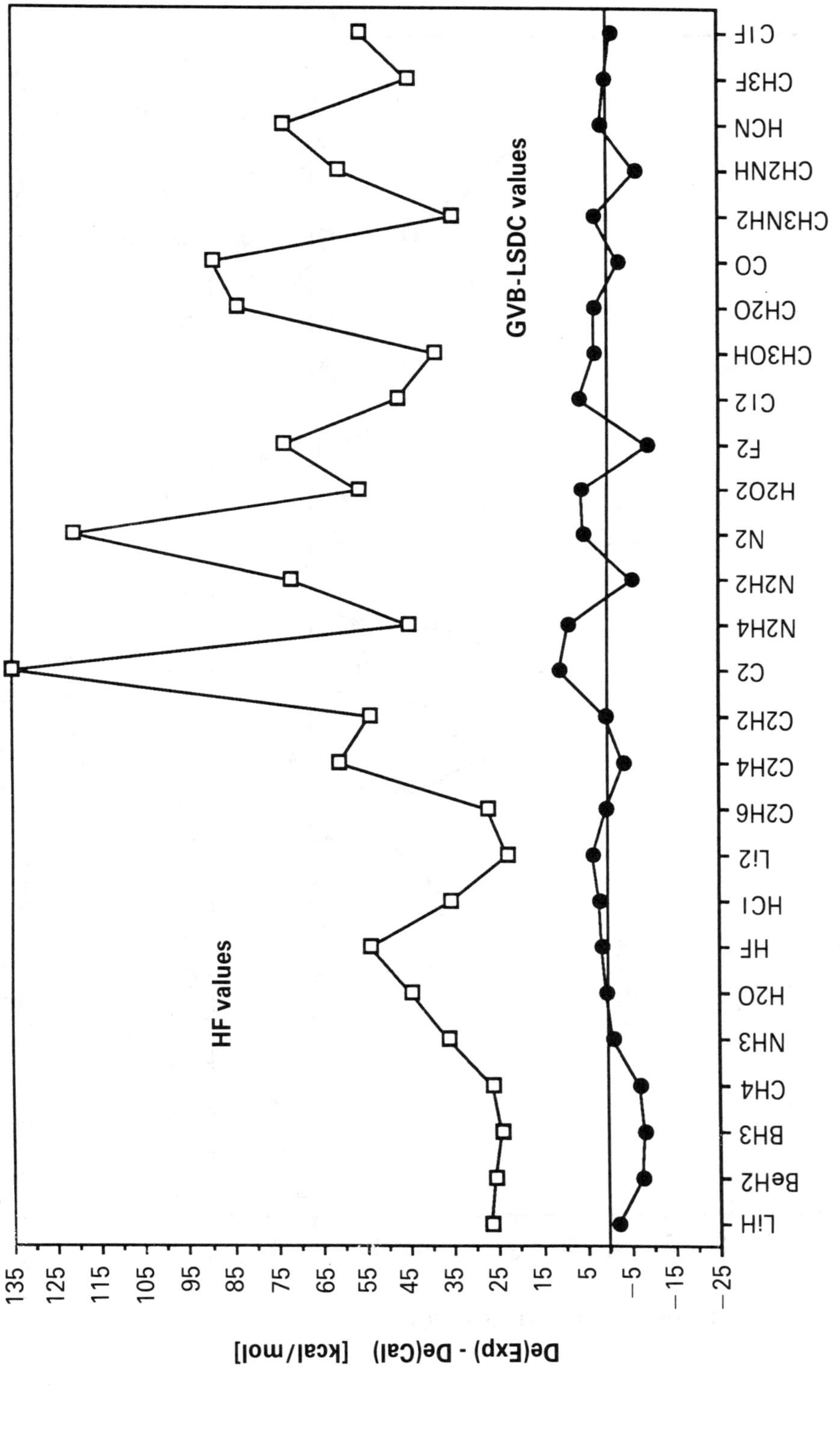

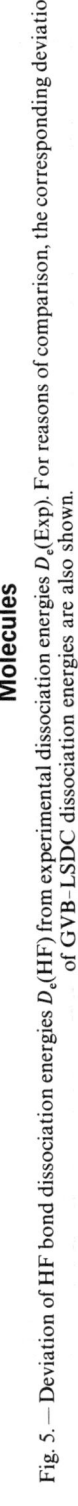

Fig. 5. — Deviation of HF bond dissociation energies D_e(HF) from experimental dissociation energies D_e(Exp). For reasons of comparison, the corresponding deviations of GVB–LSDC dissociation energies are also shown.

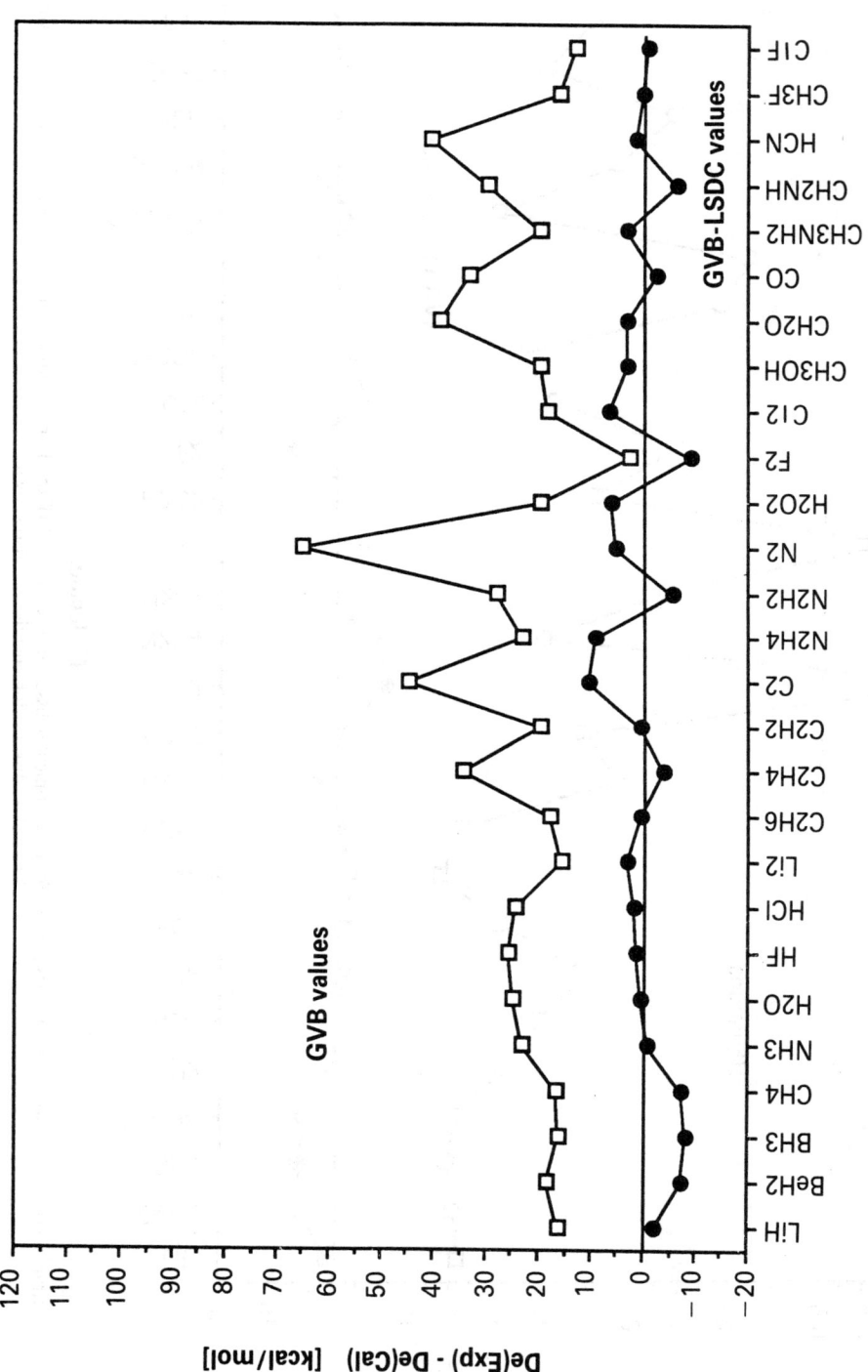

Fig. 6. — Deviation of HF–LSDC bond dissociation energies D_e(HF–LSDC) from experimental dissociation energies D_e(Exp). For reasons of comparison, the corresponding deviations of GVB–LSDC dissociation energies are also shown.

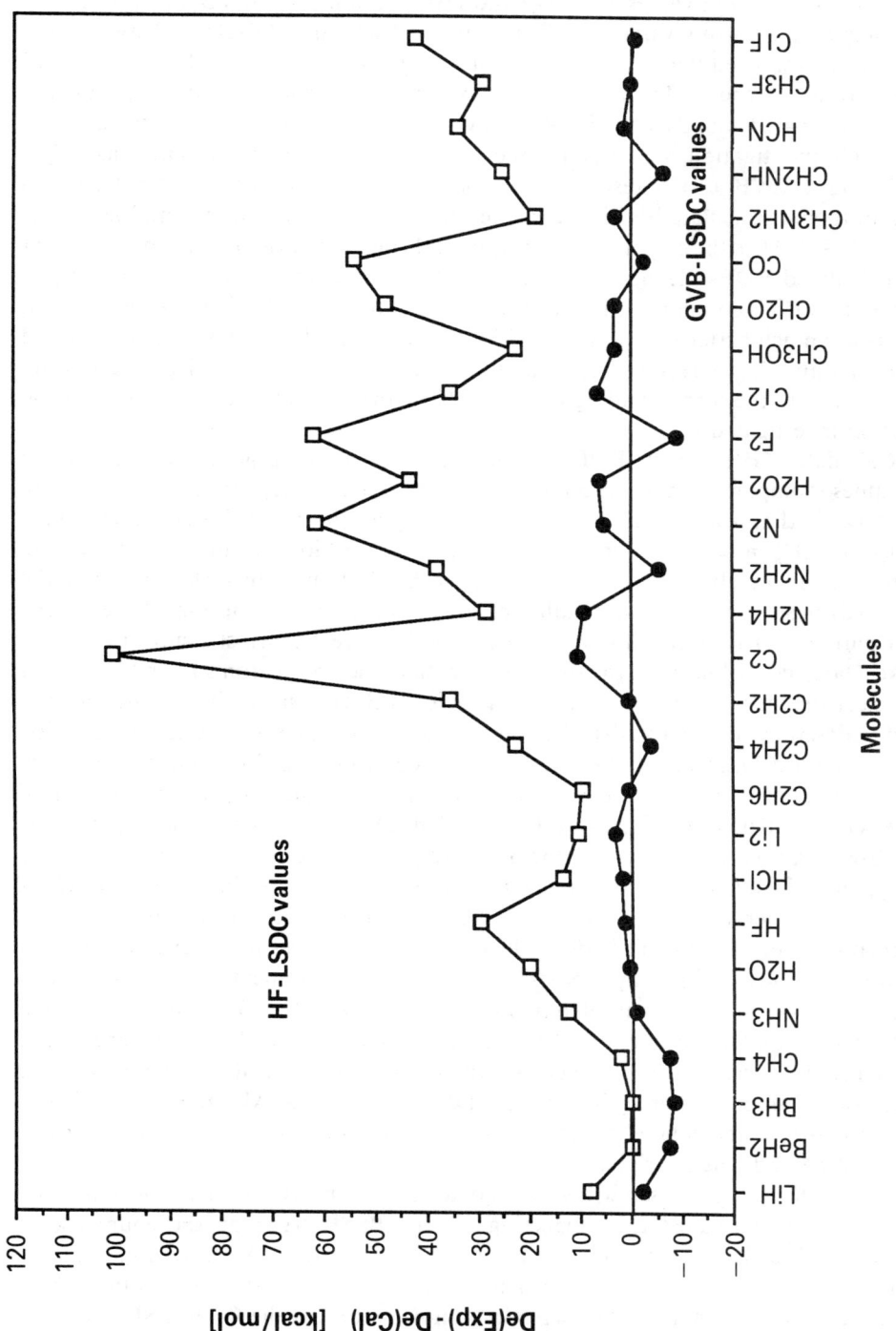

Fig. 7. — Deviation of GVB bond dissociation energies D_e(GVB) from experimental dissociation energies D_e(Exp). For reasons of comparison, the corresponding deviations of GVB–LSDC dissociation energies are also shown.

basis set. For example, these authors obtained for homolytic CC dissociation of C_2H_6, C_2H_4 and C_2H_2 BDE values of 76, 147 and 180 kcal/mol at the GVB/DZ + P level to be compared with our values of 78, 145 and 215 kcal/mol (Table 2). Hay and co-workers also showed that GVB dissociation energies can be improved at the GVB–CI level owing to the inclusion of dynamic correlation effects. Unfortunately, the computational demands of GVB–CI prevented the authors from carrying out a thorough investigation of a large number of homolytic dissociation reactions [57].

In Fig. 4, GVB–LSDC results on homolytic BDEs are shown. Obviously, at this level of theory, calculated BDEs correlate satisfactorily with experimental D_e values ($R^2 = 0.993$). Also, the correlation line is going through the origin thus indicating that the calculated BDEs are no longer too small. The average error is 4 kcal/mol (see Table 3), which is comparable to the error in experimental heats of formation. Clearly, results are much better than both HF–LSDC and GVB BDEs. This is also suggested by Figs 6 and 7, where HF–LSDC and GVB errors in calculated BDE values are compared with the corresponding GVB–LSDC errors for each homolytic dissociation reaction investigated.

Calculated GVB–LSDC BDEs can be both smaller and larger than experimental D_e values (compare with Table 3 and Fig. 7). It is particularly interesting to analyse those cases that lead to an overestimate of D_e at the GVB–LSDC level. These comprise AH_n molecules with a relatively electropositive central atom A and, therefore, a relatively high electronic charge at the H atom(s). In such a situation, the 6–31G(d) basis is certainly not sufficient for an adequate description of the charge distribution. For example, if one considers the ionic resonance structure Li^+H^- to make a large contribution to the LiH wavefunction, then both atoms Li and H possess two electrons in the LiH molecule. With the 6–31G(d) basis, 15 basis functions are used to describe the electron distribution at Li while only two basis functions describe that at H. Clearly, this leads to an unbalanced description of LiH and its homolytic dissociation reaction. Use of the 6–31G(d,p) basis would lead to a better description of the electronic structure of LiH, and this would also be true for the other AH_n systems.

However, it is easy to see that the 6–31G(d,p) basis causes a lowering of the LiH (AH_n) energy while leaving that of Li and H unaltered. Accordingly, HF and GVB dissociation energies become larger rather than lower. As noted above, LSDC correlation energies show little dependence on small changes in the electron density distribution caused either by method or basis set improvements. Test calculations carried out in this work reveal that in most cases E_c(LSDC) decreases by some mHartrees upon improvement of the basis set from minimal to DZ or better quality. Therefore, an improvement of the basis from 6–31G(d) to 6–31G(d,p) quality leads to larger rather than smaller GVB–LSDC BDE values for the AH_n molecules. Hence, overestimation of calculated D_e values for AH_n molecules such as LiH, BeH_2, BH_3, etc. must be of different origin.

An insufficient basis set will also lead to a large basis set superposition error (BSSE) in calculated BDEs. Correcting for the BSSE error by the counterpoise method [58] leads to a decrease in the calculated D_e values. We have done this for some of the AH_n molecules and, indeed, have obtained improved values for the BDEs.

Another possible reason for overestimating D_e values at the GVB–LSDC level of theory has to do with electron lone pair participation at the GVB level. For example, in the case of F_2, the inclusion of the 4 $p\pi$ electron lone pairs leads to an increase in the

D_e value from 23 to 47 kcal/mol. This, of course, is a simplification of lone pair participation since it may not only be stabilizing but also destabilizing owing to electron pair–electron pair repulsion. At the GVB level, the pairs are correlated independently of each other, thus exaggerating the stabilizing effect. On the other hand, destabilizing repulsion of electron pairs at vicinal atoms will decrease with the distance. Hence, it is not surprising that pπ lone pair inclusion at the GVB–LSDC level does not lead to the same exaggeration in D_e in the case of ClF and Cl_2 (see Tables 2 and 3).

Finally, there are multiply bonded molecules such as CH_2CH_2, CH_2NH, HNNH and CO with GVB–LSDC BDEs that are too large. This may be partially due to inaccurate or missing experimental BDEs in the case of CH_2NH and HNNH. The heat of formation of CH_2NH is not known and, therefore, it had to be estimated from experimental group additivity increments and calculated *ab initio* energies [59]. For HNNH, it is unclear whether the measured heat of formation corresponds to the *trans* configuration or to the less stable *cis* form [60]. If the latter were the case, the experimental D_e value for the *trans* form, which has been considered in this work, would be close to the calculated BDE given in Table 2. Other reasons for an overestimation of calculated D_e values may again be the number of lone pairs correlated at the GVB level (HNNH, CH_2NH, CO) or an inadequate description of the dissociation products at the LSDC level (CH_2 and NH).

The largest underestimation of D_e at the GVB–LSDC level is found for C_2 (-10.5 kcal/mol, Table 2). This is not surprising since the correct description of homolytic C_2 dissociation requires a larger MCSCF expansion than that provided by GVB. Lie and Clementi used nine configurations in their MCSCF–LDF approach, but obtained a BDE for C_2 that was 26 kcal/mol too small [35]. Future work has to show whether improvement of MCSCF, LSD or basis set will lead to a more accurate BDE.

CONCLUSIONS AND OUTLOOK

We have found that the GVB–LSDC approach provides an economic and reasonable description of the energies of homolytic dissociation reactions. This is due to an adequate treatment of static and dynamic correlation effects which are important in order to obtain a proper description of bond dissociation. GVB–LSDC results are better than both GVB and HF–LSDC BDEs. For 27 dissociation reactions investigated we obtained an average error for calculated D_e values of 4 kcal/mol.

It is obvious that future investigations have to add further proof for the usefulness of the GVB–LSDC approach. Test examples have to be extended to the dissociation of open-shell compounds and to second-row molecules. Also, it will be interesting to see whether calculated atomization and bond energies are as satisfactory as GVB–LSDC BDE values. Work is in progress to investigate this question.

The method presented is intriguing owing to its modest computational requirements. Because of this, calculations with larger basis sets than the one used in this work will be feasible. Future work has to reveal whether the use of larger basis sets will lead to a significant improvement of calculated BDEs. In this connection, it will be necessary to test the usefulness of bond functions and the impact of the BSSE on calculated values. Yet another problem to be investigated is the question of whether the fact that GVB is not size-consistent has an impact on the accuracy of D_e. Finally, it will be interesting to see whether more recent density functionals than the VWN functional [39] will lead to an improvement in theoretical BDE values.

The GVB–LSDC method presented here may be considered to be in the mainstream of theoretical approaches that attempt to cover adequately both static and dynamic correlation effects. One has early realized that even large MCSCF treatments lack an appropriate coverage of dynamic correlation effects. Therefore, MRD–CI or CASSCF–CI methods have been designed to obtain highly accurate values of reaction energies and molecular properties. As pointed out above, these methods are often too costly. Therefore, alternatives have been developed such as the GVB–MP2 method [61] or the CASSCF–MP2 approach [62]. These or similar methods will prove their calculational usefulness in the future. Owing to the physical nature of the approximations used in the application of the LSDC functional, the method proposed here cannot, with respect to the possibility of systematic improvement, compete with the *ab initio* methods just mentioned. However, its range of applicability will be much larger owing to its low computational cost.

ACKNOWLEDGEMENT

EK acknowledges the technical assistance of G. Kemister in rewriting the LSDC programs. All calculations have been carried out with the CRAY XMP/48 of the Nationellt Superdator Centrum (NSC) in Linköping, Sweden. EK and DC thank the NSC for the generous allotment of computer time, and the NFR, Stockholm, for financial support of this work.

REFERENCES

[1] The classic description of the role of dissociation energies, atomization energies, and bond energies for the understanding of chemical bonding is given by L. Pauling, *The Nature of the Chemical Bond*, 3rd edn, Cornell University Press, Ithaca, NY, 1960.

[2] J.G. Calvert and J.N. Pitts, *Photochemistry*, Wiley, New York, 1966.

[3] J.D. Cox and G. Pilcher, *Thermochemistry of Organic and Organometallic Compounds*, Academic Press, New York, 1970.

[4] S.W. Benson, *Thermochemical Kinetics*, 2nd edn, Wiley, New York, 1976.

[5] S.W. Benson, *Chem. Rev.* **78**, 23 (1978).

[6] R.T. Sanderson, *Chemical Bonds and Bond Energy*, Academic Press, New York, 1976.

[7] R.T. Sanderson, *Polar Covalence*, Academic Press, New York, 1983.

[8] CODATA Task Group, *J. Chem. Thermodyn.* **10**, 903 (1978).

[9] G. Herzberg and K.P. Huber, *Molecular Spectra and Molecular Structure 4: Constants of Diatomic Molecules*, Van Nostrand, Princeton, NJ, 1979.

[10] D.F. McMillen and D.M. Golden, *Ann. Rev. Phys. Chem.* **33**, 493 (1982).

[11] M.W. Chase, J.L. Curnutt, J.R. Downey, R.A. MacDonald, A.N. Syverud and E.A. Valenzuela, *JANAF Thermochemical Tables*, 1982. Supplement, *J. Phys. Chem. Ref. Data* **11** 695 (1982).

[12] M.W. Chase, C.A. Davies, J.R. Downey, D.J. Frurip, R.A. McDonald and A.N. Syverud, *JANAF Thermochemical Tables*, 3rd edn, *J. Phys. Chem. Ref. Data* **14**, Suppl. 1, 1 (1985).

[13] J.B. Pedley, R.D. Naylor, S.P. Kirby, *Thermochemical Data of Organic Compounds*, 2nd edn, Chapman and Hall, New York, 1986.

[14] For a compilation of relevant references, see (a) K. Ohno and K. Morokuma, Quantum Chemistry Literature Data Base, *Bibliography of ab initio Calculations for 1978–1980*, Elsevier, New York, 1982 and following supplements. (b) W.G. Richards, P.R. Scott, V. Sackwild and S.A. Robins, *A Bibliography of ab initio Molecular Wave Functions*, Supplement for 1978–1980, Clarendon Press, Oxford, 1981, and following supplements.

[15] For reviews, see (a) R.S. Mulliken and W.C. Ermler, *Diatomic Molecules, Results of Ab Initio Calculations*, Academic Press, New York, 1977. (b) R.S. Mulliken and W.C. Ermler, *Polyatomic Molecules, Results of Ab Initio Calculations*, Academic Press, New York, 1981. Some selected examples are given in references 16 to 20.

[16] W. Butscher, S.-K. Shih, R.J. Buenker and S.D. Peyerimhoff, *Chem. Phys. Lett.* **52**, 457 (1977) and references cited therein.

[17] J.S. Wright and J.R. Buenker, *Chem. Phys. Lett.* **106**, 570 (1984).

[18] C.W. Bauschlicher, *Chem. Phys. Lett.* **122**, 572 (1985).

[19] M.V. Rama Krishna and K.D. Jordan, *Chem. Phys.* **115**, 405 (1987).

[20] E.A. Carter and W.A. Goddard III, *J. Chem. Phys.* **88**, 3132 (1988).

[21] See, for example, J.A. Pople, J.S. Binkley and R. Seeger, *Int. J. Quant. Chem. Symp.* **10**, 1 (1975).

[22] See, for example J.A. Pople, R. Seeger and R. Krishnan, *Int. J. Quant. Chem.* **11**, 149 (1977).

[23] P. Mach and O. Kysel, *J. Comp. Chem.* **6**, 312 (1985).

[24] J.M.L. Martin, J.P. Francois and R. Gijbels, *J. Comp. Chem.* **10**, 152 (1989).

[25] J.M.L. Martin, J.P. Francois and R. Gijbels, *J. Comp. Chem.* **10**, 875 (1989).

[26] J.M.L. Martin, J.P. Francois and R. Gijbels, *Theoret. Chim. Acta* **76**, 195 (1989).

[27] J.M.L. Martin, J.P. Francois and R. Gijbels, *Chem. Phys. Lett.* **163**, 387 (1989).

[28] J.A. Pople, B.T. Luke, M.J. Frisch and J.S. Binkley, *J. Phys. Chem.* **89**, 2198 (1985).

[29] J.A. Pople and L.A. Curtiss, *J. Phys. Chem.* **91**, 155 (1987).

[30] L.A. Curtiss and J.A. Pople, *J. Phys. Chem.* **92**, 894 (1988).

[31] P.C. Hariharan and J.A. Pople, *Theoret. Chim. Acta* **28**, 213 (1973).

[32] R.G. Parr and W. Yang, *Density-Functional Theory of Atoms and Molecules*, Oxford University Press, Oxford, 1989, and references cited therein.

[33] R.O. Jones, in *Ab initio Methods in Quantum Chemistry, Part 1*, K.P. Lawley, Ed, *Adv. Chem. Phys.* **67**, 413 (1987).

[34] G.C. Lie and E. Clementi, *J. Chem. Phys.* **60**, 1275 (1974).

[35] G.C. Lie and E. Clementi, *J. Chem. Phys.* **60**, 1288 (1974).

[36] H. Stoll, C.M.E. Pavlidou and H. Preuss, *Theoret. Chim. Acta* **49**, 143 (1978).

[37] H. Stoll, E. Golka and H. Preuss, *Theoret. Chim. Acta* **55**, 29 (1980).

[38] H. Stoll and A. Savin, in *Density Functional Methods in Physics*, Eds. R.M. Dreizler and J. da Providencia, Plenum-Press, New York, 1985, p. 177.

[39] S.H. Vosko, I. Wilk and M. Nusair, *Can. J. Phys.* **58**, 1200 (1980).

[40] D.M. Ceperley and B.J. Alder, *Phys. Rev. Lett.* **45**, 566 (1980).

[41] G. Kemister and S. Nordholm, *Chem. Phys. Lett.* **133**, 121 (1987).

[42] G. Kemister and S. Nordholm, *J. Chem. Phys.* **83**, 5163 (1985).

[43] U. von Barth and L. Hedin, *J. Phys. C* **5**, 1629 (1972).

[44] G. Kemister, Thesis, University of Sydney, Sydney, 1984.

[45] *COLOGNE*90, J. Gauss, E. Kraka, F. Reichel and D. Cremer, Göteborg, 1990.

[46] E. Kraka, to be published.

[47] (a) W.J. Hunt, T.H. Dunning and W.A. Goddard III, *Chem. Phys. Lett.* **3**, 606 (1969) (b) W.J. Hunt, P.J. Hay and W.A. Goddard III, *J. Chem. Phys.* **57**, 738 (1972).

[48] F.W. Bobrowicz and W.A. Goddard III, in *Methods of Electronic Structure Theory*, H.F. Schaefer, Ed., Plenum Press, New York, 1977, p. 79.

[49] D. Cremer, *J. Chem. Phys.* **69**, 4440 (1978).

[50] D. Cremer, in *The Chemistry of Functional Groups, Peroxides*, S. Patai, Ed, Wiley, New York, 1983, p. 1.

[51] See compilation in W.J. Hehre, L. Radom, P.v.R. Schleyer and J.A. Pople, *Ab Initio Molecular Orbital Theory*, Wiley, New York, 1986.

[52] (a) T. Shimanouchi, *Tables of Molecular Vibrational Frequencies*, Vol 1, NSRDS-NBS-39, National Bureau of Standards, Washington, 1972. (b) T. Shimanouchi, *J. Phys. Chem. Ref. Data* **6**, 993 (1977).

[53] M.E. Jacox, *J. Phys. Chem. Ref. Data*. **13**, 945 (1984).

[54] V.E. Bondybey and J.H. English, *J. Chem. Phys.* **80**, 568 (1984).

[55] B.J. Moss, F.W. Bobrowicz, W.A. Goddard III, *J. Chem. Phys.* **63**, 4632 (1975).

[56] K. Jankowski, R. Becherer, P. Scharf, H. Schiffer and R. Ahlrichs, *J. Chem. Phys.* **82**, 1413 (1985) and references cited therein.

[57] P.J. Hay, W.J. Hunt and W.A. Goddard III, *J. Am. Chem. Soc.* **94**, 8293 (1972).

[58] S. F. Boys and F. Bernardi, *Mol. Phys.* **19**, 553 (1970).

[59] (a) D. Cremer, *J. Comp. Chem.* **3**, 165 (1982). (b) D. Cremer, Habilitation Thesis, University of Köln, Köln, 1979.

[60] N. Wiberg, G. Fischer and H. Bachhuber, *Z. Naturforsch.* **34b**, 1385 (1979).

[61] (a) K. Wolinski, H.L. Sellers and P. Pulay, *Chem. Phys. Lett.* **140**, 225 (1987). (b) K. Wolinski and P. Pulay, *J. Chem. Phys.* **90**, 3647 (1988).

[62] J.J.W. McDouall, K. Peasley and M.A. Robb, *Chem. Phys. Lett.* **148**, 183 (1988).

20

Characteristic electronic processes in etching reaction of silicon

Akitomo Tachibana[a], Susumu Kawauchi[a,b,*], Tokio Yamabe[a,b] and Kenichi Fukui[b]
[a]Kyoto University, Kyoto, Japan
[b]Institute for Fundamental Chemistry, Kyoto, Japan

INTRODUCTION

Molecular orbital calculation has been widely used as a major tool in the study of molecular problems of structure, stability, and reaction mechanism [1]. It has been anticipated that studies using *ab initio* quantum chemical techniques aided by recent developments in computers will provide a theoretical framework not only for the molecular design of novel materials but also for the design of new synthetic routes.

Recently, we reported a series of theoretical studies on the etching reaction of silicon by using *ab initio* quantum chemical techniques [2–4]. Plasma etching of silicon by fluorine is a crucial step in the fabrication of integrated circuits [5]. Since an understanding of the mechanism of the etching reaction will help us to obtain a finer resolution or new etching techniques, there are may theoretical studies [2–4, 6–9] on the etching reaction of silicon by fluorine as well as experimental studies [5, 10, 11]. However, it is difficult to model the silicon etching accurately [12], because the etching reaction is a solid–surface reaction. One possible way of overcoming these difficulties is that the local environment of the reaction site is simulated by finite clusters of atoms or ions that can be treated by *ab initio* methods [2–4, 8, 9]. Such a 'local model' approach is particularly suited for examining local phenomena that may occur as the surface reaction. Indeed, Yarmoff and McFeely [10l] suggested that the chemical mechanism participates in at least part of the mechanism of the doping effects on the silicon etching by fluorine, rather than a physical mechanism such as a purely electrostatic model [10j,k]. Therefore, in this chapter, we first introduce our local model studies and proposed chemical mechanisms on the silicon etching by using *ab initio* quantum chemical techniques [2–4], together with other theoretical studies [9]. We then examine such characteristic chemical mechanisms in terms of

* On leave from Kawasaki Plastics Laboratory, Showa Denko K.K., Kawasaki 210, Japan.

frontier orbital interactions. For details of the calculating methods, the reader is referred to the original articles.

LOCAL MODEL OF SILICON SURFACE

In order to apply the quantum chemical approach to the silicon etching reaction by using *ab initio* molecular orbital calculations, the local environment of the silicon surface is simulated by a small finite cluster of silicon, with H atoms used to saturate the remaining valences, as would occur in solid silicon as well as in amorphous silicon [2-4, 9, 12]. Fig. 1 shows the undoped silicon surface model. The calculated Si—Si bond length in H_3SiSiH_3, 2.342 Å at the HF/3-21G$^{(*)}$ level [2], is in good agreement with the experimental one for solid silicon, 2.35 Å [13].

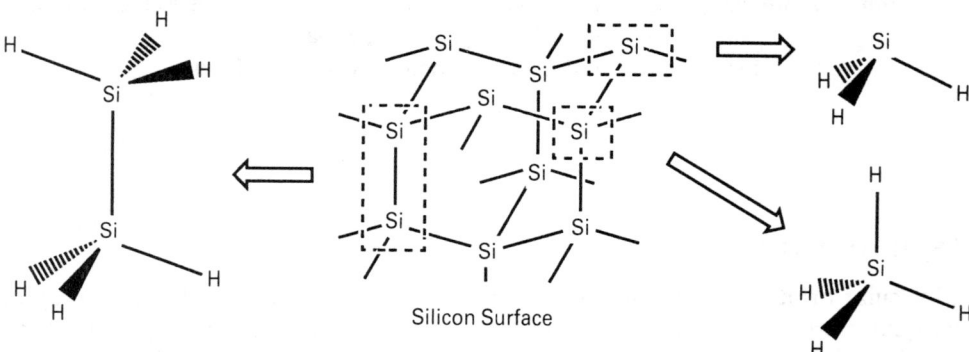

Silicon Surface

Fig. 1 — Local models for silicon surface.

Recently, Horowitz and Goddard [14] discussed one of the mysteries about the local model by using generalized valence bond (GVB) theory. The average Si—Si bond energy for the solid (54 kcal/mol) is far below that in H_3SiSiH_3 (72 kcal/mol) [14]. In comparison, the C—C bond energy in diamond (85 kcal/mol) is fairly close to that of ethane (89.8 kcal-mol) [14]. Horowitz and Goddard showed that the average of a series of Si—Si bond strengths of H_3SiSiH_3 (72 kcal/mol), H_3SiSiH_2 (51 kcal/mol), H_3SiSiH (55 kcal/mol), and H_3SiSi (49 kcal/mol) is 57 kcal/mol, which is in good agreement with the average Si—Si bond energy in solid silicon (54 kcal/mol).

For doped silicon, we assumed a local reactive site on which the electron or hole density is larger than that in bulk silicon. We called the localized electron-rich center in *n*-doped silicon the 'anion center' [4], and the localized hole-rich center in *p*-doped silicon the 'cation center' [3]. The local environment of the cation center and the anion center is simulated by positively charged and negatively charged small finite clusters of silicon, with H atoms used to saturate the remaining valences, respectively [3, 4]. The smallest of these models that we used in our study are, for example, SiH_4^+ or $H_3SiSiH_3^+$ for the cation center and SiH_4^- or $H_3SiSiH_3^-$ for the anion center, which are all stable species according to the calculation. The Si—Si bond lengths of $H_3SiSiH_3^+$ (2.720 Å) and $H_3SiSiH_3^-$ (2.641 Å) are longer than that of H_3SiSiH_3 (2.342 Å) at the HF/3-21G$^{(*)}$ level [3, 4] as shown in Table 1. The calculated Si—Si bond strengths of $H_3SiSiH_3^+$ (33.7 kcal/mol at the MP3/6-31G**//HF/3-21G$^{(*)}$ level with zero-point energy (ZPE) correction) and $H_3SiSiH_3^-$ (27.7 kcal/mol at the MP3/ 6-31 + G**//HF/3-21G$^{(*)}$ level with ZPE correction) are smaller than that of H_3SiSiH_3 (69.4 kcal/mol at the MP3/6-31G**//HF/3-21G$^{(*)}$ level with ZPE correction) as is also shown in Table 1. Therefore, the Si—Si bonds of the cation center in p-doped silicon and the anion center in n-doped silicon may be somewhat longer and weaker than that of intrinsic silicon [3, 4]. The basic idea of these models is justified experimentally [10], because the influence of dopants was small or nonexistent when the dopants were not electronically active [10b,g]. Indeed, the large etch rate for heavily n-doped Si occurred independently of whether the dopant atoms were arsenic or phosphorus [10b,g]. These results therefore suggest that the origin of the doping effect depends on the electronic structure of the surface, i.e. the number of holes or free electrons [10].

Table 1 — Calculated Si—Si bond lengths $(r_{Si-Si})^a$ and dissociation energies $(D_{Si-Si})^b$ of H_3SiSiH_3,^{c} $H_3SiSiH_3^+$,^{d} and $H_3SiSiH_3^-$ ^{e}

	r_{Si-Si}	D_{Si-Si}	
		MP3	CISD + QC
H_3SiSiH_3	2.342	69.4	68.3
$H_3SiSiH_3^+$	2.720	33.7	33.3
$H_3SiSiH_3^-$	2.641	27.7	

^{a}At the HF/3-21G$^{(*)}$ optimized geometries, in angstrom.
^{b}With ZPE corrections at the HF/3-21G$^{(*)}$ optimized geometries, in kcal/mol.
^{c}Dissociation energies at the MP3/6-31G**//HF/ 3-21G$^{(*)}$ and the CISD + QC/6-31G**//HF/ 3-21G$^{(*)}$, from ref. [2].
^{d}Diissociation energies at the MP3/6-31G**//HF/ 3-21G$^{(*)}$ and the CISD + QC/6-31G**//HF/ 3-21G$^{(*)}$, from ref. [3].
^{e}Dissociation energy at the MP3/6-31 + G**//HF/ 3-21G$^{(*)}$, from ref. [4].

CHEMICAL MECHANISM ON ETCHING REACTION OF UNDOPED SILICON

Even SiH_4, the simplest tetrahedron having only one silicon atom, maintains intrinsic chemical characteristics of the silicon compounds [15] and serves to elucidate the chemical mechanisms of the etching reaction. In order to make such a chemical characteristic more clear, SiH_4 is compared with CH_4. Fig. 2 shows the molecular orbital (MO) level diagram of SiH_4 and CH_4. There is a difference between the lowest unoccupied MO (LUMO) of SiH_4 and CH_4. Since SiH_4 has a lower LUMO, SiH_4 can act as an acceptor, with a stronger affinity for nucleophiles than CH_4. Moreover, the LUMO of SiH_4 is triply degenerated $2t_2$ orbitals, while that of CH_4 is a $2a_1$ orbital. The $2a_1$ orbital and one of the $2t_2$ orbitals are represented schematically in Fig. 3. We note here that one of the triply degenerate orbitals, $2t_{2z}$, which is also

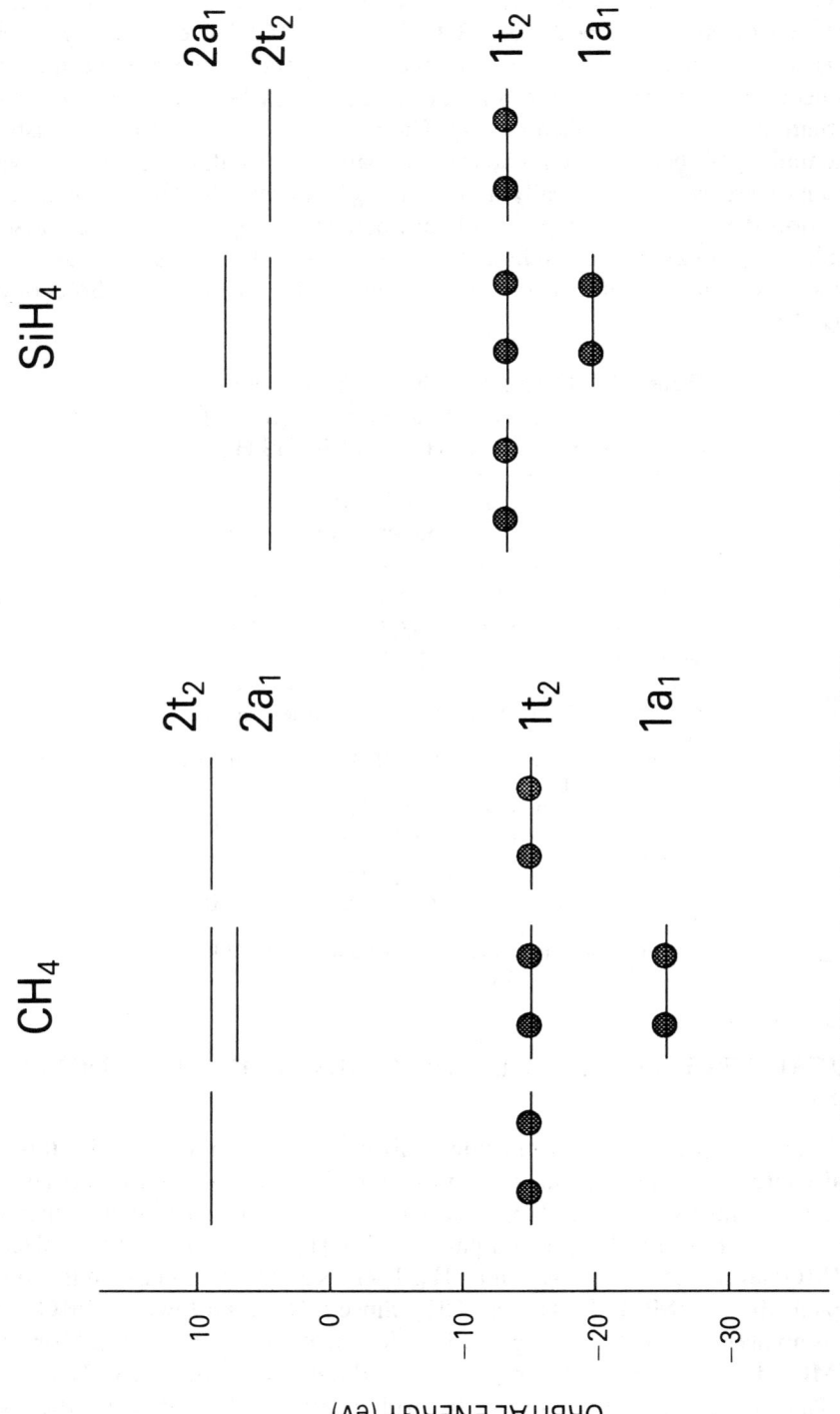

Fig. 2 — MO energy level diagram of SiH_4 and CH_4 at the HF/6–31G** level.

(a)   (b)   (c)

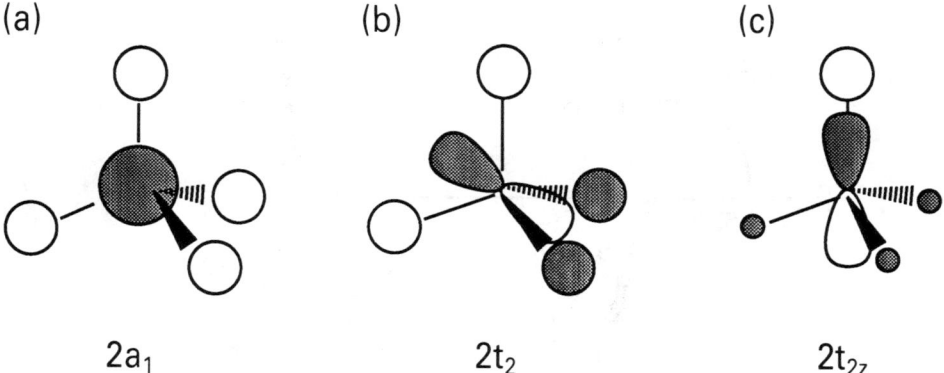

$2a_1$ $2t_2$ $2t_{2z}$

Fig. 3 — MO patterns of (a) $2a_1$, (b) $2t_2$, and (c) $2t_{2z}$.

shown in Fig. 3, can be expressed equivalently as linear combinations of three sets of $2t_2$ orbitals [16]. Such $2t_2$ and $2t_{2z}$ orbitals might contribute to stabilizing the transition-state energy, because of their more favorable orbital phase relation with nucleophiles than the $2a_1$ orbital [15a].

The $2t_{2z}$ orbital is responsible for the hypervalency of silicon compounds [15a]. This explains the etching mechanism proposed for undoped silicon by Garrison and Goddard [9] and proposed for n-doped silicon by us [4], which will be discussed in the next section. On the other hand, the $2t_2$ orbital is closely related to the F-accumulation process proposed for undoped silicon by us [2]. The F-accumulation process is related to the migration process of the adsorbed-F atoms to other Si atoms for the thermal etching reaction of silicon and explains the fluorosilyl surface layer experimentally observed [11] under the etching reaction, i.e. the adsorbed-fluorine atoms tend to accumulate locally in the bulk silicon, and the formation of the volatile main etching product SiF_4 is somewhat suppressed.

HYPERVALENCY OF SILICON ON ETCHING REACTION OF UNDOPED AND n-DOPED SILICON

The $2t_{2z}$ orbital is responsible for the hypervalency of silicon compounds [15a], which explains the etching mechanism proposed for undoped silicon by Garrison and Goddard [9] and proposed for n-doped silicon by us [4]. A pentavalent anion $FH_3SiSiH_3^-$ is a stable complex between F and the anion center $H_3SiSiH_3^-$ with exothermicity of 130.2 kcal/mol at the MP3/6-31 + G**//HF/3-21G$^{(*)}$ level [4]:

$$F + H_3SiSiH_3^- \rightarrow FH_3SiSiH_3^- \tag{1}$$

The optimized geometry of the $FH_3SiSiH_3^-$ at the HF/3-21G$^{(*)}$ level is a trigonal-bipyramid structure with apical ligands, F and SiH_3, as shown in Fig. 4. The newly formed anion center $FH_3SiSiH_3^-$ is also stable against the decomposition into F^- and H_3SiSiH_3 by endothermicity of 36.8 kcal/mol at the MP3/6-31 + G**//HF/3-21G$^{(*)}$ level [4]:

$$FH_3SiSiH_3^- \rightarrow F^- + H_3SiSiH_3 \tag{2}$$

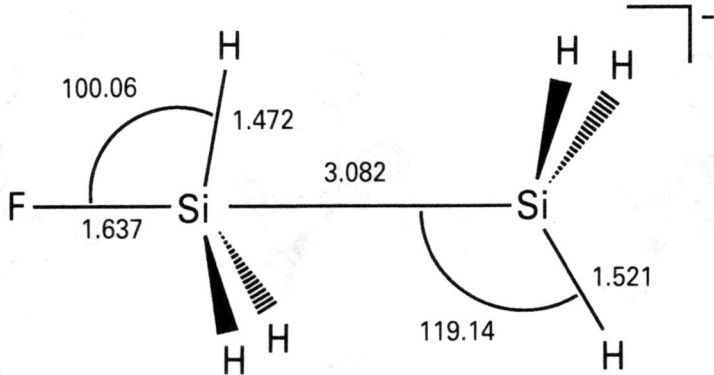

Fig. 4 — Optimized geometry of $FH_3SiSiH_3^-$

The Si—Si bond length is 0.441 Å longer than that of $H_3SiSiH_3^-$ (2.641 Å). $FH_3SiSiH_3^-$ has all positive vibrational frequencies, but two of them are fairly small values. They are an SiH_3 rotational mode (A_2 symmetry, 70 cm^{-1}) and a dissociation mode into $FSiH_3$ and SiH_3^- (A_1 symmetry, 74 cm^{-1}). Fig. 5 shows the MO level diagram of $FH_3SiSiH_3^-$ at the HF/3-21G$^{(*)}$ level. The MO of the transition state of the undoped etching reaction, FH_3SiSiH_3, is obtained by taking one electron from the HOMO of $FH_3SiSiH_3^-$. The HOMO of $FH_3SiSiH_3^-$ is an Si—Si bonding orbital. Therefore, the Si—Si bond in FH_3SiSiH_3 is weakened, and this is the reason why FH_3SiSiH_3 is the transition state. Furthermore, this is confirmed by using Bader-Pearson's perturbation theory [17], i.e. the transition density between singly occupied MO (SOMO) and LUMO of FH_3SiSiH_3 gives the symmetry corresponding to the A_1 Si—Si bond-breaking mode as shown in Fig. 6. Since $FH_3SiSiH_3^-$ is at the energy minimum, FH_3SiSiH_3 should have a lower activation barrier. Indeed, Garrison and Goddard [9] recently applied the GVB and dissociation-consistent configuration-interaction method for the reaction between the F radical and H_3SiSiH_3 as a plasma-etching reaction model for undoped silicon in which an F radical attacks the Si—Si bond from the rear:

$$F + H_3SiSiH_3 \rightarrow FSiH_3 + SiH_3 \tag{3}$$

They obtained an exothermicity of 82.6 kcal/mol, and a very low activation barrier of 6.2 kcal/mol, as expected from the analysis given above. The activation barrier was in good agreement with the experimental value (2.48 kcal/mol) [5]. These results indicated that Si—Si bond cleavage by the F radical occurs spontaneously to form an F—Si bond and a dangling bond. The dangling bond may then serve as a reactive center for further attack by the F atom.

CHEMICAL MECHANISM OF ETCHING REACTION FOR DOPED SILICON

In the etching reaction of n-doped silicon, the F radical may further react with the pentavalent anion center $FH_3SiSiH_3^-$. The insertion into the Si—Si bond of $FH_3SiSiH_3^-$ is most favourable, because of their SOMO-LUMO in-phase overlap as

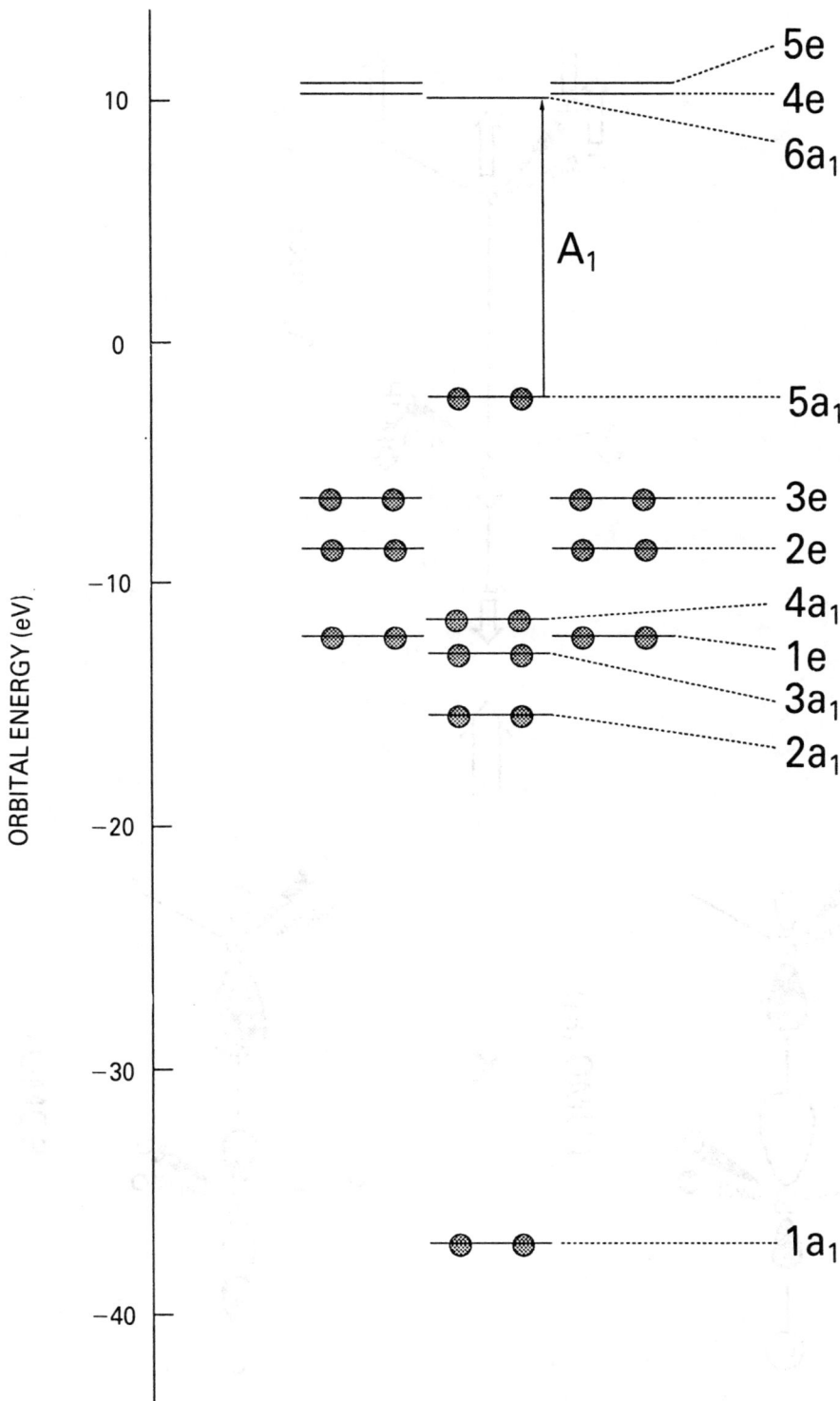

Fig. 5 — MO energy level diagram of $FH_3SiSiH_3^-$ at the RHF/3–21G$^{(*)}$ level.

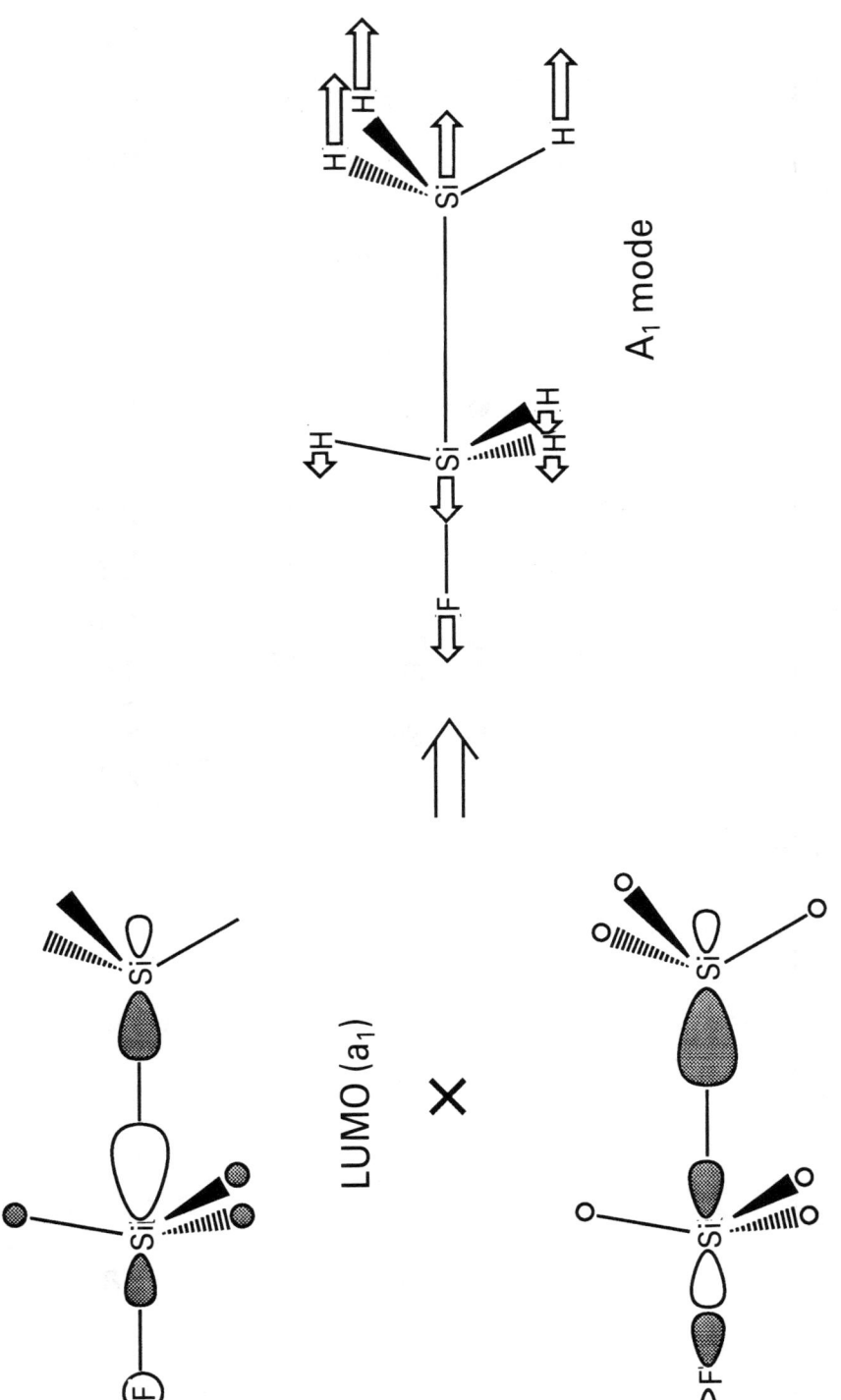

Fig. 6 — SOMO and LUMO and A_1 vibrational mode of FH_3SiSiH_3.

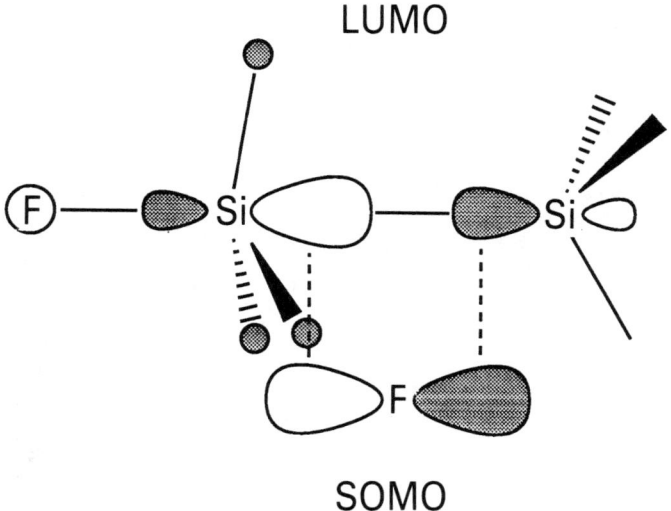

Fig. 7 — SOMO–LUMO interaction between F and $FH_3SiSiH_3^-$.

shown in Fig. 7. This leads to a doubly fluorinated pentavalent anion center and a dangling bond with exothermicity of -72.4 kcal/mol at the MP3/6-31 + $G^{**}//HF/3$-21$G^{(*)}$ level [4]:

$$F + FH_3SiSiH_3^- \rightarrow F_2SiH_3^- + SiH_3 \qquad (4)$$

The dangling bond serves as a reaction center for further attack by the F atom.

A schematic energy profile and a schematic illustration of the reactions between F and anion centers are shown in Figs. 8 and 9, respectively. At the initial stage of the reaction, an F-attached pentavalent Si anion center is formed. The apical Si—Si bond of the F-attached pentavalent Si anion center is easy for F-insertion and cleavage. The F-attached pentavalent Si anion center is a reaction intermediate which accelerates the etching reaction. When an F-attached pentavalent Si anion center is formed, the etching reaction proceeds spontaneously until F_4SiH^- is formed. Then F_4SiH^- is decomposed into volatile SiF_4 and H^- (serving as a tiny model of a trivalent anion center).

In order to compare the chemical mechanisms of the etching reactions for undoped, n-doped, and p-doped silicon, theoretical results are summarized in Fig. 10. This figure shows the schematic energy profile for the reaction between F and $H_3SiSiH_3^q$ ($q = 0$, -1, and $+1$ for undoped, n-doped, and p-doped silicon, respectively). While the undoped silicon etching reaction has a transition state [9], the n-doped and p-doped silicon etching reaction has no transition state. In addition, the exothermicities of the heat of reaction for the n-doped and p-doped silicon etching are both larger than that for undoped silicon etching. As a result, the Si—Si bond of doped silicon is easily broken by the F atom.

For the etching reaction of p-doped silicon, controversial experimental results [10] are known, i.e. its thick fluorinated surface layer and suppressed etch rate. Contrary to the conventional interpretation of these experimental results that the reactivity of F with p-doped silicon is low, we can state from Fig. 10 that there exists a

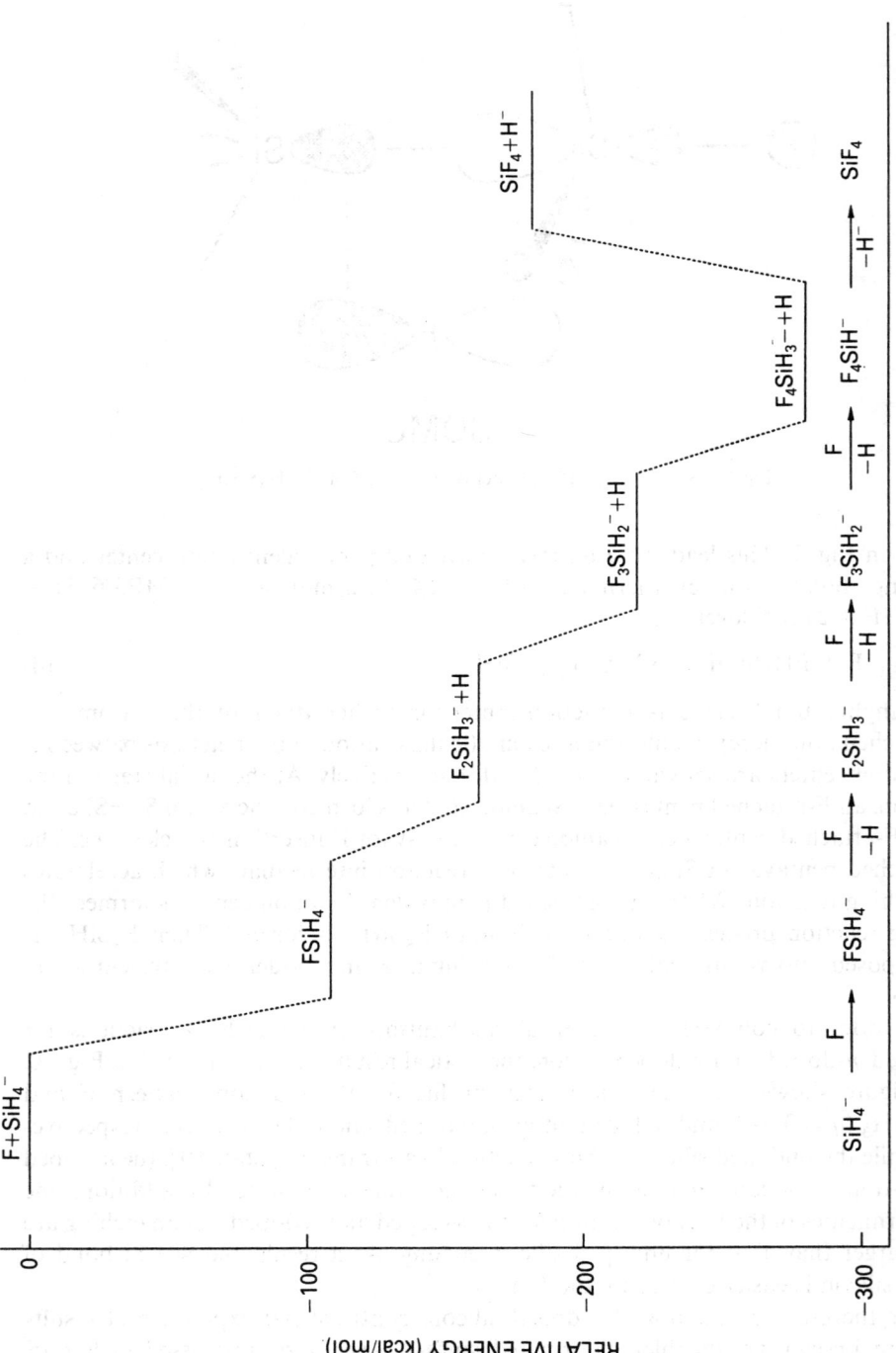

Fig. 8 — Energy profile of reaction between F and SiH_4^-.

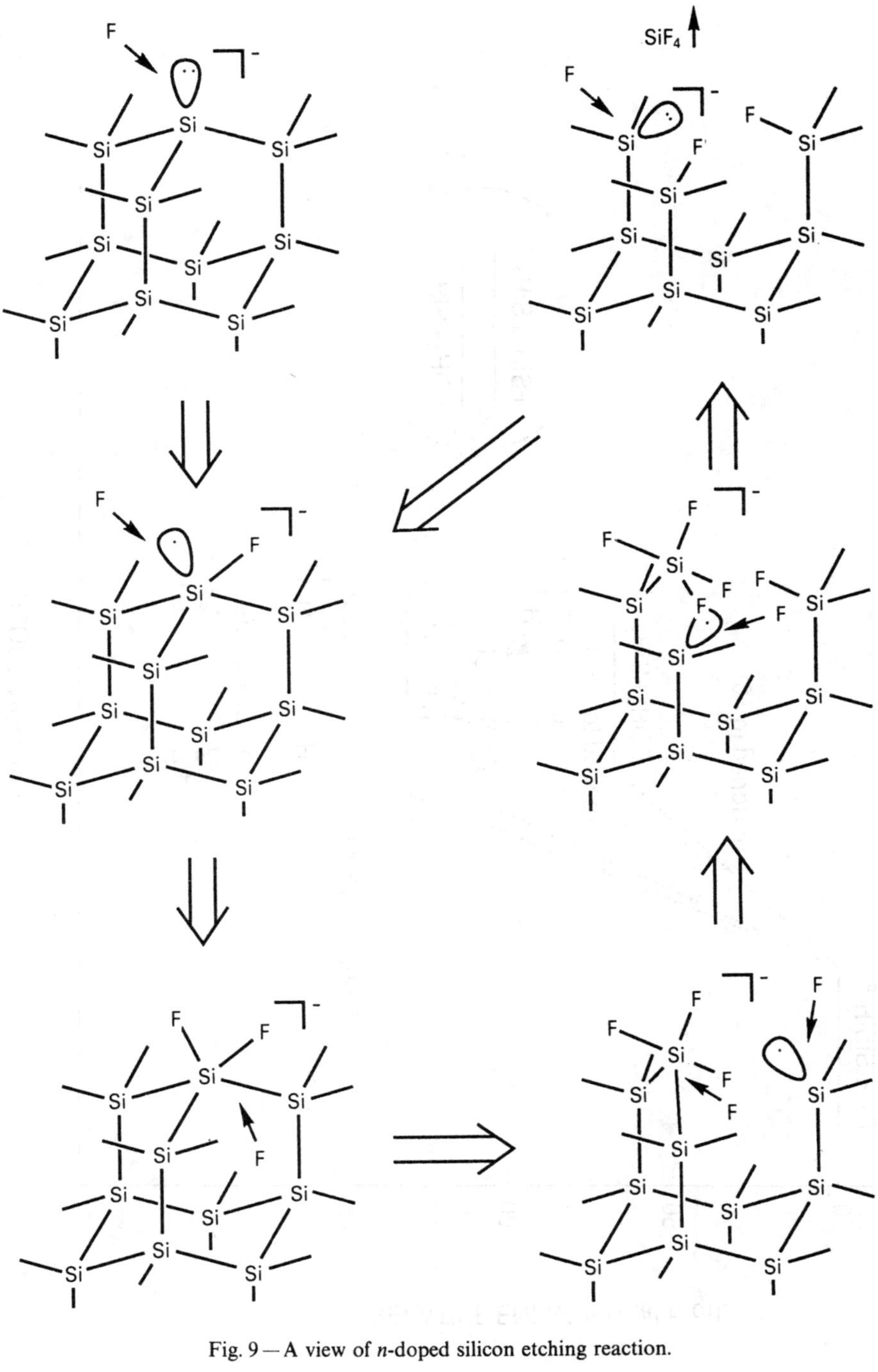

Fig. 9 — A view of n-doped silicon etching reaction.

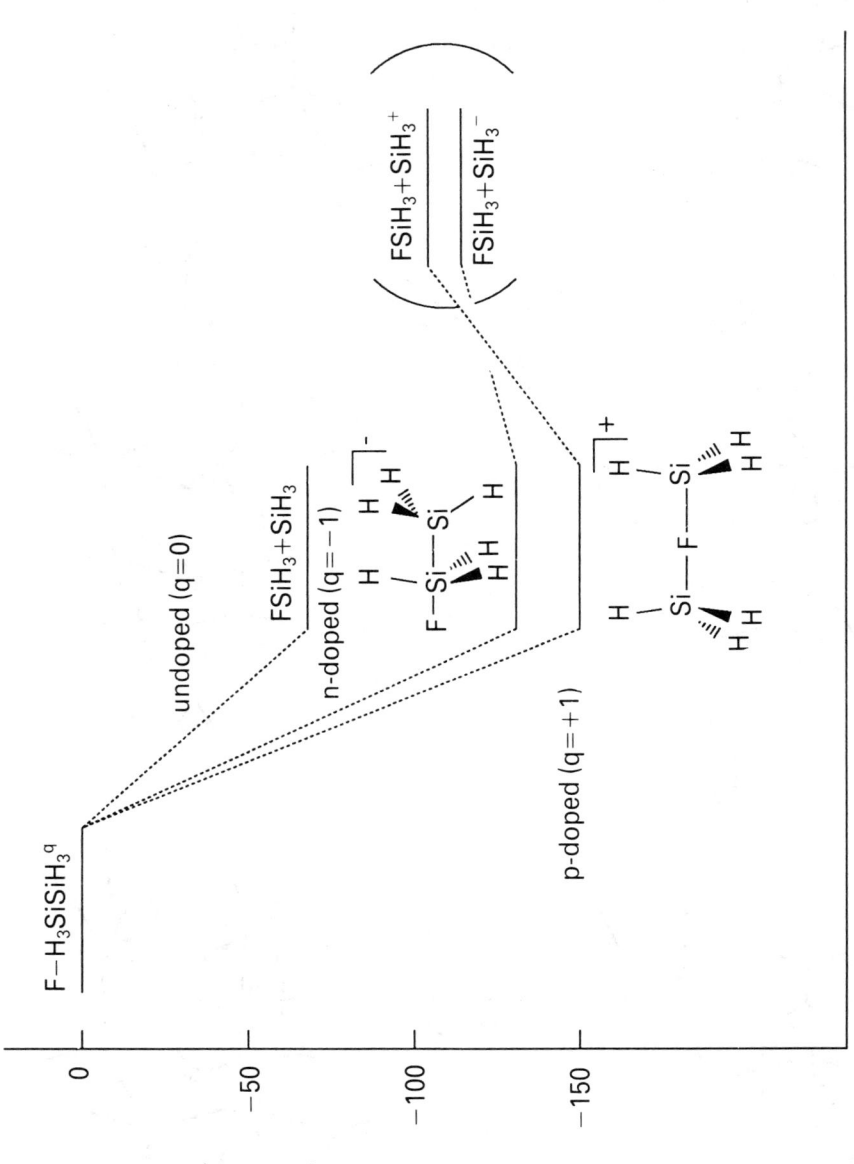

REACTION COORDINATE

Fig. 10 — Energy profile of the silicon etching reaction for undoped ($q = 0$), p-doped ($q = +1$), and n-doped ($q = 1$) silicon.

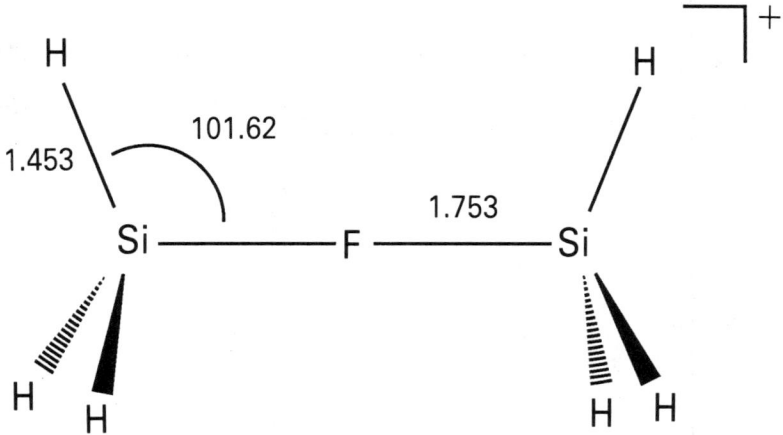

Fig. 11 — Optimized geometry of $H_3Si\text{—}F\text{—}SiH_3^+$.

high reactivity of F with p-doped silicon toward $H_3Si\text{—}F\text{—}SiH_3^+$ complex formation [3]. The reaction between F and $H_3SiSiH_3^+$ is highly exothermic (103.9 kcal/mol at the CISD + QC/6-31G**//HF/3-21G$^{(*)}$ level) to form $FSiH_3$ and SiH_3^+, and then $FSiH_3$ and SiH_3^+ can further react together to form an $H_3Si\text{—}F\text{—}SiH_3^+$ complex with exothermicity of 44.9 kcal/mol at the CISD + QC/6-31G**//HF/3-21G$^{(*)}$ level [3]:

$$F + H_3SiSiH_3^+ \rightarrow FSiH_3 + SiH_3^+ \tag{5}$$

$$\rightarrow H_3Si\text{—}F\text{—}SiH_3^+ \tag{6}$$

Fig. 11 shows the optimized geometry of *eclipsed* $H_3Si\text{—}F\text{—}SiH_3^+$ at the HF/3-21G$^{(*)}$ level. The two equivalent interacting Si—F bond lengths are 1.753 Å in both *staggered* and *eclipsed* forms, and are longer than the normal Si—F distance (1.593 Å in $FSiH_3$). *Staggered* $H_3Si\text{—}F\text{—}SiH_3^+$ has one small imaginary frequency for conversion to the stable *eclipsed* form. Nevertheless, there is no energy difference between *staggered* and *eclipsed* $H_3Si\text{—}F\text{—}SiH_3^+$. The SiH_3 groups in $H_3Si\text{—}F\text{—}SiH_3^+$ are almost free rotators.

The stability of *eclipsed* $H_3Si\text{—}F\text{—}SiH_3^+$ contrary to neutral $H_3Si\text{—}F\text{—}SiH_3$ can be understood from Bader–Pearson's perturbation theory [17]. Figs. 12 and 13 show the MO energy level diagrams of *eclipsed* $H_3Si\text{—}F\text{—}SiH_3^+$ and *eclipsed* $H_3Si\text{—}F\text{—}SiH_3$ at the HF/3-21G$^{(*)}$ level, respectively. The lowest energy transition of neutral $H_3Si\text{—}F\text{—}SiH_3$ is from SOMO $3a_1'$ to LUMO $3a_2''$. This gives a transition density of A_2'' symmetry. The SOMO and LUMO for neutral $H_3Si\text{—}F\text{—}SiH_3$ are shown in Fig. 14. The A_2'' transition density corresponds to decomposition into SiH_3 and SiH_4. This is consistent with the result of vibrational analysis from which neutral $H_3Si\text{—}F\text{—}SiH_3$ has one imaginary A_2'' frequency. The A_2'' normal mode is also shown in Fig. 14. For $H_3Si\text{—}F\text{—}SiH_3^+$, the corresponding A_2'' transition energy is not the lowest, as shown in Fig. 12. Furthermore, this is supported by their structural difference. The two equivalent Si—F bonds (1.855 Å) of *eclipsed* $H_3Si\text{—}F\text{—}SiH_3$ are longer than those of $H_3Si\text{—}F\text{—}SiH_3^+$ (1.753 Å).

Fluorine-attached silicon ($FSiH_3$) can also attack the cation center ($XSiH_3^+$, X = H, SiH_3) as well as the tri-coordinated Si cation (SiH_3^+). After various reactions

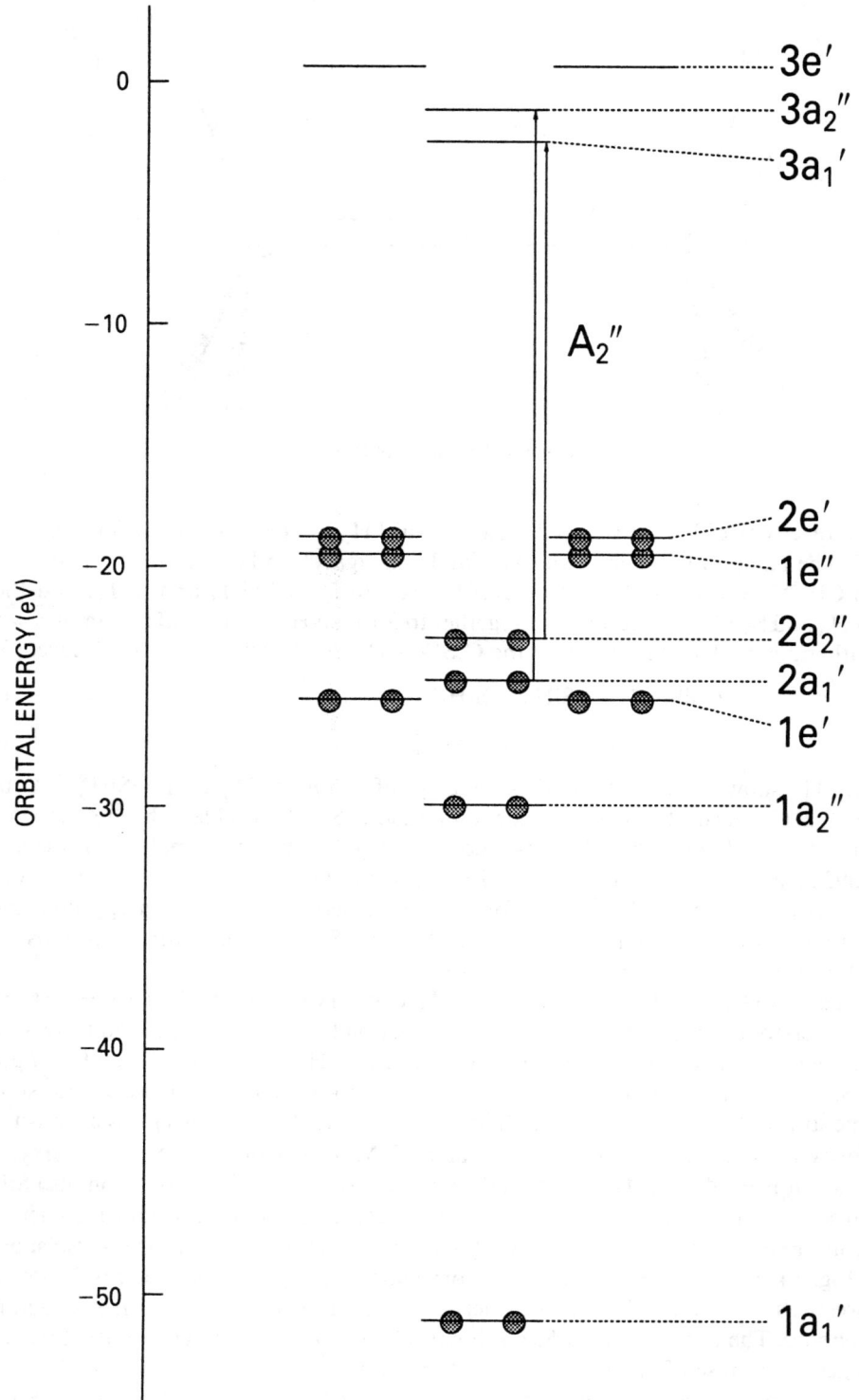

Fig. 12 — MO energy level diagrams of H_3Si—F—SiH_3^+ at the RHF/3–21G$^{(*)}$ level.

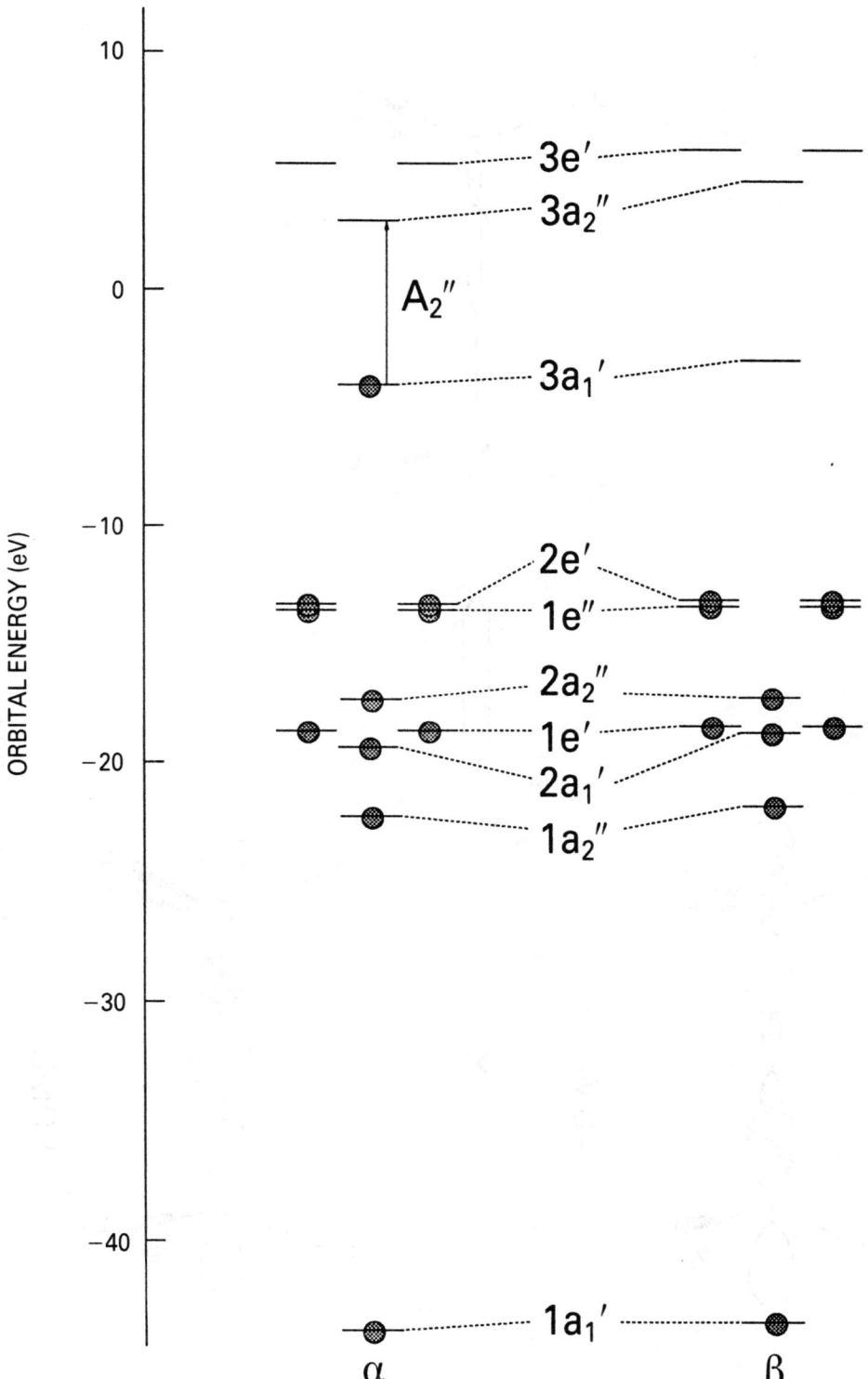

Fig. 13 — MO energy level diagram of H_3Si—F—SiH_3 at the UHF/3–21G$^{(*)}$ level.

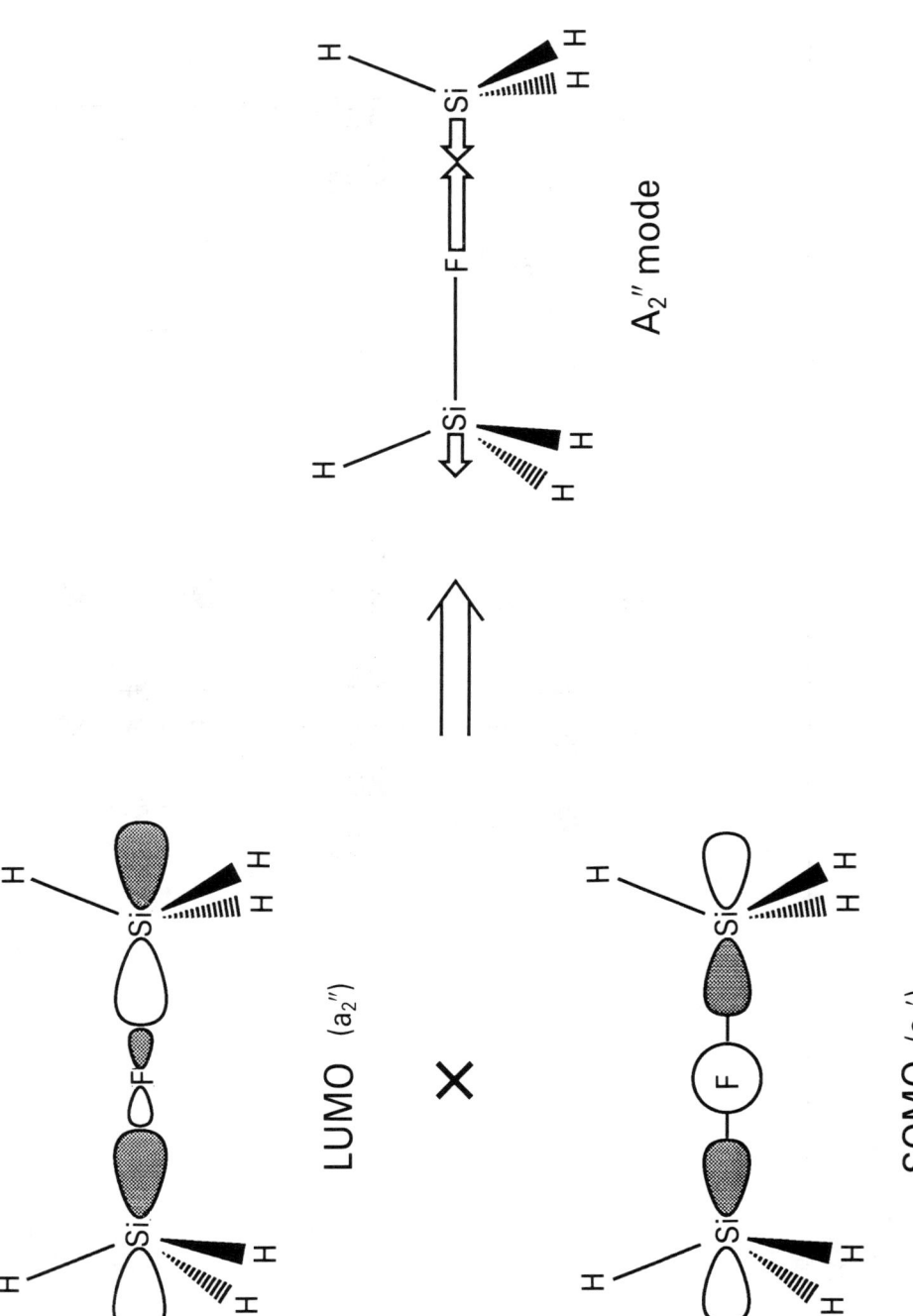

Fig. 14 — SOMO and LUMO and A_z'' vibrational mode of H_3Si—F—SiH_3.

were considered, we found the novel reaction between the cation center, $XSiH_3^+$, and $FSiH_3$ proceeds without barrier and an exothermicity of 43.9 kcal/mol for X = H and 11.6 kcal/mol for X = SiH_3 at the CISD + QC/6-31G**//HF/3-21G$^{(*)}$ level [3]:

$$XSiH_3^+ + FSiH_3 \rightarrow X\cdot + eclipsed\ H_3Si—F—SiH_3^+ \qquad (7)$$

In this reaction, the fluorine-bridged cation center and radical X· are produced. This reaction is different from previous F-migration models for undoped silicon [2], in which fluorine is migrated from fluorine-attached silicon to another silicon.

From these results, we can illustrate the following picture for the p-doping effect on plasma etching as shown in Fig. 15. At the initial stage for the p-doped silicon etching, Si—Si bond cleavage by the fluorine radical occurs more easily than that for undoped silicon etching, but subsequently the product, F-attached Si and tri-coordinated Si cation, may be recombined to form a fluorine-bridged cation center (Si—F—Si$^+$). The adsorbed-F atom also reacts with the cation center to form a stable fluorine-bridged cation center (Si—F—Si$^+$) and a dangling bond. The dangling bond may then serve as a reactive center for further attack by the F atom. As a result, a thick fluorinated layer is formed and the etch rate is suppressed for heavily p-doped silicon.

CONCLUDING REMARKS

The chemical mechanisms [2–4] of the etching reaction of silicon by fluorine are consistent with experimental results [5, 10, 11]. Although our models are rather small, it is clear that plasma etching is goverened mainly by chemical reactivity of silicon rather than by a physical mechanism. Such chemical reactions are highly localized in nature and all the essential aspects of the reactions can be modeled with fairly small compounds where accurate calculations can be performed [18].

ACKNOWLEDGMENTS

This work was supported by a Grant-in-Aid for Scientific Research from the Ministry of Education, Science and Culture of Japan. The molecular orbital calculations were carried out at the Data Processing Center of Kyoto University and the Computer Center of the Institute for Molecular Science (IMS), and we thank them for their generous permission to use the FACOM M-780 and VP-400, and HITAC M-680H and S-820 computer systems, respectively.

REFERENCES

[1] See, for example, W.J. Hehre, L. Radom, P.v.R. Schleyer and J.A. Pople, in *Ab Initio Molecular Orbital Theory*, Wiley, New York, 1986.

[2] A. Tachibana, Y. Kurosaki, S. Kawauchi and T. Yamabe, *J. Phys. Chem.* (1991), **95**, 1716.

[3] A. Tachibana, S. Kawauchi and T. Yamabe, *J. Phys. Chem.*, (1991), **95**, 2471.

[4] S. Kawauchi, A. Tachibana and T. Yamabe, *J. Phys. Chem.*, in press.

[5] D.L. Flamm, in *Silicon Chemistry*, Eds J.Y. Corey, E.R. Corey and P.P. Gaspar, Ellis Horwood, Chichester, 1988.

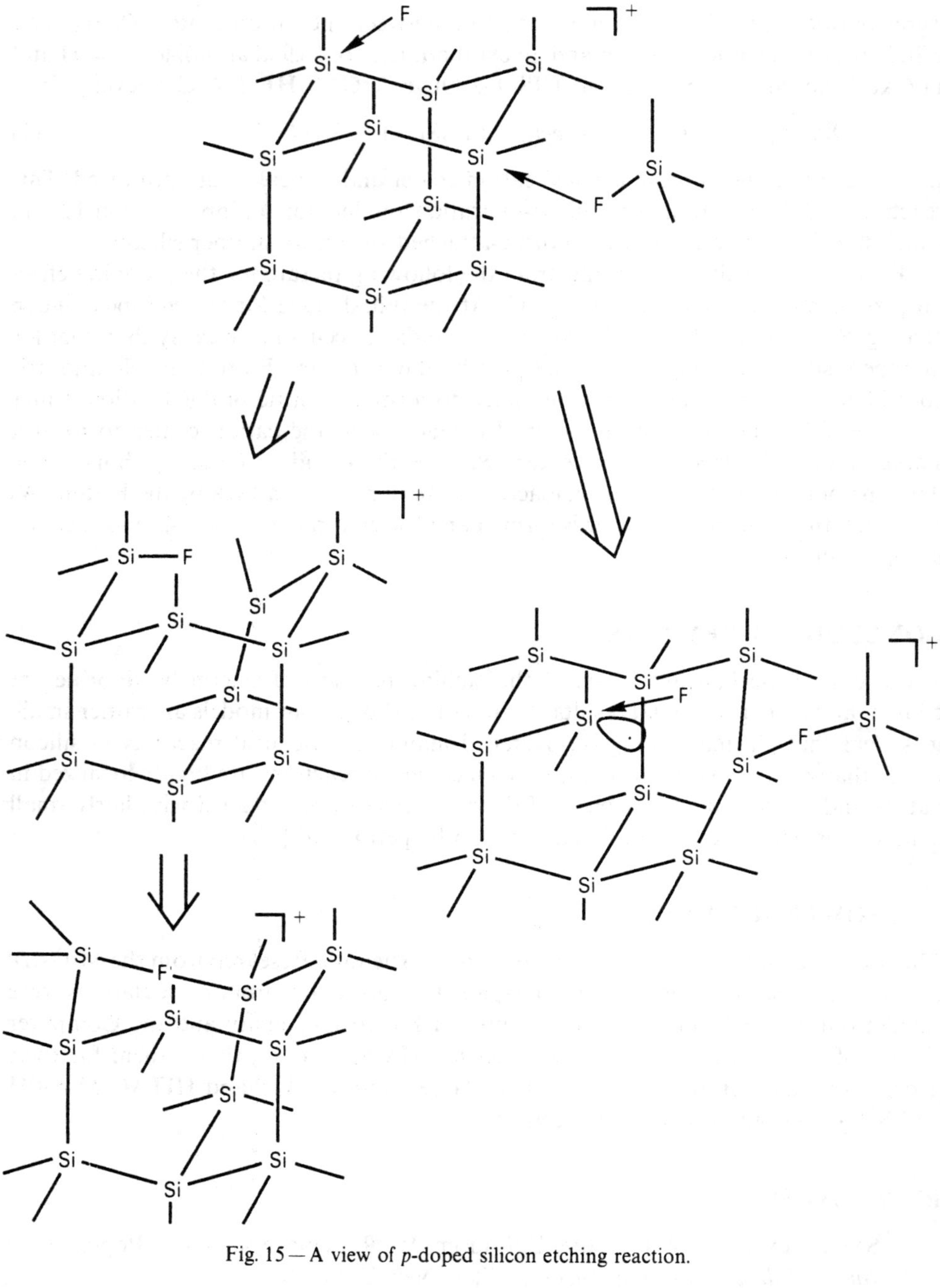

Fig. 15 — A view of p-doped silicon etching reaction.

[6] M. Seel and P.S. Bagus, *Phys. Rev. B.* (1983), **28**, 2023.

[7] (a) C.G. Van de Walle, Y. Bar-Yam, F.R. McFeely and S.T. Pantelides, *J. Vac. Sci. Technol., A* (1988), **6**, 1973. (b) C.G. Van de Walle, F.R. McFeely and S.T. Pantelides, *Phys. Rev. Lett.* (1988), **61**, 1867.

[8] B.J. Garrison and W.A. Goddard III, *J. Chem. Phys.* (1987), **87**, 1307.

[9] B.J. Garrison and W.A. Goddard III, *Phys. Rev. B* (1987), **36**, 9805.

[10] (a) K. Jinno, H. Kinoshita and Y. Matsumoto, *J. Electrochem. Soc.* (1978), **125**, 827. (b) C.J. Mogab and J. Levinstein, *J. Vac. Sci. Technol.* (1980), **17**, 721. (c) T. Makino, H. Nakamura and M. Asano, *J. Electrochem. Soc.* (1981) **128**, 103. (d) G.C. Schwartz and P.M. Schaible, *J. Electrochem. Soc.* (1983), **130**, 1898. (e) Y.H. Lee, M.-M. Chen and A.A. Bright, *Appl. Phys. Lett.* (1985), **46**, 260. (f) E. Ikawa and Y. Kurogi, *Nucl. Instrum. Methods B* (1985), **7/8**, 820. (g) L. Baldi and D. Beardo, *J. Appl. Phys.* (1985), **57**, 2221. (h) Y.H. Lee and M.-M. Chen, *J. Vac. Sci. Technol., B* (1986) **4**, 468. (i) L. Baldi and D. Beardo, *J. Electrochem. Soc.* (1986), **133**, 2202. (j) F.A. Houle, *J. Appl. Phys.* (1986), **60**, 3018(1986). (k) H.F. Winters and D. Haarer, *Phys. Rev. B* (1987), **36**, 6613. (l) J.A. Yarmoff and F.R. McFeely, *Phys. Rev. B.* (1988), **38**, 2057.

[11] (a) H.F. Winters, *J. Appl. Phys.* (1978), **49**, 5165. (b) V.M. Donnelly and D.L. Flamm, *J. Appl. Phys.* (1980), **51**, 5273. (c) D.L. Flamm, V.M. Donnelly and J.A. Mucha, *J. Appl. Phys.* (1981), **52**, 3633. (d) H.F. Winters and J.W. Coburn, *J. Vac. Sci. Technol., B* (1985), **3**, 1376. (e) J.A. Dagata, D.W. Squire, C.S. Dulcey, D.S.Y. Hsu and M.C. Lin, *Chem. Phys. Lett.* (1987), **134**, 151. (f) D.W. Squire, J.A. Dagata, D.S.Y. Hsu, C.S. Dulcey and M.C. Lin, *J. Phys. Chem.* (1988), **92**, 2827. (g) T.J. Chung, *J. Appl. Phys.* (1980), **51**, 2614. (h) H.F. Winters and F.A. Houle, *J. Appl. Phys.* (1983), **54**, 1218. (i) D.E. Ibbotson, D.L. Flamm, J.A. Mucha and V.M. Donnelly, *Appl. Phys. Lett.* (1984), **44**, 1129. (j) F.R. McFeely, J.F. Morar and F.J. Himpsel, *Surf. Sci.* (1986), **165**, 277. (k) F.A. Houle, *J. Chem Phys.* (1987), **87**, 1866. (l) J.A. Dagata, D.W. Squire, C.S. Dulcey, D.S.Y. Hsu and M.C. Lin, *J. Vac. Sci. Technol., B* (1987), **5**, 1495.

[12] J. Sauer, *Chem. Rev.* (1989), **89**, 199.

[13] R. Dovesi, M. Causa and G. Angonoa, *Phys. Rev. B: Condens. Matter* (1981), **24**, 4177.

[14] D.S. Horowitz and W.A. Goddard III, *J. Mol. Struct. (Theochem.)* (1988), **163**, 207.

[15] (a) A. Tachibana, S. Kawauchi and T. Yamabe, *Rev. Heteroatom. Chem.*, in press. (b) Y. Apeloig, in *The Chemistry of Organic Silicon Compounds*, Eds S. Patai and Z. Rappoport, Wiley, New York, 1989; Vol. 1. Chapter 2.

[16] T.A. Albright, J.K. Burdett and M.H. Whangbo, *Orbital Interactions in Chemistry*, Wiley, New York, 1985.

[17] R.G. Pearson, *Symmetry Rules for Chemical Reactions*, Wiley, New York, 1976.

[18] G.W. Trucks, K. Raghavachari, G.S. Higashi and Y.J. Chabal, *Phys. Rev. Lett.* (1990), **65**, 504.

21

Systematics of electron-rich polyhedral molecules

Paul D. Lyne and D. Michael P. Mingos,
University of Oxford, Oxford, UK

INTRODUCTION

The polyhedral skeletal electron pair theory developed primarily by Williams, Wade, Mingos and Rudolph [1–4] nearly 20 years ago has proved to be extraordinarily useful for describing the relationships between the polyhedral geometries of cluster compounds and the total number of valence electrons associated with the cluster (the polyhedral electron count, p.e.c.). It has been widely applied to metal clusters [5], main group polyhedral molecules [3b] and 'inorganometallic' clusters [6] containing both main group and transition metal atoms at the vertices. It has also been extended to capped and condensed clusters [7,8] and shown to be applicable to high-nuclearity clusters [9]. The theoretical basis of the polyhedral skeletal electron pair theory has been underpinned by numerous molecular orbital calculations on specific molecules [10] and more generally by Stone's tensor surface harmonic theory [11].

In one of the initial papers developing the relationship between polyhedral structure and the valence electron count [3] it was proposed that there is a class of electron-rich clusters whose structures were related to three-connected polyhedral molecules by the scission of an edge bond for each electron pair in excess of that required for a three-connected polyhedral molecule. A series of molecules derived from the simplest three-connected polyhedral molecules which illustrates this principle is depicted in Fig. 1. The related electron-rich molecules are derived from the polyhedra by the successive cleavage of non-adjacent bonds of the polyhedra. The last member of the series is a ring compound with a total of $6n$ valence electrons.

As the nuclearity of the three-connected parent polyhedral molecule increases, the number of molecules which may be derived by such bond cleavage processes ($n/2$) increases, and the choice of which bonds are cleaved becomes an important one. Furthermore, when heteroatoms are introduced into the cluster, questions arise about the preference for bond-breaking processes and whether such effects are the result of electronegativity or overlap considerations. This chapter attempts to address these questions in a systematic fashion. As with previous papers dealing with geometric questions of this type, the arguments will be developed from semi-empirical molecular orbital calculations [12] and analysed using perturbation theory arguments.

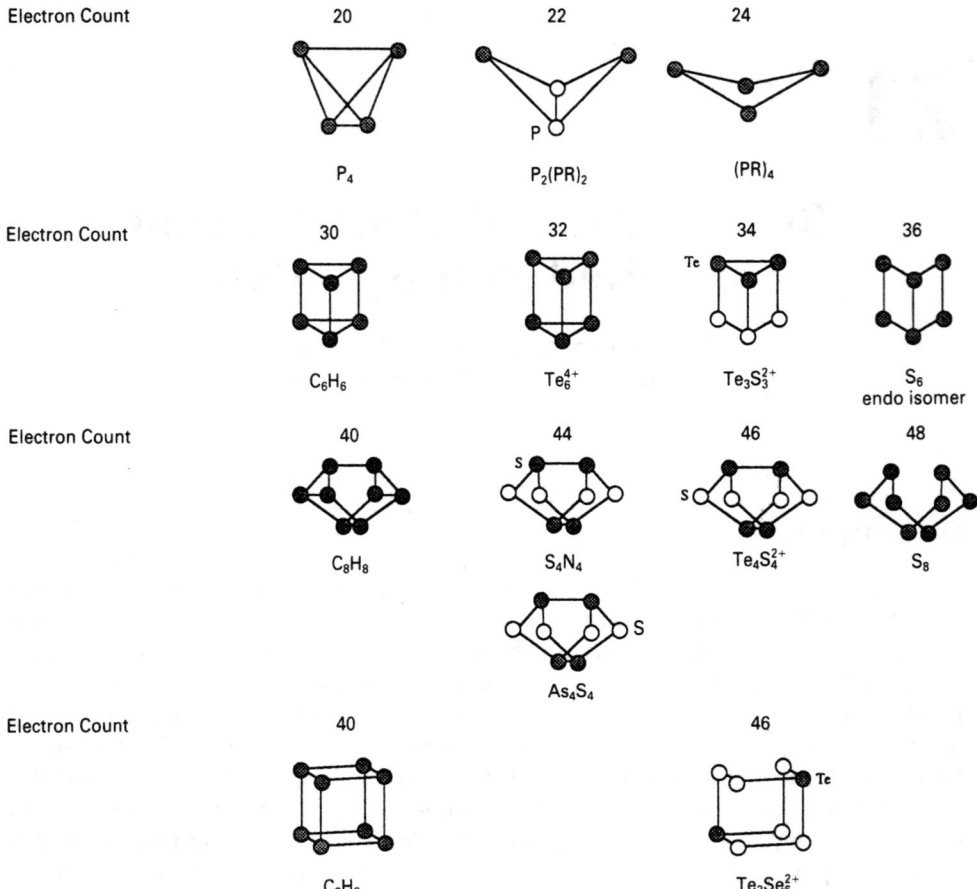

Electron Count   20   22   24

P₄   P₂(PR)₂   (PR)₄

Electron Count   30   32   34   36

C₆H₆   Te₆⁴⁺   Te₃S₃²⁺   S₆
                                                endo isomer

Electron Count   40   44   46   48

C₈H₈   S₄N₄   Te₄S₄²⁺   S₈

As₄S₄

Electron Count   40   46

C₈H₈   Te₂Se₆²⁺

Fig. 1. — Electron-rich molecules with structures that are intermediate between a cage and a ring.

ELECTRON-RICH CLUSTERS BASED ON THE TETRAHEDRON

The molecular orbital description for tetrahedral P_4 is well established [13] and the molecule has also been the subject of photoelectron spectral studies [14]. The highest occupied molecular orbitals have the symmetries a_1, t_2 and e. The e molecular orbitals are non-bonding as far as the skeleton is concerned, and in Stone's tensor surface harmonic theory are designated [15] as $(D^\pi/\bar{D}^\pi)_{\pm 1}$. The a_1 and t_2 molecular orbitals match in symmetry the lower lying a_1 and t_2 bonding skeletal molecular orbitals, and correspond approximately to the 'lone pair' orbitals on phosphorus. These molecular orbitals are illustrated schematically in Fig. 2.

The lowest unoccupied molecular orbitals of P_4 have t_1 symmetry and are skeletal antibonding. These molecular orbitals are also illustrated in Fig. 2. The influence of alternative distortion modes on the energies of these frontier orbitals can be seen in Fig. 2. Specifically, the distortions correspond to the lengthening of either one edge leading to a butterfly geometry, the lengthening of two opposite edges yielding a

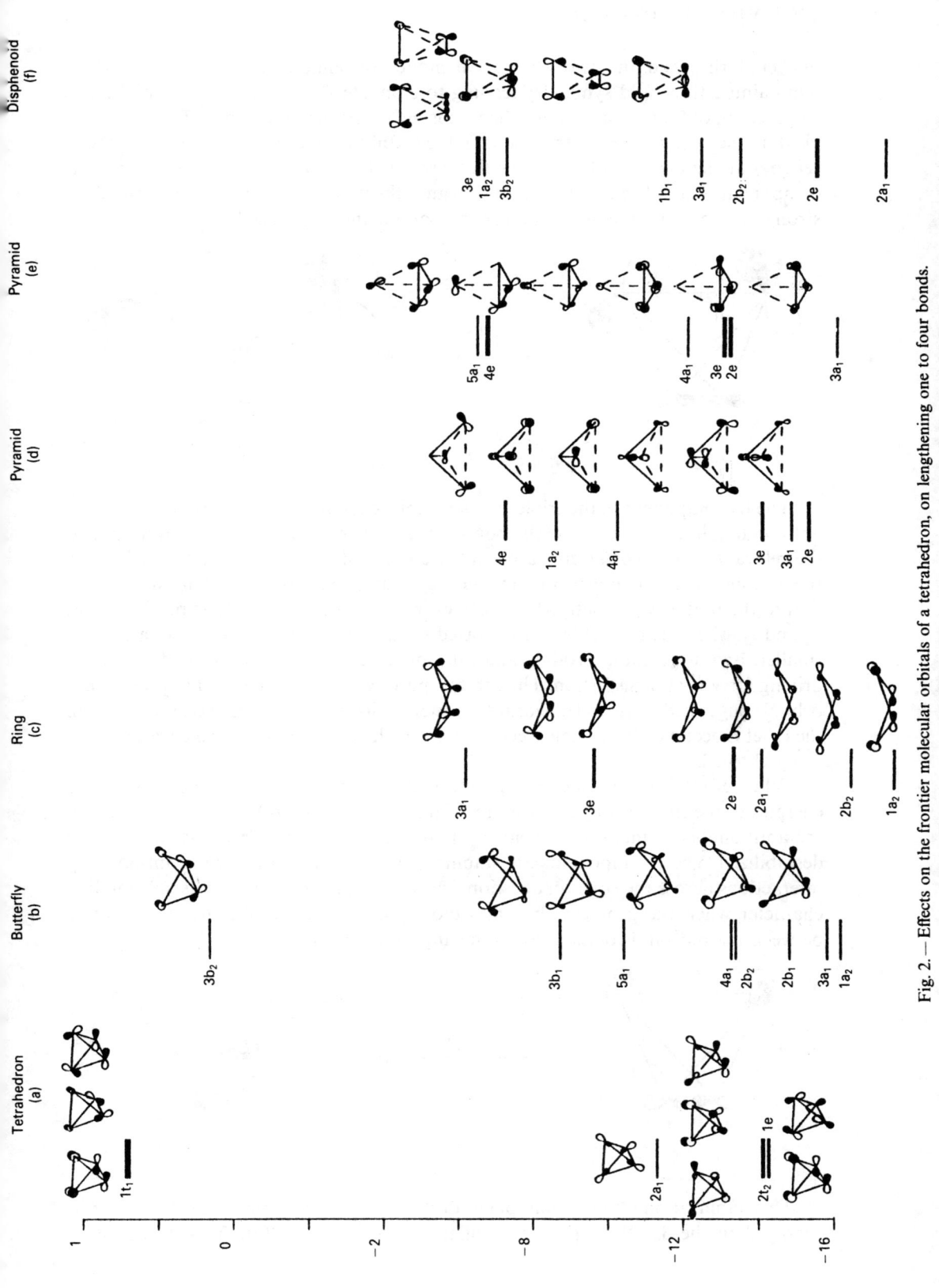

Fig. 2. — Effects on the frontier molecular orbitals of a tetrahedron, on lengthening one to four bonds.

puckered ring and the two alternative modes of lengthening three edges whilst maintaining three-fold symmetry leading to elongated or contracted pyramids. The lengthening of four edges to a disphenoid is also considered in Fig. 2. The distortion yielding the butterfly geometry results in the stabilization of one component of the t_1 set (b_1) at the expense of the other two orbitals (a_2, b_1), as can be seen in **1**. This component is stabilized preferentially because there is a substantial reduction of the strong antibonding interaction across the bond being lengthened.

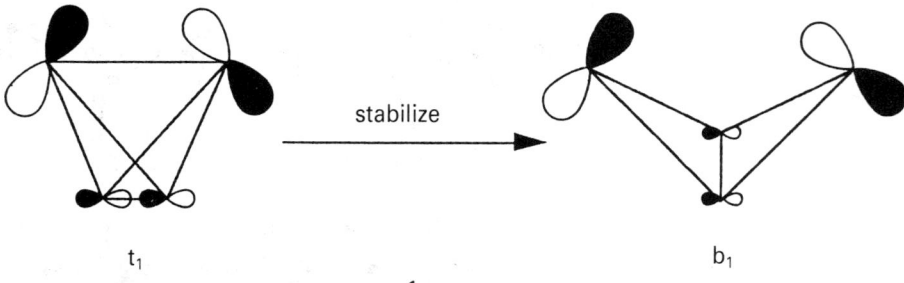

1

The b_2 component of the triplet is also stabilized, but to a much lesser extent. The antibonding interaction across the bond being broken involves a p_π–p_π overlap, and the relevant atomic coefficients are smaller compared to the b_1 component. Finally, the a_2 component undergoes minimal energy change on distortion from the tetrahedral to the butterfly geometry. The changes in the energies of the 'lone pair' orbitals, a_1 and t_2, when the tetrahedron is distorted towards the butterfly geometry are much smaller, and since these orbitals and fully occupied they do not provide a strong driving force for the distortion. The orbital splittings associated with the t_1 molecular orbitals suggest that the tetrahedral-to-butterfly distortion is most favourable when the t_1 set is occupied by a single electron pair, i.e. the total number of valence electrons is equal to 22.

When two opposite bonds of the tetrahedron are lengthened, the effects on the energies of the frontier orbitals are those illustrated in Fig. 2(c). Two components of the antibonding t_1 molecular orbitals (e) are greatly stabilized and the third (a_2) is destabilized. The e components are stabilized because they have strong antibonding interactions across the edges in question. The a_2 component increases its antibonding character when the geometry becomes more planar, because the overlap integrals between the p orbitals of adjacent atoms increase (see **2**).

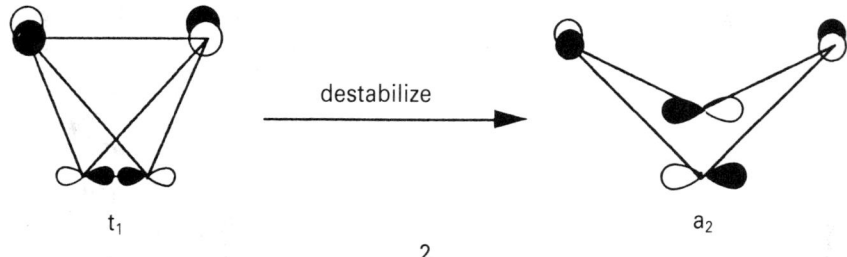

2

The stabilization of the e component of the set is so large that it now lies lower in energy than the $3a_1$ 'lone pair' molecular orbital (see Fig. 2(c)). This distortion is favoured when four electrons ocupy the t_1 set and the cluster has a total of 24 valence electrons.

There are two possible ways to lengthen three bonds of a tetrahedron while maintaining three-fold symmetry; they produce either an elongated or a contracted pyramid.

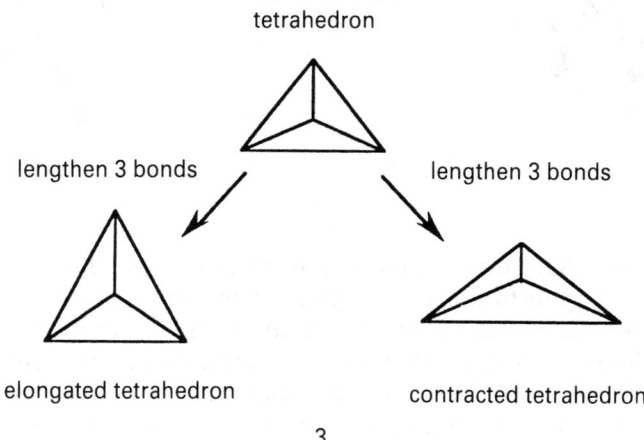

tetrahedron

lengthen 3 bonds   lengthen 3 bonds

elongated tetrahedron   contracted tetrahedron

3

The effects in each case on the frontier molecular orbitals are shown in Figs. 2(d) and (e). Elongation of three bonds of the tetrahedron results in a large stabilization of the energies of the e set of the t_1 level. The remaining component (a_2) undergoes a small destabilization in comparison to the former set of orbitals. The a_1 level from the higher lying t_2 set has been stabilized on elongation of the tetrahedron, resulting in three orbitals of similar energy in the frontier region. Contraction of the tetrahedron results in stabilization of the energies of the e set from the antibonding t_1 level. However, now the a_2 level has also been stabilized by the geometrical distortion — indeed, to a greater extent than the e levels. Both modes of lengthening three bonds produce an additional set of three orbitals in the frontier region, and the distortion is favoured by fully occupying the antibonding t_1 level, i.e. when the valence electron count is 26.

Finally, lengthening four bonds of the tetrahedron to produce the disphenoid structure has the following effects on the frontier orbitals of the tetrahedron (Fig.2(f)). All the components of the t_1 antibonding level, a_2 and e are strongly stabilized by lengthening four edges. As before, the origin of stabilization is the reduction of strong out-of-phase interactions along the appropriate edges. Lengthening of four edges brings into play the highest lying set of molecular orbitals, the t_2 set. The b_2 level of this set undergoes a dramatic stabilization, lying lower than the e and a_2 levels in the final disphenoid structure. The b_2 component in the tetrahedron (4) is a strong skeletal antibonding orbital, with inwardly pointing sp hybrid orbitals. Lengthening the four edges produces the observed large energy slope for this orbital.

The a_1 and t_2 levels are slightly stabilized while the e set undergoes no net stabilization. Distortion from the tetrahedron to the disphenoid is favoured by occupation of the t_2 level of the tetrahedron with a single pair of electrons, i.e. for tetrahedral clusters with 28 valence electrons.

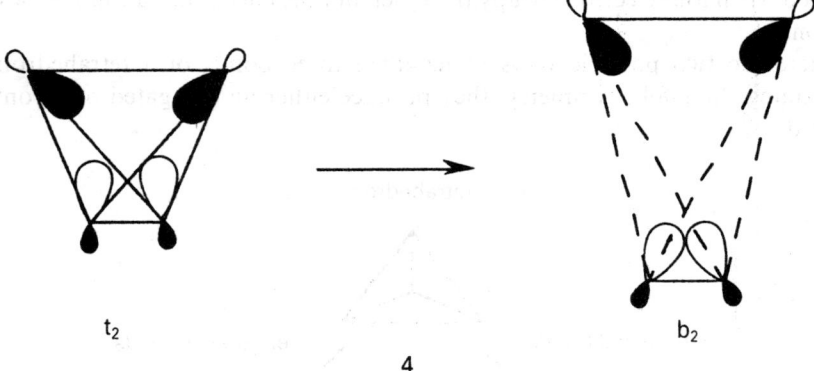

4

The conclusion derived above, that the number of antibonding skeletal molecular orbitals stabilized by distortions of the parent tetrahedron is directly related to the number of edges, can be related to a localized description of the bonding in three-connected molecules. The symmetries of the bonding skeletal molecular orbitals of a three-connected cluster may be derived by using the edges of the cluster as the basis for the irreducible representations [16]. Specifically for the tetrahedron they transform as $a_1 + t_2 + e$. The corresponding antibonding skeletal molecular orbitals may be derived by locating a single noded function along the edges of the cluster as the basis for the irreducible representations. For the tetrahedron these antibonding skeletal molecular orbitals transform as $t_1 + t_2$. In a distorted three-connected polyhedron, those edges which have lengths which may be considered as non-bonding must have pairs of edge localized bonding and antibonding molecular orbitals occupied, whereas the bonded edges have only the edge localized bonding molecular orbitals occupied. This provides an elegant method for deriving the symmetries of the occupied skeletal molecular orbitals in electron-rich clusters derived from a three-connected polyhedron. Moreover, the symmetries of the additional molecular orbitals occupied in electron-rich clusters and derived from the t_1 and t_2 antibonding skeletal molecular orbitals may be obtained by using noded functions only along the long edges as bases for the irreducible representations. These results are summarized for electron-rich clusters derived from the tetrahedron in Table 1.

Table 1 — Derived symmetries of the orbitals stabilized by various geometric distortions to the tetrahedron

Geometry	Number of lengthened bonds	Symmetries of additional molecular orbitals derived from t_1 and t_2
Butterfly	1	b_2
Ring	2	e
Elongated pyramid	3	$a_1 + e$
Compressed pyramid	3	$a_2 + e$
Disphenoid	4	$a_2 + b_2 + e_2$

These molecular orbitals correspond precisely to the molecular orbitals derived from the Walsh diagrams in Fig. 2.

Charge distributions in distorted tetrahedra

The calculated charges for the distorted tetrahedra discussed above are illustrated in **5**. The results indicate a consistent pattern involving a build-up of negative charge at those atoms which lie at the ends of the lengthened bonds. This pattern has several implications for heteroatomic clusters, and the results may be analysed using first-order perturbation theory. Specifically, more electronegative substituents may be anticipated to preferentially occupy the more negatively charged sites. From a molecular orbital viewpoint, the t_1 level of the tetrahedron is no longer degenerate, and the more stable component is that which is localized on the more electronegative atoms. Therefore, addition of an electron pair will preferentially occupy that molecular orbital, thus producing a butterfly with the electronegative atoms at the wing tips.

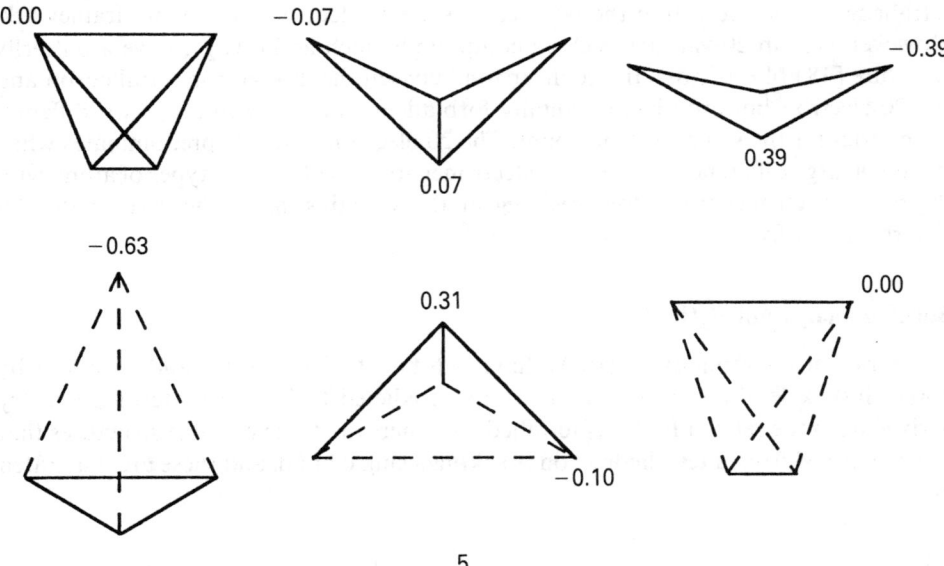

5

Although the electronegativity effects described above are important, an additional valency effect is often also present. Specifically, the replacement of an atom not only changes the electronegativity, but may also change the number of valence electrons contributed. For example, sulphur not only has a higher electronegativity than phosphorus, but also an additional valence electron. Within a one-electron approximation, these effects may be understood by keeping the atomic orbital energies constant and varying the number of electrons associated with each atom. The results of such a calculation are summarized in **6**, where M^V and M^{VI} represent identical equal electronegativity atoms with five and six valence electrons respectively. The calculated charges indicate a smaller charge separation when the valency of the atom coincides with the connectivity of the polyhedral fragment, i.e. when M^{VI} is in the two-connected site and M^V is in the three-connected site. In general, these effects

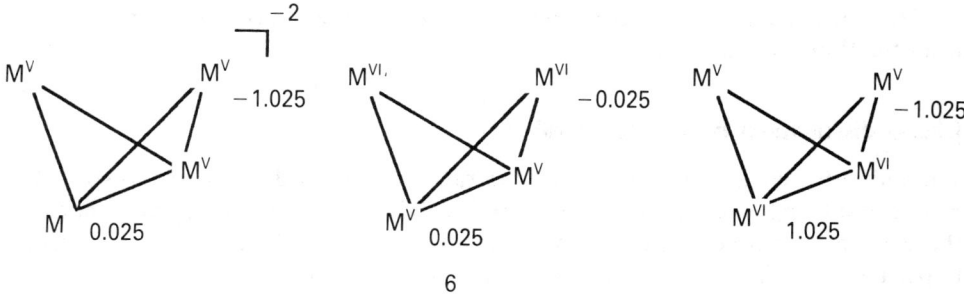

6

follow the electronegativity effects since the atom with the larger number of valence electrons is the more electronegative and prefers to occupy the lower connectivity site. However, exceptions may occur when atoms belonging to different rows of the periodic table are combined within one cluster, e.g. nitrogen and sulphur.

Eisenstein and co-workers [17] have recently performed a theoretical study on X_2Y_2 Zintl ions with 20 valence electrons. This electron count is characteristic of a tetrahedral structure within the polyhedral skeletal electron pair theory framework. However, certain 20-valence-electron compounds, such as $Tl_2Te_2^{2-}$, have a butterfly structure [18] (4 σ bonds). Interconversion between the 20-electron tetrahedron and the 20-electron butterfly is an orbitally forbidden process, permitting two different geometries for the same electron count. The 20-electron butterfly predominates when there is a large difference between the electronegativities of the two types of atom, with the more electronegative atom residing at the wingtips, as in the case of the 22-electron butterfly.

Bonding analysis of $P_2(PH)_2$

The series of compounds, $P_2(PR)_2$, have a total of 22 electrons, and therefore by polyhedral skeletal electron pair theory are predicted to have a butterfly geometry derived from the related $P_2(PR)_2$ tetrahedron. There are three possible structures that can be derived from a tetrahedron on breaking a single bond, and these are illustrated in 7.

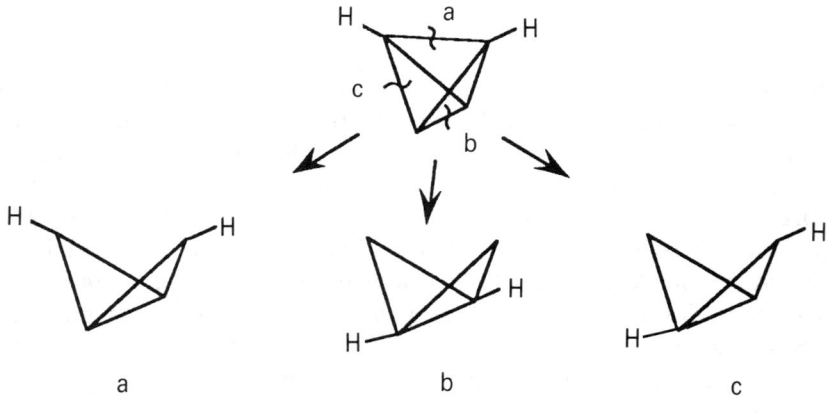

7

To understand the relative stabilities of these respective structures, calculations were performed for the interaction of a H$\cdots$H fragment with the orbitals of a P_4 butterfly derived above. The relevant interaction diagrams for the three possible structures are shown in Fig. 3. The important orbitals of this system may be identified as the b_2, b_1 and a_1 frontier orbitals of the butterfly fragment, together with the orbitals of the H_2 system. For all three structures the HOMO of the butterfly, a_1, is stabilized on interaction with the H_2 fragment. Referring to Fig. 3, it can be seen that for structures (i)–(iii), the LUMO of the butterfly (b_1) becomes the HOMO of the $P_2(PH)_2$ molecule. For structure (i) the energy of the b_1 orbital has been considerably stabilized (8.7–11.1 eV) on interaction with a lower lying b_1 orbital, together with the b_1 orbital of the H_2 fragment. The b_2 orbital of the P_4 unit, which becomes the LUMO for the $P_2(PH)_2$ molecule, does not interact with the H_2 fragment for symmetry reasons, and therefore undergoes no change in energy. For structure (ii) we see that the b_1 orbital of the P_4 unit no longer has a symmetrically compatible orbital in the H_2 fragment, and therefore is not stabilized. In this case it is a b_2 fragment orbital in the H_2 system which stabilizes the energy of the b_2 orbital of the butterfly fragment. Finally, for structure (iii), both the b_1 and the b_2 orbital energies are stabilized on interaction with the H_2 fragment, the larger stability being conferred on the b_2 orbital.

To summarize, the LUMO ($3b_1$) of the butterfly transforms as the HOMO in the $P_2(PH)_2$ molecule. For structure (i), this orbital's energy has been considerably stabilized with respect to the naked P_4 unit, while for structures (ii) and (iii) it has remained unchanged in energy and undergone a slight stabilization respectively. The b_2 orbital of the butterfly becomes the LUMO of the $P_2(PH)_2$ molecule. For structure (i) this has undergone no net energy change with respect to the P_4 unit, while for both structures (ii) and (iii) it has been stabilized. Opposite orbital energy changes are observed for structure (i) and the other two structures. This results in the former having a larger HOMO/LUMO gap, and lower total energy.

ELECTRON-RICH CLUSTERS BASED ON THE PRISM

The molecular orbital description of the skeletal orbitals of a prism is well established [19], the frontier orbitals of this system being, e', e'', a'_2, a''_2. The e' and e'' orbitals are skeletal non-bonding and are designated [16] as $D^\pi_{\pm 1}$, $\bar{D}^\pi_{\pm 1}$. It is instructive to investigate the effects on the frontier orbitals of geometrical perturbations on the prism. The resultant molecular orbital diagrams for lengthening one and two bonds of a prism are shown in Fig. 4. Unlike the previous example of the tetrahedron, the prism is not a regular polyhedron, and there are several alternative ways to lengthen either one or two bonds, while maintaining at least double connectivity for each atom. There are two options available for cleavage of a single bond, namely lengthening an intra-triangular bond, or lengthening one of the bonds parallel to the three-fold axis. In each case, one of the molecular orbitals of the prism is stabilized in energy. The stabilized orbitals have antibonding interactions across the bonds lengthened in the original prism.

For structure (ii) in Fig. 4, the orbital that has been stabilized in energy is derived from the a'_2 orbital of the prism, which has a π^* interaction between the atoms in question. Opening out one of the inter-triangular bonds to produce a boat structure

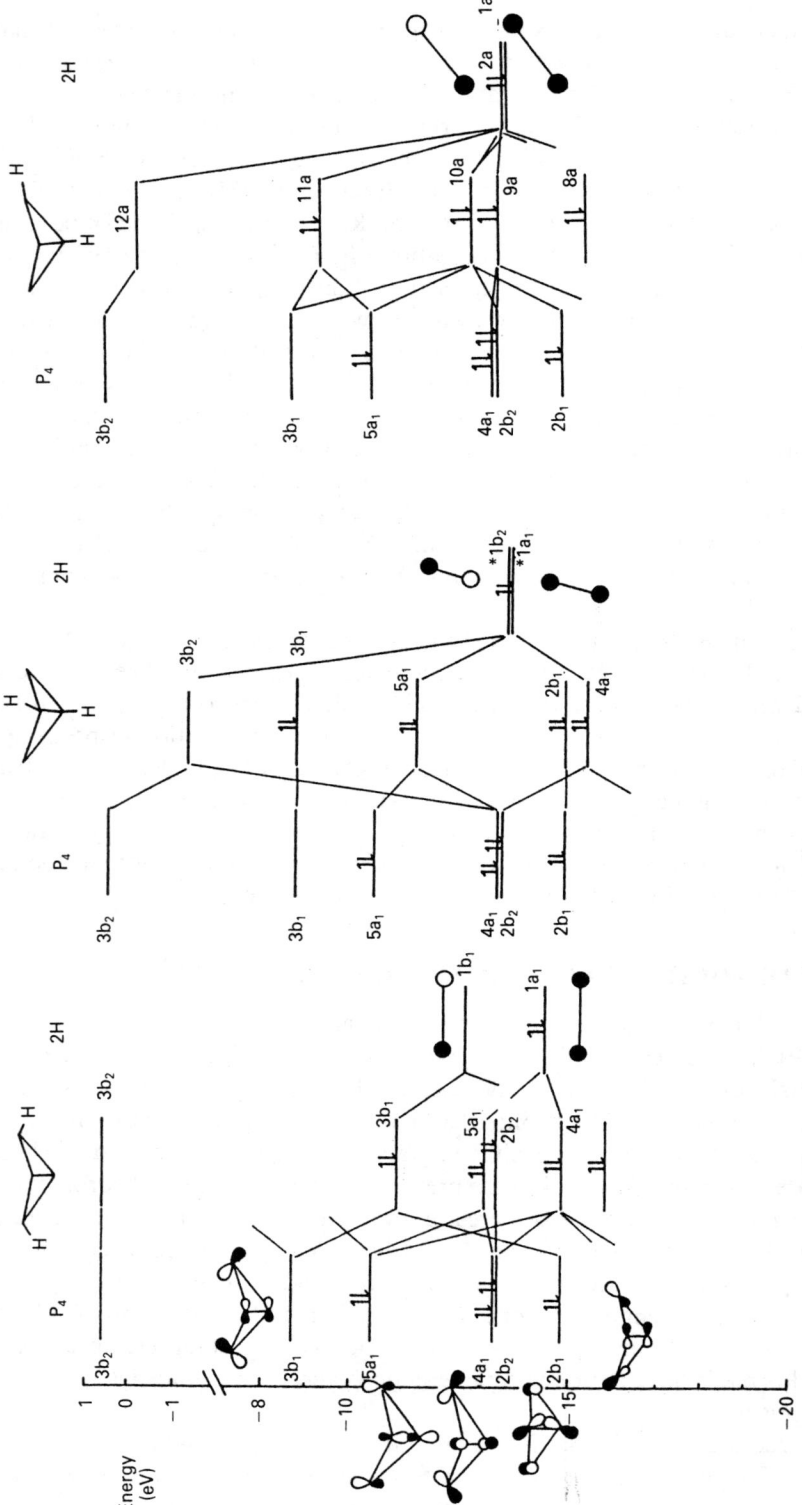

Fig. 3. — Fragment molecular orbital analysis for the three possible structures of $P_2(PH)_2$.

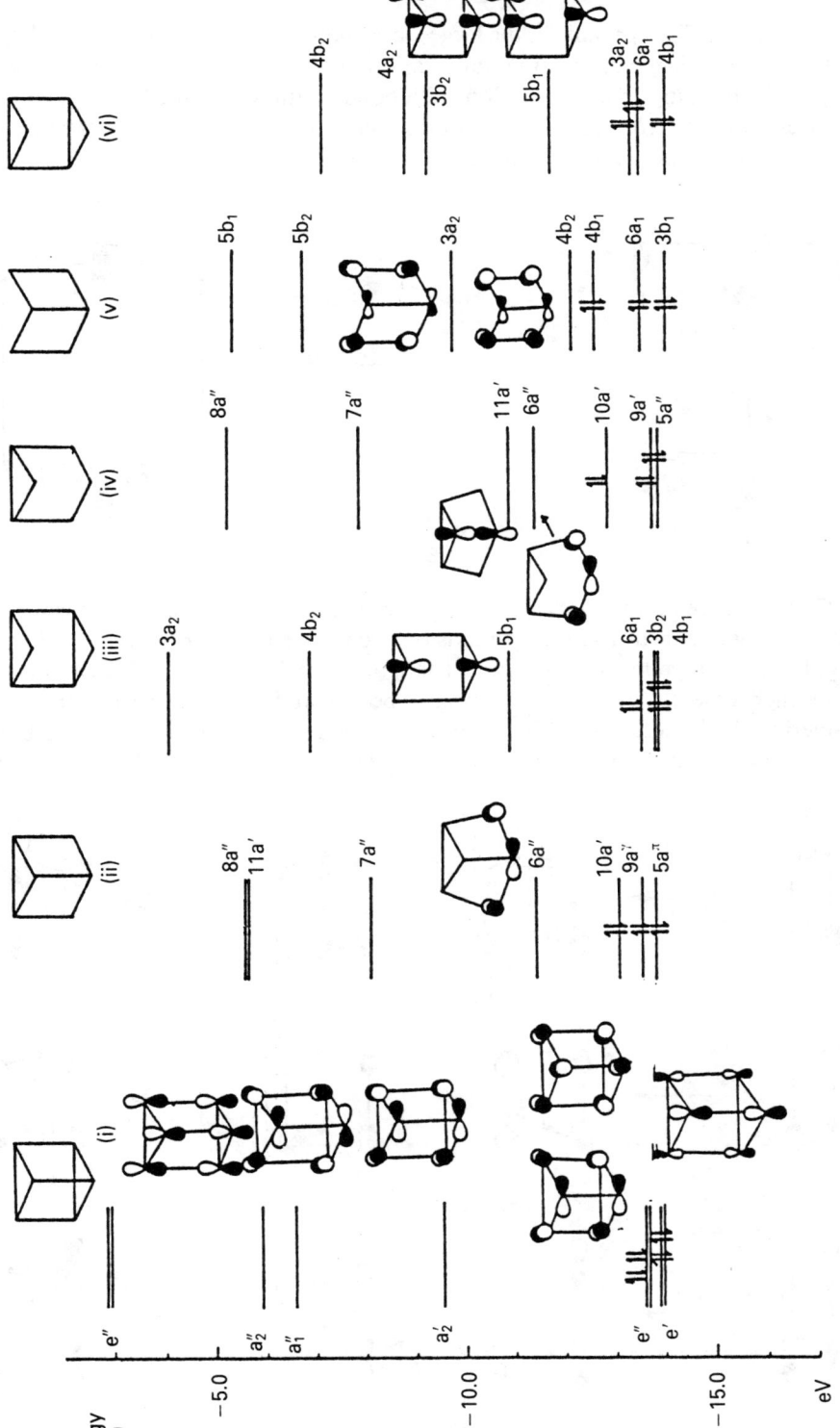

Fig. 4. — Effects on the frontier molecular orbitals of a prism on lengthening one and two bonds.

(structure (iii) in Fig. 4) results in the large stabilization of the energy of the high lying a_2'' orbital of the prism. Such a large stabilization can be attributed to the lessening of a strong σ antibonding interaction along the bond in question (see **8**). Both of these structures, resulting from the cleavage of a single bond of the prism, are favoured by 32-valence-electron systems, but the boat structure is the more stable.

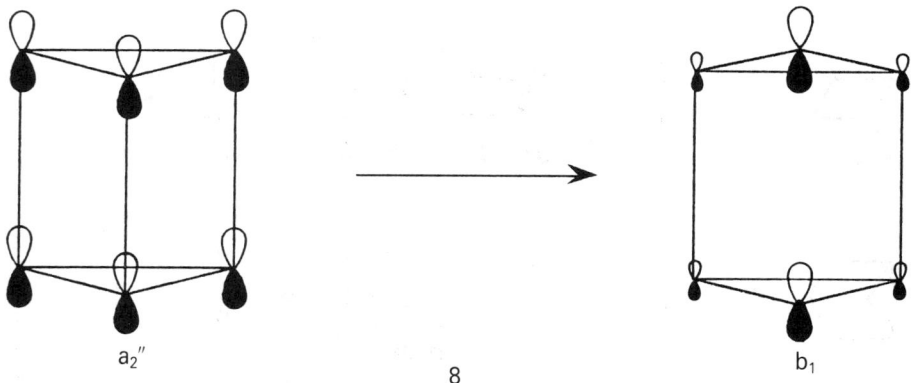

a_2'' **8** b_1

There are three options for lengthening two bonds of the prism. These correspond to lengthening two intra-triangular bonds or two inter-triangular bonds or one inter-triangular bond together with an intra-triangular bond. From Fig. 4 it can be seen that in each case two extra orbitals have been stabilized in the frontier region. Considering the structure resulting from the cleavage of two intra-triangular bonds (v), it can be seen that this results in the stabilization of the a_2' and a_1'' orbitals of the

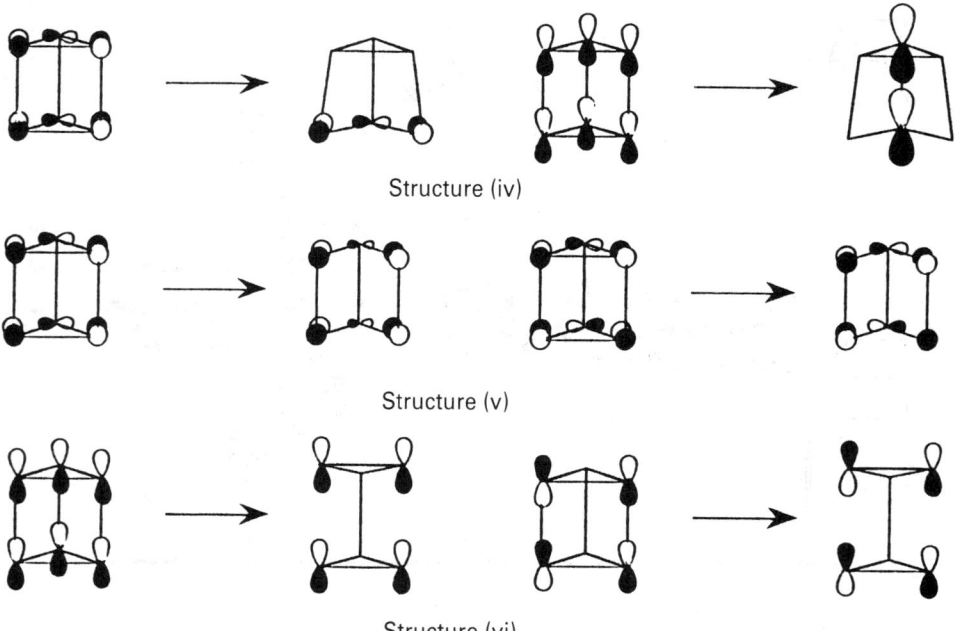

Structure (iv)

Structure (v)

Structure (vi)

9

prism, which are noded in an antibonding fashion along the edges in question. Structure (iv) in Fig. 4, resulting from lengthening an intra- and inter-triangular bond simultaneously, has stabilized the energies of the a'_2 and a''_2 orbitals of the prism. Again this can be rationalized by consideration of the nodal properties of the orbitals along the lengthened bonds. Finally, lengthening two inter-triangular bonds (vi) results in stabilization of the a'_2 and a component of the high lying e'' set. Both of these orbitals are σ antibonding between the triangular planes of the prism, and relief of this antibonding character via a geometric distortion results in a large stabilization of their respective energies. The orbitals stabilized in the three cases outlined above are depicted in **9**. These three structures, derived by lengthening two bonds of a prism, are favoured by 34-valence-electron species, with structure (vi) being the most stable structure.

The symmetries of the orbitals stabilized by each of the geometrical distortions may be derived by using the treatment developed previously for the tetrahedron; namely, using edges of the prism as bases for the irreducible representations. The results are summarized in Table 2.

Charge distributions in distorted prisms

The effects of lengthening bonds in the prism are to localize electrons on the atoms that were previously bonded to each other. This polarization of the structure plays a determining role in the positions of heteroatoms in these distorted prisms. The charge distributions for the structures resulting from lengthening one and two bonds simultaneously are shown in **10** for the Te_6^{2+} molecule.

It can be seen above that the more stable alternative on lengthening one bond of a prism is the boat (structure (c) in **10**). This structure will be favoured by a 32-valence-electron molecule. There are three alternatives available for lengthening two bonds of the prism, and the most stable geometry is that of structure (e). This will be favoured by a 34-valence-electron molecule. The $Te_3S_3^{2+}$ molecule [20] has 34 valence electrons and therefore according to polyhedral skeletal electron pair theory will have a structure derived from a prism with two bonds broken. From the calculations above, the $Te_3S_3^{2+}$ molecule may adopt either of the structures shown in **11**.

Table 2 — Derived symmetries of the orbitals stabilized by various geometric distortions to the prism. The relevant geometries, (ii)–(vi), are displayed in Fig. 4

Geometry	No. of bonds lengthened	Symmetry of stabilized molecular orbitals
(ii)	1	a''
(iii)	1	b_1
(iv)	2	$a' + a''$
(v)	2	$a_2 + b_2$
(vi)	2	$a_2 + b_1$

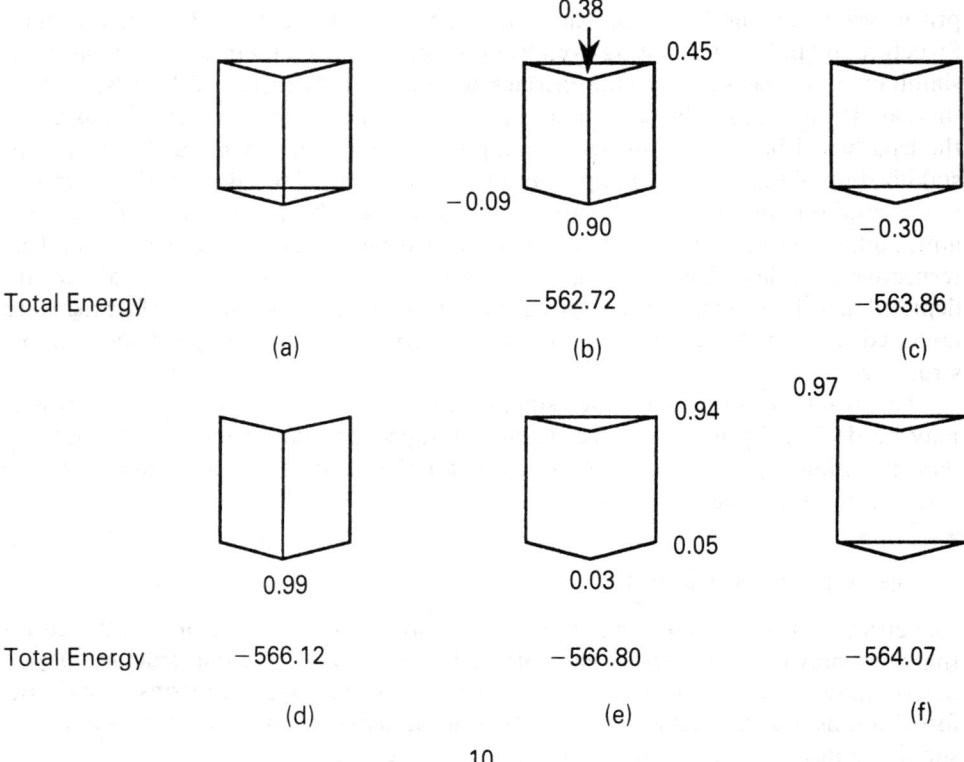

(a)   (b)   (c)

(d)   (e)   (f)

10

Consideration of the charges in **10** indicates that the more stable structure will have the more electronegative atom at the two-connected sites, and therefore structure (i) in **11** is preferred. This is also the experimentally observed structure for this compound.

An interesting pair of related compounds, $Se_2I_4^{2+}$ and $S_2I_4^{2+}$, has been synthesized by Passmore *et al.* [21, 22]. These molecules may be viewed as high electron rich systems (38 valence electrons) whose structures are based on a prism. There are four

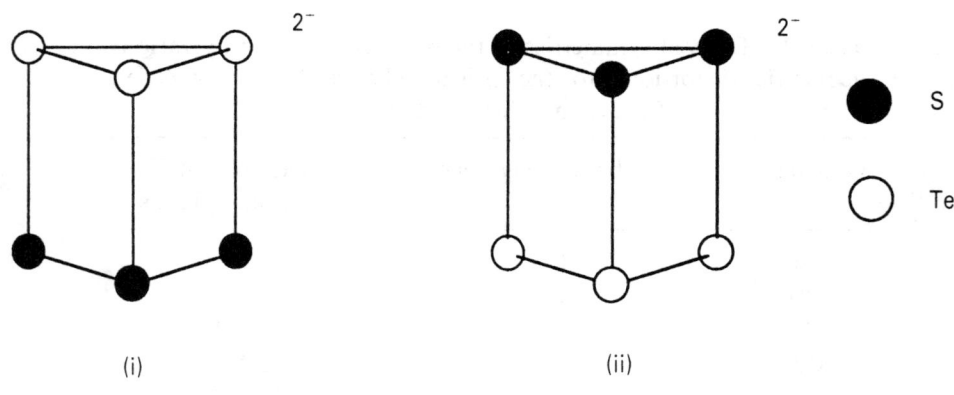

(i)   (ii)

11

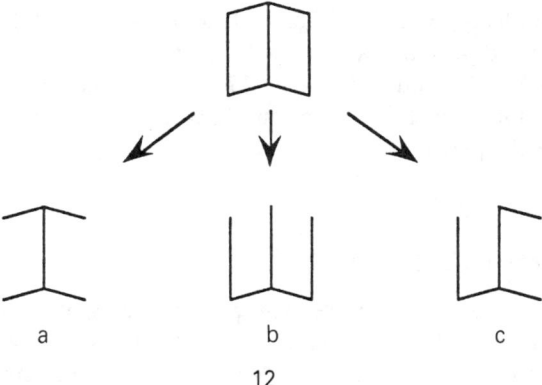

12

excess pairs of electrons, and therefore the basic structure of the molecule is equivalent to a prism with four bonds broken. These structures are shown in **12**.

Of the possibilities for a 38-electron structure derived from a prism, the most stable one involves cleavage of one bond in each of the triangles together with two inter-triangular to produce a C_{2v} structure exemplified by $Se_2I_4^{2+}$ and its isoelectronic analogues. A calculation on a 38-valence-electron prism with a C_{2v} structure was performed. As before, the charge distributions of this system show that the greatest negative charge has been located on the one-connected sites, and electronegativity arguments predict that the iodine atoms will reside in these positions. This is also confirmed by considering the frontier orbitals of the 38-electron prism. These orbitals, also depicted in **13**, are predominantly localized on the one-connected sites. Therefore, from perturbational molecular orbital theory, it is energetically more stable to have iodine, the more electronegative atom of the molecules, occupy these sites. This rationale adequately explains the structure adopted by $Se_2I_4^{2+}$ and $S_2I_4^{2+}$. However,

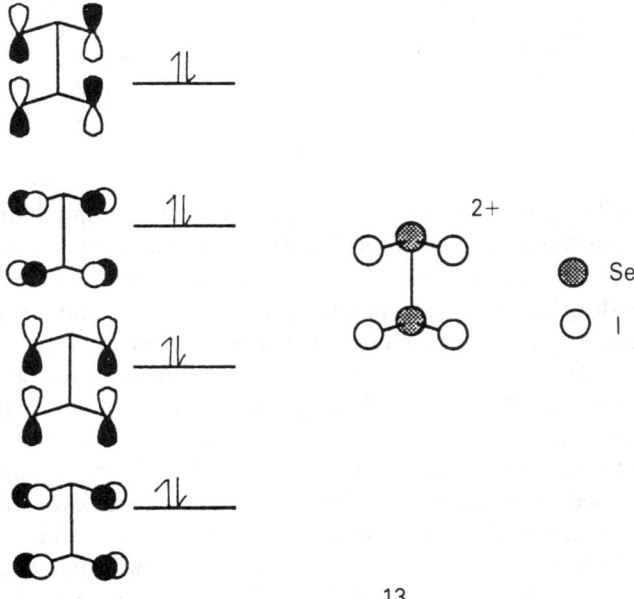

13

the isoelectronic molecule P_2I_4 [23] has a structure similar to the cations, with the exception that the iodine pairs are *trans* to each other across the P—P bond as opposed to *cis* in $Se_2I_4^{2+}$ and $S_2I_4^{2+}$ (see **12**). A detailed *ab initio* study has been performed by Passmore on these compounds which offers an explanation for this discrepancy in the P_2I_4 structure.

An interesting exception: Te_6^{4+}

Up to now, the introduction of an electron pair in excess of the polyhedral electron count for a three-connected cluster has resulted in the scission of a bond in the cluster. The Te_6^{4+} ion reported by Gillespie *et al.* [24] has 32 valence electrons and is therefore expected to have a structure derived from a prism with one bond broken: specifically, a boat structure of C_{2v} symmetry. However, it was found that the cluster maintains its D_{3h} structure, accommodating the excess electron pair not by cleavage of a single bond, but by partial lengthening of all three bonds between the triangular faces (see **14**).

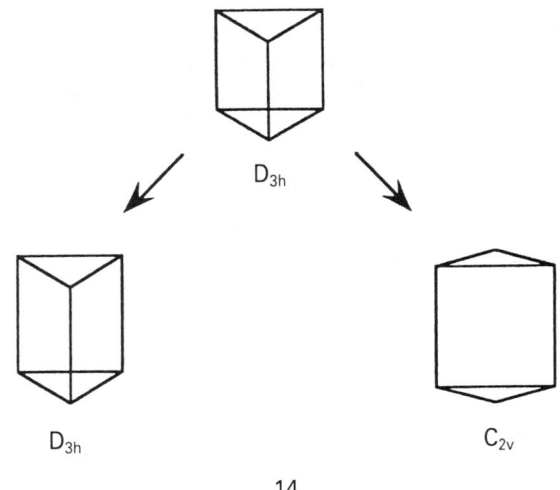

14

This unusual response to an excess of electron pairs warranted further study. The molecular orbitals for a Te_6^{4+} prism with three bonds lengthened, and the alternative boat structure, are shown together with the molecular orbitals of the regular prism in Fig. 5. The bonding between the triangular faces for the e″, a′$_2$ and the a″$_1$ orbitals of the frontier region is π in nature, and σ bonding between the planes for e′; and lengthening a single bond or three bonds simultaneously results in slight energy shifts for these orbitals. However, the a″$_2$ orbital is strongly σ antibonding in nature between the two triangular faces.

Interestingly, the highest occupied orbital is antibonding within the triangular planes, as opposed to between the triangular planes. Therefore occupation of this orbital would be expected to cause the intra-triangular bonds to lengthen. However, the effect on the frontier orbitals of lengthening inter-triangular bonds is shown in Fig. 5. This figure illustrates that lengthening of the inter-triangular bonds results in a

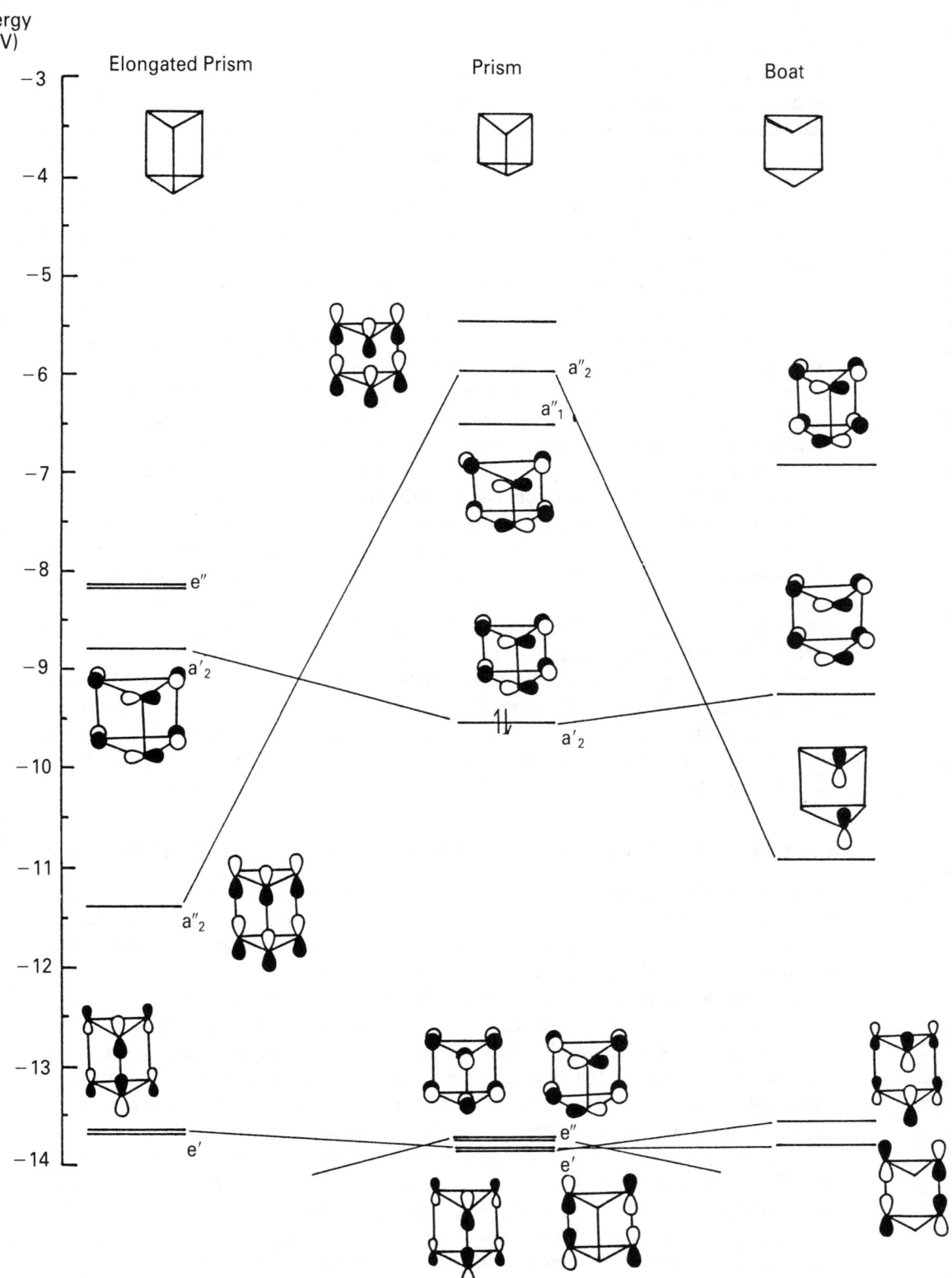

Fig. 5. — Frontier molecular orbitals of the prism, boat and elongated prism geometries.

large stabilization for the a_2'' orbital in both cases. The key to the problem must reside in the stabilization of this orbital. Therefore, this stabilization was studied in greater detail along the alternative bond-lengthening coordinates. The appropriate Walsh diagrams are shown in Fig. 6. For the resultant D_{3h} and C_{2v} structures, the energy of the a_2'' (b_1 in C_{2v} symmetry) orbital decreases steeply when the bond is lengthened. This process is formally symmetry forbidden [25] since it involves the crossing of an unoccupied and an occupied orbital. For any given bond length, the energy of the a_2'' orbital is lower for the D_{3h} geometry. This is to be expected since relief from the antibonding interaction is being accommodated by lengthening three bonds as opposed to one bond for the C_{2v} structure. The other orbitals in the frontier region are stabilized and destabilized depending on their nodal character across the vertical bonds (i.e. the bonds parallel to the three-fold axis), with the magnitude of energy being greater for the D_{3h} geometry for any given bond length.

In spite of this, the total energies of the respective geometries along the path of distortion indicate that it is more favourable to break one bond alone, rather than lengthening three bonds simultaneously. This is in contradiction with the experimental structure reported for Te_6^{4+}, and therefore within the scope of the Extended Hückel calculations it is not possible to offer an explanation for the unexpected structure that is found experimentally. Further studies on this problem were performed using Fenske–Hall calculations. It was not computationally possible to study Te_6^{4+}, so the analogous, as yet hypothetical, Se_6^{4+} molecule was studied. The energy level ordering of the frontier orbitals for both geometries was found to agree with that calculated within the Extended Hückel framework. However, the results were again inconclusive with regard to the total energies of the two geometries. In an attempt to resolve this problem, a geometrical optimization calculation was performed on the Te_6^{4+} molecule, using density functional theory. Essentially, the Te_6^{4+} molecule has an optimized structure, at the local density approximation [26] level of theory, of a prism with the three bonds parallel to the C_3 symmetry axis lengthened. The full details of these calculations are contained in a separate publication currently in preparation [27].

ELECTRON-RICH CLUSTERS BASED ON THE CUNEANE STRUCTURE

In common with the prism, the cradle structure adopted by cuneane is not a regular polyhedron. As a consequence, cleavage of one or more bonds can be accomplished in a number of ways, resulting in geometries with different symmetries. Some of the possibilities for breaking one and two bonds are shown in 15.

The primary concern is to identify which bonds, when broken, produce the more stable structure, and then to extend the analysis to site preferences of a heteroatom introduced to these systems. The skeletal frontier orbitals for the cuneane geometry together with those of the distorted structures are shown in Fig. 7. The frontier orbitals for two of the structures formed on lengthening one bond of the cuneane skeleton are shown in Fig. 7 (structure (b) and (c)). It can be seen that in each case one orbital's energy has been stabilized to the frontier region. For structures (b) and (c) the stabilized orbitals are of b_1 and b_2 symmetry respectively. These orbitals are seen to be antibonding in nature across the bond being lengthened, thus explaining the stabilization of these orbitals on scission of the bonds in question. Both structures (b) and (c) are favoured by systems with 42 valence electrons, with structure (c) being the

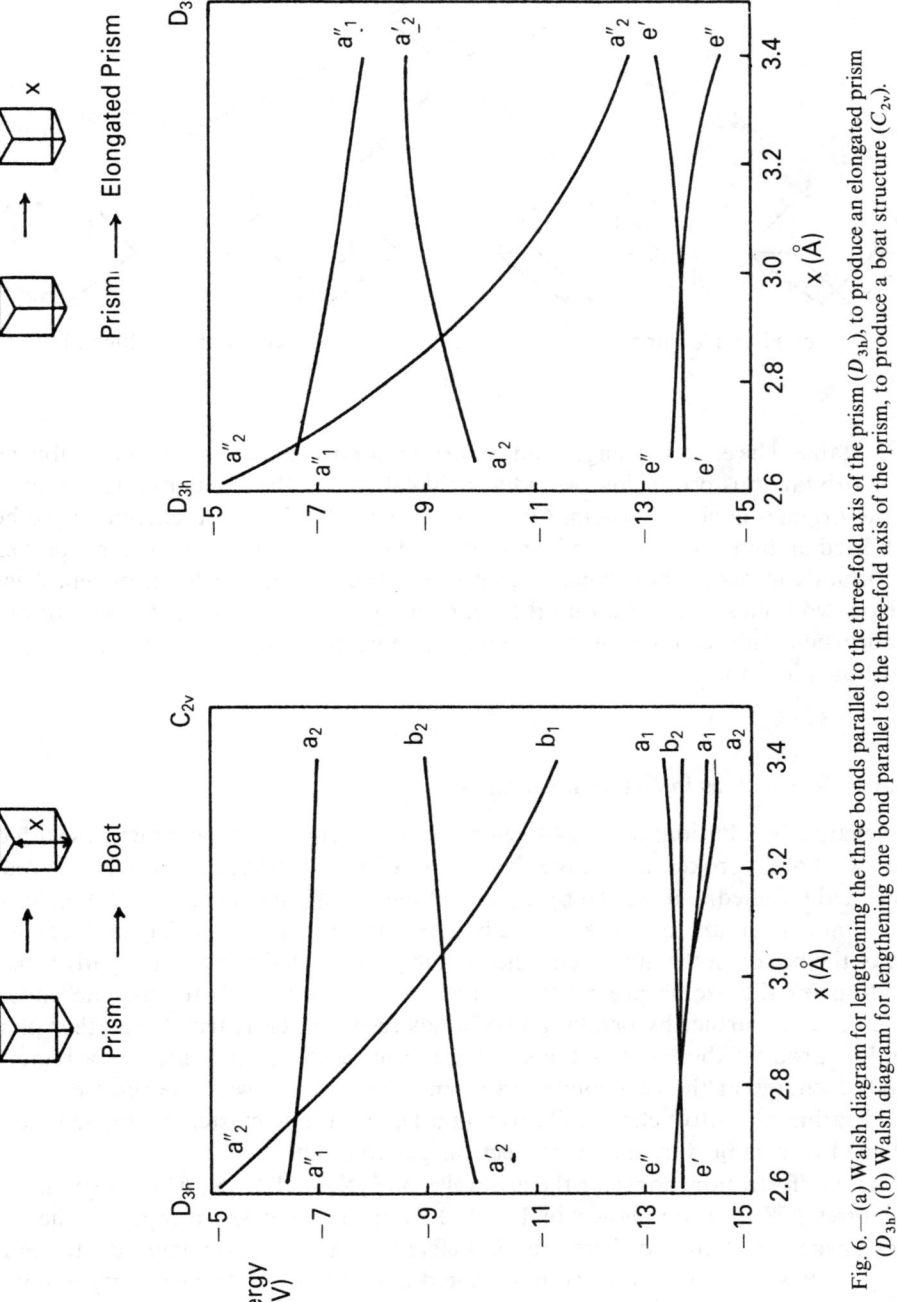

Fig. 6. — (a) Walsh diagram for lengthening the three bonds parallel to the three-fold axis of the prism (D_{3h}), to produce an elongated prism (D_{3h}). (b) Walsh diagram for lengthening one bond parallel to the three-fold axis of the prism, to produce a boat structure (C_{2v}).

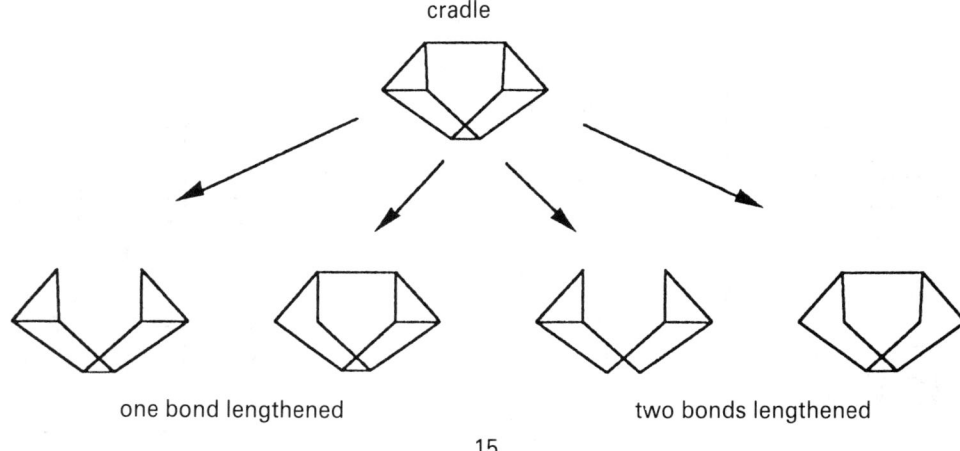

cradle

one bond lengthened   two bonds lengthened

15

more stable. The effects of lengthening a further bond in these systems is to stabilize an extra orbital, thus producing two additional orbitals in the frontier region compared to the original cuneane system. Once again, the orbitals whose energies have been stabilized on bond cleavage can be rationalized by considering the nodal properties of the orbitals across the bonds being lengthened. The stabilized orbitals on lengthening one or two bonds, derived group theoretically by using singly noded functions along the broken bonds as bases for the irreducible representations for the C_{2v} point group, are shown in Table 3.

Charge distributions in distorted cuneanes

The charge distributions for the 44-valence-electron cuneane-type structures with one and two bonds broken are shown below in **16**. These structures may be viewed as elongated tetrahedra, dissected by a plane of four atoms, arranged as a rectangle. The more stable structure for breaking one bond is shown to be that which has lengthened one of the bonds in the aforementioned plane. As expected there is a negative charge build-up on the atoms previously bonded to each other. Distorting the cuneane structure even further by breaking two bonds results in cleavage of the other planar bond to produce the more stable structure. The resulting structure has a region of negative charge at the two connected atoms of the plane, with the three-connected sites bearing a positive charge. Clearly, in a heteroatomic cluster the two-connected sites will be occupied by the more electronegative atom.

As an illustration, consider the examples of S_4N_4 and As_4S_4 whose experimental structures [28, 29] are shown in Fig. 1. In the former case, nitrogen is the more electronegative atom and therefore its preference for the two-connected sites can be easily understood. However, we have noted above that electronegativity is not the only influencing factor in site preferences for electron-rich clusters, but that valency also has a role to play. For this example, purely on valency considerations, sulphur would be predicted to occupy the two-connected sites. This is the first example that we have studied so far where valency and electronegativity factors have opposed each

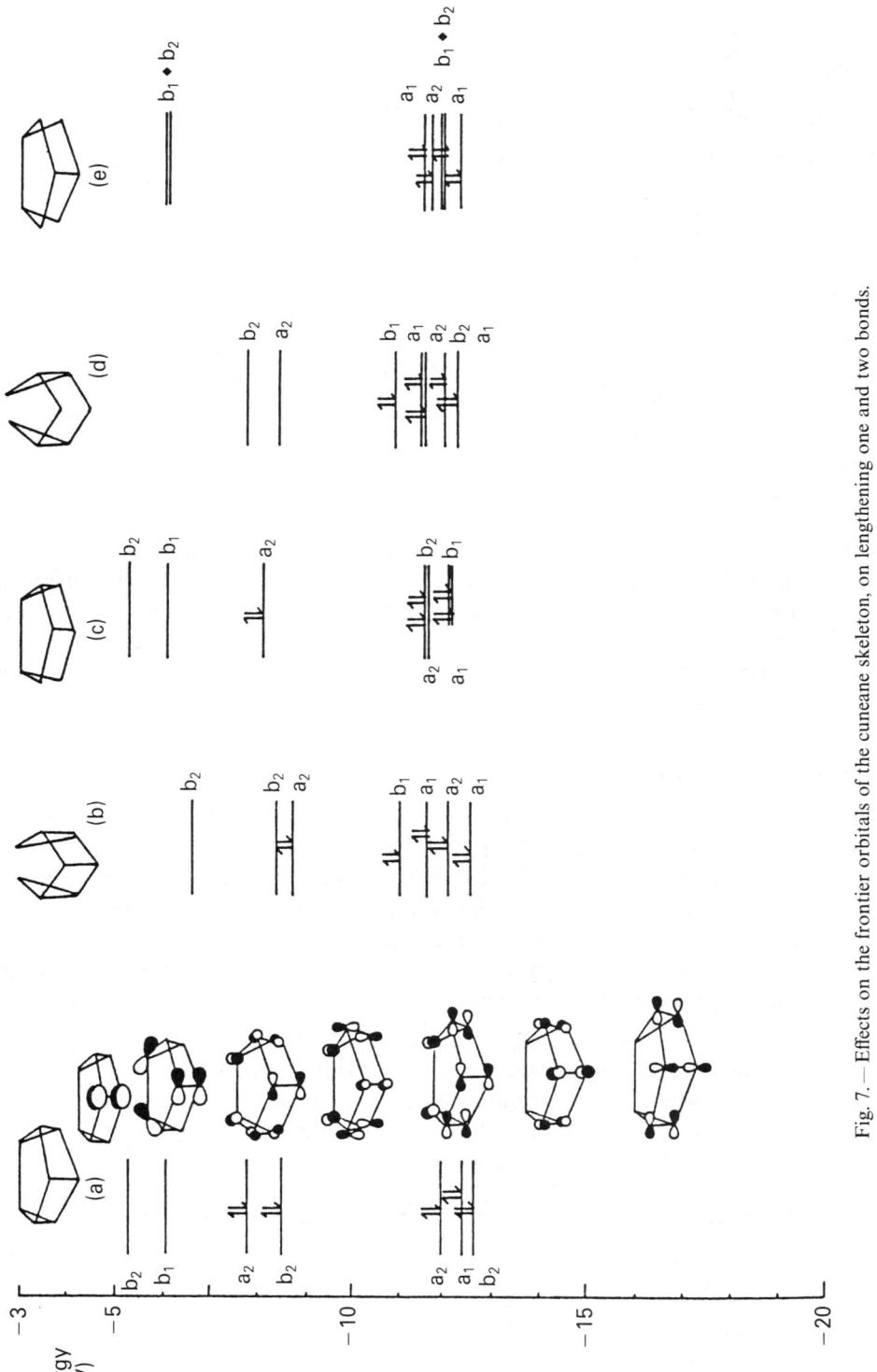

Fig. 7. — Effects on the frontier orbitals of the cuneane skeleton, on lengthening one and two bonds.

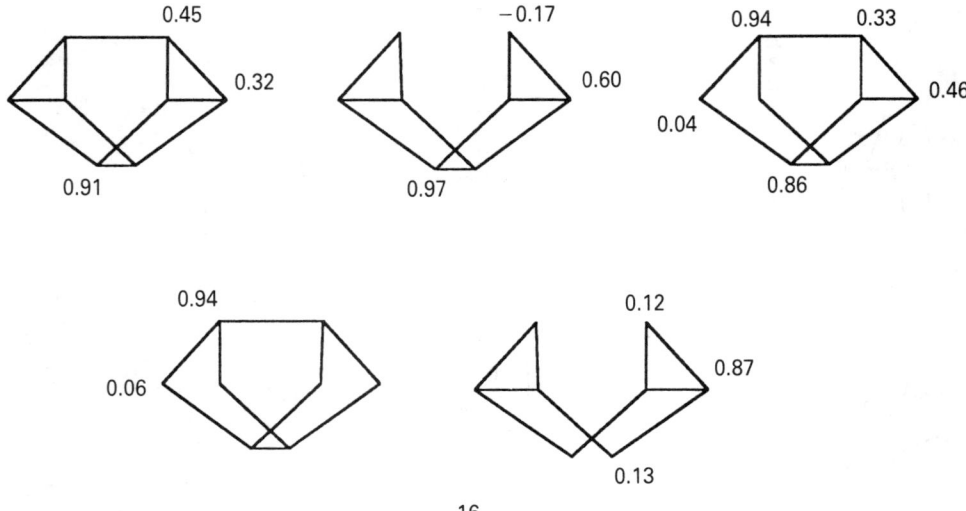

16

other. In general this conflict of preferences will only happen when the atoms concerned are from different rows of the periodic table. It is clear from this example that electronegativity has by far the greater influence on site preferences for electron-rich heteroatomic clusters. For As_4S_4, both electronegativity and valency consider-ations predict that sulphur will occupy the two-coordinate site in accordance with the experimental structure.

Molecular orbital analysis

Extended Hückel calculations were performed on S_4N_4 and As_4S_4 together with their respective alternative structures. There has already been an extensive study of the molecular and electronic structure of S_4N_4 [30, 31] and its analogues. Here, the aim of the molecular orbital study is to determine the reasons for the observed site preferences of the molecules, from a bonding viewpoint.

Calculations show the S_4N_4 geometry (S at the three-connected sites) to be the more stable structure. One of the main features of the interaction diagrams is the

Table 3—Derived symmetries of the orbitals in the D_{2d} point group stabilized by various geometric distortions to the cradle (cuneane-type structure). The relevant geometries (b)–(e) are dis-played in Fig. 7

Geometry	No. of bonds lengthened	Symmetry of stabilized molecular orbitals
(b)	1	b_1
(c)	1	b_2
(d)	2	$b_1 + b_2$
(e)	2	$a_2 + b_2$

difference in energies of the HOMO for the two structures. In the case of S_4N_4, the $4b_2$ orbital has an energy of -11.8 eV, while for N_4S_4 the $4b_2$ orbital has an energy of -9.4 eV. This difference may be considered to be a major contributor to the relative energies of the two geometries. As mentioned previously, each system may be considered as an elongated tetrahedron of one element, bisected by a square of the other element. This provides a natural division of the molecule into two fragments for an appropriate f.m.o. analysis. The interactions between the two fragments to produce the HOMO for both geometries are illustrated in Fig. 8, along with other pertinent interactions. In each case the $4b_2$ molecular orbital is constructed from the $3b_2$ orbital of the tetrahedron and the $2b_2$ orbital of the square fragment.

For S_4N_4 the $4b_2$ orbital is localized on the nitrogen fragment. However, for N_4S_4 the molecular orbital is not found to be localized on the square (sulphur) fragment, but once again on the nitrogen (tetrahedral) fragment. Table 4 shows the major overlap populations between the fragment orbitals. The $3b_2/2b_2$ pair of orbitals is shown to have a reasonable overlap for S_4N_4, while being almost non-existent for N_4S_4 (approximately five times greater for S_4N_4 than for N_4S_4). This reduction in overlap results in localization of the HOMO on the nitrogen fragment in N_4S_4, with the consequence that the HOMO is less stable than for S_4N_4.

The reason for this reduced overlap can easily be seen in **17**, which shows the orbital positions with all atoms projected onto the xy plane. Introduction of a N—N bond of 1.45 Å causes the respective orbitals to be no longer perfectly aligned as in S_4N_4. This 'skewness' results in reduced overlap between the fragment orbitals, with the consequences mentioned above. This instability that has been introduced on transforming from S_4N_4 is purely a geometric one.

The electronegativity effect on the $4b_2$ orbital can easily be illustrated by considering the analogue of this orbital in the S_8^{2+} molecule (**18**). The atomic coefficients of this molecular orbital are larger at the two-connected sites. Therefore, from first-order perturbation theory, a more electronegative atom will reside at the two-connected site to produce the more stable form of this orbital. Thus, nitrogen is predicted to occupy the two-connected sites in S_4N_4.

The frontier molecular orbital diagrams for As_4S_4 and S_4As_4 are given in Fig. 9. Unlike the previous example, the As—As and S—S bond distances are reasonably similar to facilitate the study of these two systems using the same cage geometry. This provides a more satisfactory basis for illustrating the effects of electronegativity on site preferences in the electron-rich cuneanes. Focusing on the HOMOs for the moment, it is noted that their symmetries are not the same for both structures, as was the case with S_4N_4. Here, we have the $4b_2$ molecular orbital as the HOMO for As_4S_4, while S_4As_4 the $2a_2$ orbital has been substantially raised in energy relative to As_4S_4, to become the HOMO for this structure. The effects of electronegativity can be seen in this figure. Exchanging the positions of the elements in As_4S_4 causes the tetrahedral fragment to be lowered in energy, while the square fragment orbitals are raised in energy. As a result, molecular orbitals localized on the square fragment are destabilized relative to the original As_4S_4 structure. This accounts for the high lying $2a_2$ orbital, which is totally localized on the square fragment. Also, recalling the atomic coefficients for the $4b_2$ orbital of S_8^{2+}, first-order perturbation theory predicts that the As atoms will assume the three-connected sites for the more stable structure.

Another striking difference between the two structures may be seen in the overlap

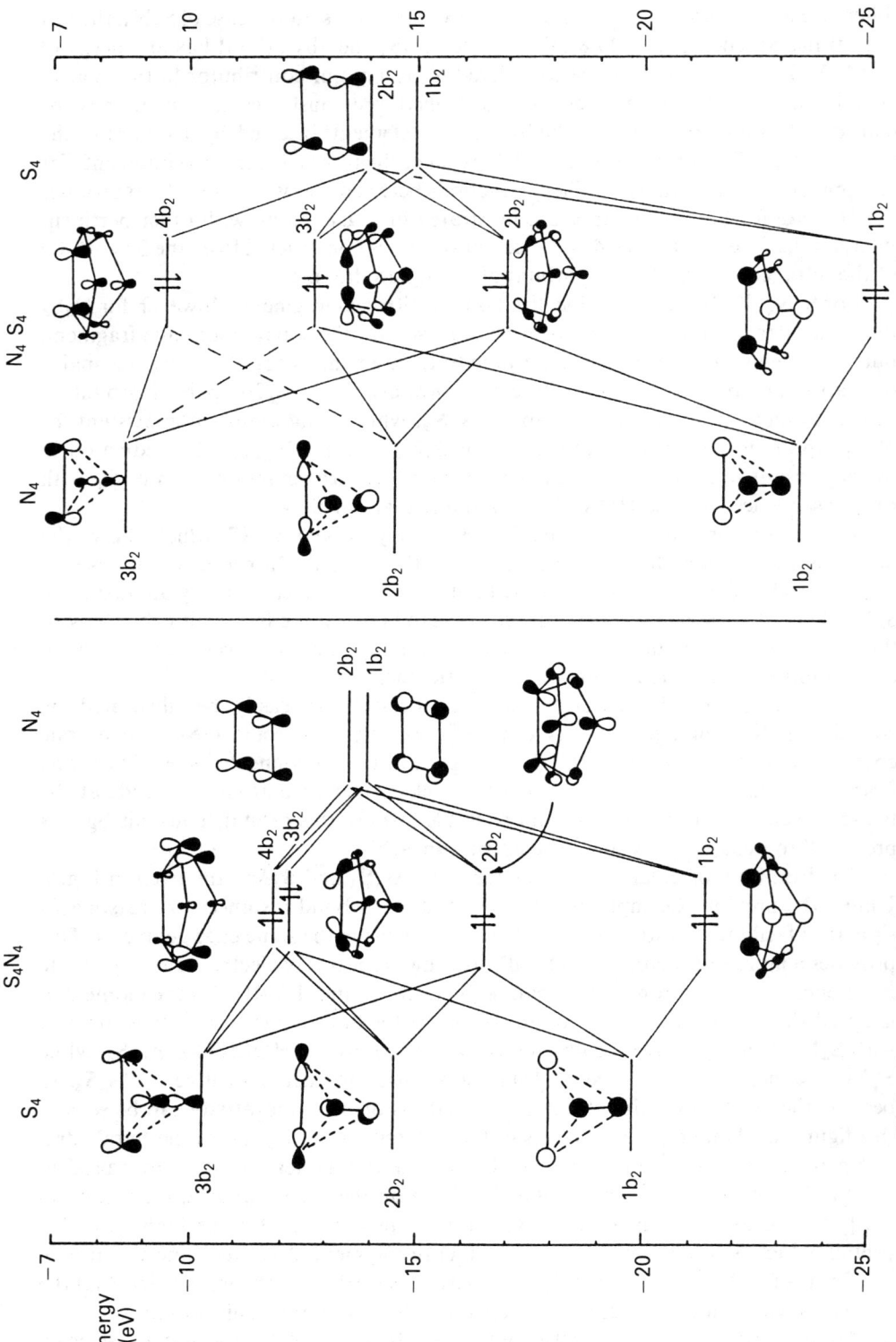

Fig. 8. — Interactions of fragment orbitals of b_2 symmetry in the region of the HOMO for S_4N_4 and N_4S_4.

Table 4 — Major overlap populations between the tetrahedral and square fragment orbitals (f.o.) of S_4N_4 and N_4S_4

S_4N_4		N_4S_4	
S_4 f.o. + N_4 f.o.	Overlap population	N_4 f.o. + S_4 f.o.	Overlap population
$1a_2 + 1a_2$	0.538	$1b_1 + 2b_1$	0.649
$1b_1 + 1b_1$	0.483	$1a_2 + 1a_2$	0.415
$3b_2 + 2b_2$	0.282	$2e + 4e$	0.171

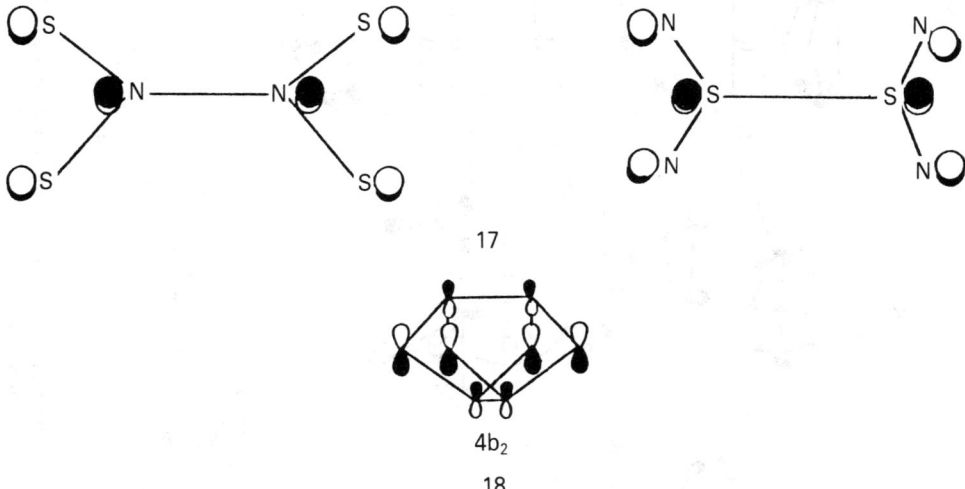

17

$4b_2$

18

populations (Table 5) between the $1b_1$ orbital of the tetrahedral fragment and the $1b_1$ and $2b_1$ orbitals of the square fragment. These interactions for both cases are shown in Fig. 10. For As_4S_4, both of these overlaps are equally strong, while for S_4As_4 the $1b_1/1b_1$ overlap population has increased at the expense of the $1b_1/2b_1$ overlap.

The effect of changing the elements from S_4As_4 to As_4S_4 on this three-orbital interaction is clear. The square fragment orbitals are stabilized to a large extent, while the tetrahedral fragment orbital has been slightly destabilized. The result of these

Table 5 — Major overlap populations between the tetrahedral and square fragment orbitals (f.o.) of As_4S_4 and S_4As_4

As_4S_4		S_4As_4	
Symmetry of f.o.	Overlap population	Symmetry of f.o.	Overlap population
$1a_2 + 1a_2$	0.5818	$1b_1 + 1b_1$	0.5737
$1b_1 + 2b_1$	0.2916	$1a_2 + 1a_2$	0.4563
$3e + 3e$	0.2743	$1b_2 + 2b_1$	0.1782
$1b_1 + 1b_1$	0.2187	$1b_1 + a_2$	0.1383

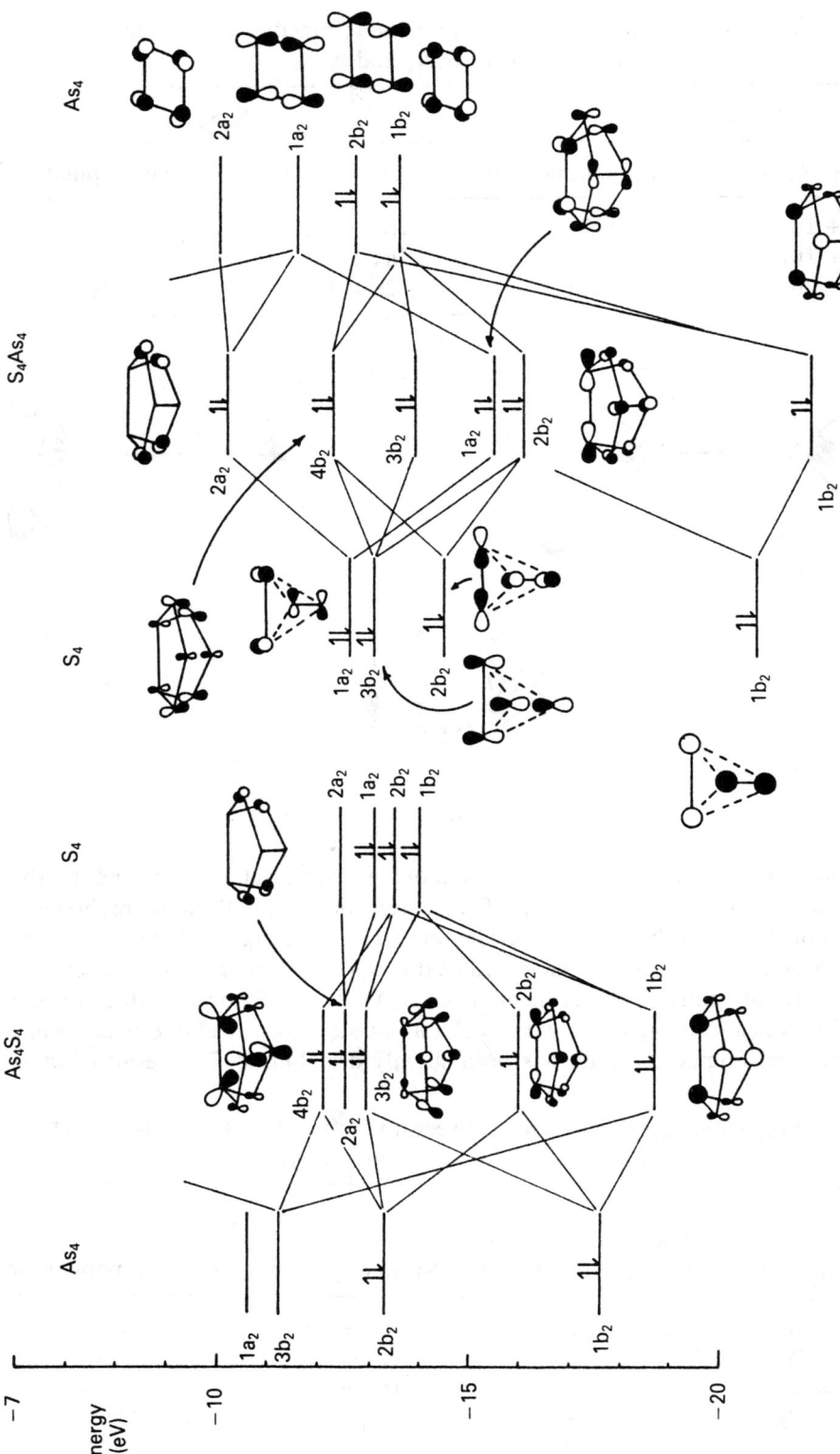

Fig. 9.— Frontier fragment molecular orbital analysis of As_4S_4 (sulphur occupying the two-connected site) and S_4As_4 (arsenic occupying the two-connected site).

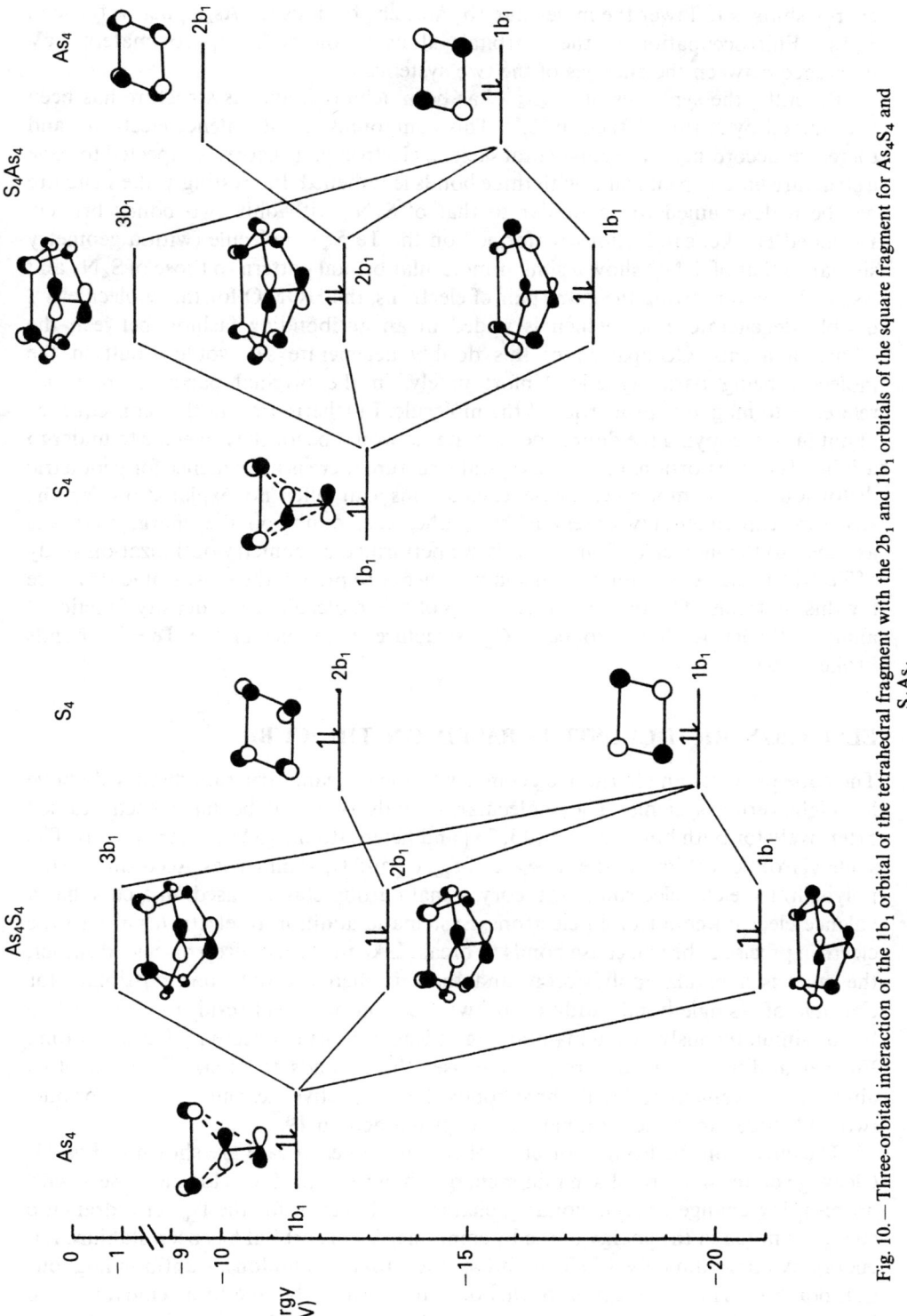

Fig. 10. — Three-orbital interaction of the $1b_1$ orbital of the tetrahedral fragment with the $2b_1$ and $1b_1$ orbitals of the square fragment for As_4S_4 and S_4As_4.

energy shifts is to lower the molecular $1b_1$ and $2b_1$ orbitals for As_4S_4 below those of S_4As_4. Full occupation of these orbitals alone accounts for approximately 4 eV difference between the energies of the two systems.

Recently, the synthesis of $Te_4S_4^{2+}$ has been achieved, and its structure has been determined by X-ray diffraction [32]. This compound has 46 valence electrons, and therefore, according to the polyhedral skeletal electron pair theory, is expected to have a structure based on cuneane, with three bonds lengthened. Interestingly, the structure has been determined to be similar to that of S_4N_4 with only two bonds broken. Extended Hückel calculations performed on the $Te_4S_4^{2+}$ molecule (with a geometry similar to that of S_4N_4) show a similar molecular orbital pattern to those of S_4N_4 and As_4S_4. However, owing the extra pair of electrons, the HOMO for this molecule is a doubly degenerate e set, which is noded in an antibonding fashion between the tellurium atoms. Occupation of this doubly degenerate set would result in the molecule being paramagnetic. Unfortunately, in the original paper, there is no reference to magnetic properties of the molecule. Furthermore, another consequence of not fully occupying the doubly degenerate set would be for the molecule to undergo a Jahn–Teller distortion. From the crystal structure there is no evidence for geometric distortion in the molecule. These calculations can offer no explanation for the structure experimentally observed, but rather cast doubt on the charge that was assigned to the molecule. Consequently we performed a geometry optimization study of $Te_4S_4^{2+}$ using density functional analysis theory to predict the most stable structure for this molecule. The optimized geometry of this molecule using density functional analysis theory is shown to be a C_{2v} structure, with one of the Te—Te bonds broken [29].

ELECTRON RICH CLUSTERS BASED ON THE CUBE

The cube provides an alternative geometry to the cuneane structure discussed above for eight-vertex systems. The molecular orbitals of the cube have been studied extensively for both homoatomic [13, 33] and heteroatomic [34] cluster systems. The skeletal frontier orbitals of the cube are of t_{2u}, t_{2g} and t_{1u} symmetries. According to the polyhedral skeletal electron pair theory, a main group cluster based on a cube has a valence electron count of 40 electrons. Systematic addition of electron pairs to the electron precise cube will cause bonds to break. Like the tetrahedron discussed earlier, the cube is a regular polyhedron, and there is therefore only one possibility for cleavage of a single bond. Addition of two electron pairs will result in cleaving two bonds simultaneously, while maintaining at least double connectivity for each atom. Further addition of an electron pair causes three bonds to break. There are two alternatives available to break three bonds. The respective cleavage patterns for one, two and three excess electron pairs are depicted here in **19**.

The effects on the frontier orbitals of the cube in each case are shown in Fig. 11. Cleavage of one bond results in minor energy changes to the levels for the t_{2g} sets, with no resulting change in their bonding character. However, for the t_{1u} set a dramatic change is noted in the energy of the b_1 component. This orbital has been stabilized in energy by approximately 4.5 eV, with the result that it is no longer antibonding, but non-bonding. The stabilization of this orbital is seen to be due to a removal of the large antibonding interaction across the bond that has been lengthened. Lengthening

cube

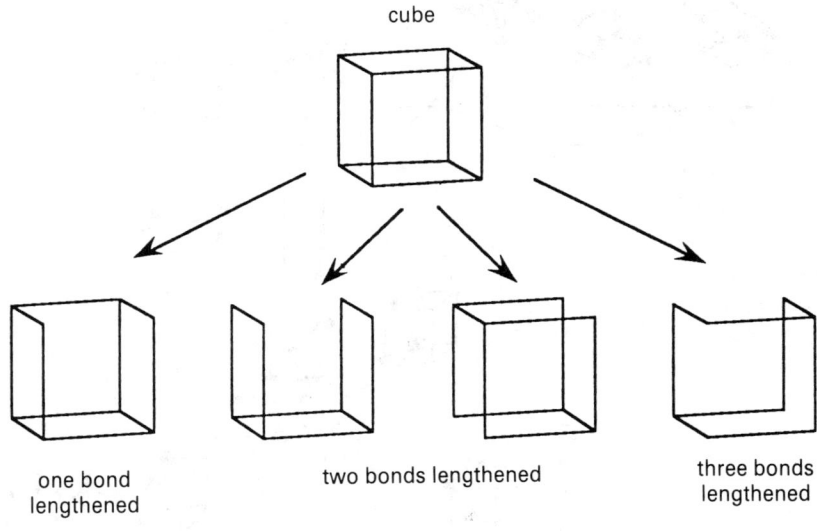

one bond
lengthened

two bonds lengthened

three bonds
lengthened

19

two bonds results in stabilization of the b_{3g} and b_{3u} orbitals for structure (c), and b_1 and b_2 orbitals of structure (d). Again, consideration of the component being stabilized reveals that weakening an antibonding interaction across the bond being lengthened is the driving force for the stabilization. Finally, lengthening three bonds results in stabilization of all the t_{1u} (a_1 + e in C_{3v} symmetry) orbitals, producing three additional orbitals in the frontier region. Once again, stabilization of the orbital energies results from weakening antibonding interactions along the lengthened bonds of the triply degenerate set.

The familiar pattern, noted previously, has once again emerged. Lengthening one to three bonds results in a corresponding number of orbitals being stabilized. As before, the symmetries of these orbitals may be derived group theoretically by using singly noded functions along the broken bonds of the cube as the bases for the irreducible representations. The symmetries of the orbitals stabilized on bond lengthening, derived in this manner, are summarized in Table 6.

Charge distributions in distorted cubes

The charge distributions for the distorted Se_8^{2+} cubes are shown in **20**. As before, there is a localization of charge on atoms previously bonded to each other.

The most stable geometry for 44 valence electrons is seen to be that of structure (c) in **20**. From an electronegativity viewpoint, the more electronegative atoms in an electron-rich heteroatomic cluster are predicted to reside at the two-connected positions. This is exemplified by the experimentally observed 46-valence-electron $Se_6Te_2^{2+}$ cube [35], which has the less electronegative tellurium atoms at the three-connected sites of structure (e) in **20**.

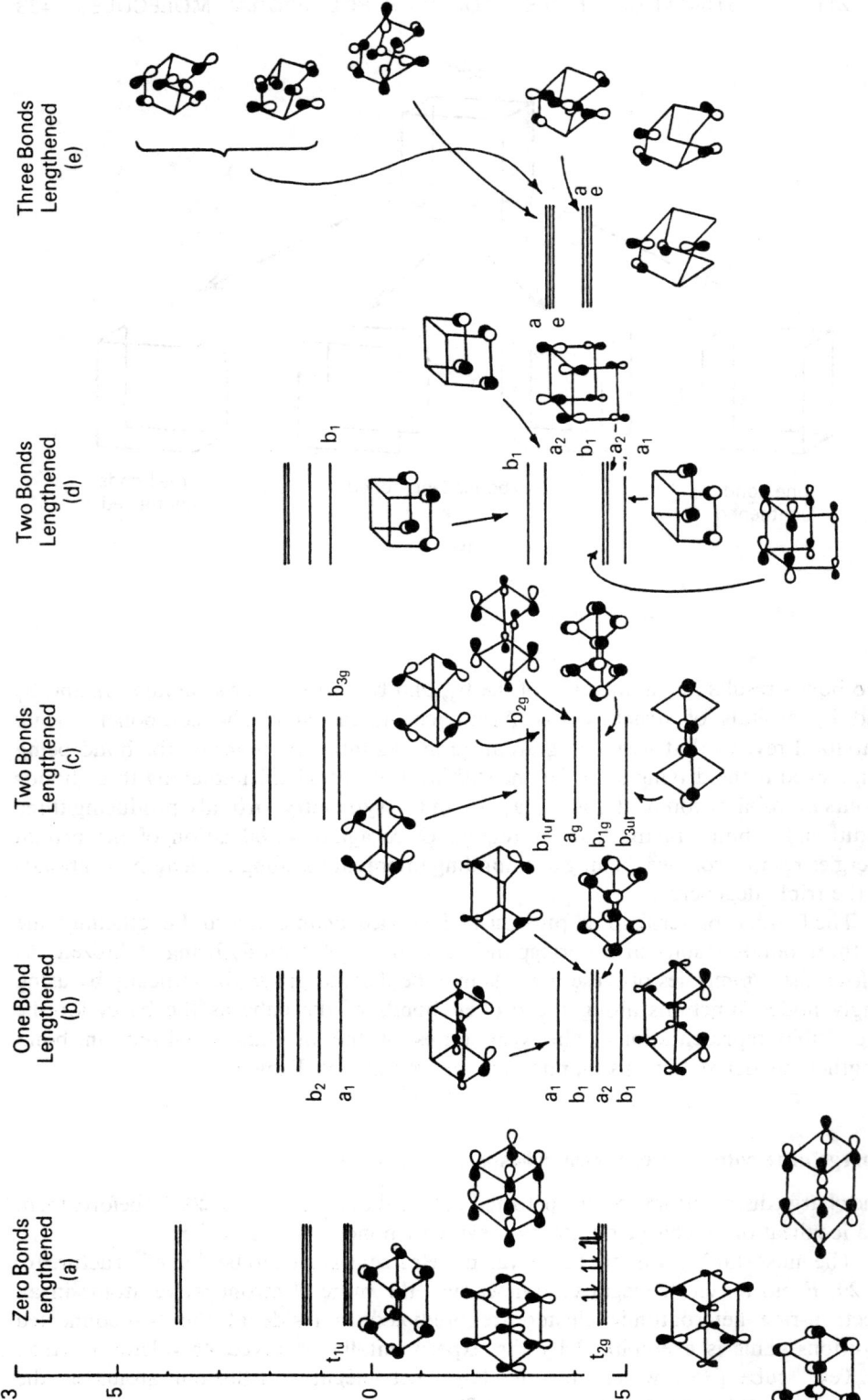

Fig. 11. — Effects on the frontier orbitals of the cube on lengthening one. two and three bonds.

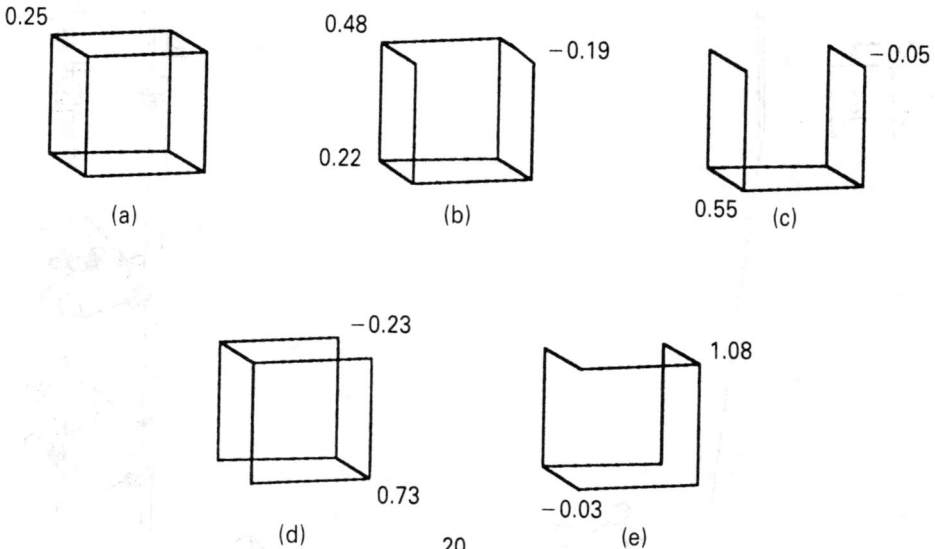

Molecular orbital analysis of $Se_6Te_2^{2+}$

For this particular example the predominant influence on site preferences must be electronegativity, since the valencies of the atoms are the same. The effect of electronegativity on the frontier molecular orbitals of Se_8^{2+}, on introduction of two tellurium atoms, is shown in Fig. 12. The highest occupied orbitals of the Se_8^{2+} molecule are of a and e symmetry. The a orbital has major contributions from all the atoms; however, the e set of orbitals has contributions only from those atoms at the two-connected sites. Therefore, from first-order perturbation theory arguments, placing the tellurium atoms at the two-connected sites destabilizes these orbitals more than if tellurium were to be placed at the three-connected sites. This can be seen to be so in Fig. 12, where introducing tellurium at the three-connected sites has caused a small destabilization in the energy of the a and e orbitals. Introducing tellurium at the two-connected sites has caused a destabilization of the e set by approximately 1 eV.

Table 6 — Derived symmetries of the orbitals stabilized by various geometric distortions to the cube. The relevant geometries (b)–(e) are displayed in Fig. 11

Geometry	No. of bonds lengthened	Symmetry of the orbitals stabilized: derived from t_{1u}
(b)	1	b_1
(c)	2	$b_{3g} + b_{3u}$
(d)	2	$b_1 + b_2$
(e)	3	$a_1 + e$

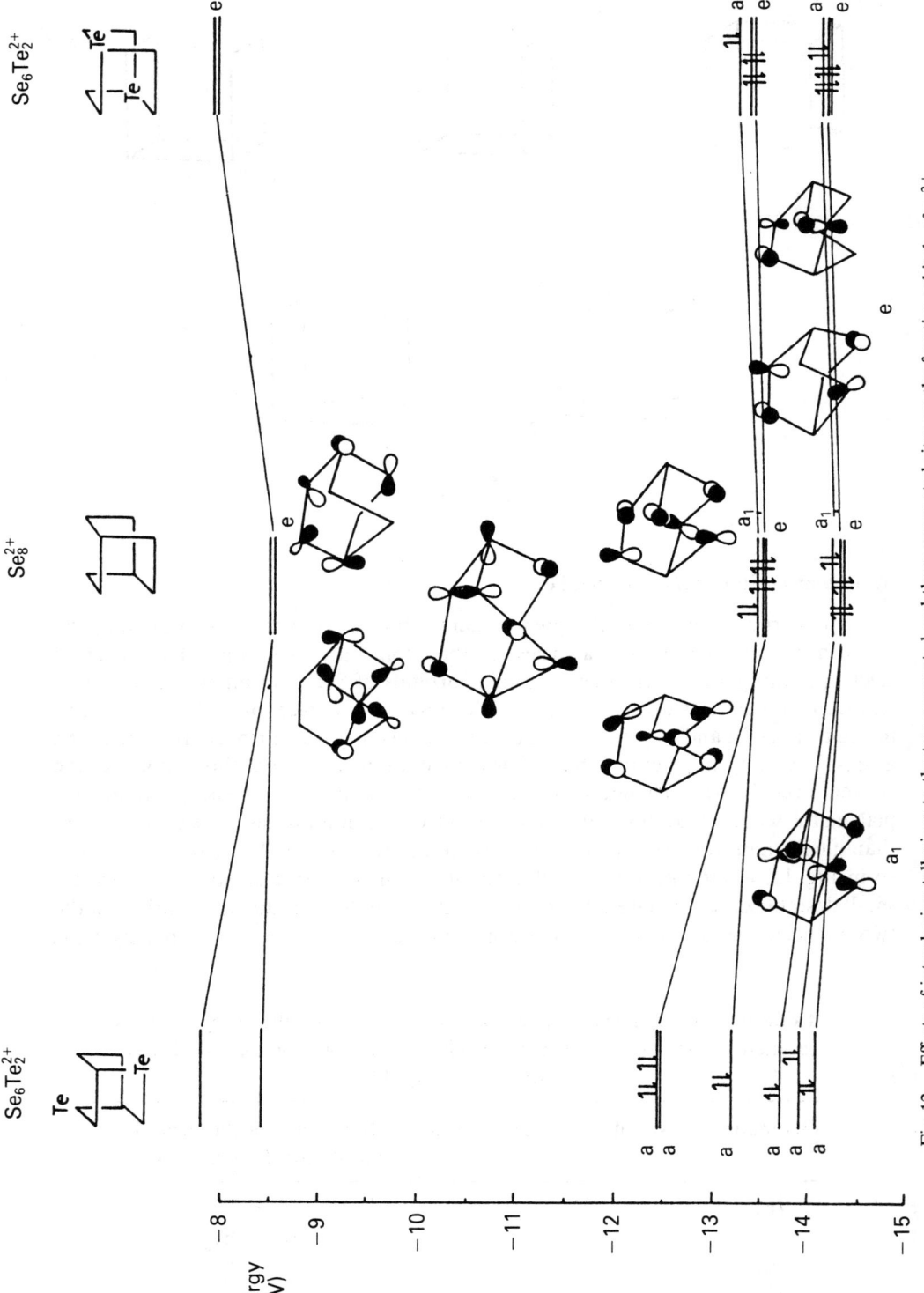

Fig. 12.— Effects of introducing tellurium to the two-connected and three-connected sites on the frontier orbitals of Se_8^{2+}.

For the lowest unoccupied set of orbitals, of e symmetry, the three-connected sites have a contribution in both components. Thus, introducing tellurium at either the two- or the three-connected sites will destabilize the LUMO relative to the LUMO of Se_8^{2+}. Therefore, the overall analysis indicates that, on perturbation theory grounds, it is more favourable to have tellurium occupy the three-connected sites of the $Se_6Te_2^{2+}$, since this causes less of a destabilization to the occupied frontier orbitals of the system.

CONCLUSION

The above analyses have demonstrated that there is a set of common factors governing the site preferences of atoms in electron-rich, three-connected main group heteroatomic clusters. The number of valence electrons of the cluster, x, provides information as to which electron-precise three-connected structure the electron-rich cluster structure is related. The actual structure adopted by the electron-rich cluster will have $(x - 5n)/2$ bonds less than the electron-precise analogue, where n is the number of atoms in the cluster. The first influencing factor is the relative electronegativities of the heteroatoms. The more electronegative atoms will reside at sites that were formerly bonded in the electron-precise structure. This has been borne out in the calculations by a consideration of the charge distributions in these electron-rich clusters, together with perturbational molecular orbital theory arguments used to assess the various effects on the frontier orbitals of the systems. A secondary factor is the valency of the atoms involved. This was highlighted above by considering a hypothetical M_4 butterfly cluster, where the electronegativities of the atoms are identical, but the respective valencies of two pairs of the atoms are different. Consideration of the charge distributions for the various structural alternatives leads to the prediction that the most preferred structure is that which had the lower valent atoms at the wingtips (two-connected sites) of the butterfly, and the higher valent atoms occupy the backbone (three-connected sites).

ACKNOWLEDGEMENTS

We thank the AFOSR for financial support, and Prof. Tom Ziegler for density functional calculations, and Dr. Catherine Housecroft for Fenske–Hall calculations.

APPENDIX: COMPUTATIONAL DETAILS

(ii) Extended Hückel calculations

The extended Hückel calculations were performed using the method developed by Hoffmann et al. [36]. The relevant atomic orbital parameters are tabulated here.

Atomic orbital	H_{ii}/eV	Slater exponent
As 4s[a]	−16.22	2.230
4p	−12.16	1.890
H 1s[b]	−13.60	1.300
N 2s[b]	−26.00	1.950
2p	−13.40	1.950
P 2s[c]	−18.60	1.750
2p	−14.00	1.300
S 3s[d]	−20.00	2.122
3p	−13.00	1.827
Se 4s[e]	−20.50	2.440
4p	−14.40	2.070
Te 5s[f]	−20.80	2.510
5p	−13.20	2.160

[a] Underwood, D.J., Nowak, M. and Hoffman, R. *J. Am. Chem. Soc.* **106**, 2837 (1984).
[b] Hoffmann, R. *J. Chem. Phys.* **39**, 1397 (1963).
[c] Summerville, R.H. and Hoffman, R. *J. Am. Chem. Soc.* **98**, 7240 (1976).
[d] Conan, F., Sala-Pala, J., Guerchais, J.E., Li, J., Hoffmann, R., Mealli, C., Mercier, R. and Toupet, L. *Organometallics* **8**, 1929 (1989).
[e] Hoffmann, R., Shaik, S., Scott, S.C., Whangbo, M.-H. and Foshee, M.J. *J. Solid State Chem.* **34**, 263 (1980).
[f] Canadell, E., Mathey, Y. and Whangbo, M.-H. *J. Am. Chem. Soc.* **110**, 104 (1988).
H_{ii} Coulomb Energy.

The relevant bond distances used in the calculations are given here:

P—P = 2.21 Å; P—H = 1.45 Å; Te—Te = 2.675 Å; Te—S = 2.45 Å;
S—S = 2.05 Å†; S—N = 1.62 Å.
As—As = 2.70 Å; As—S = 2.25 Å.

† The S—S bond distance for S_4N_4 was 2.60 Å.

REFERENCES

[1] (a) R.E. Williams, *Inorg. Chem.* **10** (1971) 210. (b) R.E. Williams, *Prog. Inorg. Chem. Radiochem.* **18** (1976) 1.

[2] (a) K. Wade, *J. Chem. Soc., Chem. Comm.* (1971) 792. (b) K. Wade, *J. Adv. Inorg. Chem. Radiochem.* **18** (1976) 1.

[3] (a) D.M.P. Mingos, *Nature Phys. Sci.* **236** (1972) 99. (b) D.M.P. Mingos, *Acc. Chem. Res.* **17** (1984) 311.

[4] (a) R.W. Rudolph and W.R. Pretzer, *Inorg. Chem.* **11** (1972) 1974. (b) R.W. Rudolph, *Acc. Chem. Res.* **9** 446 (1976) 446.

[5] D.M.P. Mingos, *Chem. Soc. Rev.* **15** (1986) 31.

[6] D.M.P. Mingos, *Inorganometallics* Ed,. T.P. Fehlner, Plenum Press, New York (1991).

[7] M.I. Forsyth and D.M.P. Mingos, *J. Chem. Soc., Dalton Trans.* (1977) 610.

[8] D.M.P. Mingos, *J. Chem. Soc., Chem. Comm.* (1983) 706.

[9] D.M.P. Mingos and L. Zhenyang, *J. Chem. Soc., Dalton Trans.* (1988), 1657.

[10] J.W. Lauher, *J. Am. Chem. Soc.* **100** (1978) 5305.

[11] (a) A.J. Stone, *Inorg. Chem.* **20** (1981) 503. (b) A.J. Stone, *Mol. Phys.* **41** (1980) 1339.

[12] J. Howell, A. Rossi, D. Wallace, K. Haraki and R. Hoffmann, *Quant. Chem. Program Exchange* **10** (1977) 344.

[13] (a) J.M. Schulman and T.J. Venanzi, *J. Am. Chem. Soc.* **96** (1974) 4739. (b) E. Heilbronner, T.B. Jones, A. Kreks, K.-D. Malsch, G. Maier, J. Pocklington and A. Schmelzer, *J. Am. Chem. Soc.* **102** (1980) 564.

[14] H. Bock and H. Müller, *Inorg. Chem.* **23** (1984) 4365 and references therein.

[15] R.L. Johnston and D.M.P. Mingos, *J. Organomet. Chem.* **280** (1985) 407.

[16] D.M.P. Mingos and L. Zhenyang. *New J. Chem.* **12** (1988) 787.

[17] R.J. Cave, E.R. Davidson, P. Sautet, E. Canadell and O. Eisenstein, *J. Am. Chem. Soc.* **111** (1989) 8105.

[18] R.C. Burns and J.D. Corbett, *J. Am. Chem. Soc.* **103** (1981) 2627.

[19] M.D. Newton, J.M. Schulman and M.M. Manus, *J. Am. Chem. Soc.* **96** (1974) 17.

[20] G.J. Schrobilgen, R.C. Burns and P. Granger, *J. Chem. Soc., Chem. Comm.* (1978) 951.

[21] W.A. Shantha Nandana, J. Passmore, P.S. White and C.-M. Wong, *Inorg. Chem.* **29** (1990) 3529.

[22] J. Passmore, G.W. Sutherland, T.K. Whidden and P.S. White, *J. Chem. Soc., Chem. Comm.* (1980) 290.

[23] Y.C. Leung and J. Waser, *J. Phys. Chem.* **60** (1956) 539.

[24] R.C. Burns, R.J. Gillespie, W.-C. Luk and D.R. Slim, *Inorg. Chem.* **18** (1979) 3086.

[25] R.B. Woodward and R. Hoffmann, *The Conservation of Orbital Symmetry*, Academic Press (1969).

[26] (a) O. Gunnarsson and I. Lundquist, *Phys. Rev.* **B10** (1974) 1319. (b) O. Gunnarsson and I. Lundquist, *Phys. Rev.* **B13** (1976) 4274. (c) O. Gunnarsson, M. Johnson and I. Lundquist, *Phys. Rev.* **B20** (1979) 3136.

[27] P.D. Lyne, D.M.P. Mingos and T. Ziegler, *in preparation.*

[28] D. Clark, *J. Chem. Soc.* (1952) 1615.

[29] H.J. Whitfield, *J. Chem. Soc.* **(A)** (1970) 1800.
[30] R. Gleiter, *Angew. Chem.* (Int. Ed.) **20** (1981) 444.
[31] K.A. Schugart, *Polyhedron* **9** (1990) 1935.
[32] R. Faggiani, R.J. Gillespie and J.E. Vekris, *J. Chem. Soc., Chem. Comm.* (1988) 902.
[33] (a) W. Schubert, M. Yoshimine and J. Pacansky, *J. Phys. Chem.* **85** (1981) 1340. (b) J.M. Schulman, C. Rutherford Fisher, P. Solomon and T.J. Venzani, *J. Am. Chem. Soc.*, **100** (1978) 2949 and references therein.
[34] R. Gleiter, K-H. Pfeifer, M. Baulder, G. Scholz, T. Wettling and M. Regitz, *Chem. Ber.* **123** (1990) 757.
[35] M.J. Collins, R.J. Gillespie and J.F. Sawyer, *Inorg. Chem.* (1987) 1476.
[36] (a) R. Hoffmann and W.N. Lipscomb, *J. Chem. Phys.* **36** (1962) 2179. (b) R. Hoffmann, *Ibid* **39** (1963) 1397.

III. Biochemistry and Biomedicine

22

DNH deoxyribonucleohelicates: self-assembly of oligonucleosidic double-helical metal complexes

Ulrich Koert, Margaret M. Harding and Jean-Marie Lehn
Université Louis Pasteur, Strasbourg, France

Nucleic acids, because of their key biological role, are prime targets for the design of either analogues that may mimic some of their features or of complementary ligands that may selectively bind to and react with them for regulation or reaction. Whereas there has been much work on the latter topic since the elucidation of the double-helical structure of DNA [1, 2], comparatively little has been done on structural and/or functional models, probably owing to the lack of self-organizing molecular systems. Here we present a class of artificial systems, the nucleohelicates, which are of interest from both points of view because they combine the double-helical structure of the double-stranded metal complexes, the helicates [2, 3], with the selective interaction features of nucleic-acid bases. These functionalized species allow the study of structural effects on the formation of the double helix and on the binding to other entities, in particular to nucleic acids.

We have shown previously that oligobipyridine molecules containing a string of 2,2′-bipyridine (bipy) groups separated by a suitable spacer unit undergo spontaneous self-organization into oligonuclear metal complexes with a double-helical structure (Fig. 1). In these helicates two molecular strands are wrapped around each other and held together by Cu(I) ions that provide the driving force for organization by imposing a (distorted) tetrahedral coordination geometry at each site. Di- to pentahelicates have thus been obtained [2, 3]. The trihelicate $[Cu_3(3-H)]^{3+}$, (4-H), schematically represented by structure $\mathscr{H}_3$, is formed by the binding of three Cu(I) ions by two tris-bipyridine molecules (3-H); it has been characterized by crystal structure determination [2]. These compounds provide a molecular framework for the spatial organization of substituent groups attached to the periphery of the double helix and projecting outwards. Exploitation of this possibility requires the introduction of functional groups on the bipyridine subunits that will allow the attachment of various molecular groups.

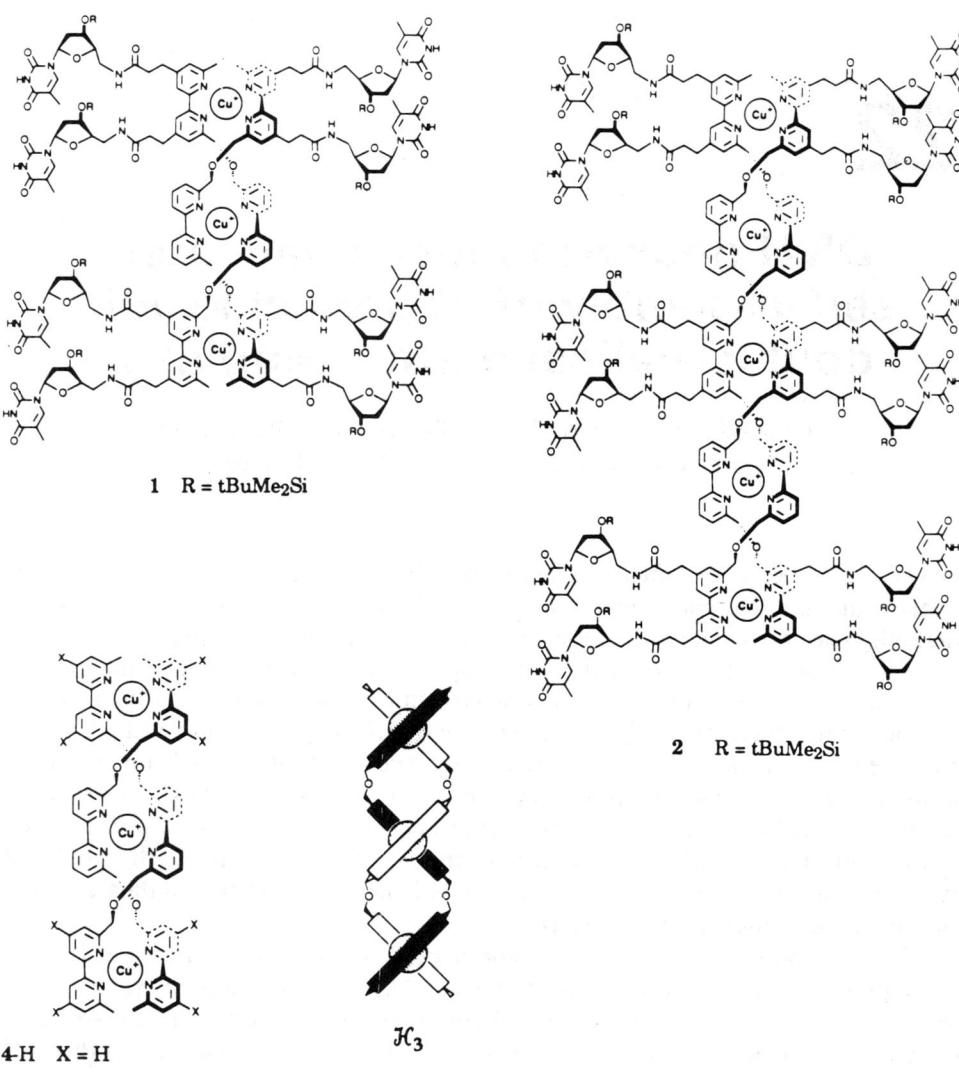

1 R = tBuMe₂Si

2 R = tBuMe₂Si

4-H   X = H

4-(C₂E)₈   X = CH₂CH₂CO₂tBu

$\mathcal{H}_3$

Fig. 1—Structure of the deoxyribonucleohelicates **1** and **2** and of the trihelicates **4-H** and **4-(C₂E)₈**; the two bipyridine groups at each Cu(I) centre are almost perpendicular to one another, as schematically represented in $\mathcal{H}_3$ for **4-H**.

common solvents except dimethyl sulphoxide (DMSO); its solubility could be increased by acylation of the free hydroxyl function to the hexa-acetyl, -butyryl-methoxyacetyl and -decanoyl derivatives. All new compounds had spectral (NMR, FAB⁺ mass) and elemental analysis data in agreement with their structure.

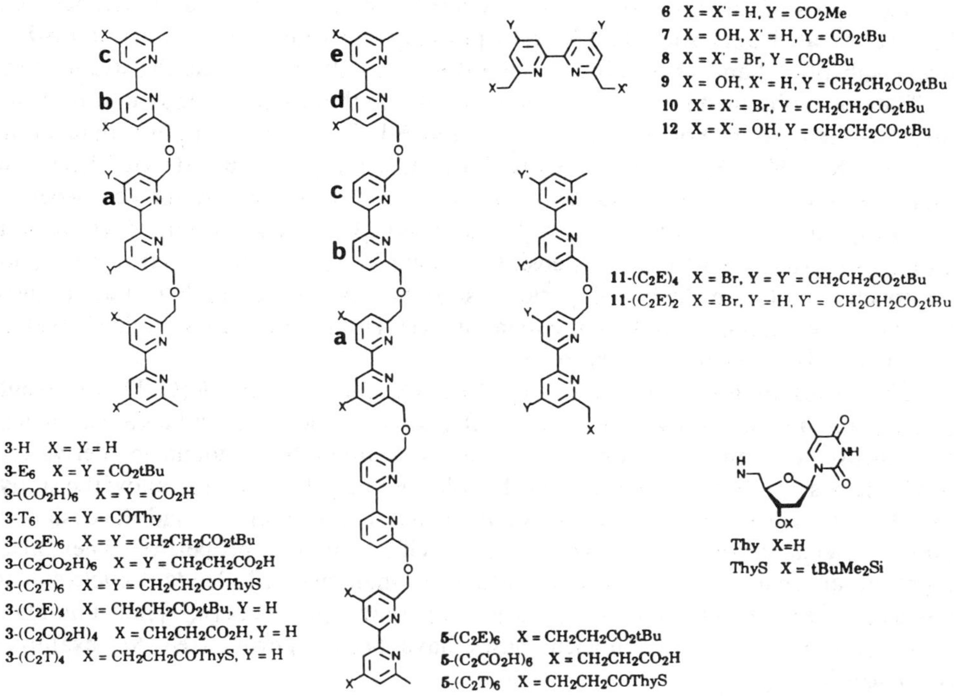

6　$X = X' = H, Y = CO_2Me$
7　$X = OH, X' = H, Y = CO_2tBu$
8　$X = X' = Br, Y = CO_2tBu$
9　$X = OH, X' = H, Y = CH_2CH_2CO_2tBu$
10　$X = X' = Br, Y = CH_2CH_2CO_2tBu$
12　$X = X' = OH, Y = CH_2CH_2CO_2tBu$

$11\text{-}(C_2E)_4$　$X = Br, Y = Y' = CH_2CH_2CO_2tBu$
$11\text{-}(C_2E)_2$　$X = Br, Y = H, Y' = CH_2CH_2CO_2tBu$

$3\text{-}H$　$X = Y = H$
$3\text{-}E_6$　$X = Y = CO_2tBu$
$3\text{-}(CO_2H)_6$　$X = Y = CO_2H$
$3\text{-}T_6$　$X = Y = COThy$
$3\text{-}(C_2E)_6$　$X = Y = CH_2CH_2CO_2tBu$
$3\text{-}(C_2CO_2H)_6$　$X = Y = CH_2CH_2CO_2H$
$3\text{-}(C_2T)_6$　$X = Y = CH_2CH_2COThyS$
$3\text{-}(C_2E)_4$　$X = CH_2CH_2CO_2tBu, Y = H$
$3\text{-}(C_2CO_2H)_4$　$X = CH_2CH_2CO_2H, Y = H$
$3\text{-}(C_2T)_4$　$X = CH_2CH_2COThyS, Y = H$

$5\text{-}(C_2E)_6$　$X = CH_2CH_2CO_2tBu$
$5\text{-}(C_2CO_2H)_6$　$X = CH_2CH_2CO_2H$
$5\text{-}(C_2T)_6$　$X = CH_2CH_2COThyS$

Thy　$X = H$
ThyS　$X = tBuMe_2Si$

Fig. 2 — Oligonucleosidic strands $3\text{-}T_6$, $3\text{-}(C_2T)_6$, $3\text{-}(C_2T)_4$ and $5\text{-}(C_2T)_6$ and synthetic intermediates.

Method. The key reaction was the condensation of compounds **7** and **8**. *t*-BuOK (2.4 eq) was added to a solution of compound **8** (2 eq) in *t*-BuOH and stirred at room temperature. After 1 h a solution of compound **7** (1 eq) in CH_2Cl_2 was added and the mixture was refluxed for 4 h. Together with the presence of the *t*-Bu groups in the ester functions, a bulky base and solvent were used to minimize side-reactions. The trimeric hexaester $3\text{-}E_6$ was isolated and purified by chromatography on silica (m.p. 135–136°C, 50% yield). Condensation of 2 eq of the monoalcohol **9** with the dibromide **10** using *t*-BuOK (1 : 1 *t*-BuOH: tetrahydrofuran (THF), 20°C 24 h) gave elongated hexaester $3\text{-}(C_2E)_6$ (m.p. 76°C, 31% yield) together with the bis-bipy monobromide $11\text{-}(C_2E)_4$ (oil, 23% yield), which were separated by column chromatography (alumina; eluent, 1 : 4 AcOEt: hexane). Condensation of 2 eq of the monoalcohol **9** with 6,6′-bis-bromomethyl bipy [20] using *t*-BuOK (1 : 1 *t*-BuOH:THF, 20°C, 4 h) gave a mixture of the mono-substituted dimeric species $11\text{-}(C_2E)_2$ (m.p. 88°C, 57% yield) and the alternating tris-bipy tetraester $3\text{-}(C_2E)_4$ (m.p. 48°C, 30% yield), which were separated by column chromatography (alumina; eluent, 4 : 1 hexane:AcOEt). Condensation of 2 eq of the bis-bipy monobromide $11\text{-}(C_2E)_2$ with the dialcohol **12** (*t*-BuOK; 1 : 1*t*-BuOH:CH_2Cl_2, room temperature 12 h) gave the alternating penta-bipy elongated hexaester $5\text{-}(C_2E)_6$ which was isolated and purified by chromatography (alumina; eluent, 2 : 1 hexane; AcOEt, m.p. 108°C, 47% yield). 6′-Amino-3′-deoxy-thymidine (H-Thy) and its 3′-*t*-butyldimethyl silyl ether (H-ThyS) were prepared from thymidine following a reported reaction sequence [21]. The polyacids $3\text{-}(CO_2H)_6$, $3\text{-}(C_2CO_2H)_6$, $3\text{-}(C_2CO_2H)_4$ and $5\text{-}(C_2CO_2H)_6$ were obtained by acid-catalysed cleavage of the corresponding *t*-butyl esters (CF_3CO_2H, CH_2Cl_2, room temperature 12 h; >95% yield) and transformed into the corresponding acid chlorides ($SOCl_2$, reflux, 1–2 h), which were used directly for condensation with either H-Thy or H-ThyS (2 eq per acid chloride function, 4 eq Et_3N, $CHCl_3$, 20°C, 2 h) to give the corresponding oligonucleoside compounds $3\text{-}T_6$ (43% yield), $3\text{-}(C_2T)_6$ (143°C, 48% yield), $3\text{-}(C_2T)_4$(167–168°C, 70% yield) and $5\text{-}(C_2T)_6$(162–163°C, 54% yield). The silylated compounds $3\text{-}(C_2T)_6$, $3\text{-}(C_2T)_4$ and $5\text{-}(C_2T)_6$, which were readily soluble in polar organic solvents (such as methanol, chloroform) were purified by column chromatography (silica, THF eluent). On the other hand $3\text{-}T_6$ had very low solubility in most

Two types of functionalized oligobipyridine compounds have been synthesized: fully substituted ones such as $3\text{-}X_6$ and partially substituted ones, $3\text{-}X_4$ and $5\text{-}X_6$, where substituted and unsubstituted bipyridine units alternate along the strand (Fig. 2). They were obtained following a scheme similar to that used for the synthesis of the unsubstituted parent compounds 3-H [2] and 5-H [3] from the appropriate basic units (X, X' = H, OH or Br; Y = CO_2t-Bu or $CH_2CH_2CO_2t$-Bu) derived from the diester 6 (X = X' = H, Y = CO_2Me) [4] using reactions to be described elsewhere.

Compounds $3\text{-}E_6$, $3\text{-}(C_2E)_6$, $3\text{-}C_2E)_4$ and $5\text{-}(C_2E)_6$ allow the introduction of a variety of groups (such as nucleosides, amino acids, sugars, electroactive and photoactive units) on the oligobipyridine strands. Only one case, that of an amino-deoxynucleoside, 6'-amino-3'-deoxythymidine (H-Thy), used as its soluble silylvated derivative (H-ThyS), is described here.

The corresponding derivatives $3\text{-}T_6$, $3\text{-}(C_2T)_6$, $3\text{-}(C_2T)_4$ and $5\text{-}(C_2T)_6$ represent artificial oligonucleosidic strands in which four or six nucleoside residues are attached to a backbone of bipyridine units. The analogous aminodeoxy-adenosine derivative $5\text{-}(C_2A)_6$ has also been synthesized. Such molecules may, for instance, allow the study of base-pairing features in model arrays of nucleic acid components and the design of novel systems presenting recognition and self-assembling properties. They also represent alternatives to flexible oligonucleoside analogues [5]. Pendant nucleic-acid bases have been introduced on the backbone of organic polymers [6, 7] and a number of systems containing two nucleic bases have been reported (see, for example, references [8–14]).

Complexation of Cu(I) by the unsubstituted oligobipyridine compounds 3-H and 5-H has been shown to give, respectively, the tri- and pentahelicates [2, 3]. Similarly, the substituted derivatives described here gave the corresponding substituted helicates when the polyesters $3\text{-}E_6$, $3\text{-}(C_2E)_4$, $3\text{-}(C_2E)_6$ and $5\text{-}(C_2E)_6$ were treated with copper(II) trifluoro $CuTf_2$ (Tf, CF_3SO_3) (2 eq) and hydrazine ($N_2H_4 \cdot H_2O$) in CH_3CN or in $CH_3CN\text{-}CHCl_3$; these results will be described elsewhere. The 200-MHz ^{1}H-NMR and FAB$^+$ (fast-atom bombardment) mass spectra of the octaester and dodecaester helicates $\{Cu_3[3\text{-}(C_2E)_4]_2\}Tf_3$, $4\text{-}(C_2E)_8$ and $\{Cu_5[5\text{-}(C_2E)_6]_2\}Tf_5$ are shown in Figs 3a, c, 4a and 5a; they agree with the structures, as do the elemental analyses.

A deep-red colour characteristic of Cu(I)-bipyridine complexes was also obtained on treatment of $3\text{-}T_6$ with the same reagents in 1:1 aqueous acetonitrile. The NMR and mass spectral data did not confirm the formation of only the dodecathymidino-trihelicate, but indicated that a mixture of complexes was obtained. This may be due to the destabilizing effect of both the carbonyl groups conjugated with the pyridine rings and steric hindrance between bulky substituent groups. Accordingly, we introduced a CH_2CH_2 segment into the side chains on alternating bipyridine units, as in $3\text{-}(C_2T)_4$ and $5\text{-}(C_2T)_6$, to reduce both adverse effects.

The following typical procedure was followed for preparing the complexes. 30.0 mg (1.35×10^{-5} mol) of $3\text{-}(C_2T)_4$ were dissolved in 15 ml $CHCl_3$ at room temperature and a solution of 14.7 mg (4.06×10^{-5} mol, 3 eq) of copper(II) trifluoro $CuTf_2$ in 5 ml CH_3CN added. The blue-green solution of the Cu(II) complex thus obtained was stirred for 10 min. Then, addition of 30 μl (6.18×10^{-4} mol, 15 eq) of $N_2H_4 \cdot H_2O$ reduced Cu(II) to Cu(I) and gave the expected colour change to red. The reaction mixture was stirred for 15 min. After evaporation of the solvent, the residue was taken up in 20 ml $CHCl_3$ and filtered through cotton wool. By addition of 30 ml

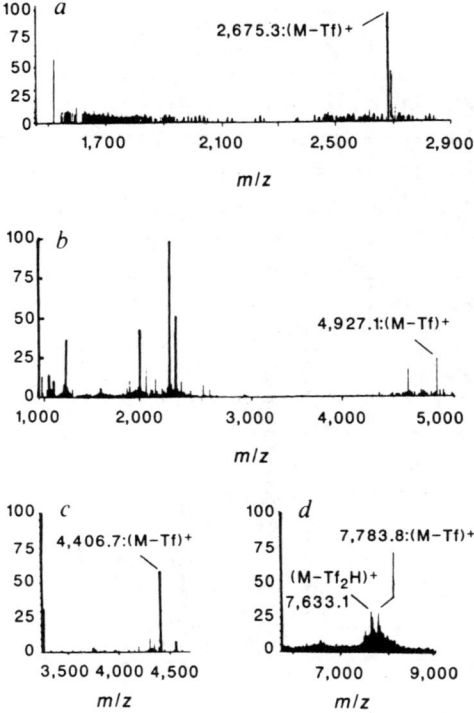

Fig. 3 — Partial FAB$^+$ mass spectra of a Cu$_3$[3-(C$_2$E)$_4$]$_2$Tf$_3$: (M–Tf)$^+$ found at 2675.4 (calc. 2675.3) (NBA, o-nitrobenzylalcohol); b, Cu$_3$[3-(C$_2$T)$_4$]$_2$Tf$_3$, 1: (M–Tf)$^+$ found at 4927.1 (calc. 4926.5) (DMSO); c, Cu$_5$[5-(C$_2$E)$_6$]$_2$Tf$_5$: (M–Tf)$^+$ found at 4406.7 (calc. 4406.3) (NBA); d, Cu$_5$[5-(C$_2$T)$_6$]$_2$Tf$_5$, 2: (M–Tf)$^+$ found at 7783.8 (calc. 7782.8) (DMSO).

of ether, the product was isolated as a red precipitate (29.9 mg, 5.89 × 10^{-6} mol, 87%; m.p. 180–181°C). Following the same procedure, treatment of 26.0 mg of 5-(C$_2$T)$_6$ with 5 eq of CuTf$_2$ and 20 eq of N$_2$H$_4$·H$_2$O in CH$_3$CN–CHCl$_3$ gave 25.1 mg of a red-orange solid (m.p. 178°C; 88% yield).

The elemental analyses (C, H, N, Cu) and ultraviolet absorption data of these solids were in agreement with the compositions [3-(C$_2$T)$_4$]$_2$[CuTf]$_3$ and [5-(C$_2$T)$_6$]$_2$[CuTf]$_5$.

The FAB$^+$ mass spectra, with peaks corresponding to {[3-(C$_2$T)$_4$]$_2$Cu$_3$Tf$_2$}$^+$ (Fig. 3b) and {[5-(C$_2$T)$_6$]$_2$Cu$_5$Tf$_4$}$^+$ (Fig. 3d), proved the presence of two ligands and, respectively, three and five copper(I) ions.

The ^{1}H-NMR spectra of the polynucleosidic trihelicate and pentahelicate were consistent with the assigned structures. High-field two-dimensional NMR techniques were used to analyse the spectra. In principle, complexation should lead to a mixture of the two diastereomers that differ in the handedness of the double helix. The presence of the chiral nucleosides might lead to a preferential induction of one helical sense, but this has not yet been proven. Fig. 4b (left) shows the NMR and TOCSY (total correlation spectroscopy) spectra of {Cu$_3$[3-(C$_2$T)$_4$]$_2$} Tf$_3$, compound 1. Changes in the ligand chemical shifts upon copper complexation were consistent with those observed earlier [2] on formation of the unsubstituted trihelicate 4-H from 3-H.

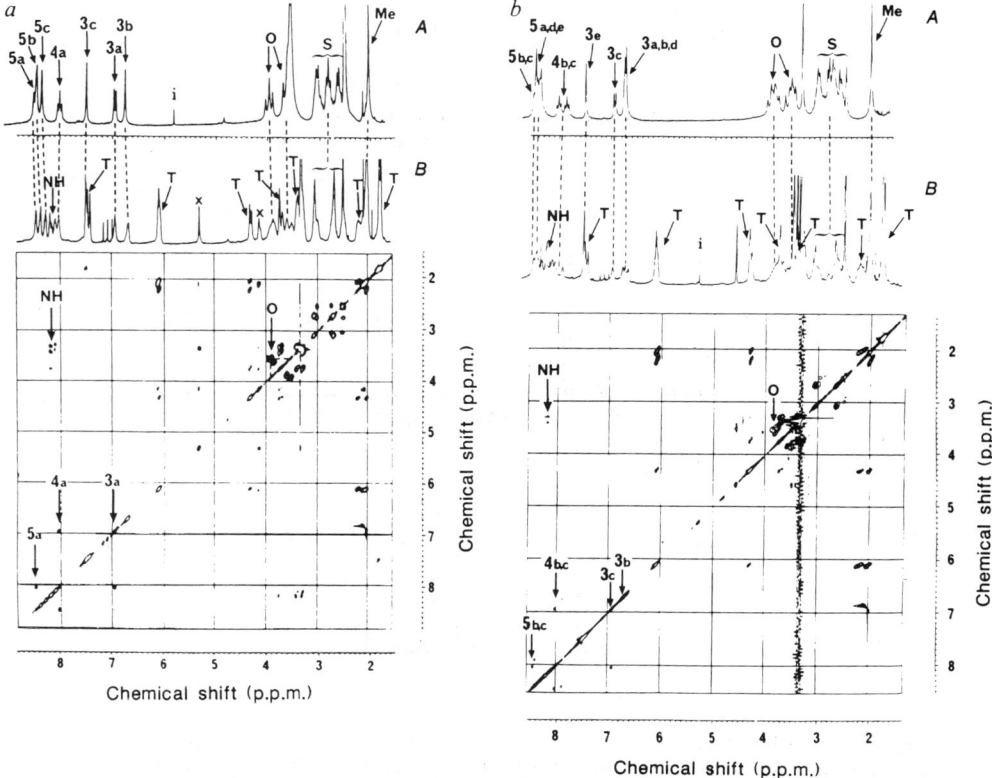

Fig. 4 — ¹H-NMR spectra (200 MHz, 293 K) between 1.5 and 8.5 p.p.m. of the octaester trihelicate $(Cu_3(3-C_2E_4)_2)Tf_3$ in d_6-DMSO (aA) and the dodecaester pentahelicate $(Cu_5(5-(C_2E)_6)_2)Tf_5$ in d_6-DMSO (bA) showing the aromatic region (3, 4, 5), the methylene oxy bridge (O), the two-carbon spacer (S) and the bipyridine methyl group (Me) signals. ¹H-NMR (600 MHz, 293 K) and TOCSY spectra of the octathymidine trihelicate **1** in d_6-DMSO (aB) and the dodecathymidine pentahelicate **2** in d_6-DMSO (bB). Helicate signals having the same chemical shift as in aA and bA are indicated by dashed lines, thymidine signals by T, and propionamide NH by NH. Cross-peaks are marked by arrows for protons 3, 4, 5 on the bipyridine ring systems a–e, the propionamide NH and the methylene oxy bridge AB systems, x, thymidine monomer impurity; triplet at 7.1 p.p.m., NH_4^+; solvent peaks at 2.5 (DMSO) and 3.5 or 3.3 p.p.m. (H_2O); i, solvent impurity; a–e refer to the different pyridine rings (see Fig. 2).

The first change is a marked upfield shift and splitting of the CH_2OCH_2 singlets into two multiplets (O) corresponding to two AB systems in 4-H. The cross-peak indicated in the TOCSY spectrum (O) results from the coupling in the two AB systems. The second change is an upfield shift of the two bipyridine protons (3a, 3b) and of the bipyridine methyl group (Me); it may be attributed to shielding by pyridine units of the other strand. The interpretation was confirmed by comparison with the ¹H-NMR spectrum of the octaester helicate 4-$(C_2E)_8$ (Fig. 4a, left), which shows similar chemical shifts and for which the same changes are observed on helicate formation.

In an analogous manner the ^{1}H-NMR spectrum (NMR and TOCSY) of [5-(C$_2$T)$_6$]$_2$[CuTf]$_5$ (Fig. 4*b*, right) is consistent with a dodecathymidino-pentahelicate {Cu$_5$[5-(C$_2$T)$_6$]$_2$}Tf$_5$, **2**. The most important spectral features are again the AB multiplets (O) (cross-peak O in the TOCSY spectrum). Comparison with the spectrum of the corresponding dodecaester helicate {Cu$_5$[5-(C$_2$E)$_6$]$_2$}Tf$_5$ (Fig. 4*a*, right) again confirms the structural and spectral assignments. Water-soluble derivatives of compounds **1** and **2** (R = H, sulphate salts) have been obtained from the corresponding desilylated strands.

In view of its features this new class of substances may be termed deoxyribonucleohelicate, DNH. The DNH complexes, such as compounds **1** and **2**, are double-helical oligonucleosides formed spontaneously by self-assembling of two suitably designed ligand strands and copper(I) ions. In contrast to DNA they have positive charges located inside the strands whereas the information-bearing nucleic-acid bases are on the outer spine of the double helix, pointing away from its axis; this structure recalls the inside-out model once proposed for DNA [15]. They should be able to interact with other species both electrostatically and by hydrogen bonding through the thymidines (in a metallonucleate exoreceptor [16] fashion). One possible target is the negatively charged double-stranded DNA; binding in its major groove would lead to a composite natural–artificial quadruple-helix entity which might present thymine · adenine · thymine interactions (of the triple helix type [17–19]) with adenine–thymine regions in the DNA. Such possibilities are being explored.

ACKNOWLEDGEMENTS

We thank A. Rigault for her contribution to the synthesis of compound **3-E$_6$**, A. Van Dorsselaer for the FAB$^+$ studies, C. Brevard, A. Pagelot and Spectrospin AG for the 600-MHz NMR spectra and E. Westhof for generating the computer graphical representations. This work was supported by the Centre National de la Recherche Scientifique (France) and by the Fonds der Chemischen Industrie (UK) and the Collège de France (MMH).

REFERENCES

[1] Watson, J. D. & Crick, F. H. C. *Nature* **171**, 737–738 (1953).

[2] Lehn, J.-M. *et al. Proc. natn. Acad. Sci. U.S.A.* **84**, 2565–2569 (1987)

[3] Lehn, J.-M. & Rigault, A. *Angew. Chem. int. Edn Engl.* **27**, 1095–1097 (1988).

[4] Alpha, B., Anklam, E., Deschenaux, R., Lehn, J.-M. & Pietraskiewicz, M. *Helv. chim. Acta* **71**, 1042–1052 (1988).

[5] Schneider, K. C. & Benner, S. A. *J. Am. chem. Soc.* **112**, 453–455 (1990).

[6] Hoffmann, S., Witkowski, W. & Schubert, H. *Z. Chem.* **14**, 154 (1974).

[7] Takenvoto, K., Kawabuko, F. & Kondo, K. *Makromol. Chem.* **148**, 131–134 (1971).

[8] Browne, D. T., Eisinger, J. & Leonard, N. J. *J. Am. chem. Soc.* **90**, 7302–7323 (1968).

[9] Leonard, N. J. *Acc. chem. Res.* **12**, 423–429 (1979).

[10] Golankiewicz, K. & Celewicz, L. *Polish J. Chem.* **52**, 1035–1038 (1978).

[11] Sasaki, I., Dufour, M.-N. & Gaudemer, A. *Nouv. J. Chem.* **6**, 341–344 (1982).

[12] Saito, I., Sugiyama, H., Matsuura, T. & Fukuyama, K. *Tetrahedron Lett.* 4467–4470 (1985).

[13] Kim, M. & Gokel, G. *JCS Chem. Commun.* 1686–1688 (1987).

[14] Sessler, J. L., Magdal, D. & Hugdall, J. *J. Inclus. Phen. molec. Recogn.* **7**, 19–26 (1989).

[15] Pauling, L. & Corey, R. B. *Nature* **171**, 346 (1953).

[16] Lehn, J.-M. *Angew. Chem., int. Edn. Engl.* **27**, 89–112 (1988).

[17] Felsenfeld, G., Danies, D. R. & Rich, A. *J. Am. chem. Soc.* **79**, 2023–2024 (1957).

[18] Moser, H. E. & Dervan, P. E. *Science* **238**, 645–650 (1987).

[19] François, J.-C., Saison-Behmoaras, T. & Hélène, C. *Nucleid Acids Res.* **16**, 11431–11440 (1988).

[20] Rodriguez-Ubis, J.-C., Alpha, B., Plancherel, D. & Lehn, J.-M. *Helv. chim. Acta* **67**, 2264–2269 (1984).

[21] Banwarth, W. *Helv. chim. Acta* **71**, 1517–1527 (1988).

23

Non-covalent interactions in biologically relevant complexes. Importance of electrostatics and the atomic charge concept

J. Köhler, R. Kikuno and A. Plückthun
Ludwig-Maximilians-Universität, Munich, FRG

INTRODUCTION

Few scientists have influenced the field of biophysics from so many different sides as has Linus Pauling. His interest has left out hardly any important topic, and many of his ideas and concepts are now called 'classic'. His views on the particular importance of the three-dimensional structures and the underlying physical forces as the basis of biochemical processes have influenced the development of the whole field. The enormous progress in this field demonstrates the fertility of his concepts. Linus Pauling has contributed the classic book on the nature of the chemical bond [1] and he has been especially aware of the nature of forces between large molecules of biological interest. His concepts about the secondary structure of proteins [2, 3] and of molecular architecture for biological catalysis [4] are of particular importance. The analysis of electrostatic forces in proteins [5, 6] and the study of their aqueous environment [7-9] might lead to a deeper understanding of protein structure, function and dynamics [10].

Over the last two decades, several empirical force field methods [11-32] have been developed, mainly for the application of molecular mechanics and molecular dynamics simulations of biological molecules [33] *in vacuo*, in solution or in the crystalline state. Typical examples are the well known program packages AMBER, CHARMM, DISCOVER, GROMOS and X-PLOR (see listing at the end). An overview of biomolecular applications has been given [32].

Here, we show what an analysis of the electrostatic interactions in molecular complexes during a molecular dynamics trajectory reveals. The system investigated is the antibody/antigen complex of McPC603/phosphorylcholine and two related antibodies, M167 and T15, which also bind to phosphorylcholine, but are derived from different light chain genes.

THE MOLECULAR DYNAMICS METHOD

The general idea of a molecular dynamics simulation [34] is to solve Newton's equations of motion for a system of interacting atoms. Currently, time scales from femto- to nanoseconds can be reached for biological molecules. The system can be investigated *in vacuo*, in solution or in the crystalline state. The interacting particles may then be only the solute or the solute plus solvent (and possibly ions). At the start of the simulation, initial velocities are assigned to the atoms corresponding to a Maxwell–Boltzmann distribution at a given temperature. All atoms then move according to the forces they create between themselves as defined by the interaction energies from the force field.

Newton's equations are integrated in small time intervals (usually 0.5 to 2 femtoseconds), and at each time step, the forces, velocities and positions of each atom are calculated to give a so-called trajectory.

$$\frac{\mathrm{d}^2 \bar{r}_i(t)}{\mathrm{d}t^2} = \frac{\bar{F}_i(\bar{r}_1, \ldots, \bar{r}_N)}{m_i} \qquad i = 1, \ldots, N$$

Different algorithms may have advantages for specific applications. These algorithms [35–40] are either computationally very fast or very accurate [41]. The simulations can be quite time consuming: a trajectory of one picosecond of a protein with 235 amino acids in about 4000 water molecules and 20 ions (about 17 000 atoms in total) takes about 1 h of Cray YMP cpu time. If the system contains only one protein molecule, one should have trajectories of up to 100 or 200 ps to sample enough conformational space so that the statistical error becomes small enough.

In the current molecular dynamics approaches [42–45], the atoms are considered to be particles with mass, van der Waals radius and electrical atomic charge. Thus, electrons are not considered explicitly. The force field is a conservative one, meaning that the forces are dependent upon the distances between the atoms only. The potential is a pair potential, making it computationally fast.

The principal form of the current potential energy may be written as

$$V(\bar{r}_1, \ldots, \bar{r}_N) = \sum_{l=1}^{N_b} \frac{1}{2} K_{b_l}(b_l - b_{0_l})^2 + \sum_{l=1}^{N_\theta} \frac{1}{2} K_{\theta_l}(\theta_l - \theta_{0_l})^2$$

$$+ \sum_{l=1}^{N_\xi} \frac{1}{2} K_{\xi_l}(\xi_l - \xi_{0_l})^2 + \sum_{l=1}^{N_\phi} K_{\phi_l}(1 + \cos(n_l \phi_l - \delta_l))$$

$$+ \sum_{i<j}^{N} \left(\frac{C_{12}(ij)}{r_{ij}} \right)^{12} - \left(\frac{C_6(ij)}{r_{ij}} \right)^6 + \frac{q_i q_j}{4\pi\varepsilon_0\varepsilon_r r_{ij}}$$

The first term concerns the covalent bonds between two atoms: as soon as the atoms move to the distance b_l and away from their equilibrium distance b_{0_l}, a force with the force constant K_{b_l} drives them back. The second term describes the bond angles. The fourth and fifth terms model the dihedral angles, which can undergo limited (i.e. in rings) or full 360° rotations, respectively. In the last three terms, the interactions between atoms that are not covalently bound to each other are described. The repulsion between atoms has a $1/r^{12}$ dependency, the attractive (or dispersion) part is proportional to $1/r^6$, and together they model the van der Waals forces in a

Lennard–Jones-type potential. The last term contains the electrostatic interactions between all atom pairs. It is actually the first term of the infinite multipole expansion. This means that all terms of higher order (dipoles, quadrupoles, etc. and their interactions) are not taken into account explicitly. The force field is an 'effective' one, meaning that all parameters such as the force constants, Lennard–Jones parameters, atomic charges, etc. are scaled such that they reproduce the experimental observables of molecular ensembles well.

There are also somewhat different terms in use. Moreover, additional terms may be introduced: for example, Morse potentials can be taken for the covalent bonds, in order to have a better description of the bond-lengths, or cross-terms can be applied as in the DISCOVER force field [46] to model the coupling between two terms (used for calculating vibrational spectra). Some force fields contain extra hydrogen bond terms of van der Waals (Lennard–Jones)-type (the AMBER force field [20, 21]).

The computation of the non-bonded interactions in the above formula between an atom and all its neighbors takes about 80% of the total simulation time, mostly because the search algorithms for neighbors are time consuming.

WHAT CAN BE LEARNED FROM MOLECULAR DYNAMICS TRAJECTORIES?

The method at the current state of the art is designed mainly for trajectories on the picosecond time scale. A few examples that have been studied in detail may serve to illustrate some of the characteristics: the agreement between the neutron diffraction structures of cyclodextrin crystals and the averaged simulated structures at room temperature (293 K) is rather high [47, 48]. The experimental positions of non-hydrogen glucose backbone atoms were reproduced within 0.34 Å. This value is smaller than the root-mean-square (rms) atomic fluctuation of 0.41 Å as derived from the crystallographic B-factors. The simulation showed an overall rms atomic fluctuation of 0.49 Å, slightly larger than the experimental value.

At low temperature (120 K) [48] the experimental positions were reproduced to within 0.46 Å. This value is larger than the rms atomic fluctuation of 0.19 Å (experimental) or 0.22 Å (simulation). Both experiment and simulation showed a reduction by a factor 2 of the atomic mobility when the temperature was lowered from 293 to 120 K. The larger deviation of the atomic positions from the experimental ones at low temperature might be due to the fact that the empirical interatomic potential function has been designed as an effective force field at room temperature. The flexibilities of molecules may be understood from the rms deviations during the trajectories, giving an impression of how extended the conformational space is at a given temperature.

On the picosecond time scale, the dynamic behavior of hydrogen bond phenomena occurs [49–51] and can be observed in the simulations. Forming and breaking of two-, three- or multiple-center hydrogen bonds [52] as well as flip-flop hydrogen bonds [53, 54] were observed in trajectories of cyclodextrin crystals, with an accuracy of about 70 to 80% agreement compared to neutron diffraction data [55, 56].

From a simulation of an α-cyclodextrin molecule in solution [57] one can see that the molecule clearly explores a larger part of the conformational space in solution than in the crystal as determined in the neutron diffraction structure.

CALCULATION OF DIFFERENCES IN FREE ENERGY WITH MOLECULAR DYNAMICS

The possibility of calculating differences in free energies between related molecular systems has given the molecular dynamics method a broad range of applications in the field of protein–ligand interaction [58] as a defined theoretical method. The principles of this method have been described by several groups [31], and computational details have been discussed for ions in solution [59, 60], or applications for biological molecules [58, 61].

The methodological concepts for the calculation of differences in free energies with the molecular dynamics method have been explained in review articles by Kollman and van Gunsteren [62], and by several other authors [29, 31, 63–65]. Some applications of this method and a somewhat different approach to calculate free reaction energies have been reported [58]. (For a general overview of biochemical thermodynamics, see ref. 66.)

The calculated differences between the free energies of two systems can be directly compared to the experimentally determined binding constants using the following thermodynamic cycle:

$$
\begin{array}{ccc}
 & \Delta G_1 & \\
Enzyme + Substrate_1 & \rightleftharpoons & Enzyme - Substrate_1 - Complex \\
\Delta G_3 \updownarrow & & \Delta G_4 \updownarrow \\
Enzyme + Substrate_2 & \rightleftharpoons & Enzyme - Substrate_2 - Complex \\
 & \Delta G_2 &
\end{array}
$$

In this example, the values for ΔG_1 and ΔG_2 may be obtained from experiment (binding constants K_1 and K_2 of the two different substrates; $\Delta G = RT \ln K$). Values for ΔG_3 and ΔG_4 come from molecular dynamics calculations. Since the G-function is a state-function, the sum of all ΔG must be zero. The way in which the vertical processes are simulated is without physical reality: the force field parameters are changed from the ones that belong to substrate$_1$ into the ones that belong to substrate$_2$ during the simulation. This can be done [31] by redefining the Hamiltonian $H(p, r)$ in potential energy $V(r)$ and momentum p space as a function of the coupling parameter λ. Now the starting state A is described by $H(p, r, \lambda_A)$ and the final state B by $H(p, r, \lambda_B)$:

$$
H(p, r, \lambda) = \sum_{i=1}^{N} \frac{p_i^2}{2m_i(\lambda)} + V(r_1, \ldots, r_N, \lambda)
$$

During such a reversible process, the molecules adopt many configurations for which the enthalpic and entropic contributions to the total energy are calculated. The value for the free energy is evaluated by integrating:

$$
\Delta F_{BA} = \int_{\lambda_A}^{\lambda_B} \left\langle \frac{\partial H(p, r, \lambda)}{\partial \lambda} \right\rangle_\lambda d\lambda
$$

For an isobaric ensemble, a similar formula is valid [67]. The evaluation of the entropy is also based on the ensemble average:

$$S = -\frac{\partial F}{\partial T}$$

so that finally the difference $\Delta G = \Delta H - T\Delta S$ in free energy between states A and B can be obtained from the simulation. This fact makes molecular dynamics simulations a very powerful tool for comparing theoretical and experimental data, although it is computationally very intensive.

The calculated data are in good agreement with experimental data as long as the two structures are closely related, or at least do not differ in their absolute charge values. Here again, the importance of the charges becomes obvious. An example for a calculation of the relative change in binding free energy of a protein–inhibitor complex is the case of the enzyme thermolysin with a pair of phosphonamidate and phosphonate ester inhibitors. The calculated difference was 4.21 ± 0.54 kcal/mol, in nice agreement with the experimental value of 4.1 kcal/mol [68].

BIOCHEMICAL RELEVANCE: ANTIBODY/ANTIGEN COMPLEXES

Antibodies are a class of proteins with which most of the problems relevant in protein–ligand interactions can be investigated. Even chemical rate accelerations can be observed and studied with antibodies. Since antibodies can in principle be made against any substance, this offers exciting prospects of developing new catalysts.

Once again, this concept can be traced back to Linus Pauling. In an article in *Chemical and Engineering News* [4], he explained his picture of enzyme catalysis. It was clear to him that the surface of the enzyme must be complementary not to the substrate molecule itself but rather to a 'strained configuration', corresponding to the activated complex. This way, part of the intrinsic binding energy of the substrate can be used to bring the substrate closer to the transition state. At this time, of course, no three-dimensional structure of an enzyme was known, but today, this concept has been verified by crystallography [69, 70]. In 1969, Jencks [71] then first proposed raising antibodies against analogs of the transition state to test whether they can cause catalysis. This has been tried several times soon thereafter, but true success only came after the invention of monoclonal antibodies, allowing one to reach the high fairly molar concentration of a pure antibody necessary to observe significant rate accelerations. A whole number of reactions have now been catalysed with this concept [72–75].

An example shall illustrate this concept. The antibody McPC603 binds phosphorylcholine, and its three-dimensional structure is known [76, 77]. During the hydrolysis of an appropriate ester, the intermediate is tetrahedral [78a, b], and therefore structurally similar to the tetrahedral arrangement of the phosphate (Fig. 1). Hammond's postulate suggests that the two transition states leading to the tetrahedral intermediate and away from it will be very similar in structure to the intermediate. Indeed, this antibody McPC603 [79–81] as well as the related antibodies T15 and M167 [82, 83] all catalyze the hydrolysis of p-NO$_2$-phenyl-choline-carbonate, albeit moderately.

Unfortunately, the structures of the antibodies T15 and M167 are not known.

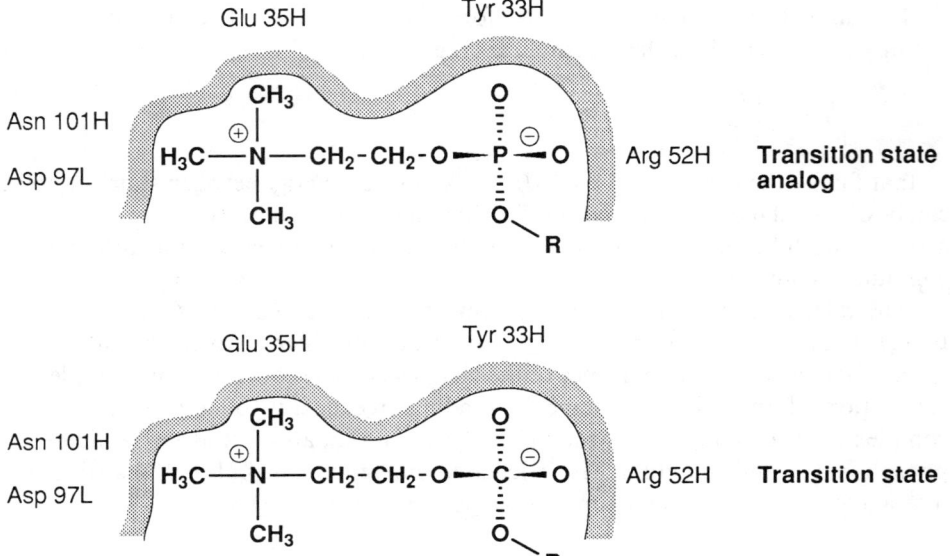

Fig. 1.—Schematic representation of the analogy of transition state binding and binding of the transition state analog. Residues from the antibody McPC603 are indicated schematically.

Recently, progress in the production of engineered antibodies from bacteria [84–86] has allowed the investigation of the effect of contributions from single amino acids [79, 80] on binding and catalysis, as well as renewed attempts of crystallizing antigen binding fragments of the antibodies T15 and M167.

The desire for a theoeretical understanding of this experimentally well-characterized system has been the impetus of some of the work described here. Binding constants between similar antigens or mutants differ only by a few kilocalories, and thus demand rather accurate methods for any useful predictions.

The first step was to model the structures of T15 and M167 starting from the known structure of McPC603 by replacing the different amino acids. This is complicated by the fact that the loops making up the binding pocket have different lengths. Models constructed with a distance matrix of the Brookhaven-Protein-Data-Base were used as the starting structures for the molecular dynamics simulations. The calculated systems contained one molecule of either McPC603, T15 or M167 (the Fv-fragments only, which is the heterodimer of the variable domains of the heavy and light chain, together about 235 amino acids), and some 4000 water molecules and ions (Fig. 2). After energy minimization, a trajectory of 30 ps at 300 K was calculated for each system (Figs 3(a), (b), (c)). The agreement between the crystal structure and the calculated trajectory of McPC603 can be seen in Figs 4(a), (b), where the B-factors, rms positional fluctuations and structural elements of Cα atoms [87] of the Fv-fragment are displayed. The atomic fluctuations in the crystal are significantly lower than in the simulated solution structure. There is a general agreement between experiment and simulation concerning the flexibility of certain areas (peaks in Figs 4(a), (b)).

Whereas the trajectory of McPC603 equilibrated, the ones for T15 and M167 did not. This can be seen from the analysis of the contributions of single amino acids to

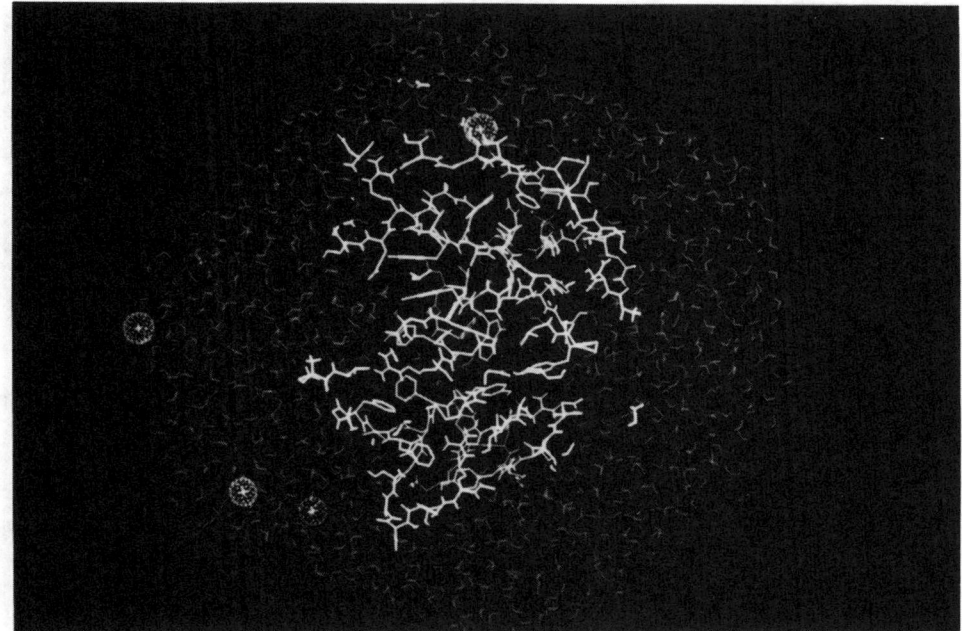

Fig. 2. — The Fv-fragment (of McPC603) and the bound antigen phosphorylcholine in a shell (6 Å) of water molecules including Na^+ and Cl^- ions. This system was simulated over 30 ps without any constraints (GROMOS87).

A

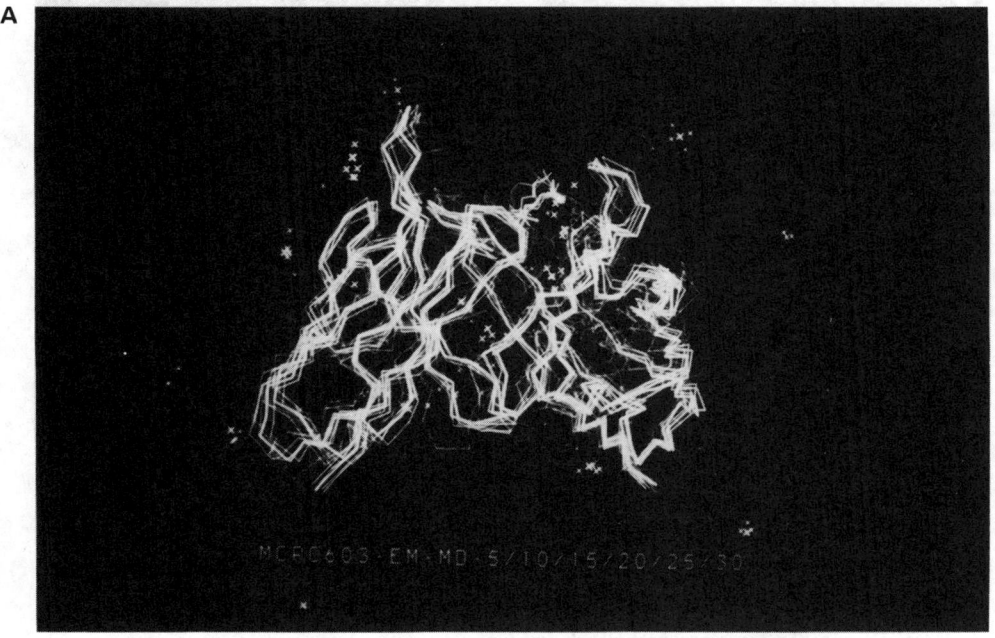

Fig. 3. — The backbone atoms of the Fv-fragments of McPC603 (A),

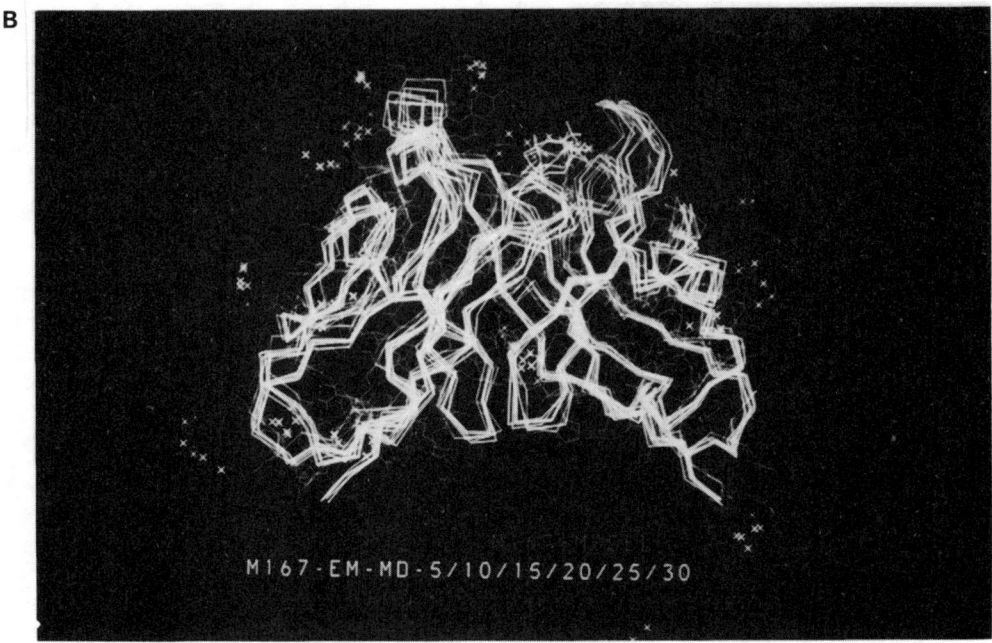

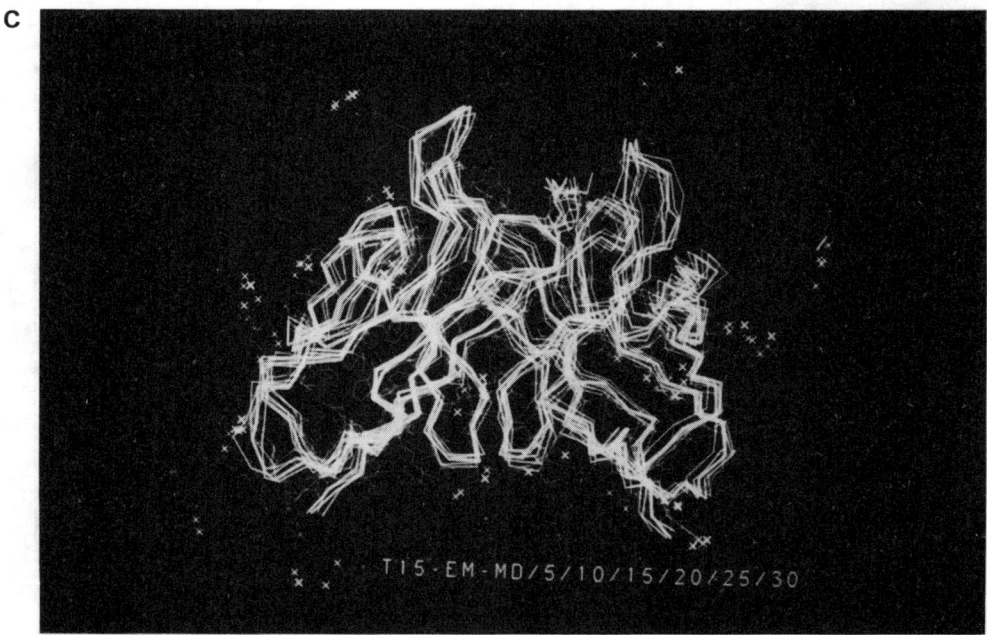

M167 (B) and T15 (C) during the molecular dynamics simulations of 30 ps each. In dark blue the starting structures including the side chain atoms are shown.

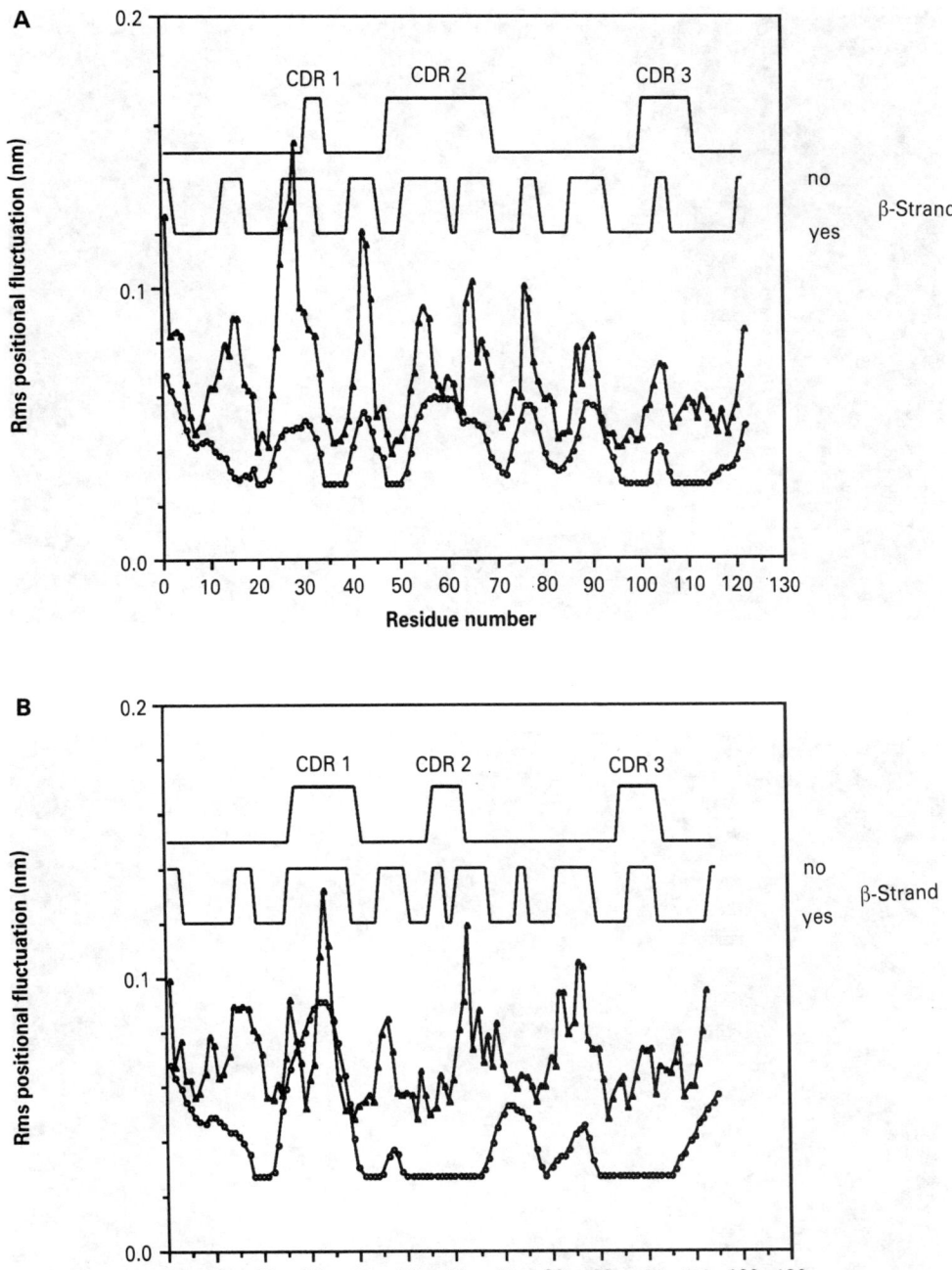

Fig. 4. — Rms fluctuations derived from crystallographic B-factors of Cα atoms (open circles) and rms positional fluctuations from a 30 ps trajectory in solution at 297 K (triangles) are shown. Both refer to the McPC603-phosphorylcholine complex [76], and (A) is the analysis for the heavy chain, (B) for the light chain. The β-strands are indicated (defined as in ref. 87), as are the complementarily determining regions (CDR) of each domain, defined by genetic variability. These are the antigen binding loops.

A

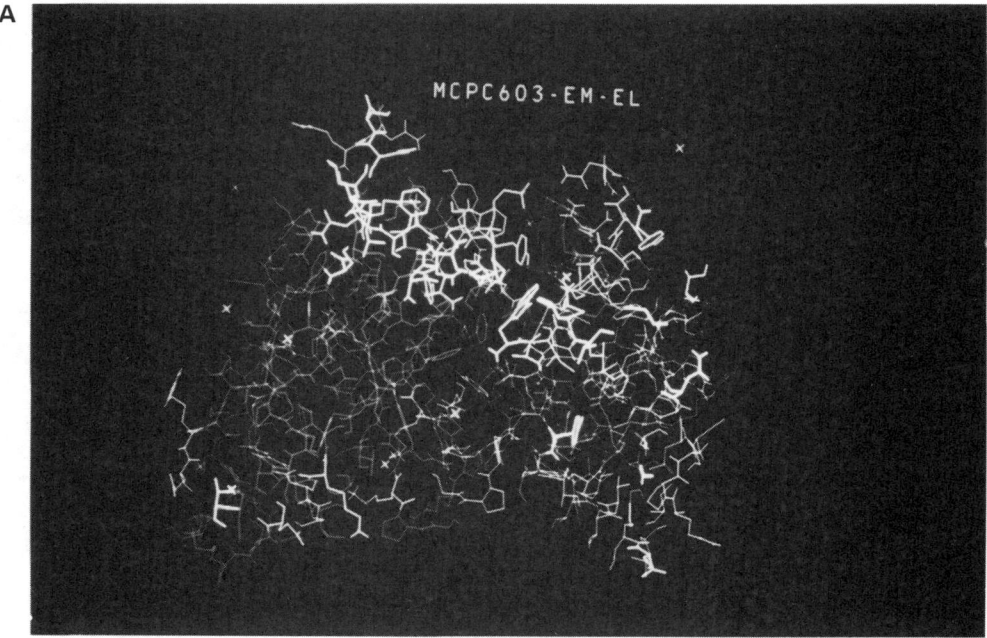

B

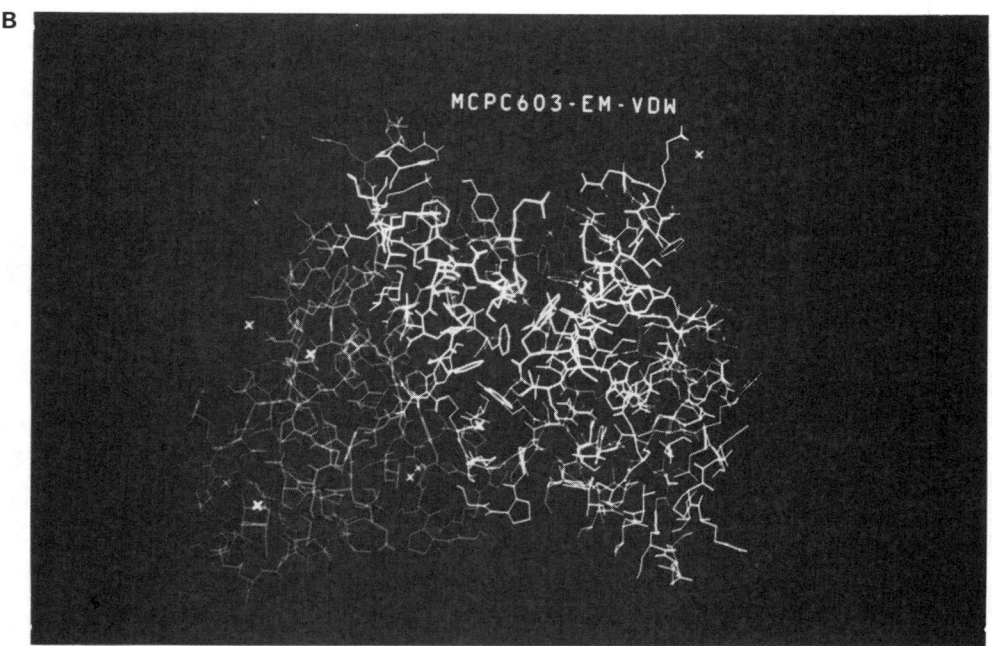

Fig. 5. — The energy-minimized structure of McPC603 with all amino acids colored according to their interaction energies towards phosphorylcholine. Red: strongest interaction calculated with the program PROELW; yellow, cyan, white, increasingly weaker interactions. (A) electrostatic, (B) van der Waals, (C) sum of (A) and (B).

C

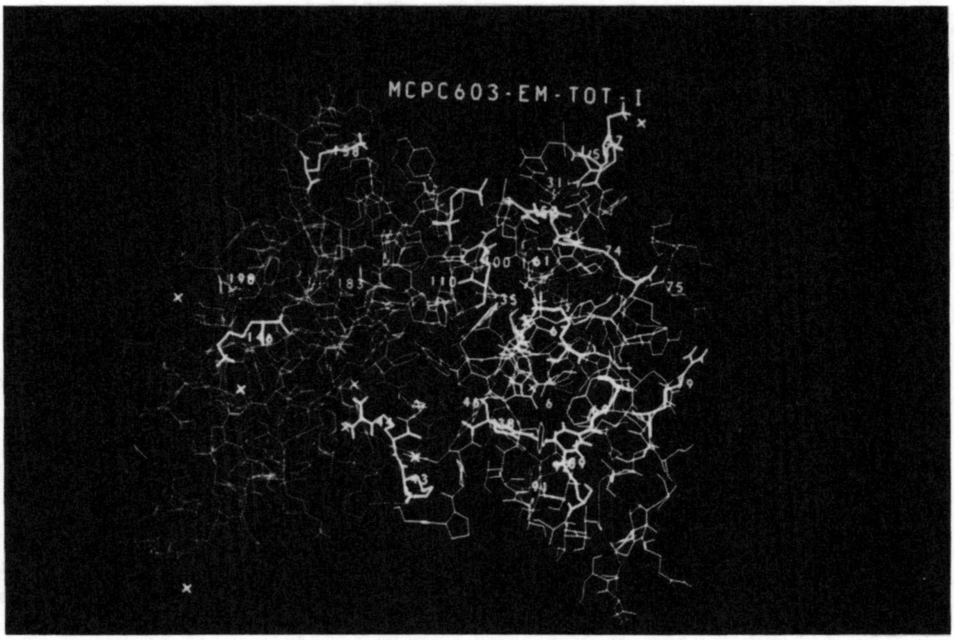

Fig. 5 *Contd.*

electrostatic and van der Waals energies and the sum of both (Figs 5(a), (b), (c)) over time [88]. The fluctuations of these electrostatic energies (Figs 6(a), (b), (c)) and van der Waals energies (Figs 7(a), (b), (c)) or the sum of both (Figs 8(a), (b), (c)) between amino acids of the antibodies and phosphorylcholine in the simulation of the modeled structures of M167 and T15 are much larger than in McPC603, which could be started from its known crystal structure. It is clear, therefore, that convergence is fairly slow under these conditions. It is hoped that the continued trajectories will converge and their results might be compared to experimental structures of M167 and T15 in the near future.

It should be noted that the comparison of potential energies, like electrostatic and van der Waals energies, must be used with care for any interpretation of substrate binding, as the real binding reaction contains entropic terms. Instead, free energy calculations have to be performed on derivatives of the protein and the ligand [89]. These calculated differences in free energies can be compared directly to the experimental binding constants.

Since in the case of the antibodies McPC603, M167 and T15 the binding constants to phosphorylcholine and its derivatives have been measured to be in the narrow range from 3 to 7 kcal/mol, a very high accuracy of the atomic charges used in the simulations is needed. This is of even higher importance since the substrates are molecules with net charges, and some have zwitterionic structures. In order to obtain atomic charges of reasonably equal quality for all derivatives we will use the SCALCHA approach [90, 91] in the future. The method performs a projection of quantum mechanically or semiempirically determined charges onto atomic charges as used in current molecular dynamics force fields.

A

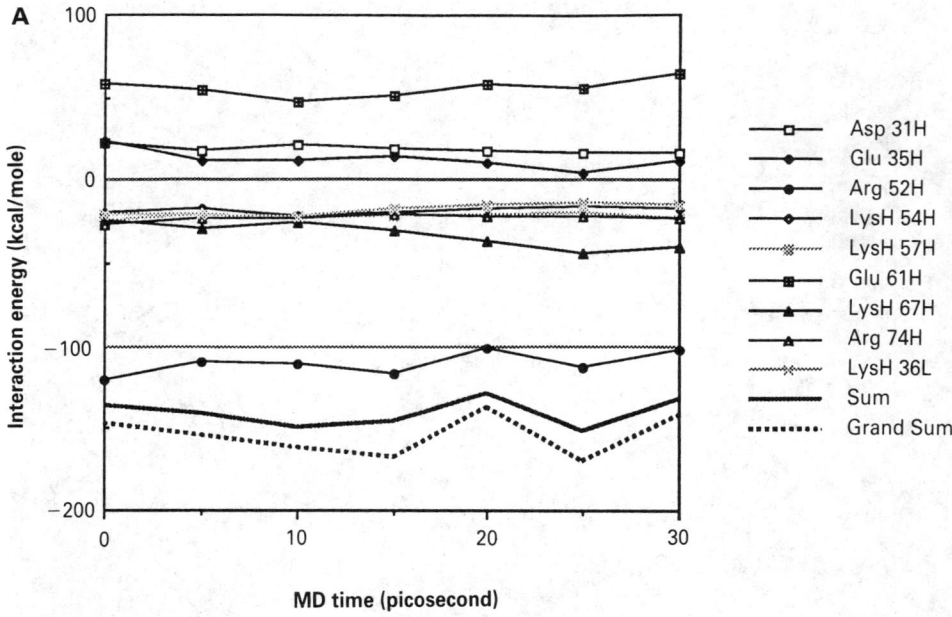

B

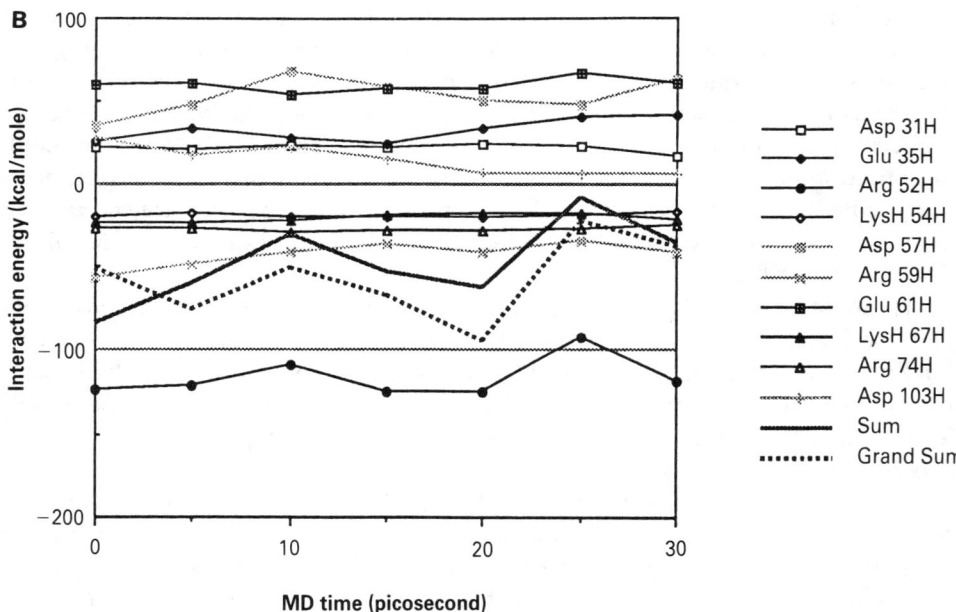

Fig. 6. — The electrostatic interaction energy (kcal/mol) for the three antibodies McPC603 (A), M167 (B) and T15 (C), each bound to (deprotonated) phosphorylcholine calculated from the structures in the trajectories at 5 ps intervals. The amino acids with the highest interaction energies with phosphorylcholine are shown. The sum is the sum of these amino acids. The grand sum is the sum over all amino acids in the whole calculated system. LysH denotes the protonated form of Lys in the GROMOS force field.

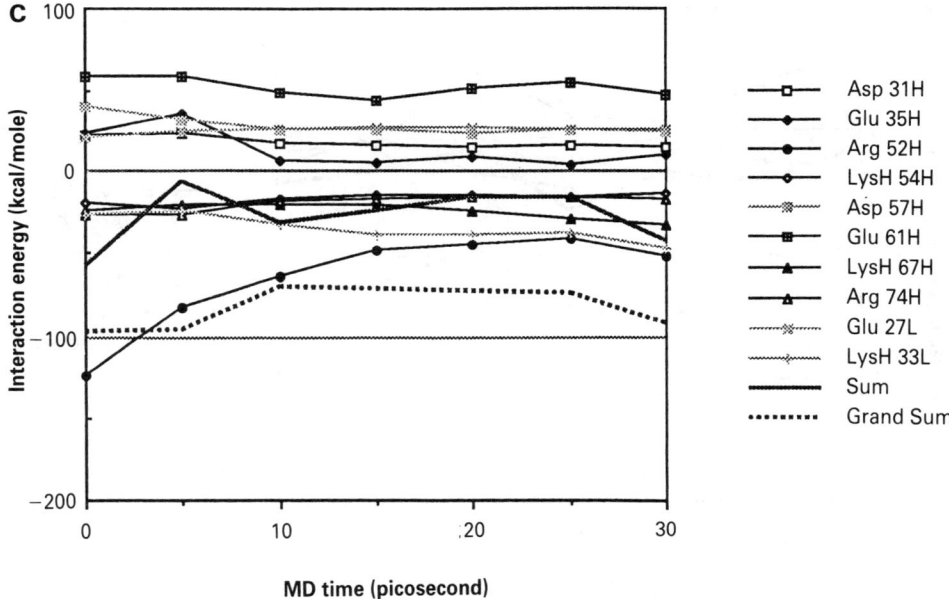

Fig. 6 *Contd.*

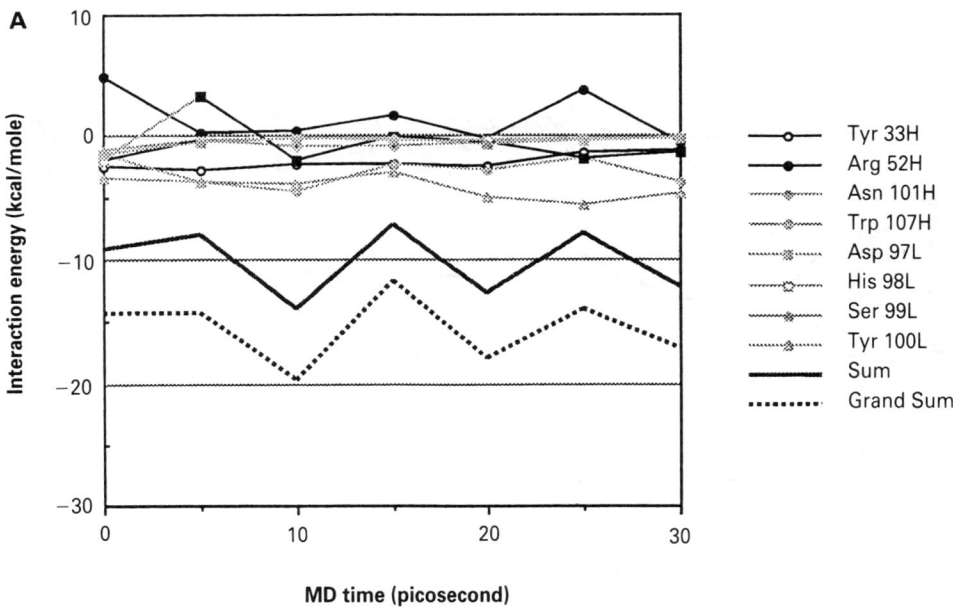

Fig. 7. — Van der Waals interaction energy (kcal/mol) for the three antibodies McPC603 (A). All parameters are as Fig. 6.

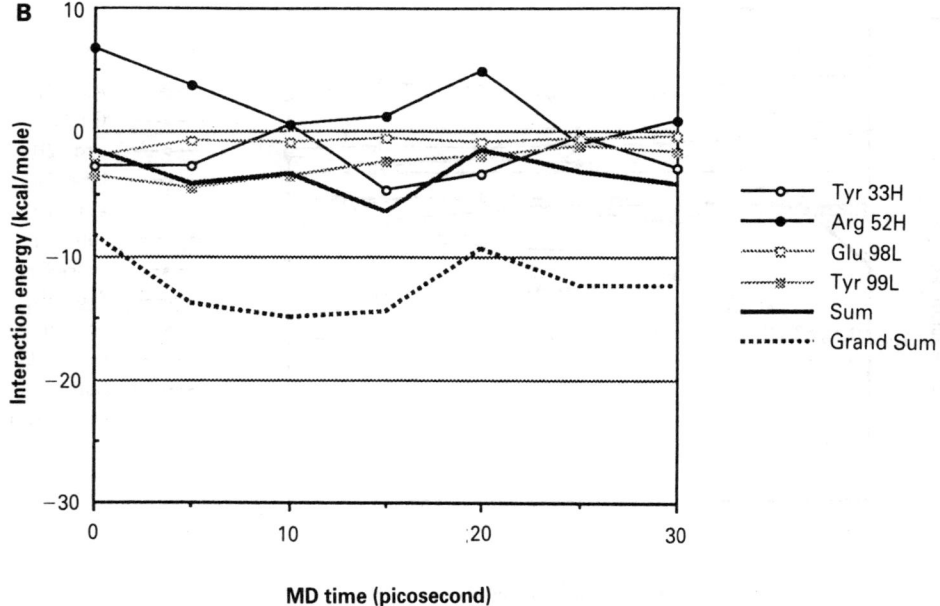

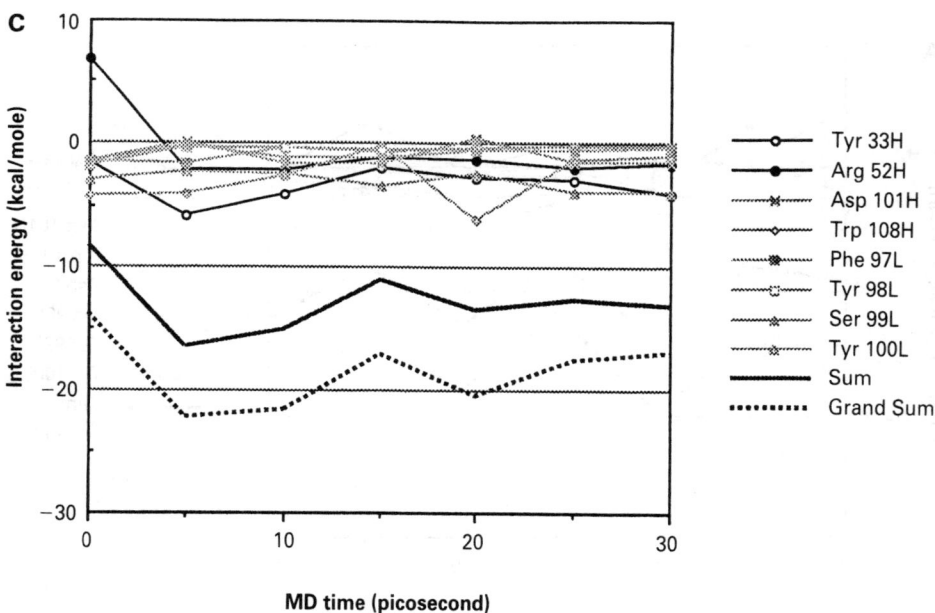

Fig. 7 *Contd.* — M167 (B) and T15 (C), each bound to (deprotonated) phosphorylcholine for the structures from the trajectories at 5 ps intervals. All parameters are as in Fig. 6.

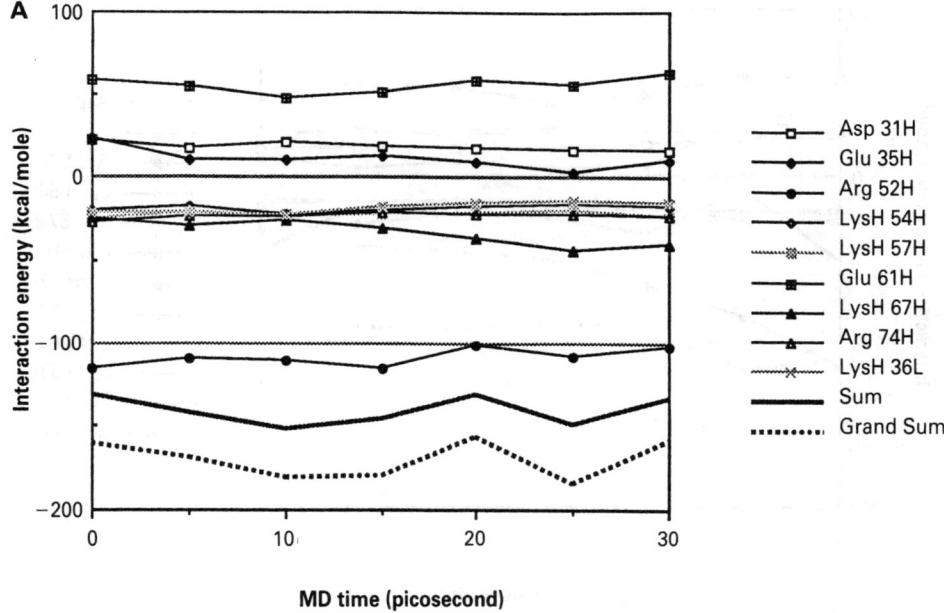

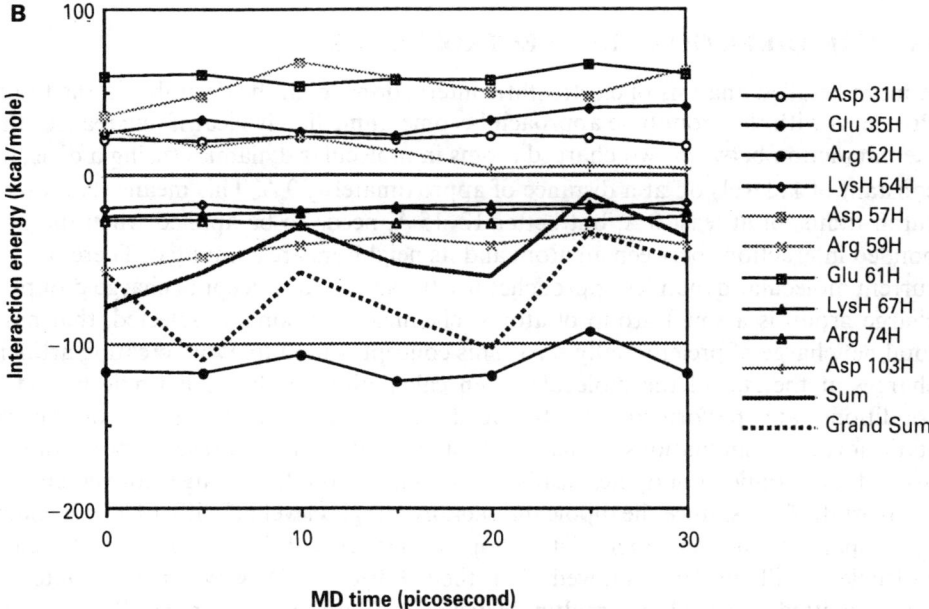

Fig. 8. — Total interaction energy (kcal/mol), which is the sum of the van der Waals and electrostatic energies for the three antibodies McPC603 (A), M167 (B) and T15. All parameters are as in Figs 6 and 7.

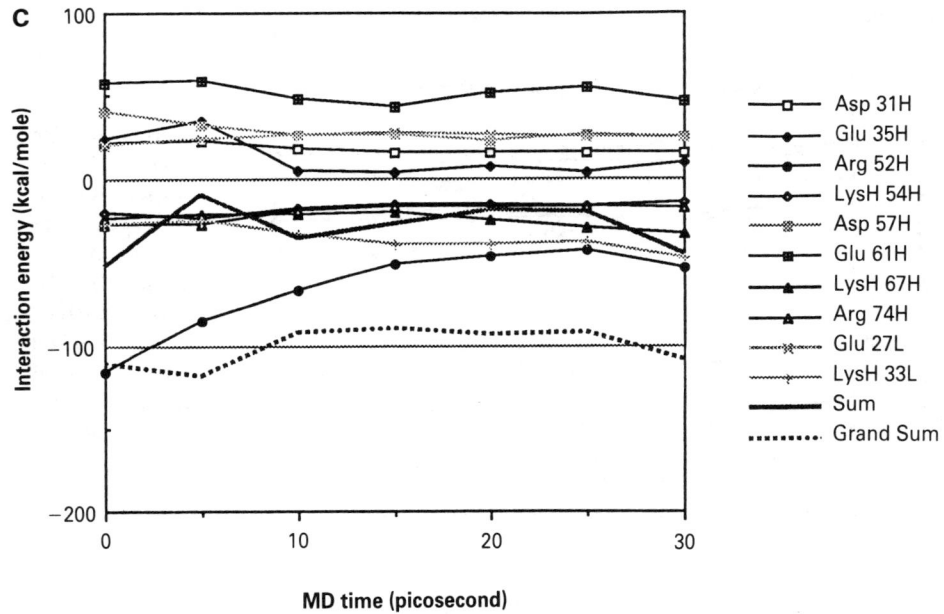

Fig. 8 *Contd.* — Total interaction energy (kcal/mol), which is the sum of the van der Waals and electrostatic energies for the three antibodies McPC603. (C), each bound to (deprotonated) phosphorylcholine for the structures from the trajectories at 5 ps intervals. All parameters are as in Figs 6 and 7.

FUTURE DIRECTIONS IN ELECTROSTATICS

A more detailed analysis of electrostatic interactions might be desirable in the future. Problems with the monopole approach become immediately clear if one realizes that the interaction between two charged atoms in molecular dynamics (using a dielectric constant of 1) levels off at a distance of approximately 30 Å. This means that a large cutoff radius of at least 8 Å, but better 12-15 Å, needs to be applied when the non-bonded interactions between an atom and its neighbors are computed. Therefore, the current molecular dynamics approaches use the so-called concept of charge groups. A charge group is a small group of atoms, chemically reasonably selected, that has a total net charge of preferentially zero. This concept is used to avoid creating artificial charges at the rim of the molecule when calculations with the periodic boundary conditions are performed. It also leads to a faster convergence of electro-static forces in simulations. It has been shown [92] that the use of bond dipoles instead of atomic monopoles leads to a quite good long-range convergence of electrostatic forces, since the dipole interaction energies level off with a $1/r^3$ dependen-cy compared to the convergence of monopole energies with $1/r$. For a set of 24 organic molecules, Williams [92] showed that their HF/6-31G** wavefunction potential could be fitted very well by a multipole distribution up to quadrupoles. If monopoles and dipoles were chosen, the fit was still very good. Moreover, it could be shown that the concept of introducing restricted bond dipole moments leads to a reproduction of the electric potential that was comparable to the pure monopole model.

Another aspect related to electrical charges is the polarizability of atoms and molecules. The effects of anisotropy and anharmonicity in crystallographic refinement and their relation to molecular dynamics have been studied [93]. The importance of including electronic polarization (e.g. electron correlation effects [94]) into model force fields has been discussed by van Duijnen and Rullman [95]. They applied their 'direct reaction field method' to the active site of the protease papain. The polarization influences molecular structures [96] and the overall results are better described by anisotropic atom–atom models. These models reproduce the behavior of π and lone-pair electrons in a better way than pure isotropic models [97]. A comparison of results for the SO_2 molecule from calculated intermolecular potentials of Lennard–Jones type and distributed multipole moments (Monopoles, Dipoles and Quadrupoles) for the electrostatic part leads to the conclusion that the second virial coefficient criterion is a necessary but not sufficient condition for a selection of atomic multipoles [98]. Water, especially with its high dielectric constant, is difficult to model in an efficient and computationally fast way.

High-speed algorithms for the calculation of atomic charges based upon the concept of orbital electronegativities have been proposed by several authors [99–102]. Dipole moments from partial equalization of orbital electronegativities have been derived [103] and a quantification of effective polarizabilities was proposed [104]. The introduction of these slightly modified concepts into the MM2 force field has been tested for 40 compounds and found to be in good agreement with experimental data [105].

As a first suggestion we think that one should clearly distinguish between improvements to analyze static structures and improvements that are related to the calculations during the trajectories. One could develop much more accurate programs including multipole terms etc. for the analysis of static structures obtained from experiment (X-ray) or from simulated trajectories. On the other hand, it might not be necessary to include the same high level of accuracy during the simulation of trajectories, since it has been shown that the most important structural features can be reproduced without. Here it might be more interesting to think about concepts and algorithms such that longer time scales might be accessible. Of course, if there should be enough computing time available, there might be a chance to also include the high-level concepts during the simulation.

CURRENT FRONTIERS OF RESEARCH FOR MOLECULAR DYNAMICS IN GENERAL

1. Attempts to exceed the picosecond time scales

One step foward is the attempt to use stochastic molecular dynamics in order to describe molecular processes that exceed the picosecond time scale [106]. Statistics can also be improved if several molecules are simulated simultaneously. An ingenious implementation has been used for studying the separation of CO from myoglobin [107]. In this work, a time-dependent Hartree approach was implemented into the molecular dynamics algorithm, so that within one simulation a system of one myoglobin molecule and about 60 CO molecules was investigated at one time. This method solves 60 'normal' trajectories in one, since the CO molecules were defined as having no interactions between themselves.

In another example the time of the molecular process exceeds by far the period that can be simulated nowadays: the question of how proteins fold cannot be answered by simulations yet, because the folding process takes place in the range of seconds or even longer. Some trials on protein folding with the Monte Carlo method [7] have been performed recently [108], however using very much simplified molecular models.

Simulated annealing methods are under development [109-112], to solve three-dimensional structures based on NMR NOE distance constraints [113]. One of the basic problems is that there is no easy way to decide where the global minimum on a multidimensional energy-hypersurface is [114].

2. Improvements in analyzing multiconfigurational space

A combined molecular dynamics and X-ray refinement [115-119] improves on the understanding of the multiconfigurational problems (tyrosine flip etc.) occurring in crystal structures.

3. Attempts to describe polarization

The polarizability of molecules [120], especially the development of a polarizable water model, is of high interest [121-125]. Cooperative effects lead to extended network water structures [50, 126-130], which makes it hard to model water with pure pair potentials unless they are properly scaled to give effective models, as mentioned above.

4. Consistent force fields

A consistent set of atomic properties might be desirable: the need of an overall parametrization procedure for molecular dynamics force fields is becoming more and more important [131, 132]. The extension of parameters also to very heavy atoms has been investigated in the shell structure approach for neutral atoms from hydrogen to uranium and for several mono-charged positive ions from helium to barium and lutetium to radium by means of non-relativistic SCF wavefunctions and the resulting spherically averaged charge densities [133]. This survey is interesting in that a more general charge density approach covering almost the whole periodic system is discussed.

5. Attempts to model chemical reactions

Since in the current force fields no electrons are considered explictly, electron transfer reactions [134] or organic reactions involving covalent bond breaking and forming [135] as well as photochemical processes cannot be described directly [136, 137], although some attempts have been reported in the direction of including quantum mechanics into molecular dynamics [63, 138, 139]. Recently, the interest in non-equilibrium molecular dynamics has been increasing [140], since the unsolved problems of finding possible structures on a folding pathway are of general interest in macromolecular systems.

6. New very fast computers with different architectures

The first step for speeding up computers was their vectorization. A Brownian dynamics simulation program has been written especially for a vector computer [141], and a polarizable water model can be computed on a vector computer Cyber 205 [123]. The development of very fast computers with new (parallel) architectures [142] and new programming languages [143] might help to calculate molecular systems [144] that are much larger or that need much longer simulation times than can now be handled.

CONCLUSION

Molecular dynamics trajectories display different behavior for molecular structures such as antibody/antigen complexes when started from model-built structures rather than from an X-ray structure.

In the case of McPC603/phosphorylcholine there was an X-ray structure available, whilst none was available for the two related complexes M167 and T15 bound to phosphorylcholine — only their sequences. We replaced individual amino acids and loop structures in the X-ray structure of McPC603 to obtain starting structures for M167 and T15. We performed one trajectory of 30 ps at 293 K for each of the three Fv-fragments in aqueous solution and analyzed the electrostatic interactions between all atoms of the most important amino acids of the proteins with respect to phosphorylcholine binding in the trajectory. The simulated structure of McPC603 was more or less equilibrated after 30 ps. For M167 and T15 we found larger fluctuations of the sensitive electrostatic energies and concluded that after 30 ps these structures were not in equilibrium. Clearly, one has to be careful with model structures built from existing X-ray structures. The change of even few amino acids might cause larger structural shifts of the protein than can be easily predicted today.

ACKNOWLEDGMENTS

Special thanks go to the authors (W.F. van Gunsteren, H.J.C. Berendsen) of the program package GROMOS 87, which we used to perform the molecular dynamics simulations. We thank B. Steipe for fruitful discussions and for providing the data from the protein structure analysis and the secondary structure.

We acknowledge the support of Cray Research Munich (D. Labrenz), the DLR Oberpfaffenhofen (H. Herchenbach, S. Leisen) and the computer center of the MPI Biochemie Martinsried, (W. Steigemann, G. Franz, H. Hanewinkel), for providing the necessary computing facilities, and Professor E. L. Winnacker and Hoechst Japan Ltd. for financial support.

REFERENCES

[1] L. Pauling, *The Nature of the Chemical Bond.* 3rd edn, Cornell University Press, Ithaca, New York (1960).

[2] L. Pauling, *Nature* **161** (1948) 707.

[3] L. Pauling and R.B. Corey, *Proc. Natl. Acad. Sci. USA* **37** (1951) 235.

[4] L. Pauling, *Chem. Eng. News* **24** (1946) 1375.

[5] D. van Belle, I. Couplet, S. Prevost and S. Wodak, *J. Mol. Biol.* **198** (1987).

[6] B. Jayaram, K.A. Sharp and B. Honig, *Biopolymers* **28** (1989) 975.

[7] W.L. Jorgensen, *J. Phys. Chem.* **87** (1983) 5304.

[8] M.M. Karelson, A.R. Katritzky and M.C. Zerner, *Int. J. Quant. Chem.* **20** (1986) 521.

[9] I. Rips, J. Klafter and J. Jortner, *J. Chem. Phys.* **88** (1988) 3246.

[10] W.S. Bennett and R. Huber, *CRC Crit. Rev. Biochem.* **15** (1984) 291.

[11] H.A. Scheraga, *Adv. Phys. Org. Chem.* **6** (1968) 103.

[12] N.L. Allinger and J.T. Sprague, *J. Am. Chem. Soc.* **94** (1972) 5734.

[13] N.L. Allinger and J.T. Sprague, *J. Am. Chem. Soc.* **95** (1973) 3983.

[14] N.L. Allinger and J.C. Graham, *J. Am. Chem. Soc.* **95** (1973) 2523.

[15] A.T. Hagler, E. Huler and S. Lifson, *J. Am. Chem. Soc.* **96** (1974) 5319.

[16] M. Karplus and J.A. McCammon, *CRC Crit. Rev. Biochem.* **9** (1981) 293.

[17] P.K. Weiner and P.A. Kollman, *J. Comp. Chem.* **2** (1981) 287.

[18] J. Hermans, H.J.C. Berendsen W.F. van Gunsteren and J.P.M. Postma, *Biopolymers* **23** (1984) 1513.

[19] B.R. Brooks, R.E. Bruccoleri, B.D. Olafson, D.J. States, S. Swaminathan and M. Karplus, *J. Comp. Chem.* **4** (1983) 187.

[20] S.J. Weiner, P.A. Kollman, D.A. Case, U.C. Singh, C. Ghio, G. Alagona, S. Profeta and P. Weiner, *J. Am. Chem. Soc.* **106** (1984) 765.

[21] S.J. Weiner, P.A. Kollman, D.T. Nguyen and D.A. Case, *J. Comp. Chem.* **7** (1986) 230.

[22] P.A. Kollman, *Acc. Chem. Res.* **18** (1985) 105.

[23] P.A. Kollman, *Methods in Enzymol.* **154** (1987) 430.

[24] D.L. Beveridge and W.L. Jorgensen (Eds), Computer Simulation of Chemical and Biomolecular Systems. *Ann. N.Y. Acad. Sci.* **482** (1986) 1.

[25] M. Karplus and J.A. McCammon, *Sci. Am.* **254** April (1986) 42.

[26] L. Nilson and M. Karplus, *J. Comp. Chem.* **7** (1986) 591.

[27] J.A. McCammon and S.C. Harvey, *Dynamics of Proteins and Nucleic Acids.* Cambridge University Press (1987).

[28] C.L. Brooks, M. Karplus and B.M. Pettitt (Eds), *Proteins: a Theoretical Perspective of Dynamics, Structure and Thermodynamics. Advances in Chemical Physics* **LXXI**, John Wiley & Sons (1988).

[29] W.F. van Gunsteren, *Protein Engineering* **2** (1988) 5.

[30] W.L. Jorgensen and J. Tirado-Rives, *J. Am. Chem. Soc.* **110** (1988) 1657.

[31] W.F. van Gunsteren and P.K. Weiner (Eds), Computer Simulations of Biomolecular Systems. Theoeretical and Experimental Applications. *ESCOM*, Leiden, The Netherlands (1989).

[32] W.F. van Gunsteren and H.J.C. Berendsen, *Angew. Chem.* **29** (1990) 992.

[33] J.M. Goodfellow (Ed.), *Molecular Dynamics. Applications in Molecular Biology.* The Macmillan Press, Basingstoke, UK (1991).

[34] D.A. McQuarrie, *Statistical Mechanics*. Harper & Row, Harper's Chemistry Series, S.A. Rice (Ed.) (1976).

[35] L. Verlet, *Phys. Rev.* **159** (1967) 98.

[36] C.W. Gear, *Numerical Initial Value Problems in Ordinary Differential Equations*. Prentice Hall, Englewood Cliffs, N.J. (1971).

[37] W.F. van Gunsteren and H.J.C. Berendsen, *Mol. Phys.* **34** (1977) 1311.

[38] J.P. Ryckaert, G. Ciccotti and H.J.C. Berendsen, *J. Comp. Phys.* **23** (1977) 327.

[39] H.J.C. Berendsen, J.P.M. Postma, W.F. van Gunsteren, A. DiNola and J.R. Haak, *J. Chem. Phys.* **81** (1984) 3684.

[40] H.J.C. Berendsen and W.F. van Gunsteren, Practical Algorithms for Dynamic Simulations. In: G. Ciccotti and W.G. Hoover. (Eds). *Proc. Enrico Fermi School of Physics*, Varenna, North Holland Physics Publishing (1986) 43.

[41] R.W. Pastor, B.R. Brooks and A. Szabo, *Mol. Phys.* **65** (1988) 1409.

[42] H.J.C. Berendsen, J.P.M. Postma, W.F. van Gunsteren and J. Hermans, In: Pullman B, (Ed.). *Intermolecular Forces*. D. Reidel Publishing Company (1981) 331.

[43] H.J.C. Berendsen and W.F. van Gunsteren, Molecular Dynamics Simulations: Techniques and Approaches. Molecular Liquids–Dynamics and Interactions. In: A.J. Barnes *et al.* (Eds). *NATO–ASI Series*, **C135** (1984) 475.

[44] W.F. van Gunsteren and H.J.C. Berendsen, *Molec. Simul.* **1** (1988) 173.

[45] W.L. Jorgensen, J. Chandrasekhar, J.D. Madura, R.W. Impey and M.L. Klein, *J. Chem. Phys.* **79** (1983) 926.

[46] A.T. Hagler, J.R. Maple, T.S. Thacher, G.B. Fitzgerald and U. Dinur, Potential Energy Functions for Organic and Biomolecular Systems. In: *Computer Simulation of Biomolecular Systems*, W.F. van Gunsteren and P.K. Weiner (Eds), *ESCOM*, Leiden (1989).

[47] J.E.H. Koehler, W. Saenger and W.F. van Gunsteren, *Eur. Biophys. J.* **15** (1987) 197.

[48] J.E.H. Koehler, W. Saenger and W.F. van Gunsteren, *Eur. Biophys. J.* **15** (1987) 211.

[49] P.A. Kollman and L.C. Allen, *Chem. Rev.* **72** (1972) 283.

[50] P. Barnes, J.L. Finney, J.D. Niocholas and J.E. Quinn, *Nature* **282** (1979) 459.

[51] J.E.H. Koehler, B. Lesyng and W. Saenger, *J. Comp. Chem.* **8** (1987) 1090.

[52] G.A. Jeffrey and W. Saenger, *Hydrogen Bonding in Biological Structures*. Springer Verlag, Berlin, Heidelberg, New York (1990).

[53] W. Saenger, *Nature* **279** (1979) 343.

[54] W. Saenger, C. Betzel, B. Hingerty and G.M. Brown, *Nature* **296** (1982) 581.

[55] J.E.H. Koehler, W. Saenger and W.F. van Gunsteren, *Eur. Biophys. J.* **16** (1988) 153.

[56] J.E.H. Koehler, W. Saenger and W.F. van Gunsteren, *J. Biomol. Struc. Dyn.* **6** (1988) 181.

[57] J.E.H. Koehler, W. Saenger and W.F. van Gunsteren, *J. Mol. Biol.* **203** (1988) 241.

[58] A. Warshel, F. Sussman and J.K. Hwang, *J. Mol. Biol.* **201** (1988) 139.

[59] T.P. Straatsma and H.J.C. Berendsen, *J. Chem. Phys.* **89** (1988) 5876.

[60] G. Ciccotti, M. Ferrario, J.T. Hynes and R. Kapral, *J. de Chimie Physique* **85** (1988) 925.

[61] A. Warshel, G. Naray-Szabo, F. Sussman and J.K. Hwang, *Biochemistry* **28** (1989) 3629.

[62] P.A. Kollman and W.F. van Gunsteren, *Methods in Enzymol.* **154** (1987) 430.

[63] P.A. Bash, M.J. Field and M. Karplus, *J. Am. Chem. Soc.* **109** (1987) 8092.

[64] U.C. Singh, F.K. Brown, P.A. Bash and P.A. Kollman, *J. Am. Chem. Soc.* **109** (1987) 1607.

[65] P. Cieplak and P.A. Kollman, *J. Am. Chem. Soc.* **110** (1988) 3734.

[66] M.N. Jones (Ed.). *Biochemical Thermodynamics.* Elsevier, Amsterdam-Oxford-New York-Tokyo (1988).

[67] W.F. van Gunsteren and H.J.C. Berendsen, *J. Comput.-Aided Mol. Design* **1** (1987) 171.

[68] P.A. Bash, U.C. Singh, F.K. Brown, R. Langridge and P.A. Kollman, *Science* **235** (1987) 574.

[69] E. Lolis and G.A. Petsko, *Ann. Rev. Biochem.* **59** (1990) 597.

[70] R. Wolfenden, *Ann. Rev. Biophys. Bioeng.* **5** (1976) 271.

[71] W.P. Jencks, *Catalysis in Chemistry and Enzymology.* McGraw-Hill, New York (1969).

[72] P.G. Schultz, *Angew. Chem. Int. Ed. Engl.* **28** (1989) 1283.

[73] K.M. Shokat and P.G. Schultz, *Ann. Rev. Immunol.* **8** (1990) 335.

[74] R.A. Lerner and S.J. Bencovic, *Bioessays* **9** (1988) 107.

[75] G.M. Blackburn, A.S. Kang, G.A. Kingsbury and D.R. Burton, *Biochem. J.* **262** (1989) 381.

[76] D.M. Segal, E.A. Padlan, G.H. Cohen, S. Rudikoff, M. Potter and D.R. Davies, *Proc. Natl. Acad. Sci. USA* **71** (1974) 4298.

[77] Y. Satow, G.H. Cohen, E.A. Padlan and D.R. Davies, *J. Mol. Biol.* **190** (1986) 593.

[78a] Euranto, E.K., Esterification and Ester Hydrolysis. In: S. Patai (Ed.), The Chemistry of Carboxylic Acids and Esters. The Chemistry of Functional Groups (series), Interscience Publishers, London, 1969.

[78b] U. Holzgrabe and G. Bejeuhr, *Pharmazie in unserer Zeit* **6** (1990) 247.

[79] J. Stadlmüller and A. Plückthun, (1991), in preparation.

[80] R. Glockshuber, J. Stadlmüller and A. Plückthun, *Biochemistry* **30** (1990) in press.

[81] A. Plückthun, R. Glockshuber, A. Skerra and J. Stadmüller, *Behring Inst. Mitt.* **87** (1990) 48.

[82] S.J. Pollack, J.W. Jacobs and P.G. Schultz, *Science* **234** (1986) 1570.

[83] S.J. Pollack and P.G. Schultz, *Cold Spring Harbor Symp. Quant. Biol.* **52** (1987) 97.

[84] A. Skerra and A Plückthun, *Science* **240** (1988) 1038.

[85] A. Plückthun, *Nature* **347** (1990) 497.

[86] A. Plückthun, *Biotechnology* **9** (1991) 545.

[87] W. Kabsch and C. Sander, *Biopolymers* **22** (1983) 2577.

[88] R. Kikuno and J.E.H. Koehler, PROELW program (1990).

[89] J. Novotny, R.E. Bruccoleri and F.A. Saul, *Biochemistry* **28** (1989) 4735.

[90] C. Köhler, Diploma Thesis, Physics Dept. TU Munich, (1990).

[91] J. Köhler, C. Köhler, V. Helms and K. Adelhard, *Croat. Chem. Acta.* (1991) in press.

[92] D.E. Williams, *J. Comp. Chem.* **9** (1988) 745.

[93] J. Kuriyan, G.A. Petsko, R.M. Levy and M. Karplus, *J. Mol. Biol.* **190** (1986) 227

[94] H.J. Werner and W. Meyer, *Molecular Physics* **31** (1979) 855.

[95] P.T. van Duijnen and J.A.C. Rullman, *Int. J. Quant. Chem.* **XXXVIII** (1990) 181

[96] S. Wolfe, *Acc. Chem. Res.* **5** (1972) 102.

[97] S. Price, *Mol. Simulat.* **1** (1988) 135.

[98] M. Pavlovic, F. Sokolic and Z.B. Maksic, *J. Mol. Struc. (Theochem.)* **202** (1989) 265.

[99] J. Gasteiger and M. Marsili, *Tetrahedron* **36** (1980) 3219.

[100] W.J. Mortier, K. van Genechten and J. Gasteiger, *J. Am. Chem. Soc.* **107** (1985) 829.

[101] L. Baumer, G. Sala and G. Sello, *Tetrahedron Comp. Methodol.* **2** (1989) 37.

[102] J. Mullay, *J. Comp. Chem.* **12** (1991) 369.

[103] J. Gasteiger and M.D. Guillen, *J. Chem. Res. (S)*, (1983) 304.

[104] J. Gasteiger and M.G. Hutchings, *J. Chem. Soc. Perkin Trans.* **II** (1984) 559.

[105] L.G. Hammarstroem, T. Liljefors and J. Gasteiger, *J. Comp. Chem.* **9** (1988) 424.

[106] B. Cartling, *J. Chem. Phys.* **91** (1989) 427.

[107] R. Elber and M. Karplus, *J. Am. Chem. Soc.* **112** (1990) 9161.

[108] J. Skolnick and A. Kolinski, *Science* **250** (1990) 1121.

[109] A.T. Bruenger, *J. Mol. Biol.* **203** (1988) 803.

[110] M. Nilges, G.M. Clore and A.M. Gronenborn, *FEBS Lett.* **239** (1988) 129.

[111] M. Nilges, G.M. Clore and A.M. Gronenborn, *FEBS Lett.* **239** (1988) 317.

[112] A.T. Bruenger, *Acta Cryst. A* **45** (1989) 42.

[113] G.M. Clore, A.T. Brunger, M. Karplus and A.M. Gronenborn, *J. Mol. Biol.* **191** (1986) 523.

[114] A. Howard and P.A. Kollman, *J. Med. Chem.* **31** (1988) 1669.

[115] A.T. Bruenger, M. Karplus and G.A. Petsko, *Acta Cryst. A* **45** (1989) 50.

[116] M. Fujinaga, P. Gros and W.F. van Gunsteren, *J. Appl. Cryst.* **22** (1989) 1.

[117] A.E. Torda, R.M. Scheek and W.F. van Gunsteren, *Chem. Phys. Lett.* **157** (1989) 289.

[118] A.E. Torda, R.M. Scheek and W.F. van Gunsteren, *J. Mol. Biol.* **214** (1990) 223.

[119] P. Gros, W.F. van Gunsteren and W.G.J. Hol, *Science* **249** (1990) 1149.

[120] B.T. Thole, *Chem. Phys.* **59** (1981) 341.

[121] F.H. Stillinger and C.W. David, *J. Chem. Phys.* **69** (1978) 1473.

[122] P. Ahlstrom, A. Wallqvist, S. Engstrom and B. Jonsson, *Mol. Phys.* **68** (1989) 563.

[123] J.C. Sauniere, T.P. Lybrand, J.A. McCammon and L.D. Pyle, *Computers Chem.* **13** (1989) 313.

[124] J. Caldwell, L.X. Dang and P.A. Kollman, *J. Am. Chem. Soc.* **112** (1990) 9144.

[125] F.A. Webster, J. Schnitker, M. Friedrichs, R.A. Friesner and P.J. Rossky, *Phys. Rev. Letters* **66** (1991) 3172.

[126] A.H. Narten and H.A. Levy, *J. Chem. Phys.* **55** (1971) 2263.

[127] L.L. Shipman and H.A. Scheraga, *J. Phys. Chem.* **78** (1974) 909.

[128] M. Mezei and D.L. Beveridge, *J. Chem. Phys.* **76** (1982) 593.

[129] R.J. Speedy, J.D. Madura and W.L. Jorgensen, *J. Phys. Chem.* **91** (1987) 909.

[130] F. Franks (Ed.), *Water Science Review 3*, Cambridge University Press (1988).

[131] M.J. Field, P.A. Bash and M. Karplus, *J. Comp. Chem.* **11** (1990) 700.

[132] K. Palmö, L.O. Pietilä and S. Krimm, *J. Comp. Chem.* **12** (1991) 385.

[133] R.P. Sagar, A.C.T. Ku and V.H. Smith, *J. Chem. Phys.* **88** (1988) 4367.

[134] D.A. Zichi, G. Ciccotti, J.T. Hynes and M. Ferrario, *J. Phys. Chem.* **93** (1989) 6261.

[135] W.L. Jorgensen, *Acc. Chem. Res.* **22** (1989) 184.

[136] H. Treutlein, K. Schulten, J. Deisenhofer, H. Michl, A. Bruenger and M. Karplus. In: *The Photosynthetic Bacterial Reaction Center: Structure and Dynamics.* J. Breton and A. Vermeglio (Eds.), Plenum Press, London (1987) 139.

[137] H. Treutlein, K. Schulten, C. Niedermeyer, J. Deisenhofer, H. Michl and D. de Vault. In: *The Photosynthetic Bacterial Reaction Center: Structure and Dynamics.* J. Breton and A. Vermeglio (Eds), Plenum Press, London (1987) 369.

[138] C. Zheng, C.F. Wong, J.A. McCammon and P.G. Wolynes, *Nature* **334** (1988) 726.

[139] R. Kosloff, *J. Phys. Chem.* **92** (1988) 2087.

[140] C. Massobrio, V. Pontikis and G. Ciccotti, *Phys. Rev. B* **39** (1989) 2640.

[141] C. Elvingson, *J. Comp. Chem.* **12** (1991) 71.

[142] P.J. Denning and W.F. Tichy, *Science* **250** (1990) 1217.

[143] A. Burns, *Programming in OCCAM 2.* Addison Wesley (1988).

[144] J.M. Goodfellow and F. Vovelle, *Eur. Biophys. J.* **17** (1989) 167.

Programs

AM1-Program, AMPAC Program Package, v 2.1. Dewar, M.J.S., Zoebisch, E.G., Healy, E.F. and Stewart, J.J.P. (1985). AM1: A New General Purpose Quantum Mechanical Molecular Model. *J. Am. Chem. Soc.* **107**, 3902–3909.

AMBER-Program, Version 3.0 (1987). Singh, U.C., Weiner, P.K., Caldwell, J. and Kollman, P.A., Dept. of Pharmaceutical Chemistry, School of Pharmacy, University of California, San Francisco, CA 94143.

CHARMM. Karplus, M., Harvard University, Dept. of Chemistry, 12 Oxford Street, Cambridge, Mass. 02138, USA.

CHELP. Chirlian, L.E. and Francl, M.M. (1985). Quantum Chemistry Program Exchange, QCPE 524

DISCOVER, published by Biosym Technologies, Inc., 9605 Scranton Rd., Suite 101, San Diego, CA 92121.

GAUSSIAN 86. Frisch, M.J., Binkley, J.S., Schelgel, H.B., Raghavachari, K., Melius, C.F., Martin, L., Stewart, J.J.P., Bobrowicz, F.W., Rohlfing, C.M., Kahn, L.R., Defrees, D.J., Seeger, R., Whiteside, R.A., Fox, D.J., Fleuder, E.M. and Pople, J.A. (1984). *Carnegie Mellon Quantum Chemistry Publishing Unit*, Pittsburgh PA.

GROMOS. Biomos, University of Groningen, 9747 AG Groningen, Nijenborgh 16, The Netherlands; Biomos, ETH-Zentrum, 8092 Zürich, Switzerland.

OCCAM2. Product Definition (preliminary). *INMOS, Bristol* and *Prentice Hall.* Hemel Hempstead, England.

INSIGHT. Biosym Technologies, 10065 Barnes Canyon Road, San Diego, CA 92121. PM3-Program. Stewart, J.J.P. (1983). MOPAC Program Package, QCPE Nr. 455.

PROBE. Hagler *et al.* (1991). Biosym Technologies, 10065 Barnes Canyon Road, San Diego, CA 92121.

QUEST. Singh, U.C. and Kollman, P.A. (1984). An Approach to Computing Electrostatic Charges for Molecules. *J. Comp. Chem.* **2**, 129–145.

X-PLOR. Bruenger, A. Howard Hughes Medical Institute, Dept. Mol. Biophys. Biochem., Yale Univ., New Haven, CT 06511, USA.

24

Orthomolecular medicine

A. Hoffer
Victoria, BC, Canada

INTRODUCTION

I began to practice orthomolecular medicine in 1951, but I did not know I was doing so until after Dr Linus Pauling published his classic 1968 paper in *Science*, on Orthomolecular Psychiatry. Many physicians practice limited aspects of orthomolecular medicine, also without being aware this is what they are doing. Curiously, they remain devoted skeptics because their education in medicine almost totally ignores the vast importance of nutrition and nutrients in the maintenance of good health. Following Dr Pauling's report it became inevitable that one day medicine would become orthomolecular.

Dr Pauling defined orthomolecular medicine as, '... the preservation of good health and the treatment of disease by varying the concentrations in the human body of substances that are normally present in the body and are required for health,' [1, 2].

Dr Pauling's interest in the molecular basis of disease began with his investigations of the structure of hemoglobin about 55 years ago. This led to studies of the structure of proteins and to the description of the helical structure of protein. The discovery of the double helix was within his grasp, but he did not have Watson and Crick's access to X-ray photographs of DNA taken by Rosalind Franklin.

Studies with hemoglobin led to his finding a disease, sickle cell anemia, that was caused by an abnormal hemoglobin molecule. He now was able to think about the molecular basis of disease. Additional data became available to him when we in Saskatchewan began to publish our therapeutic studies of schizophrenia, using very large doses of a vitamin, B-3, either niacin or niacinamide. Until then, the vitamin deficiency theory of disease rigorously prevented examination of the role of vitamins in treating diseases known not to be due to a vitamin deficiency.

According to the vitamin theory of disease, vitamins are needed in very small quantities such as are available in the foods we eat, and single nutrients such as vitamins are only required as supplements for vitamin deficiencies. One would use vitamin C for scurvy, B-1 for beriberi, B-3 for pellagra, and so on. This theory also prevented the use of *larger than vitamin (small) doses,* since these were not needed for the pure deficiency diseases, and if these diseases were not present, there was no medical indication for using vitamin supplements. Not only were there no indications for their use, but it was considered poor medical practice to use them. A colleague of mine in Saskatchewan lost his medical license. One of the charges against him was that he gave patients ascorbate by intravenous injection.

Dr Humphry Osmond and I began to treat schizophrenic patients with large doses of vitamin B-3 in 1951. Only one physician, Dr W. Kaufman [3–5], had used such doses before us, for the treatment of arthritis. The largest dose used by other authors was about 1.5 g daily. We used vitamin B-3 for a number of reasons. Our adrenochrome hypothesis of schizophrenia [6] suggested that a powerful methyl group acceptor like vitamin B-3 could decrease the production of our hypothetical schizophrenic toxin adrenochrome. We knew the pellagra psychosis was very similar to schizophrenia. I discovered later that about 100 years before, the differential diagnosis for dementia praecox (now schizophrenia) included pellagra, scurvy, and general paresis of the insane (tertiary syphilis). Water soluble vitamins are extraordinarily safe because the body excretes what is not needed. We were also aware of a few published studies which showed that 1 g doses of vitamin B-3 were helpful to a few patients. We therefore decided to try vitamin B-3, using 3 g per day. We believed smaller doses would be futile; had small doses been effective for the schizophrenias, this would certainly have been published by the pellagrologists in the mid-Thirties and mid-Forties.

Beginning in 1951, we treated schizophrenic patients who had failed to respond to electroconvulsive therapy or other treatment. The tranquilizers were just being introduced in France. The first eight patients we treated at two of our psychiatric wards in Saskatchewan responded, and stayed well. My first patient went off niacin on three occasions, relapsing each time, and then recovered when the vitamin was resumed. We then completed four prospective double-blind controlled experiments which showed that with the addition of niacin or niacinamide, we could double our two-year recovery rate from 35 percent to 75, compared to placebo. Our vitamin treatment of schizophrenia was described in 'How to Live with Schizophrenia [7]. Dr Linus Pauling was surprised by the large doses we used, and their safety. About the same time, he became familiar with Dr Irwin Stone's research with ascorbic acid [8]. Dr Stone concluded that we all suffer from a genetic disease which he called hypoascorbemia. No one could enjoy optimum health without taking optimum quantities. With less than optimum quantities, we would all have to be content with suboptimal health and repeated episodes of illness. Dr Stone's conclusions intrigued Dr Pauling, and he began to take ascorbic acid. His health improved. Since that time he has remained on ascorbic acid, increasing it over 25 years to 18 g per day. His report to Science [1], is a classic. There he showed how a species which was able to obtain enough of a nutrient, say vitamin C, from its food could, as it evolved, find it advantageous not to have to make its own. But there would be no advantage if the nutrient became unavailable in food. Hypoascorbemia is such a genetic fault. Humans have lost the ability to make vitamin C, which most animal species still have.

By 1968 a substantial number of psychiatrists and general practitioners in the United States and Canada had corroborated our early claims. We all concluded that vitamin B-3 was a very important component in the treatment of early schizophrenics suffering either their first, second, or third relapse (or more). It did not work nearly as well when given to chronic patients who had had no periods of improvement. We also had started to use large doses of pyridoxine and moderate doses of essential minerals. Nutrition in general had become much more important in our practice, as had the use of elimination diets to remove foods responsible for maintaining the psychoses. When you have seen a severely ill psychotic patient become normal after a four-day water fast, you no longer doubt the relevance of diet to disease. About that time, 1971, there appeared to be a split between our colleagues who were vitamin enthusiasts, those who favored minerals, and those who emphasized the role of environmental chemicals (clinical ecology).

Dr Pauling's term, orthomolecular, came at a most appropriate time because it defined for all of us what we were doing. A number of us therefore defined our practice as orthomolecular psychiatry or medicine, and began to identify with it at every opportunity. *Orthomolecular Psychiatry* [9] represented the state of our art. Dr Pauling's recent book, *How to Live Longer and Feel Better* [2], represents how far we have come in nearly 20 years.

HOW IT IS PRACTISED

Orthomolecular physicians work with nutrients already identified as playing a major role in health and disease, in large doses. We do not use single nutrients to treat single diseases, having given up the one-disease–one-drug concept. We try to determine the causes of the various syndromes and then remove these causes, using nutrition to provide what it can and supplementing this with vitamins, minerals, amino acids, and essential fatty acids, as needed. We take seriously Dr Pauling's definition, i.e. varying the concentrations of substances; all the substances we can control.

I can best illustrate what we do by outlining how I practice. Of course, I do not speak for all orthomolecular physicians — we all use the program with differing emphasis on the use of nutrients and diet. It is a field which is developing rapidly and has not become a rigid system, as have many treatments in medicine. We use all the methods we have been taught in medical and graduate schools, combining them with orthomolecular principles. We try to select the best from each, for each patient.

The first step in treatment is to establish a causal diagnosis. If my patient is schizophrenic, I must determine which schizophrenic syndrome is most likely present. There are several major syndromes: (a) the allergic group, (b) the vitamin-dependent group, (c) the mineral deficiency group. The vitamin-dependent group may be divided into those who are B-3 dependent, and those who are more dependent on vitamin B-6 [10–12]. The allergy group is suspected from a history of allergies and reactions of foods, and is confirmed by tests such as elimination diets and/or laboratory tests. The vitamin-dependent group is less apt to have allergies, and may show its need for vitamins by lab tests, but is finally established by therapeutic trial.

If allergy is suspected or confirmed, patients are started on a diet which avoids what is harmful. If not, they are simply advised to avoid additives, including sugar, to reestablish a diet to which evolution has adapted us [1, 13]. Supplements are not used to compensate for inadequate foods; they are really supplements to make the total dietary intake optimum when food alone cannot do so.

SUPPLEMENTS

I will describe only the ones used most often. The rest have not been studied as thoroughly but will undoubtedly find a place. There are many patients who are sick because they require these in larger than vitamin quantities, but have not yet had this need identified.

(a) Ascorbic acid

Everyone needs ascorbic acid. The main debate still rages between physicians who have not learned to use vitamins, and orthomocular physicians. The first group is content to accept nutritionists' claims that the average well-balanced diet has enough vitamin C. The RDA is around 50 mg daily. We, following the leads of Pauling [2] and Stone [8], and our own experiences, use much larger doses. Vitamin, or small, doses prevent scurvy, but do not prevent suclinical scurvy. They provide a minimal level of health which is still too close to scurvy, a terminal disease. Optimum levels vary with age, stress, and disease, from several grams per day to much more. Dr Cathcart [14] uses the effect on bowel habits as a measure. The dose is increased until the stools are loose. If more is taken, there may be too much fluidity. The dose is then adjusted downward. I have given patients up to 40 g daily. I know other people who routinely use 75 g per day. Usually if more is required it is best given by intravenous infusion using ascorbates. For treating cancer, up to 200 g may be given daily, using 60 g of sodium ascorbate per liter of fluid.

There are few side effects. A small number of people are allergic, probably to the preparation and traces of chemicals still present from their synthesis. Vitamin C does not cause kidney stones, does not cause pernicious anemia, and does not cause liver damage. These are myths perpetuated by a few physicians [2].

(b) Vitamin B-3

Niacin is used for decreasing cholesterol levels and elevating high-density lipoprotein (HDL) cholesterol, thus elevating the ratio of total cholesterol over HDL; the dose ranges from 2 to 6 g per day. Patients must be advised about the intense flush which follows the first dose. Fortunately, with succeeding doses the flush becomes less intense and may disappear with continued use after several weeks. For all other conditions, either niacin or niacinamide can be used. The upper limit is that dose which causes nausea and later vomiting if the dose is not reduced: about 6 g per day for niacinamide, much higher for niacin. Vitamin B-3 is safe if the dose is kept below this nauseant level. There are a few side effects, most of little consequence except for jaundice, which may affect one in 2000. This estimate is based upon my experience using vitamin B-3 since 1952. Most of my patients who developed jaundice were alcoholic or schizophrenic, and they are more susceptible to liver problems than other types of patient. Usually the jaundice clears when the vitamin in stopped. Every major study of niacin for lowering cholesterol has found it to be safe [15]. My usual starting dose is 3 g daily, in three divided doses [16, 17].

(c) Vitamin B-6 (pyridoxine)

The dose is below 1000 mg per day; usually 250 or 500 mg daily is given. It may sometimes make children more hyperactive unless they are also given magnesium.

(d) Minerals

Zinc is the main one [10]. Usually 30 to 50 mg per day of zinc is used, either as sulfate, citrate or gluconate.

Manganese is useful in preventing and treating tardive dyskinesia [18].

(e) Amino acids

L-tryptophan and *l*-tyrosine have been studied the most. The first is used late in the day to treat insomnia, and in 3 to 6 g doses to supplement lithium or to replace it for manic–depressive patients. I have found it very helpful for patients who cannot tolerate lithium. Slagle [19] combines it with *l*-tyrosine in the morning, in combination with vitamins, to treat depression; see also Braverman and Pfeiffer [20].

DRUGS

Drugs are used as indicated. The dose may be small compared to doses needed when orthomolecular treatment is not used. The objective is to get patients off all drugs or down to a very low level, as soon as possible, because on the usual doses of tranquilizers, patients cannot recover. I would not want my surgeon to operate on me if I knew he or she were on tranquilizers, nor my accountant to advise me. Patients can recover with the aid of tranquilizers, but must be relieved of their need as soon as possible. The orthomolecular approach makes this possible.

A schizophrenic patient might well leave my office on a sugar-free, milk-free diet with niacinamide 3 g, ascorbic acid 3 g, vitamin B-6 250 mg and zinc 50 mg daily, with or without tranquilizers.

RESULTS OF TREATING A FEW MAJOR DISEASES

Medicine will become orthomolecular. In the future, physicians who do not take nutrition and supplements into account will face malpractice suits. To illustrate the value of orthomolecular medicine, I will describe five major diseases; see also references [2, 6, 9–13, 21–64]. The five diseases are viral infections, the schizophrenias, the saccharine disease, children with learning and behavioral disorders, and cancer. Dr Pauling's work is associated with these conditions for his review of the literature and his study of vitamin C. His fierce support of the use of vitamin C in large doses has established it securely as a vital nutrient, and it is now well established in the public consciousness, even if physicians are still suspicious of it. A large proportion of cancer patients have heard about its use, and many are taking it, though usually in low doses. Vitamin C is a major component of all alternative cancer programs.

The common cold and influenza

The impact of Dr Pauling's books was amazing. The public loved them, and sales of ascorbic acid soared. The medical profession hated them and did its best to discourage patients from using vitamin C. Doctors criticized Dr Pauling because he does not have a medical degree. In fact, Dr Pauling's arguments were based upon his review of the medical literature. One of the first to use very large doses of vitamin C was Dr Fred Klenner [65–67]. Stone [8] and Pauling [68] have reviewed the literature.

Pauling made it a worldwide issue. His review of the literature and his personal experiences convinced him vitamin C was valuable and, even more, would prevent these virus infections. Every orthomolecular practitioner using vitamin C in these large doses will confirm this. From my own experiences using vitamin C over 35 years, I am convinced it is the treatment of choice, not only for the common cold and/or flu, but for all virus infections including AIDS [69, 70]. Its efficacy depends on the dose: higher doses offer more protection than lower doses. To achieve almost 100 percent protection, the dose must be optimal or sublaxative. I can recall many families where the one member I treated remained free of colds thereafter, while other family members continued to suffer their usual frequency of colds and flu.

Cancer

Dr Pauling's introduction of ascorbic acid into cancer treatment is more important clinically, for it represents a major new idea in a field which has made relatively little progress in 25 years [2, 71]. I will not describe how Dr Pauling became interested, because he has already done so in a way which cannot be improved upon, in his Chapter 19 on cancer [2]. Cameron and Pauling's [72] book, *Cancer and Vitamin C*, is one of the most important books published in medicine in the past two decades. I suggest the reader go over this chapter again before continuing with my account of our current studies of vitamin C and cancer.

I became interested in the vitamin treatment of cancer in 1960 when a psychotic patient, dying from bronchogenic cancer, became mentally normal after three days on niacin 3 g and ascorbic acid 3 g per day; he had been psychotic for three months. He continued to take both vitamins and remained well for nearly 30 months. The cancer, easily seen by X-ray at the onset, was gone 12 months later. I assumed the major therapeutic agent was the niacin. My second case was a teenage girl I saw in 1961 with osteosarcoma of her arm. She was to be treated by amputation. She started on niacinamide 3 g and ascorbic acid 3 g per day, instead. She was cured and is still well. Again I concluded vitamin B-3 was the therapeutic agent.

By 1977, when a woman with cancer of the head of the pancreas was referred to me, I was familiar with Cameron and Pauling's exceptional work. On her own, this woman started to take 10 g of ascorbic acid per day. I increased the dose to 40 g daily, adding vitamin B-3 and other vitamins. Within six months her tumor could no longer be seen on CT scan. She is still well. This time I gave credit to the ascorbic acid.

This patient changed my practice. She began to tell all her friends about her remarkable recovery, and within a few years, cancer referrals began to increase from one every two months to four to five per week. They were patients with a variety of tumors who had all ready been treated with radiation, chemotherapy or surgery, or some combination. They were almost all terminal in a matter of months or one or two years. Only a handful came early in the course of their disease. This has now changed. By 1990 many more patients were being referred in the early stages of their illness.

All the patients I see are referred to me by their general practitioners, who establish the diagnoses in consultation with the cancer clinics. These are government-operated agencies staffed by competent physicians and oncologists, and having available to them the latest advances in diagnostics and treatment; I have nothing to do with diagnosis or their treatment.

When patients present themselves in my office I try to ease their anxiety and pain, provide some hope, and offer them an orthomolecular program which will enhance their chances for recovery. Patients are not promised a cure, but are advised that from all I have read and seen, they would be better off with the program. They are then advised to improve their diet by eliminating fat and sugar. The easiest way to eliminate fats is to eliminate all dairy products, cut back on meat and avoid processed foods. They are advised that the most important nutrient is ascorbic acid, and are started on 4 g after each meal. Most patients prefer the crystalline preparation. The level of ascorbic acid is increased until they reach bowel tolerance or sublaxative levels. Patients are also given 1.5 g of vitamin B-3 daily, and smaller amounts of a B-complex preparation. To this I add vitamin E, beta carotene, selenium, zinc, calcium and magnesium.

I did not look upon this as a research study. My intention was, and is, to maximize patients' potential for recovery by using those nutrients which enhance immune system function and which have anti free radical properties. Patients are seen again in a month to answer any questions. Thereafter they are followed by their general practitioners.

After several years, it became clear that patients who followed the vitamin–mineral program lived longer than did the smaller group who did not follow the program. Every patient was advised to follow this program, but for several reasons some were not able to do so. Some patients were nauseated from chemotherapy and could not keep any medication down. A few could not obtain support from their families or physicians and became discouraged, and others believed they would die as had been predicted and so did not follow my advice. One example was a physician who was told he would die within a month. He came to see me because his girlfriend persuaded him to do so, but after listening to me he decided he would not take the program. He died in a month. A few patients died within two months of beginning to take the vitamins, before the program had a chance to really start working. Patients who did take vitamin C and the rest of the program lived much longer. The first patient in the series, the woman with pancreatic cancer, is still alive after 12 years.

There has also been a change in the attitude of the referring physicians. Before 1985, most of the referrals were patient initiated, but after that many more were physician initiated. In 1985 I examined the survival data of the first 41 patients who had consulted me [47]. Eleven patients did not follow the program. They survived an average of 4.8 months, none was alive at follow-up. Four patients took only 3 g of vitamin C per day. They survived an average of 113 months, and one was still alive. Of the 26 who did follow the program, their mean survival was 34.5 months and in 1985, 18 were still alive.

About two years later I discussed this data with Linus Pauling. He encouraged me to re-examine the results of treatment with a much larger series. Each year there has been an increase in the number of referrals. I examined the state of 134 patients seen between July 1978 and 15 April 1989, as of 1 January 1990. From this group, 33 had not followed the program. I considered that any patient not on the program for at least two months had not followed the program. This condition applied to a small proportion of this group. The rest would not or could not do so. They are a control group but are not to be considered equivalent to a control group in the usual prospective double-blind controlled experiments. I had not intended to have any controls, expecting that each patient's previous history of failure after all modern treatment would be a control. If a patient's progress is steadily downhill in spite of the best possible treatment until a new treatment is introduced, and then changes its course by markedly prolonging life, then one can make the scientific assumption that the new treatment influenced the course of that patient's illness. This may occur by chance in a few, but with each additional patient, the odds this is a chance result decrease until it is beyond the range of probability it was chance, i.e. it is a real phenomenon. I had hoped every patient would follow the program, and had no idea how many would benefit.

I contacted patients, their relatives, and their family doctors. Their clinical records received from the referring physician were available. In this way I was able to trace every patient. I used survival data only, beginning with the day I first saw them, until the day they died. Having collated this material I sent it to Dr Pauling. He had developed a novel, accurate way of measuring the significance of survival data on cancer cohorts [73]. We reported the results of our investigation [74].

The mean survival of the control group was 5.7 months. None was alive at follow-up time. The group who took vitamin C and the other nutrients had a mean survival of 92.0 months or 16 times as long, and 48 were still alive at follow-up. Another follow-up study one year later (1 January 1991) showed that survival had increased by another 5 months. The death rate from the survivors is low. Most of them probably have been cured of their cancer and are dying from other causes due to their age. One of the patients, considered terminal in 1977, died late in 1989 at age 80.

This data provides corroboration of the Cameron and Pauling studies that ascorbic acid given to cancer patients increased longevity. But it does more than that — many of these patients and their families told me how much better they felt on treatment, for which they were very grateful, again confirming Cameron and Pauling.

I have no doubt ascorbic acid is the main therapeutic variable and that the other nutrients I used added to their general good health. I cannot conclude which of the other nutrients had a synergistic effect with vitamin C. It will take more comphrehensive, carefully controlled studies on larger groups to tease out the role played by these other nutrients. I am also convinced that had each patient started on the nutrient program at the onset of their illness along with standard therapy, the survival data would have been better. At the onset of the disease, patients are stronger, have more reserves of energy and find it easier to change their lifestyle, i.e. to change their diet. If they are started on vitamin C, they will suffer fewer side effects from chemotherapy or radiation, and they will heal faster from these and from surgery. Many of these patients were already exhausted and dispirited by the time I first saw them, and they were unable to follow the program. It was especially difficult for unmarried patients who had no family to back them up or to help them in following their vitamin program. I am convinced every patient with cancer should be on optimum doses of vitamin C as soon as they have symptoms to suggest they may have cancer.

The schizophrenias

Orthomolecular psychiatric treatment of the schizophrenias yields results much superior to those using modern psychiatric tranquilizers only. These drugs are very

family and community, and engaged in the same type of activity
engaged in had they not become ill; in short, for many it means bei
taxes. I know 18 men and women who recovered from schizophrenia wh
their teens, who became physicians and are still well. One became presid
psychiatric association, one is a research psychiatrist, a third is chairman
department at a medical college. In sharp contrast, of 42 physicians who
while in practice and given only tranquilizers, only 12 went back to their pra
these, six did so with major assistance from their wives, who were their office
Only six were able to resume full-time activity equivalent to their pre-illness

The ideal treatment for the schizophrenias is to combine the best of all prog
The tranquilizers will provide rapid control by cooling hot symptoms. The orth
lecular treatment will slowly remove the illness, i.e. all the symptoms. As the treat
becomes more effective for the patients, the drugs are slowly reduced until they are
longer necessary, while patients remain on their diet and the appropriate nutrien
Until a different class of drugs comes along which removes all the symptoms, there
will be no further progress with tranquilizers. But this is highly unlikely since
schizophrenia is a biological disease, not a tranquilizer deficiency, and will only be
cured by natural means — by orthomolecular therapy.

The main vitamins are vitamin B-3, vitamin B-6 and ascorbic acid. However,
ascorbic acid may be much more important than I had considered it to be. Originally,
in 1951, we had planned to test both vitamin B-3 ascorbic acid, but our double-blind
consultants advised us not to use ascorbic acid, as it would create many problems in
design and interpretation. We therefore selected vitamin B-3, as it seemed more
appropriate. Since then I have used small orthomolecular doses, usually 3 g per day.
In a few cases I went much higher. In 1952 a schizophrenic woman was admitted soon
after a radical mastectomy. Her psychiatrist planned to give her electroconvulsive
therapy (ECT). I persuaded him to wait, as I wanted to test ascorbic acid alone. I had
planned on giving her 3 g per day. Her psychiatrist said he would only delay ECT for
3 days, and I therefore decided to give her 1 g per hour. Over Saturday and Sunday
she received 45 g. On Monday her psychiatrist canceled the ECT because she was now
normal. Also, her surgical wound, which had not healed, began to heal. She remained
mentally normal until her death from cancer one half-year later. Did she have a
scorbutic psychosis?

My next case was in 1960. I had a patient who was terribly agitated and restless.
There was no drug which would help her, as she did not respond to tranquilizers. Her
other schizophrenic symptoms were controlled by vitamin B-3. I started her on
ascorbic acid 10 g per day, and within a few days her agitation and tension were gone.

My last case was a patient in Victoria, in 1980. She was a chronic patient whose
main complaint was that half her brain was dead. On orthomolecular treatment alone
she was better, but not well. Only a series of six ECTs would remove this symptom for
a few months. She would then return, again and again for further treatment.
Eventually I gave her ascorbic acid 10 g per day instead of ECT. Within two weeks her
major symptom was gone and she has remained well. She has not required any more
ECT.

Ascorbic acid in doses up to sublaxative levels will have to be tested rigorously as a
treatment for schizophrenia, with and without vitamin B-3, to determine if there are
many ascorbic-acid-dependent patients similar to the three I have described.

Table 1 — response of schizophrenic patients to tranquilizers and to ortho-molecular treatment

Pat	Tranquilizer		Orthomolecular	
	Duration of treatment	Response	Duration of treatment	Response
any fter ell	One year	<25%	One year	90% recovery
years	2 years	<10%	2–5 years	50–75% recovery
2–5 years	2–5 years	<10%	2–5 years	25–50% recovery

The results of orthomolecular treatment compared with standard treatment are given in Table 1.

The tranquilizer recovery rates are from the vast tranquilizer literature. The orthomolecular rates are based upon my practice between 1960 and 1990 on several thousand patients I have treated. I have overestimated tranquilizer recoveries, as I have seen very few, and have underestimated orthomolecular recoveries. These estimates are also based upon the experience of my colleagues who, all together, have treated over 100 000 schizophrenic patients.

The saccharine disease

According to Cleave [75] and Cleave and Campbell [76], a large number of diseases which are considered separate are in reality symptoms of one major disease, the saccharine disease. It is caused by the high-tech diet of the western world, too rich in sugar, too rich in fat, and too low in fiber. A doughnut, made from white flour, oil and sugar, is a good example of this diet. The diseases considered variants of the saccharine disease are diabetes mellitus, cardiovascular disease, chronic constipation and colitis, appendicities, and cancer of the bowel. The evidence which supports the saccharine disease concept is powerful, and it is growing. Recently, *Time Magazine*, 14 January 1991, published data from the *Journal of the National Cancer Institute*. They showed a linear relationship between the percentage of fat in the diet and the breast cancer rate. This is shown in Fig. 1. The relationship appears to be linear, but there are too few points below 20 percent fat in the diet. Extrapolating to a 0% fat diet, it appears that breast cancer at this level would vanish. A 10 percent fat diet would be practical. There are no 0% fat diets.

Children with learning and behavioral disorders

In my opinion, orthomolecular therapy is the treatment of choice for these children. The common psychiatric nomenclature does not help very much because it is not based upon causes and is descriptive of symptoms, rather than diseases. As with the

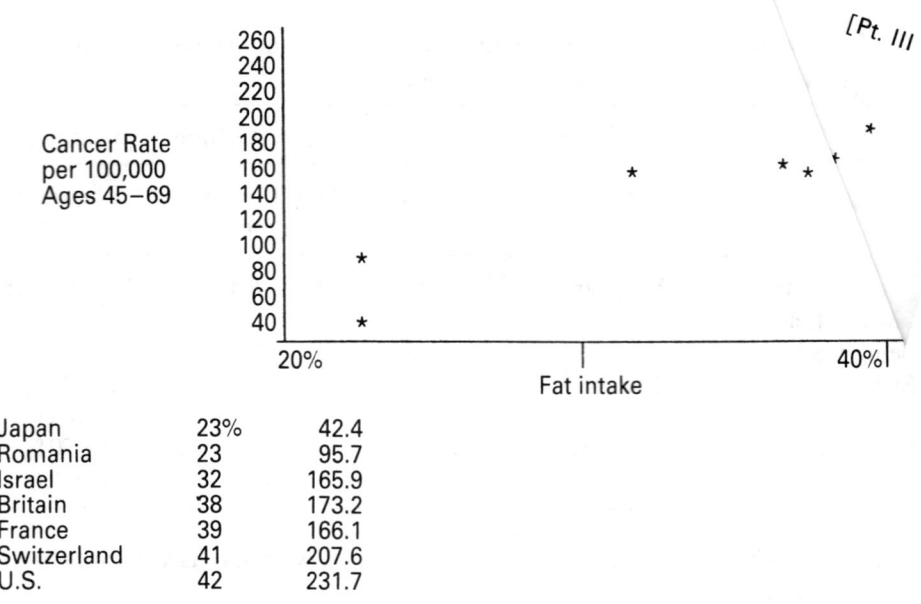

Relation between fat intake and breast cancer rate by country

Fig. 1.

Japan	23%	42.4
Romania	23	95.7
Israel	32	165.9
Britain	38	173.2
France	39	166.1
Switzerland	41	207.6
U.S.	42	231.7

schizophrenias, the main factors are dietary, including allergic reactions to foods and the need for supplements. I have concluded this from having treated over 1500 children since 1954, and from reports by Cott, Crook [77, 78], and others. Vitamin B-6 has been shown to be effective as part of the treatment for infantile autism [79–82].

About 40 percent of the young patients I have seen suffered from one or more allergies, usually to foods. The most common offenders are dairy products, sugar, and wheat. The offending foods are identified by history and elimination diets. When these foods are eliminated, the children become well in a few months. Last year a young mother brought her son, age 7, to see me. I could find nothing wrong with him. Then she told me he had been autistic until a few months before. He had been diagnosed by a Victoria clinic which specializes in diagnosing these children. Against the advice of these experts, his mother placed him on a diet free of dairy products, sugar and wheat. She had read about the connection between food and behavior. One month later he was normal. This is only one example out of hundreds I have seen.

About 40 percent of children with learning and behavioral disorders need supplementation with vitamin B-3, either niacin or niacinamide. Children can tolerate adult doses and often have to be given dosages of 2 to 3 g per day. A common program for this group includes vitamin B-3, 2 to 3 g daily, vitamin B-6, 250 mg daily, ascorbic acid, 2 to 3 g per day, and some zinc, up to 30 mg per day. Patients with infantile autism will need less vitamin B-3 and more vitamin B-6. Often these children will need additional magnesium to prevent increased irritability, which may occur if vitamin B-6 is used alone.

There is frequently an overlap between groups. Many allergic children will also need vitamin supplementation, and many who primarily need vitamins should also be placed upon a sugar-free, additive-free dietary program.

for 60 years or more after we have corrected our diet — unless, of course, we use the correct vitamin supplements in large doses. Orthomolecular nutrition will save us from our own greatest folly, but only if its principles are accepted and acted upon. I believe these high-tech diets create nutrient deficiencies, of which chronic pellagra is one example. Attempts are under way to restore our diets, but they are only partial attempts, such as diets with a moderate decrease in fats, which should decrease the incidence of cancer. *Time Magazine*, 14 January 1991, outlines the connection between fat and breast cancer.

WHAT IS WRONG WITH OUR DIET?

Adaptation is a process by which living species have developed apparatus which allows them to eat, digest and metabolize food available to them. This adaptation may be very narrow — for example, anteaters are adapted to eating only ants, and koala bears have adapted to eating the leaves of one species of tree — or it may be wide, as exemplified by omnivores. Animals that are adapted to vegetation are herbivores. They have a digestive apparatus which can chew and grind harsh foods, ferment them and make use of the products. They will not do well on very low-fiber foods. Meat eaters, carnivores, are hunters and live chiefly on the animals they hunt. They will not survive on grass and lettuce. Omnivores can digest both vegetables and meats. Humans are omnivores, as are bears, for example.

But even though we are omnivores, we must consume those foods we can digest and metabolize. We must have these foods, just as much as carnivores need meat and herbivores need vegetation. A few years ago, in Cuba, cattle were fed on high doses of a solution of raw sugar left over from the manufacture of cane sugar. The fiber content was so low they became sick. Modern zoos interested in maintaining the health of their animals understand this, and do try to provide the foods to which the animals have been adapted. Our modern diet would make them sick. Zoos caution visitors not to feed the animals with our junk food. To review, foods to which we have adapted will keep us well, and foods to which we have not adapted will make us sick.

WHAT HAVE WE ADAPTED TO?

I have described our food supply by using key adjectives which accurately portray what they are like. The foods we had adapted to and which healthy animals in the wild and in zoos still consume are:

1. *Whole* — Herbivores eat all the edible parts of plants, usually grass, seeds and fruit, often bark; the seeds are whole, not hulled. Carnivores eat all their prey including edible bones.
2. *Fresh* — Our ancestors could store some foods such as grains and nuts, but other food was consumed fresh. Animals eat living animals, occasionally carrion. Fresh foods contain all the nutrients, whereas stored foods lose essential vitamins and are subject to spoilage by fungi and bacteria.
3. *Varied* — Before agriculture there were no huge stores of grains as staple foods.

Everything edible was consumed, depending on time of day and season. Foods varied over the year. A variable diet is less apt to cause food allergies.

4. *Indigenous* — Animals and people depended upon local sources of food, i.e. cold-adapted plants and animals, rich in Omega-3 unsaturated fatty acids, were consumed by cold-adapted people.

5. *Non-toxic* — By experience, animals learned which varieties of plants and animals were safe. Foods were not adulterated by large numbers of additives.

6. *Scarce* — Before agriculture, overconsumption of calories was difficult. Native peoples, who until recently gathered and hunted as did our ancestors, are lean, and seldom fat.

Any diet which can be described by the preceding adjectives is healthy, the type of diet to which we have adapted.

Our modern high-tech diet can also be described by six adjectives:

1. *Artifact* — Only parts of foods are consumed. Wheat is milled and the most nutritious portions, the germ and the bran, are eliminated from our food. Bran is removed from rice, and sugar is extracted from sugar cane and beet. Oil is expressed from seeds and is made toxic by heat and chemical treatment. This is the primary method by which we create the saccharine disease.

2. *Stale* — Much of modern food must be stored, up to several years. Foods are canned, frozen and preserved. These are vital but food quality is compromised. With storage, a host of problems develop which destroy food quality.

3. *Monotonous* — Our diets are characterized by large quantities of a few staples: sugar, white flour, white rice, and the oils (fats). These are what I call a monotonous diet. The 100 different types of breakfast cereal do not denote a large variety of different foods, as they are mainly concocted from corn, oats, white flour and sugar, with flavoring.

4. *Exogenous* — Foods can be shipped thousands of miles. We consume tropical foods in cold areas. In this hemisphere we must be cold-acclimatized and so should not depend on heat-acclimatized foods.

5. *Toxic* — Modern foods contain a large array of additives — from sugar, the major one, to oils and chemicals — which are used to enhance stability and palatability of foods.

6. *Too Abundant* — Obesity has become a major problem in high-tech societies. We have a surplus of calories combined with a deficiency of essential nutrients. Our ancestors went through periods of famine, and humans have adapted to this by storing extra calories, when available, as fat. Fasting is beneficial and has been used therapeutically for thousands of years, and is referred to in the Bible about seventy times. Fasting is now so rare that many people believe a one-day fast will kill them. I believe that, as we had to fast historically by necessity, it is good for us to occasionally test and keep in repair our fasting mechanisms. I do not recommend rapid-weight-loss diets, which can be danerous and are usually ineffective.

The solution to most of our problems with disease is simple: return to the diet to which we had adapted (an orthomolecular diet), and supplement those nutrients which the diet can no longer provide in adequate quantities.

IF ORTHOMOLECULAR MEDICINE IS SO GOOD, WHY DOESN'T EVERY PHYSICIAN PRACTISE IT?

This is a criticism voiced by many physicians, who assume that every physician promptly begins using all the best treatments. This has never been the case in medicine, and usually there has been a 40-year lag. The main reason for this delay is that the modern paradigm for medical theory and practice is too well entrenched, and is perpetuated by medical schools. Modern or high-tech medicine is superb for emergencies or for what Dr J. Beasley has termed horizontal medicine. It can deal very effectively with most infections, most surgical crises, most acute disease and most severe hormonal anomalies, but it has failed to come to grips with the chronic diseases, such as arthritis, schizophrenia, multiple sclerosis, and many more, i.e. with vertical medicine — with patients still able to work but suffering from a poor state of health. The modern medical paradigm has ignored nutrition. This is beginning to change, but slowly and reluctantly. Medicine still embraces the old one-vitamin–one-disease concept developed when very little was known about vitamins, and it rejects orthomolecular medicine.

This view was brought home to me vividly as I was watching a television story about *Doctors without Borders*. These doctors give up their high-tech practices in their own countries to treat patients in third world countries. One physician described why she did so. She said she got immense satisfaction in seeing her patients recover, and by knowing they would not sue her. What she implied was that practicing high-tech medicine at home left most of her patients dissatisfied, and therefore more apt to sue her. This is a good summary of what has happened; modern medicine is ideal for third world countries who still suffer from diseases virtually unknown in high-tech cultures, such as beriberi, scurvy, severe infections, congenital diseases. These are easily treated. But having passed through these stages, modern society has moved into an era marked by chronic malnutrition, which can only be treated nutritionally, by orthomolecular medicine. Modern physicians, having no training in clinical nutrition, are unable to cope with these conditions. It has been repeated over and over that modern generals prepare to fight the most recent war. Modern medical schools train their students to deal with diseases they had to cope with 30 years ago, with the exception of modern surgery and anesthesia. In contrast, orthomolecular physicians are attuned to the current plethora of diseases caused by malnutrition, by a diet to which we are not adapted. There is no longer any harmony between our food and our bodies. Orthomolecular medicine tries to restore that harmony.

To answer the question: 'Why do physicians not use orthomolecular methods?', I reply that they will once they have been given a chance to see the results. Over the past 35 years, at least 50 physicians have visited me and spent a day or longer observing the way I practice. Every one has become an orthomolecular physician. No physician who has practiced orthomolecular medicine for at least six months has voluntarily given it up. Several have been forced to do so by pressure from their medical association and colleagues, or by losing their license to practice.

Medical journals are very reluctant to publish papers which describe new treatments for diseases using nutrients. A recent example is the refusal of the *New England Journal of Medicine* to publish a paper which showed that ascorbic acid cured idiopathic thrombocytopenic purpura (ITP).

ITP is a disease for which there is no treatment using drugs. Perhaps it is not a disease at all, but a unique expression of subclinical scurvy in certain individuals. ITP is quite rare. I have seen only one patient, who was referred to me several years ago. Since then she has been well except for one autumn when, due to overconfidence in her health, she stopped taking vitamin C. Within a couple of months her platelet count dropped below 100. On resuming her vitamin C, 3 g per day, her count returned to its normal level.

The discovery that ITP is a case of subclinical scurvy was due to the willingness of a clinician to follow seriously a serendipitous discovery made by his patent. Dr Brox advised his first patient there was no treatment for her ITP. A month or so later, on a subsequent visit, he found she was very much better. She told him she had started to take 1 g of ascorbic acid daily on the advice of a clerk in a health food store. Almost every other doctor would have ignored or dismissed the observation as trivial, irrelevant, or as a spontaneous recovery, denying it could have been the vitamin. But a second patient, and later a third, also responded. In Brox *et al.*'s report [87] 11 patients were treated with ascorbic acid, and seven responded. It would seem that ascorbate improved the platelet count and intravascular survival of platelets. Would their discovery have occurred if Dr Linus Pauling had not championed the use of large doses of ascorbic acid?

DR LINUS PAULING'S ROLE

From the few diseases I have described, it should be clear that medicine must inevitably become orthomolecular. More frequently over the past ten years, do we find reports dealing with clinical nutrition in establishment journals. Even the *New England Journal of Medicine*, a most determined anti-vitamin journal, published an excellent account of vitamin C several years ago. Niacin is in for hypercholesterolemia, vitamin C is coming in for its anti-viral and anti-cancer properties. Vitamin E is accepted for its anti-oxidant properties, with a number of indications. Thiamine is in for Wernicke-Korsakoff syndrome.

Dr Pauling's role in the rapid development of orthomolecular medicine is predominant. He has singlehandedly forced the world to pay attention to vitamin C. His conclusions that it has anti-viral properties are correct. His research with cancer and vitamin C are innovative, and powerfully demonstrate vitamin C's usefulness.

Linus Pauling could have retired 25 years ago to rest on his laurels, but he felt his duty to mankind overrode any wish to retire, and he has continued to fight for his views with determination and flair. All patients who have benefitted from orthomolecular medicine owe Dr Pauling for their good health.

For my part, and on behalf of every clinical orthomolecular physician, I thank Dr Pauling: not only because we are able to help so many patients, but because our practices have become more exciting. The greatest satisfaction we physicians have is to see our patients get well, and orthomolecular physicians have this satisfaction, which they did not have while practicing orthodox medicine.

The greatest tribute to Dr Linus Pauling will come when the word 'orthomolecular' is no longer needed, for then all of medicine will be orthomolecular. Usually a major paradigm shift like this requires about 40 years. I suspect this paradigm shift will occur much more quickly, perhaps only 30 years from Pauling's first major orthomolecular report to *Science* [1], but medical schools, as usual, will lag some distance behind.

LITERATURE CITED

[1] L. Pauling, Orthomolecular Psychiatry. *Science* **160**, 265–271 (1968).

[2] L. Pauling, *How to Live Longer and Feel Better*. W.H. Freeman & Co., New York, 1986.

[3] W. Kaufman, *Common Form of Niacinamide Deficiency Disease: Aniacinamidosis*. Yale University Press, New Haven, CT, 1943.

[4] W. Kaufman, *The Common Form of Joint Dysfunction: Its Incidence and Treatment*. E.L. Hildreth & Co., Brattleboro, VT, 1949.

[5] W. Kaufman, Niacinamide: a most neglected vitamin. In: *New Dimensions of Preventive Medicine*, International Academy of Preventive Medicine, Lincoln, NB, 1981.

[6] A. Hoffer and H. Osmond, *The Hallucinogens*. Academic Press, New York, 1967.

[7] A. Hoffer and H. Osmond, *How to Live With Schizophrenia*. University Books, Inc., New York, 1966, 1978.

[8] I. Stone, *The Healing Factor: Vitamin C against Disease*. Grosset and Dunlap, New York, 1972.

[9] D.R. Hawkins and L. Pauling, *Orthomolecular Psychiatry*. W.H. Freeman, San Francisco, 1973.

[10] C.C. Pfeiffer, *Mental Illness and Schizophrenia: the Nutrition Connection*. Thorsons, Wellingborough, Northamptonshire and Rochester, VT, 1987.

[11] C.C. Pfeiffer, R. Mailloux and L. Forsythe, *The Schizophrenias: Ours to Conquer*. Biocommunications Press, Wichita, KS, January 1988.

[12] C.C. Pfeiffer, J. Ward, M. El-Meligi and A. Cott, *The Schizophrenias: Yours and Mine*. Pyramid Books, New York, 1970.

[13] A. Hoffer, Orthomolecular nutrition at the zoo. *J. Ortho. Psychiat.* **12**, 116–128 (1983).

[14] R.F. Cathcart, Vitamin C: the non toxic, non rate-limited, antioxidant free radical scavenger. *Med. Hypothesis* **18**, 61–77 (1985).

[15] P.L. Canner, K.G. Berge, N.K. Wenger, J. Stamler, L. Friedman, R.J. Prineas and W. Friedewald, Fifteen year mortality in Coronary Drug Project patients: long-term benefit with niacin. *J. Am. Coll. Cardiology* **8** 1245–1255 (1986).

[16] R. Altschul, *Niacin in Vascular Disorders and Hyperlipidemia*. C.C. Thomas, Springfield, IL, 1964.

[17] R. Altschul, A. Hoffer and J.D. Stephen, Influence of nicotinic acid on serum cholesterol in man. *Arch. Biochemistry and Biophysics* **54**, 558–559 (1955).

[18] R.A. Kunin, Manganese and niacin in the treatment of drug-induced dyskinesias. *J. Ortho. Psychiat.* **5**, 4–27 (1976).

[19] P. Slagle, *The Way Up from Down*. Random House, New York, 1987.

[20] E.R. Braverman with C.C. Pfeiffer, *The Healing Nutrients Within — Facts, Findings and New Research on Amino Acids*. Keats Publishing, Inc., New Canaan, CT, 1987.

[21] A.A. Cott, Treatment of ambulant schizophrenics with vitamin B-3 and relative hypoglycemic diet. *J. Schizophrenia* **1**, 189–196 (1967).

[22] A.A. Cott, Treatment of schizophrenic children. *Schizophrenia* **1**, 44–59 (1969).

[23] A.A Cott, Orthomolecular approach to the treatment of learning disabilities. *J. Ortho. Psychiat.* **3**, 95–105 (1971).

[24] A.A. Cott, Orthomolecular approach to the treatment of children with behavioral disorders and learning disabilities. *J. Applied Nutrition* **25**, 15–24 (1973).

[25] A.A. Cott, *Dr. Cott's Help for Your Learning Disabled Child.* Times Books, New York, 1985.

[26] A. Hoffer, *Niacin Therapy in Psychiatry.* C.C. Thomas, Springfield, IL, 1962.

[27] A. Hoffer, Nicotinic acid: an adjunct in the treatment of schizophrenia. *Am. J. Psychiatry* **120**, 171–173 (1963).

[28] A. Hoffer, Treatment of organic psychosis with nicotinic acid. *Dis. Nervous System* **26**, 358–360 (1965).

[29] A. Hoffer, Enzymology of hallucinogens. Reprinted in *Enzymes in Mental Health.* J.B. Lippincott, Philadelphia, 1966, pp. 43–55.

[30] A. Hoffer, The effect of nicotinic acid on the frequency and duration of rehospitalization of schizophrenic patients: a controlled comparison study. *Int. J. Neuropsychiatry* **2**, 234–240 (1966).

[31] A. Hoffer, Enzymology of hallucinogens. *Proceedings of the Carl Neuberg Society Meeting.* J.B. Lippincott, Philadelphia, 1966.

[32] A. Hoffer, Five California schizophrenics. *J. Schizophrenia* **1**, 209–220 (1967).

[33] A. Hoffer, Safety, side effects and relative lack of toxicity of nicotinic acid and nicotinamide. *Schizophrenia* **1**, 78–87 (1969).

[34] A. Hoffer, Childhood schizophrenia: a case treated with nicotinic acid and nicotinamide. *Schizophrenia* **2**, 43–53 (1970).

[35] A. Hoffer, Pellagra and schizophrenia. *Psychosomatics* **11**, 522–525 (1970).

[36] A. Hoffer, Megavitamin B-3 therapy for schizophrenia. *Can. Psychiat. Assoc. J.* **16**, 499–504 (1971).

[37] A. Hoffer, Vitamin B-3 dependent child. *Schizophrenia* **3**, 107–113 (1971).

[38] A. Hoffer, A vitamin B-3 dependent family. *Schizophrenia* **3**, 41–46 (1971).

[39] A. Hoffer, Treatment of hyperkinetic children with nicotinamide and pyridoxine. *Can. Med. Assoc. J.* **107**, 111–112 (1972).

[40] A. Hoffer, Orthomolecular treatment of schizophrenia. *J. Ortho. Psychiat.* **1**, 46–55 (1972).

[41] A. Hoffer, Mechanism of action of nicotinic acid and nicotinamide in the treatment of schizophrenia. In: *Orthomolecular Psychiatry*, Eds. D. Hawkins and L. Pauling. W.H. Freeman & Co., New York, 1973.

[42] A. Hoffer, Nutrition and schizophrenia. *The Can. Family Phys.* **21**, 78–81 (1975).

[43] A. Hoffer, Natural history and treatment of thirteen pairs of identical twins: schizophrenic and schizophrenic–spectrum conditions. *J. Ortho. Psychiat.* **5**, 101–122 (1976).

[44] A. Hoffer, Megavitamin therapy for different cases *J. Ortho. Psychiat.* **5**, 169–182 (1976).

[45] A. Hoffer, Behavioral nutrition. *J. Ortho. Psychiat.* **8**, 169–175 (1979).

[46] A. Hoffer, *Niacin, Coronary Disease and Longevity.* Keats Publishing, Inc., New Canaan, CT, 1986.

[47] A. Hoffer, *Orthomolecular Medicine for Physicians.* Keats Publishing, Inc., New Canaan, CT, 1989.

[48] A. Hoffer and H. Osmond, *The Chemical Basis of Clinical Psychiatry*. C.C. Thomas, Springfield, IL, 1960.

[49] A. Hoffer and H. Osmond, Treatment of schizophrenia with nicotinic acid — a ten year follow-up. *Acta Psychiat. Scand.* **40**, 171–189 (1964).

[50] A. Hoffer, H. Osmond, M.J. Callbeck and I. Kahan, Treatment of schizophrenia with nicotinic acid and nicotinamide. *J. Clin. Exper. Psychopath.* **18**, 131–158 (1957).

[51] A. Hoffer and M. Walker, *Orthomolecular Nutrition*. Keats Publishing, Inc., New Canaan, CT, 1978.

[52] A. Hoffer and M. Walker, *Nutrients to Age without Senility*. Keats Publishing, Inc., New Canaan, CT, 1980.

[53] J. Hoffer, The controversy over orthomolecular therapy. *J. Ortho. Psychiat.* **3**, 167–185 (1974).

[54] R.P. Huemer, *The Roots of Molecular Medicine—a Tribute to Linus Pauling*. W.H. Freeman & Co., New York, 1986.

[55] C.C. Pfeiffer, *Mental and Elemental Nutrients*. Keats Publishing, Inc., New Canaan, CT, 1975.

[56] C.C. Pfeiffer, *Zinc and Other Micro-Nutrients*. Keats Publishing, Inc., New Canaan, CT, 1978.

[57] M.E. Rosenbaum and D. Bosco, *Super Fitness Beyond Vitamins: the Bible of Super Supplements*. New American Library, NAL Penguin, Inc., New York, 1987.

[58] H. Ross, *Fighting Depression*. Larchmont Books, New York, 1975.

[59] J. Roth and H. Ross, *The Executive Success Diet*. McGraw-Hill, New York, 1986.

[60] D.O. Rudin, The major psychoses and neuroses as Omega − 3 essential fatty acid deficiency syndrome: substrate pellagra. *Biological Psychiat.* **16**, 837–850 (1981).

[61] D.O. Rudin, The dominant diseases of modernized societies as Omega-3 essential fatty acid deficiency syndrome: substrate beri beri. *Med. Hypothesis* **8**, 17–47 (1982).

[62] D.O. Rudin and C. Felix, *The Omega-3 Phenomenon: the Nutrition Breakthrough of the '80s*. Rawson Associates, New York, 1987.

[63] M.R. Werbach, *Third Line Medicine—Modern Treatment for Persistent Symptoms*. Arkana Paperback, Routledge & Kegan Paul, New York, 1986.

[64] M. Lesser, *Nutrition and Vitamin Therapy*. Grove Press, New York, 1980.

[65] F.R. Klenner, The treatment of poliomyelitis and other virus diseases with Vitamin C. *Southern Medicine and Surgery* **111**, 209–214 (1949).

[66] F.R. Klenner, Observations on the dose and administration of ascorbic acid when employed beyond the range of a vitamin in human pathology. *J. Applied. Nutr.* **23**, 61–88 (1971).

[67] F.R. Klenner, Response of peripheral and central nerve pathology to mega doses of the vitamin B complex and other metabolites. *J. Applied. Nutr.* **25**, 16–40 (1973).

[68] L. Pauling, *Vitamin C and the Common Cold*. W.H. Freeman & Co., San Francisco, 1970.

[69] R.F. Cathcart, Vitamin C in the treatment of acquired immune deficiency syndrome (AIDS). *Medical Hypothesis* **14**, 423–433 (1984).

[70] I. Brighthope, *You Can Knock Out AIDS with Vitamin C and Immune Nutrients.* Biocenters, 18 Ripon Grove, Elsternwick, Australia, 1987.

[71] J. Heimlich, *What Your Doctor Won't Tell You.* Harper Perennial, New York, 1990.

[72] E. Cameron and L. Pauling, *Cancer and Vitamin C.* W.W. Norton, New York, 1979.

[73] Pauling (1989)

[74] A. Hoffer and L. Pauling, Hardin Jones biostatistical analysis of mortality data for cohorts of cancer patients with a large fraction surviving at the termination of the study and a comparison of survival times of cancer patients receiving large regular oral doses of Vitamin C and other nutrients with similar patients not receiving those doses. *J. Orthomolecular Medicine* **5**, 143–155 (1990).

[75] T.L. Cleave, *The Saccharine Disease.* Keats Publishing, Inc., New Canaan, CT, 1975.

[76] T.L. Cleave and G.D. Campbell, *Diabetes, Coronary Thrombosis, and the Saccharine Disease*, 1st edn, John Wright and Sons, Bristol, 1966.

[77] W.G. Crook, *Solving the Puzzle of Your Hard-to-Raise Child.* Random House, New York, 1987.

[78] W.G. Crook, *Detecting Your Hidden Allergies.* Professional Books, Jackson, TN, 1988.

[79] B. Rimland, *Infantile Autism: the Syndrome and Its Implications for a Neural Theory of Behavior.* Appleton-Century-Crofts, New York, 1964.

[80] B. Rimland, High dosage levels of certain vitamins in the treatment of children with severe mental disorder. In: *Orthomolecular Psychiatry*, Eds D. Hawkins and L. Pauling. W.H. Freeman & Co., San Francisco, 1973, pp. 513–539.

[81] B. Rimland, Risks and benefits in the treatment of autistic children. *J. Autism and Childhood Schizophrenia* **8**, 100–104 (1978).

[82] B. Rimland and G.E. Larson, Nutritional and ecological approaches to the reduction of criminality, delinquency and violence. *J. Applied Nutr.* **33**, 116–137 (1981).

[83] B. Rimland, E. Callaway and P. Dreyfus, The effect of high doses of Vitamin B-6 on autistic children: a double-blind crossover study. *Am. J. Psychiat.* **135**, 472–475 (1978).

[84] D. Burkitt, Hiatus hernia. In: *Refined Carbohydrate Foods and Disease*, Eds D.P. Burkitt and H.C. Trowell. Academic Press, New York, 1975, pp. 161–172.

[85] D.P. Burkitt and H.C. Trowell, *Refined Carbohydrate Foods and Disease. Some Implications of Dietary Fibre.* Academic Press, London, 1975.

[86] J.R. Galler, *Human Nutrition.* Plenum Press, New York, 1984.

[87] A.G. Brox, K. Howson-Jan and A.A. Fauser, Treatment of idiopathic thrombocytopenic purpura with ascorbate. *Br. J. Haematology* **70**, 341–344 (1988).

BIBLIOGRAPHY

Dr Linus Pauling is the foremost orthomolecular nutritionist, even though he does not treat patients. Over the years, his interest in the structure of proteins sensitized him to the need to further examine the role of individual nutrients, especially ascorbic acid and vitamin B-3, in promoting optimum health. From this he has progressed, until now his interests encompass the role of foods and diet, and the roles of all the nutrients. His recent book, *How to Live Longer and Feel Better*, is unsurpassed for accuracy and its coverage of the entire field. This book defines orthomolecular medicine. Medical schools would go far in atoning for their disregard of clinical nutrition if they made this book a compulsory text for their students, and the basis of their first-year course in clinical nutrition and orthomolecular medicine.

Here I will list all of Dr Pauling's orthomolecular papers. They have had an enormous impact on the world of science, and will have a greater influence on medicine and psychiatry. Orthomolecular theory and practice has taken much of psychiatric disease away from psychiatry and has placed it within the grasp of medicine, much as the discovery of vitamin B-3 removed pellagra psychosis from psychiatry, and penicillin removed general paresis of the insane from neuropsychiatry.

Cameron, E. and Pauling, L., Ascorbic acid and the glycosaminoglycans: an orthomolecular approach to cancer and other diseases. *Oncology* **27**, 181–192 (1973).

Cameron, E. and Pauling, L., The orthomolecular treatment of cancer. I. The role of ascorbic acid in host resistance. *Chemical–Biological Interactions* **9**, 273–283 (1974).

Cameron, E. and Pauling, L., Supplemental ascorbate in the supportive treatment of cancer: prolongation of survival times in terminal human cancer. *Proc. Nat. Academy of Sciences USA* **73**, 3685–3689 (1976).

Cameron, E. and Pauling, L., Supplemental ascorbate in the supportive treatment of cancer: reevaluation of prolongation of survival times in terminal human cancer. *Proc. Nat. Academy of Sciences USA* **75**, 4538–4542 (1978).

Cameron, E. and Pauling, L., Experimental studies designed to evaluate the management of patients with incurable cancer. *Proc. Nat. Academy of Sciences USA* **75**, 6252 (1978).

Cameron, E. and Pauling, L., Ascorbate and cancer. *Proc. American Philosophical Society* **123**, 117–123 (1979).

Cameron, E. and Pauling, L., *Cancer and Vitamin C*. Linus Pauling Institute of Science and Medicine, Palo Alto, CA, 1979.

Cameron, E., Pauling, L. and Leibovitz, B., Ascorbic acid and cancer: a review. *Cancer Res.* **39**, 663–681 (1979).

Hawkins, D. and Pauling, L., *Orthomolecular Psychiatry*. W.H. Freeman & Co., San Francisco, 1973.

Hoffer, A. and Pauling, L., Hardin Jones biostatistical analysis of mortality data for cohorts of cancer patients with a large fraction surviving at the termination of the study and a comparison of survival times of cancer patients receiving large regular oral doses of Vitamin C and other nutrients with similar patients not receiving those doses. *J. Orthomolecular Med.* **5**, 143–155 (1990).

Pauling, L., Protein interactions: aggregation of globular proteins. *Faraday Society Discussion*, 1953, pp. 170–176.

Pauling, L., The relation between longevity and obesity in human beings. *Proc. Nat. Academy of Sciences USA* **44**, 619–622 (1958).

Pauling, L., Observations on aging and death. *Engineering and Science Magazine*, California Institute of Technology, Pasadena, CA, May issue, 1960.

Pauling, L., A molecular theory of general anesthesia. *Science* **134**, 15–21 (1961).

Pauling, L., Orthomolecular psychiatry. *Science* **160**, 265–271 (1968).

Pauling, L., Orthomolecular somatic and psychiatric medicine. *J. Vital Substances and Diseases of Civilization* **14**, 1–3 (1968).

Pauling, L., *Vitamin C and the Common Cold*. W.H. Freeman & Co., San Francisco, CA, 1970.

Pauling, L., Evolution and the need for ascorbic acid. *Proc. Nat. Academy of Sciences USA* **67**, 1643–1648 (1970).

Pauling, L., *Vitamin C and the Common Cold*, revised edition. Bantam Books, New York, 1971.

Pauling, L., That man ... Pauling! *Nutrition Today*, March–April 1971, pp. 21–24.

Pauling, L., Vitamin C and the common cold. *J. American Med. Assoc.* **216**, 332 (1971).

Pauling, L., Vitamin C and colds. *New York Times*, 17 January 1971.

Pauling, L., Preventive nutrition. *Medicine on the Midway* **27**, 15–17 (1972).

Pauling, L., *et al.*, Results of loading test of ascorbic acid, niacinamide, and pyridoxine in schizophrenic subjects and controls. In: *Orthomolecular Psychiatry: Treatment of Schizophrenia*, Eds D. Hawkins and L. Pauling, W.H. Freeman & Co., San Francisco, 1973.

Pauling, L., *Vitamin C and the Common Cold*, abridged edition. Bantam Books, New York, 1973.

Pauling, L., Early evidence about Vitamin C and the common cold. *J. Ortho. Psychiat.* **3**, 139–151 (1974).

Pauling, L., On the orthomolecular environment of the mind: orthomolecular theory. *American J. Psychiat.* **131**, 1251–1257 (1974).

Pauling, L., Are recommended daily allowances for vitamin C adequate? *Proc. Nat. Academy of Sciences USA* **71**, 4442–4446 (1974).

Pauling, L., On fighting swine flu. *New York Times*, 5 June 1976.

Pauling, L., Ascorbic acid and the common cold: evaluation of its efficacy and toxicity. *Med. Tribune*, 24 March 1976.

Pauling, L., The case for Vitamin C in maintaining health and preventing disease. *Modern Med.*, July 1976, pp. 68–72.

Pauling, L., *Vitamin C, the Common Cold, and the Flu*. W.H. Freeman & Co., San Francisco, 1976.

Pauling, L., Robert Fulton Cathcart, III, M.D., an orthomolecular physician. *Newsletter* 1, no. 4, fall issue. The Linus Pauling Institute of Science and Medicine, Palo Alto, CA, 1978.

Pauling, L., Sensory neuropathy from pyridoxine abuse. *New Eng. J. Med.* **310**, 197 (1984).

Pauling, L., *How to Live Longer and Feel Better*. W.H. Freeman & Co., New York, 1986.

Pauling, L., Willoughby, R., Reynolds, R., Blaisdell, B.E. and Lawson, S., Incidence of squamous cell carcinoma in hairless mice irradiated with ultraviolet light in relation to intake of ascorbic acid (Vitamin C) and of D,L-alpha-tocopherol acetate (Vitamin E). In: *Vitamin C: New Clinical Applications in Immunology, Lipid Metabolism, and Cancer*, Ed. A. Hanck. Hans Huber, Bern, 1982, pp. 53–82.

Pauling, L., Nixon, J.C., Stitt, F., Marcuson, R., Dunham, W.B., Barth, R., Bensch, K., Herman, Z.S., Blaisdell, E., Tsao, C., Prender, M., Andrews, V. Willoughby, R. and

Zuckerkandl, E., Effect of ascorbic acid on the incidence of spontaneous mammary tumors in RIII mice. *Proc. Nat. Academy of Sciences USA*, **82**, 5185–5189
(1985).

Tsao, C.S., Salimi, S.L. and Pauling, L., Lack of effect of ascorbic acid on calcium
excretion. *IRCS Medical Science* **10**, 738 (1982).

Zuckerkandle, E. and Pauling, L., Molecular disease, evolution, and genic heterogen-
eity. In: *Horizons in Biochemistry*, Eds M. Kasha and B. Pullman. Academic Press,
New York, 1962, pp. 189–225.

25

Response of human tumors to ascorbic acid and copper in the subrenal capsule assay in mice

Constance S. Tsao and Ping, Y. Leung

Linus Pauling Institute of Science and Medicine, Palo Alto, CA, USA

INTRODUCTION

Several epidemiologic studies have indicated that ascorbic acid (vitamin C) may modify the risk of cancer [1, 2]. Low baseline levels of vitamin C in body fluids were significant factors for the incidence of stomach cancer and overall cancers [2]. Another study showed that fruit consumption was inversely related to 25-year lung cancer mortality and that plasma ascorbic acid levels were significantly related to citrus fruit intake [1]. Patients with malignant disease have been reported to have a very low level of tissue stores of ascorbic acid and frequently are associated with physical signs similar to subclinical scurvy [3]. There has also been evidence that patients with cancer, particularly metastatic cancer, have increased requirements for the vitamin [4] and that ascorbic acid in excess of the normal requirements may be useful in the management of patients with malignant disease [5, 6]. High doses of sodium ascorbate (10 g/day) increased the survival time — by 3–20-fold — of patients with untreatable advanced neoplasms [6, 7] compared with the survival time of patients not receiving ascorbate treatment. However, researchers from a different hospital have failed to obtain any beneficial effect of high doses of ascorbic acid on patients with cancer [8, 9].

Experiments with laboratory animals indicated that ascorbic acid had antitumor activities. Large quantities (2–8% in the diet) of dietary ascorbic acid decreased the incidence of, and delayed the first appearance of, spontaneous mammary tumors in RIII/Imr mice [10]. When hairless mice were irradiated with ultraviolet light, dietary ascorbic acid delayed the formation of skin lesions, and lesions in mice fed with ascorbic acid were consistently smaller than those in mice fed with an ascorbic acid-free diet [11, 12]. Ascorbic acid inhibited the growth of solid sarcoma-180 in Swiss mice [13], of Ehrlich ascites carcinoma in Swiss or CF_1 mice [14, 15] and of mammary carcinoma in Balb/C_f/Had/Se mice [16]. When DBA/2 mice were inoculated with S91 Cloudman melanoma, the apperance of visible tumors in the ascorbic-acid-treated mice was delayed and the survival time of these mice was longer than that of the controls [17]. Ascorbic acid also increased the survival in C3H/HEJ mice with C3HBA mouse mammary adenocarcinoma [18] and in Swiss mice with sarcoma-180 [13].

Contrary to these observations, the results of some animal studies indicated that ascorbic acid can increase tumor growth. The administration of ascorbic acid increased the growth of a transplanted solid tumor in mice [19]. Although ascorbic acid in high doses inhibited tumor growth in mice, ascorbic acid in low doses accelerated tumor growth [10–12, 16]. Injection of ascorbate stimulated the growth of melanosarcoma S39 and sarcoma-180 in mice and the growth of Watts sarcoma in rats [20]. When a tumor was induced in guinea pigs by injection of 20-methylcholanthrene, tumor regression occurred in animals maintained on a small amount of ascorbic acid (0.3 mg/kg body weight/day), but tumors in animals receiving a very large amount of ascorbic acid (1 g/kg body weight/day) grew without any sign of retardation [21]. A transplanted fibrosarcoma grew slower in guinea pigs on a scorbutic diet than in those on an adequate diet [22].

The antitumor activity of ascorbic acid in laboratory animals or in cultured cells was increased in the presence of cupric ion, a catalyst for the oxidation of ascorbic acid [23–25]; and, partly oxidized ascorbic acid was more effective than freshly prepared ascorbic acid [16]. These findings suggest that certain ascorbic acid oxidation or degradation products may be effective in controlling tumor growth. When administered intraperitoneally, the ascorbic acid oxidation products, dehydroascorbic acid and 2,3-diketogulonic acid, had higher antitumor activity in inhibiting the growth of implanted mouse sarcoma-180 in mice than ascorbic acid itself [26–28]. When mice were treated with dehydroascorbic acid, mitotic activity of ascites tumor cells was reduced [29]. There is considerable evidence that dehydroascorbic acid has a function in the control of mitotic activity [30].

Administration of ascorbic acid in the drinking water inhibited tumor growth in mice [16, 23]. However, when ascorbic acid was administered in the dry diet or when freshly prepared ascorbic acid solution was administered directly into the throat of the animal, no such effect was observed [16]. These observations also suggest that the inhibiting effect of ascorbic acid was caused by oxidized ascorbic acid or its oxidation or degradation products, which were formed in solution on standing.

We have carried out experiments in mice to test the effect of ascorbic acid, copper and a combination of ascorbic acid and copper on the growth of human tumor xenografts using the 6-day subrenal capsule assay technique [31, 32]. This assay method permits the observation of changes in the size of tumor fragments implanted beneath the renal capsule of normal immunocompetent mice. Because it is completed in 6 days, this method permits the examination of an antitumor agent before the immune system can have much effect. This separation of parameters may contribute to the understanding of the mechanisms involved.

In these assays, the growth of three commonly used human tumors and their response to the various treatments were studied. In a series of experiments with a human mammary tunor, ascorbic acid was administered in the mouse drinking water or in the diet, with or without the addition of cupric sulfate. In another series of experiments, tumors originating from the human colon, lung and breast were tested. To compare the antitumor activity of ascorbic acid and oxidized ascorbic acid, freshly prepared sodium ascorbate solutions in the presence and absence of cupric sulfate, a catalyst for the oxidation of ascorbate, were injected intraperitoneally into tumor-bearing mice. A stereoisomer of ascorbic acid, D-isoascorbic acid, was also tested as a parallel study. Ascorbic and D-isoascorbic acid are almost identical in their physical and chemical properties, but differ from each other in biological properties. A comparison of these two compounds may in turn help to elucidate the mechanisms of the action of ascorbic acid on tumor growth.

MATERIALS AND METHODS

Subrenal capsule tumor implant assay

Human colon adenocarcinoma CX-1, lung carinoma LX-1 and mammary carcinoma MX-1 were obtained from Mason Research Institute, Worcester, MA. The tumors were maintained by serial subcutaneous transplantation in immunodeficient athymic nude mice (Simonsen Laboratories, Gilroy, CA) in our laboratory. Shortly before the subrenal capsule implantation, the tumor was removed from the athymic mouse, placed in a pre-cooled petri dish containig Eagle's minimal essential medium with 10% calf serum (Gibco Laboratories, Santa Clara, CA). The tumor specimens were cleared of necrotic tissue and fat, and diced with a scalpel into 1-mm^3 pieces. The tumor fragments were kept in a ice-cold medium until implantation.

Female young adult BDF$_1$ mice, each 19–21 g, were used as xenograft recipients and were provided by Simonsen Laboratories, Gilroy, CA. After tumor implantation (described later), the mice were housed six to a cage in constant-temperature quarters. Feed and water were offered *ad libitum*. The disappearance of feed and water was measured at the end of the experiment.

The procedure for the subrenal capsule implant method has been described previous [31, 32]. The mice were anesthetized by injection of chloral hydrate solution, shaved, weighed, numbered and the left kidney was partly exteriorized. A 1-mm^3 fragment of tumor was implanted beneath the renal capsule with an 18-gage trocar. The size of the fragment was then measured with the aid of a microscope fitted with an ocular micrometer. The average diameter, (maximum width + maximum length)/2, was recorded in ocular micrometer units (OMU:1 OMU = 0.1 mm). The kidney was returned to the abdominal cavity and the incision was closed with surgical clips. A minimum of six mice were randomly assigned to each test group. On the sixth day, the mice were killed, weighed and the tumor-bearing kidneys excised. The final tumor size was measured. An assay was considered to be not evaluable if the implanted tumor grafts in the control group did not increase in size, as this could be due to the poor quality of the tumor sample. The difference between the initial and final tumor size (ΔOMU) was used as the indicator of tumor growth or growth inhibition. When the growth of the tumors in the control and test animals was compared, the implanted pieces were always from the same tumor.

Treatment

Diet.

Immediately following tumor implantation, the mice in the experimental groups were fed diets or drinking water with added ascorbic acid or cupric sulfate, or both. A nonpurified diet (Purina Certified Rodent Diet No. 5002, powdered form of Purina Diet No. 5001) was obtained from Ralston Purina Co., St. Louis, MO. This diet contained crude protein (20.8 %), crude fat (6.04 %), crude fiber (4.06 %), ash (7 %) and added minerals (2.5 %). Appropriate quantities of vitaminc C crystals (Cat. #51, Bronson Pharmaceuticals, La Cañada, CA; 1 kg of the vitamin C contains 540 g sodium ascorbate and 460 g ascorbic acid) were blended with the dry powdered diet. The mixtures, in 4 kg batches, were moistened by distilled water and pelleted with a pellet mill (California Pellet Mill Co., San Francisco, CA). The pellets were air-dried at room temperature and stored at $-40°C$. Ascorbic acid levels of the diets were analyzed after pelleting. One to five percent of ascorbic acid was destroyed during processing, but no detectable change was found during the course of the experiment. The control diet was prepared in the same manner, except that the addition of ascorbic acid was omitted. No ascorbic acid activity was found in the control diet.

Drinking water.

The drinking water for the experimental mice was prepared in the feeding bottle using a mixture of ascorbic acid–sodium ascorbate, described in the section 'Diet'. A cupric sulfate ($CuSO_4·5H_2O$, Sigma Chemical Co., St. Louis, MO) solution was prepared in the same manner. The drinking water was renewed every 48 h. The control mice were given twice-distilled water.

Injection.

Starting 24 h after tumor implantation, the test mice were injected intraperitoneally with sodium ascorbate or isoascorbate, or with one of these agents and cupric sulfate, in 0.2 ml distilled water for five days. The control animals received an equal volume of saline solution. The injection solutions were prepared immediately before injection by dissolving appropriate amounts of the test agents in distilled water. The solutions were then filtered through a 0.2 μm Acrodisc 25-mm filter assembly (Gelman, Ann Arbor, MI).

Analysis of ascorbic acid and its oxidation products

The total ascorbic acid (ascorbic acid plus dehydroascorbic acid) and dehydroascorbic acid contents of the diet, or of the solution for injection or for drinking were measured by the method of Roe and Keuther with modification [33]. Any 2,3-diketogulonic acid present would be expected to be included in these analyses. Ascorbic acid solution was diluted with 5 % trichloroacetic acid. The diet was homogenized in a blender with 0.5 % oxalic acid. The solution was then diluted with trichloroacetic acid, shaken with Norit (Sigma Chemical, St. Louis, MO), filtered and incubated with 2,4-dinitrophenylhydrazine at 37°C for 4 h. After the addition of sulfuric acid and a waiting period of 30 min at room temperature, the absorbance of each sample was read against a blank at 515 nm with a spectrophotometer. Mixtures of ascorbic acid and its oxidation products were analyzed using a gas chromatographic method (Pierce Handbook, *Silylation Reagents, Methods*, p. 144, 1988).

Materials

All chemicals were of reagent grade. Ascorbic acid was purchased from Bronson Pharmaceuticals, La Cañada, CA, and D-isoascorbic acid and cupric sulfate $\cdot 5H_2O$ from Sigma Chemical Co., St. Louis, MO.

RESULTS AND DISCUSSION

The mice had no apparent difficulty in adapting to the diets and drinking water containing ascorbic acid or copper or both. The injection of ascorbate, D-isoascorbate or copper did not show any adverse effect. The body weights of the mice showed no consistent differences between control and experimental groups. Since weight loss was not observed in any of the test groups, the administration of ascorbic acid and copper at the levels studied was apparently not toxic to these mice.

Tumors originating from human colon, lung and mammary glands have been maintained in nude mice in our laboratory for many years. The growth rates of these three tumors are very similar. Fragments of these tumors grow rapidly beneath the mouse renal capsule. The increase in tumor size during the 6-day experimental period was 50 to 100% of initial tumor size. The growth of tumor fragments implanted beneath the mouse renal capsule is illustrated in the tables. The change in tumor size (ΔOMU) is the final size of the tumor minus the initial size of the tumor.

A mammary tumor was used in a series of subrenal capsule tumor implant assay experiments in which ascorbic acid was administered in the mouse drinking water and in the diet with or without the addition of cupric sulfate. The results of these experiments are presented in Tables 1 and 2. The data indicate that the supplementation of ascorbic acid (1 or 5 g/l, corresponding to 5.68 or 28.4 mM) in the drinking water resulted in inhibition of tumor growth in the mice when compared with the controls drinking distilled water (Table 1). When ascorbic acid was added to the mouse diets (1, 5 or 50 g/kg), tumor inhibition was not observed (Table 2). However, when the mice were fed a diet containing a very large amount of ascorbic acid (50 g/kg diet) together with cupric sulfate (18 or 90 mg/l in the drinking water), the growth of the tumor fragments in these mice was depressed.

The ascorbic acid intake in a diet containing 1 or 5 g/kg added ascorbic acid is about 4 to 20 mg/day/mouse; the ascorbic acid intake in the drinking water containing 1 or 5 g/l added ascorbic acid is also about 4 or 20 mg/day/mouse. Although the amount of ascorbic acid ingested in the diet and in the drinking water was practically identical, the response of tumor growth to the administration of ascorbic acid in the diet differed from that in the drinking water. The data in Tables 1 and 2 show that an ingested solution of ascorbic acid caused an inhibition of tumor growth, but a similar dose or larger dose (50 g/kg diet) of ascorbic acid in the dry diet had no effect. A plausible explanation is as follows. Ascorbic acid solution is very unstable. The decomposition of aqueous ascorbic acid in the presence of oxygen at room temperature is rapid and autocatalytic. Many oxidation and degradation products of ascorbic acid have been detected in aqueous solutions [34–37]. In the mouse drinking water, ascorbic acid is readily oxidized to produce oxidation and degradation products, among which are tumor-growth inhibitors.

Fig. 1 shows the curves of ascorbic acid recovery in various aqueous solutions with

Table 1 — Effect of ascorbic acid (AA) and cupric sulfate in drinking water on the growth of human mammary carcinoma xenografts implanted beneath the renal capsule of BDF_1 female mice[a]

Control	Test agents: mg/kg body weight/day		
	AA: 200[c]	AA: 200 CuSO$_4$: 3.6[e]	CuSO$_4$: 3.6[e]
5.2 ± 1.1	2.9 ± 1.4* (56)	3.7 ± 0.9* (71)	5.6 ± 2.4 (108)
	AA: 200	AA: 200 CuSO$_4$: 18[f]	CuSO$_4$: 18
10.8 ± 3.0	5.4 ± 3.8* (50)	6.4 ± 2.2* (59)	11.9 ± 2.8[b] (110)
	AA: 1000[d]	AA: 1000 CuSO$_4$: 3.6	CuSO$_4$: 3.6
5.5 ± 1.0	2.5 ± 2.0** (46)	2.9 ± 1.9* (53)	6.2 ± 3.3[b] (113)
	AA: 1000	AA: 1000 CuSO$_4$: 18	CuSO$_4$: 18
9.6 ± 2.4	5.5 ± 3.3* (57)	4.9 ± 3.0* (51)	11.2 ± 2.4 (117)

[a]Values are the growth of the tumor fragment (mean ± S.D.), expressed as ΔOMU (see section 'Materials and methods'); all means are for $n = 6$ except: [b]$n = 5$. Values in parentheses are percent of control.
Significance of the difference between control and experimental values: *$P < 0.05$; **$P < 0.01$.
Dosages were calculated from water consumption and concentrations of ascorbic acid: [c]1 g/l, [d]5 g/l; and of cupric sulfate ($CuSO_4 \cdot 5H_2O$): [e]18 mg/l, [f]90 mg/l.

and without copper ion. At an initial concentration of 5.7 or 28.4 mM, corresponding to 1 or 5 g ascorbate per liter, after 48 h about 30% of the ascorbic acid was oxidized. In the presence of cupric sulfate (either 0.11 or 0.56 mM), when the concentration of ascorbic acid was 28.4 mM, more than 60% of the ascorbic acid was oxidized; when the concentration was 5.7 mM, it was completely degraded. Gas chromatographic analyses of the drinking water indicated that the ascorbic acid was converted to dehydroascorbic acid plus many other oxidation and degradation products. The chromatogram contained ascorbic acid, dehydroascorbic acid and 100–200 small measurable peaks, among which were compounds detected in other laboratories [34–37] and many other unidentified compounds.

In contrast with the ascorbic acid in the drinking water, the ascorbic acid in the mouse dry diet is relatively stable. The amounts of ascorbic acid oxidation and degradation products in the dry diet are negligibly small. Mice fed diets containing ascorbic acid during the short experimental period of six days probably attained in the body relatively low levels of these oxidation and degradation products. However, when very large quantities of ascorbic acid were consumed with cupric ion, a catalyst for the odixation of ascorbic acid, considerable amounts of these oxidation products

Table 2—Effect of ascorbic acid (AA) and cupric sulfate on the growth of human mammary carcinoma xenografts implanted beneath the renal capsule of BDF$_1$ female mice[a]

Control	Test		
	AA: g/kg body weight/day CuSO$_4$: mg/kg body weight/day		
	AA: 0.2[c]	AA: 0.2 CuSO$_4$: 3.6[f]	CuSO$_4$: 3.6
11.2 ± 3.2	13.2 ± 2.9 (118)	9.5 ± 0.9 (85)	13.7 ± 3.0 (122)
	AA: 0.2	AA: 0.2 CuSO$_4$: 18[g]	CuSO$_4$: 18
8.4 ± 2.2	9.5 ± 0.9 (113)	7.7 ± 5.2[b] (92)	6.7 ± 4.2 (80)
	AA: 1[d]	AA: 1 CuSO$_4$: 3.6	CuSO$_4$: 3.6
5.0 ± 2.7	8.7 ± 4.4 (174)	7.2 ± 3.7 (144)	6.6 ± 3.0 (132)
	AA: 1	AA: 1 CuSO$_4$: 18	CuSO$_4$: 18
13.3 ± 3.2	14.1 ± 4.2 (106)	12.9 ± 2.8 (97)	11.4 ± 4.8 (86)
	AA: 10[e]	AA: 10 CuSO$_4$: 3.6	CuSO$_4$: 3.6
9.5 ± 1.2	10.2 ± 1.5 (107)	6.6 ± 2.0[b]** (70)	9.8 ± 3.5 (103)
		AA: 10 CuSO$_4$: 18	CuSO$_4$: 18
		6.8 ± 2.4* (72)	10.3 ± 2.4 (108)
	AA: 10	AA: 10 CuSO$_4$: 18	CuSO$_4$: 18
6.2 ± 1.4	5.8 ± 2.9 (94)	3.7 ± 2.2* (60)	9.1 ± 2.8* (147)

[a] Values are the growth of the tumor fragment (mean ± S.D.), expressed as ΔOMU (see section 'Materials and methods'); all means are for $n = 6$ except [b] $n = 5$.
Values in parentheses are percent of control.
Significance of the difference between control and experimental values: *$P < 0.05$; **$P < 0.02$.
Ascorbic acid was administered in the mouse diet; cupric sulfate was in the drinking water.
Dosages were calculated from diet and water consumption and concentrations of ascorbic acid: [c] 1 g/kg, [d] 5 g/kg, [e] 50 g/kg diet; and of CuSO$_4 \cdot$ 5H$_2$O, [f] 18 mg/l, [g] 90 mg/l.

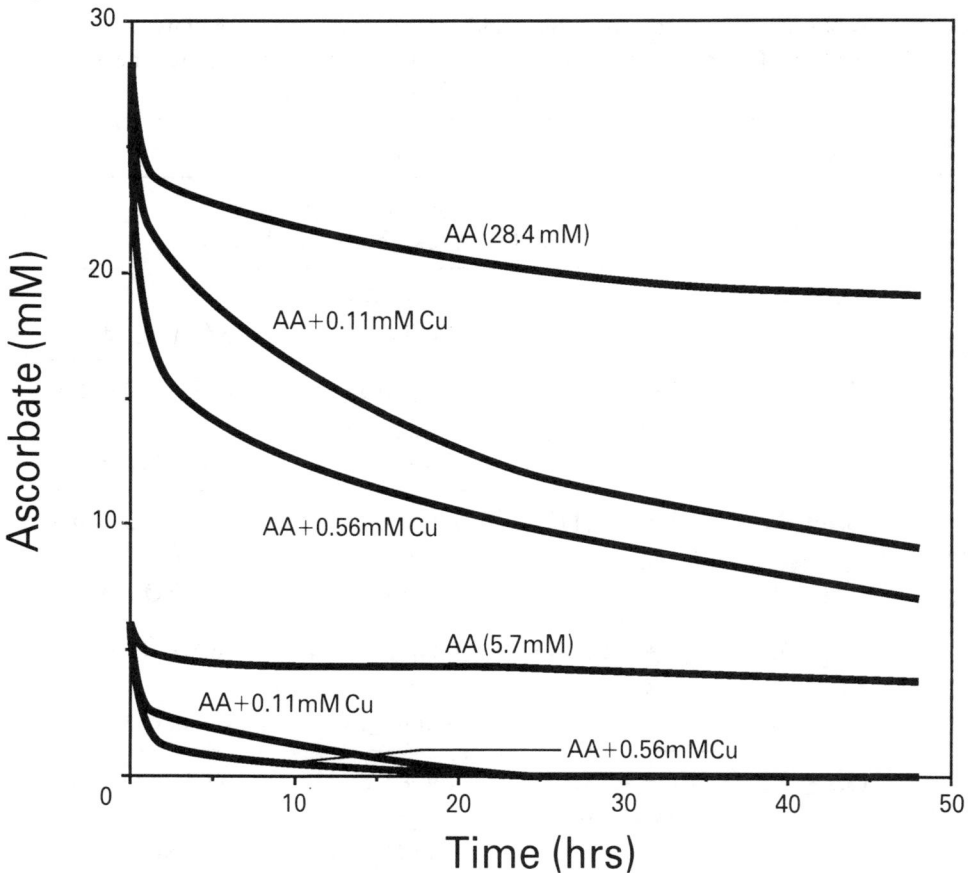

Fig. 1. — Ascorbic acid degradation.

may have been produced in the mouth and gastrointestinal tract. The results in Tables 1 and 2 suggest that the inhibiting effect of ascorbic acid was probably attributable to its oxidation products.

The data in Table 1 show that the ascorbic acid in the drinking water exhibited an inhibitory effect on tumor growth and that ascorbic acid in the absence of copper ion was as effective as ascorbic acid in the presence of copper ion. This might suggest that copper ion was not required for the formation of these effective oxidation products in a dilute solution. However, that cupric sulfate solution alone somewhat increased tumor growth. Cupric sulfate acts as a catalyst for the oxidation of ascorbic acid, thereby enhancing the antitumor activities by increasing the amounts of odixation products. On the other hand, the growth promotion effect of cupric sulfate itself may counteract the influence of the increased levels of effective oxidation products. Furthermore, when ascorbic acid was administered in the drinking water, the inhibiting effect of the ascorbic acid was not dose-responsive. Analysis of ascorbic acid recovery indicates that ascorbic acid in a dilute solution degraded to a larger extent than in a concentrated solution. Therefore, the amounts of oxidation and degradation products in a dilute solution (1 g/l) may be very similar to those in a stronger solution (5 g/l).

In a second series of subrenal capsule tumor implant assay experiments, in addition to the human mammary tumor MX-1, two more human tumors, colon adenocarcinoma CX-1 and lung carcinoma LX-1, were used to test the antitumor activity of ascorbic acid and partly oxidized ascorbic acid. Sodium ascorbate with or without cupric sulfate was administered by intraperitoneal injection. The ascorbate solution (85 mM) for injection was prepared immediately before administration to minimize oxidation or deterioriation. The results in Table 3 indicate that injection of freshly prepared ascorbate (150 mg/kg body weight/day) did not affect the growth of the three tumors during the 6-day experimental period. However, when cupric sulfate ($CuSO_4 \cdot 5H_2O$, 3 mg/kg body weight/day) was added to the injection fluids, tumor growth was depressed. Injection of cupric sulfate solution alone was ineffective or

Table 3 — Human tumor growth and response to ascorbate (AA), D-isoascorbate (IAA) and cupric sulfate in the subrenal capsule implant assay in mice[a]

Tumor	Control	Test		
		AA	IAA	
Colon adenocarcinoma	8.4 ± 3.1	8.9 ± 2.2 (106)	9.6 ± 1.9 (114)	
	10.7 ± 4.2	10.3 ± 3.4 (96)	11.8 ± 4.4[b] (110)	
		AA, $CuSO_4$	IAA, $CuSO_4$	$CuSO_4$
	11.4 ± 2.2[c]	5.7 ± 2.6*** (50)	6.4 ± 2.4*** (56)	12.0 ± 3.2 (105)
	7.8 ± 1.9	4.8 ± 2.6* (62)	4.0 ± 2.2** (51)	8.2 ± 2.9 (105)
		AA	IAA	
Lung carcinoma	6.9 ± 2.1	6.7 ± 3.8 (97)	8.2 ± 2.3[b] (119)	
	9.4 ± 3.7[c]	10.0 ± 4.5 (106)	9.9 ± 3.9 (105)	
		AA, $CuSO_4$	IAA, $CuSO_4$	$CuSO_4$
	13.4 ± 3.7	9.5 ± 2.1* (71)	9.2 ± 3.0 (69)	13.5 ± 4.0 (101)
	12.3 ± 4.1	8.4 ± 2.2* (68)	7.0 ± 1.8* (57)	14.0 ± 3.9 (114)
		AA	IAA	
Mammary carcinoma	7.3 ± 2.6[c]	7.9 ± 1.7 (108)	8.8 ± 3.8 (121)	
	7.1 ± 3.5	6.8 ± 2.7 (96)	8.3 ± 1.5[b] (117)	
		AA, $CuSO_4$	IAA, $CuSO_4$	$CuSO_4$
	6.4 ± 2.2	2.8 ± 1.7** (44)	1.9 ± 1.7*** (30)	7.1 ± 3.2 (111)
	5.5 ± 1.8	3.3 ± 1.5* (67)	2.6 ± 2.4* (47)	5.8 ± 2.2 (106)

[a] Values are means ± S.D. for the growth of tumor fragments, expressed as ΔOMU (see 'Materials and methods'); all means are for $n = 6$ except [b] $n = 5$; [c] $n = 7$. The Student's two-tailed t-test was used to determine statistical differences between the control and experimental values. Significance of the difference: *$P < 0.05$; **$P < 0.01$; ***$P < 0.005$.
Values in parentheses are percent of control.
Test agents were administered by intraperitoneal injection. The dosage of sodium ascorbate or D-isoascorbate was 150 mg/kg body weight/day; the dosage of cupric sulfate ($CuSO_4 \cdot 5H_2O$) was 3 mg/kg body weight/day.

increased growth. The concentration of ascorbate in the injection fluid was relatively high. In the absence of copper ion, ascorbate at high concentration is quite stable. In the presence of copper ion (2 mM), however, 14% of the ascorbate was oxidized. Analyses of the injection fluids by gas chromatography showed a mixture of ascorbic acid oxidation and degradation products. These results are consistent with the hypothesis that certain derivates of ascorbic acid formed *in vitro* are active antitumor agents.

D-Isoascorbic acid, a stereoisomer of ascorbic acid, was also tested in these experiments with the three human tumors. The activity of D-isoascorbic acid on the growth of the human colon, lung or mammary tumor xenografts was similar to that of ascorbic acid notwithstanding that D-isoascorbic acid has only 5% of the antiscorbutic potency of ascorbic acid and very high turnover rate [38, 39]. Analyses of the injection fluids containing D-isoascorbic acid indicated that many oxidation products of D-isoascorbic acid were formed and that the pattern of the gas chromatograms of D-isoascorbic acid were similar to those of ascorbic acid. This suggests that the antitumor activity of ascorbic acid observed in these experiments was not due to the metabolism of ascorbic acid as a vitamin, but was due to its chemical properties or its ability to produce active oxidation and degradation products.

The data in Tables 2 and 3 indicate that when the mice received a relatively low amount (for mice) of ascorbic acid in the diet (200 mg/kg body weight/day) or by injection (150 mg/kg body weight/day), tumor growth at times was increased. This phenomenon had been observed previously in long-term experiments with mice [10–12, 16, 20–22]. Studies on the effect of exogenous ascorbic acid intake on ascorbic acid metabolism in mice indicate that exogenous ascorbic acid intake leads to a decrease in the rate of biosynthesis of ascorbic acid, to an increase in the rate of catabolism and to a decreased level of ascorbic acid in various tissues in mice [40–42]. These observations are probably attributable to an over-reacting feedback control mechanism that regulates ascorbic acid metabolism. Although we know that ascorbic acid metabolism is affected by exogenous ascorbic acid intake, the relationship between altered ascorbic acid metabolism and tumor metabolism remains unknown.

When ascorbic acid is administered to animals, a considerable fraction is converted in the body into metabolites [34, 43–45]. Studies on the urine of monkeys, rats and guinea pigs given radioactive ascorbic acid show that there are many metabolic products originating from the administration of ascorbic acid. In the present study, the ascorbic acid in the dry diet or in the freshly prepared injection fluid undoubtedly would be metabolized to many metabolic products in the mice, as in the other animals. Yet, administration of ascorbic acid was effective only when it had been partly oxidized. This observation suggests that the constituents of the ascorbic acid derivatives formed *in vivo* are different from those formed *in vitro*. Chemicals react differently in different surroundings. For example, in the presence of free transition metal catalysts, ascorbic acid oxidation can yield some highly reactive breakdown products capable of initiating oxidation or free radical reactions [46]. However, these breakdown products are not formed *in vivo*, because in healthy animals, most transition metal ions are attached to binding proteins and are not available to participate in chemical reactions outside the proteins [47–49]. Therefore, derivatives of ascorbic acid formed *in vivo*, whether formed enzymatically or not, may be different from those formed *in vitro* and have different chemical properties. Since ascorbic acid itself or its derivatives formed *in vivo*, if any, did not exhibit any activity, the antitumor activity observed in this study was not the metabolic effect of the vitamin.

The mechanisms for the action of ascorbic acid on tumor cells are probably very complex. Large amounts of ascorbic acid intake can change the levels of certain amino acids in body fluids [50, 51] and may deplete the bioavailability of lysine and cysteine, which are required for rapidly growing tumors [4]. It is not known whether this mechanism was operating in the present experiments. It has been postulated that ascorbic acid may kill certain tumor cells, may increase the cell-killing or cell-retarding effect of certain tumor therapeutic agents and may increase the resistance of the host by promoting optimum function of the immune system [52]. In the present study, the mice were treated with ascorbic acid for a short period of five days after tumor implantation. There was not enough time for the immune system to act upon the implanted tumor. Therefore, the inhibiting effect observed in these experiments was probably a direct cell-retarding action of certain ascorbic acid derivatives. Ascorbate caused cleavage in nucleic acids in solution [23, 53]. Certain ascorbic acid oxidation and degradation products, including a group of furan-type compounds [35, 36], are known to be capable of adduct formation with amino and hydroxyl groups of proteins [54]. Experiments using tissue homogenates indicated that ascorbate, metal and oxygen induced structural changes in animal proteins [54, 55].

Although certain ascorbic acid breakdown products are toxic to tumor cells, they are far less harmful to normal cells. Ascorbic acid (0.2 to 0.5 mM) in cell culture medium was cytotoxic for several human malignant melanoma cell lines, and a trace amount of cupric ion increased the cytotoxicity of ascorbic acid, but nonmalignant cells were 10–20 times less sensitive to ascorbic acid than malignant cells [25]. In our laboratory, similar results were observed with cultured leukemia cells and peripheral blood cells (data not shown). Moreover, when healthy mice and guinea pigs were treated with drinking water containing ascorbic acid and cupric sulfate for a period of several months, no apparent adverse effect was detected. Thus, it seems that the growth-inhibiting effect of ascorbic acid derivatives is significant with rapidly dividing neoplastic cells but is of little importance to normal organisms with homeostatic mechanisms of their functional system.

SUMMARY

The 6-day subrenal capsule assay method has been used to investigate the effect of ascorbic acid and cupric sulfate on the growth of human mammary, colon and lung tumor xenografts in mice. The method permits the observation of changes in the size of tumor fragments implanted beneath the renal capsule of mice, without the effective intervention of the immune system.

The results indicate that ascorbic acid (1 or 5 g/l), administered in the drinking water, significantly inhibits the growth of a human mammary tumor implanted beneath the renal capsule of immunocompetent mice. The results agree with work in other laboratories carried out in animal experiments with animal tumors. Administration of ascorbic acid in the mouse diet does not affect tumor growth during the 6-day experimental period. Tumor growth, however, is inhibited when mice are fed a diet containing ascorbic acid (50 g/kg diet) together with cupric sulfate (18 or 90 mg/l of drinking water).

Intraperitoneal injection of a mixture of ascorbate (150 mg/kg body weight/day) and cupric sulfate (3 mg/kg body weight/day) significantlly inhibits the growth of the three human tumors (colon, lung and mammary). Injection of ascorbate or cupric sulfate alone has no effect. The results support the hypothesis that certain oxidation or degradation products of ascorbic acid are active antineoplastic agents for these tumors. The activity of D-isoascorbic acid, an isomer of ascorbic acid, with 5% of the antiscorbutic potency and very high turnover rate, is similar to that of ascorbate. This suggests that the antitumor activity of ascorbic acid is not due to the metabolism of ascorbic acid as a vitamin, but due to its chemical properties. The tumor-inhibiting effect seems to be a direct growth-depression action of certain degradation products of ascorbic or D-isoascorbic acid, which are formed *in vitro* and are not formed by the metabolism of ascorbic acid.

ACKNOWLEDGMENTS

We thank Professor Linus Pauling, who initiated this project, for his advice, encouragement and support. This work was supported in part by a grant from the Foundation for Nutritional Advancement. We thank M. Prender and V. Andrews for their technical assistance.

REFERENCES

[1] K.F. Gey, G.B. Brubacher and H.B. Stähelin, Plasma levels of antioxidant vitamins in relation to ischemic heart disease and cancer. *Am. J. Clin. Nutr.* (1987) **45**, 1368–1377.

[2] D. Kromhout, Essential micronutrients in relation to carcinogenesis. *Am. J. Clin. Nutr.* (1987) **45**, 1361–1367.

[3] N. Krasner and I.W. Dymock, Ascorbic acid deficiency in malignant diseases: a clinical and biochemical study. *Br. J. Cancer* (1974) **30**, 142–145.

[4] T.K. Basu, Possible role of vitamin C in cancer therapy. In: *Vitamin C: Recent Advances and Aspects in Virus Diseases, Cancer and in Lipid Metabolism*, A. Hanck and G. Ritzel (Eds). Bern: Hans Huber, 1979, 95-1-2.

[5] E. Cameron and L. Pauling, The orthomolecular treatment of cancer: I. The role of ascorbic acid in host resistance. *Chem.–Biol. Interact.* (1974) **9**, 273–283.

[6] E. Cameron and A. Campbell, The orthomolecular treatment of cancer: II. Clinical trial of high-dose ascorbic acid supplements in advanced human cancer. *Chem.–Biol. Interact.* (1974), **9**, 285–315.

[7] E. Cameron and L. Pauling, Supplemental ascorbate in the supportive treatment of cancer: prolongation of survival times in terminal human cancer. *Proc. Natl. Acad. Sci. USA* (1976) **73**, 3685–3689.

[8] E.T. Creagan, C.G. Moertel, J.R. O'Fallon, A.J. Schutt, M.J. O'Connell, J. Rubin and S. Frytak, Failure of high-dose vitamin C (ascorbic acid) therapy to benefit patients with advanced cancer. *N. Eng. J. Med.* (1979) **301**, 687–690.

[9] C.G. Moertel, T.R. Flemming, E.T. Creagan, J. Rubin, M.J. O'Connell and M.M. Ames, *N. Engl. J. Med.* (1985) **312**, 137–141.

[10] L. Pauling, J.C. Nixon, F. Stitt, R. Marcuson, W.B. Dunham, R. Barth, K. Bensch, Z.S. Herman, B.D. Blaisdell, C. Tsao, M. Prender, V. Andrews, R.

Willoughby and E. Zuckerkandl, Effect of dietary ascorbic acid on the incidence of spontaneous mammary tumors in RIII mice. *Proc. Natl. Acad. Sci. USA* (1985) **82**, 5185–5189.

[11] W.D. Dunham, E. Zuckerkandl, R. Reynolds, R. Willoughby, R. Marcuson, R. Barth and L. Pauling, Effects of intake of L-ascorbic acid on the incidence of dermal neoplasms induced in mice by ultraviolet light. *Proc. Natl. Acad. Sci. USA* (1982) **79**, 7532–7536.

[12] I. Pauling, R. Willoughby, R. Reynolds, B.E. Blaisdell and S. Lawson, Incidence of squamous cell carcinoma in hairless mice irradiated with ultraviolet light in relation to intake of (vitamin C) and of D,L-α-tocopherol acetate (vitamin E). In *New Clinical Applications in Immunology, Lipid Metabolism and Cancer*, A. Hanck (Ed). Bern: Hans Huber, 1982.

[13] R.N. Chakrabarti and P.S. Dasgupta, Effects of ascorbic acid on survival and cell-mediated immunity in tumor bearing mice. *IRCS Med. Sci.* (1984) **12**, 1147–1148.

[14] F.A. Tewfik, H.H. Tewfik and E.F. Riley, The influence of ascorbic acid on the growth of solid tumors in mice and on tumor control by X-irradiation. *Int. J. Vit. Nutr. Res.* (1982) suppl **23**, 257–263.

[15] H.E. Gruber, H.H. Tewfik and F.A. Tewfik, Cytoarchitecture of Ehrlich ascites carcinoma implanted in the hind limb of ascorbic acid-supplemented mice *Europ. J. Cancer* (1980) **16**, 441–448.

[16] F.S. Liotti and V. Talesa, Ascorbic acid and tumor growth. *Convivia Medica* (1982) **3**, suppl: 81–96.

[17] J.M. Varga and L. Arioldi, Inhibition of transplantable melanoma tumor development in mice by prophylactic administration of calcium ascorbate. *Life Sci.* (1983) **32**, 1559–1564.

[18] T.C. Frazier and M.E. McGinn, The influence of magnesium, calcium and vitamin C on tumor growth in mice with breast cancer. *J. Surg. Res.* (1979) **27**, 318–320.

[19] F.S. Liotti, V. Talesa and A.R. Menghini, Absence of accumulation phenomena in normal and tumoral tissues of mice treated with ascorbic acid. *Internat. J. Vit. Nutr. Res.* (1983) **53**, 251–257.

[20] A. Brunschwig, Vitamin C and tumor growth. *Cancer Res.* (1943) **3**, 550–553.

[21] J.A. Migliozzi, Effect of ascorbic acid on tumor growth. *Br. J. Cancer* (1977) **35**, 448–453.

[22] W.V.B. Robertson, A.J. Dalton and W.E. Heston, Changes in a transplanted fibrosarcoma associated with ascorbic acid deficiency. *J. Nat. Cancer Inst.* (1949) **10**, 53–67.

[23] C.S. Tsao, W.B. Dunham and P.Y. Leung, In vivo neoplastic activity of ascorbic acid for human mammary tumor. *In Vivo* (1988) **2**, 147–150.

[24] M.-C. De Pauw-Gillet, B. Siwek, G. Pozzi, E. Sabbioni and R.J.B. Bassleer, Control of B16 melanoma cells differentiation and proliferation by $CuSO_4$ and vitamin C. *Anticancer Res.* (1990) **10**, 391–396.

[25] S. Bram, P. Froussard, M. Guichard, C. Jasmin, Y. Augery, F. Sinoussi-Barre and W. Wray, Vitamin C preferential toxicity for malignant melanoma cells. *Nature* (1980) **284**, 629–631.

[26] K. Yamafuji, Y. Nakamura, H. Omura, T. Soeda and K. Gyotoku, Antitumor

potency of ascorbic, dehydroascorbic or 2,3-diketogulonic acid and their action on doxyribonucleic acid. *Z. Krebsforsh* (1971) **76**, 1–7.

[27] Y. Nakamura and K. Yamafuji, Antitumor activities of oxidized products of ascorbic acid. *Sci. Bull. Fac. Agr. Kyushu Univ.* (1968) **23**(3), 119–125.

[28] H. Omura, Y. Tomita, N. Yasuhiko and H Murakami, Antitumor potentiality of some ascorbate derivatives. *J. Fac. Agric. Kyushu Univ.* (1974) **18**, 181–189.

[29] M.E. Poydock, D. Reikert, J. Rice and L. Aleandri, Inhibiting effect of dehydroascorbic acid on cell division in ascites tumors in mice. *Expl. Cell Biol.* (1982) **50**, 34–38.

[30] J.A. Edgar, Dehydroascorbic acid and cell division. *Nature (London)* (1970) **227**, 24–26.

[31] A.E. Bogden, W.R. Cobb, D.J. LePage, P.M. Haskell, T.A. Gulkin, A. Ward, D.E. Kelton and H.J. Esber, Chemotherapy responsiveness of human tumors as first transplant generation xenografts in the normal mouse. *Cancer* (1981) **48**, 10–20.

[32] J.A. Stratton, The murine subrenal capsule human tumor implant assay. *J. Clin. Lab. Analy.* (1987) **1**, 62–66.

[33] J.H. Roe and C.A. Kuether, The determination of ascorbic acid in whole blood and urine through the 2,4-dinitrophenylhydrazine derivative of dehydroascorbic acid. *J. Biol. Chem.* (1943) **147**, 399–407.

[34] B.M. Tolbert, M. Downing, R.W. Carlson, M.K. Knight and E.M. Baker, Chemistry and metabolism of ascorbic acid and ascorbate sulfate. *Ann. NY Acad. Sci.* (1975) **258**, 48–69.

[35] J.H. Tatum, P.E. Shaw and R.E. Berry, Degradation products from ascorbic acid. *J. Agr. Food Chem.* (1969) **17**, 38–40.

[36] J. Velisek, J. Davidek, V. Kubelka, Z. Zelinkova and J. Pokorney, Volatile degradation products of L-dehydroascorbic acid. *Z. Lebensm. Unters-Forsch* (1976) **162**, 285–290.

[37] K. Niemela, Oxidative and non-oxidative alkali-catalysed degradation of L-ascorbic acid. *J. Chromatog.* (1987) **399**, 235–243.

[38] J.M. Rivers, E.D. Huang and M.L. Dodds, Human metabolism of L-ascorbic acid and erythorbic acid *J. Nutr.* (1963) **81**, 163–168.

[39] O. Pelletier, Turnover rates of D-isoascorbic acid and L-ascorbic acid in guinea pig organs. *Can. J. Physiol. Pharmacol.* (1969) **47**, 993–997.

[40] C.S. Tsao and M. Young, Effect of exogenous ascorbic acid intake on biosynthesis of ascorbic acid in mice. *Life Sci.* (1989) **45**, 1553–1557.

[41] C.S. Tsao and P.Y. Leung, Effect of ascorbic acid intake on tissue dehydroascorbic acid in mice. *Nutr. Res.* (1989) **9**, 1371–1379.

[42] C.S. Tsao, P.Y. Leung and M. Young, Effect of dietary ascorbic acid intake on tissue vitamin C in mice. *J. Nutr.* (1987) **117**, 291–297.

[43] M. Tolbert, Ascorbic acid metabolism and physiological function. In *Vitamin C: Recent Advances and Aspects of Virus Diseases, Cancer and in Lipid Metabolism*, A. Hanck and G. Ritzel (Eds). Bern: Hans Huber, 1979, 127–142.

[44] G. Ashwell, J. Kanfer, J.D. Smiley and J.J. Burns, Metabolism of ascorbic acid and related uronic acids, aldonic acids and pentoses. *Ann. NY Acad. Sci.* (1961) **92**, 105–114.

[45] P.C. Chan, R.R. Becker and C.G. King, Metabolic product of L-ascorbic acid. *J. Biol. Chem.* (1958) **231**, 231–240.

[46] K. Yamamoto, M. Takahashi and E. Niki, Role of iron and ascorbic acid in the oxidation of methyl linoleate micelles. *Chem. Lett.* (1987) 1149–1152.

[47] J.M.C. Gutteridge, R. Richmond and B. Halliwell, Oxygen free-radicals and lipid peroxidation: inhibition by the protein caeruloplasmin. *FEBS Lett.* (1980) **112**, 269–272.

[48] M. Levine, New concepts in biology and biochemistry of ascorbic acid. *N. Engl. J. Med.* (1986) **314**, 892–902.

[49] B. Frei, L. England and B.N. Ames, Ascorbate is an outstanding antioxidant in human blood plasma. *Proc. Natl. Acad. Sci. USA* (1989) **86**, 6377–6381.

[50] C.S. Tsao and K. Miyashita, Effects of high intake of ascorbic acid on plasma levels of amino acids. *IRCS Med. Sci.* (1984) **12**, 1052–1053.

[51] C.S. Tsao and K. Miyashsita, Effect of large intake of ascorbic acid on the urinary excretion of amino acids and related compounds. *IRCS Med. Sci.* (1985) **13**, 855–856.

[52] K.N. Prasad, Minireview: Modulation of the effects of tumor therapeutic agents by Vitamin C. *Life Sci.* (1980) **27**, 275–280.

[53] H. Omura, Y. Tomita, H. Fujiki, K. Shinohara and H. Murakami, Breaking action of reductones related to ascorbic acid on nucleic acids. *J. Nutr. Sci. Vitaminol.* (1978) **24**, 263–270.

[54] K.G. Bensch, J.E. Fleming and W. Lohmann, The role of ascorbic acid in senile cataract. *Proc. Natl. Acad. Sci. USA* (1985) **82**, 7193–7196.

[55] D. Garland, J.S. Zigler and J. Kinoshita, Structural changes in bovine lens crystallins induced by ascorbate, metal and oxygen. *Arch. Biochem. Biophysics* (1986) **251**, 771–776.

26

Effect of ascorbic acid and its derivatives on different tumors *in vivo* and *in vitro*

M. Eckert-Maksić, I. Kovaček, Z.B. Maksić, M. Osmak and K. Pavelić

Rudjer Bošković Institute, Zagreb, Croatia, Yugoslavia

INTRODUCTION

A possible positive role of vitamin C (L-ascorbic acid, AA) in homeostasis was pointed out by Jorissen and Belinfante in the early 1930s [1]. Since then, considerable research efforts have been devoted to clarify this question, with somewhat controversial results [2, 3]. It seems to be beyond doubt, however, that L-ascorbic acid has beneficial effects in strengthening the immune system [3] in general and that it provides good prophylactic protection against reactive radicals [4] and other carcinogenic substances [2]. That is to say, there is a growing body of evidence that AA is a dominant factor in controlling and enhancing many aspects of host resistance to neoplastic development and invasions [3, 5, 6]. For instance, epidemiological evidence unequivocally shows that cancer incidence in large population groups is inversely related to the average daily intake of vitamin C [7]. There is also pervasive evidence that patients with cancer possess a very low tissue level of vitamin C (in, for example, leukocytes) and frequently show symptoms of subclinical scurvy [8]. The level of ascorbic acid is dramatically lower in those cancer patients who have suffered surgical treatment [6]. It is therefore not surprising that tumor patients have an increased demand for vitamin C, in particular if a metastatic phase has taken place [9]. In this connection it should be stressed that the average survival time for terminal cancer patients was the longer if 10 g or more of sodium ascorbate was administered per day in addition to chemotherapeutic treatment [10–12]. It is also noteworthy that the quality of life and well being was increased in all patients with vitamin C intake [3]. This finding was questioned by some other researchers, and the whole topic of ascorbic acid megadosis in treating malignant tumors in man is somewhat controversial [13, 14]. However, there have been spectacular recoveries through vitamin C mega-treatments, and in particular cases, terminal cancer patients were able to control their status for a very long time [3, 12]. Hence Cameron's and Pauling's suggestion that vitamin C is one of the most important orthomolecular [15, 16] substances which can diminish the risk of cancer and to increase the rate of recovery deserves full attention.

The influence of ascorbic acid on tumor cells has also been extensively studied *in vivo* on laboratory animals [17–22]. Most of these studies revealed that AA possesses significant antitumor activity. Contrary to these observations, no effect was observed in some other experiments [23, 24], whereas some animal studies have revealed that ascorbic acid can even increase the growth of some tumors [25–27]. This is not surprising since vitamin C is not expected to act similarly on all types of tumor. As a matter of fact, some selectivity is even desirable.

Let us finally mention that considerable attention has recently been devoted to the examination of the antitumor potential of vitamin C supplemented with other agents [28–35]. There are several sound reasons derived primarily by *in vitro* [28, 29, 34] and animal [31, 35] studies but also by studies in man [30, 32, 33], that antitumor activity of AA can be considerably enhanced by synergistic action with other certain drugs. Its joint action with other vitamins (predominantly A and E) seems to be beneficial too [3, 36, 37].

In this chapter we present an overview of the results obtained in investigating the antitumor activity of L-ascorbic acid and its derivates at the 'Rudjer Bošković Institute in Zagreb over the past several years [38–40].

This brief review starts with the results of our recent study related to the intracellular mechanism of the antitumor activity of L-ascorbic acid [38, 39]. The following section addresses the question of modifications of L-ascorbic acid by substituent(s) and their influence on antitumor activity. This will be illustrated by using 6-bromo-6-deoxy ascorbic acid (6-Br-AA) as an example [40]. Finally, in the concluding section, some preliminary results of *in vitro* experiments designed to investigate the antitumor potential of AA and its 6-bromo derivate supplemented with β-carotene will be presented. The latter is of considerable interest in view of the extensively discussed role of β-carotene and other retinoids in preventing and curing benign and malignant tumors [41–45].

ANTITUMOR ACTIVITY OF VITAMIN C: INTRACELLULAR MECHANISMS OF ACTION

The time- and dose-dependent activity of AA was examined on different human and murine tumors or normal cell lines (*in vitro*). Subsequently, *in vivo* investigations on human tumors transplanted in athymic 'nude' mice were performed.

Treatment of human neuroblastoma cells in culture with L-ascorbic acid resulted in significant dose-dependent decreases in [³H]thymidine, [³H]uridine and [³H]leucine incorporation into cells (Fig. 1). AA dramatically decreased the viability of cells after 1 h incubation without affecting normal neuronal cells (Fig. 1, inset).

The treatment of human mammary or kidney carcinoma cells and murine tumor cell lines with L-ascorbic acid resulted in dose- and time-dependent decreases of [³H]thymidine incorporation into DNA (Figs 2 and 3). Moreover, AA inhibited DNA protein synthesis in human mammary carcinoma cells (Fig. 2, inset).

The inhibitory effect of vitamin C on DNA synthesis was much less pronounced in normal murine T-lymphocytes, bovine corneal endothelial cells and human fibroblasts (Fig. 3).

The data in Fig. 3 indicate that vitamin C inhibited [³H]thymidine incorporation by transformed cells more efficiently than by normal cells. The possibility of induction

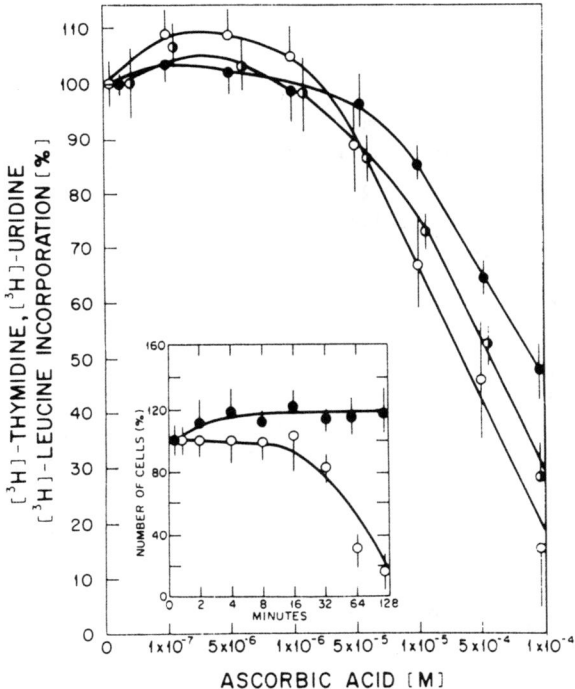

Fig. 1—Effects of different concentrations of L-ascorbic acid on [³H]thymidine (○),
[³H]uridine (●) and [³H]leucine (◐) incorporation into human neuroblastoma cells. AA was
added for 2 h at the indicated concentration. The basal number of 75,000 ± 9820 neuroblastoma
cells spontaneously incorporated [³H]thymidine which decayed at the rate of 5232 ± 620 cpm,
[³H]uridine at the rate of 4997 ± 212 cpm and [³H]leucine at the rate of 5351 ± 821 cpm. Error
bars represent standard deviations. Inset: effect of 1×10^{-4} M AA on number of neuroblastoma
(○) and normal mouse neuronal cells (●) after different incubation times (see ref. [38] for
experimental details).

of DNA strand breakage by L-ascorbic acid was investigated by treating human
mammary carcinoma cells with vitamin C for 2 h followed by analysis of DNA
integrity by the alkaline elution method. The treatment of cells *in vitro* with L-ascorbic
acid increased rates of DNA-elution in a concentration-dependent manner, indicating
vitamin-C-induced DNA strand breaks (Fig. 4(a)).

To determine whether vitamin C could induce DNA protein cross links, cells were
irradiated with 300 rad of γ-rays from a ^{130}Cs source after 2 h of treatment with
vitamin C. The formation of either DNA–DNA or DNA–protein cross links increases
the retention of DNA as compared with that of irradiated control DNA (Fig. 4(b)).
Since DNA–protein cross links could be removed by proteinase K, proteinase-
induced increases in DNA elution rates indicated the presence of DNA–protein cross
links. Additionally, DNA retention was reduced by proteolytic digestion with
proteinase K.

To summarize, then, our results with alkaline elution suggest that strand breaks
appear after treatment with AA. It suggests that vitamin C is effective either with or
without proteinase treatment. The very steep line in Fig. 4(a) is an irradiated line,

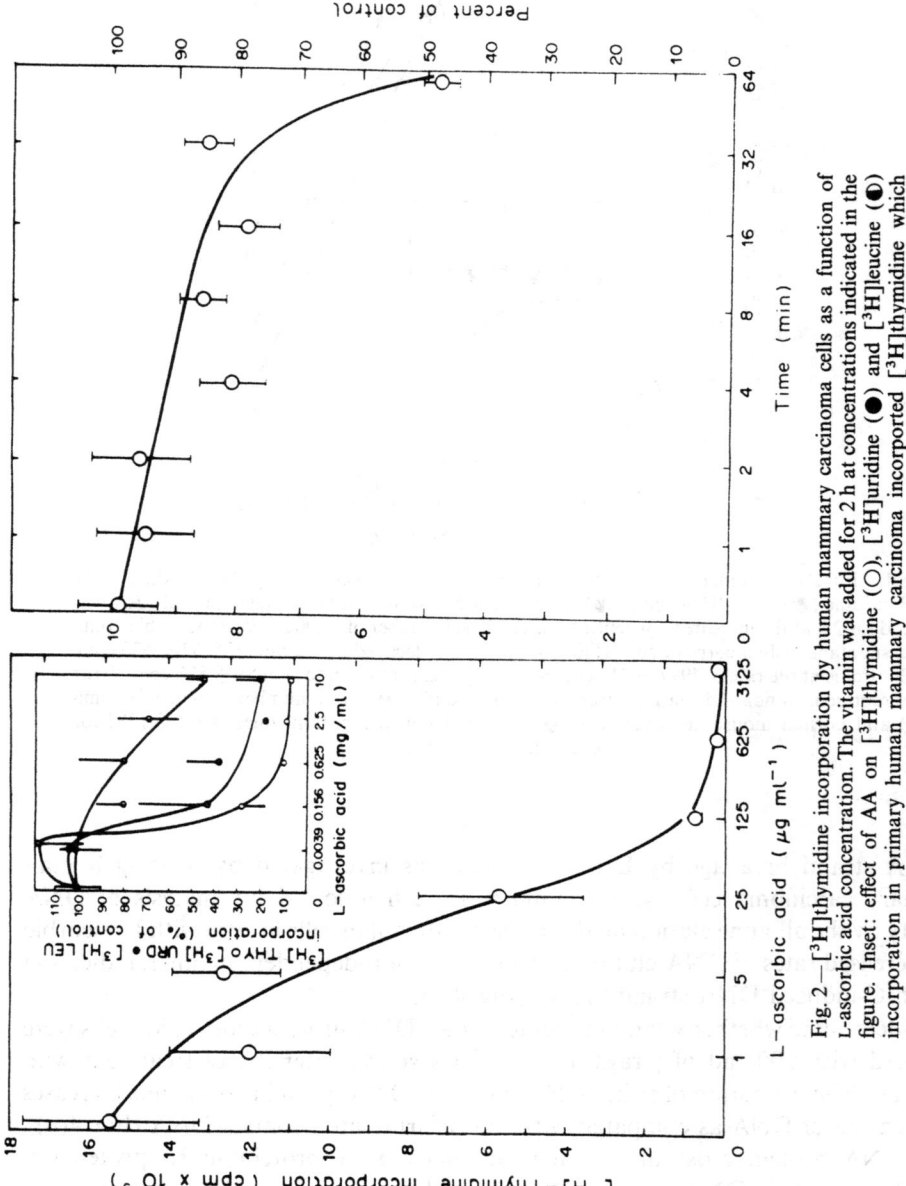

Fig. 2 — [³H]thymidine incorporation by human mammary carcinoma cells as a function of L-ascorbic acid concentration. The vitamin was added for 2 h at concentrations indicated in the figure. Inset: effect of AA on [³H]thymidine (○), [³H]uridine (●) and [³H]leucine (●) incorporation in primary human mammary carcinoma incorporated [³H]thymidine which decayed at the rate of 14,614 ± 2074 cpm, [³H]uridine at the rate of 2620 ± 274 and [³H]leucine at the rate of 689 ± 61 cpm. Time dependence of [³H]thymidine incorporation into carcinoma cells in presence of 100 µg/ml AA (right panel). Error bars represent standard deviations of four parallel samples (see ref. [39] for experimental details).

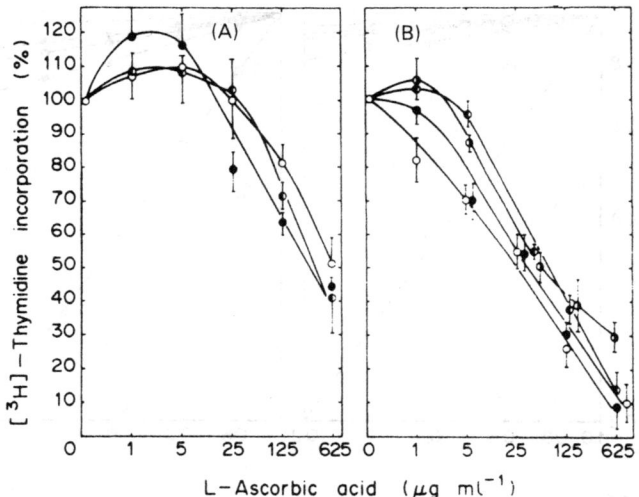

Fig. 3 — Effects of AA on [³H]thymidine incorporation normal (a) and transformed cells: (a) ○, bovine corneal endothelial cells; ●, murine T-lymphocytes; ◐, normal human fibroblasts derived from prostate, (b) ●, murine Lewis lung carcinoma cells; ◑, murine melanoma B16BL6 cells; ○, human primary kidney carcinoma cells; ◐, murine L1210 leukemia. Vitamin C was added for 2 h at the indicated concentrations. The basal number of 10,000 cells spontaneously incorporated [³H]thymidine which decayed at 11,298 ± 1177 (L1210), 5139 ± 198 (B16BL6), 9546 ± 779 (Lewis lung carcinoma) 4758 ± 230 (kidney carcinoma) and 6693 ± 1515 (human fibroblasts) (see ref. [39] for experimental details.

whereas the lines with shallower slopes are those exposed to the vitamin. In fact, based on drugs that are known to introduce a significant number of strand breaks, the extent of strand breakage is less pronounced, particularly in relation to the high concentrations used in the experiments.

The second series of experiments deals with the introduction of the removal of some DNA-protein cross links by the proteinase cross links. These data suggest that AA-induced suppression of transformed cell proliferation could be caused by DNA damage.

Fig. 5 displays the effect of AA on the tumor volume in athymic nude mice. Starting from day 1 after s.c. transplantation of the 5×10^6 neuroblastoma cells from the culture, 'nude mice' were given 500 mg/kg of L-ascorbic acid for 10 days, and the changes in tumor size were noted. It appears that in the treated nude mice, tumors grow slower and mice survive longer (43.3 ± 8.0 in controls vs 67.2 ± 7.8 in the AA-treated group, $p < 0.01$).

ANTITUMOR ACTIVITY OF 6-BROMO-6-DEOXY ASCORBIC ACID

Since its synthesis [49] in 1980, the 6-bromo-6-deoxy derivate of L-ascorbic acid aroused considerable interest concerning its physicochemical properties, which in many respects resemble those of the parent compound [49, 50]. For instance, both substances undergo Michael reaction [50–54] with a number of α, β-unsaturated carbonyl compounds and exhibit similar reactivity towards oxidizing radicals like HO˙, Br₂˙⁻, RS˙, etc. [55].

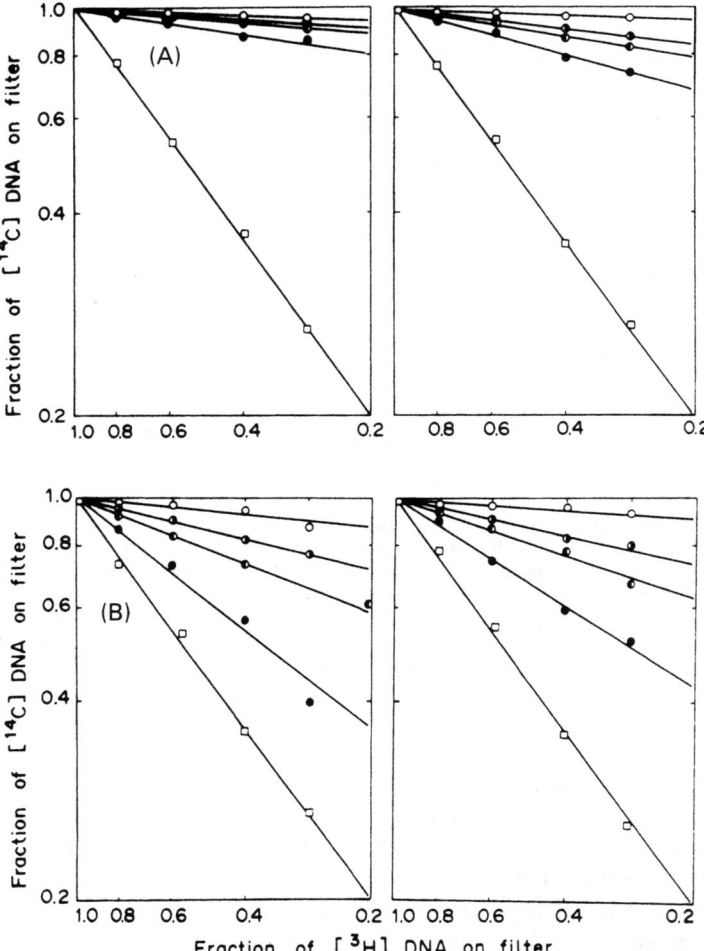

Fig. 4 — (a) Induction of DNA strand breaks in human mammary carcinoma treated with L-ascorbic acid. The cells were treated for 2 h with different AA concentrations: ◑, 0.1 mg/ml; ◐, 1 mg/ml; ●, 10 mg/ml. The cell sample indicated by (□) received 300 rad of γ-irradiation only. ○, control cells without any treatment. Data presented in the left panel represented DNA eluted after proteinase K treatment. (b) DNA cross links in human mammary carcinoma treated with AA: ●, 0.1 mg/ml; ◑, 1 mg/ml; ◐, 10 mg/ml; ○, 20 mg/ml; □, cells that did not receive drug treatment. All cells received ·300 rad of γ-irradiation before alkaline elution. The effects of proteinase K treatment are shown in the left panel. Alkaline elution of Kohn *et al.* [46,47] adapted by King *et al.* [48].

Recently, a new synthetic route for its preparation based on the use PPh$_3$—CBr$_4$ as brominating agent was reported [56]. The crystal structure, which gives important clues for molecular properties, has also become available [57]. Its molecular and electronic structure were studied theoretically by the MNDO semiempirical quantum-chemical procedure [58]. The same calculation method was applied to its radical anion, which was also characterized by ESR spectroscopy [55, 58]. It was found that replacement of the terminal hydroxyl group by bromine results in lowering the

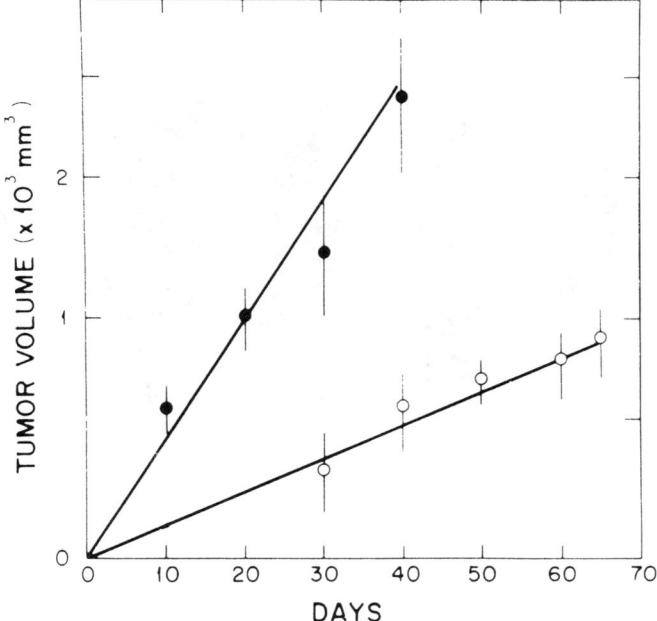

Fig. 5 — Human neuroblastoma volume ± standard deviation in 'nude' mice as a function of time after treatment with AA (500 mg/kg/day) was given i.p. for the first 10 days after tumor cell inoculation in mice (○). Control group (●) received physiological saline. All the groups consisted of 5 mice. The differences between vitamin-treated and non-treated groups are statistically significant ($p < 0.001$) (see ref. [38] for experimental details).

stability of both the neutral molecule and its radical anion by c. 50 kcal/mol. It is, however, interesting to point out that the aforementioned substitution does not affect either the energy of the highest occupied molecular orbitals (-9.4 eV vs. -9.5 eV in AA and 6-Br-AA, respectively) or the electron distribution within the lactonic ring. This finding indicates that these two substances might be equally reactive in certain chemical reactions such as, for example, oxidation processes.

Our interest in the potential biological activity of 6-Br-AA, and particularly in its possible antitumor effects, is a result of the idea that modified orthomolecular substances may have some specific useful features while still being similar to parent compounds which have been selected by evolution over millions of years. This is compatible with the recent proposition of Tsao *et al.* that the antitumor activity of ascorbic acid might be due to the chemical properties of the substance and not to its metabolism as a vitamin [35].

The activity of AA and 6-Br-AA was examined first on mouse fibroblasts (L929) as a model of normal cells and two tumor cell lines: mouse melamoma (MeL-B16) and human cervical carcinoma (HeLa) cells *in vitro* [40]. Subsequently, *in vivo* investigations on mice were performed [40]. We give here the most interesting outcome of these studies only. Let us commence with results obtained for mouse fibroblast cells. These are displayed in Fig. 6. We observe that both substances only slightly inhibit the proliferative capacity of this type of cells, with 6-Br-AA being somewhat more toxic. This holds for the whole range of doses and of the incubation period involved.

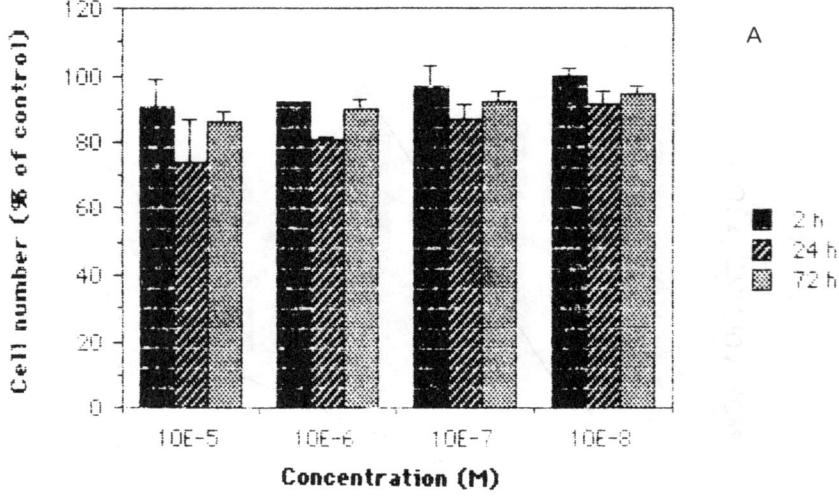

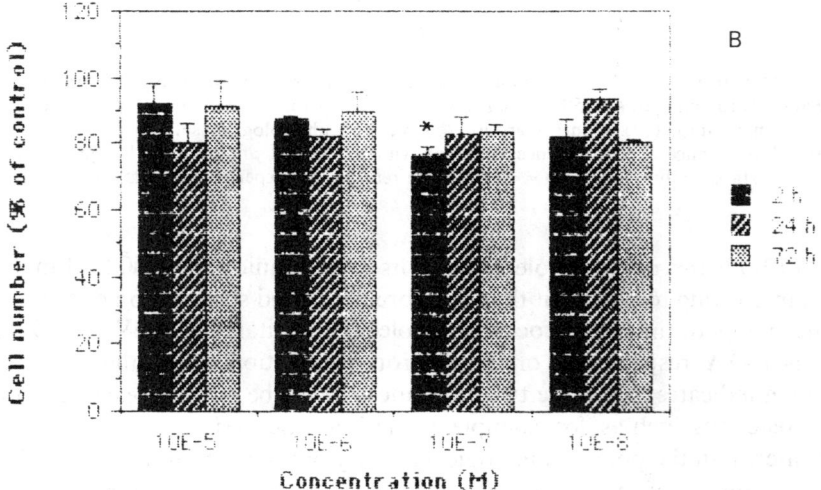

Fig. 6 — Effect of AA (a) and 6-Br-AA (b) on proliferative capacity of mouse L929 cells examined in the exponentially growing cell population. A ratio (in %) between a number of cells treated by AA or 6-Br-AA and controls (untreated sample) is given. The absolute number of cells in the control samples was $4.03 \pm 0.31 \times 10^5$. The asterisk indicates values statistically different from the control ($p < 0.05$) (see ref. [40] for experimental details).

On the contrary, results obtained with tumor Mel–B16 cells (Fig. 7) indicate a considerable difference in activity of the two agents. Whereas AA suppresses proliferation only slightly, as in the case of L929 fibroblasts, its derivative 6-Br-AA exhibits a pronounced inhibitory activity. It also appears that the extent of inhibition depends strongly on the concentration and incubation period. The best results from experiment described by Fig. 7 were achieved with a 72 h incubation and 10^{-5}M solution, which caused a decrease of the number of viable cells by about 60%. It is noteworthy

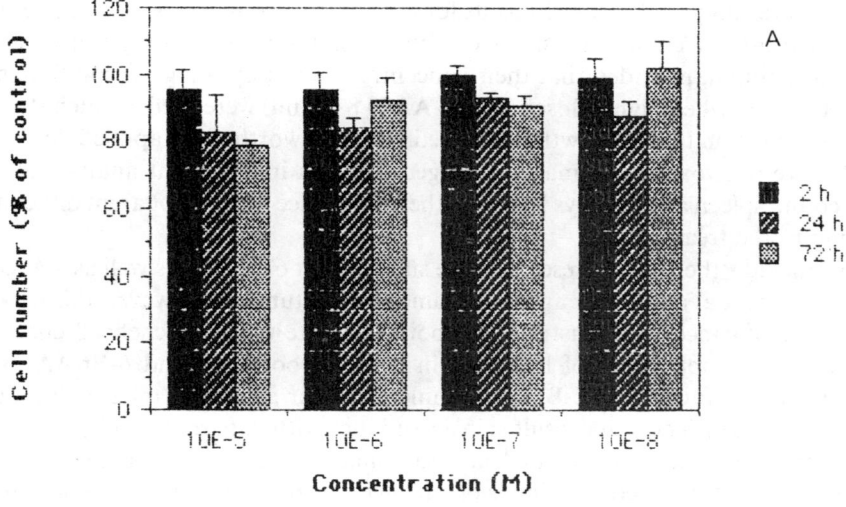

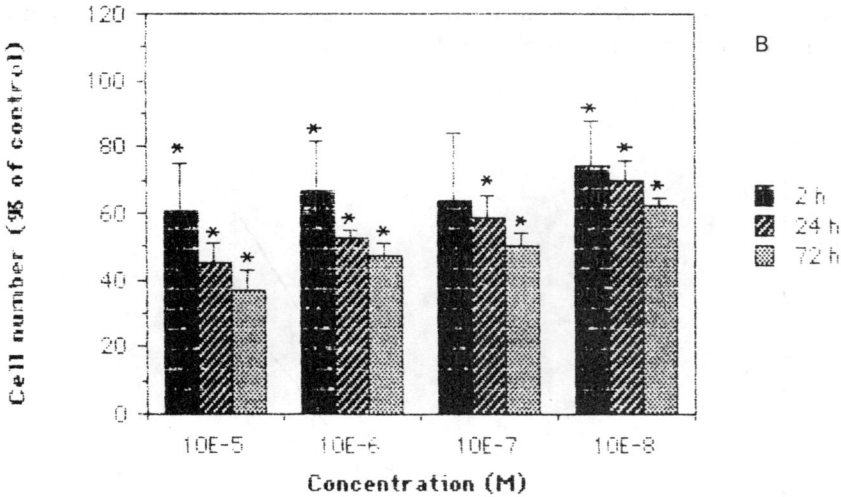

Fig. 7 — Effect of AA (a) and 6-Br-AA (b) on the proliferative capacity of mouse melanoma B16 cells examined in the exponentially growing cell population. A ratio (in %) between a number of cells treated by AA or 6-Br-AA and controls (untreated samples) is given. The absolute number of cells in the control samples was $1.09 \pm 0.12 \times 10^5$. The asterisks indicate values statistically different from the control ($p < 0.05$) (see ref. [40] for experimental details).

that even stronger activity was achieved by further increase of the 6-Br-AA dose. More detailed discussion is presented in ref. [40]. Let us mention in passing that the results for Mel-B16 were also confirmed by the [³H]thymidine incorporation test. Since this test measures the rate of DNA synthesis, it implies that 6-Br-AA, similarly to ascorbic acid, induces damage in treated cells.

An inhibitory effect of 6-Br-AA was also confirmed by *in vivo* experiment. The results of this experiment together with those obtained by treating animals with vitamin C are compared with the results for the control group in Fig. 8. Their scrutiny reveals that 6-Br-AA exhibits a slightly stronger inhibition of tumor growth than the parent compound, provided that their concentration in physiological solution is the same. However, when higher doses of 6-Br-AA (4.8 mg ml) were used, a much stronger inhibiting effect on tumor growth is attained. It is noteworthy that applied doses of 6-Br-AA were not toxic to animals, as judged from an independent analysis of their blood count, spleen and kidneys [59]. Furthermore, mice had no apparent difficulty in adapting to the treatment.

To conclude, the results presented here suggest that 6-Br-AA, as well as AA, could serve as a potential antitumor agent for some specific tumors. However, this depends on the type of cancer. To illustrate this point we note that no beneficial effect was observed in the treatments of leukemia in mice by both AA and 6-Br-AA. More importantly, it is found that 6-Br-AA stimulates tumor growth in mice with induced fibrosarcoma [59]. The latter result is in accordance with data reported by Robertson *et al.*, who found that transplanted fibrosarcoma grew slower in guinea pigs on a scorbutic diet as compared to animals on an adequate diet [60]. Hence, some care has to be exercised in applying AA derivates. Nevertheless, selective inhibition of mouse melanoma B-16 cell growth by the 6-Br-AA treatment makes further *in vitro* and *in vivo* experiments warranted. In particular, studies directed toward elucidation of the mechanism of antitumor action of 6-Br-AA are desirable.

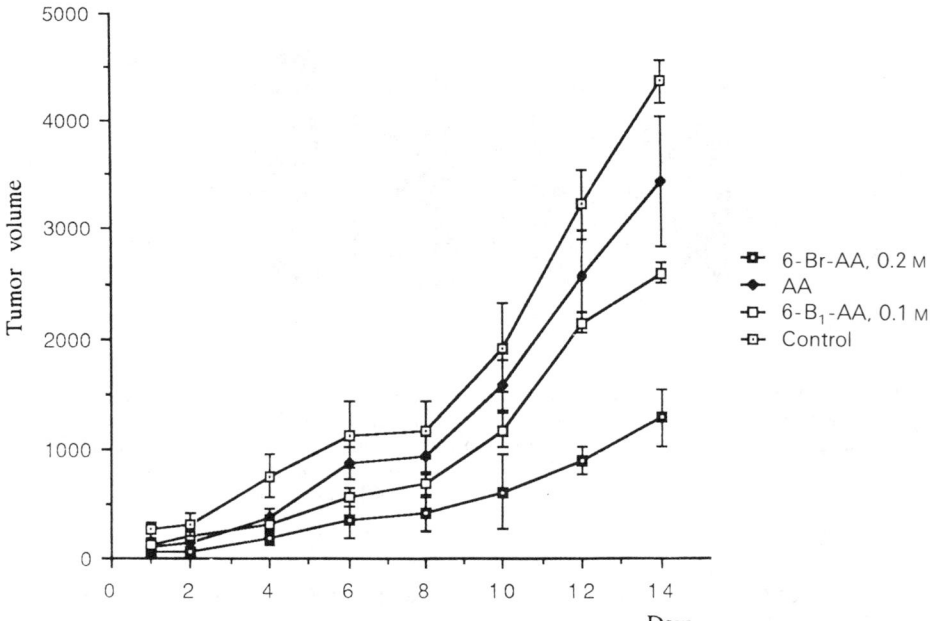

Fig. 8 — Value of solid melanoma B16 volume ± SD as a function of time after transplantation for AA: (-♦-), 6-Br-AA-treated (-□-)and (□) and control mice treated with physiological saline (□). The daily dose of AA (3 × 90 mg/kg of body mass) and 6-Br-AA (3 × 120 (□) and 3 × 240 (⊥) mg/kg of body mass) was administered i.p. in three partions. All the groups consisted of 11 animals. The differences in tumor volumes between the 6-Br-AA-treated or AA-treated and untreated groups are statistically significant ($p < 0.001$)

VITAMIN C AND β-CAROTENE

Most experiments aiming to elucidate the anticancer activity of vitamins involve studies of single substances and not their various combinations. In other words, little attention has been focused on maintaining the antitumor potential of two or more vitamins, despite conclusive evidence that certain combinations of these nutrients provide considerable prophylactic protection. Hence this topic deserves to be examined.

Recently, we tackled this problem by exploring the effect of ascorbic acid supplemented with β-carotene (a precursor of vitamin A) on the proliferative capacity of several normal and tumor cells. Additionally, the effect of 6-Br-AA combined with β-carotene was examined. Here we present results obtained by *in vitro* studies with mouse fibroblasts (L929) (Fig. 9) and with mouse melanoma (Mel-B16) cells (Fig. 10).

The results displayed in Fig. 9(a) reveal that β-carotene (contrary to AA and 6-Br-AA — see previous section) inhibits the growth of L929 cells, although to a relatively small extent (about 20% for the highest concentration used). If, however, β-carotene is given together with either AA (Fig. 9(b) or 6-Br-AA (Fig. 9(c) the extent of inhibition becomes insignificant, which is a favorable result.

Surprisingly enough, the effect of β-carotene on the proliferation of Mel-B16 cells is even less pronounced than for the L929 cell line (Fig. 10(a)). Further, a combined regime involving both AA and β-carotene leads to an insignificant increase in the number of viable cells for incubation periods of 2 and 24 h. However, if the incubation period is extended to 72 h then the number of viable cells becomes equal to that in the control samples (Fig. 10(b)).

On the other hand, it is remarkable that a joint application of β-carotene and 6-Br-AA (in concentrations of 10^{-8}–10^{-5} M) leads to a considerable decrease in the proliferation of cancer cells (Fig. 10(c)). Moreover, the inhibitory effect for the highest concentration of 6-Br-AA and β-carotene used seems to be slightly more pronounced than for this substance alone. (Fig. 7). The contrary holds for lower concentrations of 6-Br-AA. Finally, comparison of the results shown in Figs 7 and 10 suggests that the antitumor activity of 6-Br-AA and β-carotene strongly resembles the pattern obtained by the former compound itself. Hence, we conclude that β-carotene contributes little to the antitumor activity of the 6-Br-AA derivative of vitamin C. This finding is disappointing, but a possible synergistic effect of β-carotene and L-ascorbic acid should be carefully examined in other tumors before a final conclusion is drawn.

CONCLUDING REMARKS

Three different topics related to the antitumor activity of L-ascorbic acid have been discussed. The mechanism of antitumor activity of AA was considered first. The results presented suggest that L-AA, if implemented into cell cultures in high concentration, decreases DNA, RNA and protein synthesis, as well as mitosis of transformed cells, causing both DNA strand breaks and DNA cross links.

The second question addressed is related to the 6-bromo-6-deoxy derivative of ascorbic acid. Investigations performed suggest that appropriate substitution of ascorbic acid moiety might represent an important way of modifying and enhancing antitumor activity of the parent substance against specific tumors. On the other hand,

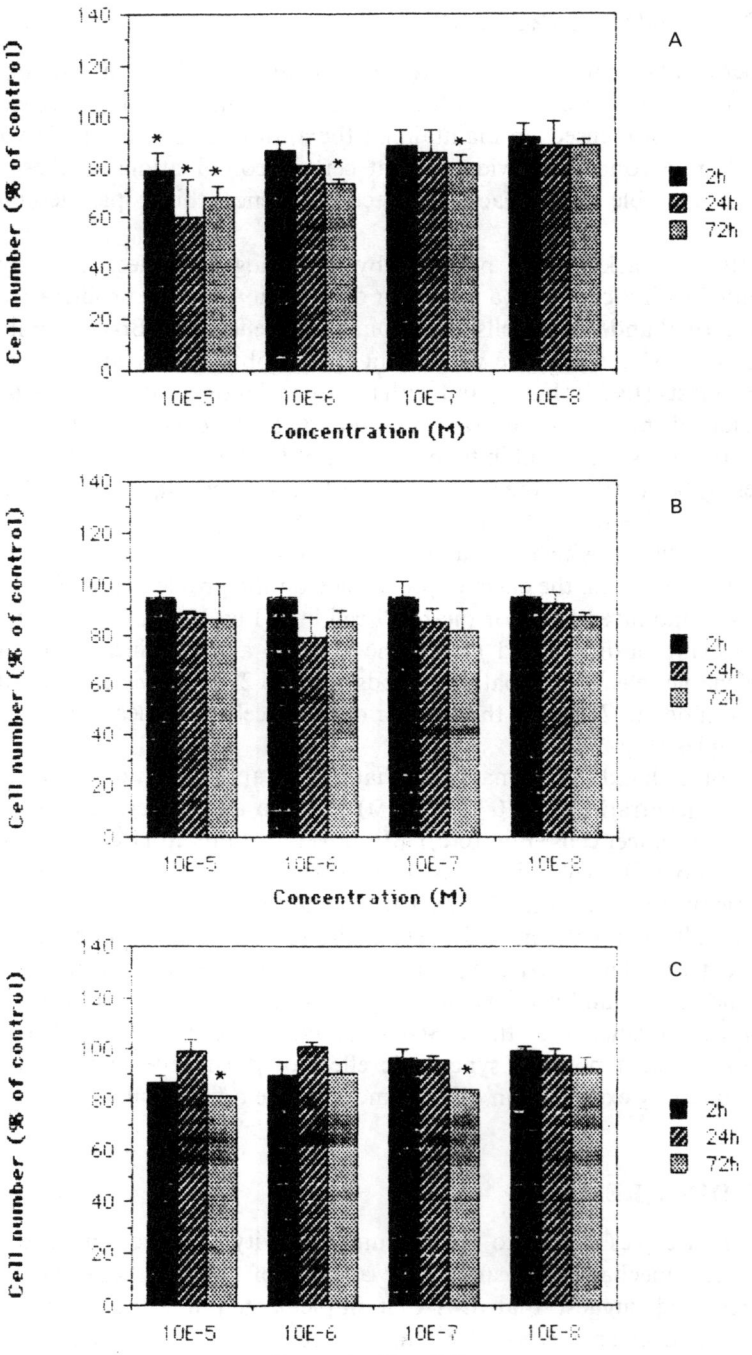

Fig. 9 — Effect of β-carotene (a), β-carotene + AA (b) or β-carotene + 6-Br-AA (c) on the proliferative capacity of mouse L929 cells examined in exponentially growing cell population. A ratio between a number of treated and control cells is given. The absolute number of cells in control samples was 3.81 ± 0.29. Asterisks indicate values statistically different from control $(p < 0.05) \times 10^5$. (see ref. [40] for experimental details).

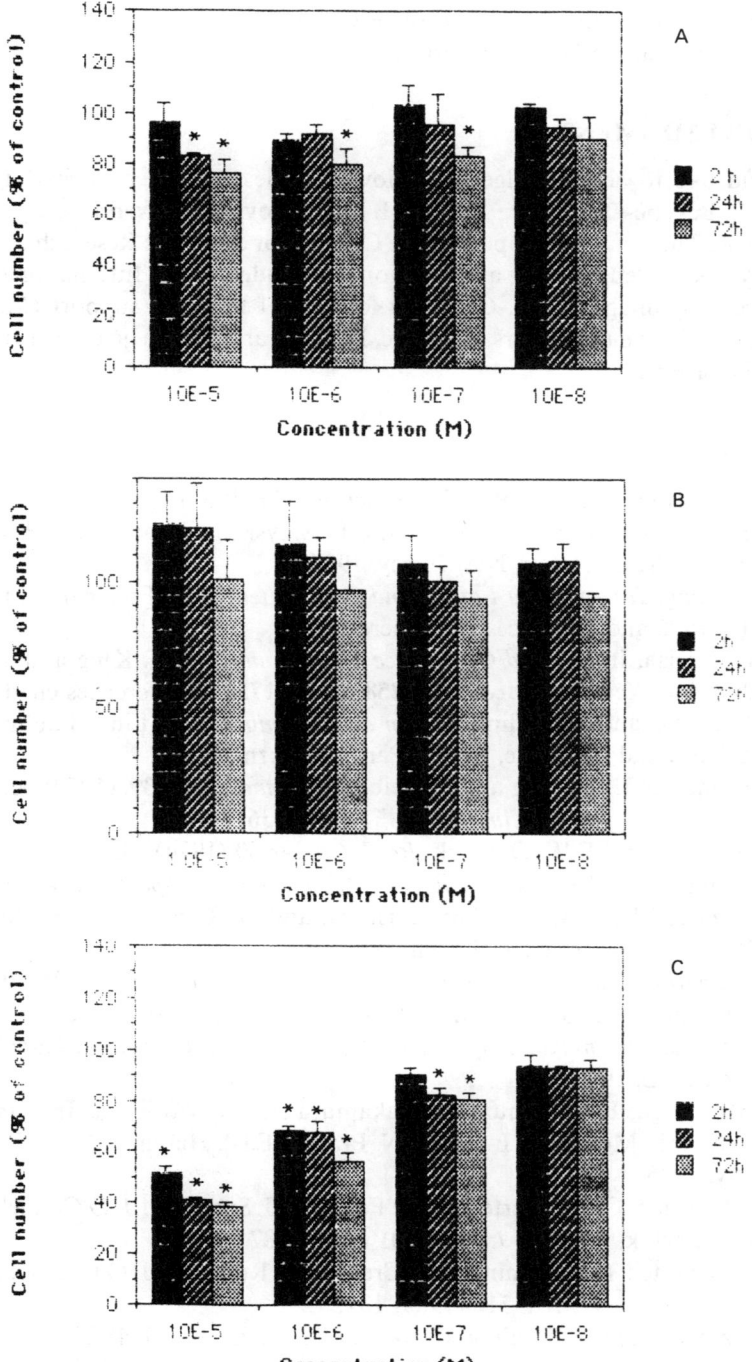

Fig. 10—Effect of β-carotene (a), β-carotene + AA (b) or β-carotene + 6-Br-AA (c) on the proliferative capacity of mouse melanoma Mel-B16 cells examined in exponentially growing cell population. A ratio between a number of treated and control cells is given. The absolute number of cells in control samples was $2.06 \pm 0.17 \times 10^5$. Asterisks indicate values statistically different from control ($p < 0.05$) (see ref. [40] for experimental details).

a combination of vitamin C and β-carotene does not seem to play an important role in this respect, at least in *in vitro* experiments.

ACKNOWLEDGMENT

We would like to acknowledge the following colleagues for participating in these studies: L. Beketić-Orešković, Z. Kos, I. Ljubenkov, R. Spaventi, Š. Spaventi and B. Šušković. The financial support of the Council for Scientific Research of Croatia is gratefully acknowledged. Our appreciation is extended to the International Office of Kernforschungsanlage Jülich, Germany for partial financial support (M.E.M. and Z.B.M.). Finally, we thank Mrs M. Fiolić, Lj. Krajcar and N. Ujčić for their excellent technical assistance.

REFERENCES

[1] W.P. Jorissen and A.H. Belinfante, *Science*, **79** (1934) 13.

[2] A. Hanck, In: *Vitamins and Cancer*, F.L. Meyskens, Jr. and K.N. Prasad (Eds), Humana Press, Clifton, New Jersey, 1986.

[3] L. Pauling, *How to Live Longer and Feel Better*, W.H. Freeman and Co., New York, 1986 and references cited therein.

[4] S.S. Mirvish, In: *Second Conference on Vitamin C*, C.G. King and Y.Y. Burns (Eds), *Ann. New York Acad. Sci.* **258** (1975) 175 and references cited therein.

[5] E. Cameron and L. Pauling, *Cancer and Vitamin C*, The Linus Pauling Institute of Science and Medicine, Menlo Park, California, 1979.

[6] E. Cameron, L. Pauling and B. Leibovitz, *Cancer Res.* **39**, (1979) 663.

[7] D. Kromhout, *Am. J. Clin. Nutr.* **45** (1987) 1361.

[8] N. Krasner and E.W. Dymock, *Br. J. Cancer* **30** (1974) 142.

[9] T.K. Basu, In: *Vitamin C: Recent Advances and Aspects in Virus Diseases, Cancer and Lipid Metabolism*, A. Hanck and G. Ritzel (Eds), H. Huber, Bern, 1979 and references cited therein.

[10] E. Cameron and L. Pauling, *Proc. Natl. Acad. Sci. USA* **73** (1976) 3685.

[11] A. Murata, F. Morishige and H. Yamaguchi, In: *Vitamin C: New Clinical Applications in Immunology, Lipid Metabolism and Cancer*, A. Hanck (Ed), H. Huber, Bern, 1982.

[12] F. Morishige, T. Nakamura, N. Nakamura and N. Morishige, In: *Vitamins and Cancer*, F.L. Meyskens, Jr. and K.N. Prasad (Eds), Humana Press, Clifton, New Jersey, 1986.

[13] E.T. Creagan, C.G. Moertel, J.R. O'Fallon, A.J. Schutt, M.J. O'Connell, J. Rubin and S. Frytak, *N. Engl. J. Med.* **301** (1979) 687.

[14] C.G. Moertel, T.R. Fleming, E.T. Creagan, J. Rubin, M.J. O'Connell and M.M. Ames, *N. Engl. J. Med.* **312** (1985) 137.

[15] E. Cameron and L. Pauling, *Chem.–Biol. Interact.* **9**, (1974) 273.

[16] E. Cameron and A. Campbell, *Chem.–Biol. Interact.* **9** (1974) 285.

[17] T.G. Frazier and M.E. McGinn, *J. Surg. Res.* **27** (1979) 318.

[18] H.E. Gruber, H.H. Tewfik and F.A. Tewfik, *Europ. J. Cancer* **16** (1980) 441.

[19] F.A. Tewfik, H.H. Tewfik and E.F. Riley, *Int. J. Vit. Nutr. Res.* (1982); Suppl. 23, 257.

[20] F.S. Liotti and V. Talesa, *Convivia Medica* (1982); 3 suppl, 81.

[21] J.M. Varga and L. Arioldi, *Life Sci.* **32** (1983) 1559.

[22] R.N. Chakrabarti and P.S. Dasquipta, *IRCS Med. Sci.* **12** (1984) 1147.

[23] L. Pauling, J.C. Nixon, F. Stitt, R. Marcuson, W.B. Dunham, R. Barth, K. Bensch, Z.S. Herman, B.D. Blaisdell, C. Tsao, M. Prender, V. Andrews, R. Willoughby and E. Zuckerkandell, *Proc. Natl. Acad. Sci. USA* **82** (1985) 5185.

[24] J.A. Migliozzi, *Br. J. Cancer* **35** (1977) 448.

[25] A. Brunschwig, *Cancer Res.* **3** (1943) 550.

[26] L. Pauling, R. Willoughby, R. Reynolds, B.E. Blaisdell and S. Lawson, In: *New Clinical Applications in Immunology, Lipid Metabolism and Cancer*, A. Hanck (Ed.), H. Huber, Bern, 1982 and references cited therein.

[27] F.S. Liotti, V. Talesa and A.R. Menghini, *Int. J. Vit. Nutr. Res.* **53** (1983) 251.

[28] S. Bram, P. Froussard, M. Guichard, C. Jasmin, Y. Augery, F. Sinoussi-Barre and W. Wray, *Nature* **284** (1980) 629.

[29] K.N. Prasad, P.K. Sinha, M. Ramanujam and A. Sakamoto, *Proc. Natl. Acad. Sci.* **76** (1979) 829.

[30] W.R. Waddel and R.E. Gerner, *J. Surg. Oncol.* **15** (1980) 85.

[31] E. Poydock, *IRCS Med. Sci.* **12** (1984) 813.

[32] C.F. van der Merwe, *Sa Med. J.* **65** (1984) 712.

[33] C.S. Tsao, W.B. Dunham and P.Y. Leung, *In Vivo* **2** (1988) 147.

[34] M.C. de Pauw-Gillet, B. Siwek, G. Pozzi, E. Sabbioni and R.J.B. Bassler, *Anticancer Res.* **10** (1990) 391.

[35] C.S. Tsao and P.Y. Leung, Chapter 25 in this volume.

[36] W.R. Bruce and P.W. Dion, *Am. J. Clin. Nutr.* **33** (1980) 2511.

[37] E. Bright-See and H.L. Newmark, In: *Modulation and Mediation of Cancer by Vitamins*, F.L. Meyskens and K.N. Prasad (Eds), Karger, Basel, 1983.

[38] K. Pavelić, *Brain Res.* **342** (1985) 369.

[39] K. Pavelić, Z. Kos and S. Spaventi, *Int. J. Biochem.* **21** (1989) 931.

[40] M. Osmak, M. Eckert-Maksić, K. Pavelić, Z.B. Maksić, R. Spaventi, L. Beketić, I. Kovaček and B. Šušković, *Res. Exp. Med.* **190** (1990) 443.

[41] R. Lotan, *Biochem. Biophys. Acta* **605** (1980) 33.

[42] W. Bollag, In: *Protective Agents in Cancer*, D.C.H. McBrien and T.F. Slate (Eds), Academic Press, 1983.

[43] M.M. Mathews-Roth, *Pure & Appl. Chem.* **57** (1985) 717.

[44] R.C. Moon, D.L. McCormick and R.G. Mehta, In: *Vitamins and Cancer*, F.L. Meyskens, Jr. and K.N. Prasad (Eds), Humana Press, New Jersey, 1986 and references cited therein.

[45] A. Kornhauser, W. Wamer and A. Giles, Jr., In: *Antimutagenesis and Anticarcinogenesis Mechanism* (D.M. Shankel, P.E. Hartman, T. Kada and A. Hollaender (Eds), Plenum, 1986, p. 465.

[46] K.W. Kohn, R.A.G. Friedman and Z.M. Iqbal, *Biochemistry* **13** (1974) 4134.

[47] K.W. Kohn, R.A.G. Ewing, W. Rouse and F. Reigler, In: *A Laboratory Manual of Research Procedures*, E. Friedberg and P. Hanawolt (Eds), Marcel Dekker, New York, 1981, p. 379.

[48] C.L. King, W.N. Hittelman and R.L. Loo, *Cancer Res.* **44** (1984) 5634.

[49] K. Kiss, K.P. Berg, A. Dirscherl, W.E. Oberhansli and W. Arnold, *Helv. Chim. Acta* **63** (1980) 237.

[50] I. Ljubenkov, M.Sc. Thesis, University of Zagreb, 1991; M. Eckert-Maksić and I. Ljubenkov, to be published.

[51] G. Fodor, R. Arnold, T. Mohacsi, I. Karle and J. Flippen-Anderson, *Tetrahedron* **39** (1983) 2137.

[52] R. Arnold, G. Fodor, H. Mathelier, I. Monasci, A. Szent-Györgyi and R.W. Vetri, In: *Protective Agents in Cancer*, D.C.H. McBrien and T.F. Slater (Eds), Academic Press, New York, 1983.

[53] K. Sussangkarn, G. Fodor, I. Karle and C. George, *Tetrahedron* **44** (1988) 7047.

[54] K. Eger, G. Folkers, W. Zimmermann, R. Schmidt and W. Hiller, *J. Chem. Res. (M)* (1987) 2352.

[55] M. Bonifačić, I. Ljubenkov and M. Eckert-Maksić, manuscript in preparation.

[56] B.A. Šušković, *Croat. Chem. Acta* **58** (1985) 231.

[57] Ž. Ružić-Toroš, B. Kojić-Prodić, D. Horvatić and B. Šušković, *Acta Cryst.* **C45** (1989) 269.

[58] M. Eckert-Maksić, P. Bischof and Z.B. Maksić, *Croat. Chem. Acta* **58** (1985) 179.

[59] K. Pavelić, I. Kovaček, R. Spaventi, M. Eckert-Maksić and Z.B. Maksić, unpublished results.

[60] W.V.B. Robertson, A.J. Dalton and W.E. Heston, *J. Natl. Cancer Inst.* **10** (1949) 53.

27

Nephroprotective effect of L-ascorbic acid and its 6-bromo-6-deoxy derivative

M. Radačić,[a] M. Eckert-Maksić,[a] V. Šverko,[a] J. Jerčić,[b] E. Suchanek,[c] M. Golić[a] and Z. B. Maksić[a,d]

[a]Rudjer Bošković Institute, Zagreb, Croatia, Yugoslavia
[b]Veterinarian Faculty, Zagreb, Croatia, Yugoslavia
[c]Clinic for Gynaecology and Obstetrics, University of Zagreb, Croatia, Yugoslavia
[d]Faculty of Natural Sciences and Mathematics, University of Zagreb, Croatia, Yugoslavia

INTRODUCTION

Since its discovery in 1969 [1], cis-dichlorodiamineplatinum (II) (DDP) has proved to be one of the most effective agents against a variety of experimental and human neoplasms [2–5]. However, this drug has certain shortcomings, such as narrow range of responsive tumors, severe host toxicity and the development of resistance in tumor cells [6, 7]. Furthermore, the administration of cis-DDP to humans can produce several undesirable side-effects—such as nausea and vomiting, ototoxicity, neurotoxicity and nephrotoxicity, the last being most often considered as dose-limiting [8–10]. It is thus not surprising that substantial attention has been focused on the development of procedures for prevention or at least diminishing the nephrotoxicity of cis-DDP. Of these, diuresis [12] and hypersalination [13] have been most commonly used. Alternative approaches involve (a) application of newly developed coordination complexes [13–15] and (b) combined treatment involving cis-DDP and various chemical agents [16–20]. Typical examples of the latter are provided by diethyldithio-carbamate [16], probenecid [17], WR2721 [18], sodium thiosulfate and thiourea [19], 4,5,6,7-tetrahydro-3-oxo-2H-indazole-5,5-dicarboxylic acid, etc. [20].

The present study was stimulated by a recent observation that several platinum complexes of L-ascorbic acid (AA) show reduced nephrotoxicity [13, 14]. A question arises as to whether the same effect could be achieved if the L-ascorbic acid and platinum agent were given separately. To answer this question we performed a series of experiments designed to examine the influence of L-ascorbic acid and its 6-bromo-6-deoxy derivative (6-Br-AA) [21] administered together with cis-DDP on (a) kidneys and (b) on the overall body-status in a mouse model. The 6-bromo-6-

deoxy derivative of L-ascorbic acid is particularly interesting owing to its physicochemical properties which in many respects resemble those of the parent acid [21–23].
This substance is similar to AA in that it exhibits pronounced antitumor activity
towards certain tumor cells *in vitro* as well as *in vivo* [24].

MATERIAL AND METHODS

Animals

Inbred male and female CBA/H2gr mice, 10–12 weeks old, 20–25 g in weight, were
obtained from our own breeding facility. The animals were housed in Macrolone
cages, 20 per large (26.5 × 38.0 cm) cage before experiment and 5 per small (12.0 ×
18.0 cm) cage during experiment. Ten animals were used in each treatment group.
Commercial food pellets and tap water were offered *ad libitum*.

Drugs

cis-Diaminodichloroplatinum (*cis*-DDP, platinol) was obtained by courtesy of
Bristol Myers (Copenhagen, Denmark). The L-ascorbic acid (AA) (p.a. purity) was
obtained from PLIVA Pharmaceutical Works, Zagreb. 6-Bromo-6-deoxy-ascorbic
acid (6-Br-AA) was prepared by reacting L-ascorbic acid with HBr-HCOOH following the procedure described by Kiss *et al* [21]. Prior to dissolution, the substance was
rerystallized from nitromethane (m.p. 176–177°C, $[x]^{20}$ D-6.5 (c 1.0, H_2O)).

Drug injections

All drugs were injected intraperitoneally and dissolved in deionized water immediately before application. The pH of the AA and 6-Br-AA solutions was adjusted to 7.0
using 1 N NaOH. *cis*-DDP (10 mg per kg of body mass) was given at the rate of
0.5 ml/mouse. The required doses of AA (330 mg/kg of body mass) and 6-Br-AA
(480 mg/kg of body mass) were given at the rate of 0.25 ml/mouse 2 h before or 15 min
after *cis*-DDP injection.

Measurements

On the fifth day after treatment all animals were anesthetized using ether, and blood
samples were collected from jugular vessels for the measurement of blood urea
nitrogen (BUN) and creatinine. Spleen and thymus weights were recorded on a
Sartorius electronic balance. The BUN plasma level was determined by the spectro-
photometric urease Berthelot method (Ames, Sera-Pack) and by the creatinine
enzymatic method (Boehringer, Mannheim) [25].

The results are presented as the mean values obtained with 10 animals ±
standard error. The significance of the differences between groups was assessed by the
Student's *t*-test. The level of significance was set at 0.05%.

RESULTS AND DISCUSSION

The results of three experiments are summarized in Tables 1 and 2.

Table 1 shows the blood urea nitrogen (BUN) and creatinine levels in animals treated with AA or 6-Br-AA 2 h before and $\frac{1}{4}$ h after injecting cis-DDP. Results of the same test obtained by treating the animals with each of the aforementioned substances

Table 1 — Effects of AA and 6-Br-AA rescue on blood urea nitrogen (BUN) and creatinine plasma levels (CPL) (means + standard error) in mice 5 days after cis-DDP (10 mg/kg i.p.)

Group[a]	Treatment	BUN (mmol/dm^3) (X + S.E.)	CPL (mmol/dm^3) (X + S.E.)
I	AA (−2 h)[b]	29.4 ± 5.1	38.3 ± 5.9[c]
II	AA (+$\frac{1}{4}$ h)[b]	70.7 ± 7.2	130.2 ± 16.4
III	6-Br-AA (−2 h)[b]	19.3 ± 4.0[c]	62.0 ± 14.6
IV	6-Br-AA (+$\frac{1}{4}$ h)[b]	68.9 ± 11.0	109.6 ± 17.2
V	cis-DDP	41.1 ± 7.7	71.2 ± 12.6
VI	AA	5.7 ± 0.3	—
VII	6-Br-AA	6.5 ± 0.2	—
VIII	Control	7.2 ± 0.6	56.5 ± 10.2

[a] Each group consisted of 10 animals.
[b] Combined treatment; protective substance given before (−) or after (+) cis-DDP injection.
[c] Significantly different from group V ($p < 0.05$ or better).

Table 2 — Protective effect of AA and 6-Br-AA treatment on BUN plasma level, the weight of spleen (SW) and thymus (TW) (means + standard error) in mice 5 days after cis-DDP (10 mg/kg i.p.)

Group[a]	Treatment	BUN (mmol/dm^3) ($X \pm$ S.E.)	SW ($X \pm$ S.E.)	TW ($X \pm$ S.E.)
Experiment no. 2				
I[b]	cis-DDP + AA	6.0 ± 0.6[c]	60.8 ± 2.0	20.4 ± 2.1
II[b]	cis-DDP + 6-Br-AA	5.4 ± 0.4[c]	61.0 ± 3.9	24.6 ± 3.1
III	cis-DDP	38.0 ± 10.2	54.0 ± 5.4	19.7 ± 2.2
IV	Control	5.4 ± 0.4	78.0 ± 3.0	45.0 ± 2.3
Experiment no. 3				
V	cis-DDP + 6-Br-AA	20.3 ± 3.5[d]	60.0 ± 3.1	10.0 ± 0.1
VI	cis-DDP	50.7 ± 8.7	43.0 ± 3.0	10.0 ± 0.1
VII	Control	6.6 ± 0.3	78.3 ± 3.1	33.0 ± 3.3

[a] Each group consisted of 10 animals.
[b] Protective substance given 2 h before cis-DDP injection.
[c] Significantly different from group III ($p < 0.05$ or better).
[d] Significantly different from group VI ($p < 0.01$).

separately and those obtained for the untreated control group are given for comparison. Perusal of the relevant numbers reveals several interesting features. Firstly, it is important to note that the BUN values in the latter three groups are of the same order of magnitude. This strongly indicates that AA and 6-Br-AA do not exert any renal damage. Application of AA or 6-Br-AA 2 h before cis-DDP injection has had a significant protective effect on renal function, whereby 6-Br-AA was apparently more effective. The corresponding BUN values are 19.3 and 29.4 mmol/dm^3, respectively, as compared to 41.1 mmol/dm^3 found in animals treated with cis-DDP only. In contrast, the BUN levels in animals treated with AA or 6-Br-AA $\frac{1}{4}$ h after cis-DDP were even higher than in animals treated by cis-DDP alone. It should be noted, however, that results for each of the treated groups exhibit considerable scattering, as indicated by the rather high standard deviations. Hence, our conclusions are more qualitative than quantitative in nature.

The creatinine plasma levels (CPLS) in animals protected by AA 2 h before cis-DDP as compared to those treated with cis-DDP only were significantly lower. Specifically, the CPL value dropped from 71.2 ± 12.6 to 38.3 ± 5.9 mmol/dm^3. The latter value is even lower than in the control group (56.5 ± 10.2 mmol/dm^3). In 6-Br-AA protected animals the CPL values changed only slightly, in contrast with the results of the BUN test. The reason for this discrepancy is not clear at the moment.

Since the BUN level is generally considered to be a more reliable test of kidney damage than the CPL value [26], it follows that 6-Br-AA is a more potent protective agent than AA. Consequently, only BUN values were measured in subsequent experiments (Table 2). The protective substances (AA or 6-Br-AA) were given 2 h before cis-DDP. As in experiment no. 1 (Table 1), all AA or 6-Br-AA pretreated animals have had significantly lower BUN values than animals given cis-DDP only. There is a pronounced difference between values measured in different experiments. In experiment no. 2 (Table 2), the BUN values were considerably lower than those obtained in experiment no. 1 (Table 1) and no. 3 (Table 2), respectively. Part of the discrepancies can be ascribed to different samples of cis-DDP used in various experiments and to a natural variation in biochemical individualities of the treated animals. Moreover, crudeness in the testing procedure should be taken into account.

In order to get an insight into the overall status of experimental animals we have also measured the total body weight (not shown) and the weights of the spleen and the thymus of the treated mice.

All of the treated animals lost body weight, but the loss was smaller in animals protected by AA or 6-Br-AA than in mice treated with cis-DDP only. The same holds for the spleen. The changes in thymus weight are negligible.

CONCLUDING REMARKS

The results, albeit qualitative, show that L-ascorbic acid, as well as its 6-bromo-6-deoxy derivative, reduces renal toxicity caused by cis-DDP if administered before it. Both substances also reduce the loss of body weight and spleen atrophy caused by cis-DDP, thus exerting beneficial effects.

The present results do not, however, allow any definite conclusion to be drawn about specific site or the mechanism by which these effects take place. It is conceivable, however, that these substances (or their metabolic products) compete with *cis*-DDP for the same receptors in kidney tubules. Another possibility would be that AA and its derivatives [27], owing to the strong complexing ability, chelate *cis*-DDP. Additional studies are necessary in order to clarify the precise mechanism(s) of protection.

Finally, it would be of considerable interest to test the antitumor activity of *cis*-DDP given together with AA or 6-Br-AA and to see whether these substances act synergistically or antagonistically with *cis*-DDP in this respect, in addition to their nephroprotective ability.

ACKNOWLEDGMENT

The financial support of the Council for Scientific Research of Croatia is gratefully acknowledged. Our appreciation is extended to the International Office of Kernforschungsanlage Jülich, Germany, for their partial financial support (M.E.M. and Z.B.M.).

REFERENCES

[1] B. Rosenberg, L. Van Cowp, J.F. Trosko and V.H. Mansour, *Nature (Lond.)* **222** (1969) 385.
[2] C. Grau and J. Overgard, *Raiothes. Oncol.* **13** (1988) 301.
[3] J.F. Holland, H.W. Bruckner, C.J. Cohen, R.C. Walleck, C.B. Brusberth, E.M. Greenspan and J. Goldberg, *Cisplatin Therapy of Ovarian Cancer*, Academic Press, New York, 1980, p. 308.
[4] P.J. Loehrer and L.rH. Einborn, Drugs five year later: Cisplatin, *Ann. intern. Med.* **100** (1984) 704.
[5] S.D. Williams and L.H. Einborn, In: *Cisplatin: Current Status and New Developments*, A.W. Prestayko, S.T. Crooke and S.K. Carter (Eds), Academic Press, New York, 1980, p. 323.
[6] A. Eastman and E. Bresnick, *Biochem. Pharmacol.* **30** (1981) 2721.
[7] J. Burchenal, K. Kalahar, K. Dew, L. Lokys and G. Gale, *Biochem. (Paris)* **60** (1978) 691.
[8] I.H. Krokoff, *Cancer Treat. Rep.* **63** (1979) 1523.
[9] D.D. Van Hoff, R. Schilsky and C.M. Reichert, *Cancer Treat. Rep.* **63** (1979) 1439.
[10] J.B. Vermorken and H.M. Pinedo, *Neth. J. Med.* **25** (1982) 270.
[11] D.M. Hayes, E. Critković and R.B. Golbey, *Cancer (Philo)* **39** (1977) 1372.
[12] R.F. Ozols, B.J. Corden, J. Collins and R.C. Young, In: *Platinum Coordination Complexes in Cancer Therapy*, M.P. Hacker, E.B. Douple and I.H. Krakoff (Eds), Mortinus Nijhoff, Boston 1984, pp. 321–329.
[13] S.L. Hollis, A.R. Amundsen and E.W. Stern, *J.Am. Chem. Soc.* **107** (1984) 274.
[14] M.P. Hacker, A.R. Khokkar, D.B. Brown, J.J. McCormack and I.H. Krakoff, *Cancer Res.* **45** (1985) 4748.
[15] W.K. Anderson, D.A. Quagliato, R.D. Haugwitz, V.L. Narayanan and M.K. Wolpert-DeFilippes, *Cancer Treat. Rep.* **70** (1986) 997.
[16] R.F. Borch and M.E. Pleasants, *Proc. Nat. Acad. Sci. USA* **76** (1979) 6611.
[17] D.A. Ross and G.R. Gale, *Cancer Treat. Rep.* **63** (1979) 781.

[18] J.M. Yuhas and F. Čulo, *Cancer Treat, Rep.* **64** (1980) 57.

[19] M. Ishizawa, S. Toniguchi and T. Baba, *Jap. J. Pharmacol.* **31** (1981) 883.

[20] M. Radačič, M. Boranić, D. Škarić, V. Škarić, M. Mihalić, V. Gajšak, J. Jerčić and P. Lelieveld, *Oncology* **44** (1987) 34.

[21] K. Kiss, K.P. Berg, A. Dirscherl, W.E. Oberhansli and W. Arnold, *Helv. Chim. Acta* **63** (1980) 1728.

[22] M. Bonifačić, I Ljubenkov and M. Eckert-Maksić, manuscript in preparation.

[23] I. Ljubenkov, M.Sc. Thesis, University of Zagreb, 1991, M. Eckert-Maksić and I. Ljubenkov, to be published.

[24] M. Osmak, M. Eckert-Maksić, K. Pavelić, Z.B. Maksić, R. Spaventi, L. Beketić, I. Kovaček and B. Šušković, *Res. Exp. Med.* **190** (1990) 443.

[25] P.S. Whitton, D. Orešković, B. Jernej, M Radačić, F. Plavšić, and M. Bulat, *Biomed. Pharmacother.* **40** (1986) 191.

[26] R.J. Kociba and S.D. Sleight, *Cancer Chemother. Rep.* **55** (1971) 1.

[27] W. Jabs and W. Gaube, *Wiss Z. Ernst-Moritz-Arndt-Univ., Greifswald. Math.-nat. wis. Reihe* **34** (1985) 25 and references cited therein.